# Jellyfish Blooms: Ecological and Societal Importance

# Developments in Hydrobiology 155

*Series editor*
H. J. Dumont

# Jellyfish Blooms: Ecological and Societal Importance

Proceedings of the International Conference on Jellyfish Blooms, held in Gulf Shores, Alabama, 12–14 January 2000

*Edited by*

## J. E. Purcell[1], W. M. Graham[2] & H. J. Dumont[3]

[1] *University of Maryland, Center for Environmental Science, Horn Point Lab., Cambridge, MD, U.S.A.*
[2] *Dauphin Island Sea Lab., Dauphin Island, AL, U.S.A.*
[3] *University of Ghent, Institute of Animal Ecology, Belgium*

*Reprinted from Hydrobiologia, volume 451 (2001)*

SPRINGER SCIENCE+BUSINESS MEDIA, B.V.

**Library of Congress Cataloging-in-Publication Data**

Jellyfish blooms : ecological and societal importance / edited by J.E. Purcell, W.M.
Graham and H.J. Dumont.
    p. cm. -- (Developments in hydrobiology ; 155)
  This volume developed from a conference held in Gulf Shores, Alabama in January 2000.
  Includes bibliographical references (p. ).
  ISBN 978-0-7923-6964-6     ISBN 978-94-010-0722-1 (eBook)
  DOI 10.1007/978-94-010-0722-1
   1. Jellyfishes--Congresses. I. Purcell, J. E. II. Graham, W. M. III. Dumont, H. J.
(Henry J.) IV. Series.

QL377.S4 J48 2001
593.5'3--dc21

2001029532

ISBN 978-0-7923-6964-6

**Cover illustration:** *Chrysaora fuscescens* aggregation in Monterey Bay, California. Photo by Dave Wrobel.

# TABLE OF CONTENTS

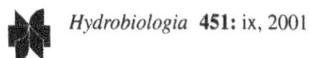*Hydrobiologia* **451:** ix, 2001.

# International Conference on Jellyfish Blooms

**Gulf Shores, Alabama**
**12–14 January 2000**

Conference Organizers:   Dr W. M. Graham (Dauphin Island Sea Lab, Alabama)

                             Dr J. E. Purcell (Horn Point Laboratory, Maryland)

Conference Sponsors:     Mississippi–Alabama Sea Grant Consortium

                             Dauphin Island Sea Lab

The conference logo depicting the 'global jellyfish aggregation' was conceived by Dr Graham, and graciously placed on paper by Mr R. Dixon.

The photograph on the cover of *Chrysaora fuscescens* aggregation in Monterey Bay, California, is by David Wrobel.

Bibliographic references to works from this volume should be in the following form:

*For the entire volume:*
Purcell, J. E., W. M. Graham & H. J. Dumont (eds), 2001. Jellyfish Blooms: Ecological and Societal Importance. Developments in Hydrobiology 155. Kluwer Academic Publishers, Dordrecht: xviii + 334 pp. Reprinted from Hydrobiologia 451.

*For individual papers:*
Burnett, J. W., 2001. Medical aspects of jellyfish envenomation: pathogenesis, case reporting and therapy. Hydrobiologia 451 (Dev. Hydrobiol. 155): 1–9.

Participants of the International Conference on Jellyfish Blooms held in Gulf Shores, Alabama, 12–14 January 2000.

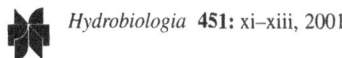*Hydrobiologia* **451**: xi–xiii, 2001.

# Introduction

Scientific awareness of gelatinous zooplankton, which includes medusae, siphonophores, cteno-phores, salps, and larvaceans, has increased markedly in the past three decades. At the forefront of this research field is the possibility that ecological change, forced by a variety of factors, may be enhancing gelatinous zooplankton populations in estuarine and marine environments. To address ecological and societal issues of jellyfish blooms around the world, the International Conference on Jellyfish Blooms was held in Gulf Shores, Alabama, on the coast of the northern Gulf of Mexico, in January 2000. This conference convened many of the world's current and future investigators of gelatinous zooplankton ecology to discuss the past, present and future of our field. The Conference was hugely successful in assembling scholars that ranged from patriarchs and matriarchs in the field to graduate and undergraduate students. In total, more than 70 scientists representing 13 countries from 5 continents attended the three-day conference hosted by the Dauphin Island Sea Lab. and the Mississippi–Alabama Sea Grant Consortium. This group of scientists represents the majority of jellyfish researchers worldwide. A total of 53 oral and poster presentations were contributed.

The International Conference on Jellyfish Blooms was a natural progression in the lineage of jellyfish ecology across scales of understanding. We can even link historical events to major transitions in how scientists have approached ecological studies of jellyfish. The first major transition occurred in the late 1960s following the United States' passage of a legislative bill entitled 'The Jellyfish Act', which released the single largest research allocation at that time to study the regulation of jellyfish populations, primarily *Chrysaora quinquecirrha* (Desor) in the Chesapeake Bay. The 'Jellyfish Act' served as a significant transition of emphasis from the organism to the. population. We would like to especially acknowledge the important contributions of Mr David Cargo (e.g., Cargo & Schultz, 1966, 1967; Cargo & King, 1990), who, regrettably, passed away before the Gulf Shores conference. The anticipated product of the Jellyfish Act, which was the ability for humans to regulate jellyfish populations, never materialized. Nevertheless, Jellyfish Act-funded research produced a wealth of information on feeding, growth, reproduction, development, and behavior of jellyfish.

The next transition in jellyfish ecology research was from the population level to the ecosystem level in the 1970s and 1980s. A major factor contributing to this transition was the development of an action plan to address pollution issues in the Mediterranean Sea. As part of the Mediterranean Action Plan (MAP), workshops were held in 1983 (Athens, Greece) and in 1986 and 1987 (Trieste, Italy) to understand the possible linkages between environmental pollution and jellyfish blooms, principally of *Pelagia noctiluca* (Forsskål), in the Mediterranean Sea. These workshops and their subsequent scientific reports (UNEP 1984, 1991) highlighted the complex relationships between jellyfish blooms, environmental variability, and human activity.

It was not until the late 1980s that the 'human dimension' to jellyfish blooms could be fully appreciated. It was at this transition point when ecosystem level research was linked to socio-economics and politics as a consequence of the major ecological shift that took place in the

Black Sea. As discussed in this volume, the introduction of the ctenophore *Mnemiopsis leidyi* A. Agassiz in the Black Sea following extreme environmental changes (including over-harvesting of fish, cultural eutrophication and freshwater diversion) was implicated in the collapse of the valuable Black Sea anchovy fishery (e.g., GESAMP, 1997). In the wake of the Black Sea 'environmental disaster', the scientific community has made significant advances into understanding the complex nature of the relationships of coastal marine ecosystems and the functional role of gelatinous zooplankton. The naïve belief in the 1960s that jellyfish were the 'problem' has been replaced today with the sobering realization that jellyfish blooms more likely are symptoms of environmental problems such as cultural eutrophication and associated effects like hypoxia, commercial over-harvesting of fish and invertebrates, habitat modification, freshwater diversion, species introductions, and global climate change.

Unfortunately, not since the MAP workshops (UNEP 1984, 1991) has the international community of gelatinous zooplankton scientists come together with the specific intent to address jellyfish blooms (but see GESAMP, 1997). The International Conference on Jellyfish Blooms clearly reaffirmed the high degree of collegiality and devotion among the scientists of this research community. It is this international collegiality that will ultimately foster the next major advances in understanding global dynamics of jellyfish blooms. The Conference also recognized a unified voice that will give direction to this rapidly expanding field. The attendees realized that a greater impact could be achieved with a collective voice, and called for future meetings to be held every 5–6 years. It was recommended that increased efforts be made to globalize our research, and discussions on specific mechanisms to accomplish this resulted in the conceptual foundation of the 'JellyWatch' network of global jellyfish bloom dynamics. The research community also called for increased public awareness, expansion of interdisciplinary activities, larger scale studies, and increased funding potential. This volume serves as an extension of the Conference as a message to the larger scientific community and to the world.

While this volume was being compiled, the northern Gulf of Mexico provided striking examples of the socio-economic effects of jellyfish and our lack of understanding of most jellyfish populations. During the summer of 2000, public attention was drawn to 'invasions' of two large, previously unseen jellyfish species. The first was identified as *Phyllorhiza punctata* (von Lendenfeld), which is native to the tropical western Pacific Ocean. Millions of these large medusae (50–60 cm bell diameter) accumulated in the coastal waters of the northern Gulf. An immediate economic impact on shrimp harvesting was felt in the region due to fouling of nets, however, longer-term ecological effects due to predation are unknown. The second highly unusual species, the large cyaneid medusa, *Drymonema dalmatinum* (Haeckel), occurred over a large area of the northern Gulf from Florida to Louisiana. Individuals measuring up to 75 cm in bell diameter, with tentacles of 20 m in length were observed preying heavily on local populations of *Aurelia aurita* medusae. Whether these species will persist in the northern Gulf of Mexico is unknown at this time. This example illustrates that we need to better understand why jellyfish often suddenly appear in profusion, and their effects on the coastal ecosystems.

This volume is comprised of a subset of papers from the Conference, including both reviews and research papers. The Conference was designed to integrate various aspects of jellyfish blooms as they impact societies around the world. In the first section, Jellyfish and Human Enterprise: Fisheries and Tourism, topics range from the medical aspects of jellyfish stings (Burnett), jellyfish as food (Hsieh et al.) and jellyfish fisheries (Omori & Nakano), to interactions between jellies and fish (Purcell & Arai; Mianzan et al.). In the second section, Jellyfish and Changing Ecosystems, papers consider whether jellyfish blooms have increased in recent years (Mills; Arai; Graham; Sullivan et al.), the effects of environmental factors such as dissolved oxygen, temperature and salinity on jellyfish population size (Condon et al.; Graham; Sullivan et al.; Raskoff; Dawson et al.), and the effects of introductions of ctenophores in the Black Sea region (Purcell et al.; Finenko et al.; Shiganova et al.). The third section, Physical and Hydrodynamic Interactions with Jellyfish, pertains to the role that physical processes play in mediating aggregations and distributions of jellyfish (Graham et al.; Johnson & Perry) and feeding (D'Ambra et al.). The fourth section comprises a review of reproduction and life history strategies (Lucas), asexual reproduction rates (Watanabe & Ishii), and molecular phylogenetics (Dawson & Martin). The final section encompasses a variety of topics related to morphology (Moss et al.), epibionts and parasites of jellies (Moss et al.; Martorelli), and feeding (Ishii & Tanaka; Youngbluth & Båmstedt). All papers were subjected to the full peer review procedure of the journal *Hydrobiologia*, in which these manuscripts also are published.

To those of us at the Conference, the importance of jellies is obvious, but to most other aquatic scientists, their importance often goes unnoticed. We believe that this volume will be a key reference for the experts in our field. We hope that it also will stimulate the interest of other scientists and provide valuable references to those who may enter the field.

W. M. GRAHAM &
J. E. PURCELL
*Organizers and Editors*

## Acknowledgments

The International Conference on Jellyfish Blooms was generously funded primarily by the Mississippi–Alabama Sea Grant Consortium, Dr B. Costa-Pierce, Director. Additional support was graciously provided by the Dauphin Island Sea Lab, the Maryland Sea Grant College and the Horn Point Laboratory of the University of Maryland Center for Environmental Science.

## References

Cargo, D. G. & D. R. King, 1990. Forecasting the abundance of the sea nettle, *Chrysaora quinquecirrha*, in the Chesapeake Bay. Estuaries 13: 486–491.

Cargo, D. G. & L. P. Schultz, 1966. Notes on the biology of the sea nettle, *Chrysaora quinquecirrha*, in Chesapeake Bay. Chesapeake Sci. 7: 95–100.

Cargo, D. G. & L. P. Schultz, 1967. Further observations on the biology of the sea nettle and jellyfishes in the Chesapeake Bay. Chesapeake Sci. 8: 209–220.

GESAMP (IMO/FAO/UNESCO-IOC/WMO/WHO/IAEA/UN/UNEP Joint Group of Experts on the Scientific Aspects of Marine Environmental Protection), 1997. Opportunistic settlers and the problem of the ctenophore *Mnemiopsis leidyi* invasion in the Black Sea. Rep. Stud. GESAMP 58: 84 pp.

UNEP (United Nations Environmental Programme), 1984. Workshop on Jellyfish Blooms in the Mediterranean. Athens, 31 October–4 November 1983: 221 pp.

UNEP, 1991. Jellyfish blooms in the Mediterranean, Proc. of II Workshop on Jellyfish in the Mediterranean Sea. Mediterranean Action Plan Technical Reports Series 47: 320 pp.

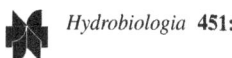 *Hydrobiologia* **451**: xv–xviii, 2001.

# A Complete Listing of Conference Presentations

## ORAL PRESENTATIONS

**Jellyfish and Global Change: Population Increases and Decreases in Response to Changing Conditions in the Sea.** C. E. Mills. Friday Harbor Laboratories and Department of Zoology, University of Washington, 620 University Road, Friday Harbor, WA 98250, U.S.A.

**Pelagic Coelenterates and Eutrophication: A Review.** M. N. Arai. Pacific Biological Station,Nanaimo, British Columbia, V9R 5K6, Canada and Department of Biological Sciences, University of Calgary, 2500 University Drive NW, Calgary, Alberta, T2N 1N4, Canada.

**Does Low Dissolved Oxygen Favor Gelatinous Zooplankton in Eutrophic Estuaries?** D. L. Breitburg. The Academy of Natural Sciences, Estuarine Research Center, St. Leonard, MD 20685, U.S.A.

**Effects of Low Dissolved Oxygen on Predation of Zooplankton by the Ctenophore *Mnemiopsis leidyi*.** M. B. Decker[1], D. L. Breitburg[2] and J. E. Purcell[1]. [1]Horn Point Laboratory, University of Maryland Center for Environmental Science, Cambridge, MD 21613-0775, U.S.A. [2]The Academy of Natural Sciences, Estuarine Research Center 10545 Mackall Road, St. Leonard, MD 20685, U.S.A.

**The Influence of Hypoxia on the Ctenophore *Mnemiopsis leidyi*: Behavior and Predation on *Gobiosoma bosc* Larvae and *Anchoa mitchilli* Eggs and Yolk Sac Larvae.** S. E. Kolesar[1], D. L. Breitburg[2], K. A. Rose[3] and J. H. Cowan, Jr[4]. [1]Chesapeake Biological Laboratory, Solomons, MD 20688, U.S.A. [2]The Academy of Natural Sciences Estuarine Research Center, St. Leonard, MD 20685, U.S.A. [3]Louisiana State University, Baton Rouge, LA 70803, U.S.A. [4]University of South Alabama, Mobile, AL 36688, U.S.A.

**Historical and Current Trends in Jellyfish Population Abundance in Chesapeake Bay, and a Tribute to Dave Cargo.** J. E. Purcell[1], K. Mountford[2] and [1]M. B. Decker. [1]Horn Point Laboratory, University of Maryland Center for Environmental Science, Cambridge, MD 21613, U.S.A. [2]U.S. Environmental Protection Agency, Chesapeake Bay Program, Annapolis, MD 21403, U.S.A.

**Are Jellyfish Blooms in the Northern Gulf of Mexico on the Increase? An Analysis of 10 Years of Fishery Survey By-Catch Data.** W. M. Graham, Dauphin Island Sea Lab, Marine Environmental Sciences Consortium, 101 Bienville Boulevard, Dauphin Island, AL 36528, U.S.A.

**Distribution of *Aurelia aurita* (Cnidaria), *Pleurobrachia pileus* and *Mnemiopsis leidyi* (Ctenophora) in the Southern Black Sea During 1995–1998.** A. E. Kideys[1] and Z. Romanova[2]. [1]Institute of Marine Sciences, P.O. Box 28, Erdemli, 33731, Turkey. [2]Institute of Biology of the Southern Seas, Nachimov Avenue 2, Sevastopol, Ukraine.

**Spatio-Temporal Distributions of Jellyfish (Hydromedusae and Scyphomedusae) Along the West Coast of South Africa: Difficulties of Forecasting Their Fluctuations.** E. Buecher[1,2] and M. J. Gibbons[1]. [1]Zoology Department, University of the Western Cape, Private Bag X17, Bellville, 7535, South Africa. [2]Laboratoire d'Océanographie Biologique et Ecologie du Plancton Marin, Station Zoologique, B.P. 28, F-06234, Villefranche-sur-Mer, France.

**Distribution Patterns of *Chrysaora hysoscella* and *Aequorea aequorea* in the Northern Benguela Ecosystem.** C. Sparks[1] and M. Gibbons[2]. [1]School of Life Sciences, Cape Technikon, Box 652, Cape Town, 8000, South Africa, and Zoology Department, University of the Western Cape, Private Bag X17, Bellville, 7535, South Africa. [2]Zoology Department, University of the Western Cape, Private Bag X17, Bellville, 7535, South Africa.

**Jellyfish Swarms, Tourists, and "The Christchild".** M. N Dawson[1,2], L. E. Martin[1,2] and L. K. Penland[2]. [1]Department of Organismic Biology, Ecology and Evolution, University of California, Los Angeles, CA 90095-1606, U.S.A. [2]Coral Reef Research Foundation, Box 1765, Koror, PQ 96940, Republic of Palau.

**Is there a need for a Jellywatch Network?** L. P. Madin[1] and P. Kremer[2]. [1]Woods Hole Oceanographic Institution, Woods Hole, MA 02543,

**els.** J. H. Cowan, Jr.[1] and K. A. Rose[2]. [1]University of South Alabama, Dauphin Island Sea Lab, 101 Bienville Boulevard, Dauphin Island, AL 36528, U.S.A. [2]Louisiana State University, Baton Rouge, Louisiana, U.S.A

**Up Close, *In situ* Investigations of the Midwater Medusa *Periphylla periphylla*.** M. J. Youngbluth[1] and U. Båmstedt[2]. [1]Harbor Branch Oceanographic Institution, Fort Pierce, FL 34946, U.S.A. [2]Department of Fisheries and Marine Biology, University of Bergen, N-5020 Bergen, Norway.

**Functional roles of the lobate ctenophore *Bolinopsis mikado* in the marine coastal ecosystem of Japan.** S. Uye[1] and T. Kasuya[2]. [1]Faculty of Applied Biological Science, Hiroshima University, Higashi-Hiroshima, 739-8528, Japan. [2]Tokyo University of Fisheries, Tokyo, 108-8477, Japan.

**The Impact of El Niño Events on Blooms of Midwater Hydromedusae.** K. A. Raskoff. Monterey Bay Aquarium Research Institute, Moss Landing, CA 93950, U.S.A.

**Developing Jellyfish Strategy Hypotheses Using Circulation Models.** D. R. Johnson[1], H. M. Perry[2] and W. D. Burke[2]. [1]NRL Code 7332, Stennis Space Center, MS 39529, U.S.A. [2]Gulf Coast Research Laboratory, Ocean Springs, MS 39564, U.S.A.

**A Physical Context for Jellyfish Aggregations: A Review.** F. Pagès[1], W. M. Graham[2] and W. M. Hamner[3]. [1]Institut de Ciències del Mar (CSIC), Plaça del Mar s/n, Barcelona, Catalunya, 08039, Spain. [2]University of South Alabama and Dauphin Island Sea Lab, 101 Bienville Boulevard, Dauphin Island, AL 36528, U.S.A. [3]University of California, Los Angeles, Deparment OBEE, Box 951606, Los Angeles, CA 90095-1606, U.S.A.

**Population Dynamics and Sexual Reproduction of the Scyphomedusa *Aurelia aurita* from Two Contrasting Ecosystems.** C. Lucas. School of Ocean & Earth Science, University of Southampton, Southampton Oceanography Centre, European Way, Southampton SO14 3ZH, United Kingdom.

**Population Structure, Growth and Age Determination in the Cubozoan *Chiropsalmus* Sp.** M. Gordon, C. Hatcher and J. Seymour. James Cook University, Cairns, Queensland, 4878, Australia.

**First bloom of the invader *Beroë ovata* in the Black Sea.** T. A. Shiganova, P. Yu. Sorokin and Yu. F. Luksahev. P. P. Shirshov Institute of Oceanology RAS, 36 Nakhimovsky Pr., Moscow, 117851, Russia.

## POSTER PRESENTATIONS

**The Relationship Between Temperature, Size and Metabolic Rate in the Cubozoan *Chiropsalmus* Sp.** M. Gordon, A. Krockenberger and J. Seymour. James Cook University, Cairns, Queensland, 4878, Australia.

**The Effects of Low Dissolved Oxygen on the Survival and Development of Sea Nettle Polyps (*Chrysaora quinquecirrha*).** R. H. Condon[1], M. B. Decker[2] and J. E. Purcell[2]. [1]2/32 Stephensons Road, Mount Waverley, VIC 3149, Australia. [2]Horn Point Laboratory, University of Maryland Center for Environmental Science, Cambridge, MD 21613, U.S.A.

**Food Groove Morphology and Function in the Lobate Ctenophore, *Mnemiopsis*.** R. C. Rapoza, A. G. Moss and L. Muellner. Woods Hole Oceanographic Institution, Woods Hole, MA 02543, U.S.A. and Auburn University, Auburn, AL 36849, U.S.A.

**Effect of Medusa Size on Prey Selectivity in the Jellyfish *Aurelia aurita* in the Northern Gulf of Mexico.** R. M. Kroutil and W. M. Graham, Dauphin Island Sea Lab, 101 Bienville Boulevard, Dauphin Island, AL 36528, USA.

**Parasites of Jellyfishes and Ctenophores of the Southern Atlantic.** S. R. Martorelli. Gulf Coast Research Laboratory, 703 East Beach Drive, Ocean Springs, MS 39566-7000, U.S.A. Visiting Research Scholar (CONICET-Argentina).

**Flow and Prey Capture by the Scyphomedusa *Phyllohriza punctata* von Lendenfeld 1884.** I. D'Ambra[1], F. Bentivegna[1] and J. H. Costello[2]. [1]Naples Aquarium, Stazione Zoologica "A. Dohrn", Villa Communale, 80121 Naples, Italy. [2]Biology Department, Providence College, Providence, RI 02918-0001, U.S.A.

**Partial Purification and Characterization of a Hemolysin (CAI) From Hawaiian Box Jellfish (*Carybdea alata*) Venom.** J. Chung, J. Kuroiwa, L. Ratnapala, I. M. Cooke and A. A. Yanagihara. Békésy Laboratory of Neurobiology, Pacific Biomedical Research Center, University of Hawaii, Honolulu, HI 96822, U.S.A.

**Temporal Variability in the Unfished Abundance and Biomass of the Commercially Harversted Jellyfish, *Catostylus mosaicus* (Rhizostomeae) in New South Wales, Australia.** K. A. Pitt and M. J. Kingsford. University of Sydney, Sydney, New South Wales, 2006, Australia.

*Hydrobiologia* **451**: 1–9, 2001.
© 2001 *Kluwer Academic Publishers.*

1

# Medical aspects of jellyfish envenomation: pathogenesis, case reporting and therapy

Joseph W. Burnett

*Department of Dermatology University of Maryland School of Medicine 405 W. Redwood Street, 6th floor, Baltimore, MD 21201, U.S.A. E-mail: jburnett@umaryland.edu*

*Key words:* jellyfish, sting, envenomation, venom, Portuguese-man-o' war, sea nettle

## Abstract

With larger human population numbers and their need for recreation, contact between humans and jellyfish is increasing. The pathogenesis of cnidarian stings is discussed here and some of the factors influencing the variability in adverse reactions they produce are mentioned. The pharmakinetics of venom delivery determines the organ site of damage and the extent of abnormality. Since venoms can injure man by allergic or toxic reactions, the differences between these processes is elucidated. Toxic reactions predominate and allergic ones are unusual. A more complete list of disease entities caused by jellyfish stings has been compiled. These sting reactions may be local, systemic, chronic or fatal. Most follow cutaneous stinging but some occur after stings to the eye or following ingestion. Increased case loads and experience has also lead to a more comprehensive understanding of sting pathogenesis and treatment. Accordingly, the principles of prevention and first aid therapy are outlined. Finally, some recommendations for more complete recording of adverse stings are suggested.

The purpose of this paper is to increase the understanding of marine biologists about the medical aspects of jellyfish stingings by: (I) stressing the importance of pathogenesis including the process of envenomation, the pharmokinetics of the stinging, the role of allergy and toxicity in the disease state, and the importance of prior exposure to other animal venoms, (II) suggesting a scheme for adequate history taking and case recording, (III) cataloguing the jellyfish envenomation syndromes, and (IV) outlining the best therapeutic and prevention techniques.

## Pathogenesis of envenomation

### Process of envenomation

Jellyfish are pelagic cnidarians found worldwide. Some have a stinging apparatus injurious to humans. Many species are cited as nuisances or health hazards and can appear individually or in swarms. Each species has a distinctive armamentarium of cnidocytes, cells with cytoplasmic nematocysts containing a coiled, toxin-delivering tubule (Halstead, 1965).

These cnidocytes are arranged in clustered batteries along the surfaces of the tentacles. In some jellyfish, the nematocysts are located in the outermost layer when in the 'fire ready' position or are withdrawn by a fibrillar network to an unexposed position (Rifkin & Endean, 1988). Nematocyst firing is triggered by combined chemical and tactile stimuli produced by the victim. The movement of nearby prey is an important factor in controlling jellyfish nematocyst discharge (Watson & Hessinger, 1989). Discharge of the nematocyst into human tissues involves the eversion of the tubule into our epidermis and its subjacent vascular and nerve-rich dermis.

### Pharmacokinetics of the sting

Along their trajectory into the tissue, the nematocyst threads cross the epidermis, penetrate the dermis and enter, transfix or penetrate lymphatics, nerves and capillaries. Venom present on the outside of the thread is deposited in all these areas of skin. Thus, some venom has been injected intraepidermally, intradermally, intravascularly and probably in some cases subcutaneously. Each of these injection routes has an established circulation time, which is also influenced by the molecular size of the venom, the health of the patient, the movement of the injured part, the perfusion ability of the patient and the location of the injury.

While venom dose will depend upon the number of nematocysts fired, the rapidity of the symptoms and their duration will vary with each case depending upon several factors that control the circulation time and the interval lapse between injection and the onset of symptoms. To make the problem more complicated, each injurious molecule attaches to different receptor sites at different organ systems. All these require a variation in reaction time before the pathogenic episodes appear. Thus, a spectrum of time interval or 'incubation' is necessary before disease is established.

*The role of allergy/toxicity in the disease state*

It is important, now, to emphasize that most of the bathers' disease from envenomation is due to a toxic reaction to the venom rather than an allergic one (allergy can be defined as an exaggeration of either of the body's immune systems, see below next paragraph). This critical distinction means that the bather has little chance of exacerbating his reaction with succeeding stings. Toxic reactions imply that all stung people are injured and that both supersensitivity and resistance do not occur. The fact that a jellyfish sting is a toxic reaction is usually overlooked because the rash is red and may contain lesions with hive-like components. Even very skilled emergency care workers will diagnose this type of rash as allergy-induced when it is associated with hypotension, difficult breathing and cold clammy skin. Unfortunately, they forget that many toxins produce the same clinical picture.

Jellyfish stings initiate the expected immune defense reactions in bathers that any other foreign proteins produce. These are not associated with disease or abnormality. Accordingly, there is a stimulation of both serum antibody (humoral immunity) and reactive lymphocytes or monocytes (cell mediated immunity). These events can be used to test for the patient's exposure to that certain venom (serodiagnosis). It is now thought that an exaggerated stimulation of the cell mediated system (a form of allergy) may be responsible for some of the later symptoms (sequellae) patients will experience after their stings (Burnett et al., 1986). These sequellae, then induced by allergy, stand in contrast to the usual immediate response to the jellyfish sting which is 'toxic' in nature. The exaggerated or

persistent immunological reactions (allergy) are rarely found in the envenomated population and some stung bathers may be even supersusceptible. No person exhibits these unusual allergic clinical reactions after his initial sting since prior exposure to the venom is necessary for their appearance. Modulation of the immune response by external agents such as drugs or ultraviolet light at the time of the sting should theoretically alter the swimmer's response to the venom, but no illustrative case of this phenomenon has been reported (Miura et al., 1996).

*The relationship between different animal venoms and its importance in sting pathogenesis*

Because the clinical course of human injuries produced by different animal venoms appears similar, some misconceptions about the deleterious or beneficial effect of prior exposure to diverse venoms exists. The chemical and immunological relationships between different animal venoms have been extensively studied. Thus, jellyfish, bees, wasps and some snakes contain related components (i.e. peptidases, hyaluronidase) which might induce similar toxic reactions to account for some of the local redness and swelling. Although immunological cross reactivity between proteins of different animal venoms has been demontrated *in vitro*, this relationship does not appear to be clinically important (Olson et al., 1985; Togias et al., 1985). The frequency of immunological cross reactivity between venoms increases as the poisonous animals are more closely related phylogenetically. Just as in jellyfish evenomations, most bee, wasp or snake injuries are toxic in nature (everyone is hurt with the first contact, no one is naturally resistant, the severity of the reaction is directly dose-related) and allergy (anaphylaxis) is rare. Additionally, humans do not acquire clinical resistance or susceptibility to bee stings by exposure to cnidarian venoms (or vice versa).

**Adequate history taking and case recording**

Many readers are employed at biological laboratories located at remote seasides or on board ship-locations where jellyfish stings occur but medical personnel are sparse. Therefore, a guide for facilitating adequate information collecting is offered in Table 1. In addition, both the offending animal and the resulting rash can be photographed or sketched and the skin can be scraped with a blade or stripped with cellulose tape to obtain released cells for microscopic examination and nematocyst identification. A good description of the rash would include three elements: distribution on the body (localized or widespread), configuration or arrangement of the component lesions (symmetrical, confluent, linear or circular) and a word about the initial individual component or 'spot' of the rash (macule = flat spot; papule = gooseflesh bump; nodule = larger bump; vesicle/bulla = small to large blister; pustule = pus filled blister). The above description should be only settled upon after examining the skin and hearing also what the bather saw as his disorder progressed.

The compilation of this information will allow future scientists to know exactly which animal produces what disease and geographically where the more serious stings occur. This information will be invaluable as therapy becomes more advanced with species specific immunodiagnostic (serology) and immunotherapeutic (antivenoms) techniques develop.

## Catalogue of syndromes

Jellyfish envenomation syndromes can be separated on their temporal (delayed or immediate) and spatial (local or systemic) extents (Table 2).

### Fatal reactions

Death produced by jellyfish stings may occur by either toxic or allergic mechanisms. Anaphylaxis, the most serious allergic reaction, rarely occurs from jellyfish stings (Togais et al., 1985). This type of reaction, which occurs within a few minutes to an hour, is characterized by cardiovascular collapse, fainting, hypotension, tachycardia, cold, clammy skin and respiratory distress with varying degrees of severity. There has been no known fatality from anaphylaxis or allergy-induced disease following a jellyfish sting.

A similar, rapidly progressive syndrome produced by a toxic reaction to the sting is more common. Death can occur from toxic reactions to the jellyfish by four different mechanisms depending upon the dose of absorbed venom. Large doses affect the heart by producing varying degrees of cardiac block, ventricular arrhythmia, disturbance to the Purkinje fiber network and coronary artery vasoconstriction (Kleinhaus et al., 1973; Burnett & Calton, 1987; Lin et al., 1988). This cardiotoxic action is thought to be mediated by alterations in calcium and/or sodium ionic transfer (Burnett & Calton, 1977, 1983, 1987; Cobbs et al., 1983;

Dubois et al., 1983). Moderate doses of venom depress respiration by a central nervous system effect several minutes to hours after the sting. Thus, if the patient survives the cardiotoxic action of the venom, death can occur by respiratory toxicity (Burnett & Calton, 1987). Finally, still lower doses of venom may prove fatal later by producing acute renal failure with acute tubular necrosis (Guess et al., 1982) or by inducing cellular necrosis leading to liver failure (Garcia et al., 1994).

### Systemic reactions

The systemic reactions produced by jellyfish envenomation include nausea, vomiting, muscle cramps, diarrhea, dizziness, diaphoresis, fainting, coma, convulsions, muscular spasms and (if the sting is on the thorax) respiratory acidosis (Kizer & Piel, 1982). A specific symptom complex, the 'Irukandji' reaction, occurs after envenomation by *Physalia physalis* (Linnaeus) (Portuguese man-o'war) or any of several carybdeids, especially *Carukia barnesi* Southcott (Irukandji boxjellyfish). This reaction usually follows a minor painful sting, which may be anywhere on the body and is either unnoticed or its symptoms disappear within minutes. After a latent period ranging from 5 to 40 min, a deep pain begins on the sacrum or abdomen and spreads quickly into the chest and thighs. Intercostal tightness and waves of excruciatingly severe muscle pain intensify over a few minutes before fading then recurring cyclically. There are increased 'sighing' respirations, restlessness, tremor, anxiety, headache, localized or generalized piloerection, sweating, pallor, cyanosis, low urine output, tachycardia, nausea, hypertension, pulmonary edema with left ventricular dilatation and feelings of imminent death. Fortunately this syndrome, which lasts 1–2 days, is self resolving (Fenner et al., 1988; Gunawardane & Murtha, 1988).

Stings by other jellyfish have been accompanied by symptoms in other body regions. Psychiatric symptoms have followed *Stomolophus nomurai* (Kishinoouye) stings with psychosis, convulsions, coma and stupor (Zhang & Li, 1988). Fever and muscle spasms can occur after *Chrysaora quinquecirrha* (Desor) (sea nettle), *Physalia physalis* and box-jellyfish envenomations.

### Local reactions

Local reactions produced by jellyfish envenomation include painful, linear, red, hive-like lesions. Pain is perceived instantly, is maximal within 5 min and

*Table 1.* Pertinent historical questions and observations recommended for jellyfish sting patients

---

**Details of the sting**

    Date, time, precise geographical location

    Patient activity at the time (entering water, swimming, sunbathng)

    Weather, tide, water conditions, amount of sun exposure

    Jellyfish species-drawing, photograph

    Length of tentacle contact with skin

    How long did adherent tentacles persist?

    Description of marks on the skin from tentacles

    Part of body stung

**Symptoms**

| Dizziness | Nausea | Vomiting |
|---|---|---|
| Fever | Faintness | Coma |
| Local or generalized sweating | Shortness of breath | |
| Muscle spasms | Piloerection ('gooseflesh') | |
| Increased heart rate | | |

**Past medical history**

    Significant medical diseases

    Drugs taken at the time of the sting

    Allergy history - asthma, hay fever, eczema

    Treatment

**First aid**

    Time of arrival at hospital

    Condition of the patient

    Initial hospital treatment

    Microscopy of skin scrapings for nemaocyst identification

    Skin biopsy

---

dissipates over the next few hours. In some patients, the cutaneous eruptions may be dramatically swollen, may be delayed for several hours, may persist up to several months or be recurrent without further stings (Mansson et al., 1985; Maretic, 1986; Burnett & Calton, 1987). During the recurrences, some patients have significant diarrhea. The recurrent reactions may be similar in severity but will be accompanied by itching, not pain. The recurrent eruption can appear at any interval between 4 and 30 days and can last anywhere between 1 and 7 days. Red, swollen, hive-like eruptions may also develop distant to the site of sting (Matusow, 1980). Two patients with papular urticarial lesions located on the body away from the envenomation site have been reported (Burnett & Calton, 1987). Contact dermatitis to jellyfish tentacles has also been recorded in humans (Yaffee, 1968; Kokelj et al., 1995).

Ophthalmological complications of jellyfish stings can occur. Corneal ulcerations follow tentacle-orbital contact (Rapoza et al., 1986). Additional more serious local reactions were reported with cases of eye pain, blurred or diminished vision and sensitivity to bright light (Burnett & Burnett, 1990, Glasser et al., 1992). One of the patients with inflammation of the anterior eye chamber developed spontaneously remitting increased ophthalmic pressure and persistant glaucoma-requiring local therapy for over 4 years. Presumably, local skin damage from stings can trigger injury to nearby or underlying tissues. Symptoms in these patients were delayed 24 h for unknown reasons.

Other examples of injury to tissues adjacent to sting areas include the following. A patient with an inflamed superficial chest wall vein (Mondor's disease) after a local sting was noted as has a patient with a deep venous clotting inflammation of the leg

*Table 2.* Syndromes resulting from jellyfish envenomations

| Jellyfish envenomation syndromes | Causative species |
|---|---|
| **Fatal Reactions** | |
| Toxin-induced | |
| Immediate cardiac arrest | CF, P |
| Rapid respiratory arrest | CF |
| Delayed renal failure | P |
| Liver destruction* | A |
| Allergy-induced | |
| Anaphylaxis** | |
| **Systemic Reactions** | |
| Toxin-induced | CQ, CC, P, CF |
| Irukandji reaction | CB, P, CF |
| Respiratory acidosis | P |
| Pulmonary edema | CB |
| Arthritis* | P |
| **Local Reactions** | |
| Toxin-induced to skin, mucosa, and cornea | CQ, P, CF, PN, CC |
| Exaggerated local reaction (angioedema) | A |
| Recurrent reactions up to four episodes | P, L, CO |
| Delayed persistent reactions up to several months | PN, CA |
| Distant site reactions | P |
| Local lymphadenopathy | P |
| Seabathers' dermatitis | L |
| **Chronic Reactions** | |
| Keloids | CF |
| Pigmentation | PN |
| Fat atrophy* | |
| Contractions | P, CF |
| Gangrene | CF |
| Ulceration | |
| Vascular spasm | P, CQ, CF |
| Mononeuritis | P, CQ, CF |
| Autonomic nerve paralysis* | unknown box jellyfish |
| Ataxia | unknown box jellyfish |
| Increased ocular pressure | CQ |
| Mondor's disease (thrombophlebitis)* | |
| Gullian Barre* | PN |
| Limb bluing | |
| Deep venous thrombosis | |
| Blurred vision | CQ, L |
| Acquired cold urticaria* | P |
| Facial swelling | A |
| **Post-episode Dermatitis** | |
| Herpes simplex | CQ |
| Granuloma annulare | P, CH |
| **Reactions from Jellyfish Ingestion** | |
| Gastrointestinal symptoms | |
| Urticaria | R, P, L |
| Ciguatera* | R |

Key: A–anemone; CA–*Carybdea alata*; CF–*Chironex fleckeri* Southcott; CQ–*Chrysaora quinquecirrha* (sea nettle); CB–*Carukia barnesi* (Irukandji jellyfish); CC–*Cyanea capillata* (lion's mane jellyfish); CO–Corals; CH–*Chrysoaora hysocella, Eschscholtz*; L–*Linuche unguiculata* (thimble jellyfish); P–*Physalia physalis* (Portuguese man-o'war); PN–*Pelagia noctiluca* (Mauve baubler); R–*Rhizostoma*; S–*Stomolophus nomurai*; **=theoretical, no case verified. *-only 1 case reported-uncorroborated.

(Ingram et al., 1992; Al-ibrahim et al., 1995). Diffuse facial swelling, and acquired cold urticaria (long lasting propensity to hives following exposure to cold) have been reported in others (Mathelier-Fusade & Leynadier, 1993). Another patient, in whom an attack of Guillian-Barre syndrome (an unknown ascending spinal cord paralysis) (Pang & Schwartz, 1993) was observed, dated the onset of that disorder from exposure to *Pelagia noctiluca* (mauve baubler).

The clinical syndrome occurring from the sting of *Linuche unguiculata* Schwartz (thimble jellyfish) (all life stages) is known as seabather's eruption (Segura-Puertas et al., 2001). Four to 24 h after exposure, a prickling sensation may or may not be accompanied by local urticarial or macular lesions under the swimsuit from entrapped larvae or on open skin if large ephyrae or medusae are the culprits. The eruptions persist for 7–12 days but secondary recurrences or persistent hives can occur. Therapy is not effective so sedation and topical menthol aimed only at symptom relief is recommended (Wong et al., 1994). An indentical clinical syndrome caused by anemone larvae has occurred off Long Island waters (Freudenthal & Joseph, 1993).

## Chronic reactions

Chronic reactions at the envenomated site include persistent nodules (Reed et al., 1984), hyperpigmentation (Kokelj & Burnett, 1990), keloids, contractions, fat atrophy and vasospasm with gangrene (Gunn, 1947; Adiga, 1984; Williamson et al., 1988). Local neurological complications may be seen. Several cases of damage to one or two peripheral nerves have been reported subsequent to cnidarian envenomations (Filing-Katz, 1984; Moats, 1992). The affected nerves were near the stung area. Another patient had autonomic nerve paralysis with an extended abdomen and urinary bladder. Impotency, and inability to lacrimate or urinate were noted. This patient had electrocardiographic abnormalities unaffected by breathing or cartoid sinus massage (Chand & Sellah, 1984). A similar patient had 48 h of low grade fever, bladder atony, diplopia and ataxia (Burnett & Calton, 1987). Monoarticular arthraligia and reactive arthritis can postdate *Physalia physalis* stings (Weinberg, 1988).

## Post-episode dermatitis

Local stings may be followed by recurrent herpes simplex or granuloma annulare (dermal necrosis with hard circular nodules of unknown cause) at the site of wounds induced by *Chrysaora quinquecirrha* and the unitentacular form of *Physalia physalis*, respectively (Burnett & Calton, 1987; Mandojana, 1990).

## Disorders following ingestion

In certain areas of the world, ingestion of jellyfish as condiments or *hors d'oeuvres* is common (Hsieh et al., 2001). Occasional gastrointestinal complaints such as cramping and abdominal pain can be produced by these delicacies. One woman developed gastrointestinal symptoms and persistent urticaria after ingestion of dry jellyfish (Yaffee, 1968), and another had a ciguatera-like outbreak after eating reconstituted frozen jellyfish products flown from the Orient (Zlotnick et al., 1993).

# Therapy and prevention of stings

## Therapy

Knowledge of the chronological symptoms and their pathogenesis (toxic vs. allergic) should dictate first aid therapy. As in all injuries, maintenance of the vital signs (pulse, respiration and blood pressure) is the essential initial maneuver (Williamson et al., 1996). The stung subject should be reassured, kept warm, quiet and recumbent.

Analgesia should be provided once the patient is stabilized. This therapy has to be systemic (oral or parenterally) because the topical route of drug-containing ointment penetrating the skin is too slow or ineffective. The painful stimulus caused by the venom coated nematocyst thread was delivered instantly through the epidermis into the dermis. Topically administered agents such as anti-inflammatory, anesthetics or corticosteroids penetrate down to that site too slowly and are not effective-even when mixed with penetrants (dimethylsulfoxide) or occluded under plastic wrap. The type of analgesic and its route of administration depends upon the clinical situation but oral agents (ibuprofen, acetylsalicylic aid) are sufficient for most American jellyfish injuries.

Consideration can now be given to removing tentacle debris, presumably containing unfired nemaocysts, from the bather's skin. With the possible exception of *Carukia barnesi* and Irukandji, it is doubtful that removal of unfired nematocysts will reduce the pain of the sting already in progress. However, if properly performed, this maneuver cannot be damaging to the patient. Thus, dousing the skin with copious amounts of various solutions for over 30 s has been advocated. In the U.S., vinegar (a weak acid) has been found to be useful for all jellyfish forms except *Cyanea capillata* (Linnaeus)(lion's mane jellyfish) and *Chrysaora quinquecirrha* (Burnett et al., 1983; Fenner & Fitzpatrick, 1986). For the latter species, a 50:50 baking soda: sea water slurry (an alkali) has been recommended. Once adherent tentacles with unfired nematocysts have been removed, additional pain relief can be achieved by cold packs or warm compresses. Cold appears to alleviate *Physalia physalis* and Australian box-jellyfish stings (Exton, 1988) whereas warmth aids Hawaiian *Carybdea alata* Reynaud injuries and either temperature change can alleviate American *Chrysaora quinquecirrha* injuries. Care must be taken not to over-chill a small patient nor to apply sufficient heat that vasodilatation opens extra avenues for venom entry to the core of the body. Additionally, pain relief can be accomplished non-specifically in many cases with placebos or counter irritation from an abraident. The latter is often used against human cancer pain with transepidermal nerve stimulation (TENS) units which non-specifically stimulate one cutaneous region to 'offset' pain reception from other areas. The efficacy of placebos and counter irritation are mentioned to suggest an explanation for reported anecdotal remedies such as meat tenderizer, sand and papain.

Specific antivenom serum therapy is in its infancy. None are available for jellyfish in American waters. An antivenom is manufactured for stings delivered by one Australian species, but its efficacy and standardization are questioned (Endean & Sizemore, 1988; Sutherland, 1994). Experimental intravenous L verapamil is effective in animals against the cardiotoxicity of several jellyfish venoms but no human data have been obtained (Burnett & Calton, 1983).

Treatment for the long term complications of cnidarian stings is difficult. Special surgery may be required for fat atrophy, gangrene, contractions, ulceration or vascular spasm. Keloids may require corticosteroid injections and pigmentation, topical hydroquinone. The neurological disorders usually remit spontaneously after many months. Irkuandji patients receive narcotics and alpha adrenergic blockers (phentolamine) parenterally. Swelling and venous thrombosis are aided by bed rest, sedation and warm compresses.

*Prevention*

Stings can be prevented by prohibiting tentacles from contacting the skin. Barrier clothing (lycra suits) known as 'stinger suits' are effective; gloves, masks and full wet suits should be worn by divers (Williamson et al., 1988). Barrier creams are theoretically possible. Large thick layers of topical petrolatum (Vaseline R) were helpful in one study (Burnett et al., 1968). Several attempts to add another nematocyst inactivating compound to the topical base have been attempted (Heeger et al., 1992). These investigations have yielded varying degrees of success and are difficult because they must be conducted in environments appropriate for the animal with adequate controls and significant numbers of subjects. One protocol outlining the necessary studies to prove efficacy of such barrier preparations has been published (Burnett et al., 1999). A major difficulty is that there are no data on the human pain threshold for jellyfish stings. That is, what number of nematocysts are required to induce pain? Is it necessary to totally block all stinging to lessen and make the pain tolerable? Can the bather perceive greater pain with greater numbers of nematocysts firing?

Mechanical barriers to jellyfish at the beach are effective. Both nets and air- and water-jet screens have been used in different parts of the world (Schultz & Cargo, 1969). Moveable fine mesh metal fences hanging on floats are successful in Australia (Williamson et al., 1996; Moss & Stark, 1998).

Hopefully by expanding research on barrier nets, clothing and topical preparations, most stings can be prevented. Then, by better understanding the pathogeneses of the jellyfish injuries and, by knowing which animal causes what disorders, definite treatment can be improved to reduce the nuisance and dangers of jellyfish stings.

## References

Adiga, K. M., 1984. Brachial artery spasm as a result of a sting. Med. J. Aust. 1: 181–182.

Al-ibrahim, M. Z., M. Z. Tahir, M. Rustom & H. Shafei, 1995. Jellyfish-venom-induced deep venous thrombosis. Angiology 46: 449–451.

Burnett, H.W. & J. W. Burnett, 1990. Prolonged blurred vision following coelenterate envenomation. Toxicon 28: 731–733.

Burnett, J. W. & G. J. Calton, 1977. The chemistry and toxicology of some venomous pelagic coelenterates. Toxicon 15: 177–196.

Burnett, J. W. & G. J. Calton, 1983. Response of the box-jellyfish (*Chironex fleckeri*) cardiotoxin to intravenous administration of verapamil. Med. J. Aust. 2: 192–194.

8

Burnett, J. W. & G. J. Calton, 1987. Jellyfish envenomation syndromes updated. Ann. Emerg. Med. 9: 1000–1005.

Burnett, J. W., K. P. Hepper & L. Aurelian, 1986. Lymphokine activity in coelenterate envenomation. Toxicon 24: 104–107.

Burnett, J. W., H. Rubinstein & G. J. Calton, 1983. First aid for jellyfish envenomation. Southern Med. J. 76: 870–872.

Burnett, J. W., L. J. Pierce Jr., U. Nawachinda & J. H. Stone, 1968. Studies on sea nettle stings. Arch. Derm. 98: 587–589.

Burnett, J. W., J. E. Purcell, D. B. Learn & T. Meyers, 1999. A protocol to investigate blockade of jellyfish nematocysts by topical agents. Contact Dermatitis 40: 55–56.

Chand, R. P. & K. Sellah, 1984. Reversible parasympathetic dysautonomia following stings contributing to the box-jellyfish (*Chironex fleckeri*). Aust. N. Z. J. Med. 14: 673–675.

Cobbs, C. S., P. K. Gaur, A. J. Russo, J. E. Warnick, G. J. Calton & J. W. Burnett, 1983. Immunosorbent chromatography of sea nettle (*Chrysaora quinquecirrha*) venom and characterization of toxins. Toxicon 21: 385–391.

Dubois, J. W., J. Tanguy & J. W. Burnett, 1983. Ionic channels induced by sea nettle toxin in the nodal membrane. Biophys. J. 42: 199–202.

Endean, R. & D. J. Sizemore, 1988. The effectivenes of antivenom in countering the actions of box-jellyfish (*Chironex fleckeri*) nematocyst toxins in mice. Toxicon 26: 425–431.

Exton, D. R., 1988. Treatment of *Physalia physalis* envenomation. Med. J. Aust. 149: 54.

Fenner, P. J. & P. J. Fitzpatrick, 1986. Experiments with the nematocysts of *Cyanea capillata*. Med. J. Aust. 145: 174.

Fenner, P. J., J. A. Williamson, J. W. Burnett, D. M. Colquhoun, S. Godfrey, K. Gunawardane & W. Murtha, 1988. The 'Irukandji syndrome' and acute pulmonary oedema. Med. J. Aust. 149: 150–156.

Filing-Katz, M. R., 1984. Mononueritis multiplex following jellyfish stings. Ann. Neurol. 15: 213.

Freudenthal, A. R. & P. R. Joseph, 1993. Seabather's eruption. N. Eng. J. Med. 329: 542–544.

Garcia, P. J., R .M. H. Schein, & J. W. Burnett, 1994. Fuliminat hepatic failure from a sea anemone sting. Ann. Int. Med. 120: 665–666.

Glasser, D. B., M. J. Noell , J. W. Burnett, S. S. Kathuria & M. M. Rodrigues, 1992. Ocular jellyfish stings. Ophthalmol 99: 1414–1418.

Guess, H. A., P. L. Saveteer & C. R. Morris, 1982. Hemolysis and acute renal failure following a Portuguese man-o'-war sting. Pediatrics 70: 979–981.

Gunawardane, K. & W. Murtha, 1988. The 'Irunkandji syndrome' and acute pulmonary oedema. Med. J. Aust. 149: 150–154.

Gunn, M. A., 1947. Localized fat atrophy after jellyfish sting. Br. Med. J. 2: 687.

Halstead, B. W., 1965. Phylum Coelenterata. In Poisonous and Venomous Marine Animals of the World, 1. U.S. Government Printing Office, Washington, DC: 297–536.

Heeger, T. H. & U. Mrowietz, 1992. Protection of human skin against jellyfish (*Cyanea capillata*) stings. Mar. Biol. 113: 669–678.

Hsieh, Y-H. P., F-M. Leong & J. Rudloe, 2001. Jellyfish as food. Hydrobiologia 451 (Dev. Hydrobiol. 155): 11–17.

Ingram, D. M., H. J. Sheiner & A. M. Ginsberg, 1992. Mondor's disese of the breast resulting from a jellyfish sting. Med. J. Aust. 157: 836–837.

Kizer, K. W. & M. Piel, 1982. Arterial blood gas changes with bluebottle envenomation. Hawaii Med. J. 41: 193–194.

Kleinhaus, A. L., P. F. Cranefield & J. W. Burnett, 1973. The effects on canine cardiac purkinje fibers of *Chrysaora quinquecirrha* (sea nettle) toxin. Toxicon 11: 341–349.

Kokelj, F. & J. W. Burnett, 1990. Treatment of a pigmented lesion induced by a *Pelagia notiluca* sting. Cutis 46: 62–64.

Kokelj, F., G. Stinco, M. Avian, H. Mianzan & J. W. Burnett, 1995. Cell-mediated sensitization to jellyfish antigens confirmed by positive patch tests to *Olindias sambaquiensis* preparations. J. Amer. Acad. Dermatol. 33: 307–308.

Lin, W. W., C. Y. Lee & J. W. Burnett, 1988. Effect of sea nettle (*Chrysaora quinquecirrha*) venom on isolated rat aorta. Toxicon 26: 1209–1212.

Mandojana, R. M., 1990. Granuloma annulare following a bluebottle jellyfish (*Physalia utriculus*) sting. J. Wilderness Med. 1: 220–224.

Mansson, T., H. Randall, R. M. Mandojana, G.J. Calton & J.W. Burnett, 1985. Recurrent cutaneous jellyfish eruptions without envenomation. Acta derm.-venereol. 65: 72–75.

Maretic, Z., 1986. Cnidarismus nudorum: a new epidemiological and clinical entity. Dermatologica 172: 123–126.

Mathelier-Fusade P. & P. F. Leynadier, 1993. Acquired cold urticaria after jellyfish sting. Contact Dermatitis 29: 273.

Matusow, R. J., 1980. Oral inflammatory responses to a sting from a Portuguese man- o'war. J. Am. Dental Assoc. 100: 73–75.

Miura S., J. W. Burnett & L. Aurelian, 1996. Ultraviolet light and immunity to coelenterate venom. Cutis 57: 201–204.

Moats, W. E., 1992. Case report: fire coral envenomation. J Wilderness Med. 3: 284–287.

Moss, K. & K. Stark, 1988. Stinger resistant enclosures. James Cook University of North Queensland, Department of Civil Engineering Publication, Townsville, Q, Australia.

Olson, C. E., M. G. Heard, C. J. Calton & J. W. Burnett, 1985. Interrelationships between toxins: studies on the cross-reactivity between bacterial or animal toxins and monoclonal antibodies to two jellyfish venoms. Toxicon 23: 307–316.

Pang, K. A. & M. C. Schwartz, 1993. Gullian-Barre syndrome following jellyfish stings (*Pelagia noctiluca*). J. Neurology, Neurosurgery & Psychiatry 56: 1133.

Rapoza, P. A., S. K. West, H. S. Newland & H. R. Taylor, 1986. Ocular jellyfish sting in Chesapeake Bay watermen. Am. J. Ophthalmol. 102: 536–537.

Reed, K. M., B. R. Bronstein & H. P. Baden, 1984. Delayed and persistent cutaneous reactions to coelenterates. J. Am. Acad. Dermatol. 10: 462–466.

Rifkin, J. F. & R. Endean, 1988. Arrangement of accessory cells and nematocysts bearing mastigophores in the tentacle of the cubozoan *Chironex fleckeri*. J. Morph. 195: 103–115.

Schultz, L. P. & D. G. Cargo, 1969. Sea nettle barriers for bathing beaches in upper Chesapeake Bay. Natural Resources Institute University of Maryland, Reference no. 69–58, Solomons, Md: 1–16.

Segura Puertas, L., M. E. Ramos, C. Aramburo, E. P. Heimer de la Cotera & J. W. Burnett, 2001. One *Linuche* mystery solved: all three stages of the coronate schyphomedusa *Linuche ungiuculata* cause seabather's eruption. J. Amer. Acad. Dermatol. 44: 624–628.

Sutherland, S. K., 1994. Antivenom research in Australia. Med. J. Aust. 161: 48–50.

Togias, A. J., J. W. Burnett, A. Kagey-Sobatka & L. M. Lichtenstein, 1985. Anaphylaxis after contact with a jellyfish. J. Allergy Clin. Immunol. 75: 672–675.

Watson, G. M. & D. A. Hesssinger, 1989. Cnidocyte mechanoreceptors are tuned to the movements of swimming prey by chemoreceptors. Science, N.Y. 243: 1589–1591.

Weinberg, S. R., 1988. Reactive arthritis following a sting by a Portuguese man-o'war. J. Fla. Med. Assn. 75: 280–281.

Williamson, J. A., P. J. Fenner, J. W. Burnett & J. Rifkin, 1996. Venomous and Poisonous Marine Animals. Univ. N.S. Wales Press, Sydney.

Williamson, J. A., J. W. Burnett, P. J. Fenner, V. Hach-Wunderle, L. Y. Hoe & K. M. Adiga, 1988a. Acute regional vascular insufficiency after jellyfish envenomation. Med. J. Aust. 149: 697–701.

Wong, D. E., T. L. Meinking, L. B. Rosen, D. Taplan, D. J. Hogan & J. W. Burnett, 1994. Seabather's eruption: clinical, histological and immunological features. J. Am. Acad. Derm. 30: 399–406.

Yaffee, H. S., 1968. A delayed cutaneous reaction following contact with jellyfish. Dermatology International (April–June): 75–77.

Zhang, M.-L. & M. Li, 1988. Study on the jellyfish *Stomolophus nomurai* stings in Beidehei, Med. J. China 68: 449.

Zlotnick, B. A., S. Kintz, D. L. Park & P. S. Auerbach, 1993. Ciguatera poisoning after ingestion of imported jellyfish: diagnostic application of serum immunoassay. Wilderness Environment Med. 6: 288–294.

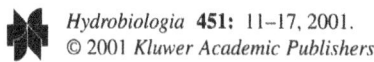
*Hydrobiologia* **451**: 11–17, 2001.
© 2001 *Kluwer Academic Publishers.*

# Jellyfish as food

Y-H. Peggy Hsieh[1], Fui-Ming Leong[2] & Jack Rudloe[3]
[1]*Department of Nutrition & Food Science, Auburn University, Auburn, AL 36849, U.S.A.*
*E-mail: hsiehyp@auburn.edu*
[2]*Perseco Asia Pacific Singapore, Blk 511, Kampong Bahru Road, Unit No 05-03, Keppel Distripark, Singapore 099447*
[3]*Gulf Specimen Marine Laboratories, Inc. P.O. Box 237, Panacea, FL 32346, U.S.A.*

*Key words:* jellyfish, fishery, food, health, arthritis

## Abstract

Jellyfish have been exploited commercially by Chinese as an important food for more than a thousand years. Semi-dried jellyfish represent a multi-million dollar seafood business in Asia. Traditional processing methods involve a multi-phase processing procedure using a mixture of salt (NaCl) and alum ($AlK[SO_4]_2 \cdot 12 H_2O$) to reduce the water content, decrease the pH, and firm the texture. Processed jellyfish have a special crunchy and crispy texture. They are then desalted in water before preparing for consumption. Interest in utilizing *Stomolophus meleagris* L. Agassiz, cannonball jellyfish, from the U. S. as food has increased recently because of high consumer demand in Asia. Desalted ready-to-use (RTU) cannonball jellyfish consists of approximately 95% water and 4–5% protein, which provides a very low caloric value. Cannonball jellyfish collagen has shown a suppressing effect on antigen-induced arthritis in laboratory rats. With the great abundance of cannonball jellyfish in the U. S. coastal waters, turning this jellyfish into value-added products could have tremendous environmental and economicbenefits.

## Introduction

While jellyfish are shunned by swimmers in most places, several species of scyphozoan jellyfish with mild stings are edible. Edible jellyfish are largely estuarine in nature, aggregating around river mouth drainages and primarily caught from the Indian, Northwest Pacific, and Western Central Pacific Oceans by several countries including Thailand, Indonesia, Malaysia, the Philippines and China (Huang, 1988). Among the edible species *Rhopilema esculentum* Kishinouye is the most abundant and important species in the Asian jellyfish fishery which represents a multi million-dollar seafood business in Asia (Omori & Nakano, 2001). China is the first country to process jellyfish for human consumption (Morikawa, 1984). Although the Chinese have been eating jellyfish for more than a thousand years, the jellyfish industry only recently has become a commercial fishery. Other jellyfish producing countries learned the traditional processing techniques from the northern Chinese with slight modific-

ation. Processing jellyfish in Asia is a low-cost operation that requires little capital but is labor intensive.

Distributions of the jellyfish populations are sporadic and seemingly unpredictable in nature. Meterological conditions, currents, water temperature, pressure, salinity and predation may play a significant role in determining the population size (Suelo, 1986). Asian countries are actively developing fisheries management plans in an effort to conserve jellyfish. In both China and Thailand, the government fisheries departments control the jellyfish season (Rudloe, 1992). During the last several weeks of the season, the governments do not allow catching because the jellyfish are largest and are reproducing. Recently, Australia, India and the U. S. began utilizing their available species to produce jellyfish products for export. Jellyfish enters into commerce from all over the world. However, little information has been published on utilization of jellyfish for human consumption. In the present paper, the processing of edible jellyfish, the nutritional and potential medicinal value of cannonball jellyfish, and the future of jellyfish fishery in the U. S. are briefly addressed.

## Jellyfish processing

Fresh jellyfish readily spoil at ambient temperature. Therefore, processing of jellyfish is carried out preferably within a few hours of capture while the animals are still alive. The body of jellyfish consists of a hemispherical transparent umbrella. The mouth is on the undersurface of the umbrella and is protected by fused oral arms, commonly known as 'legs'. The umbrella and oral arms of jellyfish are separated immediately after catching. Jellyfish are cleaned with sea water, scraped to remove mucus membranes and gonadal material. Both umbrella and oral arms are used in processing. Traditional methods of processing involve a step-wise reduction of the water content using salt and alum. A salt mix containing about 10% alum is used for initial salting of jellyfish using about 1 kg salt-alum mix for 8–10 kg of jellyfish (Subasinghe, 1992). Salted jellyfish are then left in the brine for 3–4 days, followed by several transfers to another container salted with fresh mix containing a smaller amount of alum. The salted jellyfish can then be heaped and left dry on a draining rack at room temperature for 2 days, and the heap is turned upside down several times during that period to allow excess water to drain out through compression by its own weight. The entire process requires 20–40 days to produce a salted final product with 60–70% moisture and 16–25% salt (Huang, 1988; Subasinghe, 1992). The processed jellyfish has a yield of about 7–10% of the raw weight depending on the species and processing formula.

Preservation of jellyfish in a mixture of salt and alum is necessary to obtain products of desirable structure and texture. Alum reduces pH, acts as a disinfectant and a hardening agent, giving and maintaining a firm texture by precipitating protein (Huang, 1988). Salt aids in reducing the water content and in keeping the product microbially stable. Salt or alum used singularly in the processing of jellyfish does not produce a product of satisfactory properties (Wootton et al., 1982). Extensive liquidation of the tissue occurs in the absence of salt, while disagreeable odors develop in the absence of alum. In Malaysia and Thailand, a small amount of soda is often added in addition to salt and alum. Addition of soda facilitates water dehydration in the curing solution and increases the crispiness of the cured jellyfish. In China, soda is omitted. Because there are large variations among species, and even with different batches of the same types of jellyfish, processors vary the amounts of alum and soda in the salt from one batch to the next to achieve standardization of product. Processing jellyfish is more of an art than an exact science, hence Asians employ Jellyfish Masters who make adjustments in the amount of salt, alum, soaking periods and compression to obtain the right quality of product (Rudloe, 1992). Exact procedures are often kept as trade secrets. Processing reduces the pH from about 6.6 in fresh cannoball jellyfish to 4.5–4.8 (Table 1). The lowered pH greatly reduces the possibility of microbial growth and enhances the shelf life of the product. The quality of salt affects the ash content in processed jellyfish. Low grade curing chemicals and processing equipment can be the source of heavy metal contamination (Hsieh et al., 1996). Refined curing agents should be used for jellyfish processing.

Cured jellyfish has a special crunchy and crispy texture that makes it unique. The price also depends on the quality of the processed jellyfish as measured by its texture, a combination of tender, elastic and crunchy characteristics, and its colour. The colour of freshly processed jellyfish is creamy white, but gradually turns yellowish as the product ages. In the Asian jellyfish market, a whiter coloured product has higher retail value. Yellow, but not a brown colour, is acceptable. The longer the sample remains around, the darker it becomes. If it remains too long, it turns brown and the product is unacceptable.

The salted jellyfish has a stable shelf life up to 1 year at room temperature. The shelf life can be increased to more than 2 years if the product is kept cool, however, freezing spoils the product, which dries out completely and becomes covered with wrinkles. Prolonged storage at warm ambient temperature may cause a loss of crispness or spoilage of the product. The market price varies considerably depending on the size and condition. The larger the jellyfish, the better is the price. Oral arms product has a lower market value than umbrella due to the irregular shape. A premium Grade A jellyfish, with a wholesale price of $10.00–12.00 per pound in Asia, must be 18 inches in diameter, have a white to creamy colour, and have a crispy texture and be tender at the same time.

*Table 1.* Chemical composition, caloric value, and pH of fresh cannonball jelly-fish, *Stomolophus meleagris*, desalted ready-to-use (RTU) processed cannonball products, and a RTU commercial Malaysian *Rhopilema* product. Numbers are means (± one standard deviation) of 3 determinations from composite samples

| Composition | Fresh cannonball umbrella | RTU Malaysian umbrella | RTU cannonball umbrella | RTU cannonball leg |
|---|---|---|---|---|
| Moisture (%) | 96.10 (0.06) | 95.63 (0.01) | 95.04 (0.04) | 94.08 (0.02) |
| Ash (%) | 1.25 (0.16) | 0.69 (0.00) | 0.33 (0.00) | 0.34 (0.01) |
| Protein (%) | 2.92 (0.04) | 4.13 (0.01) | 4.69 (0.03) | 5.60 (0.02) |
| Fat (%) | < 0.01 | < 0.01 | < 0.01 | < 0.01 |
| Total (%) | 100.27 | 100.45 | 100.06 | 100.02 |
| Cal 100 $g^{-1}$ | 11.68 | 16.52 | 17.84 | 22.4 |
| PH | 6.67 (0.01) | 4.64 (0.01) | 4.46 (0.01) | 4.46 (0.01) |

## Preparation of jellyfish dishes

Jellyfish is more than a gourmet delicacy; it is a tradition. A Chinese wedding or formal banquet is rarely completed without a jellyfish salad. The processed jellyfish should be desalted and rehydrated in water for several hours to overnight before preparing the jellyfish dishes. The desalted ready-to-use (RTU) products have little flavour but are served with sauces or as part of more elaborate dishes. The Chinese have various methods for preparing jellyfish, cooked or un-cooked. They usually are shredded, scalded and served in a dressing composed of oil, soy sauce, vinegar and sugar. Sliced jellyfish may be eaten as a salad mixed with shredded vegetables and/or thinly sliced meat. Japanese prepare cured jellyfish by rinsing in fresh water, cutting into thin strips and served with vinegar as an appetizer (Firth, 1969). In Thailand, cured jelly-fish are sliced into small threads like noodles, washed several times to remove salt and dipped in hot water; the dipped jellyfish is ready for use in various recipes (Soonthonvipat, 1976).

The overnight desalting procedure and preparation of jellyfish dish may become a barrier for modern consumers with busy life styles. This has been overcome by developing shredded RTU products with varieties of flavor and sauces. Recently, shredded jellyfish have appeared on the Japanese market packaged together with condiments such as wasabi or mustard as a convenient ready-to-eat food. The preparation of shredded RTU products will also eliminates the problems related to the size of the jellyfish and increase the utilization of their oral arms.

## Nutrient composition

Not only are they delectable, but jellyfish are a natural diet food. We have analyzed the chemical composition of RTU cannonball jellyfish products and compared with that of a commercial Malaysian *Rhopilema* product and a fresh cannonball jellyfish sample. Results of the chemical composition, pH and calculated caloric value per 100 g of serving are shown in Table 1. RTU jellyfish mainly consists of water and protein. They are low in calories, with no detectable crude fat and cholesterol, and trace amounts of sugar. The average moisture content of RTU samples is in a range of 94.1–95.6% . The tissue can bind a large quantity of water; yet maintain a non-watery, crunchy texture.

Fresh unprocessed jellyfish are rich in minerals such as Na, Ca, K and Mg, but the processed product is depleted of salts after desalting in fresh water (Hsieh et al., 1996). Salt (NaCl) can be completely removed by soaking in clean water. The hardness of water used and the number of water changes during desalting could affect the residual amount of these elements. However, RTU jellyfish contain a significantly higher amount of aluminum than fresh jellyfish (Hsieh et al., 1996). Apparently, aluminum is contributed from the curing agent, alum and remains in the tissues in a bound form. Processing time, temperature and the amount of alum used affect the retention of aluminum in jellyfish tissues (unpublished data).

Except in relatively well-developed gonads during the reproductive cycle, jellyfish contain no visible lipid deposits (Joseph, 1979). Hooper & Ackman (1973) report that the lipid content of whole jellyfish range from 0.0046 to 0.2% on a wet weight basis. In our study, crude fat content in RTU cannonball jellyfish is less than 0.01%. This may be due to the complete removal of mucus and gonads, which contain trace amounts of lipids. According to Hsieh & Rudloe (1994), the cholesterol content of whole fresh jellyfish calculated on a wet weight basis is less than 0.35 mg 100 g$^{-1}$ based on four species of jellyfish, thus, jellyfish can be declared as a fat-free and cholesterol-free food.

Carbohydrate is the other macronutrient that contributes to the caloric value of food. In jellyfish tissue, a trace amount of carbohydrate in the form of sugar is bound to protein as glycoproteins (Kimura et al., 1983). The sum of the percentages of water, lipid, protein, and ash is approximately 100%; therefore, the carbohydrate content in jellyfish is negligible for calorie calculations (Table 1). The calculated caloric value for a normal serving (100 g) of RTU is less than 20 Kcal. Such low caloric value makes jellyfish a natural diet food from the sea.

Protein analysis of salted jellyfish indicate that a Malaysian commercial product and a Chinese product contain an average of 5.5 and 6.8 g of protein per 100 g of salted product, respectively (Huang, 1988). However, the salted product is not the form for consumption. Our results show that percent crude proteins in RTU cannonball jellyfish ranged from 4.7% in umbrellas to 5.6% in legs, values that are higher than the RTU Malaysian commercial product (Table 1). When the combined moisture and ash content increases, protein content decreases. Amino acid analysis shows that tryptophan, a limiting amino acid, is either not detectable or is found in small amounts in jellyfish tissue (Kimura et al., 1983). Thus, the nutritional quality of jellyfish protein quality is low. Glycine accounts for one third of the total amino acid residues with a high proportion of hydroxyproline and hydroxylysine indicating that jellyfish protein is mainly collagen (Barzansky et al., 1975).

## Medicinal value of jellyfish

Jellyfish have long been recognized for medicinal value (Omori, 1981; Hsieh & Rudloe, 1994). It is believed to be an effective cure for arthritis, hypertension, back pain and ulcers, while softening skin and improving digestion. Jellyfish is also alleged to remedy fatigue and exhaustion, stimulate blood flow during the menstrual cycle of women, and ease any type of swelling. Most of these claims regarding the medicinal value of jellyfish are described in non-scientific publications in Chinese. In Korea, jellyfish are promoted on television and in women?s magazines as an aid for weight loss and beautiful skin. Australian aboriginal shamans have prescribed dried jellyfish powder as a treatment for burns (Hsieh & Rudloe, 1994). However, no scientific research has been carried out to document the medicinal efficacy of jellyfish. Collagen has been hypothesized to be the ingredient in jellyfish contributing to the beneficial health effects because collagen is the essential building material of muscle tissue, cartilage and bone, and has a great medicinal promise (Hsieh & Rudloe, 1994).

Recently, we conducted a small scale animal study investigating the health effect of jellyfish collagen on arthritis in rats. Laboratory rats were divided equally into 5 experimental groups with 5 animals in each group. Three groups were fed daily with 10 $\mu$g of jellyfish collagen per rat for 6 days before (pre-JC), after (post-JC), and before and after (pre & post-JC) injection of the arthritis inducing reagent, bovine type II collagen (Wood et al., 1969). A positive control group was induced with arthritis without collagen treatment and a negative control group consisted of normal healthy rats. All animals were examined daily for joint thickness and swelling to evaluate the onset and severity of the disease. An arthritis score was derived by grading the severity of involvement of each paw from 0 to 4 (0=normal, 1=redness, 2=redness plus mild swelling. 3=severe swelling, 4=joint deformity) according to Trentham et al. (1977). Delayed-type hypersensitivity and serum anti-CII antibody levels were also monitored according to Yoshino et al. (1995). The average onset of arthritis in the positive control group was 19 days after induction of the disease, while in the jellyfish-fed groups was delayed to 24–29 days. The average arthritis scores for all jellyfish fed groups were lower than the positive control group. Our results demonstrated that laboratory rats fed with low doses of jellyfish collagen had significantly ($p < 0.05$) reduced incidence, onset and severity of antigen-induced arthritis, a model that shares clinical, histological, immunological and genetic features with human rheumatoid arthritis. Detailed methods and complete results of the study will be published elsewhere.

## Cannonball jellyfish - A new fishery in the United States

*Stomolophus meleagris* or 'cannonball jellyfish' is an

edible rhizostome jellyfish species. Cannonball jelly-fish are small in size; about 127 mm high and 180 mm wide. They are hemispherical, thick, tough and milky-bluish or yellowish in colour, with or without a pale-spotted brown band around the margin. Cannonball jellyfish have no tentacles. Rather, they have 16 short, forked fused oral arms. Jellyfish of this species are of no significant public hazard because the toxin of their stinging cells (nematocysts) is relatively innocuous to humans (Toom & Chan, 1972).

Cannonball jellyfish are frequently seen in coastal waters from North Carolina to Florida and in the north-ern Gulf of Mexico (Mayer, 1910). Each year, from August to December, ranging from Chesapeake Bay to Texas, millions of cannonball jellyfish can be ob-served. When abundant, they become a nuisance to shrimp fishermen by clogging the nets and crushing the shrimp. They sometimes severely damage fishing nets due to their huge volumes and weights (Huang, 1988). One swarm observed at Port Arkansas, Texas, was estimated drifting through the channel at a rate of 2 million per hour (Meinkoth, 1981). Kraeuter & Setzler (1975) reported finding jellyfish of this species from March through October off the coast of Geor-gia. Burke (1976) noted that *S. meleagris* was almost always present in the Mississippi Sound. In South Car-olina, they occur throughout the year in abundance (Calder & Hester, 1978). It has also been recorded from southern California to Ecuador in Pacific and from the Sea of Japan to the South China Sea in the western Pacific (Kramp, 1961; Omori, 1978). The species has also been reported from New England to Brazil in the western Atlantic Ocean (Kramp, 1961; Larson, 1976).

Because cannonball jellyfish are so abundant in the coastal waters of the U.S., they have been con-sidered a nuisance. Turning cannonball jellyfish into value-added products has tremendous environmental and economic benefits. Interest in utilizing cannon-ball jellyfish from the U. S. has increased recently because of high consumer demand in Asia. A fishery in Florida has initiated the first venture of processing cannonball jellyfish since 1992 (Rudloe, 1992). They use the Asian method to clean and cure the jellyfish. Since cannonball jellyfish are much smaller than the Asian species, reduced-time processing methods were developed to produce the salted product (Huang, 1988; Hsieh et al., 1996).

A sensory study was conducted at Auburn Uni-versity to compare the colour, texture and overall preference of laboratory processed cannonball um-

Table 2. Sensory evaluation of jellyfish products from 35 experi-enced panelists who had consumed jellyfish before, and 16 inexperi-enced panelists who had not consumed jellyfish before. A structured 8-point hedonnic scale was used to evaluate the lightness of colour, crunchiness of the texture and overall preference of unflavoured jellyfish product. The higher the score indicates the lighter in colour, crunchier in texture, and more preferred product. Means within rows followed by the same letter are not significantly different at $p < 0.05$

| Sample | Malaysian umbrella | Cannonball umbrella | Cannonball leg |
|---|---|---|---|
| Experienced panelists | | | |
| Colour, lightness | $2.80^b$ | $5.31^a$ | $5.74^a$ |
| Crunchiness | $4.97^b$ | $7.03^a$ | $5.63^b$ |
| Overall preference | $5.06^b$ | $5.80^a$ | $4.97^b$ |
| Inexperienced panelists | | | |
| Colour, lightness | $3.00^b$ | $5.19^a$ | $5.56^a$ |
| Crunchiness | $6.25^a$ | $6.44^a$ | $5.75^a$ |
| Overall preference | $4.88^a$ | $4.88^a$ | $4.06^a$ |

brella and leg products with a commercial Malaysian product (Leong, 1995). The study involved 51 pan-elists including 35 who had consumed jellyfish at least once before, and 16 inexperienced panelists. The three jellyfish samples were desalted overnight, sliced, coded with 3-digit random numbers and presented to the panalists on a white plate at the ambient temperat-ure. An eight-point structured hedonic scale was used to evaluated the lightness of colour, crunchiness and preference of the these unflavoured jellyfish products. Based on the sensory scores, cannonball products that had been stored in a refrigerator for 1 year presented a whiter colour and crunchier texture than the com-mercial one tested (Table 2). Significant differences ($p \leq 0.05$) were found between the cannonball products and the Malaysian sample on the attribute of colour. Both cannonball products were rated lighter in col-our than the commercial product. It is possible that the cannonball products were more recently processed than the commercial product. The low temperature of refrigerated storage also keeps the light colour longer. In terms of texture and overall preference, experienced panelists rated cannonball umbrella as a crunchier product than the Malaysian sample and gave higher score of preference to the cannonball products. Results from inexperienced panelists showed no signi-ficant difference in crunchiness and overall preference among all samples tested. Leg tissue was accepted as well as the umbrella parts of the cannonball jellyfish.

16

Processed jellyfish are in high demand in Asia (Omori & Nakano, 2001). For example, Japan is one of the leading jellyfish consuming countries. More than 10 000 tons of raw jellyfish are caught in 1978 and 1979 in the Ariake Bay; however, domestic production is too small to meet the high consumer demand. In 1980, Japanese imports from countries like Malaysia, China, Indonesia, Burma and the Philippines amounted to more than $40 million (Omori, 1981). On the other hand, jellyfish populations have been unstable or declining in Asian waters due to pollution, over fishing or changing climate (personal comunication), causing Asian dealers to explore new sources of jellyfish. Although the U. S. cannonball jellyfish industry is in its infancy, its future looks promising due to the abundance of these jellyfish in U. S. waters, and the increasing Asian demand. The utilization of this valuable marine resource will potentially reduce the interference of this species with other fisheries and tourists and offer a great economical opportunity for the deprived fishing industry along the Gulf of Mexico (Rudloe, 1992). In the long run, jellyfish may be a valuable health food and collagen source for a world clientele. Greater recognition of the value of jellyfish will enable the U. S. to develop the jellyfish industry for penetration into the world market.

## Future product developments

In spite of their wide commercial availability, jellyfish processing and utilization are not sufficiently studied and reported in the literature. Intensive labor harvesting and traditional manual processing are still used in Malaysia, China, Indonesia and elsewhere. With the trend towards globalization of the jellyfish industry, a cost-effective harvesting design and automated processing are needed to reduce the labor cost and improve production. The establishment of a standardized production for optimum quality from each species will facilitate the quality control of the jellyfish product.

Even though cured jellyfish is a delicacy with high demand in Asia, Westerners are repulsed by the idea of eating them. The jellyfish product is a great seafood alternative that can be sprinkled on salads for extra crunch, prepared as seafood salad, or displayed on the sushi bar. This low calorie seafood product with the potential of being used for treating rheumatoid arthritis or providing other health benefits, could eventually become welcome by Westerners.

The myths of the medicinal value of eating jellyfish should be unveiled by conducting carefully-controlled studies on animal models and human subjects. The preventive and/or therapeutic effects of jellyfish collagen on arthritis needs to be confirmed, and the effective dose range, treatment duration and mechanisms of the suppressing effect should be further investigated in animal models. Carefully controlled preclinical studies would be essential before use of jellyfish collagen as a treatment for rheumatoid arthritis patients. Because jellyfish are a huge untapped resource of collagen, they may find a special niche in the near future for food, clinical and industrial applications.

16

## References

16

Barzansky, B., H. M. Lenhoff & H. Bode, 1975. *Hydra* mesoglea: similarity of its amino acid and neutral sugar composition to that of vertebrate basal lamina. Comp. Biochem. Physiol. 50B: 419–424.

Burke, W. D., 1976. Biology and distribution of the macrocoelenterates of Mississippi Sound and adjacent waters. Gulf Res. Rep. 5: 17–28.

Calder, D. R. & B. S. Hester, 1978. Phylum Cnidaria. In Zingmark, R. G. (ed.), An Annotated Checklist of the Biota of the Coastal Zone of South Carolina. Univ. South Carolina Press, Columbia: 87–93.

Firth, F. E., 1969. The Encyclopedia of Marine Resources. Van Nostrand Reinhold Co., New York: 324–325.

Hooper, S. N. & R. G. Ackman, 1973. Distribution of trans-6-hexadecenoic acid, 7-methyl-7-hexadecenoic acid and common fatty acids in lipids of the ocean sunfish *Mola mola*. Lipids 8: 509–516.

Hsieh, Y-H. P. & J. Rudloe, 1994. Potential of utilizing jellyfish as food in Western countries. Trends Food Sci. Tech. 5: 225–229.

Hsieh, Y-H. P., F-M. Leong & K.W. Barnes, 1996. Inorganic constituents in fresh and processed cannonball jellyfish (*Stomolophus meleagris*). J. Agric. Food Chem. 44: 3117–3119.

Huang, Y. W., 1988. Cannonball jellyfish, *Stomolophus meleagris* as a food resource. J. Food Sci. 53: 341–343.

Joseph, J. D., 1979. Lipid composition of marine and estuarine invertebrates. Porifera and Cnidaria. Prog. Lipid Res. 18: 1–30.

Kimura, S., S. Miura & Y. H. Park, 1983. Collagen as the major edible component of jellyfish (*Stomolophus nomurai*). J. Food Sci. 48: 1758–1760.

Kraeuter, J. N. & E. M. Setzler, 1975. The seasonal cycle of Scyphozoa and Cubozoa in Georgia estuaries. Bull. mar. Sci. 25: 66–74.

Kramp, P. L., 1961. Synopsis of the medusae of the world. J. mar. biol. Ass. U. K. 40: 1–469.

Larson, R. J., 1976. Marine flora and fauna of the northeastern United States. Cnidaria: Scyphozoa. NOAA Tech. Rep. NMFS Circ. 397: 17.

Leong, F-M., 1995. Processing, chemical composition, and quality evaluation of cannonball jellyfish. M.S. Thesis. Auburn University, Alabama, U.S.A.

Mayer, A. G., 1910. Medusae of the World. Volume 3. 'The Scyphomedusae'. Carnegie Institution of Washington, Washington, DC: 711 pp.

Meinkoth, N. A., 1981. The Audubon Society Field Guide to North American Seashore Creatures. Alfred A. Knopf, New York: 364 pp.

Morikawa, T., 1984. Jellyfish. FAO INFOFISH Marketing Digest 1: 37–39.

Omori, M., 1978. Zooplankton fisheries of the world: a review. Mar. Biol. 48: 199–205.

Omori, M., 1981. Edible jellyfish (Scyphomedusae: Rhizostomeae) in Far East waters. A brief review of the biology and fishery. Bull. Plankton Soc. Japan 28: 1–11 (in Japanese).

Omori, M. & E. Nakano, 2001. Jellyfish fisheries in southeast Asia. Hydrobiologia 451 (Dev. Hydrobiol. 155): 19–26.

Rudloe, J., 1992. Jellyfish: A new fishery for the Florida panhandle. A report to the U. S. Department of Commerce Economic Development Administration. EDA Project no. 04-06-03801: 35 pp.

Soonthonvipat, V., 1976. Dried jellyfish. In Tieros, K. (ed.), Fisheries Resources and their Management in South-east Asia. Proc. Int'l. Seminar Nov-Dec, 1974. German Foundation for Int'l. Dev. Bonn: 149–151.

Subasinghe, S., 1992. Jellyfish processing. INFOFISH Int. 4: 63–65.

Suelo, L.G. 1986. Utilization of the Australian jellyfish *Catostylus* sp. as a food product. Ph.D. Thesis. University of New South Wales, Sydney, Australia.

Toom , P. M. & D. S. Chan, 1972. Preliminary studies on nematocysts from the jellyfish *Stomolophus meleagris*. Toxicon 10: 605–610.

Trentham, D. E., A. S. Townes & A. H. Kang, 1977. Autoimmunity to type II collagen: an experimental model of arthritis. J. exp. Med. 146: 857–868.

Wood, F. D., C. M. Pearson & A. Tanaka, 1969. Capacity of mycobacterial wax D and its subfractions to induce adjuvant arthritis in rats. Int. Arch. Allergy Appl. Immunol. 35: 456–467.

Wootton, M., K. A. Buckle & D. Martin, 1982. Studies on the preservation of Australian jellyfish (*Catostylus* spp.). Food Tech. Aust. 34: 398–400.

Yoshino, S., E. Quattrocchi & H. L. Weiner, 1995. Suppression of antigen-induced arthritis in Lewis rats by oral administration of type II collagen. Arthritis & Rheum. 38: 1092–1096.

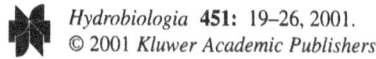
*Hydrobiologia* **451**: 19–26, 2001.
© 2001 *Kluwer Academic Publishers.*

19

# Jellyfish fisheries in southeast Asia

Makoto Omori[1,3] & Eiji Nakano[2]

[1]*Tokyo University of Fisheries, Tokyo 108-8477, Japan*
*E-mail: makomori@amsl.or.jp*
[2]*Sanko Kaisanbutsu Co. Ltd., Kobe 652-0842, Japan*
[3]*Present address: Akajima Marine Science Laboratory 179 Aka, Zamamison, Okinawa 901-3311, Japan*

*Key words:* Rhizostomeae, edible jellyfish, Southeast Asian fisheries, Chinese cooking

## Abstract

A few large jellyfish species in the order Rhizostomeae constitute an important food in Chinese cooking. For more than 1700 years, they have been exploited along the coasts of China. Such jellyfish became an important fishery commodity of Southeast Asian countries in the 1970s with increasing demand from the Japanese market. Recently, Japan has imported 5400–10 000 tons of jellyfish products per year, valued at about 25.5 million US dollars, annually from the Philippines, Vietnam, Thailand, Malaysia, Indonesia, Singapore and Myanmar. Judging from the type names at market and the external appearance of the semi-dried products, the edible jellyfish harvest in Southeast Asia is composed of more than 8 species. They are caught by various kinds of fishing gear including set-nets, drift-nets, hand-nets, scoop-nets, beach-seines and hooks. The fishery is characterized by large fluctuations of the annual catch and a short fishing season that is restricted from two to four months. The average annual catch of jellyfish between 1988 and 1999 in Southeast Asia is estimated to be about 169 000 metric tons in wet weight and the worldwide catch is approximately 321 000 metric tons. Needs for future study on the biology of rhizostome jellyfish are discussed as they relate to understanding population fluctuations.

## Introduction

Some species of large jellyfish are considered to be delicacies for Chinese cooking. Their medicinal value has also been recognized for a long time (South China Sea Institute of Oceanology, 1978; Hsieh & Rudloe, 1994). For over a thousand years, jellyfish have been exploited along the coasts of China. The Chinese philosopher Zhang Hua (232–300 A.D.) described the use of jellyfish as food in the old literature 'Natural History' during the Tsin Dynasty (Wu, 1955). Today, the Japanese are the leading consumers of jellyfish. Since the 1970s, with increasing demand from the Japanese market, jellyfish fishing has become popular in Southeast Asia. More recently, small-scale exploitation of jellyfish has also commenced in other countries such as Australia, India, Mexico, Turkey and the U.S.A.

In Southeast Asia, jellyfish are fished in the Philippines, Vietnam, Malaysia, Thailand, Indonesia, Singapore and Myanmar. One of the present authors (E. N.) has engaged in trade of jellyfish commodity for 27 years. He estimates that about two thirds of the products are exported to Japan and the remainders are sold to South Korea, Taiwan, Singapore, Hong Kong and recently the U.S.A.

In spite of its importance as a commodity, only a little is known about the biology and fishery of edible jellyfish. This is particularly so in Southeast Asia where scientific studies cannot catch up with the rapid development of exploitation. The fishery is characterized by considerable fluctuation in the catch and the good season is restricted to a few months, which vary by locality. This circumstance causes instability of the fishery. The present review compiles available information on the fishery of edible jellyfish in Southeast Asia, in order to call some attention to the ecology of the animals and perhaps to stimulate further development of the jellyfish fishing industry in the region.

*Table 1.* Identified species of edible jellyfish in the world

**Cepheidae**

    *Cephea cephea* (Forskål, 1775)

**Catostylidae**

    *Catostylus mosaicus* (Quoy & Gaimard, 1824)

    *Crambione mastigophora* Maas 1903

    *Crambionella orsisi* (Vanhöffen, 1888)

**Lobonematidae**

    *Lobonema smithii* Mayer, 1910

    *Lobonemoides gracilis* Light, 1914

**Rhizostomatidae**

    *Rhizostoma pulmo* (Macri, 1778)

    *Rhopilema esculentum* Kishinouye, 1891

    *Rhopilema hispidum* (Vanhöffen, 1888)

    *Neopilema nomurai* Kishinouye, 1922

**Stomolophidae**

    *Stomolophus meleagris* L. Agassiz, 1862

## Species

Edible jellyfish all belong to the order Rhizostomeae, in the Scyphomedusae. The bodies of these jellyfish are large, and considerably tough and rigid, with a thick umbrella. At least 11 species in 5 families, i.e. Cepheidae, Catostylidae, Lobonematidae, Rhizostomatidae and Stomolophidae, are known to be exploited worldwide (Table 1). Taxonomic characteristics and geographic distribution of each species can be found in Kramp (1961).

*Cephea cephea* is distributed widely in the Indo-West Pacific from the Red Sea to Touamotu Archipelago. *Catostylus mosaicus, Crambione mastigophora* and *Crambionella orsini* were added rather recently to the list of marketable species. *Catostylus mosaicus* is distributed in the Philippines, New Guinea and west coasts of Australia, and is exploited in New South Wales, Australia. *Crambione mastigophora* occurs in the Malay Archipelago, Java and Truk Island, whereas *Crambionella orsini* is found in the Red Sea, Iranian Gulf and Bengal Bay. *Lobonema smithii* and *Lobonemoides gracilis* are restricted to tropical waters in the Indo-West Pacific. Morphologically these two species are quite similar, and Dr P. Cornelius (pers. comm.) considers that all 'species' of *Lobonema* and *Lobonemoides* to be just one species *Lobonema*

*smithii. Rhizostoma pulmo* is distributed in the Mediterranean, Bay of Biscay, North Sea and Black Sea. According to Dr A. Kideys (pers. comm.), a small amount of this species is commercially fished in Turkish coasts of the Sea of Marmara and Black Sea. *Rhopilema esculentum*, the most expensive species at market, is distributed in the western part of Japan, Po Hai, Yellow Sea and the East and South China Sea (Hon et al., 1978). On the other hand, *Rhopilema hispidum* occurs in warmer waters in the Indo-west Pacific, from the southern part of Japan, southern coasts of China, Philippines, Malaysia and Indonesia to the Indian Ocean and the Red Sea. *Stomolophus meleagris* has been recorded in the southeast Atlantic coast of the U.S.A., Gulf of Mexico, off Baja California and off Panama. On the other hand, *Nemopilema nomurai* is found in the marginal seas of the northwestern Pacific. This species was described by Kishinouye (1922) as a distinct species, but was mistakenly placed in the genus *Stomolophus* by Uchida (1954), and moreover, it was included under the species *S. meleagris* (Kramp, 1961). However, they are not the same species at all, and we propose to refer *S. nomurai* to its original genus (Omori, Kitamura & Cornelius, unpublished). It grows to an enormous size, being sometimes greater than 1 m in diameter and as much as 150 kg. Heavy occurrence of the species along the Japanese coast of the Sea of Japan was reported in 1920, 1958 and 1995 (Kishinouye, 1922; Shimomura, 1959; Yasuda, 1995). In addition to these 11 species, Dr T. Heeger (pers. comm.) recently informed us that he saw fresh *Cassiopea ndrosia* at the market in Carmen, north of Cebu, Philippines, for local consumption.

Because of their large size and difficulties in preservation for taxonomic study, taxonomic specialists have not yet had opportunities to study many specimens of edible jellyfish. Therefore, some species from Southeast Asia have still not been properly identified. Variations in morphological features, size and coloration are considerable, and the taxonomy remains somewhat confused.

Jellyfish are processed with a mixture of salt and alum, and the semi-dried products are marketed as a commodity (Hon et al., 1978; Omori, 1981; Rumpet, 1991; Hsieh & Rudloe, 1994; Hsieh et al., 2001). Judging from the shape of these commodities, the edible jellyfish harvest in Southeast Asia is composed of more than 8 species. Dealers and merchants call the jellyfish at market simply by the following 8 types based on the color, form, texture and size of the semi-dried products.

(1) Red type or China type. Slightly reddish with smooth exumbrella. Umbrella 300-600 mm in diameter. Probably *Rhopilema esculentum*.

(2) White type. Exumbrella whitish with numerous 1-3 cm long, pointed papillae. Umbrella up to 500 mm in diameter. Most certainly *Lobonema smithii*.

(3) Sand type. Exumbrella whitish with numerous small projections and brown spots. Umbrella reaches over 500 mm in diameter. Probably *Rhopilema hispidum*.

(4) River type. Whitish or slightly brown, small; ex-umbrella not smooth. Diameter up to 200 mm. This is often fished in estuaries and near the mouths of rivers.

(5) Cilacap type. Lilac colored when alive; diameter up to 250 mm. There are fine peripheral radial ridges on the exumbrella. This type occurs mainly in Cilacap, East Java.

(6) Prigi type. Reddish purple when alive; diameter up to 400 mm. Peripheral radial ridges are not very clear. This type occurs abundantly in Prigi Bay and Muncar, East Java, and identified in Muncar as *Crambione mastigophora* by Dr M. Toyokawa (pers. comm.).

(7) Semi-China type. Light yellowish; exumbrella smooth. Diameter up to 150 mm. This type is similar to the Red type but smaller.

(8) Ball type. Slightly brownish; umbrella thick and rigid. The wide marginal ridges and grooves of exumbrella are themselves distinctive. Dr P. Cornelius (pers. comm.) identified as *Crambionella orsi*.

## Fishery in Southeast Asia

### History

Chinese immigrants probably first introduced the jellyfish fishery to a few places in Southeast Asia. But, the fishery remained at a small scale before 1970, as Japan imported most of its semi-dried jellyfish from China. Due to instability of production and a rapid rise of the price in China, however, Japanese merchants in Thailand, Malaysia and Indonesia emphasized jellyfish fishing and technical development of the processing in the 1970s. After that, the jellyfish fishery was expanded to sites in the Philippines, Vietnam and Myanmar.

### Fishing methods and behavior of jellyfish

The fishing gear used includes various set-nets, drift-nets, push-nets (scoop-nets), hand nets, beach-seines and weirs. A typical set-net with rectangular mouth is set at a depth of 2–10 m across a tidal current. Drift-nets are also used across the current flow with a system of floats and sinkers. Hand nets, long poles with two-edged iron hooks on top and trawling nets are also commonly used for incidental catch. There is no distinctive difference in fishing methodology applied to different species or locality.

In general, the entire body of the Red and Semi-China types are brought back, but for the White and Sand types, fishermen often cut off the 'leg' portion (mouth-arms) while at the sea and only umbrellas of the medusae are loaded into the boat.

Fishing is carried out during the daytime only, as fishermen search for jellyfish when they appear at the surface of the water. Naturally, weather conditions and tide affect such fishing operations, as jellyfish aggregate at the water surface only when the sea is calm. They occur near the seashore at high tide, and are transported to offshore as the tide starts to recede.

The River type is found mainly in brackish water near the mouth of rivers, whereas the White type occurs in more offshore waters. The Sand type is found in both environments. According to fishermen, the distance of horizontal migration of the Red type is generally greater than other types.

### Fishing grounds and fishing season

Figure 1 shows the main fishing grounds of jellyfish in Southeast Asia. Based on data collected mainly by the authors and some information from previous reports, the fishing season for various types of jellyfish in each location is indicated on Table 2. In Thailand, fishing is seen along the Gulf of Thailand from Rayong to Songkla and around Ranong in the Andaman Sea (Soonthonvipat, 1976). In Malaysia, the fishing grounds are at Telok-Anson and neighbouring waters in the Strait of Malacca and around Kuching and Ulu Kuala Matu in Sarawak (Rumpet, 1991).

There is an inverse relationship between the fishing season and the monsoon. The main fishing season of jellyfish is between March and May and August and November. Fishermen stay at home during the dry season (December and February) when northeast or northwest winds and rough seas prevail. The fishing is also intermittent in some areas during the southwest monsoon (rainy season).

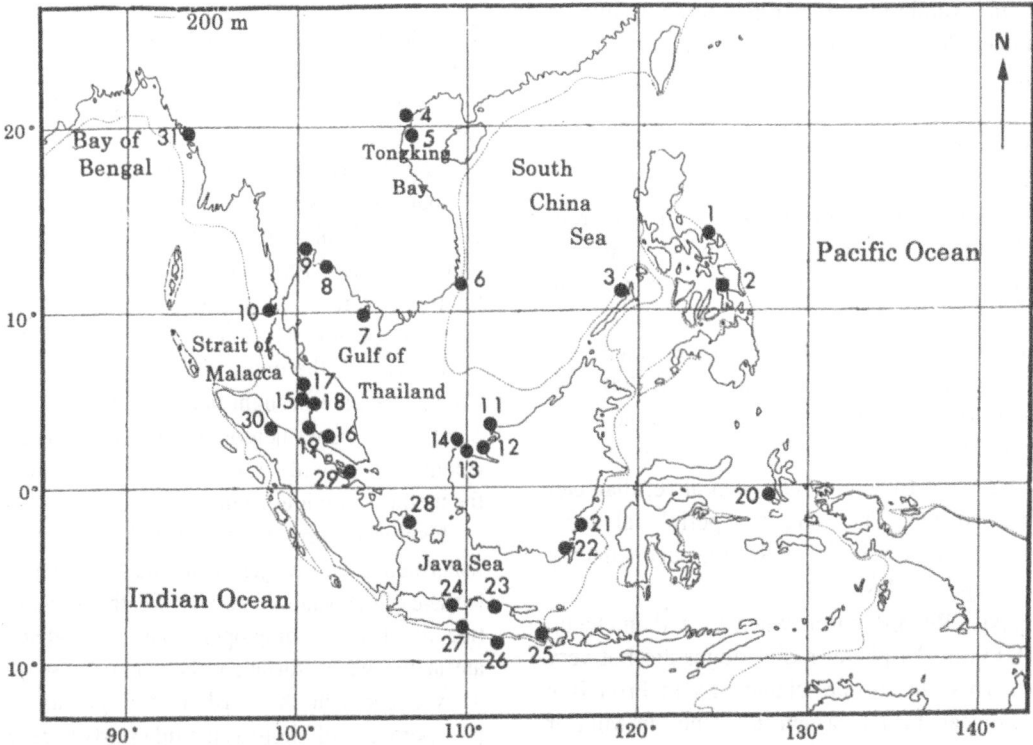

*Figure 1.* Main fishing grounds of jellyfish in Southeast Asia. 1. San Miguel Bay, 2. Carigara Bay, 3. Malampaya Sound, 4. Haiphon, 5. Tongking Bay, 6. Cam Ranh, 7. Phu Quoc Island, 8. Rayong, 9. Samut Sakhon, 10. Ranong, 11. Ulu Kuala Matu, 12. Kabong, 13. Kuching, 14. Sematan, 15. Ipoh, 16. Kuala Lumpur, 17. Penang, 18. Pangkor, 19. Telok Anson, 20. Bacan Island, 21. Balikpepan, 22. Kotabalu, 23. Tuban, 24. Cirebon, 25. Muncar, 26. Prigi, 27. Cilacap, 28. Bangka Island, 29. Tanjung Balai, 30. Medan, 31. Sittwe.

In order to make the analysis simple, we have assumed the peak season of each fishing ground in Table 2 to be the first even-numbered month after fishing starts. Then, 35% of the peaks were in August, 26% in April, and 18% in October. Figure 2 shows comparison between locations of fishing grounds with peak season and the general pattern of the offshore surface currents (Wyrtki, 1961). Most of the fishing grounds are found where water currents have velocities of 12 or 25 cm s$^{-1}$. In the fishing grounds located very near shore, the current velocity must be much lower. The life history of jellyfish and other factors such as fishing method and size of fishing boat may be also responsible for setting the fishing season. In some regions, jellyfish fishing is carried out only during periods when the main offshore fishing for another fish is suspended by unfavorable weather conditions.

The places where great numbers of edible jellyfish occur are characterized by having a large tidal range, shallow depth, semi-enclosed water mass, freshwater inflow through river systems and development of mangrove swamps. Such factors apparently create favourable conditions for settling of polyps and recruitment. In Thailand and Malaysia, fishermen say that the jellyfish catch has decreased after extensive coastal development and cutting of mangrove trees.

**Production and catches**

According to records of the Tokyo Customs House, the amount of semi-dried jellyfish commodities imported annually from Southeast Asia to Japan varied from 5369 to 10 084 metric tons (average 7874 tons per year) during the period from 1988 to 1999 (Table 3). As the total amount imported from China during the same 12-year period averaged 2933 tons, the figures show the current importance of jellyfish production in Southeast Asia. A number of dealers and merchants estimate that the actual catch of jellyfish in Southeast Asia may be approximately 1.5 times larger than the amount exported to Japan. In this connection, for example, the amount of semi-dried products exported from Sarawak varied from 358 to 1510 tons annually

*Table 2.* Main fishing grounds and fishing season of various types of jellyfish in Southeast Asia

| Country | Fishing ground | No. in Figure 1 | Type | Fishing season |
|---|---|---|---|---|
| Philippines | San Miguel Bay (Luzon) | 1 | White | Feb–May |
| | Carigara Bay (Samar and Leyte) | 2 | White | Feb–May |
| | Malampaya Sound and Port Barton (Palawan) | 3 | White | Apr–May |
| Vietnam | Haiphon and Tongking Bay | 4,5 | Sand | July–Sept |
| | Cam Ranh (South China Sea) | 6 | White | Feb–Apr & July–Oct |
| | Phu Quoc Island (Gulf of Thailand) | 7 | White | Oct–Mar |
| Thailand | Rayong and Samut Sakhon (Gulf of Thailand) | 8,9 | White & Sand | May–July |
| | Ranong (Andaman Sea) | 10 | White | Dec–Mar |
| Malaysia | Matu (Sarawak) | 11 | Red | Feb–Apr & Aug–Dec |
| | Kabong, Kuching, and Sematan (South China Sea) | 12,13,14 | White | Mar–Oct |
| | Ipoh and Kuala Lumpur (Strait of Malacca) | 15,16 | White, Red & River | July–Sept |
| | Penang, Pangkor, and Telok Anson (Strait of Malacca) | 17,18,19 | Semi-China | July–Sept |
| Indonesia | Bacan Island (Halmahera) | 20 | Semi-China | July–Aug |
| | Balikpepan and Kotabalu (East Kalimantan, Makassar Strait) | 21,22 | White | Aug–Nov |
| | Tuban (East Java, Java Sea) | 23 | White & Sand | Mar–May & Sept–Nov |
| | Cirebon (West Java, Indian Ocean) | 24 | Sand | Aug–Nov |
| | Muncar and Prigi (West Java, Indian Ocean) | 25,26 | Prigi | July–Nov |
| | Cilacap (West Java, Indian Ocean) | 27 | Cilacap | Aug–Nov |
| | Bangka Island (South Sumatra, Java Sea) | 28 | White | May–Nov |
| | Tanjung Balai and southern coasts (Strait of Malacca) | 29 | River | May–Aug |
| | Medan (North Sumatra) | 30 | White | June–Dec |
| Myanmar | Sittwe (Arakan, Bay of Bengal) | 31 | White & Ball | Mar–May & June–Sept |

during the period from 1980 to 1987 and of that 59–96% (average 83%) was for Japan (Rumpet, 1991). Water content of edible jellyfish is about 95% (measured on *Rhopilema esculantum,* Omori, unpublished) and production yield from fresh jellyfish to the marketed commodity is about 7% of the original weight (Omori, 1981). Thus, we can estimate average annual catch of jellyfish in Southeast Asia to be approximately 169 000 tons in wet weight. This estimate is larger than the statistics indicated by Food and Agriculture Organization of the United Nations (FAO) on their website (1999). In the FAO statistics, the average annual jellyfish catch in Southeast Asia during the period from 1988 to 1997 was 85 697 tons wet weight. Because of the difficulty of data collection and the processing treatments in Southeast Asia, we are convinced that our catch statistics are more realistic than those of the FAO.

According to the FAO statistics, the jellyfish catch in other areas than Southeast Asia (almost all from China) is 152 382 tons in wet weight per year during the same period. Adding up this figure with that in Southeast Asia, we estimate annual catch of jellyfish in the world to be about 321 000 tons in wet weight.

The commercial value of imported commodities from Southeast Asia to Japan varied annually from 1479 to 4113 million JPN yen (average 2733 JPN yen which is equivalent to about 25.5 million US dollars) between 1988 and 1999 (Table 4). This figure is only slightly higher than the total value of jellyfish imported from China, because the Chinese commodity consisted mainly of the most favored species, *Rhopilema esculentum.* In this connection, the price of the umbrella of *R. esculentum* is about 2400 JPN yen $kg^{-1}$, whereas that of whole Southeast Asian (mixed) species is only about 350 JPN yen $kg^{-1}$. The price of

*Table 3.* Amount of jellyfish commodity imported to Japan from 1988 to 1999 (metric tons). Data from the Information Office, Tokyo Customs House. + means less than 0.1%

| Country | 1988 | 1989 | 1990 | 1991 | 1992 | 1993 | 1994 | 1995 | 1996 | 1997 | 1998 | 1999 | Mean | % |
|---|---|---|---|---|---|---|---|---|---|---|---|---|---|---|
| China | 2279.1 | 2650.6 | 2465.3 | 2425.3 | 4805.0 | 5215.6 | 3854.7 | 3015.6 | 2675.9 | 1830.9 | 1808.4 | 2168.3 | 2932.9 | 27.7 |
| North Korea | 0 | 2.8 | 9.5 | 0 | 0 | 0 | 0 | 0 | 0 | 0 | 0 | 0 | 1.2 | + |
| South Korea | 0 | 0 | 0 | 73.5 | 1.0 | 12.0 | 0 | 0 | 0 | 0 | 0 | 0 | 8.7 | + |
| Taiwan | 0.5 | 0 | 7.5 | 0 | 0 | 0 | 0 | 0 | 0 | 14.5 | 9.5 | 0 | 2.3 | + |
| Hong Kong | 5.1 | 0.8 | 0 | 0 | 0 | 0.5 | 0 | 1.3 | 6.0 | 1.2 | 2.8 | 0 | 1.5 | + |
| | | | | | | | | | | | | | | |
| Vietnam | 50.1 | 2.0 | 1.0 | 0 | 88.8 | 40.0 | 254.8 | 25.3 | 216.5 | 103.4 | 99.5 | 78.2 | 80.0 | 0.7 |
| Thailand | 966.2 | 2808.6 | 3698.4 | 5437.0 | 3621.1 | 965.4 | 2860.7 | 1583.6 | 3465.7 | 4232.6 | 1486.8 | 4214.2 | 2945.0 | 27.1 |
| Malaysia | 2802.8 | 1985.7 | 1778.0 | 1472.6 | 1611.5 | 1667.4 | 870.6 | 959.4 | 818.7 | 1865.6 | 732.7 | 1719.6 | 1523.7 | 14.0 |
| Philippines | 151.5 | 0 | 0 | 5.0 | 349.6 | 64.8 | 92.2 | 161.2 | 0 | 65.7 | 0 | 169.9 | 88.3 | 0.8 |
| Indonesia | 5211.1 | 3765.7 | 824.9 | 768.9 | 608.8 | 2256.1 | 2023.9 | 1504.5 | 2156.8 | 2317.8 | 2603.3 | 1094.8 | 2094.8 | 19.3 |
| Myanmar | 196.1 | 290.8 | 128.8 | 196.8 | 797.3 | 2957.3 | 1600.2 | 1812.1 | 3426.3 | 1412.2 | 431.5 | 364.6 | 1134.5 | 10.5 |
| Singapore | 0 | 14.0 | 0 | 13.5 | 14.0 | 0 | 38.5 | 0 | 0 | 0 | 15.5 | 1.8 | 8.0 | + |
| | | | | | | | | | | | | | | |
| SE Asia subtotal | 9377.8 | 8866.8 | 6431.1 | 7893.8 | 7091.1 | 7951.0 | 7740.9 | 6046.1 | 10084.0 | 9997.3 | 5369.3 | 7643.1 | 7874.3 | 72.6 |
| | | | | | | | | | | | | | | |
| India | 10.0 | 10.2 | 0 | 0 | 0 | 0 | 0 | 9.2 | 0 | 26.0 | 69.3 | 135.9 | 21.7 | 0.2 |
| Turkey | 0 | 0 | 0 | 0 | 0 | 0 | 0 | 0 | 4.2 | 0 | 0 | 0 | 0.4 | + |
| Australia | 0 | 0 | 0 | 0 | 0 | 0 | 0 | 0 | 0 | 0.2 | 0 | 0 | - | + |
| U.S.A. | 0 | 0 | 0 | 0 | 0 | 0.5 | 0 | 0 | 20.3 | 4.9 | 0 | 199.9 | 18.8 | 0.2 |
| Mexico | 0 | 0 | 0 | 0 | 0 | 0 | 0 | 0 | 0 | 0 | 0 | 20.6 | 1.7 | + |
| | | | | | | | | | | | | | | |
| Total | | 11672.5 | 11531.2 | 8913.4 | 10392.6 | 11897.1 | 13179.6 | 11595.6 | 9072.2 | 12790.4 | 11875.0 | 7259.3 | 10167.8 | 10863.5 |

*Table 4.* Values of jellyfish commodity imported to Japan from 1988 to 1999 (× million JPN yen). Data from the Information Office, Tokyo Customs House

| Year | China | Southeast Asia | Others | Total |
|---|---|---|---|---|
| 1988 | 2351.0 | 4113.4 | 1.7 | 6466.1 |
| 1989 | 3519.6 | 2560.4 | 2.6 | 6082.6 |
| 1990 | 2831.3 | 2197.2 | 0.0 | 5028.5 |
| 1991 | 2624.6 | 2967.2 | 0.0 | 5591.8 |
| 1992 | 3297.7 | 3119.2 | 0.0 | 6416.9 |
| 1993 | 2278.0 | 2788.6 | 0.0 | 5066.6 |
| 1994 | 2129.0 | 2199.2 | 0.0 | 4328.2 |
| 1995 | 2130.5 | 1648.2 | 2.2 | 3780.9 |
| 1996 | 2771.3 | 3711.4 | 10.8 | 6493.5 |
| 1997 | 2061.1 | 3162.2 | 11.1 | 5234.4 |
| 1998 | 1617.3 | 1478.6 | 14.3 | 3110.2 |
| 1999 | 1766.0 | 2855.3 | 151.6 | 4770.9 |
| | | | | |
| Mean | 2448.1 | 2733.4 | 16.2 | 5197.7 |
| % | 47.1 | 52.6 | 0.3 | |

the mouth-arms is generally less than half that of the umbrella portion.

## Needs for further biological study

In spite of their importance as a fishery commodity, almost nothing is known about the biology and ecology of edible jellyfish in Southeast Asia. Many commercial jellyfish species in the region have no scientific name. At first, therefore, taxonomic study should be carried out with many specimens from various fishing grounds for proper identification of the species involved.

Secondly, we emphasize the need for life history studies. The jellyfish fishery is characterized by considerable fluctuations in catch and the fishing season is restricted to a few months of each year. Unprecedented mass occurrences of rhizostomes sometimes disturb net fishing, while on other occasions they suddenly disappear from fishing grounds. In addition to local weather conditions, certain biological factors

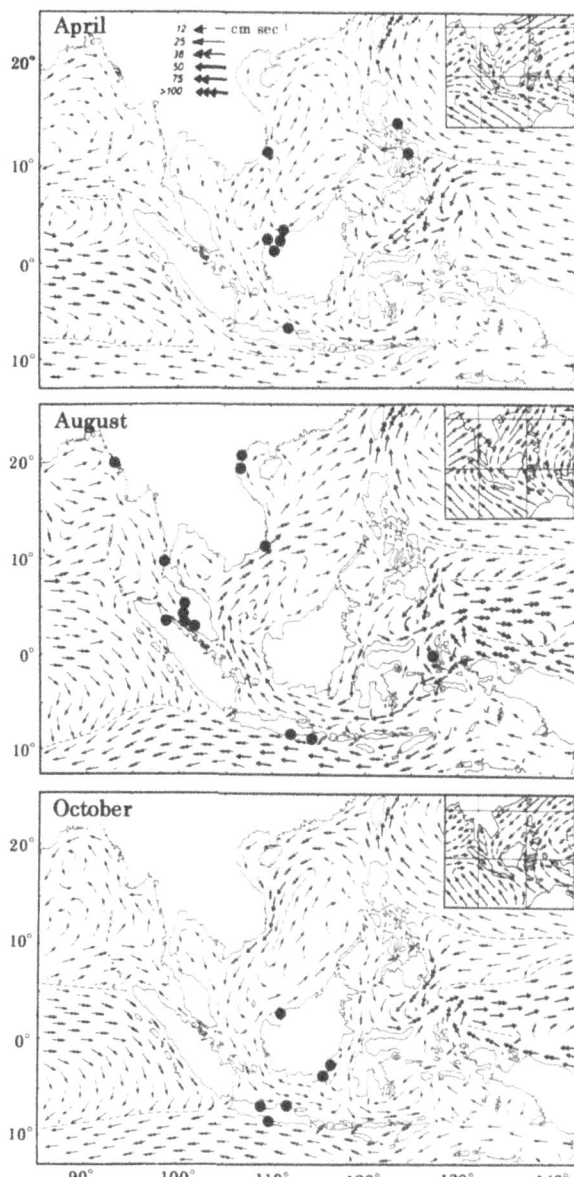

*Figure 2.* Fishing grounds of jellyfish in April, August and October and surface currents in Southeast Asia. Figures of the surface currents are derived from Wyrtki (1961). Small inset-charts on upper right corners show wind distribution over the region.

crucial factors that determine the population size of the 'harvested' medusa stage. Therefore, we particularly emphasize needs for future study on the life of the polyp stage for prediction of the fishery resources and fluctuations.

Only a few studies exist on the growth and feeding of rhizostome jellyfish. According to Ding & Chen (1981), *Rhopilema esculentum* needs 2–3 months from ephyra of 1.5–3.0 mm in diameter to maturity with umbrella 250–450 mm in diameter in the Liaodong Bay, northeast China. In the Ariake Sea, Japan, the same species completes its metagenic life cycle within one year (S. Nakano, 1980, cited by Omori, 1981). The population increased in average diameter of umbrella from 17 mm to 700 mm (from 0.61 g to 27 kg in wet weight) between the middle of May and early September. Thus, the growth exponent, $k$, is 0.09 or, expressed as a daily growth rate, 9% for a medusa of 17 mm umbrella diameter, calculated using the relationship:

$$W_t = W_0 e^{kt},$$

where $W$ is the wet weight and t is the duration at time $t$. This is the only information on growth of any edible jellyfish.

What kind of food supports such large growth efficiency of jellyfish? As rhizostome jellyfish have many suctorial mouths, each with a diameter smaller than 1 mm, they appear to be plankton feeders. Hon et al. (1978) reports that diatoms, ciliates and small planktonic crustaceans are digested extracellularly by *Rhopilema esculentum*. Larson (1991) reports that *Stomolophus meleagris* feed primarily on small zooplankton, with bivalve veligers constituting 63% of the gut contents, followed by tintinnids (9%), copepod nauplii (9%), gastropod veligers (8%) and small numbers of many other zooplankters. Additional research on growth and feeding is clearly needed to understand the biology and ecology of commercial jellyfish.

## Acknowledgements

Mr M. Matsumura gave us much information about jellyfish fisheries in Southeast Asia. Dr P. Cornelius provided useful suggestions on taxonomy of Rhizostomeae. Dr C. Mills read the manuscript and provided valuable comments. We greatly appreciate their assistance.

such as life history, growth and migrations must be involved in these phenomena. Scyphomedusae that have a polyp stage are typically tied to coastal regions where a shallow bottom or floating material can be found, onto which the planulae can settle. Then, when strobilation is completed, ephyrae are liberated and develop into young medusae. We consider that the number of polyps reproduced asexually and the number of ephyral discs liberated from the polyps are

# References

Ding, G.-W & J.-K. Chen, 1981. The life history of *Rhopilema escuenta* Kishinouye. J. Fish. China. 5: 93–102 (in Chinese).

Food and Agriculture Organization of United Nations, 1999. *http//www.fao.org/fi/statist/FISOFT/FISHPLUS.asp*

Hon, H.-C., S.-M. Chang & C.-C. Wang, 1978. Hai tsue (edible jellyfish). Science Publications, Beijing: 70 pp. (in Chinese).

Hsieh, Y-H. & J. Rudloe, 1994. Potential of utilizing jellyfish as food in western countries. Trends Food Sci. Technol. 5: 225–229.

Hsieh, Y-H.P., F.-M. Leong & J. Rudloe, 2001. Jellyfish as food. Hydrobiologia 451 (Dev. Hydrobiol. 155): 11–17.

Kishinouye, K., 1922. Echizen kurage. Doubutsugaku zasshi 34: 343–346 (In Japanese).

Kramp, P. L., 1961. Synopsis of the medusae of the world. J. mar. biol. Ass. U. K. 40: 1–469.

Larson, R. J., 1991. Diet, prey selection and daily ration of *Stomolophus meleagris*, a filter-feeding scyphomedusa from the NE Gulf of Mexico. Estuar. coast. shelf Sci. 32: 511–525.

Omori, M., 1981. Edible jellyfish (Scyphomedusae: Rhizostomeae) in the Far East waters: a brief review of the biology and fishery. Bull. Plankton Soc. Japan 28: 1–11 (in Japanese).

Rumpet, R., 1991. Some aspects of the biology and fishery of jellyfish found along the coast of Sarawak, Malaysia. Dept. of Fisheries, Ministry of Agriculture, Malaysia: 37 pp.

Shimomura, T., 1959. On the unprecedented flourishing of 'Echizen-kurage', *Stomolphus nomurai* (Kishinouye), in the Tsushima warm current regions in autumn, 1958. Bull. Japan Sea Reg. Fish. Res. Lab. 7: 85–107 (in Japanese).

Soonthonvipat, V., 1976. Dried jellyfish. In Kieros, T. (ed.), Seminar Fisheries Resources and their Management in Southeast Asia. German Foundation International Development, Bonn: 149–151.

South China Sea Institute of Oceanology, Academia Sinica, 1978. Medicinal Organisms in the South China Sea. Science Publications, Beijing, 153 pp. (in Chinese).

Uchida, T., 1954. Distribution of scyphomedusae in Japanese and adjacent waters. J. Fac. Sci., Hokkaido Univ. Ser. 6, Zool. 12: 209–219.

Wu, B-L., 1955. Hai-tsue (*Rhopilema esculentum*). Sheng Fu Xie Tong Bao [Biological Newsletter] 4: 35–40 (in Chinese).

Wyrtki, K., 1961. Physical Oceanography of the Southeast Asian Waters. Naga Rep. 2: 1–195.

Yasuda, T., 1995. Mass occurrence of *Stomolophus nomurai* was seen again. Umiushi Tsusin 9: 6–8 (in Japanese).

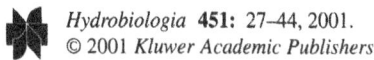
*Hydrobiologia* **451:** 27–44, 2001.
© 2001 *Kluwer Academic Publishers.*

# Interactions of pelagic cnidarians and ctenophores with fish: a review

Jennifer E. Purcell[1] & Mary N. Arai[2]
[1]*University of Maryland Center for Environmental Science, Horn Point Laboratory, P.O. Box 775, Cambridge, MD 21613, U.S.A.*
[2]*Pacific Biological Station, Nanaimo, British Columbia, V9R 5K6 Canada; Department of Biological Sciences, University of Calgary, Alberta, Canada T2N IN41*
*Present address: Shannon Point Marine Center, 1900 Shannon Point Road, Anacortes, WA 98221, U.S.A.*
*E-mail: purcell@hpl.umces.edu*

*Key words:* ichthyoplankton, predation, zooplankton, competition, parasite, commensalism

## Abstract

Medusae, siphonophores and ctenophores (here grouped as 'pelagic coelenterates') interact with fish in several ways. Some interactions are detrimental to fish populations, such as predation by gelatinous species on pelagic eggs and larvae of fish, the potential competition for prey among pelagic coelenterates and fish larvae and zooplanktivorous fish species, and pelagic coelenterates serving as intermediate hosts for fish parasites. Other interactions are positive for fish, such as predation by fish on gelatinous species and commensal associations among fish and pelagic coelenterates. The interactions range from beneficial for the gelatinous species (food, parasite removal), to negative (predation on them). We review existing information and present new data on these topics. Although such interactions have been documented frequently, the significance to either fish or pelagic coelenterate populations is poorly understood. The effects of pelagic coelenterates on fish populations are of particular interest because of the great importance of fisheries to the global economy. As fishing pressures mount, it becomes increasingly important to understand how they may influence the balance between pelagic coelenterates and fish.

## Introduction

The relationships between pelagic coelenterates and fish have been of particular interest because of the potential effects on commercially important fisheries. Although pelagic coelenterates generally are known for being deleterious to fish and fisheries (e.g. Hay et al., 1990), the interactions potentially are both negative and positive. Such interactions include predation on ichthyoplankton by pelagic coelenterates (reviewed in Möller, 1980; Purcell, 1985, 1997; Arai, 1988; Bailey & Houde, 1989), potential competition between pelagic coelenterates and zooplanktivorous fish and fish larvae for prey (op. cit.), predation by fish on medusae and ctenophores (reviewed by Arai, 1988; Ates, 1988; Harbison, 1993), parasite transmission to fish (Lauchner, 1980a,b; Arai, 1988), and commensal associations between fish and medusae (reviewed in Mansueti, 1963; Thiel, 1970, 1978).

Unfortunately, we know little about the significance of most of the above interactions for either fish or pelagic coelenterate populations. The most data exist for pelagic coelenterates as predators on ichthyoplankton. Descriptive information for the other types of interactions usually is without any understanding of the potential importance. Herein, we review existing information on the various interactions among fish and pelagic coelenterates, a term we use to include scyphomedusae, hydromedusae, cubomedusae, siphonophores and ctenophores. We discuss these interactions and the possibility of increasing pelagic coelenterate populations in the context of global changes that are occurring due to commercial fishing activities (Parsons, 1995; Pauly et al., 1998a).

## Predation by pelagic coelenterates on fish eggs and larvae

### Predator types and characteristics

Gelatinous predators of ichthyoplankton range from a few, highly specific predators, like cystonect siphonophores, to many species with broad diets (Purcell, 1997). Species that include high proportions of soft-bodied prey in their diets often eat many fish eggs and larvae when available (Purcell, 1997). The cystonect siphonophores eat mainly larval fish (94–100% of the diets), and other soft-bodied organisms (chaetognaths, small squids) are also eaten by *Physalia physalis* (Linnaeus) (see Purcell, 1984b). The diet of the hydromedusan *Aequorea victoria* (Murback & Shearer) consists almost exclusively of Pacific herring (*Clupea pallasi* Valenciennes) larvae when the larvae first hatch (Purcell, 1989; Purcell & Grover, 1990), and includes a variety of larvae of other fish families (Pleuronectidae, Cottidae, Scorpaenidae, Stichaeidae, Pholidae and Gadidae), as well as pelagic eggs from pleuronectids, in addition to gelatinous and crustacean prey (Purcell, 1989). Diets of semaeostome scyphomedusae, such as *Aurelia aurita* (Linnaeus), *Cyanea capillata* (Linnaeus) and *Chrysaora quinquecirrha* (Desor), may also contain large numbers of ichthyoplankton prey in addition to gelatinous and crustacean prey (Möller, 1980, 1984; Fancett, 1988; Purcell et al., 1994b). For example, gut contents of *C. quinquecirrha* may contain as many as 1497 eggs and 71 larvae medusa$^{-1}$.

Gelatinous predators that consume primarily crustacean prey usually contain few fish eggs or larvae at typical ichthyoplankton densities (reviewed in Purcell, 1997). For example, hydromedusae like *Sarsia tubulosa* (M. Sars), which eat mostly crustaceans (Purcell, 1990), contained few fish larvae (0 in 2100 medusae) at low larval densities ($<0.5$ m$^{-3}$; Van Der Veer, 1985), but contained several Pacific herring larvae (17 in 38 medusae) at high larval densities (mean 632 m$^{-3}$; Purcell, 1990). Some species of cubomedusae frequently eat post-larval fish, while other species eat mainly prawns or copepods, however, the information available is mostly qualitative (Larson, 1976; Hamner et al., 1995; Stewart, 1996). For example, *Carukia barnesi* Southcott eats almost exclusively fish and *Chironex fleckeri* Southcott eats prawns when small, but switches to fish when about 80 mm in bell diameter (J. Seymour, pers. comm.).

Likewise, the incidence of fish larvae in the gut contents of ctenophores is generally low, however, predation on fish eggs by lobate ctenophores can be substantial. Among cydippid species, only 0.06% of 15 000 *Pleurobrachia pileus* (O. F. Müller) contained fish larvae (Van Der Veer, 1985), about 5% of 710 *P. bachei* L. Agassiz contained herring larvae at high densities near hatching (Purcell, 1990). None of 75 lobate *Mnemiopsis leidyi* A. Agassiz from Chesapeake Bay contained bay anchovy larvae, while one-third contained 1–3 eggs (Purcell et al., 1994b). Ichthyoplankton also compose a small proportion of the prey items of ctenophores in the Black Sea. Ichthyoplankton averaged 1% of the prey items of *P. pileus* averaged over six cruises, but the highest percentage (10%) was on one cruise in July, 1992 (Mutlu & Bingle, 1999). Similarly, fish eggs or larvae averaged 0.12% of the prey items in *M. leidyi*, but reached 4% during one (August, 1993) of the six cruises (Mutlu, 1999). Nevertheless, because of their great abundance during the spawning season of pelagic schooling fish, *M. leidyi* can consume substantial proportions of those ichthyoplankton (reviewed by Purcell et al., 2001).

Selection for fish eggs and larvae by gelatinous predators has been positive for every species for which it has been calculated: scyphomedusae *Cyanea capillata, Pseudorhiza haeckeli* Haacke, *Stomolophus meleagris* L. Agassiz, *Chrysaora quinquecirrha* (in Fancett, 1988; Larson, 1991; Purcell et al., 1994b); hydromedusae *Aequorea victoria* (in Purcell, 1989); and cystonect siphonophores (Purcell, 1981b, 1984b). Fish eggs and yolk-sac larvae may be positively selected by the medusae because they have little or no escape ability, and they are large relative to most other zooplankton, thereby increasing encounter rates.

### Factors affecting pelagic coelenterate encounter and capture of ichthyoplankton

Feeding rates of gelatinous species on ichthyoplankton depend on several factors that affect encounter rates between the predator and prey, and subsequent capture. This has been reviewed in Purcell (1985), Arai (1988), Arai (1997) and Purcell (1997), therefore we will discuss the factors briefly as they apply to ichthyoplankton. Thickness and spacing of the tentacles affect encounters of tentaculate predators with prey (Madin, 1988; Purcell, 1997). This is reflected by the general trend of predators having numerous, closely-spaced tentacles eating smaller prey than predators with few, widely-spaced tentacles (op. cit.). This trend holds in general terms, with the first group eating mostly crustacean prey, and the second group including soft-bodied prey, like fish larvae, in the diets.

Some taxa swim while feeding ('cruising', e.g., scyphomedusae, some hydromedusae, lobate ctenophores), and others do not swim while feeding ('ambush', e.g. siphonophores, some hydromedusae, cydippid ctenophores). Cruising predators are predicted to capture predominantly small, slow prey, while ambush predators are predicted to capture mostly larger, faster prey (Gerritsen & Strickler, 1977; Greene, 1986). Those trends hold to a degree among the predators of ichthyoplankton. For example, fish eggs do not swim and are eaten in great numbers by cruising taxa like scyphomedusae, *Aequorea victoria*, and *Mnemiopsis leidyi* (Table 1), but are not eaten much by ambush taxa like siphonophores or *Pleurobrachia bachei* (Purcell, 1981a, 1990; Van Der Veer, 1985). Post-yolksac fish larvae are active swimmers, and are eaten in large numbers by some ambush taxa (cystonect siphonophores), and are not eaten much by some cruising predators (*M. leidyi*). There are obvious exceptions to these trends as well, for example, the cruising scyphomedusae and *A. victoria* capture many fish larvae.

Trends in prey capture by cnidarians are related to the types of nematocysts present, with soft-bodied prey, including fish larvae, being captured by predators such as cystonect siphonophores that have only a few types of nematocysts, which penetrate prey, and hard-bodied prey being captured by predators such as most siphonophores and hydromedusae that have several types of nematocysts, some of which adhere to prey (Purcell, 1984a; Purcell & Mills, 1988). Most of the various predators, such as scyphomedusae, have a mix of nematocysts types, and include both hard and soft-bodied prey in the diets (Purcell, 1997). Chemical stimuli from the prey may also affect nematocyst discharge and the types of prey captured (Purcell & Anderson, 1995).

Characteristics of the prey, such as size, stage and swimming speed, also affect their encounter rates with the predators and abilities to escape (reviewed by Purcell, 1985; Bailey & Houde, 1989). Fish eggs do not swim, small yolk-sac larvae swim weakly and larger post-yolksac larvae swim with increasing speed as they grow (Bailey, 1984; Purcell et al., 1987). Increased size and swimming speed would increase encounter rates with predators, but also increase escapes from the predators. For example, yolksac herring lar-

vae escaped in only 9% of contacts with *Aequorea victoria* tentacles, while post-yolksac larvae escaped from 87% of contacts (Purcell et al., 1987). Most studies show greater feeding on yolksac larvae than on post-yolksac larvae (Bailey, 1984; Purcell et al., 1987), and on small larvae than on large (Cowan & Houde, 1992, 1993), however, the opposite trend was observed for the large hydromedusa *Staurophora mertensi* Brandt feeding on capelin (*Mallotus villosus*) larvae (De Lafontaine & Leggett, 1988). Also, unfed larvae show reduced swimming compared with fed larvae, which reduces encounters and escapes (Bailey, 1984). This is reflected in greater capture of unfed larvae than fed (Bailey, 1984; Purcell et al., 1987).

Several studies incorporate characteristics of both gelatinous predators and ichthyoplankton prey in individual-based models in order to predict encounter and predation rates (Bailey & Batty, 1983; Cowan & Houde, 1992; Cowan et al., 1996; Paradis et al., 1996, 1999; Breitburg et al., 1999). All of these models have been based on the encounter theory of Gerritsen & Strickler (1977). They use experimental measurements of swimming speeds, encounter radius (size), and densities of both predator and prey organisms, and predation rates to determine encounter probabilities. Most of the studies focus on the effects of size on predation. Encounter and predation rates increase with diameter for *Aurelia aurita* medusae (5–21 mm) (Bailey & Batty, 1984). Susceptibility (the probability of capture after contact) of bay anchovy decreases rapidly with larval size (3–9 mm) to both *Mnemiopsis leidyi* ctenophores and *Chrysaora quinquecirrha* medusae, but the percent eaten is relatively constant with larval size (Cowan & Houde, 1992). Predation rate *versus* prey:predator length is maximum when prey size is about 10% of predator size, decreasing linearly at greater percentages from combined data on 5 species of medusae from 9 studies; a dome-shaped relationship resulted for *M. leidyi* ctenophores combined from 3 studies (Paradis et al., 1996). Other studies extend the models to predict survival and growth of ichthyoplankton *in situ* (Cowan et al., 1997; Breitburg et al., 1999).

*Predation rates and effects on ichthyoplankton populations*

Feeding rates of gelatinous species on ichthyoplankton show some important general trends. We use 'feeding rates' to include both 'clearance rates' (liters cleared predator$^{-1}$ d$^{-1}$) and 'predation rates' (numbers of

*Table 1.* Predation rates, clearance rates and predation effects of gelatinous predators feeding on fish eggs and larvae. Medusa sizes are bell diameters, and ctenophore sizes are in ml live volume. Prey consumed are the estimated percentages consumed daily *in situ.*

| Predator species and size | Condition | Prey type and density (No. prey m$^{-3}$) | Prey eaten (No. pred$^{-1}$ d$^{-1}$) | Clearance rates[h] (l pred$^{-1}$ d$^{-1}$) | Prey consumed (% d$^{-1}$) | Source |
|---|---|---|---|---|---|---|
| *Physalia physalis* | Field | ~0.2 larvae m$^{-3a}$ | 120 | 600 000 | 60.0 | Purcell (1984b) |
| *Rhizophysa eysenhardti* | Field | 28 larvae m$^{-3a}$ | 9 | 311 | 28.3 | Purcell (1981b) |
| *Aequorea victoria* | Field | < 10 larvae m$^{-3b}$ | 13±13 | 5650±6114 | 18±29 | Purcell (1989, 1990); Purcell & |
| 33–68 mm | | 10–100 larvae m$^{-3b}$ | 55±48 | 1357±908 | 49±35 | Grover (1990) |
| | | >100 larvae m$^{-3b}$ | 91±47 | 288±210 | 33±32 | |
| *Nemopsis bachei* | | | | | | |
| 0.6 ml | 2.2 m$^3$ | 23–90 eggs m$^{-3c}$ | 4±3 | 72.9$^i$, 40.5 | na | Cowan et al. (1992) |
| *Aurelia aurita* | | | | | | |
| 6–50 mm | Field | larvae$^d$ | 1.6 | na | 2.6–4.4 | Möller (1980) |
| 35–50 mm | 0.27 m$^3$ | larvae 96–3704 m$^{-3e}$ | na | 204±69 | na | De Lafontaine & Leggett (1987) |
| 40–80 mm | 6.3 m$^3$ | larvae 39–410 m$^{-3e}$ | na | 526±173 | na | De Lafontaine & Leggett (1987) |
| 40–85 | 5 m$^3$ | 10–80 larvae m$^{-3d}$ | na | 182 | na | Gamble & Hay (1989) |
| na | 2 m$^3$ | 50 larvae m$^{-3f}$ | na | na | 1.0–5.8$^{g,h}$ | Duffy et al. (1997) |
| *Chrysaora quinquecirrha* | | | | | | |
| 53 ml$^h$, 39 ml | Lab, 750 l | 33–133 eggs m$^{-3g}$ | 110±67 | 1383$^i$, 885 | 20–40 | Cowan & Houde (1993) |
| 100 mm–calc av | 3.0 m$^3$ | 20 eggs m$^{-3g}$ | 68±28$^i$ | 2983±788$^i$ | 20–40 | Cowan & Houde (1993) |
| 40–70 mm | Field | av. 164 eggs m$^{-3g}$ | 343±419 | 2213±1625 | 14±4 | Purcell et al. (1994b) |
| na | 2.0 m$^3$ | 50 larvae m$^{-3f}$ | na | na | 0.1–10.4$^{g,h}$ | Duffy et al. (1997) |
| 36 ml$^h$, 52.5 ml | 3.0 m$^3$ | 10 larvae m$^{-3g}$ | 6$^i$,13 | 1344$^i$, 630 | 20–40 | Cowan & Houde (1993) |
| 40–70 mm | Field | av. 43 larvae m$^{-3g}$ | 86±136 | 1818±1861 | 29±14 | Purcell et al. (1994b) |
| *Cyanea capillata* | | | | | | |
| 40 mm | Lab, 25 l | eggs and larvae$^a$ | na | 140$^i$ | 0.1-2.4 | Fancett & Jenkins (1988) |
| *Pseudorhiza haeckeli* | | | | | | |
| 40 mm | Lab, 25 l | eggs and larvae$^a$ | na | 400$^i$ | 0.1–3.8 | Fancett & Jenkins (1988) |
| *Stomolophus meleagris* | | | | | | |
| 55 mm | Field | eggs$^a$ | na | 3120 | na | Larson (1991) |
| na | 2.0 m$^3$ | 50 larvae m$^{-3f}$ | na | na | 0.3–15.8$^{g,h}$ | Duffy et al. (1997) |
| *Mnemiopsis leidyi* | | | | | | |
| 13 ml | Lab, 200 l | eggs$^g$ | na | 60–170 | 10–65 | Monteleone & Duguay (1998) |
| 15–16 ml | LAb, 750 l | 33–133 eggs m$^{-3g}$ | 10$^i$, 7 | 128$^i$, 82 | 20–40 | Monteleone & Duguay (1988) |
| 3 ml | 2.2 m$^3$ | 45 eggs m$^{-3g}$ | 2–5 | 50±46 | 38 | Cowan et al. (1992) |
| 17 ml | 3.0 m$^3$ | 20 eggs m$^{-3g}$ | 7±1$^i$ | 366±58$^i$ | 20–40 | Cowan & Houde (1993) |
| 40 ml | Field | 224±178 eggs m$^{-3g}$ | 42±33 | 128±58 | 9±14 | Purcell et al. (1994b) |
| 21 ml | 3.0 m$^3$ | 10 larvae m$^{-3g}$ | 1.7$^i$, 0.1 | 172$^i$, 15 | 20–40 | Cowan & Houde (1992, 1993) |
| 40 ml | Field | av. 43 larvae m$^{-3g}$ | 0 | 0 | 0 | Purcell et al. (1994b) |

[a]Mixed species or unidentified; [b]Pacific herring, [c]Black drum, *Pogonias cromis* (Linnaeus); [d]Atlantic herring, *Clupea harengus* Linnaeus; [e]Capelin, *Mallotus villosus* (Müller); [f]Red drum, *Sciaenops ocellatus* Linnaeus, ranges represent 1-2 d old to 12-13 d old larvae; [g]Bay anchovy; [h]calculated from data in source; [i]with no alternative prey; na = not available. Data on *Mnemiopsis leidyi* in the Black Sea are not included here, but are summarized in Purcell et al. (2001).

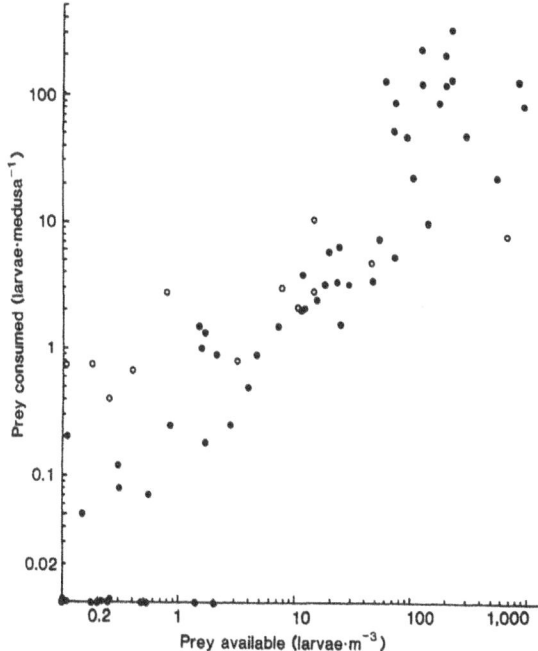

*Figure 1.* Predation rates by the hydromedusan *Aequorea victoria* relative to density of Pacific herring larvae in two locations on Vancouver Island, British Columbia, Canada. Data are compiled from 1983 to 1987. Each point represents averages for several medusae at individual stations. Daytime (○); nighttime (●). Data sources as in Table 2.

common in ocean waters, can make accurate estimates of predation on ichthyoplankton difficult *in situ*.

Experimental conditions greatly affect the feeding rates obtained. Experiments in small containers sometimes appear to overestimate predation, perhaps because the prey become trapped by the predators, and because extremely high prey densities often are used. Predation rates are negatively correlated with container size and experiment duration for medusae and ctenophores (Paradis et al., 1996). Counter to those results, small containers appear to disturb feeding by pelagic coelenterates, and predation rates sometimes increase with container size (Table 1; De Lafontaine & Leggett, 1987; Toonen & Chia, 1993). Nevertheless, those opposing trends lead to the same conclusion, that experiments in small containers do not yield realistic results. Additionally, consumption of ichthyoplankton is much greater in experiments without alternative prey than when zooplankton also is present (Table 1, Cowan & Houde, 1992, 1993). The effects of predator and prey species, sizes, prey density, container size and the presence of alternative prey, which differ among the various studies, make comparisons difficult.

It is possible to compare among predator species and prey types by using clearance rates, indicating the 'volume swept clear', which generally do not vary with prey density at natural levels (Purcell, 1997). Satiation at high larval densities may occur in laboratory experiments and in the field, however, as suggested by the decrease in clearance rates of *Aequorea victoria* with increasing density of large Pacific herring larvae *in situ* (Table 1). Field estimates show that large medusae can clear > 1 $m^3$ of ichthyoplankton daily (Table 1). Individual clearance rates of bay anchovy, *Anchoa mitchilli* (Valenciennes), eggs and larvae for the lobate ctenophore, *Mnemiopsis leidyi*, are about an order of magnitude lower than for the scyphomedusan, *Chrysaora quinquecirrha* in Chesapeake Bay (Table 1). Clearance rates standardized to predator volume are more similar, but still greater for the medusae (21.2±2.8 versus 39.0±6.7 l ml predator$^{-1}$ d$^{-1}$) (Cowan & Houde, 1993). Clearance rates differ dramatically with prey type for any given predator, for example, *C. quinquecirrha* clears only 16 l medusa$^{-1}$ d$^{-1}$ of copepods (Purcell, 1992), but clears hundreds to thousands of liters daily of ichthyoplankton (Table 1). Clearance rates (Table 1) on fish larvae are lower than for fish eggs, which have large diameters and no escape abilities.

prey eaten predator$^{-1}$ d$^{-1}$), and distinguish between the two when appropriate. Increasing size of the predator increases the predation rates (Table 1). For example, small species like *Nemopsis bachei* L. Agassiz (0.6 ml volume) consumed only 4 eggs medusa$^{-1}$ d$^{-1}$, while large species like *Chrysaora quinquecirrha* (≤100 ml) ate 100 or more eggs daily (Table 1). Within a species, feeding rates on ichthyoplankton increase with medusa size, as shown for *Aurelia aurita* (Möller, 1980; Bailey & Batty, 1984; Gamble & Hay, 1989) *Cyanea capillata* (Fancett & Jenkins, 1988), and *C. quinquecirrha* (Purcell et al., 1994b).

In addition to the sizes of pelagic coelenterates and prey, predation rates (prey eaten predator$^{-1}$ d$^{-1}$) of eggs and larvae increase directly with prey density (Bailey & Batty, 1984; De Lafontaine & Leggett, 1987; Fancett & Jenkins, 1988; Purcell et al., 1994b). Large gelatinous species feeding at high densities of ichthyoplankton may eat tens to hundreds of fish eggs and larvae daily (Table 1). For example, predation by the large hydromedusan, *Aequorea victoria*, increased over four orders of magnitude and larval herring densities similarly differed over four orders of magnitude (Fig. 1). Low prey densities, which are

*Table 2.* Consumption of newly-hatched herring larvae by the hydromedusan *Aequorea victoria* on Vancouver Island, British Columbia, Canada. Data are averages of 3 stations in both locations

| Location and date | No. of medusae examined | Larvae in each medusa (No.) | Medusae m$^{-3}$ | Larvae m$^{-3}$ | % larvae eaten d$^{-1}$ |
|---|---|---|---|---|---|
| **Kulleet Bay** | | | | | |
| 22 Mar 1983[a] | 24 | 22.0 | 0.012 | 162.1±284.6 | 0.8 |
| 9 Apr 1984[b] | 12 | 2.0 | 0.35 | 38.5 | 11.1 |
| 14–21 Apr 1985[c] | 322 | 0.2–135.3 | 1.1–5.1 | 3.5–264.1 | 40.5–72.8 |
| 31 Mar 1986[d] | 10 | 3.1 | <0.006 | 19.3 | 0.7 |
| 31 Mar–2 Apr 1987[d] | 33 | 51.3 | 0.02 | 145.6±263.6 | 4.6 |
| | | | | | |
| **Baynes Sound** | | | | | |
| 20–21 Mar 1983[d] | 201 | 0.04–2.4 | 0.01–0.43 | 0.8–29.0 | 2.5–8.9 |
| 6–7 Apr 1984[d] | 79 | 0.06–0.4 | 0.1–0.4 | 43.1–62.5 | 0.4 |
| 30 Mar 1985[d] | 12 | 0.08 | 0.5±0.4 | 3.4±3.7 | 7.7 |
| 30 Mar 1986[d] | 9 | 3.1 | 0.03 | 29.8±14.0 | 2.1 |
| 30 Mar 1987[d] | 13 | 3.2 | <0.0002 | 47.6±30.0 | <0.1 |

[a]Purcell (1989); [b]from Purcell (1990); [c]from Purcell & Grover (1990); [d]Purcell (unpublished data).

Predation effects by pelagic cnidarians on fish larvae often are substantial (>30% d$^{-1}$) in environments where predators are numerous, as for the scyphomedusan *Chrysaora quinquecirrha*, the hydromedusan *Aequorea victoria*, and the siphonophores *Rhizophysa eysenhardti* Gegenbaur and *Physalia physalis* (Purcell, 1981b, 1984b, 1992; Purcell & Grover, 1990). The magnitude of predation depends critically on predator population size, as illustrated for *A. victoria* at two locations in British Columbia, Canada, where medusa densities differed over four orders of magnitude and predation differed from < 0.1 to 73% d$^{-1}$ (Table 2). By contrast, effects of ctenophores on fish larvae are small (e.g. Van Der Veer, 1985; Purcell et al., 1994b, 2001), but lobate ctenophores may have large effects on fish eggs (10–65% d$^{-1}$, Monteleone & Duguay, 1988; 0–40% d$^{-1}$, Cowan & Houde, 1993; Purcell et al., 1994b, 2001). Intense daily predation on ichthyoplankton can have serious consequences because the spawning period of the fish may be limited (e.g. Pacific herring spawn once annually). Inverse correlations in predator and larval fish abundances have been interpreted as predation effects (e.g. Alvariño, 1980; Möller, 1984), however, there can be other explanations for such negative relationships (Frank & Leggett, 1985; Purcell, 1985). Unless supported by more direct evidence, inverse correlations should not be used as indicators of predation effects.

## Competition for food among pelagic coelenterates and fish

The diets of many species of gelatinous predators include mostly copepods, cladocerans, larvaceans and meroplanktonic invertebrate larvae in different proportions depending on predator species and prey availability (summaries in Purcell, 1981a; Purcell & Mills, 1988; Mills, 1995; Arai, 1997; Purcell, 1997), and overlap with the diets of zooplanktivorous fish, such as anchovies, herrings and sardines. Although several authors have speculated on the importance of potential competition among zooplanktivorous pelagic coelenterates and fish (e.g. Möller, 1980; Van Der Veer, 1985; Arai, 1988, 1997; Bailey & Houde, 1989; Shiganova, 1998), very few direct comparisons of diets have been made. Purcell & Grover (1990) compare the diets of post-yolksac larval Pacific herring and those of seven co-occurring hydromedusan species. The prey of larval herring (copepod nauplii and eggs, shelled protozoans and bivalve veligers) were abundant (41 l$^{-1}$) in the environment, and predation on those prey by the gelatinous species was estimated to be only 0.2% d$^{-1}$ of the standing stocks. They conclude that competition for food among the larvae and pelagic coelenterates was not occurring at that time. Consumption of mesozooplankton by the introduced ctenophore *Mnemiopsis leidyi* and zooplanktivorous fish (European anchovy, *Engraulis encrasicolus* (Lin-

*Table 3.* Percent diet similarities (Schoener, 1974) among species of jellyfish and forage fish in PWS. The similarities among mainly crustacean-eating species (top left) and among mainly larvacean-eating species (bottom right) are highlighted (Purcell & Sturdevant, 2001)

| | Percent Diet Similarity (%) | | | |
|---|---|---|---|---|
| | Walleye pollock | Pacific sandlance | Pacific herring | Pink salmon |
| *Aurelia labiata* | 67.2 | 75.1 | 73.3 | 18.7 |
| *Pleurobrachia bachei* | 41.1 | 63.4 | 62 | 5.3 |
| *Cyanea capillata* | 34.8 | 33.7 | 50.2 | 78.1 |
| *Aequorea aequorea* | 55.5 | 35.5 | 48.9 | 59.0 |

naeus), European sprat, *Sprattus sprattus* (Linnaeus) and Mediterranean horse mackerel, *Trachurus mediterraneus* (Steindachner)) in the Black Sea are estimated by Vinogradov et al. (1996). Biomass of zooplankton and fish, and prey consumption by fish were high until 1988, but decreased dramatically during the outbreak period of *M. leidyi* that began in 1989. Competition for food among the ctenophores and fish was inferred.

Purcell & Sturdevant (2001) quantify the diets of four gelatinous species and the zooplanktivorous juveniles of four fish species in Prince William Sound, Alaska. *Aurelia labiata* Chamisso & Eysenhardt and *Pleurobrachia bachei* eat mostly hard-bodied prey, such as small copepods, cladocerans and bivalve veligers, and *Cyanea capillata* and *Aequorea aequorea* var. *albida* Bigelow consume mostly larvaceans, with some crustacean prey. The juvenile fish all consume some of each zooplankton taxon, however, walleye pollock, *Theragra chalcogramma* (Pallas), Pacific sandlance, *Ammodytes hexapterus* Pallus, and Pacific herring eat mainly small copepods, but pink salmon, *Oncorhynchus gorbuscha* (Walbaum) consumed mostly larvaceans. Percent Similarity Index (PSI) values comparing the diets of the gelatinous and juvenile fish show the greatest similarities among crustacean-eating species (*A. labiata*, 67–75%; *P. bachei*, 41–62%) with walleye pollock, Pacific sandlance and Pacific herring), and among larvacean-eating species (*C. capillata*, 78%; *A. aequorea*, 59%) with pink salmon (Table 3). PSI comparisons among all pelagic coelenterate and fish species average 50%.

The potential for competition for zooplankton prey among fish and pelagic coelenterates is very difficult to assess. Determination of all the necessary parameters has not been encompassed by any study, to

our knowledge. Whether competition would occur depends on the extent of the spatial, depth and temporal co-occurrence of the various species, as well as their consumption rates of zooplankton, and the production rates of the zooplankton. In addition, demonstrating competition would require evidence that prey populations are limited by predation.

Generally, estimated predation effects of gelatinous species on copepod populations are too small to cause prey population declines (e.g. <10% d$^{-1}$; Kremer, 1979; Larson, 1987a; Kuipers et al., 1990; Purcell et al., 1994a). Other studies, however, indicate much higher predation and possible reduction of zooplankton standing stocks (e.g. >20% d$^{-1}$; Deason, 1982; Matsakis & Conover, 1991; Purcell, 1992; Olesen et al., 1994; Schneider & Behrends, 1998; Purcell et al., 2001). Inverse relationships in the abundances of gelatinous predators and zooplankton often have been interpreted as resulting from predation (e.g. Feigenbaum & Kelly, 1984; Behrends & Schneider, 1995). Correlations do not show cause and effect and may be misleading, as for mid-Chesapeake Bay, where gelatinous predators were not responsible for the spring decline of copepods in 1987 and 1988 (Purcell et al., 1994a).

## Associations among pelagic coelenterates and fish

Associations of fish species with pelagic coelenterates include no cases that are species-specific and obligate partnerships. Even in the classic example of the man-of-war fish, *Nomeus gronovii* (Gmelin), with the pleustonic siphonophore, *Physalia physalis*, the fish occur with alternative hosts, and the siphonophore with six other fish species. *N. gronovii* remains as an adult in the association, and survives with this extremely toxic host due to a variety of special characteristics (reviewed in Arai, 1988). The fish has at least one antibody to the toxins and can survive high doses, but is not completely immune. It has a complex skin structure that may reduce nematocyst penetration. The fish is very flexible due to an unusually large number of vertebrae, and actively avoids contact with the tentacles. In this association, the fish is known to feed on the siphonophore, as well as be eaten (Jenkins, 1983).

Observations of fish associated with siphonophores other than *Physalia physalis* are rare. Biggs (1976) reports juvenile fish with camouflaging pigmentation sheltered among the tentacles of *Forskalia tholoides* Haeckel, and Robison (1983) reports California smoothtongue, *Leuroglossus stilbius* Gilbert, and a myctophid, *Stenobrachius leucopsarus* (Eigenmann & Eigenmann), associated with midwater siphonophores, *Apolemia* sp.

Few observations of fish associated with ctenophores exist. Several fish species, *Chloroscombrus chrysurus* (Linnaeus), *Hemicaranx amblyrhynchus* (Cuvier), *Palinurichthys* sp., *Syngnathus* sp. and P*eprilus alepidotus* (Linnaeus), are reported with *Beroe* sp. in the Gulf of Mexico (Matthews & Shoemaker, 1952), where larval spadefish, *Chaetodipterus faber* (Broussonet), are seen laying between the comb rows of *Beroe* sp. (A. G. Moss, pers. comm.). Several unidentified myctophid fish have been observed among groups of unidentified ctenophores (Auster et al., 1992).

The most common associations are among juvenile fish and scyphomedusae (reviewed in Mansueti, 1963; Thiel, 1970, 1978; Arai, 1997; Table 4). Although at least 80 species pairs have been documented (op. cit.), those represent a small fraction of the possible combinations of species. These non-specific, facultative associations occur mainly between scyphomedusae and juvenile fish in the families Carangidae, Stromateidae and Gadidae, although some associations with species in at least six additional families are recorded. Several fish species within a genus may associate with pelagic coelenterates. Typically, a given fish species may be found consorting with more than one gelatinous species. For example, ROV observations show up to five age-0 walleye pollock swimming among tentacles of *Cyanea capillata* and up to 30 with *Chrysaora melanaster* Brandt at depths of 30–40 m during the day in Alaskan waters (Brodeur, 1998). Juveniles of several fish species may associate with a species of medusa, either under the swimming bell or near the tentacles of semaeostomes, and often schooling in front of the bell of rhizostomes. The fish seek refuge among the tentacles and oral arms or in the sub-genital pouches when frightened. Several fish typically consort with an individual medusa, however, schools of juvenile walleye pollock were associated with aggregations of *Aurelia labiata* medusae in Alaska (Purcell et al., 2000).

It generally is assumed that through associations with pelagic coelenterates, the fish gain protection from vertebrate predators. Only post-larval fish are seen consorting with pelagic coelenterates, however, presumably because larvae would be eaten. Protection from predators has been demonstrated in only one study, in which juvenile American butterfish, *Peprilus triacanthus* (Peck), were eaten by birds when displaced from their scyphomedusan host, *Cyanea capillata* (in Duffy, 1988). The potential benefits of cover in the pelagic environment are further suggested by the tendencies of small fish to gather around floating plants and debris (Kingsford, 1993).

The nature of the associations among fish and pelagic coelenterates may change as the fish grow. Juvenile harvestfish, *Peprilus alepidotus*, apparently use their medusa hosts initially for protection from their many vertebrate predators (Mansueti, 1963). As they grow, the harvestfish begin to consume parts of the medusae, and possibly to steal food from the medusae. Fish gut contents and observations indicate that many fish species utilize their pelagic coelenterate hosts for food (reviewed in Mansueti, 1963). These associations generally appear to benefit the fish, while being potentially detrimental to the pelagic coelenterates. The pelagic coelenterates may benefit by removal of amphipod parasites, as documented for whiting and *Cyanea capillata*, and by occasionally consuming fish consorts (reviewed in Mansueti, 1963; Thiel, 1970).

The importance of these associations for either the fish or pelagic coelenterates is unknown. There are few systematic, quantitative data on the frequencies or length of associations. In Mississippi Sound, three species of scyphomedusae occur with several known associated fish species, allowing comparisons among species. *Cyanea capillata* medusae associated with butterfish only, while *Chrysaora quinquecirrha* and *Stomolophus meleagris* medusae were associated with three fish species. Atlantic bumpers (*Chloroscombus chrysurus*) were the most abundant fish consorts, and constituted 82–94% of the fish associated with *C. quinquecirrha* and *S. meleagris* (Table 5). We are not aware of any data on the percentages of the daily rations obtained by either the hosts or consorts from their associates. These associations merge into predation upon the pelagic coelenterates (next section), and the effects are further complicated by the transfer of parasites from the intermediate gelatinous hosts to the definitive fish hosts (last section).

## Fish predation on pelagic coelenterates

In addition to predation by pelagic coelenterates on fish, there is also predation by fish on coelenterates. Examination of stomach contents has shown that gelatinous organisms form a portion of the diet of a

*Table 4.* Associations of fish with scyphomedusae. This table is based on new observations of these associations published since 1978. A large amount of earlier literature is reviewed in Mansueti (1963), Thiel (1970) and Thiel (1978).

| Scyphomedusan species | Fish species | Source |
|---|---|---|
| **Order Coronatae** | | |
| *Periphylla periphylla* (Péron & Lesuer) | *Icichthys (= Pseudoicichthys) australis* Haedrich | Pagès et al. (1996) |
| **Order Semaeostomeae** | | |
| *Aurelia aurita* | *Chloroscombrus chrysurus* | Tolley (1987) |
| | *Hemicaranx amblyrhynchus* | Tolley (1987) |
| | *Hippocampus zosterae* Jordan & Gilbert | Tolley (1987) |
| | *Peprilus paru* (Linnaeus) | Tolley (1987) |
| *Aurelia labiata* | *Theragra chalcogramma* | Purcell et al. (2000) |
| *Chrysaora achlyos*[a] | *Peprilus simillimus* (Ayres) | Martin & Kuck (1991) |
| *Chrysaora melanaster* | *Theragra chalcogramma* | Brodeur (1998) |
| | *Zaprora silenus* Jordan | Brodeur (1998) |
| *Chrysaora* sp. | *Pseudocaranx dentax* (Block & Schneider) | Southcott & Glover (1987) |
| *Cyanea capillata* | *Merlangius merlangus* (Linnaeus) | Hay et al. (1990) |
| | *Peprilus triacanthus* | Duffy (1988) |
| | *Theragra chalcogramma* | Brodeur (1998) |
| *Cyanea lamarcki* | *Zaprora silenus* | Brodeur (1998) |
| Péron & Lesueur | *Merlangius merlangus* | Alvariño (1985), Hay et al. (1990) |
| | *Trachurus trachurus* (Linnaeus) | Alvariño (1985) |
| *Cyanea* sp. | *Melanogrammus aeglefinus* (Linnaeus) | Koeller et al. (1986) |
| *Desmonema gaudichaudi* | *Pseudocaranx dentax* | Southcott & Glover (1987) |
| (Lesson) | *Trachurus* spp. | Kingsford (1993) |
| *Drymonema dalmatinum* | *Caranax crysos* Mitchill | Larson (1987) |
| Haeckel | *Chloroscombus chrysurus* | Larson (1987) |
| **Order Rhizostomae** | | |
| *Cassiopea andromeda* (Forsskål) | Unidentified fish | Thiel (1979) |
| *Catostylus mosaicus* | *Trachurus novaezelandiae* (Richardson) | Southcott & Glover (1987) |
| (Quoy & Gaimard) | *Trachurus* spp. | Kingsford (1993) |
| *Cotylorhiza tuberculata* (Macri) | *Trachurus* sp. | Thiel (1979) |
| *Pseudorhiza haeckeli* | *Pseudocaranx dentex* | Southcott & Glover (1987) |
| *Rhopilema nomadica* | *Alepes djedaba* (Forsskål) | Spanier & Galil (1991) |
| *Rhizostoma pulmo* (Macri) | *Merlangius merlangus* | O'Connor & McGrath (1978) |
| *Thysanostoma thysanura* Haeckel | *Caranx* sp. | Thiel (1979) |

[a] Martin, Gershwin, Burnett, Cargo & Bloom.

wide variety of fish species. Arai (1988) and Ates (1988) independently published lists of over 50 fish species that consume pelagic coelenterates. As summarized in Arai (1988), much of the earlier evidence for coelenterates in the diets of fish was circumstantial. As ichthyologists become more aware of the necessity of examining fresh fish stomachs, before most gelatinous material is digested or destroyed by preservation, the numbers of known coelenterate eaters is steadily rising. So far, no fish are known for which pelagic coelenterates are the only prey. Some are primarily coelenterate predators, but even they at least eat the hyperiid amphipods associated with the medusae and often also eat salps (Kashkina, 1986; Mianzan et al., 1997). Most fish that eat pelagic coelenterates have broad diets.

A recent study showing the proportion of fish eating pelagic coelenterates is that of Mianzan et al. (1996). They examined over 25 000 stomachs from 69 fish species on the Argentine continental shelf. The

*Table 5.* Fish species associated with scyphomedusae in the Gulf of Mexico. Numbers are the percentages of each fish species collected by dip net with each jellyfish species, unless otherwise specified (from Tables 1 and 2 in Phillips et al., 1969)

| Fish species | *Chrysaora quinquecirrha* | *Stomolophus meleagris* | *Cyanea capillata* | Totals (No.) | Fish size (mm) |
|---|---|---|---|---|---|
| *Chloroscombrus chrysus* | 88.2 | 93.8 | 0 | 150 | 11–39 |
| *Peprilus alepidotus* | 5.9 | 5.5 | 0 | 9 | 14–38 |
| *Caranx* sp. | 5.9 | 0 | 0 | 2 | 18 |
| *Peprilus burti* Fowler | 0 | 0 | 100 | 30 | 6–34 |
| *Stephanolepis hispidus* (Linnaeus) | 0 | 0.8 | 0 | 1 | 17.5 |
| Total fish (No.) | 34 | 128 | 30 | 192 | – |
| Total jellyfish (No.) | 27 | 191 | 39 | 257 | – |
| Jellyfish with fish (%) | 53 | 10 | 72 | – | – |
| Jellyfish size (mm) | 25–110 | 60–130 | 30–130 | – | – |

abundance of ctenophores, the predominant coelenterate group, was greatest during the spring. During that period, over 28% of the stomachs, belonging to 35% of the fish species, included at least some ctenophores. During the remainder of the year, these values fell to 7–17% of the stomachs, and 15–23% of the fish species. Two stromateoid fish, *Stromateus brasiliensis* Fowler and *Seriolella porosa* Guichenot, fed almost exclusively on ctenophores most of the year, and one scorpaenid fed primarily on ctenophores during the spring and summer.

Because there are no quantitative data on the passage times of gelatinous prey in fish guts, it is not possible to convert data on stomach contents to feeding rates. Nevertheless, it is possible to predict that, unless such digestion rates vary greatly with specific fish or prey, predation by the large numbers of species of fish with generalized diets is more ecologically important than the predation by the relatively small numbers of specialized fishes with primarily gelatinous diets.

Harbison (1993) points out that the suborder Stromateoidei includes the group of fishes for which we have the most extensive evidence of specialized feeding on salps or pelagic coelenterates. As in the Mianzan et al. (1996) study, species of the family Centrolophidae and the family Stromateidae are particularly well documented as extensive coelenterate feeders. The latter family includes such prized human food fishes as the butterfish *Peprilus triacanthus*.

On the other hand, pelagic coelenterates form portions of the diets of a broad range of such common and commercially important species as spiny dogfish, *Squalus acanthias* Linnaeus, chum salmon, *Onco-*

*rhynchus keta* (Walbaum), sablefish, *Anoplopoma fimbria* (Pallas), and various gadoids, scorpaenids and scombrids (Arai, 1988; Brodeur & Pearcy, 1992). Early work on the spiny dogfish rarely found pelagic coelenterates in the diet; however, when measurements were made of fresh specimens at sea, pelagic coelenterates represented 30–40% by volume of the stomach contents (Bowman et al., 1984; McFarlane et al., 1984). More recently, Brodeur & Pearcy (1992) found the By-the-Wind Sailor, *Velella velella*, (Linnaeus) to be a major food item when available off Oregon, and Ellis et al. (1996) found ctenophores in the diet of dogfish in the Irish Sea.

Salmon are the dominant intermediate level predators in the upper 50 m of the North Pacific. There is a good deal of overlap in the generalist diets of the six species, however, chum salmon differ in also utilizing gelatinous prey. As shown by a number of Russian, Japanese, Canadian and American scientists, the diet varies with fish age, and among years and localities from primarily arthropod, to primarily gelatinous. On the high seas, the gut contents of fresh caught chum juveniles is often an amorphous mass of white or red jelly that is difficult to identify but may include the hydromedusan, *Aglantha digitale* (O. F. Müller), and hyperiid amphipods, which are frequent associates of gelatinous animals (Tsuruta, 1963; Davis et al., 1998). Inshore, other medusae and ctenophores may be utilized (Black & Low, 1983; Healey, 1991). This partially gelatinous diet may reduce competition with the other salmon species (Azuma, 1992, 1995; Tadokoro et al., 1996).

In midwater, pelagic coelenterates are found in the diets of lantern fish (family Myctophidae), grenadiers (family Macrouridae), slickheads (family Alepocephalidae) and deep sea smelts (family Bathylagidae) (Arai, 1988; Carrasson & Matallanas, 1990; Beamish et al., 1999). The myctophid, *Stenobrachius leucopsarus*, is the dominant species in the midwater community of the subarctic Pacific. The stomach contents include up to 12% cnidarians and ctenophores depending on location (Beamish et al., 1999). Also in locations of the Pacific Ocean and neighboring seas, for deep-sea smelts such as *Bathylagus ochotensis* Schmidt and *Leuroglossus schmidti* Rass, pelagic coelenterates may form up to 30% and 41%, respectively, of the stomach contents (Beamish et al., 1999).

Fish may show physical adaptations to a gelatinous diet. Harbison (1993) noted that primarily medusivorous fish, such as the stromateoid species, typically have deep bodies and underslung jaws, however, fish that include a significant proportion of pelagic coelenterates in generalized diets may be much more streamlined. For example, the beautifully streamlined chub mackerel, *Scomber japonicus* Houttuyn, of the Pacific or the very similar Atlantic mackerel, *S. scombrus* Linnaeus, can feed either by filter feeding using the gill rakers, or by individual selection of organisms. In the laboratory, Atlantic mackerel select individuals of the hydromedusan, *Aglantha digitale*, in preference to filtering copepods (Runge et al., 1987).

Better correlated than body shape with a gelatinous diet is an enlarged digestive tract. Harbison (1993) summarizes work by Buhler (1930) and later workers indicating the stromateoid fishes have both exceptionally large stomachs and extremely long intestines. The lumpfish, *Cyclopterus lumpus* Linnaeus, which eats the ctenophore *Pleurobrachia* sp., has a large stomach and a narrow moderately long intestine (Eggeling, 1908).

Of interest is the variation in digestive tract structure within the salmon of the North Pacific. In keeping with their partially gelatinous diet, chum salmon have enlarged stomachs compared with the other species of *Oncorhynchus* (Welch, 1997). The stomach is a large bag-like structure that nearly fills the coelomic cavity and is formed of a thick, soft tissue lacking the muscle tone evident in the five other species. The chum stomach is capable of holding approximately 3.5-times the volume of other species at equivalent body lengths. Chum do not have an enlarged or lengthened intestine.

Fish feeding on gelatinous species also may have pharyngeal or oesophageal modifications, presumably to prevent regurgitation. *Mola mola* Linnaeus has three rows of recurved pharyngeal teeth (Suyehiro, 1942). In the stromateoid fishes, there are oesophageal sacs with denticulate papillae (Gilchrist, 1922; Isokawa et al., 1965). *Genicanthus personatus* Randall, the masked angelfish, includes hydromedusae and siphonophores in its diet and has finger-like oesophageal papillae that point posteriorly (Howe, 1993). The oesophagus of chum salmon is strongly muscular with a well-defined sphincter (Azuma, 1992).

Pelagic coelenterates display a variety of characteristics and behaviors that might be effective in reducing fish predation. The stinging nematocysts of cnidarians presumably deter some potential fish predators. Cnidae are released by the scyphomedusan, *Stomolophus meleagris*, when pinched to simulate a predator attack (Shanks & Graham, 1988). These cnidae drive off the associated juvenile fish and could potentially also affect larger fish predators. Some planktonic coelenterates respond to stimulation by 'crumpling' or escape behaviors (Mackie, 1995). It is unclear how crumpling would deter a visual predator, but a vigorous escape response such as that of *Aglantha digitale* might reduce the frequency of capture. There is no direct proof that transparency reduces predation, although its prevalence in gelatinous animals suggests that is of considerable survival value (Johnson & Widder, 1998). Conversely, the loss of transparency upon disturbance in the siphonophore, *Hippopodius hippopus* (Forsskål), has been hypothesized as a predator deterent (Mackie, 1995). Bioluminescence occurs in all known genera of ctenophores, except *Pleurobrachia*, is widespread among pelagic hydrozoans, and is found in some scyphomedusae (Widder et al., 1989; Haddock & Case, 1995; Mackie, 1995; Arai, 1997). Its prevalence and the elaborate displays suggest that bioluminescence is of great value in detering visual predation. While octocorals may contain chemicals which are unplatable or cause vomiting in fish (e.g. Gerhart & Coll, 1993), and hydroid polyps may contain metabolites that are unpalatable (Stachowicz & Lindquist, 1997), similar compounds have not yet been identified in pelagic cnidarians or ctenophores.

Since pelagic coelenterates are utilized not only by the small number of specialist fish feeders but by large numbers of species with more generalized diets, this predation is probably of ecological importance. Arai (1988) summarizes data on the caloric content of pelagic coelenterates and speculates on their dietary value to the fish. As noted above, it is not possible to calculate field based feeding rates of fish from observed stomach contents since digestion rates are lacking. In turn, without field based feeding rates, it is also not possible to quantitatively evaluate the importance of fish predation on pelagic coelenterates in marine food webs. A single paper, Oviatt & Kremer (1977), combines laboratory measurements of feeding rates by American butterfish, *Peprilus triacanthus*, on the ctenophore, *Mnemiopsis leidyi*, with population abundances in Narragansett Bay and estimated possible predation rates of 5–15% $d^{-1}$. These subjects will not advance greatly until digestion rates become available and allow calculation of field based feeding rates.

It is to be expected that if the population of a fish species that feeds on pelagic coelenterates decreases, the prey species may increase in numbers. In spite of the lack of quantitative data on predation rates as discussed above, there are instances where overfishing of a predator is believed to have contributed to increases of pelagic coelenterate populations. Historically Atlantic mackerel, *Scomber scombrus*, did not reproduce in the Black Sea, but there were massive migrations from the Sea of Marmara onto the Northwestern Shelf of the Black Sea in most years. Mackerel had disappeared as a commercial species of the Black Sea by the end of the1960s (Caddy, 1993). As noted above, mackerel are known predators of pelagic coelenterates (Scott, 1914, 1924; Runge et al., 1987). Zaitsev & Polischuk (1984) suggest that the increase of *Aurelia aurita* in the Black Sea in the 1970s and 1980s was due, at least in part, to decreased autumn predation by the mackerel (see also Zaitsev, 1992). In the Adriatic Sea, a population increase of the scyphomedusan, *Pelagia noctiluca*, was correlated with increased catch of several predator fish in the late 1970s (Avian & Rottini-Sandrini, 1988). This led to proposals that overfishing was decreasing these predators and allowing the *P. noctiluca* bloom (Avian & Rottini-Sandrini, 1988; Legović, 1991; Parsons, 1995). However, in this case, Vućetić & Alegria-Hernandez (1988), in a more extensive analysis of similar data, concluded that other factors were increasing the populations of both fish and jellyfish. Overfishing can not explain either the occurrence of a past bloom of *P. noctiluca* in the Adriatic Sea 1910–1914 or its precipitous decline in 1986 following the 1976 bloom (see references cited in Purcell et al., 1999a).

## Pelagic coelenterates as intermediate hosts for parasites of fish

Trematode, cestode and nematode larvae are widely distributed in pelagic Hydrozoa, Scyphozoa and Ctenophora (Lauchner, 1980a, b). It is not known to what extent fish acquire the parasites by eating coelenterates rather than eating other possible intermediate hosts. As knowledge expands on inclusion of gelatinous species in marine fish diets, there is also increasing awareness of the probable role of coelenterates in the transmission of helminth (metazoan worm) parasites to fish (Arai, 1988; Marcogliese, 1995).

Digenetic trematodes are the most thoroughly investigated parasites of pelagic coelenterates and fish. Larval stages of genera such as *Opechona* and *Neopechona* in the family Lepocreadiidae include cercariae, which develop into rediae in gastropods, followed by metacercariae that develop in medusae, ctenophores or other intermediate hosts. Metacercariae have been found in several hydromedusae, such as *Aglantha digitale* (see Koie, 1975; Martorelli, 1996, 2001) and in ctenophores, such as *Pleurobrachia pileus* (see Yip, 1984). The definitive hosts, where the trematode becomes sexually mature, are fish, such as mackerel, which are known to eat pelagic coelenterates. Similar life cycles with mollusc, coelenterate and fish hosts have been demonstrated in the family Hemiuridae for species of *Lecithocladium* (see Koie, 1991), and in the family Fellodistomidae for species of *Monascus* (see Girola et al., 1992; Martorelli & Cremote, 1998).

The rates of natural infections of other zooplankton intermediate hosts with fish parasites are usually very low (Marcogliese, 1995). In the few cases so far investigated, the rates of infections of hydromedusae and ctenophores with trematodes may be higher than for other zooplankton, but vary with location, season or host size. For example, in hydromedusae (1670 specimens of *Phialidium* sp. and 1892 *Liriope tetraphylla* (Chamisso & Eysenhardt)) examined from the Argentine-Uruguayan Common Fishing Zone, the prevalences of *Monascus filiformis* (Rudolphi) were 16–39% and 2–25% respectively, in three collecting zones (Girola et al., 1992). In *Pleurobrachia pileus* from Galway Bay, Ireland, prevalences of *Opechona bacillaris* Mokin showed maxima of up to 50% in early summer but no infection in mid-winter (Yip, 1984). No infections were found in *P. pileus* less than 1 mm in length.

Adult nematodes (Ascaridoidea, Anisakidae) of the genus *Hysterothylacium* are very widely distributed in marine fish. Coelenterates are among the wide variety of planktonic and benthic invertebrates that may serve as intermediate hosts. *Hysterothylacium* larvae have been recorded in hydromedusae, such as *Aglantha digitale* (see Svendsen, 1990) and in the ctenophores, *Pleurobrachia pileus* and *Mnemiopsis leidyi* (see Svendsen, 1990; Koie, 1993; Gaevskaya & Mordvinova, 1993; Mutlu & Bingle, 1999). Although cestodes have been recorded in coelenterates, the larvae are difficult to identify and no life cycles involving coelenterates have been established as yet.

## Discussion

This paper describes what is known about the interactions between pelagic coelenterates and fish. The interactions can be either positive or negative in their potential effects on the commercially important fish populations, which are of particular interest. The negative interactions include predation by pelagic coelenterates on fish eggs and larvae, for which considerable quantitative data exist, potential competition for food, and transmission of parasites from pelagic coelenterates to fish, however, the ultimate effects of these interactions on fish populations are unknown in most cases. On the other hand, fish may benefit from predation on pelagic coelenterates and from commensal relationships between young fish and medusae. Again, data are lacking to evaluate the magnitude and dietary importance of fish predation on pelagic coelenterates, or whether young fish of some species are dependent on medusae for survival.

With the poor understanding of the importance of these interactions, it obviously is difficult to assess the effects of predicted changes in climate, eutrophication or over fishing on pelagic coelenterate populations and their interactions with fish. Climate changes may have various effects. For example, release of medusae by hydroids and scyphistomae depends on light, temperature, salinity, and feeding (Arai, 1992, 1997; Purcell et al., 1999b). These environmental factors could change the timing and abundance of pelagic coelenterates and ichthyoplankton and alter the predation effects (e.g. Table 2). Eutrophication, which may decrease biodiversity but increase some holoplanktonic and estuarine species especially in coastal waters, is discussed by Arai (2001).

Over fishing has dramatically altered fish populations around the world, not only in numbers but also in trophic structure. Fisheries typically first have concentrated on the large, long-lived, piscivorous fish and gradually are turning to small, planktivorous fish (Pauly et al., 1998b). The extent of this concentration on fishing high trophic levels shows wide regional variation (Caddy et al., 1998; Pauly et al., 1998a). Also, the distinction between piscivorous and planktivorous fish is often not clear. Nevertheless, it is interesting to speculate on what effects this fishing may have on the trophic structure of marine ecosystems and particularly on the pelagic coelenterate populations.

Characteristics of gelatinous zooplankton allow them to quickly exploit new resources in changing circumstances (Alldredge, 1983). Many pelagic coelenterates increase their populations very rapidly through a combination of high growth rates, and sexual plus asexual reproduction, whereas fish lack the abilities to asexually multiply and to have several generations in one season. Greve & Parsons (1977) suggest a dichotomy in which two pathways might exist for the transfer of energy up the marine food web leading alternatively to fish or pelagic coelenterates. In waters with low productivity, nanophytoplankton (e.g. small flagellates), small zooplankton and zooplanktivorous ctenophores or medusae would predominate, and alternatively, in highly-productive waters, microphytoplankton (e.g. large diatoms), large zooplankton and zooplanktivorous fish would predominate. This oversimplified dichotomy has been criticized by Longhurst (1985) because pelagic coelenterates also utilize larger particles including fish, by Arai (1988) because pelagic coelenterates are not necessarily the top predators, but are eaten by a variety of predators from chaetognaths to birds, and by Mills (1995) because pelagic coelenterates are also abundant in areas of high productivity.

More recently, Parsons (1992, 1995) summarizes several possible effects of removal of top pelagic fish predators, and suggests that pelagic coelenterates may supplant many commercial fish species as top predators. It is unlikely that many pelagic coelenterates will replace piscivorous fish in marine food webs. Diets primarily of larval or juvenile fish are confined to a few cnidarian species such as cystonect siphonophores and some cubomedusae (Larson, 1976; Purcell, 1997). Most predation on fish eggs and larvae by pelagic coelenterates is by species with broad diets. Most pelagic coelenterates consume primarily zooplankton, and would be expected to compete with zooplanktivorous fish such as anchovies, herring and sardines. When over fishing includes those fish species, there could be significant unconsumed zooplankton, and pelagic coelenterate populations might expand (Caddy, 1993). Even then, the outcome is unclear because many pelagic coelenterate populations can be controlled by predation by other gelatinous species (Purcell, 1997; Purcell et al., 2001).

As discussed above, commercial removal of fish predators of pelagic coelenterates also could allow their populations to increase. Many prized commercial fish, such as chum salmon, mackerel and Atlantic butterfish, consume pelagic coelenterates as well as fish and zooplankton. Again, predation by other pelagic coelenterates might prevent the populations from expanding in response to the removal of fish predators.

It is not possible to predict the final ecosystem balances from over fishing, given the present shortage of information. In combination with the effects of climate change and eutrophication, ecosystem changes may vary widely among locations. It is obvious that further research is essential.

## Acknowledgements

Previously unpublished research from British Columbia was supported by a Department of Fisheries and Oceans, Canada, subventions grant to G. O. Mackie and J. E. Purcell. We also thank D. W. Welch for comments on the manuscript, the library staff at the Pacific Biological Station for assistance with literature, and J. S. Nelson for advice on ichthyological nomenclature. UMCES Contribution No. 3387.

## References

Alldredge, A. L., 1983. The quantitative significance of gelatinous zooplankton as pelagic consumers. In Fasham, M. J. R. (ed.), Flow of Enrergy and Materials in Marine Ecosystems: Theory and Practice. Plenum Press, New York: 407–433.

Alvariño, A., 1980. The relation between the distribution of zooplankton predators and anchovy larvae. CalCOFI Rep. 21: 150–160.

Alvariño, A., 1985. Predation in the plankton realm, mainly with reference to fish larvae. Invest. Mar. CICIMAR 2: 1–122.

Arai, M. N., 1988. Interactions of fish and pelagic coelenterates. Can. J. Zool. 66: 1913–1927.

Arai, M. N., 1992. Active and passive factors affecting aggregations of hydromedusae: a review. Sci. mar. 56: 99–108.

Arai, M. N., 1997. A Functional Biology of Scyphozoa. Chapman & Hall, London, 316 pp.

Arai, M. N., 2001. Pelagic coelenterates and eutrophication: a review. Hydrobiologia 451 (Dev. Hydrobiol. 155): 69–87.

Ates, R. M. L., 1988. Medusivorous fishes, a review. Zool. Med. Leiden 62: 29–42.

Auster, P. J., C. A. Griswold, M. J. Youngbluth & T. G. Bailey, 1992. Aggregations of myctophid fishes with other pelagic fauna. Envir. Biol. Fishes 35: 133–139.

Avian, M. & L. Rottini-Sandrini, 1988. Fishery and swarmings of Pelagia noctiluca in the central and northern Adriatic Sea: middle term analysis. Rapp. Comm. Int. Mer Medit. 31: 231.

Azuma, T., 1992. Diel feeding habits of sockeye and chum salmon in the Bering Sea during the summer. Nippon Suisan Gakkaishi 58: 2019–2025.

Azuma, T., 1995. Biological mechanisms enabling sympatry between salmonids with special reference to sockeye and chum in oceanic waters. Fish. Res. (Amst.) 24: 291–300.

Bailey, K. M., 1984. Comparison of laboratory rates of predation on five species of marine fish larvae by three planktonic invertebrates: effects of larval size on vulnerability. Mar. Biol. 79: 303–309.

Bailey, K. M. & R. S. Batty, 1983. A laboratory study of predation by Aurelia aurita on larval herring (Clupea harengus): experimental observations compared with model predictions. Mar. Biol. 72: 295–301.

Bailey, K. M. & R. S. Batty, 1984. Laboratory study of predation by Aurelia aurita on larvae of cod, flounder, plaice and herring: development and vulnerability to capture. Mar. Biol. 83: 287–291.

Bailey, K. M. & E. D. Houde, 1989. Predation on eggs and larvae of marine fishes and the recruitment problem. Adv. mar. Biol. 25: 1–83.

Beamish, R. J., K. D. Leask, O. A. Ivanov, A. A. Balanov, A. M. Orlov & B. Sinclair, 1999. The ecology, distribution and abundance of midwater fishes of the subarctic Pacific gyres. Prog. Oceanogr. 43: 399–442.

Behrends, G. & G. Schneider, 1995. Impact of Aurelia aurita medusae (Cnidaria, Scyphozoa) on the standing stock and community composition of mesozooplankton in the Kiel bight (western Baltic Sea). Mar. Ecol. Prog. Ser. 127: 39–45.

Biggs, D. C., 1976. Nutritional ecology of Agalma okeni and other siphonophores from the epipelagic western North Atlantic Ocean. Ph.D. Thesis, WHOI-MIT Joint Program: 141 pp.

Black, E. A. & C. J. Low, 1983. Ctenophores in salmon diets. Trans. am. Fish. Soc. 112–728.

Bowman, R., R. Eppi & M. Grosslein, 1984. Diet and consumption of spiny dogfish in the Northwest Atlantic. ICES (Int. Counc. Explor. Sea) cm 1984/G:27.

Breitburg, D. L., K. A. Rose & J. H. Cowan, Jr., 1999. Linking water quality to larval survival: predation mortality of fish larvae in an oxygen-stratified water column. Mar. Ecol. Prog. Ser. 178: 39–54.

Brodeur, R. D., 1998. In situ observations of the association between juvenile fishes and scyphomedusae in the Bering Sea. Mar. Ecol. Prog. Ser. 163: 11–20.

Brodeur, R. D. & W. G. Pearcy, 1992. Effects of environmental variability on trophic interactions and food web structure in a pelagic upwelling ecosystem. Mar. Ecol. Prog. Ser. 84: 101–110.

Bühler, H., 1930. Die Verdauungsorgane der Stromateidae (Teleost). Zeitschr. Morphol. Okolog. Tiere 19: 59–115.

Caddy, J. F., 1993. Toward a comparative evaluation of human impacts on fishery ecosystems of enclosed and semi-enclosed seas. Rev. Fish. Sci. 1: 57–95.

Caddy, J. F., J. Csirke, S. M. Garcia & R. J. R. Grainger, 1998. How pervasive is "Fishing down marine food webs"? Science 282: 1383a.

Carrasson, M. & J. Matallanas, 1990. Preliminary data about the feeding habits of some deep-sea Mediterranean fishes. J. Fish. Biol. 36: 461–463.

Cowan, J. H., Jr. & E. D. Houde, 1992. Size-dependent predation on marine fish larvae by ctenophores, scyphomedusae and planktivorous fish. Fish. Oceanogr. 1: 113–126.

Cowan, J. H., Jr. & E. D. Houde, 1993. Relative predation potentials of scyphomedusae, ctenophores and planktivorous fish on ichthyoplankton in Chesapeake Bay. Mar. Ecol. Prog. Ser. 95: 55–65.

Cowan, J. H., Jr., E. D. Houde & K. A. Rose, 1996. Size-dependent vulnerability of marine fish larvae to predation: an individual-based numerical experiment. ICES J. mar. Sci. 53: 23–37.

Cowan, J. H., Jr., K. A. Rose & E. D. Houde, 1997. Size-based foraging success and vulnerability to predation: selection of survivors in individual-based models of larval fish populations. In Chambers, R. C. & E. A. Trippel (eds), Early Life History and Recruitment in fish populations. Chapman & Hall, London: 357–386.

Cowan, J. H., Jr., R. S. Birdsong, E. D. Houde, J. S. Priest, W. C. Sharp & G. B. Mateja, 1992. Enclosure experiments on survival and growth of black drum eggs and larvae in lower Chesapeake Bay. Estuaries 15: 392–402.

Davis, N. D., K. W. Myers & Y. Ishida, 1998. Caloric values of high-seas salmon prey organisms and simulated salmon ocean growth and prey consumption. NPAFC (North Pacific Anadromous Fish Commission) Bull. 1: 146–162.

Deason, E. E., 1982. Mnemiopsis leidyi (Ctenophora) in Narragansett Bay, 1975–79: abundance, size composition and estimation of grazing. Estuar. coast. shelf Sci. 15: 121–134.

De Lafontaine, Y. & W. C. Leggett, 1987. Effect of container size on estimates of mortality and predation rates in experiments with macrozooplankton and larval fish. Can. J. Fish. aquat. Sci. 44: 1534–1543.

De Lafontaine, Y. & W. C. Leggett, 1988. Predation by jellyfish on larval fish: an experimental evaluation employing in situ enclosures. Can. J. Fish. aquat. Sci. 445: 1173–1190.

Duffy, D. C., 1988. Predator-prey interactions between common terns and butterfish. Ornis Scand 19: 160–163.

Duffy, J. T. C. E. Epifanio & L. A. Fuiman, 1997. Mortality rates imposed by three scyphozoans on red drum (Linnaeus) larvae in field enclosures. J. exp. mar. Biol. Ecol. 212: 123–131.

Eggeling, H., 1908. Dünndarmrelief und Ernährung bei Knochenfischen. Jena. Z. Naturwiss. 43: 417–529.

Ellis, J. R., M. G. Pawson & S. E. Shackley, 1996. The comparative feeding ecology of six species of shark and four species of ray (Elasmobranchii) in the North-east Atlantic. J. mar. biol. Ass. U. K. 76: 89–106.

Fancett, M. S., 1988. Diet and prey selectivity of scyphomedusae from Port Phillip Bay, Australia. Mar. Biol. 98: 503–509.

Fancett, M. S. & G. P. Jenkins, 1988. Predatory impact of scyphomedusae on ichthyoplankton and other zooplankton in Port Phillip Bay. J. exp. mar. Biol. Ecol. 116: 63–77.

Feigenbaum, D. & M. Kelly, 1984. Changes in the lower Chesapeake Bay food chain in presence of the sea nettle Chrysaora quinquecirrha (Scyphomedusa). Mar. Ecol. Prog. Ser. 19: 39–47.

Frank, K. T. & W. C. Leggett, 1985. Reciprocal oscillations in densities of larval fish and potential predators: a reflection of present or past predation? Can. J. Fish. aquat. Sci. 42: 1841–1849.

Gamble, J. C. & S. J. Hay, 1989. Predation by the scyphomedusan Aurelia aurita on herring larvae in large enclosures: effects of predator size and prey starvation. Rapp. P.-v. Réun. Cons. Int. Explor. Mer 191: 366–375.

Gaevskaya, A. V. & T. N. Mordvinova, 1993. Parasitism of nematode larvae in Mnemiopsis mccradyi in the Black Sea. Gidrobiol. Zh. 29: 104–105.

Gerhart, D. J. & J. C. Coll, 1993. Pukalide, a widely distributed octocoral diterpenoid, induces vomiting in fish. J. Chem. Ecol. 19: 2697–2704.

Gerritsen, J. & J. R. Strickler, 1977. Encounter probabilities and community structure in zooplankton: a mathematical model. J. Fish. Res. Bd Can. 34: 73–82.

Gilchrist, J. D. F., 1922. Note on the oesophageal teeth of the Stromateidae. Ann. Mag. Nat. Hist. Ser. 9 (9): 249–255.

Girola, C. V., S. R. Martorelli & N. H. Sardella, 1992. Presencia de metacercarias de Monascus filiformis (Digenea, Fellodistomidae) en hidromedusas del Océano Atlántico Sur. Rev. Chilena Hist. Nat. 65: 409–415.

Greene, C. H., 1986. Patterns of prey selection: Implications of predator foraging tactics. Am. Nat. 128: 824–839.

Greve, W. & T. R. Parsons, 1977. Photosynthesis and fish production: hypothetical effects of climatic change and pollution. Helgol. wiss. Meeresunters. 30: 666–672.

Haddock, S. H. D. & J. F. Case, 1995. Not all ctenophores are bioluminescent: Pleurobrachia. Biol. Bull. 189: 356–362.

Hamner, W. M., M. S. Jones & P. P. Hamner, 1995. Swimming, feeding, circulation and vision in the Australian blox jellyfish Chironex fleckeri (Cnidaria: Cubozoa). Aust. J. mar. Freshwat. Res. 46: 985–990.

Harbison, G. R., 1993. The potential of fishes for the control of gelatinous zooplankton. ICES (Int. Counc. Explor. Sea) CM S(1993/L:74): 1–10.

Hay, S. J., J. R. G. Hislop & A. M. Shanks, 1990. North Sea scyphomedusae: summer distribution, estimated biomass and significance particularly for 0-group gadoid fish. Neth. J. Sea Res. 25: 113–130.

Healey, M. C., 1991. Diets and feeding rates of juvenile pink, chum and sockeye salmon in Hecate Strait, British Columbia. Trans. am. Fish. Soc. 120: 303–318.

Howe, J. C., 1993. A comparative analysis of the feeding apparatus in pomacanthids, with special emphasis of oesiphageal papillae in Genicanthus personatus. J. Fish Biol. 43: 593–602.

Isokawa, S., K. Kubota, T. Kosakai, I. Satomura, M. Tsubouchi & A. Sera, 1965. Some contributions to study of esophageal sacs and teeth of fishes. J. Nihon Univ. Sch. Dent. 7: 103–111.

Johnson, S. & E. A. Widder, 1998. Transparency and visibility of gelatinous zooplankton from the Northwestern Atlantic and Gulf of Mexico. Biol. Bull. (Woods Hole) 195: 337–348.

Jenkins, R. L., 1983. Observations on the commensal relationship of Nomeus gronovi with Physalia physalis. Copeia 1983: 250–252.

Kashkina, A. A., 1986. Feeding of fishes on salps (Tunicata, Thaliacea). J. Ichthyol. (Engl. Transl. Vopr. Ikhtiol.) 26: 57–64.

Kingsford, M. J., 1993. Biotic and abiotic structure in the pelagic environment: importance to small fishes. Bull. mar. Sci. 53: 393–415.

Koie, M., 1975. On the morphology and life history of Opechona bacillaris (Molin, 1859) Looss, 1907 (Trematoda, Lepocreadiidae). Ophelia 13: 63–86.

Koie, M., 1991. Aspects of the morphology and life cycle of Lecithocladium excisum (Digenea, Hemiuridae), a parasite of Scomber spp. Int. J. Parasit. 21: 597–602.

Koie, M., 1993. Aspects of the life cycle and morphology of Hysterothylacium aduncum (Rudolphi, 1802) (Nematoda, Ascaridoidea, Anisakidae). Can. J. Zool. 71: 1289–1296.

Koeller, P. A., P. C. F. Hurley, P. Perley & J. D. Neilson, 1986. Juvenile fish surveys on the Scotian Shelf: implications for year-class size assessments. J. Cons. int. Explor. Mer. 43: 59–76.

Kremer, P., 1979. Predation by the ctenophore *Mnemiopsis leidyi* in Narragansett Bay, Rhode Island. Estuaries 2: 97–105.

Kuipers, D. R., U. Gaedke, L. Enserink & H. Witte, 1990. Effect of ctenophore predation on mesozooplankton during a spring outburst of *Pleurobrachia pileus*. Neth. J. Sea Res. 26: 111–124.

Larson, R. J., 1976. Cubomedusae: feeding-functional morphology, behavior and phylogenetic position. In Mackie, G. O. (ed.), Coelenterate Ecology and Behavior. Plenum Press, New York: 237–245.

Larson, R. J., 1987a. Daily ration and predation by medusae and ctenophores in Saanich Inlet, B.C., Canada. Neth. J. Sea Res. 21: 35–44.

Larson, R. J., 1987b. First report of the little-known scyphomedusa *Drymonema dalmatinum* in the Caribbean Sea, with notes on its biology. Bull. mar. Sci. 40: 437–441.

Larson, R. J., 1991. Diet, prey selection and daily ration of *Stomolophus meleagris*, a filter-feeding scyphomedusa from the NE Gulf of Mexico. Estuar. coast. shelf Sci. 32: 511–525.

Lauchner, G., 1980a. Diseases of Cnidaria. In Kinne, O. (ed.), Diseases of Marine Animals. Vol. 1. General Aspects, Protozoa to Gastropoda. John Wiley and Sons, New York: 167–237.

Lauchner, G., 1980b. Diseases of Ctenophora. In Kinne, O. (ed.), Diseases of Marine Animals. Vol. 1. General Aspects, Protozoa to Gastropoda. John Wiley and Sons, New York: 239–252.

Legović, T., 1991. Causes, consequences and possible control of massive occurrence of jellyfish *Pelagia noctiluca* in the Adriatic Sea. In UNEP: Jellyfish blooms in the Mediterranean. Proc. II Workshop on Jellyfish in the Mediterranean Sea. MAP Tech. Rep. Ser. 47: 128–132.

Longhurst, A. R., 1985. The structure and evolution of plankton communities. Prog. Oceanogr. 15: 1–35.

Mackie, G. O., 1995. Defensive strategies in planktonic coelenterates. Mar. Fresh. Behav. Physiol. 26: 119–129.

Madin, L. P., 1988. Feeding behavior of tentaculate predators: *in situ* observation and a conceptual model. Bull. mar. Sci. 43: 413–429.

Mansueti, R., 1963. Symbiotic behaviour between small fishes and jellyfishes, with new data on that between the stromateid, *Peprilus alepidotus*, and the scyphomedusae, *Chrysaora quinquecirrha*. Copeia 1963: 40–80.

Marcogliese, D. J., 1995. The role of zooplankton in the transmission of helminth parasites to fish. Rev. Fish Biol. Fish. 5: 336–371.

Martin, J. W. & H. G. Kuck, 1991. Faunal associates of an undescribed species of *Chrysaora* (Cnidaria, Scyphozoa) in the Southern California Bight, with notes on unusual occurrences of other warm water species in the area. Bull. South. Calif. Acad. Sci. 90: 89–101.

Martorelli, S. R., 1996. First record of encysted metacercariae in hydrozoan jellyfishes and ctenophores of the Southern Atlantic. J. Parasit. 82: 352–353.

Martorelli, S. R., 2001. Digenea parasites of jellyfish and ctenophores of the Southern Atlantic. Hydrobiologia 451 (Dev. Hydrobiol. 155): 305–310.

Martorelli, S. R. & F. Cremonte, 1998. A proposed three-host life history of *Monascus filiformis* (Rudolphi, 1819) (Digenea: Fellodistomidae) in the south west Atlantic Ocean. Can. J. Zool. 76: 1198–1203.

Matsakis, S. & R. J. Conover, 1991. Abundance and feeding of medusae and their potential impact as predators on other zooplankton in Bedford Basin (Nova Scotia, Canada) during spring. Can. J. Fish. aquat. Sci. 48: 1419–1430.

Matthews, J. E. & H. H. Shoemaker, 1952. Fishes using ctenophores for shelter. Copeia 4: 270.

McFarlane, G. A., R. J. Beamish, M. W. Saunders, M. S. Smith & T. Butler, 1984. Data for the biology and diet studies of spiny dogfish (*Squalus acanthias*) in Hecate Strait, B. C., August 1977 and June 1978. Can. Data Rep. Fish. aquat. Sci. No. 443: 1–410.

Mianzan, H. W., N. Mari, B. Prenski & F. Sanchez, 1996. Fish predation on neritic ctenophores from the Argentine continental shelf: a neglected food resource? Fish. Res. 27: 69–79.

Mianzan, H., P. M. Pájaro, A. Colombo & A. Madirolas, 2001. Feeding on survival-food: gelatinous plankton as a source of food for anchovies. Hydrobiologia, this volume.

Mianzan, H. W., M. Pajaro, L. Machinandiarena & F. Cremonte, 1997. Salps: possible vectors of toxic dinoflagellates? Fish. Res. 29: 193–197.

Mills, C. E., 1995. Medusae, siphonophores and ctenophores as planktivorous predators in changing global ecosystems. ICES J. mar. Sci. 52: 575–581.

Möller, H., 1980. Scyphomedusa as predators and food competitors of larval fish. Meeresforschung 28: 90–100.

Möller, H., 1984. Reduction of a larval herring population by jellyfish predator. Science 224: 621–622.

Monteleone, D. M. & L. E. Duguay, 1988. Laboratory studies of predation by the ctenophore *Mnemiopsis leidyi* on the early stages in the life history of the bay anchovy, *Anchoa mitchilli*. J. Plankton Res. 10: 359–372.

Mutlu, E., 1999. Distribution and abundance of ctenophores and their zooplankton food in the Black Sea. II. *Mnemiopsis leidyi*. Mar. Biol. 135: 603–613.

Mutlu, E. & F. Bingel, 1999. Distribution and abundance of ctenophores and their zooplankton food in the Black Sea. I. *Pleurobrachia pileus*. Mar. Biol. 135: 589–601.

O'Connor, B. & D. McGrath, 1978. On the occurrence of the scyphozoan *Rhizostoma octopus* (L.) around the Irish Coast in 1976. Ir. Nat. J. 19: 261–263.

Olesen, N. J., K. Frandsen & H. U. Riisgård, 1994. Population dynamics, growth and energetics of jellyfish *Aurelia aurita* in a shallow fjord. Mar. Ecol. Prog. Ser. 105: 9–18.

Oviatt, C. A. & P. M. Kremer, 1977. Predation on the ctenophore, *Mnemiopsis leidyi*, by butterfish, *Peprilus triacanthus*, in Narragansett Bay, Rhode Island. Chesapeake Sci. 18: 236–240.

Pagès, F., M. G. White & P. G. Rodhouse, 1996. Abundance of gelatinous carnivores in the nekton community of the Antarctic Polar Frontal Zone in summer 1994. Mar. Ecol. Prog. Ser. 141: 139–147.

Paradis, A. R., P. Pépin & J. A. Brown, 1996. Vulnerability of fish eggs and larvae to predation: review of the influence of the relative size of prey and predator. Can. J. Fish. aquat. Sci. 53: 1226–1235.

Paradis, A. R., M. Pépin & P. Pepin, 1999. Disentangling the effects of size-dependent encounter and susceptibility to predation with an individual-based model for fish larvae. Can. J. Fish. aquat. Sci. 56: 1562–1575.

Parsons, T. R., 1992. The removal of marine predators by fisheries and the impact of trophic structure. Mar. Pollut. Bull. 25: 51–53.

Parsons, T. R., 1995. The impact of industrial fisheries on the trophic structure of marine ecosystems. In Polis, G. A. & K. O. Winemiller (eds), Food Webs: Integration of Patterns and Dynamics. Chapman & Hall, New York: 352–357.

Pauly, D., R. Froese & V. Christensen, 1998a. How pervasive is "Fishing down marine food webs"? Response. Science 282: 1383a.

Pauly, D., V. Christensen, J. Dalsgaard, R. Froese & R. Torres, Jr., 1998b. Fishing down marine food webs. Science 279: 860–863.

Phillips, P. J., W. D. Burke & E. J. Keener, 1969. Observations on the trophic significance of jellyfishes in Mississippi Sound with quantitative data on the associative behavior of small fishes with medusae. Trans. am. Fish. Soc. 98: 703–712.

Purcell, J .E., 1981a. Dietary composition and diel feeding patterns of epipelagic siphonophores. Mar. Biol. 65: 83–90.

Purcell, J. E., 1981b. Feeding ecology of *Rhizophysa eysenhardti*, a siphonophore predator of fish larvae. Limnol. Oceanogr. 26: 424–432.

Purcell, J. E., 1984a. The functions of nematocysts in prey capture by epipelagic siphonophores (Coelenterata, Hydrozoa). Biol. Bull. 166: 310–327.

Purcell, J. E., 1984b. Predation on fish larvae by *Physalia physalis,* the Portuguese man of war. Mar. Ecol. Prog. Ser. 19: 189–191.

Purcell, J. E., 1985. Predation on fish eggs and larvae by pelagic cnidarians and ctenophores. Bull. mar. Sci. 37: 739–755.

Purcell, J. E., 1989. Predation by the hydromedusa *Aequorea victoria* on fish larvae and eggs at a herring spawning ground in British Columbia. Can. J. Fish. aquat. Sci. 46: 1415–1427.

Purcell, J. E., 1990. Soft-bodied zooplankton predators and competitors of larval herring (*Clupea harengus pallasi*) at herring spawning grounds in British Columbia. Can. J. Fish. aquat. Sci. 47: 505–515.

Purcell, J. E., 1992. Effects of predation by the scyphomedusan *Chrysaora quinquecirrha* on zooplankton populations in Chesapeake Bay. Mar. Ecol. Prog. Ser. 87: 65–76.

Purcell, J. E., 1997. Pelagic cnidarians and ctenophores as predators: Selective predation, feeding rates and effects on prey populations. Ann. Inst. Oceanogr. Paris 73: 125–137.

Purcell, J. E. & P. A. V. Anderson, 1995. Electrical responses to water-soluble components of fish mucus recorded from the cnidocytes of a fish predator, *Physalia physalis*. Mar. Fresh. Behav. Physiol. 26: 149–162.

Purcell, J. E. & J. J. Grover, 1990. Predation and food limitation as causes of mortality in larval herring at a spawning ground in British Columbia. Mar. Ecol. Prog. Ser. 59: 55–67.

Purcell, J. E. & C. E. Mills, 1988. The correlation of nematocyst types to diets in pelagic Hydrozoa. In Hessinger, D. A. & H. M. Lenhoff (eds), The Biology of Nematocysts. Academic Press, San Diego: 463–485.

Purcell, J. E. & M. V. Sturdevant, 2001. Prey selection and dietary overlap among zooplanktivorous jellyfish and juvenile fishes in Prince William Sound, Alaska. Mar. Ecol. Prog. Ser. 210: 67–83.

Purcell, J. E., A. Malej & A. Benović, 1999a. Potential links of jellyfish to eutrophication and fisheries. In Malone, T. C., A. Malej, L. W. Harding, Jr., N. Smodlaka & R. E. Turner (eds), Ecosystems at the Land–sea Margin: Drainage Basin to Coastal Sea. American Geophysical Union, Coastal and Estuarine Studies 55: 241–263.

Purcell, J. E., T. D. Siferd & J. B. Marliave, 1987. Vulnerability of larval herring (*Clupea harengus pallasi*) to capture by the jellyfish *Aequorea victoria*. Mar. Biol. 94: 157–162.

Purcell, J. E., J. R. White & M. R. Roman, 1994a. Predation by gelatinous zooplankton and resource limitation as potential controls of *Acartia tonsa* copepod populations in Chesapeake Bay. Limnol. Oceanogr. 39: 263–278

Purcell, J. E., T. Shiganova, M. B. Decker & E. D. Houde, 2001. The ctenophore *Mnemiopsis* in native and exotic habitats: U.S. estuaries *versus* the Black Sea basin. Hydrobiologia 451 (Dev. Hydrobiol. 155): 145–175.

Purcell, J. E., J. R. White, D. A. Nemazie & D. A. Wright, 1999b. Temperature, salinity and food effects on asexual reproduction and abundance of the scyphozoan *Chrysaora quinquecirrha*. Mar. Ecol. Prog. Ser. 180: 187–196.

Purcell, J. E., E. D. Brown, K. D. E. Stokesbury, L. J. Haldorson & T. C. Shirley, 2000. Aggregations of the jellyfish *Aurelia labiata*:

abundance, distribution, association with age-0 walleye pollock, and behaviors promoting aggregation in Prince William Sound, Alaska, U.S.A. Mar. Ecol. Prog. Ser. 195: 145–158.

Purcell, J. E., D. A. Nemazie, S. E. Dorsey, E. D. Houde & J. C. Gamble, 1994b. Predation mortality of bay anchovy (*Anchoa mitchilli*) eggs and larvae due to scyphomedusae and ctenophores in Chesapeake Bay. Mar. Ecol. Prog. Ser. 114: 47–58.

Robison, B. H., 1983. Midwater biological research with the WASP ADS. Mar. Tech. Soc. J. 17: 21–27.

Runge, J. A., P. Pepin & W. Silvert, 1987. Feeding behavior of the Atlantic mackerel *Scomber scombrus* on the hydromedusa *Aglantha digitale*. Mar. Biol. 94: 329–333.

Schneider, G. & G. Behrends, 1998. Top-down conrol in a neritic plankton system by *Aurelia aurita* medusae – a summary. Ophelia 48: 71–82.

Scott, A., 1914. The mackerel fishery off Walney in 1913. Proc. Trans. Liverpool Biol. Soc. 28: 109–115.

Scott, A., 1924. Food of the Irish Sea herring in 1923. Proc. Trans. Liverpool Biol. Soc. 38: 115–119.

Shanks, A. L. & W. M. Graham, 1988. Chemical defense in a scyphomedusa. Mar. Ecol. Prog. Ser. 45: 81–86.

Shiganova, T. A., 1998. Invasion of the Black Sea by the ctenophore *Mnemiopsis leidyi* and recent changes in pelagic community structure. Fish. Oceanogr. 7: 305–310.

Southcott, R. V. & C. J. M. Glover, 1987. Tthe occurrence of *Desmonema gaudichaudi* (Lesson) (Scyphozoa, Semaeostomeae) in South Australian waters with records of fish–jellyfish symbioses. Trans. r. Soc. S. Australia 111: 131–132.

Spanier, E. & B. S. Galil, 1991. Lessepsian migration: a continuous biogeographical process. Endeavour (Oxf.) 15: 102–106.

Stachowicz, J. J. & N. Lindquist, 1997. Chemical defense among hydroids on pelagic *Sargassum*: predator deterrence and absorption of solar UV radiation by secondary metabolites. Mar. Ecol. Prog. Ser. 155: 115–126.

Stewart, S. E., 1996. Field behaviour of *Tripedalia cystophora* (Class Cubozoa). Mar. Fresh. Behav. Physiol. 27: 175–188.

Suyehiro, Y., 1942. A study on the digestive system and feeding habits of fish. Jpn. J. Zool. 10: 1–299.

Svendsen, Y. S., 1990. Hosts of third stage larvae of *Hysterothylacium* sp. (Nematoda, Anisakidae) in zooplankton from outer Oslofjord, Norway. Sarsia 75: 161–167.

Tadokoro, K., Y. Ishida, N. D. Davis, S. Ueyanagi & T. Sugimoto, 1996. Change in chum salmon (*Oncorhynchus keta*) stomach contents associated with fluctuations of pink salmon (*O. gorbuscha*) abundance in the central subarctic Pacific and Bering Sea. Fish. Oceanogr. 5: 89–99.

Thiel, M. E., 1970. Das Zusammenleben von Jung- und Kleinfischen mit Rhizostomeen (Scyphomedusae). Ber. Dtsch. Wiss. Komm. Meeresforsch. 21: 444–473.

Thiel, M. E., 1978. Das Zusammenleben von Jung- und Kleinfischen mit Semaeostomen (Scyphomedusae). Mitt. Hamb. Zool. Mus. Inst. 75: 19–47.

Thiel, H., 1979. Assoziationen von Quallen und Fischen. Nat. Mus. 109: 353–360.

Tolley, S. G., 1987. Association of young *Chloroscombrus chrysurus* (Pisces: Caragidae) with the jellyfish *Aurelia aurita*. Copeia 1987: 216–219.

Toonen, R. H. & F.-S. Chia, 1993. Limitations of laboratory assessments of coelenterate predation: container effects on the prey selection of the limnomedusa, *Proboscidactyla flavicirrata* (Brandt). J. exp. mar. Biol. Ecol. 167: 215–235.

Tsuruta, A., 1963. Distribution of plankton and its characteristics in the oceanic fishing grounds with special reference to their relation to fishery. J. Shimonoseki Univ. Fish. 12: 13–214.

Van Der Veer, H. W., 1985. Impact of coelenterate predation on larval plaice *Pleuronectes platessa* and flounder *Platichthys flesus* stock in the Western Wadden Sea. Mar. Ecol. Prog. Ser. 25: 229–238.

Vinogradov, M. E., E. A. Shushkina & Yu. V. Bulgakova, 1996. Consumption of zooplankton by the comb jelly *Mnemiopsis leidyi* and pelagic fishes in the Black Sea. Oceanology 35: 523–527.

Vućetić, T. & V. Alegria-Hernandez, 1988. Trands of annual catches or stock densities of some pelagic fishes in recent '*Pelagia* years' in the Adriatic. FAO Fish Rep. 394: 133–136.

Welch, D. W., 1997. Anatomical specialization in the gut of Pacific salmon (*Oncorhynchus*): evidence for oceanic limits to salmon production? Can. J. Zool. 75: 936–942.

Widder, E. A., S. A. Bernstein, D. F. Bracher, J. F. Case, K. R. Reisenbichler & J. J. Torres, 1989. Bioluminescence in the Monterey submarine canyon: image analysis of video recordings from a midwater submersible. Mar. Biol. 100: 541–551.

Yip, S. Y., 1984. Parasites of *Pleurobrachia pileus* Müller, 1776 (Ctenophora), from Galway Bay, Western Ireland. J. Plankton Res. 6: 107–121.

Zaitsev, Yu. P., 1992. Recent changes in the trophic structure of the Black Sea. Fish. Oceanogr. 1: 180–189.

Zaitsev, Yu. P. & L. N. Polischuk, 1984. An increase in the number of *Aurelia aurita* (L.) in the Black Sea. Ekol. Morya 17: 35–46.

*Hydrobiologia* **451**: 45–53, 2001.
© 2001 *Kluwer Academic Publishers.*

# Feeding on survival-food: gelatinous plankton as a source of food for anchovies

H. Mianzan[1,2], M. Pájaro[2], G. Alvarez Colombo[2] & A. Madirolas[2]
[1]*CONICET;* [2]*INIDEP; P.O. Box 175, 7600 Mar del Plata, Argentina*
*E-mail: hermes@inidep.edu.ar*

*Key words: Engraulis anchoita*, gut contents, predation, *Iasis zonaria*, acoustics, south-western Atlantic Ocean

## Abstract

The gelatinous zooplankton, composed by members of different phyla (Cnidaria, Ctenophora, Tunicata), are usually neglected in most studies about energy transfer in the marine trophic web, and often it is assumed that such soft-bodied fauna are trophic 'dead ends' in the food webs. In recent years, however, it has been shown that many fish species feed extensively on gelatinous zooplankton, while other species may feed on them occasionally when other food is scarce. We found that anchovies, *Engraulis anchoita* Hubbs & Marini, 1935, shoaled close to the Río de la Plata surface salinity front, where dense aggregations of the salp, *Iasis zonaria* (Pallas, 1774), were detected acoustically in May, 1994. Densities of non-gelatinous zooplankton were low at this interface, and anchovies fed on the salps. In this paper, we describe the environmental and biological conditions that led a normally planktivorous filter feeder *E. anchoita* to prey on gelatinous plankton.

## Introduction

The gelatinous zooplankton, composed of members of different phyla (Cnidaria, Ctenophora, Tunicata), are usually neglected in most of the studies about energy transfer in marine trophic webs, principally because these animals are usually damaged beyond recognition when sampled with conventional plankton nets. As a consequence, their distribution and abundance patterns are poorly known. Furthermore, even though many papers have focused on their role as consumers (Alvariño, 1985; Madin & Kremer, 1995; Madin et al., 1997; Purcell, 1997), there are relatively few reports that document the predators of gelatinous zooplankton (Purcell, 1991). These discrepancies have fostered a belief that such soft-bodied fauna are trophic 'dead ends' in marine food webs. This leads to the assumption that some jelly species that can reach enormous biomass merely die, sink and decompose.

In recent years, it has been shown that many fishes feed on gelatinous zooplankton (Ates, 1988; Arai, 1988; Mianzan et al., 1996; Mianzan et al., 1997; Purcell & Arai, 2001). Although some fish species may depend heavily and be specialized to feed on gelatinous species, others utilize them only occasionally. The reasons why those fish species should accept gelatinous organisms as food are not clearly understood, but it is hypothesized that this happens when nothing better is available to fill fish stomachs. Kashkina (1986) described this behaviour as feeding on survival food.

To test this hypothesis, a multidisciplinary investigation that included acoustical monitoring, plankton net sampling and analysis of fish stomach contents was performed in order to obtain a synoptic picture. In this paper, we describe the environmental and biological conditions that led a normally planktivorous filter feeder (*Engraulis anchoita*) to prey on gelatinous plankton.

## Materials and methods

Day and night sampling was conducted on the Argentine Continental shelf (Fig. 1), with the R/V Eduardo Holmberg (INIDEP) from 15th to 28th May, 1994, focused on acoustical estimation of *Engraulis anchoita* biomass. Zooplankton was sampled using a CalVET net (200 $\mu$m), 25 cm diameter. Sixty five plankton stations were performed from Río de la Plata up to the slope front (Fig. 1). The net was towed vertically from

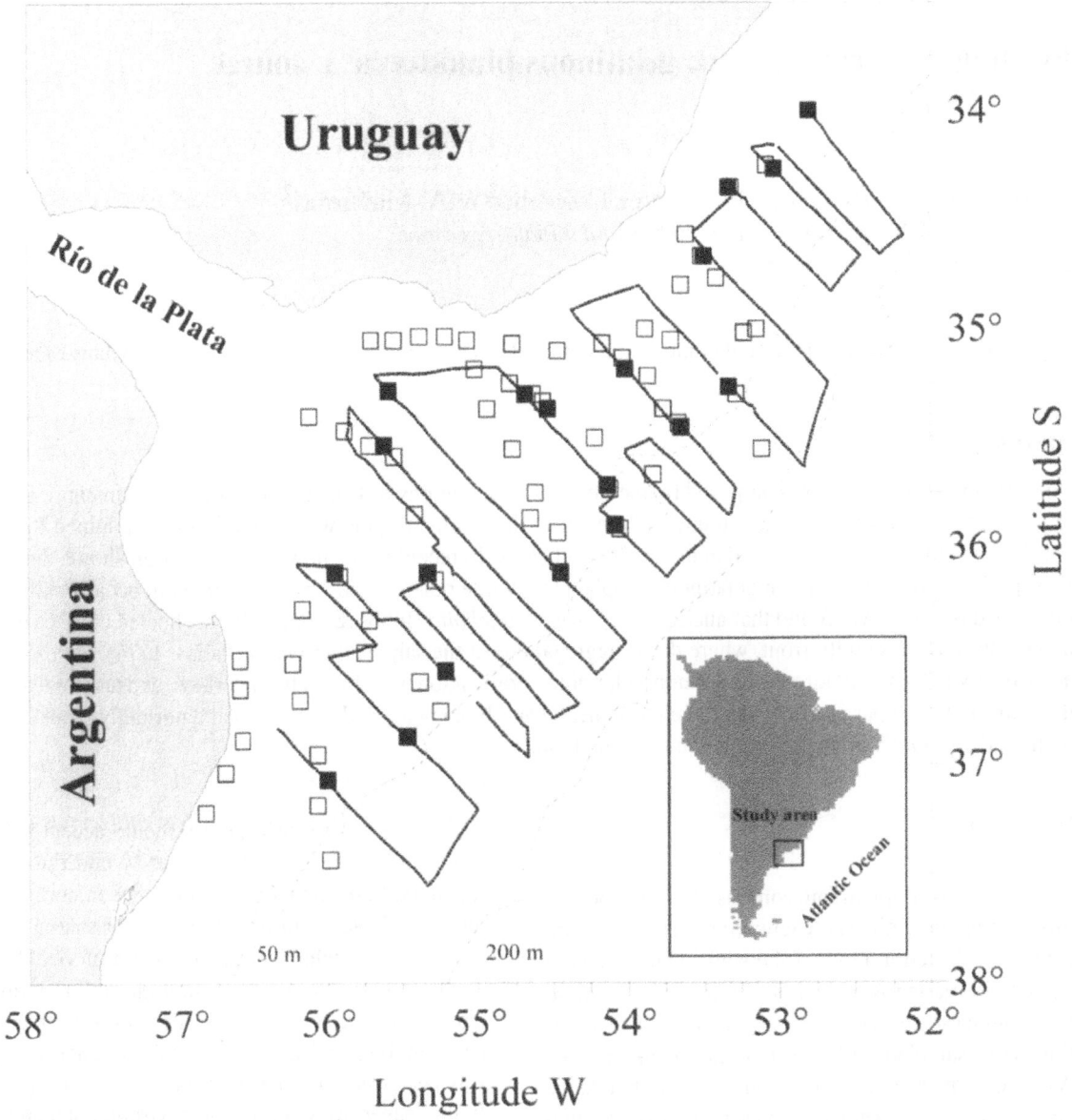

*Figure 1.* Cruise design: CTD (conductivity-temperature-depth profiler) and plankton (CalVET net) stations (□), fishing stations (Nichimo Midwater trawl) (■) and acoustic sampling (solid line) along parallel transects (SIMRAD EK500 echosounder operating at 38 kHz), performed during Autum, 1994 by the R/V Dr E. Holmberg (INIDEP).

the bottom to the surface or from 70 m up the surface. The average volume of water filtered was 1.4 m³ (range 0.4–3.2 m³). Standing stocks were estimated by converting wet weight (ww) or number of individuals to dry weight (dw) and then to organic carbon using conversion factors (Omori, 1969; Madin et al., 1981; Larson, 1986; Fernandez Araoz, 1994). Results are expressed as mg carbon m⁻³. Conductivity and temperature were measured with a SeaBird 19 CTD at a sampling rate of 2 scans per second. Data were

processed to achieve a one-meter vertical resolution. Salinity data are reported with a precision of 0.05.

Nineteen fishing stations (Fig. 1) were sampled with a Nichimo midwater trawl, with an inner mesh of 10 mm at the cod end. One hundred and twenty anchovies were sub-sampled from ten stations for analyses. Length and weight of each specimen was recorded. Gut contents were identified from preserved collected specimens either macroscopically or using a dissecting microscope (Wild M8). Salps were recognized by their muscle bands and stomachs or by the whole body of the animal. Salps were identified according to Esnal (1981) and Esnal & Daponte (1999). Each stomach was weighed with and without the stomach content in order to determine the weight of the ingested prey, expressed as the difference between both weights. The 'Stomach Repletion Index' (SRI) was calculated following the scale proposed by Angelescu (1982). This index indicates the state of satiety from 1 (an empty stomach: <0.5% of the anchovy weight) to 4 (full stomach: >6% of the anchovy weight).

Acoustic sampling was performed along parallel transects (Fig. 1). A Simrad EK500 echosounder operating with a 38 kHz split-beam transducer was employed. The processing method was echo-integration (Forbes & Nakken, 1972). The averaging interval was 1 nm. Echograms as well as integrated area scattering coefficient values ($s_a$, in units of $m^2 \, nm^{-2}$), were recorded with a color printer. Nine surface-referenced layers starting at 3 m from the transducer face, i.e. 6.5 m below sea surface, were programmed. The echosounder was calibrated during the cruise, according to the centered sphere method with standard targets (Foote et al., 1987). Echograms were analyzed in order to determine the $s_a$ fraction corresponding to the concentrations of salps.

## Results

The surface salinity contour of the Río de la Plata showed a typical autumn NNE drainage, parallel to the Uruguayan coast as a low salinity wedge 30–35 nautical miles off the coastline (Fig. 2). The central and southern sections of the water column were highly stratified. The central section had more than 200 km of estuarine surface waters. The southern one was shorter and the stratified region occupied a small portion, up to 60 km in length. Here, vertically homogeneous waters were observed at the outer sector of the section. The northern section was weakly stratified due to a Continental shelf marine waters intrusion lying over the coast.

Mesozooplankton was dominated by copepods (*Acartia tonsa* Dana, 1848, *Paracalanus parvus* (Claus, 1863), *Labidocera fluviatilis* F. Dahl, 1894, *Corycaeus* sp. Dana, 1849, *Euterpina acutifrons* (Dana, 1852)) and cladocerans (*Evadne* sp. Lovén, 1836, *Podon* sp. Lilljeborg, 1853), with mean densities of 466 copepods $m^{-3}$ (standard deviation: 710.5;

range: 0–3406) and 112 cladocerans $m^{-3}$ (standard deviation: 338.3; range: 0–2291). Many other taxa were collected: chaetognaths, appendicularians, meroplankton including bivalve, decapod and polychaete larvae, ichthyoplankton (*Engraulis anchoita* eggs and larvae), ctenophores (*Mnemiopsis* sp. L. Agassiz, 1860), and hydromedusae (*Liriope tetraphylla* (Chamisso & Eysenhardt, 1821) and *Turritopsis nutricola* McCradyi, 1859). Copepod and cladoceran biomasses reached maxima of 3.8 and 4.4 mg C $m^{-3}$, respectively. These maximum values were found along the surface salinity front. However, less than 30% of the samples showed values higher than 1 mg C $m^{-3}$. Of the rest, most of 40% showed values less than 0.1 mg C $m^{-3}$. When present, salps (*Iasis zonaria*) largely dominated zooplankton biomass with aggregations up to 277 mg C $m^{-3}$, several orders of magnitude greater than non-gelatinous zooplankton biomass. Except for the stations at which salps were abundant, the biomass of total zooplankton was very low during the survey (Fig. 3).

Stomach contents of anchovies reflected what was observed in plankton samples. Fifty five percent of the stomachs studied were empty or with very little identifiable food. No regurgitated material was observed. Stomach contents of the rest (45%) included a few copepods (calanoids and harpacticoids, 22.5%), cladocerans(2.5%), appendicularians (2%), anchovy eggs (2%) and salps (oozooids and blastozooids, 11.7%). Almost 90% of the stomachs analyzed showed SRI values of 1–2 (corresponding to up to 1% of the body weight), implying very little or no food in them (Fig. 4A). Salps were found in the remaining 11.7% of the stomachs reaching the maximum SRI values found (SRI= 3) (corresponding to up to 6% of the body weight) (Fig. 4B). More than 50 specimens of salps (mostly aggregated zooids: Fig. 5B) were found in one anchovy stomach. In one single haul, 40% of the specimens analyzed showed salps in their stomachs (Fig. 5), the rest being empty.

At the working frequency (38 kHz, widely used for fish biomass estimations) large planktonic aggregations were detected, some of them covering areas of more than 1000 square nautical miles between transects. Vertical profiles of the echotraces showed differences along the acoustic transects, either occupying most part of the water column or forming well-defined scattering layers with marked diurnal migrations. Aggregations of zooplankton were often close to anchovy shoals. The highest numbers of salps observed in plankton samples, fishing mid-water trawls and stom-

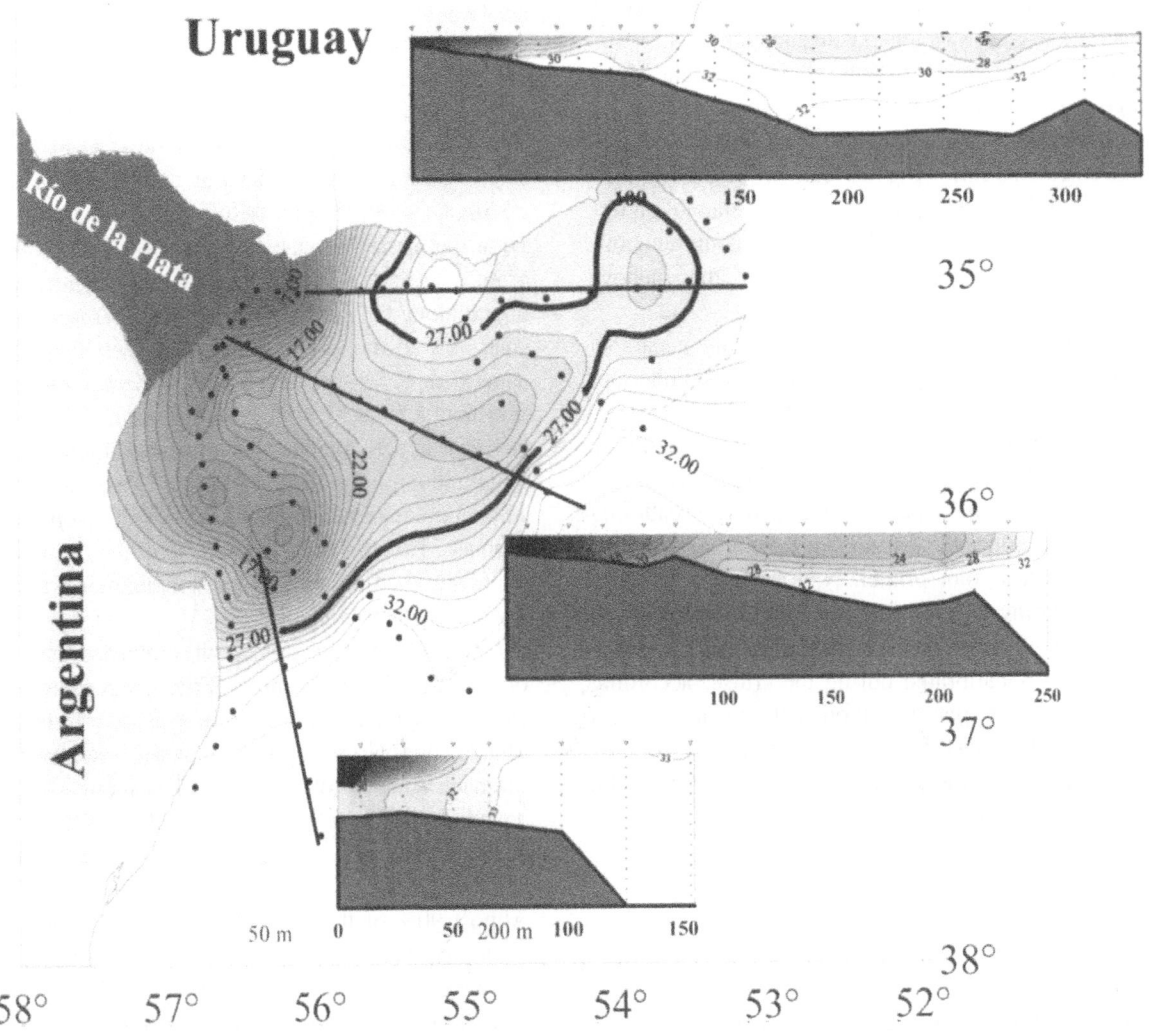

*Figure 2.* Surface salinity contours in the Río de la Plata area. Bold isoline denotes the surface halocline of 27. Sections show the pattern of vertical stratification.

ach contents were coincident with areas where high plankton concentrations were detected acoustically (Fig. 6). From this accordance between echograms and sampling results, we therefore assume that the plankton aggregations registered were primarily due to the salp, *Iasis zonaria*.

The horizontal distribution of *Iasia zonaria* aggregations closely followed the salinity contours. Aggregations occurred mainly between a salinity range of 26–33. This pattern suggested that some *I. zonaria* aggregations tended to be formed at the edge or just outside the Río de la Plata Surface Salinity Front (SSF).

## Discussion

There are very few reports on mesozooplankton densities for the study area (see Méndez et al., 1997). Biomass estimates derived from abundance data (Fernández Araoz et al., 1991; Fernández Araoz, 1994) for neighbouring areas reach values within the range of the present study. For the Río de la Plata salt wedge regime, Mianzan et al. (2001) found up to 8000 copepods $m^{-3}$ for the inner part of the salt wedge and several thousands for the outer section. In the present study, salps when present dominated plankton carbon biomass. Salps form an oceanic group (Esnal,

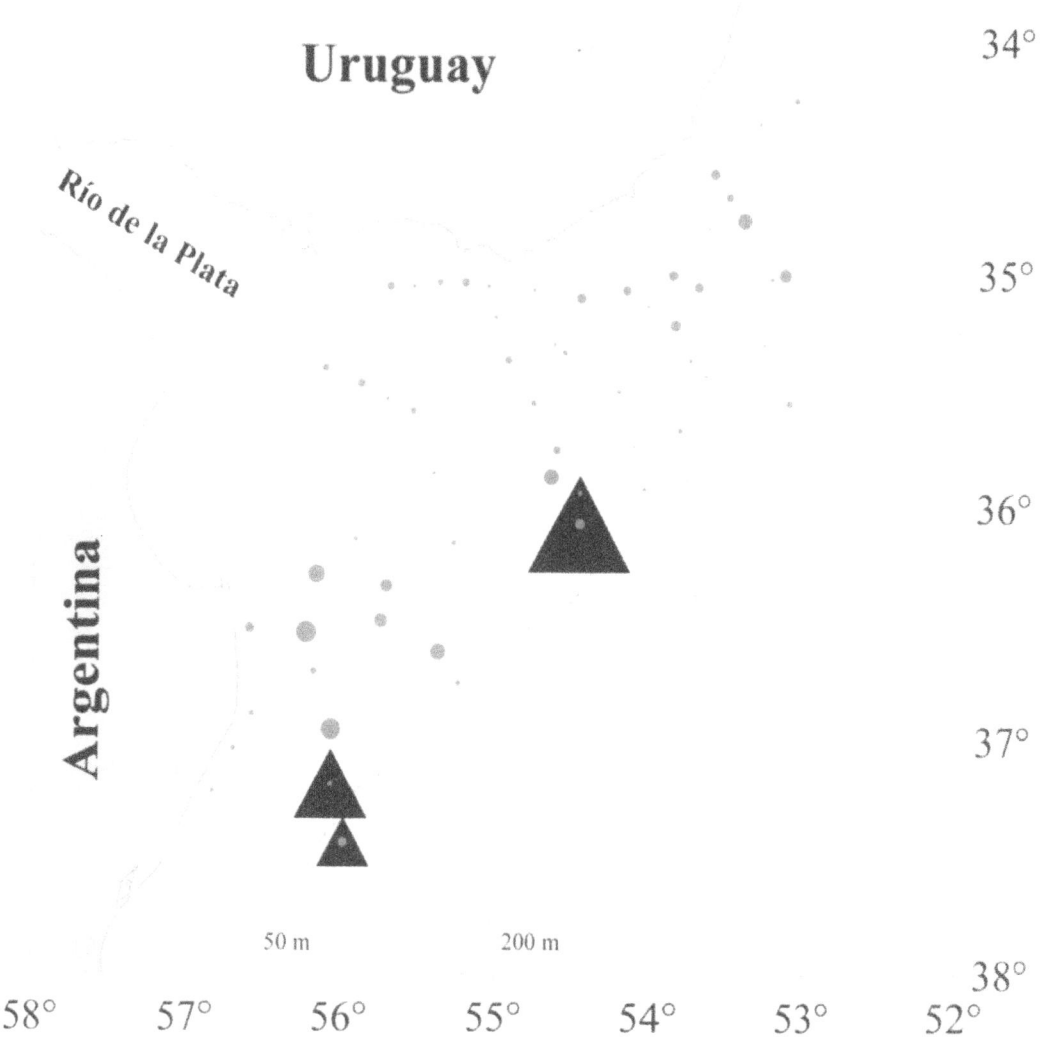

*Figure 3.* Biomass of Copepoda and Cladocera (○) and salps (▲). Maximum values: 8.2 and 277 mg C m$^{-3}$, respectively.

1981; Esnal & Daponte, 1999), however among them, *Iasis zonaria* invades large areas of the Buenos Aires Continental shelf. The ability of salps to develop large populations quickly (blooms) and their capability to aggregate in large numbers, are consistent with the enormous biomass acoustically detected, and confirmed by some of the plankton samples where salps represented more than 99% of the total plankton carbon biomass.

Salps are principally composed of gelatinous tissue. A low sound reflectivity is expected from their high water content (>95%), being considered as a fluid-like scatterer (Stanton et al., 1994). Demer & Hewitt (1994) considered that most of the scattering produced by these animals results from their spherical stomach where diatoms, copepods and other organisms are actively concentrated. However, in spite of their low scattering properties, salp aggregations were registered even at the 38 kHz frequency and their scattering measured. Stanton et al. (1996) observed that the number of salps of *Salpa aspera* Chamisso, 1819 of 2.6 cm in length, required to produce a volume scattering of −70 dB was 110 individuals per m$^3$, working at the same frequency. According to our results, a maximum density obtained from the CalVET net was 45 individuals of *Iasis zonaria* per m$^3$. Also, volume scattering (S$v$) values were as high as −53 dB. In this sense, the comparatively rigid structure of its tunic and

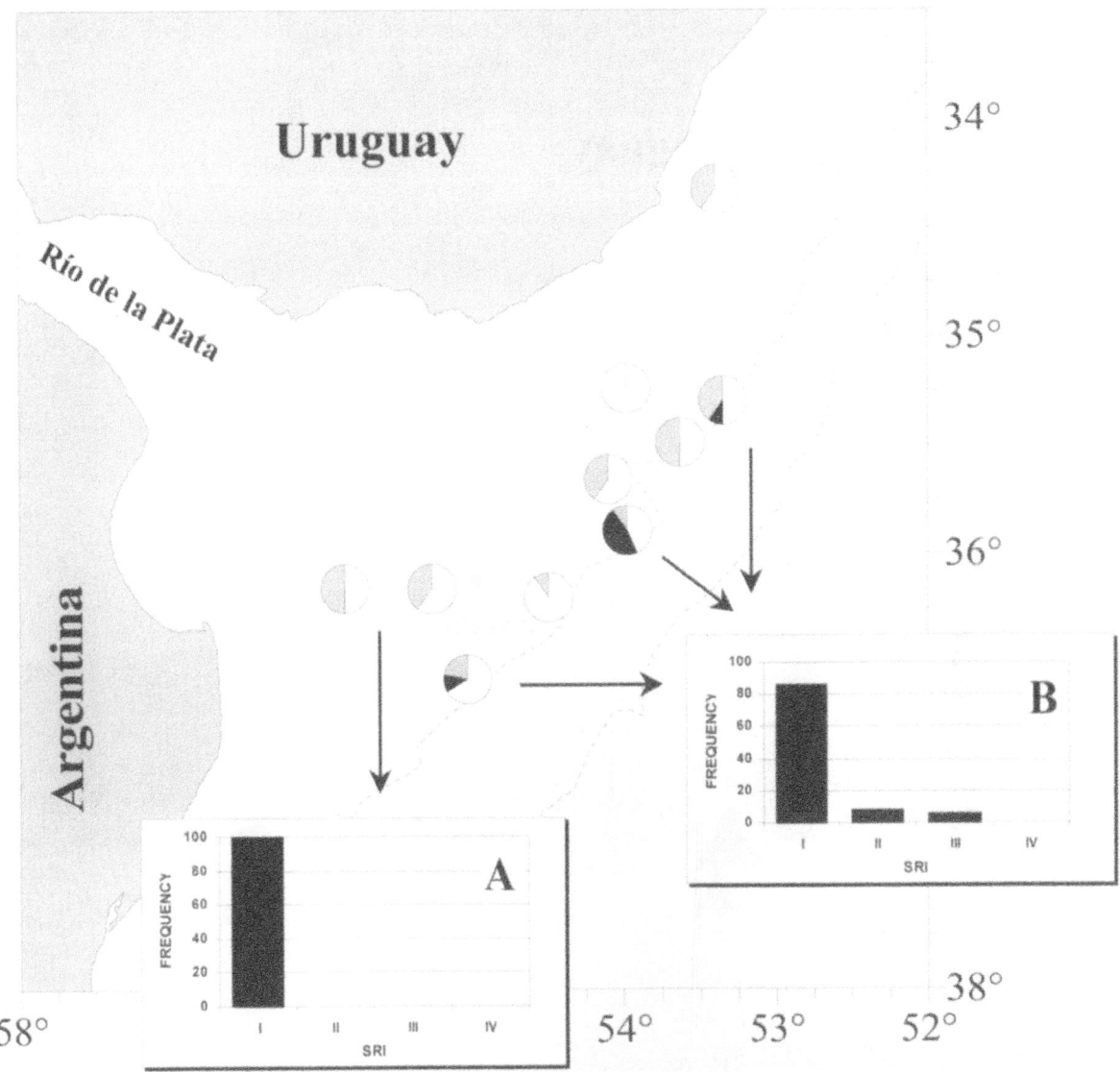

*Figure 4.* Anchovies' stomach contents. Empty (○), Salps (●) and non-gelatinous plankton (◎). Histograms indicate the stomach repletion index (SRI) for 2 groups of stomachs: (a) stomachs without salps, (b) stomachs with salps. This index indicates the state of satiety from I (an empty stomach: <0.5% of the anchovy weight) to IV (full stomach: >6% of the anchovy weight).

larger size (3–6 cm in length) could lead to a higher sound reflectivity of *I. zonaria*.

Under some special circumstances, only one or a few species will consistently dominate the scattering, as in this case with *Iasis zonaria*. The acoustic monitoring constitutes a helpful tool to assess horizontal distribution patterns and relative abundance. In this case in particular, it allowed us to link apparently unconnected results, like the co-occurrence of the presence of salps in plankton samples and the presence of salps in anchovy gut contents, with the background synoptic picture given by the acoustic recording (Fig. 6).

These identified echo traces looked like well-defined 'scattering layers' (Fig. 6) that probably were due to the strong vertical stratification of the area. Mianzan & Guerrero (2000) and Mianzan et al. (2001) reported on salp aggregations that tended to occur near the Río de la Plata Surface Salinity Front. Physical forces and behavior may cause this scale of patchiness (see Folt & Burns, 1999). Our observations suggest that the salinity gradient, which represents the boundary regime of two different water masses, enhances the vertical nutrient flux, fertilizes the frontal area (Méndez et al., 1997), and is responsible for the generation of the *Iasis zonaria* aggregations.

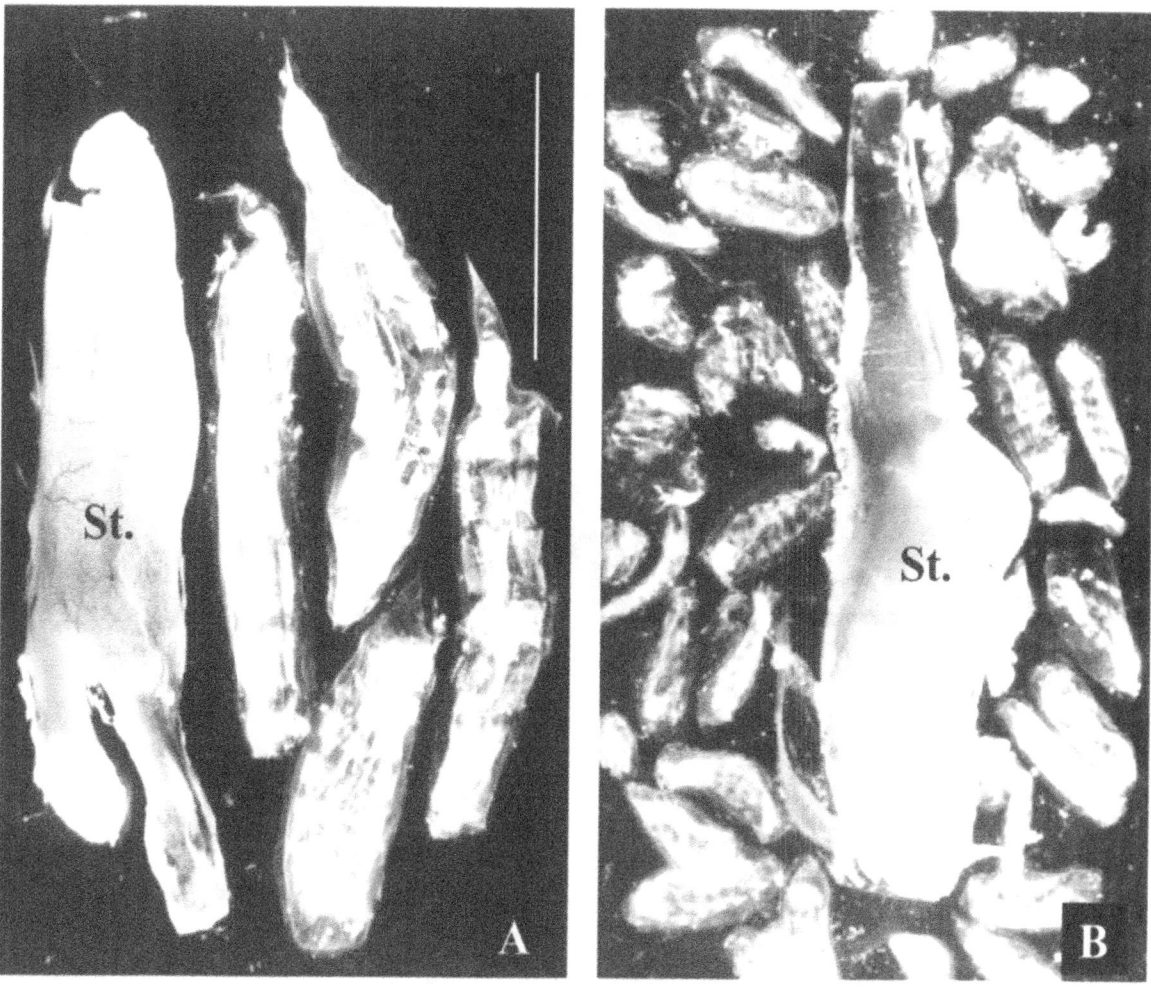

*Figure 5.* Specimens of *Iasis zonaria* found in anchovy stomachs (St.). (a) oozooids, (b) blastozooids. Scale bar = 1 cm.

Several fish species find gelatinous zooplankton an acceptable source of food. Most specialized species have developed morphological and physiological adaptations to feed on these fauna (Arai, 1988; Purcell & Arai, 2001). However, it was also shown that many species that usually are not considered as gelatinous predators, do actually feed from time to time on gelatinous fauna (Mianzan et al., 1996, 1997; Purcell & Arai, 2001). Salps in particular represent a significant proportion of the diet of several mesopelagic fish (Kashkina, 1986) and also for some demersal species (Mianzan et al., 1996).

Anchovies have not been recognized as gelatinous plankton consumers, as they typically feed on mesozooplankton and macrozooplankton. Filtration and capture mechanisms vary in accordance with the relationships between size of the prey and diameter of the anchovy's mouth, as well as prey density; microscopic food (<3 mm) is ingested by filtration, and

larger size prey are ingested by capture (Angelescu, 1981). Our results showed that under low mesozooplankton density, anchovies filled their stomachs by capture of salps. The only previous references to gelatinous prey in the gut contents of anchovies were the occasional presence of juvenile zooids of *Iasis zonaria* (Angelescu, 1982) and the consumption of appendicularians (Capitanio et al., 1997). As a possible source of error, it must be considered that some of the salps present in anchovies' guts could have been ingested in the net cod end. However, the salps were found in different digestion stages identifiable by the remaining muscle bands and nucleus. Moreover, the blastozooids (3–5 mm in length) were too small to be retained by the net cod end mesh (10 mm).

Over recent decades, man's expanding influence on the oceans is beginning to cause major changes and there is reason to think that in some regions, new blooms of jellyfish are occurring in response to

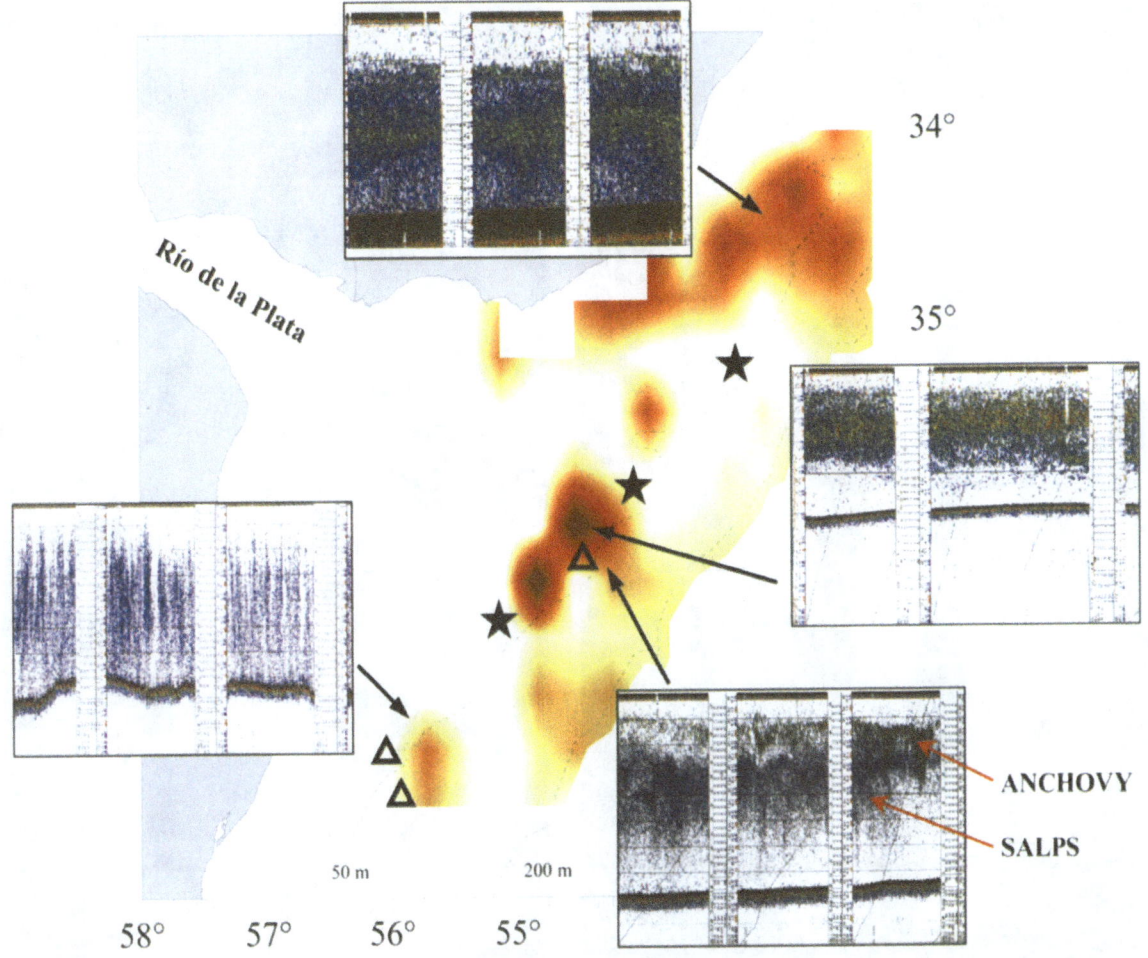

*Figure 6.* Horizontal (map) and vertical (sections) salp aggregations and anchovy shoals detected acoustically. Colour scale is proportional to square root transformation of $s_a$ (from 0 = white to 90 = brown). Density contours as well as vertical sections represent the $s_a$ fraction corresponding to the concentrations of salps (except as indicated as anchovy). (★) Salps in stomach contents; (△) salps in plankton samples.

some of the cumulative effects of these impacts (Mills, 2001). Overfishing, eutrophication, introduction of alien species and pollution were identified as major sources of change. Overfishing seems to have resulted in the increase of some medusae species (Mills, 1995; Mianzan & Cornelius, 1999). Increases in 'jellyfish' have also been the result of recent invasions of non-indigenous species into coastal ecosystems (see Mianzan & Cornelius, 1999; Purcell et al., 2001). Salp aggregations were not monitored systematically in the Argentine Sea, although they would play a complex role in marine trophic webs. As an example, mortality of mackerel, *Scomber japonicus* Houttuyn, 1872, was recently ascribed to the ingestion of salps that were feeding on a red tide bloom (Mianzan et al., 1997). Red tides blooms have increased both in time and space through the Argentine coastal ecosystem (El Busto et al., 1993) and it is possible that salps, which

could be commonly eaten by fish species, may also increase. It is not by chance that the first report on PSP (paralytic shellfish poisons) on *Engraulis anchoita* was recently published (Montoya et al., 1998) and salps may have been a possible trophic link, among other herbivorous zooplankters, from dinoflagellates to anchovies.

## Acknowledgements

We are indebted to Raúl Guerrero (Argentina) for his comments on oceanography, Fernando Ramírez and Laura Machinandiarena for their help with taxonomical identifications, and Marsh Youngbluth, Mary N. Arai and Jennifer Purcell for their critical reading and comments on the manuscript. This is INIDEP contribution no. 1176

# References

Alvariño, A., 1985. Predation in a plankton realm: mainly with reference to fish larvae. Inv. Mar. Cicimar 2: 1–122.

Angelescu, V., 1981. Ecología trófica de la anchoíta del Mar Argentino (Engraulidae, *Engraulis anchoita*). Parte I. Morfología del sistema digestivo en relación con la alimentación. In Salinas, P. J. (ed.), Zoología Neotropical. Actas VIII Congreso Latinoamericano de Zoología, Mérida: 1317–1350.

Angelescu, V., 1982. Ecología trófica de la anchoíta del Mar Argentino (Engraulidae, *Engraulis anchoita*). Parte II. Alimentación, comportamiento y relaciones tróficas en el ecosistema. Series Contribución del Instituto Nacional de Investigación y Desarrollo Pesquero, INIDEP, Mar del Plata. 409: 83 pp.

Arai, M., 1988. Interactions of fish and pelagic coelenterates. Can. J. Zool. 66: 1913–1927.

Ates, R. M., 1988. Medusivorous fishes, a review. Zool. Med. Leiden 62: 29–42.

Capitanio, F. L., M. Pajaro & G. B. Esnal, 1997. Appendicularians (Chordata, Tunicata) in the diet of anchovy (*Engraulis anchoita*) in the Argentine Sea. Sci. mar. 61: 9–15.

Demer, D. & R. Hewitt, 1994. *In situ* target strength measurements of Antarctic zooplankton (*Euphausia superba* and *Salpa thompsoni*) at 120 Khz and 200 Khz, corroboration of scattering models, and a statistical technique for delineating species. CCAMLR WG-Krill 4: 22 pp.

El Busto, C., J. I. Carreto, H. Benavides, H. Sancho, D. Cucchi Coleoni, M. Carignan & A. Fernandez, 1993. Paralytic shelfish toxicity in the Argentine Sea, 1990: an extraordinary year. In Smayda, T. J. & Y. Shimizu (eds), Toxic Phytoplankton Blooms in the Sea. Elsevier Science Publishers, Amsterdam: 229–233.

Esnal, G. B., 1981. Thaliacea: Salpidae. In Boltovskoy, D. (ed.), Atlas del Zooplancton del Atlantico Sudoccidental. Publ. Esp. Inidep: 793–808.

Esnal, G. B. & M. C. Daponte, 1999. Salpida. In Boltovskoy, D. (ed.), South Atlantic Zooplankton. Backhuys Publishers, Leiden: 1423–1444.

Fernandez Araoz, N., 1994. Estudios sobre la biomasa de Copepoda (Crustacea), con especial énfasis en Calanoida, del Atlántico Sudoccidental. Parte I y II. Ph.D. Thesis, Univ. Nac. Mar del Plata: 76 + 111 pp.

Fernandez Araoz, N., G. M. Perez Seijas, M. D. Viñas & R. Reta, 1991. Asociaciones zooplanctónicas de la zona común de pesca Argentino-uruguaya en relación con parámetros ambientales, primavera 1986. Frente Maritimo 8: 85–99.

Folt, C. L. & C. W. Burns, 1999. Biological drivers of zooplankton patchiness. Trends Ecol. Evol. 14: 300–305.

Foote, K. G., H. P. Knudsen, G. Vestnes, D. N. Maclennon & E. J. Simmonds, 1987. Calibration of acoustic instruments for fish density estimation: a practical guide. ICES Cooperative Res. Rep. 144: 69 pp.

Forbes, S. T. & O. Nakken, 1972. Manual of Methods for Fisheries Resources Surveys and Appraisal. Part 2. The Use of Acoustic Instruments for Fish Detection and Abundance Estimation. Food and Agricultural Organization, Manual in Fisheries Science 5: 138 pp.

Kashkina, A. A., 1986. Feeding of fishes on salps (Tunicata, Thaliacea). J. Ichthyol. 26: 57–64.

Larson, R. J., 1986. Water content, organic content and carbon and nitrogen composition of medusae from the northeast Pacific. J. exp. mar. Biol. Ecol. 99: 107–120.

Madin, L. P., C. M. Cetta & V. L. McAllister, 1981. Elemental and biochemical composition of salps (Tunicata, Thaliacea). Mar. Biol. 63: 217–226.

Madin, L. P. & P. Kremer, 1995. Determination of the filter-feeding rates of salps (Tunicata, Thaliacea). ICES J. mar. Sci. 52: 583–595.

Madin, L. P., J. E. Purcell & C. B. Miller, 1997. Abundance and grazing effects of *Cyclosalpa bakeri* in the subarctic Pacific. Mar. Ecol. Prog. Ser. 157: 175–183.

Méndez, S., M. Gómez & G. Ferrari, 1997. Planktonic studies of the Río de la Plata and its oceanic front. In Wells, P. G. & G. R. Daborn (eds), The Río de la Plata. An Environmental Overview. An ECOPLATA Project background Rep., Dalhousie Univ., Halifax: 85–112.

Mianzan, H. & P. F. S. Cornelius, 1999. Cubomedusae and scyphomedusae. In Boltovskoy, D. (ed.), South Atlantic Zooplankton. Backhuys Publishers, Leiden: 513–559.

Mianzan H. & R. Guerrero, 2000. Environmental patterns and biomass distribution of gelatinous macrozooplankton. Three study cases in the Southwestern Atlantic. Sci. mar. 64(suppl.1): 215–224.

Mianzan, H., N. Marí, B. Prenski & F. Sanchez, 1996. Fish predation on neritic ctenophores from the Argentine continental shelf: a neglected food resource? Fish. Res. 27: 69–79.

Mianzan, H., M. Pájaro, L. Machinandiarena & F. Cremonte, 1997. Salps: possible vectors of toxic dinoflagellates? Fish. Res. 29: 193–197.

Mianzan, H., C. Lasta, M. Acha, R. Guerrero, G. Macchi & C. Bremec, 2001. The Río de la Plata estuary, Argentina-Uruguay. In Seeliger, U., L. D de Lacerda & B. Kjerve (eds), Ecological Studies 44: Coastal Marine Ecosystems of Latin América, Springer-Verlag 13: 185–204.

Mills, C. E., 1995. Medusae, siphonophores and ctenophores as planktivorous predators in changing global ecosystems. ICES J. mar. Sci. 52: 575–581.

Mills, C. E., 2001. Jellyfish blooms: are populations increasing globally in response to changing ocean conditions? Hydrobiologia 451 (Dev. Hydrobiol. 155): 55–68.

Montoya, N., M. I. Reyero, R. Akselman, J. M. Franco & J. I. Carreto, 1998. Paralytic shelfish toxins in the anchovy *Engraulis anchoita* from the Argentinian coast. In Reguera, B., J. Blanco, M. L. Fernández & T. Wyatt (eds), Harmful Algae. Xunta de Galicia and COI (UNESCO): 72–73.

Omori, M., 1969. Weight and chemical composition of some important oceanic zooplankton in the North Pacific Ocean. Mar. Biol. 3: 4–10.

Purcell, J. E., 1991. A review of cnidarians and ctenophores feeding on competitors in the plankton. Hydrobiologia 216/217: 335–342.

Purcell, J. E., 1997. Pelagic cnidarians and ctenophores as predators: selective predation, feeding rates and effects on prey populations. Ann. Inst. Oceanogr., Paris 73: 125–137.

Purcell, J. E. & M. N. Arai, 2001. Interactions of pelagic cnidarians and ctenophores with fish: a review. Hydrobiologia 451 (Dev. Hydrobiol. 155): 27–44.

Purcell, J. E., T. A. Shiganova, M. B. Decker & E. D. Houde, 2001. The ctenophore *Mnemiopsis* in native and exotic habitats: U. S. estuaries versus the Black Sea basin. Hydrobiologia 451 (Dev. Hydrobiol. 155): 145–176.

Stanton, T. K., D. Chu & P. H. Wiebe, 1996. Acoustic scattering characteristics of several zooplankton groups. ICES J. mar. Sci. 53: 289–295.

Stanton, T. K., P. H. Wiebe, D. Chu, M. C. Benfield, L. Scanlon, L. Martin & R. L. Eastwood, 1994. On acoustic estimates of zooplankton biomass. ICES J. mar. Sci. 51: 505–512.

*Hydrobiologia* **451**: 55–68, 2001.
© 2001 *Kluwer Academic Publishers.*

# Jellyfish blooms: are populations increasing globally in response to changing ocean conditions?

Claudia E. Mills

*Friday Harbor Laboratories and Department of Zoology, University of Washington,*
*620 University Road, Friday Harbor, WA 98250, U.S.A.*
*E-mail: cemills@u.washington.edu*

*Key words:* biodiversity, Cnidaria, Ctenophora, hydromedusae, nonindigenous species, scyphomedusae, siphonophore

## Abstract

By the pulsed nature of their life cycles, gelatinous zooplankton come and go seasonally, giving rise in even the most undisturbed circumstances to summer blooms. Even holoplanktonic species like ctenophores increase in number in the spring or summer when planktonic food is available in greater abundance. Beyond that basic life cycle-driven seasonal change in numbers, several other kinds of events appear to be increasing the numbers of jellies present in some ecosystems. Over recent decades, man's expanding influence on the oceans has begun to cause real change and there is reason to think that in some regions, new blooms of jellyfish are occurring in response to some of the cumulative effects of these impacts. The issue is not simple and in most cases there are few data to support our perceptions. Some blooms appear to be long-term increases in native jellyfish populations. A different phenomenon is demonstrated by jellyfish whose populations regularly fluctuate, apparently with climate, causing periodic blooms. Perhaps the most damaging type of jellyfish increase in recent decades has been caused by populations of new, nonindigenous species gradually building-up to 'bloom' levels in some regions. Lest one conclude that the next millennium will feature only increases in jellyfish numbers worldwide, examples are also given in which populations are decreasing in heavily impacted coastal areas. Some jellyfish will undoubtedly fall subject to the ongoing species elimination processes that already portend a vast global loss of biodiversity. Knowledge about the ecology of both the medusa and the polyp phases of each life cycle is necessary if we are to understand the true causes of these increases and decreases, but in most cases where changes in medusa populations have been recognized, we know nothing about the field ecology of the polyps.

## Introduction

For the purposes of this article, the term 'jellyfish' is used in reference to medusae of the phylum Cnidaria (hydromedusae, siphonophores and scyphomedusae) and to planktonic members of the phylum Ctenophora. Though not closely related, these organisms share many characteristics including their watery or 'gelatinous' nature, and a role as higher-order carnivores in plankton communities; I also cite one example of fragments of the benthic portion of a hydrozoan that occur in high numbers up in the water column, functioning more or less like small jellyfish in terms of their diet. I will not discuss the salps or other planktonic tunicates, which also have bloom characteristics in their appearances and disappearances in the water column, but which are herbivores feeding on very small particles, and have many other quite different aspects to their life cycles.

As parts of the oceans become increasingly disturbed and overfished, there is some evidence that energy that previously went into production of fishes may be switched over to the production of pelagic Cnidaria or Ctenophora (Mills, 1995). Commercial fishing efforts continue to remove top-predator fishes throughout the world oceans (Pauly et al., 1998), and it seems reasonable to watch concomitant trends in jellyfish populations, as jellyfish typically feed on the same kinds of prey as do many either adult or larval fishes.

Increases in jellyfish populations that will be detailed in this paper include some cases where native species have increased in local or regional ecosystems. Increases in jellyfish in some other cases have been the result of recent introductions followed by population explosions of nonindigenous species into coastal ecosystems. Decreases in jellyfish populations have also been documented in local or regional ecosystems in which the habitat has been degraded, typically by increased development, industrialization and pollution, but the proximate causes of these declines may not be evident. Although environmental degradation typically leads to species loss, eutrophication can apparently also sometimes lead to increases of jellyfish in local environments; such cases typically involve only single species and may sometimes in fact be non-native species (see Arai, 2000). Decreasing levels of oxygen (hypoxia) in some bodies of water, often associated with eutrophication, may also favor increases in jellyfish populations (Purcell et al., 2001b.) Finally, there is a small amount of evidence suggesting that some jellyfish blooms may also turn out to be indicators of climate-induced regional regime shifts (Shimomura, 1959; Brodeur et al., 1999), rather than a response to anthropogenic change.

To some extent, what we interpret as a jellyfish bloom may reflect our expectations about an ecosystem. The life cycles of jellyfish lead to the transient appearance of 'blooms' in nearly all cases. Most medusae are budded from benthic polyps – that asexual reproduction process is usually seasonal, with the period of medusa budding varying from days to months long, but nearly always resulting in seasonal appearance and disappearance of medusa populations (with a few exceptions, most medusae appear to live less than one year). Even holoplanktonic species like ctenophores increase in number in the summer when planktonic food is available in greater abundance, giving rise to a form of jellyfish 'bloom'. While locally appearing and disappearing with great annual regularity, medusa and ctenophore populations also undergo interannual fluctuations, some years bringing much larger populations of each species than others. On top of the interannual population variation, there is now evidence for some species showing overall net gains or losses in numbers in certain locations over many-year periods. This last type of population trend is the one I am addressing in the present paper. Blooms resulting from aggregations enhanced by physical oceanographic processes will be reviewed elsewhere in this volume (Graham et al., 2001). Many of the species

included have both planktonic (medusa or ctenophore) and benthic (polyp) phases of their life cycles and both parts must be considered in order to understand the changes taking place. Unfortunately, in most cases, we have information about only the planktonic phase of each life cycle.

There seems to be general agreement that man's activities are having measurable effects on the oceans in many places and certainly in most coastal habitats. Jellyfish populations (hydromedusae, scyphomedusae and ctenophores) respond to these changes, yet the general awareness of these phenomena is still embryonic and few data are available. Local increases in jellyfish abundance seem to be of two sorts. In some cases, species that have always been present suddenly experience severe increases or 'blooms', often with little evidence of what caused the population increase. In other cases, introductions of nonindigenous species to an ecosystem can lead to their unchecked population growth; several recent increases of medusa and ctenophore populations can be attributed to such circumstances. The following selected examples illustrate increases of native jellyfish species and nonindigenous species, as well as decreases in some other species.

## Intrinsic increases in species native to an ecosysem

### Chrysaora, Cyanea *and* Aequorea *populations in the Bering Sea*

Scientists working on Alaskan fisheries for the U.S. National Oceanic and Atmospheric Administration (NOAA) realized about 5 years ago that there is an unprecedented biomass of large jellyfish in the Bering Sea this decade. Biomass, especially of the scyphomedusae *Chrysaora melanaster* Brandt, 1835 [combined with less abundant *Cyanea capillata* (Linnaeus, 1758) and hydromedusae *Aequorea aequorea* (Forsskål, 1775)], has been estimated in NOAA's eastern shelf trawl samples from 1975 to 1999. After remaining more or less constant throughout the 1980s, the combined medusa biomass has increased more than 10-fold over the 1980s values during the 1990s (Brodeur et al., 1999). This increase has been confirmed by other researchers and fishermen who have worked for decades in the Bering Sea and report never having seen such high numbers as in recent years (R. D. Brodeur, pers. comm.). There is very little historic mention of *C. melanaster* in the Bering Sea since its initial description from Avachinsky Bay on Kamtschatka (south of the Aleutians) more than 160 years ago, but at least in the 1990s, this has become the dominant jellyfish in the Bering Sea pelagic ecosystem.

The Bering Sea is a very productive region, accounting for up to 5% of the world's total fishery production and 56% of the U.S. fishery production of fish and shellfish (National Research Council, 1996). There is some public debate over whether or not the Bering Sea is being overfished, with no consensus in sight. Changes in numbers of pollock and other fishes are thought to be effecting an entire trophic cascade in the North Pacific, including the feeding of sea lions, and eventually seemingly causing some killer whales to shift diets to sea otters (Estes et al., 1998).

The dramatic increase in *C. melanaster* is very likely in exchange for some other member(s) of the ecosystem. It is not known at this time what process is facilitating the jellyfish increase, but data imply a correlation with a climate shift in the area that occurred about the same time (Brodeur et al., 1999). Nothing is known about the ecology of the polyp of *C. melanaster*, which is the only life cycle phase present during the colder months of the year, and which could be driving the change. Alternatively, medusae might be surviving better or growing faster, thus accounting for the huge increase in biomass.

### Chrysaora *and* Aequorea *populations in the Benguela Current, Southern Africa and Namibia*

Similar increases in populations of *Chrysaora hysoscella* (Linnaeus, 1766) and *Aequorea aequorea* medusae are implied to have taken place in the Benguela Current off the west coast of Southern Africa during the 1970s (Fearon et al., 1992). The evidence in that case is circumstantial; in fact, the increase is hypothesized only in that these prominent members of the 1980s Benguela Current plankton did not even appear in comprehensive data records from the 1950s and 1960s, and thus their populations are assumed to have previously been very low or nonexistent. High numbers seen in the 1970s have persisted through the 1980s and into the late 1990s off Namibia where both species are still abundantly present, to the point of negatively impacting the fishing industry (Sparks et al., 2001; H. Mianzan, pers. comm.).

In general, there is a long history of removing and discarding jellyfish from net plankton samples because they encumber the smaller planktonic study-species. Additionally, a net full of large scyphomedusae may tear upon recovery and be very costly to repair or replace, so large jellyfish populations are usually systematically avoided by those who study the general plankton (or fish). Such traditions make the 20-year Bering Sea data set that much more remarkable, but also lend some uncertainty to the purported mid-century absence of jellyfish data in the Benguela Current.

### Pelagia *in the Mediterranean*

*Pelagia noctiluca* (Forsskål, 1775) is a small scyphomedusa with fairly cosmopolitan distribution and is apparently endemic to the Mediterranean, as well as other locations (Kramp, 1961). *P. noctiluca* blooms in the Mediterranean have been especially noteworthy because the medusae sting and the summer blooms are considered highly offensive to summer bathers. A several-year bloom in the early 1980s stimulated two 'Jellyfish Blooms' meetings in Athens in 1983 and 1991 (Vućetić, 1983; Boero, 1991). The Mediterranean location of this bloom phenomenon provided an unusually complete simple binary (presence/absence) database covering more than two centuries from which the fluctuation pattern could be teased out (information came from research by European scientists and associated collection data at 4 nearby museums and field laboratories) (see Goy et al., 1989b).

This is perhaps the only species of jellyfish for which regular population fluctuations are known. In the western and central Mediterranean Sea, population peaks have occurred on average every 12 years between 1785 and 1985, with each peak enduring over several years (Goy et al., 1989b). The authors found eight population highs separated by seven low *P. noctiluca* periods and conclude that climatic factors between May and August including low rainfall, high temperature and high atmospheric pressure appear to correlate well with *P. noctiluca* blooms, these factors occurring during the reproductive period for this species and likely influencing it, at least indirectly. Years without large numbers of *P. noctiluca* off southern France seem to be typified by higher numbers of a variety of other species of medusae, siphonophores and ctenophores, but not apparently in such high numbers as might themselves be described as blooms (Morand & Dallot, 1985; Goy et al., 1989a; Buecher et al., 1997).

*Pelagia noctiluca* appears to have been historically much less common in the Adriatic portion of the Mediterranean, where substantial blooms have been noted only in 1910–1914 and 1976–1986 (Purcell et al., 1999b). *P. noctiluca* densities in the northern Adriatic in 1984–1985 reached typical offshore densities of about 20 per m³, with probably wind-driven nearshore accumulations of up to 600 medusae per m³ (Zavodnik, 1987).

Periodic fluctuations of *Pelagia noctiluca* have not been described from elsewhere in the world, although the species is common in warm waters world-wide. It is possible that fluctuating *P. noctiluca* populations also occur elsewhere, but have not been recorded yet.

## Stomolophus nomurai *in the Sea of Japan*

Shimomura (1959) described a very large bloom of very large rhizostome medusae, *Stomolophus nomurai* (Kishinouye, 1922), in the Sea of Japan in 1958. This species seems to be tolerant of a wide temperature range, occurring that year in temperatures from 12–28 °C, and the bloom extended from the Sea of Japan even to waters off Hokkaido. The medusae occurred from the surface to 200 m, being deeper in the day and nearer the surface at night. The bloom, which was a serious fisheries nuisance, lasted well into the winter, ending in December in the Sea of Japan and in January on the Pacific side of Japan. Individual medusae were to 200 cm in diameter, weighed up to 40 kg, and were visible every few m of surface at peak abundance. Fishermen are reported to have caught 20 000–30 000 *S. nomurai* medusae per day during the yellow tail fishery in October and November 1958. While Shimomura reported that local occurrences of this species occur most years, he also cited a bloom of similar magnitude from 20 years earlier, when hindsight indicates that it signaled a regime shift and the end of a several-year sardine peak. Another very large and unpredicted bloom of *S. nomurai* occurred in the Sea of Japan in 1995, with small numbers seen also in 1972 and 1998 (M. Omori & Y. Hirano, pers. comm.). The biology of this very impressive species is so little-known that the whereabouts of the polyps and whether or not there is a small annual production of medusae *somewhere* is not known.

## Siphonophores Muggiaea *in the German Bight and* Apolemia *off the coast of Norway*

Greve (1994) described a seemingly unprecedented invasion of the small calycophoran siphonophore, *Mug-giaea atlantica* Cunningham, 1892, into the German Bight, North Sea, where it is typically absent or found only in very low numbers. *M. atlantica* is commonly collected in the adjoining western English Channel, but its sudden presence in numbers up to 500 per m³ in waters west of Helgoland in July, 1989 seems unexplainable in terms of the understanding of local oceanographic processes. Although other populations of the pelagic ecosystem seemed unaffected by the unusual presence of all of these siphonophores, Greve (1994) pointed out that the usual dominant carnivore in the system is the ctenophore *Pleurobrachia pileus* (O. F. Müller, 1776), which is secondarily controlled by another ctenophore, the highly specific ctenophore-feeder *Beroe gracilis* Künne, 1939. If *M. atlantica* were to replace *P. pileus* as the dominant carnivore, its population would be unlikely to be preyed upon by *B. gracilis*, thus altering the balance in this pelagic ecosystem.

Similarly perhaps, Båmstedt et al. (1998) report an unusual mass occurrence of the virulant and very long siphonophore *Apolemia uvaria* (Lesueur, ?1811) along much of the Norwegian west coast beginning in November, 1997 and lasting at least into February, 1998. The primary effect reported of this invasion was killing of penned (farmed) salmon, although such high numbers of large siphonophores probably also preyed heavily on the coastal zooplankton community.

Although both the *Muggiaea atlantica* and *Apolemia uvaria* events in the North Atlantic were rare and peculiar, they very likely represent changes in local hydrography (Edwards et al., 1999), as does the nearly-annual stranding of the oceanic, neustonic, hydroid *Velella velella* (the by-the-wind-sailor), which is blown ashore in huge numbers by prevailing winds nearly every year in mid-to-late spring along most of the beaches of Washington, Oregon and California. Peculiar winds or ocean currents are certainly capable of causing the appearance of local jellyfish blooms by advecting unusual species into new areas. Whether such species remain in a system long enough to cause long-term changes in the plankton community determines to some extent our interest in the events and whether or not they are seen as 'blooms'.

## Siphonophore Nanomia *in the Gulf of Maine*

Twice in the last two decades, unusually high numbers of the siphonophore *Nanomia cara* A. Agassiz, 1865 have been reported by observers in manned submersibles in the Gulf of Maine (Rogers et al., 1978;

Mills, 1995). The 1975 observations were corroborated by fishermen whose trawl nets were being clogged by the high numbers of siphonophores. The authors respectively reported maximum densities of 1–8 siphonophores per m$^3$ in 1975–1976 and up to 50–100 per m$^3$ (concentrated near the bottom) in 1992–1993. In both cases, access to submersibles for observations was limited and follow-up counts were not performed. It is not known if such high numbers of *N. cara* occur with some regularity or if some special ecological factors in the environment, for instance the poor fishing conditions of the early 1990s (resulting from decades of overfishing), might be related. Further study of *N. cara* in the Gulf of Maine is planned for the next few years (M. Youngbluth, pers. comm.).

*Pelagic hydroid fragments in the Gulf of Maine*

Beginning in 1994, immediately following observations of high numbers of *Nanomia cara* in Wilkinson Basin in the Gulf of Maine (Mills, 1995), another team of scientists found unusually high numbers of floating bits of hydroid colonies suspended in the water column about 150 km to the southeast, on Georges Bank (Madin et al., 1996), where floating hydroids do not typically form a noticeable element in the zooplankton. In fact, from May to June 1994 the net zooplankton in the region was dominated by fragments of hydroid colonies, primarily *Clytia gracilis* (M. Sars, 1850), but also including other *Clytia* and *Obelia* species. A shore-based observer reports large numbers of hydroid polyps washing ashore most autumns since 1990 on the south shore of Nantucket Island, in the same general oceanic system (J. T. Carlton, pers. comm.). Shipboard feeding experiments in 1994 indicated that the unexpected hydroids in the water column might be eating half of the daily production of copepod eggs and 1/4 of the production of copepod nauplii, potentially affecting recruitment of fishes whose larvae normally feed on these copepods (Madin et al., 1996). A careful search of the literature and unpublished data sets by L. P. Madin revealed that similar floating hydroids were reported in the same area by Bigelow in 1913, 1914, 1916, and also recorded in 1939–1940, in the 1980s, and every year since 1994 (L. P. Madin, pers. comm.). Nevertheless, the phenomenon is not well known, and not well understood. The intriguing question of whether these bits of usually-bottom-living animals have been broken up and become resident in the water column as a result of increased trawling activities on the bottom remains un-

answered (Mlot, 1997), but it seems that the numbers of 'jellies' may have increased in recent decades in these important, but now decimated, fishing grounds.

*Scyphomedusae in the northern Gulf of Mexico*

Graham (2001) found some evidence for recent increases in large scyphomedusan jellyfish near the coasts of Alabama, Mississippi, Louisiana and Texas in the Gulf of Mexico by examining bycatch data from routine government shrimp and groundfish surveys from 1985 to 1997. He presents data that suggests localized increases in both number and distribution (increasingly offshore) of *Chrysaora quinquecirrha* (Desor, 1848) medusae in high productivity waters near the Mississippi River delta during the summer months and more general numerical increases in *Aurelia aurita* (Linnaeus, 1758) over much of the study area in the autumn months of the study period. It will likely take at least another decade for these trends to sort out and to fully understand the importance of the apparent increases in medusae in the northern Gulf of Mexico. The additional arrival of a new large jellyfish, the Indo-Pacific rhizostome *Phyllorhiza punctata* von Lendenfeld, 1884, in large numbers from coastal Alabama to Louisiana throughout summer 2000 (W. M. Graham, pers. comm.) may further change the pelagic ecosystem dynamics in this economically important fishing area.

*Pelagic Cnidaria and Ctenophora in the Southern Ocean*

Pagès (1997) suggests in a review of gelatinous zooplankton in the pelagic system of the Southern Ocean that recent several-year periods in which the pelagic ecosystem seems to have been dominated by Cnidaria and Ctenophora may alternate regionally with periods dominated by krill and/or salps. For example, in the Antarctic Polar Frontal Zone in the South Georgia sector, in summer 1994, gelatinous carnivores, together with myctophid fish, were the most abundant nektonic organisms. At the Weddell Sea ice edge in autumn 1986, salps, medusae and ctenophores accounted for 3/4 of the wet weight and 1/3 of dry weight of the micronekton/macrozooplankton in the upper 200 m. Pagès (1997) notes that in spite of reports of such high densities, no comments on the apparent importance of these animals in the pelagic system have been put forth.

In the Antarctic pelagic ecosystem, the greatest scientific effort has been on the commercially valuable krill, with some reluctant study of salps in years when krill were few and salps dominated (Loeb et al., 1997). In some years when salps were locally abundant (1991, 1993), medusae, siphonophores and/or ctenophores were also important players either in nearby regions or even different water layers than the salps. Pagès (1997) found it difficult to put together enough data to even document the apparent recent increase in carnivorous jellies and there is certainly too little data yet to understand the nature or regularity of these apparent fluctuations. Scientists who study carnivorous pelagic jellies are rarely included in studies of the Southern Ocean.

### Mesopelagic and deep-sea jellyfish

If we can barely define the extent of blooms in the visible uppermost layers of the sea, what can be said about midwater jellies? Do they form periodic blooms? Are the huge numbers of *Periphylla* present since the 1970s in a Norwegian fjord reported by Fosså (1992) an invasion of a new habitat, or a bloom of a pre-existing population? Will the 'new' population maintain at its present level, or increase, or fall off?

Jarms et al. (1999) and Youngbluth & Båmstedt (2001) report population increases of the characteristically-midwater *P. periphylla* medusae in Lurefjorden during the 1990s (relative to the data of Fosså, 1992), indicating further changes in the jellyfish population in that fjord, growing in recent years through recruitment within the fjord. Other fjords in Norway also host small numbers of *P. periphylla* medusae, but for reasons that are not clear, have not suffered the explosion that has disrupted fisheries in Lurefjorden, and remain dominated by mesopelagic fishes. Whether Lurefjorden is a special case or represents one of several possible outcomes in such isolated waters is still unknown. Eiane & Bagøien (1999) compare the jellyfish-dominated Lurefjorden with a nearby fish-dominated fjord and note that light levels below 100 m in Lurefjorden are substantially lower and may, therefore, disadvantage visual predators such as fish, giving jellyfish an advantage in exploiting the food web in that particular situation.

In the open ocean, although total biomass drops off with depth (Angel & Baker, 1982), specific biomass and species diversity of medusae and siphonophores (and probably ctenophores) apparently increases with depth at least to several thousand m (Thurston, 1977; Angel & Baker, 1982). Our very rough knowledge of the midwater pelagic ecosystem typically includes only a sketchy understanding of changes over time. Raskoff (2001) has examined a unique decade-long record of the midwater gelatinous fauna in Monterey Bay, California, and found evidence for changes in the mesopelagic jellyfish populations during two El Niño events in the 1990s that might be interpreted as short-term blooms. It is not yet clear if the influence of man's activities extends in general to the deep water column, and how or when we will be able to read the signals if it is.

Burd & Thomson (2000) report increased abundance of medusae in the water column above hydrothemal vent fields compared to the same depths in surrounding waters. Such population differences most likely relate to increased nutrient availability above vent sites and may be better interpreted as site-specific patchiness rather than as blooms.

## Increases in nonindigenous species that recently invaded an ecosystem

### Rhopilema *and other scyphomedusae in the Mediterranean*

*Rhopilema nomadica* Galil, Spanier, & Ferguson, 1990 is a large (to 80 cm diameter) scyphomedusa that has become increasingly abundant in the eastern Mediterranean over the past two decades (see below) (Lotan et al., 1992). Like *Pelagia noctiluca*, another jellyfish resident in the Mediterranean (see above), *R. nomadica*'s presence creates an environmental hazard to fishermen and bathers alike, because it has an unpleasant sting and can be present in such large numbers as to clog fishing nets. First recorded in 1976 in the Mediterranean, the origin of this new hazardous jellyfish is surprisingly unclear. Although assumed to have arrived via the Suez Canal, *R. nomadica* is rare in the Red Sea and is not known from elsewhere [it was only recently described, after its arrival to the Mediterranean (Galil et al., 1990)]. Its reproductive potential in the eastern Mediterranean appears to be very high (Lotan et al., 1992) and it has been present in large numbers off the coast of Israel every summer since 1986 (Lotan et al., 1994).

The population has so far remained in the eastern Mediterranean, where it can now be found in coastal areas from Egypt to Turkey (Kideys & Gücü, 1995; M. Fine, pers. comm.). The jellyfish blooms in Mersin Bay, Turkey, of the mid-1980s, although not identified to species by Bingel et al. (1991) were attributed to a new population of *R. nomadica* (Lotan et al., 1994). In contrast to the fluctuating population peaks demonstrated as typical of *Pelagia noctiluca*, there is no question that the 'bloom' of *Rhopilema nomadica* is simply a population explosion in a new habitat.

Two other species of scyphomedusae have recently become established in the Mediterranean Sea (and elsewhere) – these are *Phyllorhiza punctata* of the warm western Pacific and the epi-benthic, Indo-Pacific species *Cassiopea andromeda* (Forsskål, 1775) (M. Fine, pers. comm.). Little is yet known about the population dynamics of these newer populations.

### *Increasingly common new estuarine hydromedusae in San Francisco Bay and the Chesapeake Bay*

Three species of hydromedusae, all apparently indigenous to the Black Sea, *Maeotias marginata* (=*inexspectata*) Ostroumoff, 1896, *Blackfordia virginica* Mayer, 1910, and *Moerisia* sp., have now become established in both San Francisco Bay and the Chesapeake Bay in North America (Calder & Burrell, 1967, 1969; Mills & Summer, 1995; Mills & Rees, 2000; Rees & Gershwin, 2000). All three species occur in very low salinity regions of these two large estuary systems. In San Francisco Bay, such regions were not previously inhabited by (native) jellyfish (Smith & Carlton, 1975), whereas in the Chesapeake Bay the nonindigenous species join native low-salinity jellyfish populations (Purcell et al., 1999c). The impacts of these new residents (known in the Chesapeake Bay since the 1960s–70s, but only discovered in San Francisco Bay in the 1990s) are largely still unknown, and their ubiquity in these ecosystems has only recently been recognized (Purcell et al., 1999a; Rees & Gershwin, 2000).

Originally located in one tributary to San Francisco Bay (Mills & Summer, 1995), *Maeotias marginata* is now known to be present in at least 4 equally-low salinity sloughs in the region (J. T. Rees, pers. comm.). *Moerisia lyonsi* Boulenger, 1908 has become so numerous in parts of the Chesapeake Bay that it has become an accidental nuisance in experimental mesocosms (Purcell et al., 1999a); a related (or possibly the same) species is still a rarity in San Francisco Bay (Rees & Gershwin, 2000). *B. virginica* is now known to be present in two tributaries to San Francisco Bay, as well as a variety of other harbors all over the world (Mills & Sommer, 1995), but little is known about its effect in these ecosystems.

### *Medusae and ctenophores in the Black Sea*

Pollution, eutrophication and many anthropogenic alterations of the natural environment have vastly altered the Black Sea and its adjacent Sea of Azov in the past 50 years (Zaitsev & Mamaev, 1997). This system provides the most graphic example to date of a highly productive ecosystem that has converted from supporting a number of valuable commercial fisheries to having few fishes and high numbers of 'jellyfishes' – medusae and ctenophores. By the 1960s, largely due to the effects of pollution combined with over fishing, many of the native fishes in the Black Sea had become uncommon, including the jellyfish-eating mackerel *Scomber scombrus*. Perhaps directly related to the loss of this and other fishes, and to increasing eutrophication, the Black Sea has experienced severe outbreaks of three different species of 'jellyfish' in the past 3 decades (Zaitsev & Mamaev, 1997).

The first, little publicized, bloom was of the Mediterranean (and presumptively Black Sea native) scyphomedusa, *Rhizostoma pulmo* (Macri, 1778). In the late 1960s and early 1970s, this species (with bell diameters to 40 cm) reached abundances of 2–3 per m$^3$ in nearshore waters, later washing ashore and leaving 1–1.5 m high piles along beaches in late summer and early fall (Zaitsev & Mamaev, 1997).

For unexplained reasons, the *Rhizostoma pulmo* population dropped back to some lower 'non-bloom' level by the mid-1970s, but at the same time, the resident population of *Aurelia aurita* began to increase, perhaps in response to the generally increasing salinity, as large amounts of incoming fresh water were diverted for irrigation (Studenikina et al., 1991). Increasing numbers of commercial benthic and pelagic fish populations were also crashing during this period, leaving *A. aurita* as one of the top water-column predators. Its population peaked in the late 1980s, with a biomass estimated at 300–500 million tons, when it was estimated to be eating 62% of the annual production of the Black Sea zooplankton, most of which had previously been supporting fishes (Vinogradov et al., 1989; Zaitsev & Mamaev, 1997).

Perhaps because of a several year influx of additional fresh water in the 1980s, the *Aurelia aurita* population began to decline in the late 1980s when the salinity became unfavorably low (Studenikina et al., 1991), but at about the same time the Atlantic American (New England to Argentina) ctenophore, *Mnemiopsis leidyi* A. Agassiz, 1865, was accidentally introduced in the Black Sea, probably via ballast water from a grain ship. This ctenophore is more eury-haline than *A. aurita* and was apparently not adversely affected by the lowering salinity. The nonindigenous *M. leidyi* population first peaked in the late 1980s to early 1990s with an estimated biomass of over a billion tons (300–500 animals per m$^3$ observed in some regions), while in the same period the *A. aurita* population dropped to less than 1/20 of its earlier peak value (Vinogradov et al., 1989; Zaitsev, 1992; Zaitsev & Mamaev, 1997). Nearly all of the zooplankton production in the Black Sea at that time had gone from feeding fishes to feeding ctenophores, and commercial fisheries in the Black Sea became nearly non-existent.

Economic turmoil in Russia during the 1990s has interrupted a regular sampling program in the northern Black Sea, so it is not entirely clear how jelly populations have fared this decade. Both Russian and Turkish scientists are now sampling regionally to follow events in the Black Sea. Kovalev & Piontkovski (1998), Mutlu et al. (1994) and Shiganova (1998) give data that indicate continuing very high numbers of jellies in the system, but with peaks alternating this decade from *Mnemiopsis leidyi* to *Aurelia aurita* and then back to *M. leidyi* by the mid-1990s. Kideys et al. (2000) review data from the past decade including new data from the southern portion of the Black Sea, where *M. leidyi* and *A. aurita* numbers have dropped in the late 1990s. Purcell et al. (2001a) review the history and biology of *M. leidyi* in the Black Sea basin and compare it with the same species in its native North American estuaries elsewhere in this volume.

The arrival of the ctenophore *Beroe ovata* Bruguière, 1789 in the Black Sea in 1997 (Finenko et al., 2001; Shiganova et al., 2001) promises to redirect the story there yet again. This species already appears to be having some local effect on *Mnemiopsis* populations. *B. ovata* is well known in the Mediterranean, but had not previously been recorded in the Black Sea; in some ways its extension into the Black Sea might be seen as yet another nonindigenous species there.

The gradual domination of the Black Sea and Sea of Azov pelagic systems by jellies is a story that combines outbreaks of both native and introduced medusae and ctenophores. There is little doubt that extensive anthropogenic alterations over time have led to severe disruptions in the functioning pelagic ecosytem, and the absence of jellyfish predators has undoubtedly fueled these imbalances. *Mnemiopsis leidyi* was newly observed in the Caspian Sea in 1999 (Volovik, 2000, T. A. Shiganova, pers. comm.), and it is predicted that again, a highly unique pelagic ecosystem with a large number of endemic species and important fisheries resources, may be massively disrupted by the arrival of this ctenophore.

*Aurelia blooms around the world*

In addition to the 1980s bloom in the Black Sea, *Aurelia 'aurita'* populations have recently swelled to huge numbers in many coastal areas worldwide, often causing significant economic damage. Although it is usually considered to be a cosmopolitan species, I currently favor a theory of 19th and 20th century introduction of *A. aurita* to harbors throughout the world via shipping, citing the fairly recent nuisance status of this species in many areas. 'Rediscovery' of *A. labiata* Chamisso and Eysendardt, 1821, another species that is apparently endemic to the Pacific Coast of North America (L. A. Gershwin, pers. comm.) adds weight to the idea of a more restricted original range of *A. aurita*, probably in the North Atlantic. Understanding the biogeography of all of the *Aurelia* species requires serious molecular genetic study. Several researchers are undertaking aspects of the problem at this time (see Dawson & Martin, 2001), and the situation may eventually be sorted out.

*Aurelia* sp. is known to have been present in Japan at least since 1891 (Kishinouye, 1891) and was first mentioned in Tokyo Bay in 1915 (Hirasaka, 1915). The species in Japan is generally thought to be *A. aurita* and is well-known by individuals in the American aquarium display business to culture differently than the west coast of North America species, *A. labiata* (Japanese material strobilates nearly all of the time in the laboratory, wheras *A. labiata* polyps from the west coast of North America strobilate only occasionally). Whether or not *A. aurita* is indigenous to Japan is not known, but summer blooms of this species in Japanese bays have caused increasing socio-economic problems since the 1950s. Shimomura (1959) documents disruptions of fisheries in the Sea of Japan by *A. aurita* blooms as early as 1950. Matsueda (1969) describes power plant restrictions and temporary shut-downs throughout Japan due to clogging of intake screens by *A. aurita* medusae beginning in the mid-1960s as increasing numbers of power plants used seawater cooling systems. This technology highlighted the already-occurring summer *A. aurita* blooms, whose origins in time are obscure. Problems in net fishing and power plant operations in Tokyo Bay from exceedingly high *A. aurita* numbers are described by Yasuda (1988). Omori et al. (1995) note that the importance of

*A. aurita* in the pelagic ecosystem in Tokyo Bay began in the 1960s when the dominant copepods switched from *Acartia omorii* Bradford and *Paracalanus* spp. to the smaller *Oithona davisae* Ferrari & Orsi. This switch had many food web ramifications which may have included an increasingly favorable situation for *A. aurita* in Tokyo Bay (although this is not proven). Feeding and digestion by *A. aurita* in Tokyo Bay has recently been studied by Ishii & Tanaka (2001). Similar general zooplankton changes may explain the increasing importance of *A. aurita* in bays throughout Japan in the latter half of this century, but the blooms may have already occurred in some places decades earlier (M. Toyokawa, pers. comm.).

*Aurelia 'aurita'* has caused upsets in power plants throughout the world. Besides Japan, shut-downs due to medusae clogging the seawater intake screens have been reported in the Baltic region, Korea, India, Saudi Arabia, Australia and more (Möller, 1984a; Rajagopal et al., 1989; Y. Fadlallah & S. Baker, pers. comm.). Half of the Philippines lost power on December 10, 1999 when large numbers of *Aurelia* sp. were sucked into the cooling system of a power station there (The Economist, Dec. 18, 1999, pp. 36–37). The 'bloom' nature of such events can be seen at many levels. *A. aurita*, like most jellyfish, has a more or less annual cycle, so the clogging problem peaks annually during the months that medusae are largest and also most abundant. There is also some variation between years, with clogging being much more problematic in some years than others. The final issue is whether or not the entire clogging phenomenon is becoming increasingly severe over a period of several to many years.

It appears that *Aurelia aurita* may become especially abundant in highly eutrophic areas, and if so, increasing eutrophication of some harbors may increase *A. aurita* globally in coming decades. Elefsis Bay in Greece supports a uniquely high *A. aurita* population in the Mediterranean, which is assumed to correlate to the high eutrophication there (Papathanassiou et al., 1987). Sewage effluent, in this case from Athens, provides both inorganic and organic nutrients that are available to medusae both directly and indirectly through the food web (Wilkerson & Dugdale, 1984).

*Aurelia aurita* has been extensively studied in a variety of locations in the North Sea and Baltic Sea (Hay & Hislop, 1980; Möller, 1980, 1984a,b; Gröndahl, 1988; Schneider & Behrends, 1994; reviewed by Lucas, 2001), where it is implied that it is a natural endemic species. Such *A. aurita* populations fluctuate enormously throughout the year as annual medusa populations mature and die fairly synchronously, but the scientific literature contains no evidence or mention of long-term changes in these *A. aurita* populations in recent decades. Schneider & Behrends (1994) discuss large, interannual variations in the *A. aurita* medusa populations in Kiel Bight, but their data (1978–1993) and discussion gives no hint of gradual or abrupt increase in the Baltic Sea populations over time. Such lack of change in comparison with *A. aurita* populations in Japan is noteworthy and possibly indicative of longer residence time of *A. aurita* in northern Europe, or of different scenarios of anthropogenic disturbance and biological response in bays in Europe and Asia.

## Recent decreases in jellyfish populations

Decreases in either jellyfish abundance or species richness or both have been reported in a variety of locations worldwide in the past decade. Examples enumerated below come from both the community/ecosystem level and from the level of a single taxon in a fairly restricted location.

### Hydromedusae in the northern Adriatic Sea

Benović et al. (1987) report a decrease in hydromedusa abundance and species richness in the northern Adriatic, which they believe correlates with declining water quality resulting from increasingly eutrophic nearshore conditions. There is a long tradition of marine plankton work in the North Adriatic and the fauna is well-known. Since the 1960s, there has been a trend in those waters toward growing oxygen depletion in near-bottom water, while at the same time the near-surface water was becoming increasingly hypersaturated with oxygen. The authors report a substantial loss of metagenic anthomedusae (22 species) and leptomedusae (9 species), out of a total of 42 known regional species of hydromedusae, have disappeared from the northern Adriatic from 1910 to 1984. All of the affected species have benthic polyps that may have been eliminated by the low oxygen bottom water, while the holoplanktonic (without benthic hydroids) trachymedusae and narcomedusae were only slightly affected by changes in the water column. The effects on polyps are all inferred, with no actual polyp studies available. The continuation of this study for more than another decade, through 1997 (Benović et al., in press), shows that the biodiversity of hydromedusae in the northern Adriatic has remained low in spite of evidence for seasonal immigration by medusae of species previously established there. Low oxygen conditions on the bottom remain unfavorable to benthic polyps living in the northern Adriatic.

## Medusa biodiversity in St. Helena Bay, west coast of South Africa

Buecher & Gibbons (2000) examined hydromedusa, scyphomedusa and ctenophore populations within this oceanically-influenced bay in a set of 264 samples taken from 1988 to 1997. The area is important as a major recruitment center for commercial pelagic fish in the southern Benguela ecosystem. The authors identified a total of 53 species of pelagic Cnidaria and Ctenophora from the Bay, but show a decided trend towards a loss of species richness of this gelatinous fauna during the 10-year study period, with 21–24 species present each of the first 5 years, declining to only 11 or 12 species present the last 2 years. No reason for this decrease in biodiversity over the 10-year study period was proposed, and although it is likely that a decrease in sample numbers may account for loss of numerous rare species in later years, it is also possible that an undefined change in the ecosystem is recorded in this loss of biodiversity.

## Aequorea victoria in Washington State and British Columbia

Aequorea victoria (Murbach & Shearer, 1902) (sometimes locally reported as A. aequorea or A. forskalea in the literature) has been perhaps the most abundant medusa both numerically and in terms of biomass in parts of the Puget Sound/Strait of Georgia inland marine waters of Washington State, U.S.A., and British Columbia, Canada. Between the early 1960s and the mid-1990s, hundreds of thousands of these medusae were collected by various different laboratory groups in order to extract natural aequorin and green fluorescent protein (gfp), respectively luminescent and fluorescent proteins that have proved useful in biological and medical research. Annual collections varied enormously, but it is estimated that 25 000 to 150 000 Aequorea mature medusae were harvested nearly every summer during that period in and around Friday Harbor, Washington (J. F. Blinks & O. Shimomura, pers. comm.). Only the largest specimens were collected, not out of special concern for their ecology but because the protein yield per individual was more

favorable. Since the early 1990s, both the numbers and maximum sizes of Aequorea medusae in the Friday Harbor area have fallen off gradually, but continuously, so that in the late 1990s, there have not been enough animals for commercial collections (although the ability to manufacture aequorin and gfp has also now largely supplanted the need for collection). Finding even 1000 Aequorea medusae over several weeks would have been difficult during the summers of 1997, 1998, and 1999, and average size was much smaller than in earlier year (few were as large as the minimum size example painted onto remaining collecting screens from the 1970s); in summer 2000, numbers have been the lowest yet (C. E. Mills, pers. obs.).

In trying to assess this obvious population decrease in which little real population data is available, it should be noted that we also have no idea what might be the functional geographic limits of the Aequorea victoria population that is resident in and near Friday Harbor. We do not know if the decline is a slow response to nearly three decades of collections or (more likely) if it is the result of an unrelated environmental change. Furthermore, there is virtually no field data about the polyp phase of A. victoria, outside of a few isolated field collections over the decades. As in the cases of Polyorchis penicillatus (Eschscholtz, 1829), Spirocodon saltatrix (Tilesius, 1818) (below) and Chrysaora melanaster (see above), one cannot determine whether the change has been effected by the medusa or polypoid phase of the life cycle of this species. We have no sense of when, if ever, the A. victoria medusa population will rebound.

## The family Polyorchidae in the North Pacific

The hydrozoan family Polyorchidae is comprised of five species of anthomedusae that have historically inhabited many of the protected bays and inlets between about 30° and 55° N Latitude on both sides of the north Pacific Ocean (Uchida, 1927; Kramp, 1961; Rees & Larson, 1980; Y. M. Hirano, pers. comm.; C. E. Mills, unpublished). On the Asian side of the Pacific, two species have non-overlapping distributions: the medusa Spirocodon saltatrix used to be commonly found from southern Kyushu to the top of Honshu (Japan), and Polyorchis karafutoensis Kishinouye, 1910 occurs from central Hokkaido to northern Sakhalin Island (Russia). On the west coast of North America, Polyorchis penicillatus has been collected from the northern Gulf of California and San Diego to the Aleutian Islands, and is joined by Polyorchis haplus

Skogsberg, 1948 and *Scrippsia pacifica* Torrey, 1909 in California.

All of these large, easily recognized, hydromedusae seem to assume the same ecological role, spending much of their time perched on their tentacles and feeding on the bottom, but also swimming and feeding in the water column some of the time. All of them presumably have a benthic polyp phase in their life cycle, but it is not known for any species, in spite of many attempts to raise the easily-obtained planula larvae in the laboratory from field-collected medusae. The polyp could be the most vulnerable part of the polyorchid life cycle, yet we know nothing about it.

*Spirocodon saltatrix* is now uncommon or rare throughout most of its range in Japan (Y. M. Hirano, pers. comm.) and *P. penicillatus* is much less abundant in some Washington and British Columbian bays and probably throughout California than it was only a couple of decades ago (a strong showing of *P. penicillatus* in some central California bays in the winter of 1999–2000 now clouds the picture slightly). There is too little information about the remaining 3 species to speculate on the robustness of their present populations.

The problem of marine habitat loss as a result of coastal development is sadly exemplified by this family of large, showy hydromedusae. Once well-known in shallow bays along more than 1500 linear miles of coastlines on both sides of the North Pacific Ocean, these medusae are now increasingly rare. The general urbanization of many bays, accompanied by dredging and filling, and construction of marinas and tourist facilities along most of this range has all contributed to a vast degradation of their habitat. Additionally, both *S. saltatrix* and *P. penicillatus* have been favorite research animals and heavily collected from many of their previous haunts; one cannot discount the possibility that over-collection by scientists has led to their demise in some bays. These large medusae are correspondingly highly fecund, producing around 10 000 eggs per day for much of their lives; it is possible that some aspect of their ecology requires this huge egg/embryo input in order to maintain stable local populations. One wonders how long into the next millennium this family of unusual semi-benthic medusae will manage to persist.

## Conclusions

It has been said that many biologists who have observed marine communities over a period of time believe they have seen significant declines in populations of some species, but that they do not have the data to confirm or refute these impressions (Thorne-Miller & Catena, 1991). Even though relatively few scientists study pelagic medusae or ctenophores, many cases of upsets in medusa or ctenophore populations have been documented as man's influence on the oceans becomes increasingly apparent. The problem of ocean change is very real.

It is unfortunate that we have so little population and ecological data about medusae and ctenophores in the field that we usually cannot presently distinguish between natural fluctuations and long term, possibly irreversible, change. Even in the case of *Chrysaora melanaster* in the Bering Sea, with an unusual 25 year data set (1975–1999), it would require data from the preceding 20 years, when the international fin-fishing effort was considerably less, to understand if man's influence in the Bering Sea is driving the ecosystem toward a long-term increase in medusae. Seemingly enormous numbers of jellyfish are now being harvested in Southeast Asia for the global market (Omori & Nakano, 2001). We know nothing about the population biology of these species; in many cases, we do not even know the species names of the commodity-products coming to market, and certainly we do not know how these populations are responding either to to harvest pressure or to nearshore changes in recent decades. Uye & Kasuya (1999) suggest that numbers of indigenous ctenophores, especially *Bolinopsis mikado* (Moser, 1907), may be rising in some Japanese coastal waters; this situation bears following in coming years.

Although Cnidaria and Ctenophora are low on the phylogenetic tree, they generally feed high on marine food chains, directly competing in many cases with fishes for food. Massive removals of fishes from ecosystems might be expected to open up food resources for gelatinous predators, which seems in some cases to be what has happened. Further interactions between jellyfishes and fish are explored by Purcell & Arai (2001), elsewhere in this volume. Although some jellyfishes are preyed upon by fishes, others of the carnivorous jellies prey nearly exclusively other jellies, forming a somewhat independent food web named the 'jelly web' by B.H. Robison (Robison & Connor, 1999).

Largely through the aquarium industry's handsome efforts to display jellyfish, the general public is becoming much better acquainted with this group of animals at the same time that jellyfish seem to be increasing their presence on the world stage of ocean ecosystems. If I could offer one piece of advice to young scientists seeking a project on pelagic Cnidaria, it would be to study the population dynamics of some of the common and abundant species that occur in coastal regions throughout the world, whose populations must be substantially influenced by changes in their local ecosystems, and about which we know next to nothing beyond their names.

## Acknowledgements

Thanks to Larry Madin, Masaya Toyokawa, Hermes Mianzan, Francesc Pagès, Yusef Fadlallah, Yayoi Hirano, Maoz Fine, John Rees, Lisa Gershwin, John Blinks, Osamu Shimomura and Jenny Purcell for providing information that has not yet reached the literature. Thanks to Francesc Pagès, Richard Brodeur, Hermes Mianzan, Jenny Purcell, Monty Graham and Jim Carlton for offering valuable suggestions that improved the manuscript. And finally, thanks to the buzz on the Internet and by the media which caused me to buckle down and take a serious look at this subject.

## References

Angel, M. V. & A. de C. Baker, 1982. Vertical distribution of the standing crop of plankton and micronekton at three stations in the northeast Atlantic. Biol. Oceanogr. 2: 1–30.

Arai, M. N., 2001. Pelagic coelenterates and eutrophication: a review. Hydrobiologia 451 (Dev. Hydrobiol. 155): 69–87.

Båmstedt, U., J. H. Fosså, M. B. Martinussen & A. Fosshagen, 1998. Mass occurrence of the physonect siphonophore Apolemia uvaria (Lesueur) in Norwegian waters. Sarsia 83: 79–85.

Benović, A., D. Justić & A. Bender, 1987. Enigmatic changes in the hydromedusan fauna of the northern Adriatic Sea. Nature 326: 597–600.

Benović, A., D. Lučić & V. Onofri, in press. Does change in an Adriatic hydromedusan fauna indicate an early phase of marine ecosystem destruction? P.S.Z.N.: Mar. Ecol.

Bingel, F., D. Avsar & A. C. Gücu, 1991. Occurrence of jellyfish in Mersin Bay. In Jellyfish Blooms in the Mediterranean, Proceedings of the II Workshop on Jellyfish in the Mediterranean Sea. MAP Technical Reports Series No. 47, UNEP, Athens: 65–71.

Boero, F., 1991. Contribution to the understanding of blooms in the marine environment. In Jellyfish Blooms in the Mediterranean, Proceedings of the II Workshop on Jellyfish in the Mediterranean Sea. MAP Technical Reports Series No. 47, UNEP, Athens: 72–76.

Brodeur, R. D., C. E. Mills, J. E. Overland, G. E. Walters & J. D. Schumacher, 1999. Evidence for a substantial increase in gelatinous zooplankton in the Bering Sea, with possible links to climate change. Fish. Oceanogr. 8: 296–306.

Buecher, E. & M. J. Gibbons, 2000. Interannual variation in the composition of the assemblages of medusae and ctenophores in St Helena Bay, southern Benguela ecosystem. Sci. mar. 64(Supl. 1): 123–124.

Buecher, E., J. Goy, B. Planque, M. Etienne & S. Dallot, 1997. Long-term fluctuations of Liriope tetraphylla in Villefranche Bay between 1966 and 1993 compared to Pelagia noctiluca pullulations. Oceanol. Acta 20: 145–157.

Burd, B. J. & R. E. Thomson, 2000. Distribution and relative importance of jellyfish in a region of hydrothermal venting. Deep Sea Res. Pt. I: Oceanogr. Res. 47: 1703–1721.

Calder, D. R. & V. G. Burrell Jr., 1967. Occurrence of Moerisia lyonsi (Limnomedusae, Moerisiidae) in North America. Am. midl. Nat. 78: 540–541.

Calder, D. R. & V. G. Burrell Jr., 1969. Brackish water hydromedusa Maeotias inexpectata in North America. Nature 222: 694–695.

Dawson, M. N & L. E. Martin, 2001. Geographic variation and ecological adaptation in Aurelia (Scyphozoa, Semaeostomeae): some implications from molecular phylogenetics. Hydrobiologia 451 (Dev. Hydrobiol. 155): 259–273.

Edwards, M., A. W. G. John, H. G. Hunt & J. A. Lindley, 1999. Exceptional influx of oceanic species into the North Sea late 1997. J. mar. biol. Ass. U.K. 79: 737–739.

Eiane, K. & E. Bagøien, 1999. Fish or jellies – a question of visibility? Limnol. Oceanogr. 44: 1352–1357.

Estes, J. A., M. T. Tinker, T. M. Williams & D. F. Doak, 1998. Killer whale predation on sea otters linking oceanic and nearshore ecosystems. Science 282: 473–476.

Fearon, J. J., A. J. Boyd & F. H. Schülein, 1992. Views on the biomass and distribution of Chrysaora hysoscella (Linné, 1766) and Aequorea aequorea (Forskål, 1775) off Namibia, 1982–1989. Sci. mar. 56: 75–84 & 383–384 (errata).

Finenko, G. A., B. E. Anninsky, Z. A. Romanova, G. I. Abolmasova & A. E. Kideys, 2001. Chemical composition, respiration and feeding rates of the new alien ctenophore, Beroe ovata, in the Black Sea. Hydrobiologia 451 (Dev. Hydrobiol. 155): 177–186.

Fosså, J. H., 1992. Mass occurrence of Periphylla periphylla (Scyphozoa, Coronatae) in a Norwegian fjord. Sarsia 77: 237–251.

Galil, B. S., E. Spanier & W. W. Ferguson, 1990. The scyphomedusae of the Mediterranean coast of Israel, including two Lessepsian migrants new to the Mediterranean. Zool. Meded. (Leiden) 64: 95–105.

Goy, J., S. Dallot & P. Morand, 1989a. Les proliférations de la méduse Pelagia noctiluca et les modifications associées de la composition du macroplancton gélatineux. Oceanis 15: 17–23.

Goy, J., P. Morand & M. Etienne, 1989b. Long term fluctuation of Pelagia noctiluca (Cnidaria, Scyphomedusa) in the western Mediterranean Sea. Prediction by climatic variables. Deep-Sea Res. 36: 269–279.

Graham, W. M., 2001. Numerical increases and distributional shifts of Chrysaora quinquecirrha (Desor) and Aurelia aurita (Linné ) (Cnidaria: Scyphozoa) in the northern Gulf of Mexico. Hydrobiologia 451 (Dev. Hydrobiol. 155): 97–111.

Graham, W. M., Pagès, F. & W. M. Hamner, 2001 A physical context for gelatinous zooplankton aggregations: a review. Hydrobiologia 451 (Dev. Hydrobiol. 155): 199–212.

Greve, W., 1994. The 1989 German Bight invasion of Muggiaea atlantica. ICES J. mar. Sci. 51: 355–358.

Gröndahl, F., 1988. A comparative ecological study on the scyphozoans *Aurelia aurita*, *Cyanea capillata* and *C. lamarckii* in the Gullmar Fjord, western Sweden, 1982 - 1986. Mar. Biol. 97: 541–550.

Hay, S. J. & J. R. G. Hislop, 1980. The distribution and abundance of scyphomedusae in the North Sea during the summer of 1979. Int. Counc. Explor. Sea, Comm. Meet. (Oceanogr. Biol. Comm.) L: 1–7.

Hirasaka, K., 1915. Medusae of Tokyo Bay. Jap. J. Zool. 27: 164 (in Japanese)

Ishii, H. & F. Tanaka, 2001. Food and feeding of *Aurelia aurita* in Tokyo Bay with an analysis of stomach contents and a measurement of digestion times. Hydrobiologia 451 (Dev. Hydrobiol. 155): 311–320.

Jarms, G., U. Båmstedt, H. Tiemann, M. B. Martinussen & J. H. Fosså, 1999. The holoplanktonic life cycle of the deep-sea medusa *Periphylla periphylla* (Scyphozoa, Coronata). Sarsia 84: 55–65.

Kideys, A. E. & A. C. Gücü, 1995. *Rhopilema nomadica*: a lessepsian scyphomedusan new to the Mediterranean coast of Turkey. Israel J. Zool. 41: 6145–617.

Kideys, A. E., A. V. Kovalev, G. Shulman, A. Gordina & F. Bingel, 2000. A review of zooplankton investigations of the Black Sea over the last decade. J. mar. Systems 24: 355–371.

Kishinouye, K., 1891. Mizu-Kurage. Jap. J. Zool. 3: 289–291 (in Japanese).

Kovalev, A. V. & S. A. Piontkovski, 1998. Interannual changes in the biomass of the Black Sea gelatinous zooplankton. J. Plankton Res. 20: 1377–1385.

Kramp, P. L., 1961. Synopsis of the medusae of the world. J. mar. biol. Ass. U. K. 40: 1–469.

Loeb, V., V. Siegel, O. Holm-Hansen, R. Hewitt, W. Fraser, W. Trivelpiece & S. Trivelpiece, 1997. Effects of sea-ice extent and krill or salp dominance on the Antarctic food web. Nature 387: 897–900.

Lotan, A., R. Ben-Hillel & Y. Loya, 1992. Life cycle of *Rhopilema nomadica*: a new immigrant scyphomedusan in the Mediterranean. Mar. Biol. 112: 237–242.

Lotan, A., M. Fine & R. Ben-Hillel, 1994. Synchronization of the life cycle and dispersal pattern of the tropical invader scyphomedusan *Rhopilema nomadica* is temperature dependent. Mar. Ecol. Prog. Ser. 109: 59–65.

Lucas, C. H., 2001. Reproduction and life history strategies of the common jellyfish, *Aurelia aurita*, in relation to its ambient environment. Hydrobiologia 451 (Dev. Hydrobiol. 155): 229–246.

Madin, L. P., S. M. Bollens, E. Horgan, M. Butler, J. Runge, B. K. Sullivan, G. L. Klein-MacPhee, E. Durbin, A. G. Durbin, D. Van Keuren, S. Plourde, A. Bucklin & M. E. Clarke, 1996. Voracious planktonic hydroids: unexpected predatory impact on a coastal marine ecosystem. Deep-Sea Res. II 43: 1823–1829.

Matsueda, N., 1969. Presentation of *Aurelia aurita* at thermal power station. Bull. mar. biol. Sta. Asamushi 13: 187–191.

Mills, C. E., 1995. Medusae, siphonophores and ctenophores as planktivorous predators in changing global ecosystems. ICES J. mar. Sci. 52: 575–581.

Mills, C. E. & J. T. Rees, 2000. New observations and corrections concerning the trio of invasive hydromedusae *Maeotias marginata* (=*M. inexpectata*), *Blackfordia virginica* and *Moerisia* sp. in the San Francisco Estuary. Sci. mar. 64(Suppl. 1): 151–155.

Mills, C. E. & F. Sommer, 1995. Invertebrate introductions in marine habitats: two species of hydromedusae (Cnidaria) native to the Black Sea, *Maeotias inexspectata* and *Blackfordia virginica*, invade San Francisco Bay. Mar. Biol. 122: 279–288.

Mlot, C., 1997. Unusual fish threat afloat in the Atlantic. Sci. News 152: 325.

Möller, H., 1980. Population dynamics of *Aurelia aurita* medusae in Kiel Bight, Germany (FRG). Mar. Biol. 60: 123–128.

Möller, H., 1984a. Effects on jellyfish predation by fishes. Proceedings of the Workshop on Jellyfish Blooms in the Mediterranean, Athens 1983. UNEP 1984: 45–59.

Möller, H., 1984b. Reduction of a larval herring population by jellyfish predator. Science 224: 621–622.

Morand, P. & S. Dallot, 1985. Variations annuelle et pluriannuelles de quelques espèces du macroplancton cotier de la Mer Ligure (1898–1914). Rapp. Comm. Int. Mer Méd. 29: 295–297.

Mutlu, E., F. Bingel, A. C. Gücü, V. V. Melnikov, U. Nierman, N. A. Ostr & V. E. Zaika, 1994. Distribution of the new invader *Mnemiopsis* sp. and the resident *Aurelia aurita* and *Pleurobrachia pileus* populations in the Black Sea in the years 1991–1993. ICES J. mar. Sci. 51: 407–421.

National Research Council, 1996. The Bering Sea Ecosystem: Report of the Committee on the Bering Sea Ecosystem. National Academy Press, Washington, D.C.: 307 pp.

Omori, M., H. Ishii & A. Fujinaga, 1995. Life history strategy of *Aurelia aurita* (Cnidaria, Scyphomedusae) and its impact on the zooplankton community of Tokyo Bay. ICES J. mar. Sci. 52: 597–603.

Omori, M. & E. Nakano, 2001. Jellyfish fishery in southeast Asia Hydrobiologia 451 (Dev. Hydrobiol. 155): 19–26.

Pagès, F., 1997. The gelatinous zooplankton in the pelagic system of the Southern Ocean: a review. Ann. Inst. océanogr., Paris 73: 139–158.

Papathanassiou, E., P. Panayotidis & K. Anagnostaki, 1987. Notes on the biology and ecology of the jellyfish *Aurelia aurita* Lam. in Elefsis Bay (Saronikos Gulf, Greece). P. S. Z. N. I.: Mar. Ecol. 8: 49–58.

Pauly, D., V. Christensen, J. Dalsgaard, R. Froese & F. Torres Jr., 1998. Fishing down marine food webs. Science 279: 860–863.

Purcell, J. E. & M. N. Arai, 2001. Interactions of pelagic cnidarians and ctenophores with fish: a review. Hydrobiologia 451 (Dev. Hydrobiol. 155): 27–44.

Purcell, J. E., U. Båmstedt & A. Båmstedt, 1999a. Prey, feeding rates and asexual reproduction rates of the introduced oligohaline hydrozoan *Moerisia lyonsi*. Mar. Biol. 134: 317–325.

Purcell, J. E., D. L. Breitburg, M. B. Decker, W. M. Graham, M. J. Youngbluth & K. A. Raskoff, 2001b. Pelagic cnidarians and ctenophores in low dissolved oxygen environments: a review. In Rabalais, N. N. & R. E. Turner (eds), Coastal Hypoxia: Consequences for Living Resources and Ecosystems, American Geophysical Union. Coastal & Estuar. Stud. 58: 77–100.

Purcell, J. E., A. Malej & A. Benović, 1999b. Potential links of jellyfish to eutrophication and fisheries. In Malone, T. C., A. Malej, L. W. Harding Jr., N. Smodlana & R. E. Turner (eds), Ecosystems at the Land-Sea Margin: Drainage Basin to Coastal Sea. Coastal and Estuar. Stud. 55: 241–263.

Purcell, J. E., T. A. Shiganova, M. B. Decker & E. D. Houde, 2001a. The ctenophore *Mnemiopsis* in native and exotic habitats: U.S. estuaries *versus* the Black Sea basin. Hydrobiologia 451 (Dev. Hydrobiol. 155): 145–175.

Purcell, J. E., J. R. White, D. A. Nemazie & D. A. Wright, 1999c. Temperature, salinity and food effects on asexual reproduction and abundance of the scyphozoan *Chrysaora quinquecirrha*. Mar. Ecol. Prog. Ser. 180: 187–196.

Rajagopal, S., K. V. K. Nair & J. Azariah, 1989. Some observations on the problem of jelly fish ingress in a power station cooling system at Kalpakkam, east coast of India. Mahasagar Quart. J. Oceanogr., Nat. Inst. Oceanogr, Goa, India 22: 151–158.

Raskoff, K. A., 2001 The impact of El Niño events on populations of mesopelagic hydromedusae. Hydrobiologia 451 (Dev. Hydrobiol. 155): 121–129.

Rees, J. T. & L. A. Gershwin, 2000. Non-indigenous hydromedusae in California's upper San Francisco Estuary: life cycles, distribution, and potential environmental impacts. Sci. Mar. 64(Suppl. 1): 73–86.

Rees, J. T. & R. J. Larson, 1980. Morphological variation in the hydromedusa genus *Polyorchis* on the west coast of North America. Can. J. Zool. 58: 2089–2095.

Robison, B. & J. Connor, 1999. The Deep Sea. Monterey Bay Aquarium Press, Monterey, California: 80 pp.

Rogers, C. A., D. C. Biggs & R. A. Cooper, 1978. Aggregation of the siphonophore *Nanomia cara* in the Gulf of Maine: observations from a submersible. Fish. Bull. 76: 281–284.

Schneider, G. & G. Behrends, 1994. Population dynamics and the trophic role of *Aurelia aurita* medusae in the Kiel Bight and western Baltic. ICES J. mar. Sci. 51: 359–367.

Shiganova, T. A., 1998. Invasion of the Black Sea by the ctenophore *Mnemiopsis leidyi* and recent changes in pelagic community structure. Fish. Oceanogr. 7: 305–310.

Shiganova, T. A., Yu. V. Bulgakova, S. P. Volovik, Z. A. Mirzoyan & S. I. Dudkin, 2001. The new invader *Beroe ovata* Mayer, 1912 and its effect on the ecosystem in the northeastern Black Sea. Hydrobiologia 451 (Dev. Hydrobiol. 155): 187–197.

Shimomura, T., 1959. On the unprecedented flourishing of 'Echizen-Kurage', *Stomolophus nomurai* (Kishinouye), in the Tsushima Warm Current regions in autumn, 1958. Bull. Jap. Sea Reg. Fish. Res. Lab. 7: 85–107.

Smith, R. I & J. T. Carlton (eds), 1975. Light's Manual: Intertidal Invertebrates of the Central California Coast. University of California Press, Berkeley: 716 pp.

Sparks, C., E. Buecher, A. S. Brierley, B. E. Axelsen, H. Boyer & M. J. Gibbons, 2001. Observations on the distribution and relative abundance of the scyphomedusan *Chrysaora hysoscella* (Linné, 1776) and the hydrozoan *Aequorea aequorea* (Forskål, 1775) in the northern Benguela ecosystem. Hydrobiologia 451 (Dev. Hydrobiol. 155): 275–286.

Studenikina, Ye. I., S. P. Volovik, I. A. Mirzoyan & G. I. Luts, 1991. The ctenophore *Mnemiopsis leidyi* in the Sea of Azov. Oceanology 31: 722–725.

Thorne-Miller, B. & J. Catena, 1991. The Living Ocean: Understanding and Protecting Marine Biodiversity. Island Press, Washington, D.C., 180 pp.

Thurston, M. H., 1977. Depth distributions of *Hyperia spinigera*

Bovallius, 1889 (Crustacea: Amphipoda) and medusae in the North Atlantic Ocean, with notes on the associations between *Hyperia* and coelenterates. In Angel, M. (ed.), A Voyage of Discovery: George Deacon 70th Anniversary Volume. Pergamon Press, Ltd., Oxford: 499–536.

Uchida, T., 1927. Studies on Japanese Hydromedusae. 1. Anthomedusae. J. Fac. Sci. imp. Univ. Tokyo. IV. Zool. 1: 145–241, pls. 10, 11.

Uye, S. & T. Kasuya, 1999. Functional roles of ctenophores in the marine coastal ecosystem. In Okutani, T., S. Ohta & R. Ueshima (eds) Update Progress in Aquatic Invertebrate Zoology. Tokai University Press, Tokyo: 57–76 (in Japanese with English Abstract).

Vinogradov, M. Ye., E. A. Shushkina, E. I. Musayeva & P. Yu. Sorokin, 1989. A newly acclimated species in the Black Sea: the ctenophore *Mnemiopsis leidyi* (Ctenophora: Lobata). Oceanology 29: 220–224.

Volovik, S. P., (ed), 2000. Ctenophore *Mnemiopsis leidyi* (A. Agassiz) in the Azov and Black Seas: its Biology and Consequences of its Intrusion. State Unitary Enterprize Research Institute of the Azov Sea Fishery Problems (GUP AzNIIRKH), Rostov-on-Don, Russia, 497 pp. (in Russian with English Contents, Introduction and Concluding Remarks).

Vućetić, T., 1983. Some causes of the blooms and unusual distribution of the jellyfish *Pelagia noctiluca* in the Mediterranean (Adriatic). In Proceedings of the Workshop on Jellyfish Blooms in the Mediterranean, Athens 1983. UNEP 1984: 167–176.

Wilkerson, F. P. & R. C. Dugdale, 1984. Possible connections between sewage effluent, nutrient levels and jellyfish blooms. In Proceedings of the Workshop on Jellyfish Blooms in the Mediterranean, Athens 1983. UNEP 1984: 195–201.

Yasuda, T., 1988. Unusually gregarious occurrences of jellyfishes in Japanese waters. Saishu to Shiiku 50: 338–346 (in Japanese).

Youngbluth, M. & U. Båmstedt, 2001. Distribution, abundance, behavior and metabolism of *Periphylla periphylla*, a mesopelagic coronate medusa in a Norwegian fjord. Hydrobiologia 451 (Dev. Hydrobiol. 155): 321–333.

Zaitsev, Yu. P., 1992. Recent changes in the trophic structure of the Black Sea. Fish. Oceanogr. 1: 180–189.

Zaitsev, Yu. & V. Mamaev, 1997, Marine biological diversity in the Black Sea: a study of change and decline. United Nations Publications, New York: 208 pp.

Zavodnik, D., 1987. Spatial aggregations of the swarming jellyfish *Pelagia noctiluca* (Scyphozoa). Mar. Biol. 94: 265–269.

*Hydrobiologia* **451:** 69–87, 2001.
© 2001 *Kluwer Academic Publishers.*

# Pelagic coelenterates and eutrophication: a review

Mary N. Arai[1]

*Pacific Biological Station, Nanaimo, British Columbia, Canada V9R 5K6; Department of Biological Sciences, University of Calgary, 2500 University Drive N.W., Calgary, Alberta, Canada T2N 1N4. E-mail: araim@island.net*

*Key words:* nutrients, Hydrozoa, Scyphozoa, Ctenophora, *Aurelia, Cassiopea,* oxygen

## Abstract

Although eutrophication is a widespread problem in marine waters, its effects are often difficult to separate from normal fluctuations of pelagic coelenterate populations and from other anthropogenic changes due to industrial pollution, construction, introductions, global warming and overfishing. The least complex situations are in small coastal water bodies such as the Caribbean lagoons and Scandinavian fjords. Typically, the diversity of pelagic coelenterates decreases, but the biomass of a small number of species (such as the hydromedusae *Aglantha digitale* and *Rathkea octopunctata* and the scyphomedusae *Aurelia aurita* and *Cassiopea* spp.) may increase. Adaptations that may allow these species to survive under eutrophic conditions are discussed.

## Introduction

Eutrophication in the marine environment is a fairly recent concept (see history in Nixon, 1995). A number of different uses (definitions) of the term have come into common practice. For the purpose of this paper, eutrophication is defined as "the process of changing the nutritional status of a given water body by increasing the nutrient resources" (Richardson & Jorgensen, 1996). Nixon (1995) suggested that eutrophication be defined more strictly as "an increase in the rate of supply of organic carbon to an ecosystem". Nevertheless, most literature continues to describe eutrophication as increases in the input of mineral nutrients (primarily nitrogen and phosphorus) to a particular water body rather than considering other factors that might effect primary production, such as trace metals and/or growth factors.

Marine eutrophication can occur as a result of natural processes such as upwelling and river inflow, however, modern concerns are centered on 'cultural' or anthropogenic eutrophication. Nutrients are primarily increased in water bodies due to addition of sewage, to forest cutting and fertilizer use on adjoining land and to deposition of reactive nitrogen emitted to the atmosphere during fossil fuel combustion. If nutrients, rather than other factors such as light and temperature, have been limiting plant production, then production is increased. In most systems, that causes an increase in phytoplankton and/or macrophyte biomass as plant production exceeds loss.

Nixon (1995) redefined four trophic states (oligotrophic, mesotrophic, eutrophic and hypertrophic) along a quantitative cline of primary production rates. These terms are also still normally used with some degree of imprecision (Caddy, 1993; Richardson & Jorgensen, 1996). Oligotrophic systems have a low availability of nutrients due to a minor influence of upwelling, terrestrial and/or atmospheric inputs. As a result, mid-oceanic systems such as the Sargasso Sea have a low organismal biomass. Mesotrophic systems are moderately enriched with nutrients, allowing supplemented primary production. Typically, this allows a well-developed pelagic ecosystem and an oxygenated benthic system. Such moderate eutrophication may be advantageous from an anthropomorphic point of view, as fish biomass may increase at the expense of biodiversity. In eutrophic systems, a high influx of extrinsic nutrients allows very high primary production. Surplus detrital rain typically leads to seasonal anoxia of bottom water and sediments with impacts on benthic and demersal fauna. In coastal regions, increasing turbidity may cause a decline of sea grasses. In heterotrophic systems, chemical changes may occur leading to permanent, low oxygen levels in much of the system.

It is clear that some species of pelagic coelenterates are surviving and even thriving under modern eutrophic conditions. It is also well documented that eutrophic areas are increasing (Caddy, 1993; Kennish, 1997). It is less clear what effect anthropogenic eutrophication *per se* has had on coelenterate diversity and biomass. There is little quantitative historical data for most areas so that the level of previous natural variation in diversity and biomass is not known. Even where changes can be traced historically, it is also difficult to separate the direct and indirect effects of eutrophication from the effects of other anthropogenic activities, including industrial pollution, construction and dredging, overfishing, introductions and global warming. Some of these other anthropogenic activities are discussed in this volume by Mills (2001), Purcell & Arai (2001) and Sullivan et al. (2001).

The first section of this review includes a survey of world-wide literature on occurrence of pelagic coelenterates in eutrophic environments. Smallest water bodies are discussed first, followed by increasingly larger and more complex ones. A small number of species are widely present. The second section briefly considers the adaptations of these species that allow them to survive under eutrophic conditions.

## Pelagic coelenterates in eutrophic marine waters

Most of the data on pelagic coelenterates in eutrophic waters are from the coastlines of Europe, North America and eastern Asia (Table 1). Only in these areas is it possible to correlate eutrophication with changes in diversity and biomass of coelenterates. Eutrophication is proceeding elsewhere, such as in coastal waters of Africa and south-east Asia, but there has been less work on coelenterates in these regions and few publications.

In the following examples, I have attempted to distinguish the effects of eutrophication (and associated hypoxia) from the effects of other anthropogenic activities. There are no cases where eutrophication is the only factor causing changes in coelenterate populations. The simplest cases are usually in very small water bodies, although they may also be subject to construction and dredging. There may be a gradient from point sources of nutrients such as sewage but this is often accompanied by industrial pollution. Overfishing may contribute to changes in pelagic coelenterate biomass due to decreasing competition and to reduced predation by fish on the coelenterates. Blooms of introduced species may partially obscure the effects of eutrophication. Finally, all water bodies are subject to temperature changes, including the possible effects of global warming.

### Small lagoons

Localized eutrophication of coastlines is often caused by waste from resorts. Such development is usually accomplished by construction and dredging, which also have effects on the ecosystem.

Rhizostome scyphomedusae of the genus *Cassiopea* are widely distributed in tropical lagoons. The adult medusae spend much of the time inverted on the bottom allowing light to reach their zooxanthellae. They are present throughout the Caribbean, extending north to the southern tip of Florida (Wagenaar Hummelink, 1968; Zamponi et al., 1990).

One of the earliest reports of anthropogenic effects on coelenterates was the appearance of large numbers of *Cassiopea xamachana* R. P. Bigelow in the moat of Fort Jefferson, Tortugas, Florida (Mayer, 1910; Cary, 1917). This dependable population was utilized by scientists for physiological work for many years. The moat at that time had a 'weedy' bottom with a filamentous algae mat indicating probable eutrophication. By 1936, the moat had silted up and the medusae had completely disappeared (Smith, 1936).

In 1910, *Cassiopea xamachana* also occurred in many other semi-stagnant, salt lagoons along the Florida Keys as far north as Miami (Mayer, 1910). At that time, *Cassiopea frondosa* (Pallas) was also found in protected openings in mangroves in the cuts between the keys. Recently, the sea grass of lagoon habitats in the Florida Keys has been adversely affected by eutrophication (Lapointe & Clark, 1992; Lapointe et al., 1994). There are anecdotal accounts of large increases in numbers of *Cassiopea* spp. in nutrient-enriched areas (Lapointe et al., 1994; Fitt & Costley, 1998).

The international resort of Cancun in Quintana Roo, Mexican Carribbean coast was developed around the shallow Nichupte Lagoon System in the last three decades (Reyes & Merino, 1991; Merino et al., 1992). The unaltered areas of the lagoon system are oligotrophic resulting from the low-nutrient surface water of the Caribbean and the absence of runoff in the Yucatan Peninsula. The submerged aquatic vegetation is dominated by a sea grass, *Thalassia testudinum* Banks, community. The Bojorquez lagoon, which is particularly impacted by development, has an area of 2.47 km$^2$ and a mean depth of 1.7 m. It has suffered

*Table 1.* Pelagic coelenterates in eutrophic marine waters. For references see text

| Water bodies | Coelenterate responses | Notes/Other anthropogenic activities |
|---|---|---|
| **Small Lagoons** | | |
| Florida Keys and Nichupte Lagoon System, Mexico | Increased scyphomedusa *Cassiopea* with development | Construction, dredging |
| **Fjords** | | |
| Kertinge Nor | Highest recorded population scyphomedusa *Aurelia* | Records only post-eutrophication |
| Oslofjord | Decreasing diversity hydrozoa; Gradients of increasing pelagic hydromedusae *Aglantha* and *Rathkea* and decreasing epibenthic *Tesserogastria* towards sewage source | Scyphomedusae and ctenophores not monitered |
| **Bays/estuaries** | | |
| Elefsis Bay | Abundant *Aurelia* | Data 1983–1985; Industrial pollution |
| Chesapeake Bay | Abundant hydromedusae *Liriope* and *Nemopsis*, scyphomedusa *Chrysaora*, and ctenophore *Mnemiopsis*; Low diversity other hydromedusae | Abundances stable since 1960; Industrial pollution |
| Tokyo and Osaka Bays | Increased *Aurelia*; Abundant hydromedusae *Rathkea* and *Liriope*, siphonophore *Muggiaea*, ctenophore *Bolinopsis*; Low diversity other hydromedusae | Industrial pollution |

*Table 1.* Continued

| Water bodies | Coelenterate responses | Notes/Other anthropogenic activities |
|---|---|---|
| Hangzhou Bay | Fishing catch of rhizostome scyphomedusae peaked 1966 and then declined sharply | Industrial pollution; Overfishing |
| **Semi-Open Enclosed Seas** | | |
| Baltic Sea | Abundance of scyphomedusae *Aurelia* and *Cyanea* and ctenophore *Pleurobrachia* not clearly correlated with eutrophication | Strong, variable salinity gradient |
| North Sea | Distribution of pelagic coelenterates not correlated with eutrophication pattern | Variable intrusion of North Atlantic water |
| North Adriatic Sea | Abundance of scyphomedusa *Pelagia* not correlated with eutrophication; Decreased diversity of anthomedusae and leptomedusae | Industrial pollution |
| Black Sea | Increased *Aurelia* 1960s–1980s; Decreased stauromedusa *Lucernaria* | Overfishing; Industrial pollution; Introduction of ctenophores *Mnemiopsis* and *Beroe* |
| Gulf of Mexico | Increased *Chrysaora* and *Aurelia* | Industrial pollution; Oil rig construction |

dredging of over 20% of its original bottom (which released nutrients and organic compounds from the sediments), continuous sewage discharge, increasing boat traffic, and destruction and filling of surrounding mangroves. In the Bojorquez lagoon, nutrient concentrations in the water column are not significantly higher than in the rest of the lagoon system, however, a soft unconsolidated layer of organic matter covers much of the bottom and the sea grass has been largely replaced by phytoplankton and macroalgae.

In the Bojorquez lagoon, *Cassiopea xamachana* and *C. frondosa* are very abundant compared with the rest of the Nichupte lagoon system. The population observed at a control station in a neighbouring lagoon was less than one medusa m$^{-2}$. In Bojorquez, the population reached an average density of 42 medusae m$^{-2}$ in 1985–1986 (Collado Vides et al., 1988).

Even in this simple case, eutrophication may not be the cause of all population changes. It should be noted that while the Florida and Mexican data clearly show increases in numbers of *Cassiopea* spp. medusae following human activity, they do not as clearly relate those increases to eutrophication *per se* rather than construction and dredging. Although *Cassiopea xamachana* medusae have zooxanthellae, they also must obtain a portion of their nutrients from feeding or other external sources (Larson, 1997; Vodenichar, 1995 quoted in Fitt & Costley, 1998). In eutrophic conditions, it is, therefore, possible that *Cassiopea* spp. are able to utilize the increased organic matter near the bottom. On the other hand, construction or dredging may also allow *Cassiopea* spp. to expand its populations into areas where the seagrass or mangrove cover has been cleared away from the bottom. *C. xamachana* typically lives on soft, muddy and weedy bottoms, whereas *C. frondosa* lives on mud to coarser sediments in more open positions (Mayer, 1910; Larson, 1997). *C. xamachana* requires a peptide derived from the cell walls of decomposing plants such as mangrove leaves as an inducer of settling and metamorphosis (Fleck & Fitt, 1999; Fleck et al., 1999). There may also be greater concentrations of such peptides in areas where construction is disturbing the surrounding plant cover. Finally there are no data related to eutrophication or construction effects on the incidence of predators such as the nudibranchs known to eat *Cassiopea* spp. in the Caribbean (Brandon & Cutress, 1985).

Although other scyphomedusan species are present in similar lagoons of Florida and the Caribbean, there have been no reports to date of increases in these species correlated with human activity. Laguna Joyuda, a shallow lagoon in Puerto Rico, has blooms of the scyphomedusa *Phyllorhiza punctata* von Lendenfeld during the warm/wet season, (as well as smaller numbers of *Cassiopea* sp.) (Garcia & Lopez, 1989). *Aurelia aurita* (Linnaeus) and *Stomolophus meleagris* L. Agassiz are present in the shallow coastal lagoons of the southern Gulf of Mexico (Gonzalez, 1979; Gomez-Aguire, 1980).

## Fjords

Fjords such as those in Scandinavia and NW America are long, narrow inlets from the sea. They vary greatly in depth and are typically partially blocked by a sill near the mouth. The sill may separate a basin fauna from the open sea. Depending on the height and position of the sill (particularly above or below the halocline), the tidal range, and the extent of fresh water inflow water exchange in the basin also varies greatly. The deeper water of the basins may become stagnant and in extreme cases anoxic even without anthropogenic eutrophication.

The Kertinge Nor is a 5.8 km$^2$ cove connected to the Great Belt east of Denmark by the more narrow Kerteminde Fjord. The cove has a mean water depth of 2 m (maximum 3 m). The fjord has a maximum depth of 8 m and a sill at its mouth. Domestic sewage had been fed into the cove which, with a mean residence time of water of approximately 2 mo., had become eutrophic. This input of domestic sewage was stopped in 1989 and investigations of a variety of factors were conducted in 1991–1992 (Riisgard et al., 1995, 1996). By this time, Kertinge Nor had a bottom partially covered with eel grass and partially with a dense mat of filamentous green algae over anoxic sediment. The sediment was the dominant source of nutrients. Except when the algal mat floated up due to production of oxygen bubbles, the water was exceptionally clear. High populations of benthic filter feeding ascidians reduced the phytoplankton.

In the summers of 1991 and 1992, populations of up to 300 *Aurelia aurita* m$^{-3}$ controlled the populations of harpacticoid copepods and rotifers (Olesen et al., 1994; Olesen, 1995). The maximum mean umbrella diameter of the medusae was only 54 mm in 1991 and 37 mm in 1992, indicating that the medusae were food limited. By 1995, maximum population had fallen to 98 m$^{-3}$, and maximum mean umbrella diameter had risen to 73 mm (Nielsen et al., 1997).

It is unclear to what extent the properties of the *Aurelia aurita* population in Kertinge Nor are due to the eutrophic conditions. The Kertinge Nor records are the highest populations recorded for the ubiquitous *A. aurita*. Of *A. aurita* populations described in the literature, the population dynamics in Kertinge Nor most resembles that in Horsea Lake in southern England (Lucas, 1996; Lucas et al., 1997). Horsea Lake is shallow, brackish and man-made, with a limited intermittent intake of salt water from Portsmouth Harbour. Although not eutrophic, a simplified food chain supports a large population of small, food limited me-dusae there, also. The peak abundance in May 1994 was 24.9 m$^{-3}$. In neither location is it clear how the simplified food chain was established or why the pop-ulation numbers are so high. In both locations, the size at maturity is decreased, but there are no data on polyp or cyst populations.

Oslofjord is one of the most polluted fjords of Nor-way (Rosenberg et al., 1987; Beyer & Indrehus, 1995). The inner fjord, is separated from the outer south-ern segments of the fjord by the Drobak Sound and a 20 m deep sill. The inner fjord consists of two natural basins, Vestfjord and Bunnefjord, on which the city of Oslo is situated, which are 160 m and 164 m deep, respectively, and are separated by a sill at 50 m depth. In winter, oxygen-rich water flows over the Drobak sill and partially flushes the inner basins so that a complete renewal of water occurs approximately every 3 years.

Prior to 1900, there was a commercial fishery for shrimp in Bunnefjord, the innermost basin. Addition of sewage caused increased phytoplankton, decreased benthic biomass and periodic anoxia as early as 1917, which accelerated in the 1960s. By 1981, 11 sewage plants discharged treated sewage into the inner fjord, producing a gradient in diversity of bottom fauna from the heavily polluted Bunnefjord to the Drobak sill (Mirza & Gray, 1981). Since 1982, some sewage has been piped farther down the fjord and cellulose indus-trial discharge to the outer fjord has increased, but the areas outside the Drobak sill remain well oxygenated (Rosenberg et al., 1987).

There has been less examination of plankton than bottom fauna but, nevertheless, trends can be identi-fied. Sverdrup (1921) and Kramp and Damas (1925) described a number of Norwegian hydromedusae, si-phonophores and ctenophores but confined their work to the Drobak region of Oslofjord (then Kristiani-afjorden).

In June 1947, *Sarsia tubulosa* (M. Sars) hy-dromedusae were present in the 8–14 m deep ther-mocline of Bunnefjord (Hansen, 1951). Beneath the thermocline was found the hydromedusa *Aglantha di-gitale* (O. F. Müller) and 5 species of siphonophores. By 1962–1964, the hydromedusae *A. digitale* and *Rathkea octopunctata* (M. Sars) each had a horizontal gradient of increasing population from Drobak to the innermost eutrophic part of the Bunnefjord (Beyer, 1968). For *A. digitale*, there was a range from 0.3 m$^{-3}$ near the sill to 16 m$^{-3}$ at the most polluted sta-tion. *A. digitale* is holoplanktonic and *R. octopunctata* can reproduce by manubrial budding. The siphono-phore *Lensia conoidia* (Keferstein & Ehlers) was also abundant in Bunnefjord (Tveite, 1969), but the other siphonophores and *S. tubulosa* were not recorded. In 1967–1968, *A. digitale* in Bunnefjord was the subject of a detailed study of population numbers and growth in relation to season, diet, depth and concentration of oxygen and hydrogen sulphide (Smedstad, 1972). The small copepods, *Oithona* sp. and *Oncaea* sp., which were the most important prey of *A. digitale*, were also most abundant in the inner most eutrophic parts of the fjord.

The epibenthic trachymedusa, *Tesserogastria mus-culosa* Beyer, was first described from soft bottom in Oslofjord (Beyer, 1958). By 1962, this species showed a decreasing population gradient toward the source of sewage near Oslo (Beyer, 1968). Sampling in 1981–1993 showed population decreases in more extensive areas of both Bunnefjord and Vestfjord (Beyer & Indrehus, 1995).

*Bays and estuaries*

Bays and estuaries differ primarily in the degree of freshwater input. On a worldwide basis, many are highly polluted by human populations, but sewage and other nutrients producing eutrophic conditions are usually mixed with a high proportion of toxic wastes. In very few cases is there enough historical inform-ation about pelagic coelenterates to enable tracing population changes to eutrophication. Also because most bays where eutrophication is occurring are ports, there is a high incidence of introduced species.

Elefsis Bay is a small (67 km$^2$) and shallow (max-imum depth 33 m) bay separated by sills and break-waters from the inner part of the Saranikos Gulf at the southern end of the Aegean Sea. It is highly eutrophic due primarily to outflow of sewage and in-dustrial wastes such as petroleum products from the city of Athens (Friligos, 1982). In February 1985, populations of *Aurelia aurita* in the bay reached a maximum of 44 ephyrae m$^{-3}$ (Papathanassiou et al., 1986, 1987). During 1983–1985, the maximum mean umbrella diameter was approximately 200 mm (Panayotidis et al., 1985, 1986; Papathanassiou et al., 1987). The medusae carried out diel vertical migra-tions between 1 and 10 m, i.e. above the zone of oxy-gen depleted water. Unfortunately, there are no data on how the population developed as eutrophication increased.

Chesapeake Bay is the largest U.S. estuary, a 300 km long shallow system with a central mainstem

74

communicating with the Atlantic Ocean, and several major tributaries including the Susquehanna River. Historically, this bay was very productive, but fisheries have been threatened for many years by eutrophication and toxic substances from farming and urban centers such as Washington (D.C.) and Baltimore (Kennish, 1997). In addition, deposit of atmospheric nitrogen had reached approximately 40% of total nitrogen input by 1990 (Paerl, 1995). In summer, hypoxia and anoxia develop below the pycnocline in the deeper parts of the mainstream bay and tributaries. Much of the extensive research on the ecosystem has recently been summarized in Malone et al. (1999).

*Chrysaora quinquecirrha* (Desor) is the predominant species of scyphomedusa, although *Aurelia aurita, Cyanea capillata* (Linnaeus) and *Rhopilema verrilli* (Fewkes) are also present (Purcell et al., 1999a). The *C. quinquecirrha* medusae appear first in the tributaries in the spring, and continue more dense in those locations than in the main bay (Purcell, 1992). Unfortunately, little is known about coelenterate abundances prior to 1960 when the bay was already heavily impacted by human activity (Cargo & King, 1990). There has been no obvious trend in abundance of *C. quinquecirrha* since 1960, possibly because there is wide interannual variablity correlated with salinity and temperature (Purcell et al., 1999b). The effects of hypoxic bottom water on medusae and polyps in the bay have been described by i.e., Keister et al. (2000), Condon et al. (2001) and Purcell et al. (2001b). Nevertheless, Purcell et al. (1999a), reviewing the potential links of *C. quinquecirrha* in the bay to eutrophication, concluded that the interactions between physical and biological factors are very complex and without historical data "direct connections between human effects on estuarine systems and changes in jellyfish populations are difficult to make".

Hydromedusae and ctenophores are also present. The most abundant hydromedusan species is the holoplanktonic *Liriope tetraphylla* (Chamisso & Eysenhardt) in the lower bay, although the euryhaline anthomedusan, *Nemopsis bachei* L. Agassiz, which has a hydroid stage, is widely distributed in the spring (Calder, 1971; Purcell & Nemazie, 1992; Purcell et al., 1999a). The abundance and diversity of other hydromedusae is low. The most abundant ctenophore is *Mnemiopsis leidyi* A. Agassiz (Kremer, 1994; Purcell & Cowan, 1995) which is discussed elsewhere in this volume (Purcell et al., 2001a).

Eutrophication in Japan has most greatly affected Tokyo Bay, which receives the output of nutrients and industrial waste from Tokyo, Yokohama and Kawasaki. Osaka Bay, at the eastern end of the Seto Inland Sea, receives input from the next largest human population center with a group of cities including Kobe and Osaka. The nutrient load to both bays increased drastically in the 1960s and 1970s (Tatara, 1991). The rate of increase since has been partially controlled, but they remain highly eutrophic (Uye & Shimazu, 1997; Uye et al., 1999). Freshwater discharge into inner Tokyo Bay and water exchange with the outer bay has increased since 1988 (Nomura & Ishimaru, 1998). Tokyo Bay is usually stratified, but in the summer, wind-driven upwelling may bring anoxic, hydrogen sulfide-containing, water to the surface. Oxidation produces colloidal sulfur particles which cause the sea surface to become milky-blue in the 'Aishio' phenomenon (Otsubo et al., 1991). Little fishery remains in Tokyo Bay. In the Inland Sea, away from the most concentrated pollution, total fishing catch increased during the 1960s and 1970s, but the trophic level of that catch decreased (Tatara, 1991).

The most dramatic changes in pelagic coelenterate populations with eutrophication have been the blooms of *Aurelia aurita*, which first caused clogging of intakes of power plants along the coast of Tokyo Bay in 1963 (Kuwabara et al., 1969; Uye, 1994). Similar problems also appeared in the Inland Sea (Matsueda, 1969). Aggregations form in the most polluted and shallow (less than 30 m depth) inner part of Tokyo Bay (Kuwabara et al., 1969; Toyokawa & Terazaki, 1994; Omori et al., 1995; Ishii et al., 1995; Toyokawa et al., 1997; Nomura & Ishimaru, 1998). Although these aggregations are often referred to as 'dense', the maximum density in the aggregations reported in the inner harbour was 25 ind. $m^{-3}$ (Kuwabana et al., 1969), whereas the overall population is less than 1 $m^{-3}$ (Toyokawa & Terazaki,1994; Omori et al., 1995; Nomaru & Ishimaru, 1998). This should be compared with other populations cited in this paper, or with aggregations of up to 600 $m^{-3}$ and population density of 71 $m^{-3}$ in Urazoko Bay, Japan Sea (Yasuda, 1969, 1970). Ephyrae mainly originate from canals in the innermost bay where salinity is reduced (Toyokawa et al., 2000; Watanabe & Ishii, 2001). Smaller numbers of the scyphomedusa *Chrysaora melanaster* Brandt have been present since 1988 and *Pelagia noctiluca* (Forsskål) was recorded in October, 1992 (Toyokawa et al., 1997; Nomura & Ishimura, 1998).

The most abundant hydromedusae in inner Tokyo Bay 1977–1978 and 1988–1989 was the manubrial-budding *Rathkea octopunctata* (in Sugiura, 1980; Toyokawa & Terazaki, 1994). In the latter period, the holoplanktonic hydromedusa, *Liriope tetraphylla,* and the siphonophore, *Muggiaea atlantica* Cunningham, were also common. The most abundant ctenophore is *Bolinopsis mikado* (Moser) (Nomura & Ishimura, 1998; Uye & Kasuya, 1999), and since 1988, *Beroe cucumis* Fabricius has also been present (Toyokawa & Terazaki, 1994; Nomura & Ishimura, 1998).

Although eutrophication has occurred in many bays of eastern and southeastern Asia, there is little documentation of the effects on pelagic coelenterates. The most polluted bay in eastern China is Hangzhou Bay. Four rivers, including the Yangtze River, deliver almost 50% of the national discharge of nutrients, petroleum residues and heavy metals from a number of cities including Shanghai (Liu et al., 1991). The fishing output and zooplankton biomass from the Zhoushan fishing ground in the bay has dropped since the early 1970s due to pollution and overfishing. Scyphomedusae of the order Rhizostomae are important food in Chinese and Japanese cooking (Hsieh et al., 2001; Omori & Nakano, 2001). The fishery of scyphozoan jellyfish also reached a maximum in 1966 and has declined precipitously, while the diameter of the medusae was reduced by 50% by 1985 (Liu et al., 1991).

*Semi-open and enclosed seas*

As the size of the water body increases, point sources of nitrogen such as localized agricultural, municipal and industrial wastes become less important, and air borne nitrogen becomes a more significant proportion of the input. This atmospheric nitrogen, primarily as nitrogen oxides and ammonia, is generated by industrial and automobile fossil fuel combustion and discharges from agriculture such as dairy farms (Paerl, 1995). It may be borne long distances down wind to affect areas such as the eastern coastal zone of North America and the Baltic, North and Western Mediterranean Seas. In these areas, it may reach more than 50% of the total nitrogen input.

Other factors that significantly affect production over wide areas, and may be confused with eutrophication, are changes in temperature and salinity. Temperature may be affected by normal climate variations or by global warming. For example, a temperature regime shift over the northern hemisphere in 1989/90 coincided with changes in the biomass and composition of gelatinous zooplankton in the Okhotsk, Bering, North, Adriatic and Black Seas (Avian & Rottini

Sandrini, 1994; Shuntov et al., 1996; Niermann & Greve, 1997; Niermann et al., 1998; Brodeur et al., 1999). However, in the Baltic and western Mediterranean Seas, changes in salinity were not always synchronous with temperature and also affected zooplankton populations (Buecher, 1997, 1999; Buecher et al., 1997).

It is also in these larger water bodies that the effects of overfishing may become more apparent. Removal of fish may allow increase of gelatinous organisms, both by removing direct predation of fish on pelagic coelenterates and by decreasing competition for other food (Arai, 1988; Purcell & Arai, 2001).

The Baltic Sea is an enclosed basin connected to the North Sea only through narrow and shallow sounds. Eutrophication has increased nutrient concentrations in the water and organic contents in the sediments this century, but particularly since the 1970s (Schulz et al., 1992; Nehring, 1992; Bonsdorff et al., 1997). There has been increased production of phytoplankton and decreased oxygen concentrations leading to anoxic bottom water in the main basins. Although total fish catch has increased, the population of cod, the main apical species, has decreased since 1980 due to changing food chains and to poor survival of the sinking demersal eggs in hypoxic or anoxic bottom water (Elmgren, 1989; Caddy, 1993).

It is difficult to distinguish the effects of eutrophication on zooplankton and fish from temperature effects and more importantly from the effects of changes in salinity (Vuorinen & Ranta, 1988). There is a gradient in salinity through the Gulf of Bothnia (Bothnian Bay to Bothnian Sea) and the Gulf of Finland to the Baltic Sea proper and thence through the Kattegat and Skagerrak to the North Sea, and a strong halocline over most of the area. Changes in freshwater runoff and irregular intrusions of saline water through the Kattegat can affect the composition and abundance of the plankton as far north as the Bothnian Sea (Vuorinen & Ranta, 1987). From the beginning of the century to 1977, salinity rose irregularly, but in the period of less intrusion during the 1980s, the salinity dropped (Schulz et al., 1992; Hanninen et al., 2000).

No pelagic coelenterates are present in the Bothnian Bay, however, the ctenophore *Pleurobrachia pileus* (O. F. Müller) penetrates into the Bothnian Sea (Sandström & Sörlin, 1981; Vuorinen, 1987). It penetrates to salinities down to at least 6.5, but at these low salinities it becomes very fragile. Nevertheless, the species is, surprisingly, more abundant in the Bothnian Sea and the Gulf of Finland than in the Baltic Sea proper (Vuorinen & Vihersaari, 1989). Since temperature, salinity and food all increase toward the south, it is unlikely to be limited by those factors in the Baltic Sea proper. It is possible that this distribution reflects better oxygenation of the shallow northern water bodies than in the Baltic proper. It may also reflect decreased predation by predators such as cod.

In the Baltic Sea proper during 1968–1972, four coelenterates were commonly present; *Pleurobrachia pileus*, the hydromedusa *Sarsia tubulosa* only in the south, the scyphomedusa *Cyanea capillata* sparsely as far north as the Gulf of Finland, and *Aurelia aurita* (Haahtela & Lassig, 1967; Hernroth & Ackefors, 1979). *A. aurelia* was already present in 'large concentrations' in the south and had been extending its range in the north further into the Gulf of Finland as salinity increased (Segerstråle, 1951; Palmén, 1953; Hernroth & Ackefors, 1979). Against this non-quantitative picture of earlier abundance of *A. aurelia*, it is not possible to ascribe the more recent populations in the southern Baltic (Möller, 1980; Janas & Witek, 1993; Margonski & Horbowa, 1996) to the effects of eutrophication as done by Schulz (1989).

Although largely a shallow sea, the North Sea is less affected by pollution than the Baltic due to its interchange with the Norwegian Sea to the north and access to the North Atlantic via the English Channel to the south. The main currents move south along the east coast of the British Isles and north again along the European coast. The output from the European rivers is carried north so that the greatest nutrient levels, phytoplankton blooms and most oxygen depleted bottom waters occur off Denmark and Germany in the German Bight (Caddy, 1993). Eutrophication increased from 1950 through the 1980s, but has decreased in the 1990s especially with respect to phosphorus. Increasing catches of demersal fish and changes in species caught have been attributed to this eutrophication, increased fishing effort and climatic changes.

ICES surveys of gadoid fishes sampled by-catch of scyphomedusae from early summers of 1971–1986 (Hay et al., 1990). *Aurelia aurita*, *Cyanea capillata* and *Cyanea lamarckii* Péron and Lesueur were the most abundant species. While *A. aurita* was found primarily in the coastal waters, it was not more abundant off the coast of continental Europe than off the Scottish east coast. *C. capillata* was widely distributed. Only *C. lamarckii* showed a concentration off the European coast which, if only this species were examined, might be interpreted as correlated with

eutrophication. A survey by Möller in August 1978 showed a similar distribution, although *Chrysaora hysoscella* (Linnaeus), which appears later in the year, was also collected off the Netherlands and Germany (Möller, 1980). These data do not differ substantially from those collected off the Netherlands 1933–39, prior to eutrophication, by Maaden (1942) and previous literature from the North Sea reviewed by Verwey (1942).

Irregular interannual changes in biomass and distribution of zooplankton, including hydromedusan, siphonophore and ctenophore populations, in the North Sea area have been documented for many years (Russell, 1939; Fraser, 1969, 1970). These changes were early correlated with variations in transport of water from the North Atlantic with associated effects on North Sea salinity and temperature, and more recently with more widely distributed climate changes (Aebischer et al., 1990; Fransz et al., 1991; Nicholas & Frid, 1999). For example, there was an exceptional influx of southern oceanic species coincident with the temperature regime shift in 1988–1989 (Edwards et al., 1999; Reid et al., 2001). The siphonophore, *Muggiaea atlantica*, which previously had been rarely reported in the North Sea off the east coast of Scotland (Fraser, 1967), bloomed in the German Bight (Greve, 1994). Against this shifting backgound, it is not possible to evaluate the effect of eutrophication on hydromedusae, siphonophores or ctenophores in the North Sea.

The Mediterranean Sea was historically oligotrophic with evaporative losses exceeding river inflow (Caddy, 1993). The nutrient poor near-surface inflow from the Atlantic Ocean flows over the shallow sill at the Strait of Gibraltar, eastward along the African coast and then westward along southern Europe. To the east the Levantine Basin was the most oligotrophic area, whereas the Adriatic Sea had input from the Po River, and the Aegean Sea from the Black Sea. With the operation of the Aswan Dam on the Nile River in 1965, the Levantine Basin has become one of the most nutrient limited areas in the world's oceans (Azov, 1991). Under these oligotrophic conditions, the anthropogenic addition of nutrients can lead to particularly striking changes. The effects in small bays such as Elefsis Bay on the coast of Greece have already been mentioned, but in the northern Adriatic Sea, changes are more widespread.

Nutrient inputs to the shallow northern Adriatic Sea, particularly from the Po River and in the Bay of Trieste, have been increasing over the last four decades. Water from the Po River flows south along the

Italian coast creating a nutrient gradient across the Adriatic from west to east, as well as from north to south. Inflow from the southern Adriatic occurs along the eastern coast particularly in the fall and winter (Boicourt et al., 1999). Phytoplankton blooms, and since 1988, formation of gelatinous aggregates, occur over much of the northern portion of the sea, the distribution depending on the extent of the Po outflow but also on wind driven currents (Degobbis et al., 1995). Hypoxia in bottom layers, under a halocline or thermocline, has increased since the period 1955–1965 (Legovic & Justic, 1997). It was particularly widespread in summers of 1977 and 1988 (Boicourt et al., 1999). The dynamics of eutrophication, as well as comparisons with Chesapeake Bay, are reviewed in Malone et al. (1999). The over 150 year history of observations on medusae are reviewed by Purcell et al. (1999a) in that volume.

Periodic blooms of the holoplanktonic scyphomedusa, *Pelagia noctiluca,* have been documented in the Mediterranean Sea for over two centuries. In the western Mediterranean, they have appeared with a periodicity of about 12 years, which is correlated (not necessarily causally) with fluctuations in temperature and rainfall (Goy et al., 1989). In the Adriatic, population changes have been more irregular, large swarms appearing between 1910–14 and 1976–86 in both northern and southern regions of the Adriatic (UNEP, 1984, 1991; Rottini-Sandrini & Avian, 1986). Although *P. noctiluca* can grow and reproduce more rapidly in nutrient-rich environments (Purcell et al., 1999a), the populations in the northern Adriatic are more dependent on the hydrographic and climatic conditions and the extent of intrusion of water from the southern Adriatic and Ionian Sea (Vućetić, 1991). The bloom has not continued beyond 1986, in spite of continuing eutrophication.

Seven other species of Scyphozoa have been reported from the North Adriatic (Avian & Rottini-Sandrini, 1994). There is insufficient evidence to clearly correlate abundance or distribution of these species with eutrophication. *Aurelia aurita* has been reported as abundant in 1962, 1989 and 1994–1997 (Avian & Rottini-Sandrini, 1994; Purcell et al., 1999a), so its numbers may be increasing.

The number of species of hydromedusae recorded in the northern Adriatic declined from 40 to 10 between 1910 and 1984/85 (Benović et al., 1987). This was a gradual decrease in numbers first of leptomedusae and then of anthomedusae, both with benthic hydroids, which was correlated with decreasing oxy-

gen concentrations of the bottom water. Seven of these species reappeared in 1992–1993, although three Trachymedusae were no longer collected (Benovic & Lucic, 1995; Purcell et al., 1999a). Present diversity is greatest on the eastern side of the northern Adriatic where eutrophication is less extensive and species might be reintroduced by currents from the south (Benovic & Lucic, 1996).

The Black Sea is the largest body of eutrophic or hypertrophic water in the world. The sea has a broad shallow northwestern coastal shelf with input from the Danube, Dniester and Dnieper Rivers. To the northeast, the Don River drains into the Black Sea through the brackish Sea of Azov. The remainder of the Black Sea has only narrow coastal shelves. To the southwest, the sea drains into the Mediterranean Sea via the shallow Bosporus, the Sea of Marmara and the Dardanelles. The Black Sea has a permanent halocline at 100–200 m. Above the halocline, there is a cyclonically meandering rim current and two (western and eastern) central gyres. Below the halocline, thousands of years of accumulation of organic material dropping from above has led to a permanent anoxic zone (Caddy, 1993). The rate of accumulation has increased in recent years with greater input of nutrients (as well as toxic wastes) from the Danube and Dniester Rivers and resulting algal blooms (Bologa et al., 1995; Cociasu et al., 1997). At present, more than 90% of the water volume below the halocline is anoxic. The shallow coastal waters have been affected so that on the previously very productive northwestern shelf, favourable benthic environmental conditions at present remain only at the edge of the sea in depths of less than 5–6 m (Zaitsev, 1992; Zaitsev & Alexandrov, 1997; Gucu, 1997).

It is difficult to separate the effects of eutrophication on coelenterate populations from effects of overfishing, temperature and salinity changes, and especially introductions to the Black Sea. Prior to 1982, the pelagic coelenterates present included the scyphomedusae, *Aurelia aurita* and *Rhizostoma pulmo* (Macri), and the ctenophore, *Pleurobrachia pileus.* There were also a small number of native hydrozoans and early introductions (Zaitsev & Mamaev, 1997). *R. pulmo* is largely confined to coastal areas, but the other scyphomedusae and ctenophore species are more widely distributed. *P. pileus* is found in hypoxic water above the anoxic zone but below the pycnocline by day, and migrates up toward the surface at night (Vinogradov et al., 1985; Mutlu & Bingel, 1999). Small *A. aurita* medusae are found from the surface to the thermocline, and larger individuals are found just below it, with planulae concentrated just above the anoxic layer (Vinogradov & Shushkina, 1982).

There was a poorly documented bloom of *Rhizostoma pulmo* starting in the late 1960s and subsiding by 1974 (Zaitsev & Mamaev, 1997). There was also an increase of several orders of magnitude in the population of *Aurelia aurita* through the 1960s, 1970s and 1980s (Gomoiu, 1980; Shushkina & Musayeva, 1983; Shushkina & Vinogradov, 1991). Most authors have stressed eutrophication, and increased prey production as the cause of these increases (Caddy & Griffiths, 1990; Kideys, 1994; Zaitsev, 1994). In addition to eutrophication, increases may have been due at least in part to removal of predators. Mackerel, *Scomber scombrus* Linnaeus, are predators of pelagic coelenterates (Scott, 1914, 1924; Runge et al., 1987), and have been observed eating medusae in the Black Sea (Zaitsev & Polischuk, 1984). Mackerel reproduced primarily outside the Black Sea, but there were massive seasonal migrations from the Marmara Sea onto the Northwestern Shelf of the sea in most years prior to the 1960s. Mackerel had disappeared as a commercial species of the Black Sea by the end of the 1960s due to over fishing (Caddy, 1993), which may have allowed the medusae to increase (Zaitsev, 1992). Also during this time, the worsening conditions of the Romanian littoral caused a number of benthic species, including the stauromedusan *Lucernaria campanulata* Lamouroux, to became extinct or very rare (Gomoiu, 1981). It is probable that, as Roginskaya (1988) suggested, there was a decrease in predation by Black Sea nudibranchs of the genera *Trinchesia* and *Facelina,* which are known to feed on *Aurelia aurita* polyps in other waters.

Introduction of the ctenophore *Mnemiopsis leidyi* was first recorded in 1982 (Zaika & Sergeeva, 1990). The species first bloomed in 1988–9 (Vinogradov et al., 1989; Shushkina & Vinogradov, 1991). This was at the time of a major temperature regime shift upwards (Niermann & Greve, 1997; Niermann et al., 1998). Following that massive bloom of a new species in the ecosystem, it is not possible to clearly distinguish eutrophication effects on coelenterates during the 1990s. After the initial dramatic bloom, the population numbers of *M. leidyi* have decreased and fluctuated (Shiganova, 1997, 1998; Mutlu, 1999; Purcell et al., 2001a). The greatest competition is with *Aurelia aurita* which overlaps in depth range and food resources. *A. aurita* at first declined in 1989, then fluctuated at lower levels than *M. leidyi* (Shushkina

& Vinogradov, 1991; Shiganova, 1997; Kideys et al., 2000). *Pleurobrachia pileus*, remained somewhat more stable (Shiganova et al., 1998; Mutlu & Bingel, 1999; Kideys et al., 2000). The interactions between the three species have been reviewed by Mutlu et al. (1994), Kovalev & Piontkovski (1998), Kideys et al. (2000) and Purcell et al. (2001a). In addition to competition for zooplankton prey, *A. aurita* may feed on small *M. leidyi* (Weisse & Gomoiu, 2000).

The roles of the three main coelenterates as predators and competitors of fish in the 1990s have been reviewed or modelled by Kideys (1994), Vinogradov et al. (1996), Gucu (1997), Shiganova (1997, 1998), Berdnikov et al. (1999) and Shiganova & Bulgakova (2000). There has been little consideration given to the reverse possibility, that population fluctuations of fish predators may have affected the pelagic coelenterate populations. Anchovy, *Engraulis encrasicolus* (Linnaeus), and horse mackerel, *Trachurus mediterraneus* (Steindachner) both belong to genera that include species that feed on gelatinous prey (Arai, 1988; Ates, 1988; Mianzan et al., 1996; Mianzan et al., 2001). Zaika (quoted in Kideys, 1994) has noted that *T. mediterraneus* consume *Mnemiopsis leidyi* juveniles off the Bulgarian coast. Anchovy and horse mackerel populations rose during the 1970 and 1980's and collapsed at the end of the 1980s due to overfishing as well as trophic interactions (Prodanov et al., 1997). This collapse may have released the coelenterate populations from predation.

The recent bloom of the ctenophore *Beroe ovata* Bruguiére, which eats other ctenophores (Swanberg, 1974), has been recorded in the Black Sea since 1997, and has increased the predation on ctenophores *Mnemiopsis leidyi* and *Pleurobrachia pileus* (in Finenko et al., 2001; Shiganova et al., 2001). This is again shifting the balances between *M. leidyi*, *Aurelia aurita*, and their predators and prey (Shiganova & Bulgakova, 2000; Shiganova et al., 2001). The exact effects of eutrophication continue to be difficult to distinguish, although the Black Sea certainly constitutes an example of large pelagic coelenterate populations in eutrophic conditions.

In the Gulf of Mexico and Carribean Sea, the prime instance of eutrophication is the input of the Mississippi and Atchafalaya Rivers to the broad shallow coastal shelf of the northern Gulf of Mexico. The Mississippi watershed covers 41% of the continental United States including over 50% of the farmland. Although the river carries toxic waste from as far north as the Chicago industrial complex, the main pollution is by nutrients derived from fertilizer from the central United States (Rabelais et al., 1996). The freshwater discharge flows primarily westward over more saline water along the Louisiana coast to the south Texas coast. The 'Dead Zone' of hypoxic water that forms in summer below the pycnocline has increased. Since 1993, it may cover up to $18\,000$ km$^2$ of the bottom between the 5 and 60 m depth contours of the shelf (Malakoff, 1998; Rabalais & Turner, 2001).

Before the 1980s, the scyphomedusa *Chrysaora quinquecirrha* was confined to the near shore estuarine areas of the northern Gulf. Since 1985, trawl by-catch data has shown that the distribution is extending off shore as the area of hypoxic water increases (Purcell et al., 2001b; Graham, 2001). In 1987, an aerial survey of *Aurelia aurita* aggregations found them to be common east of the Mississippi River, but not to the west either spring or fall (Roden et al., 1990). Between 1987 and 1997, this species also increased, particularly in-shore (Graham, 2001). These changes may be linked to the eutrophication, however, they may also reflect up-current increased hard surface for the polyps on oil and gas structures and climate changes that influence the rate of river discharge (Graham, 2001).

## Adaptations of pelagic coelenterates

At this time, it is not possible to predict what the results of anthropogenic eutrophication will be in any given location. This is partly due to the presence of the other complicating factors that have been considered above. It is also because of our poor knowledge of the effects of eutrophication itself, not only on the adult pelagic coelenterates, but also on the developmental stages and on other fauna which may compete with, be preyed on, or prey upon the coelenterates. The most direct effect is the increased availability of dissolved organic material (DOM) for uptake by the coelenterates. Then, when increased nutrients allow increased phytoplankton production, the coelenterates are affected by decreased oxygen and light. Finally, the coelenterates are affected by changes in the abundance of prey, predators and competitors, and by their responses to turbidity and hypoxia.

The known effects of hypoxia on pelagic Cnidaria and Ctenophora have recently been extensively reviewed by Purcell et al. (2001b). Many species are able to survive by avoidance of oxygen-depleted layers. Since surface layers are well oxygenated even under eutrophic conditions, it is to be expected that near-surface holoplanktonic forms will be able to survive. This presumably applies to hydromedusae like *Aglantha digitale* and *Liriope tetraphylla,* siphonophores such as *Muggiaea atlantica* and *Lensia* spp., as well as several ctenophores. Species of Scyphozoa such as *Aurelia aurita*, with benthic stages in shallow water may also survive in this manner. To date, there are no recorded blooms in eutrophic waters of anthomedusae and leptomedusae with hydroids in the littoral zones or on floating objects. This may simply reflect the lack of knowledge of the distribution of many hydroids. For example, the anthomedusa, *Nemopsis bachei*, is abundant in eutrophic Chesapeake Bay in the spring but the hydroid, less than 1 mm in height, has not been collected in the field (Kuhl, 1962; Purcell & Nemazie, 1992).

Other species may have at least one stage able to survive hypoxic, or even temporary anoxic, conditions. The tolerance of low oxygen levels has not been extensively measured in the laboratory, however, *Chrysaora quinquecirrha* medusae and *Mnemiopsis leidyi* ctenophores have prolonged survival in dissolved oxygen concentrations $<2$ mg l$^{-1}$ (Purcell et al., 2001b). Forty percent of *C. quinquecirrha* polyps can survive 0.5 mg l$^{-1}$ dissolved oxygen for $>20$ d although mortality is less and development enhanced at higher oxygen levels (Condon et al., 2001). A large amount of field data shows that many other coelenterates are tolerant of hypoxic water (reviewed by Purcell et al., 2001b). A few, such as *Rathkea lizzioides* O'Sullivan even penetrate, at least briefly, into anoxic water (Bayly, 1986). Developmental stages may differ in tolerance. *Aurelia aurita* planulae are more tolerant than the medusae, as indicated by their presence in the Black Sea, below the medusae, at $<1.0$ mg l$^{-1}$ (Vinogradov et al., 1985). Scyphozoa such as *Cyanea* spp., produce short-term cysts (planulocysts) from the planulae, and produce podocysts from the polyps, which may remain encysted until the following year (Brewer & Feingold, 1991). There are no measurements to date on the tolerance of these presumably resistant cysts to low oxygen levels.

Anaerobic metabolism is often ignored in calculating coelenterate energy budgets but could be very important to survival in oxygen depleted water (Arai, 1997a, b). Very little is known as yet about the relevant chemical reactions. The production of lactate, as indicated by the presence of lactate dehydrogenase, occurs in the swimming muscle of several hydrozoan and scyphozoan medusae (Thuesen & Childress, 1994). The lactate is probably only temporarily produced during strenuous swimming bouts and then further metabolized. An alternate pathway for anaerobiosis of some invertebrates is production of succinate rather than pyruvate from phosphoenol pyruvate. The enzyme phosphoenolpyruvate carboxykinase, present in scyphistomae of *Aurelia aurita* and *Chrysaora quinquecirrha* indicates that this pathway may be present (Lin & Zubkoff, 1977). There have been no measurements in coelenterates of accumulation or excretion of succinate, or of the fatty acid, proprionate, which may be produced from it.

Like anaerobic metabolism, nutrition is a term of the energy budget that is not well understood (Arai, 1997a). Dissolved organic material (DOM) and translocation from symbionts may be utilized by coelenterates. To date, there are no pelagic coelenterates known to feed directly on phytoplankton, although they may ultilize microzooplankton such as ciliates. *Aurelia aurita* and *Mnemiopsis leidyi* feed on the larger aloricate ciliates rather than on tintinnids (Stoecker et al., 1987 a & b; Båmstedt, 1990). Most work has concentrated on animal prey, especially fish and arthropods.

DOM uptake is potentially especially important in eutrophic situations (Wilkerson & Dugdale, 1984). Uptake of labelled glucose or amino acids by hydromedusae, scyphomedusae and siphonophores has been observed in the laboratory (Shick, 1975; Erokhin, 1980; Ferguson, 1988; Wilkerson & Kremer, 1992). Shick (1975) demonstrated that uptake of glycine could benefit otherwise starved, strobilizing *Aurelia aurita*. Labeled DOM is incorporated into complex organic compounds and labeled carbon dioxide is released during metabolism. However, Cnidaria may also release amino acids to the environment. Until measurements are made of the net influx or loss of organic compounds, it is not possible to quantitatively evaluate the importance of DOM to the survival of coelenterates.

With eutrophication there is not only an increase in production of phytoplankton, but often a change in the type produced. Diatoms require silicate to form the external shell and their growth becomes silicate limited if the ratio of silicate to nitrogen falls below 1:1 (Justić et al., 1995; Turner et al., 1998). This may result in an increase in populations of flagellates replacing the diatoms. In turn, there may be changes in the composition of the higher trophic levels. In many cases, small copepods are favored over larger forms in eutrophic situations. For example in Tokyo Bay, the populations of larger copepods such as *Acartia omorii*

Bradford have decreased and been replaced by smaller forms such as *Oithona davisae* Ferrari & Orsi since the 1960s (Nomura & Murano, 1992; Uye, 1994). *A. omorii* feeds on diatoms whereas *O. davisae* utilizes flagellates (Uchima, 1988). It is not clear how changes in the size of available copepods would affect survival of coelenterate predators. *Aurelia aurita* preferentially decreases the abundance of *A. omorii* rather than the smaller *O. davisae* in Tokyo Bay (Omori et al., 1995), although *A. aurita* can also utilize *O. davisae* (Ishii & Tanaka, 2001). In Kiel Bight (also eutrophic) the reverse is true; in the presence of *A. aurita Oithona similis* is decreased but *Acantia* spp. are unaffected (Behrends & Schneider, 1995).

Rates of budding of Hydrozoa and strobilation of Scyphozoa, and growth depend on nutrition (see reviews in Arai, 1992, 1997b; Purcell et al., 1999b). Nevertheless, there is very little known about the relative ability of coelenterates to assimilate and utilize alternative diets for growth and reproduction. It has recently been shown that growth of ephyrae of *Cyanea capillata* and *Chrysaora quinquecirrha* is enhanced by ctenophore prey rather than alternative zooplankton prey (Olesen et al., 1996; Bamstedt et al., 1997). Further data of this type are necessary to evaluate what effects changes in population structure caused by eutrophication might have on nutrition and resulting population changes of coelenterates.

When an increase of phytoplankton biomass is associated with eutrophication, there is increased turbidity. Decreased light may have direct effects on coelenterates. Triggered by changes in light intensity, many pelagic Cnidaria carry out diel vertical migration, which may be important for contact with prey (Pagès et al., 2001). On the other hand, aside from a few cubozoans, actual feeding by coelenterates is not visual and is therefore not affected by decreasing light. Strobilation of Scyphozoa and budding of hydroids may be affected by light (reviewed in Arai, 1992, 1997b). For rhizostome scyphomedusae, such as *Cassiopea* spp., light is essential for activity of the symbionts (Arai, 1997b). In the Nichupte Lagoon System, described above, most of the organic material is deposited in a layer on the bottom of the lagoon, which allows light to reach the medusae.

The relative ability of coelenterates and their predators and prey to survive and to hunt under eutrophic conditions will influence the food webs. The relative resistance of nudibranchs and their scyphozoan and hydrozoan prey to oxygen deficiencies or turbidity is not known, however, copepod prey are more sensitive to hypoxia than are medusae and ctenophores (Purcell et al., 2001b). Breitburg and her co-workers have shown that hypoxia can increase predation by *Chrysaora quinquecirrha* on fish larvae, but decrease predation by striped bass (Breitburg et al., 1997, 1999). *Mnemiopsis leidyi* is also more tolerant of hypoxia than are fish (Purcell et al., 2001b). Turbidity also favors predation by tactile predators such as the scyphomedusa *Periphylla periphylla* (Péron and Lesueur) over that by visual predators such as fish (Eiane et al., 1997; Eiane et al., 1999).

modify the food webs. For example, Cnidaria and Ctenophora are non-visual predators and may compete more successfully with visual predators, such as fish, in turbid conditions.

## Acknowledgements

I am grateful for the assistance of the librarians of the Pacific Biological Station, Nanaimo, B. C. in obtaining much of the literature on which this review is based. I am also grateful to several correspondents who have sent pertinent foreign literature or information including U. Båmstedt, S. D. Batten, G. Harris, S. Kaartvedt, C. E. Mills, S. Morison, M. Omori, J. E. Purcell, M. Toyokama and S.-I. Uye. Z. Kabata translated critical portions of the Russian literature.

## Summary

In the above survey, the changes of pelagic coelenterate populations in particular eutrophic areas can rarely be unequivocally linked only to results of eutrophication rather than to normal fluctuations of coelenterate populations or other anthropogenic changes. Nevertheless, a general picture does begin to emerge from the accumulated data. The biomass of coelenterates usually, but not necessarily, increases as a result of eutrophication. Typically, there is a loss of species diversity. This loss is demonstrated most clearly for Hydrozoa with polyp stages. The coelenterate species that most often survive and may increase in biomass are holoplanktonic or shallow water forms. They are widely distributed and common in non-eutrophic areas as well. For example, there have been spectacular increases of the scyphomedusa *Aurelia aurita* in some eutrophic locations, however, the population may remain low as in Chesapeake Bay. Elsewhere, the biomass of *A. aurita* may also be very high in some locations with oligotrophic or mesotrophic conditions such as Horsea Lake or Wakasa Bay.

In order to understand responses to eutrophication, there will need to be a great deal more research on the effects of changed diets, and decreased oxygen and light. There are some data on the ability of adult medusae and ctenophores to avoid or tolerate low oxygen levels, however, there are very few data on developmental stages and none on scyphozoan cysts. Anaerobic metabolism, which could be very important to survival in oxygen depleted water, has not been investigated in these animals. In respect to nutrition, uptake of dissolved organic material is known to occur, but its quantitative importance is not clear. The ability to utilize different diets for growth and reproduction is little understood. Finally, the relative ability of predators and prey to adapt to the changed conditions will

## References

Aebischer, N. J., J. C. Coulson & J. M. Colebrook, 1990. Parallel long-term trends across four marine trophic levels and weather. Nature (Lond.) 347: 753–755.

Arai, M. N., 1988. Interactions of fish and pelagic coelenterates. Can. J. Zool. 66: 1913–1927.

Arai, M. N., 1992. Active and passive factors affecting aggregations of hydromedusae: a review. Sci. mar. 56: 99–108.

Arai, M. N., 1997a. Coelenterates in pelagic food webs. In Den Hartog, J. C. (ed.), Proceedings of the 6th International Conference of Coelenterate Biology, 1995. National Natuurhistorisch Museum, Leiden: 1–9.

Arai, M. N., 1997b. A Functional Biology of Scyphozoa. Chapman & Hall, London: 316 pp.

Ates, R. M. L., 1988. Medusivorous fishes, a review. Zool. Med. Leiden 62: 29–42.

Avian, M. & L. Rottini Sandrini, 1994. History of scyphomedusae in the Adriatic Sea. Boll. Soc. Adriat. Sci. Trieste 75: 5–12.

Azov, Y., 1991. Eastern Mediterranean-a marine desert? Mar. Pollut. Bull. 23: 225–232.

Båmstedt, U., 1990. Trophodynamics of the scyphomedusae *Aurelia aurita*. Predation rate in relation to abundance, size and type of prey organism. J. Plankton Res. 12: 215–229.

Båmstedt, U., H. Ishii & M. B. Martinussen, 1997. Is the scyphomedusa *Cyanea capillata* (L.) dependent on gelatinous prey for its early development? Sarsia 82: 269–273.

Bayly, I. A. E., 1986. Ecology of the zooplankton of a meromictic antarctic lagoon with special reference to *Drepanopus hispinosus* (Copepoda: Calanoida). Hydrobiologia 140: 199–231.

Behrends, G. & G. Schneider, 1995. Impact of *Aurelia aurita* medusae (Cnidaria, Scyphozoa) on the standing stock and community composition of mesozooplankton in the Kiel Bight (western Baltic Sea). Mar. Ecol. Prog. Ser. 127: 39–45.

Benović, A. & D. Lučić, 1995. Appearance of hydromedusae in the northern Adriatic Sea in 1992 and 1993. Rapp. Comm. Int. Mer. Medit. 34: 203.

Benović, A. & D. Lučić, 1996. Comparison of hydromedusae findings in the northern and southern Adriatic Sea. Sci. mar. 60: 129–135.

Benović, A., D. Justić & A. Bender, 1987. Enigmatic changes in the hydromedusan fauna of the northern Adriatic Sea. Nature (Lond.) 326: 597–600.

Berdnikov, S. V., V. V. Selyutin, V. V. Vasilchenko & J. F. Caddy, 1999. Trophodynamic model of the Black and Azov Sea pelagic ecosystem: consequences of the comb jelly, *Mnemiopsis leidyi*, invasion. Fish. Res. 42: 261–289.

Beyer, F., 1958. A new, bottom-living trachymedusa from the Oslo Fjord: description of the species, and a general discussion of the life conditions and fauna of the fjord deeps. Nytt Mag. Zool. (Oslo) 6: 121–143.

Beyer, F., 1968. Zooplankton, zoobenthos, and bottom sediments as related to pollution and water exchnage in the Oslofjord. Hellgoländer wiss. Meeresunters 17: 496–509.

Beyer, F. & J. Indrehus, 1995. Effects of pollution and deep water exchange on the fauna along the bottom of Oslofjorden, Norway, based on material collected since 1952. Rep. Norsk Inst. Vannforsk. 621: Vol. 1: 1–143: Vol. 2: 1–153.

Boicourt, W. C., M. Kuzmić & T. S. Hopkins, 1999. The Inland Sea: Circulation of Chesapeake Bay and the Northern Adriatic. Coastal Estuar. Stud. 55: 81–129.

Bologa, A. S., N. Bodeanu, A. Petran, V. Tiganus & Yu. P. Zaitsev, 1995. Major modifications of the Black Sea benthic and planktonic biota in the last three decades. Bull. Inst. Ocean. (Monaco) Special No. 15: 85–110.

Bonsdorff, E., E. M. Blomquist, J. Mattila & A. Norkko, 1997. Coastal eutrophication: Causes, consequences and perspectives in the archipelago areas of the northern Baltic Sea. Estuar. coast. shelf Sci. 44 (Suppl. A): 63–72.

Brandon, M. & C. E. Cutress, 1985. A new *Dondice* (Opisthobranchia: Favorinidae), predator of *Cassiopea* in southwest Puerto Rico. Bull. mar. Sci. 36: 139–144.

Breitburg, D. L., K. A. Rose & J. H. Cowan, 1999. Linking water quality to larval survival: predation mortality of fish larvae in an oxygen-stratified water column. Mar. Ecol. Prog. Ser. 178: 39–54.

Breitburg, D. L., T. Loher, C. A. Pacey & A. Gerstein, 1997. Varying effects of low dissolved oxygen on trophic interactions in an estuarine food web. Ecol. Monogr. 67: 489–507.

Brewer, R. H. & J. S. Feingold, 1991. The effect of temperature on the benthic stages of *Cyanea* (Cnidaria: Scyphozoa), and their seasonal distribution in the Niantic River estuary, Connecticut. J. exp. mar. Biol. Ecol. 152: 49–60.

Brodeur, R. D., C. E. Mills, J. E. Overland, G. E. Walters & J. D. Schumacher, 1999. Evidence for a substantial increase in zooplankton in the Bering Sea, with possible links to climate change. Fish. Oceanogr. 8: 296–306.

Buecher, E., 1997. Distribution and abundance of *Pleurobrachia rhodopis* (Cydippid Ctenophore) in the Bay of Villefranche-sur-Mer (Northwestern Mediterranean) studied using three different planktonic time series. Ann. Inst. oceanogr. 73: 173–184.

Buecher, E., 1999. Appearance of *Chelophyes appendiculata* and *Abylopsis tetragona* (Cnidaria, Siphonophora) in the Bay of Villefranche, northwestern Mediterranean. J. Sea Res. 41: 295–307.

Buecher, E., J. Goy, B. Planque, M. Etienne & S. Dallot, 1997. Long-term fluctuations of *Liriope tetraphylla* in Villefranche Bay between 1966 and 1993 compared to *Pelagia noctiluca* pullulations. Oceanol. Acta 20: 145–157.

Caddy, J. F., 1993. Toward a comparative evaluation of human impacts on fishery ecosystems of enclosed and semi-enclosed seas. Rev. Fish. Sci. 1: 57–95.

Caddy, J. F. & R. C. Griffiths, 1990. Recent trends in the fisheries and environment in the General Fisheries Council for the Mediterranean (GFCM) area. Gen. Fish. Counc. Mediterr. Stud. Rev. 63: 43–71.

Calder, D., 1971. Hydroids and hydromedusae of southern Chesapeake Bay. Virginia Inst. mar. Sci. Spec. Papers in mar. Sci. 1: 1–125.

Cargo, D. G. & D. R. King, 1990. Forecasting the abundance of the sea nettle, *Chrysaora quinquecirrha*, in the Chesapeake Bay. Estuaries 13: 486–491.

Cary, L. R., 1917. Studies on the physiology of the nervous system of *Cassiopea xamachana*. Carnegie Inst. Wash. Publ. 251: 121–170.

Cociasu, A., V. Diaconu, L. Popa, L. Buga, I. Nae, L. Dorogan & V. Malciu, 1997. The nutrient stock of the Romanian Shelf of the Black Sea during the last three decades. In Oszoy, E. & A. Mikaelyan (eds), Sensitivity of Change: Black Sea, Baltic Sea and North Sea. Kluwer Academic Publishers, Dordrecht, The Netherlands: 49–63.

Collado Vides, L., L. Segura Puertas & M. Merino Ibarra, 1988. Observaciones sobre dos escifomedusas del genero *Cassiopea* en la laguna de Bojorquez, Quintana Roo, Mexico. Rev. Inv. Mar. 9: 21–27.

Condon, R. H., M. B. Decker & J. E. Purcell, 2001. Effects of low dissolved oxygen on survival and asexual reproduction of scyphozoan polyps (*Chrysaora quinquecirrha*). Hydrobiologia 451 (Dev. Hydrobiol. 155): 89–95.

Degobbis, D., S. Fonda-Umani, P. Franco, A. Malej, R. Precali & N. Smodlaka, 1995. Changes in the northern Adriatic ecosystem and the hypertrophic appearance of gelatinous aggregates. Sci. Total Envir. 165: 43–58.

Edwards, M., A. W. G. John, H. G. Hunt & J. A. Lindley, 1999. Exceptional influx of oceanic species into the North Sea late 1997. J. mar. biol. Ass. U.K. 79: 737–739.

Eiane, K., D. L. Aksnes & J. Giske, 1997. The significance of optical properties in competition among visual and tactile planktivores: a theoretical study. Ecol. Model. 98: 123–136.

Eiane, K., D. L. Aksnes, E. Bagoien & S. Kaartvedt, 1999. Fish or jellies – a question of visibility? Limnol. Oceanogr. 44: 1352–1357.

Elmgren, R., 1989. Man's impact on the ecosystem of the Baltic Sea: Energy flows today and at the turn of the century. Ambio 18: 326–332.

Erokhin, V. E., 1980. Invertebrates capacity for utilizing organic substances in sea water. Ekol. Morya 2: 3–15.

Ferguson, J. C., 1988. Autoradiographic demonstration of the use of free amino acid by Sargasso Sea zooplankton. J. Plankton Res. 10: 1225–1238.

Finenko, G. A., B. E. Anninsky, Z. A. Romanova, G. I. Abolmasova & A. E. Kideys, 2001. Chemical composition, respiration and feeding rates of the new alien ctenophore, *Beroe ovata*, in the Black Sea. Hydrobiologia 451 (Dev. Hydrobiol. 155): 177–186.

Fitt, W. K. & K. Costley, 1998. The role of temperature in survival of the polyp stage of the tropical rhizostome jellyfish *Cassiopea xamachana*. J. exp. mar. Biol. Ecol. 222: 79–91.

Fleck, J. & W. K. Fitt, 1999. Degrading mangrove leaves of *Rhizophora mangle* Linne provide a natural cue for settlement and metamorphosis of the upside down jellyfish *Cassiopea xamachana* Bigelow. J. exp. mar. Biol. Ecol. 234: 83–94.

Fleck, J., W. K. Fitt & M. G. Hahn, 1999. A proline-rich peptide originating from decomposing mangrove leaves is one natural metamorphic cue of the tropical jellyfish *Cassiopea xamachana*. Mar. Ecol. Prog. Ser. 183: 115–124.

Fransz, H. G., J. M. Colebrook, J. C. Gamble & M. Krause, 1991. The zooplankton of the North Sea. Neth. J. Sea Res. 28: 1–52.

Fraser, J. H., 1967. Siphonophora in the plankton to the north and west of the British Isles. Proc. r. Soc. Edinb. Sect. B (Biol.) 70: 1–30.

Fraser, J. H., 1969. Variability in the oceanic content of plankton in the Scottish area. Prog. Oceanogr. 5: 149–159.

Fraser, J. H., 1970. The ecology of the ctenophore *Pleurobrachia pileus* in Scottish waters. J. Cons. int. Explor. Mer 33: 149–168.

Friligos, N., 1982. Some consequences of the decomposition of organic matter in the Elefsis Bay, an anoxic basin. Mar. Poll. Bull. 13: 103–106.

Garcia, J. R. & J. M. Lopez, 1989. Seasonal patterns of phytoplankton productivity, zooplankton abundance and hydrological conditions in Laguna Joyuda, Puerto Rico. Sci. mar. 53: 625–631.

Gomez-Aguirre, S., 1980. Variacion estacional de grandes medusas (Scyphozoa) en un sistema de lagunas costeras del sur del Golfo de Mexico (1977/1978). Bol. Inst. Oceanogr. 29: 183–185.

Gomoiu, M.-T., 1980. Ecological observations on the jellyfish *Aurelia aurita* (L.) populations. Cercet. Mar. 13: 91–102.

Gomoiu, M.-T., 1981. Some problems concerning actual ecological changes in the Black Sea. Cercet. Mar. 14: 109–127.

González, A. C., 1979. Contribución al conocimiento de las medusas (Coelenterata) de la Laguna de Términos, Camp. México. An. Cent. Cienc. Mar Limnol. Univ. Nac. Auton. Mex. 6: 183–188.

Goy, J., P. Morand & M. Etienne, 1989. Long-term fluctuations of *Pelagia noctiluca* (Cnidaria, Scyphomedusa) in the western Mediterranean Sea. Prediction by climatic variables. Deep Sea Res. 36: 269–279.

Graham, W. M., 2001. Numerical increases and distributional shifts of *Chrysaora quinquecirrha* (Desor) and *Aurelia aurita* (Linné) (Cnidaria: Scyphozoa) in the northern Gulf of Mexico. Hydrobiologia 451 (Dev. Hydrobiol. 155): 97–111.

Graham, W. M., Pagès, F. & W. M. Hamner, 2001. A physical context for gelatinous zooplankton aggregations: a review. Hydrobiologia 451 (Dev. Hydrobiol. 155): 199–212.

Greve, W., 1994. The 1989 German Bight invasion of *Muggiaea atlantica*. ICES (Int. Counc. Explor. Sea) J. mar. Sci. 51: 355–358.

Gucu, A. C., 1997. Role of fishing in the Black Sea ecosystem. In Ozsoy, E. & A. Mikaelyan (eds), Sensitivity to change: Black Sea, Baltic Sea and North Sea. Kluwer Academic Publishers, Dordrecht, The Netherlands: 149–162.

Haahtela, I. & J. Lassig, 1967. Records of *Cyanea capillata* (Scyphozoa) and *Hyperia galba* (Amphipoda) from the Gulf of Finland and the northern Baltic. Ann. Zool. Fenn. 4: 469–471.

Hanninen, J., I. Vuorinen & P. Hjelt, 2000. Climatic changes in the Atlantic control the oceanographic and ecological changes in the Baltic Sea. Limnol. Oceanogr. 45: 703–710.

Hansen, K. V., 1951. On the diurnal migration of zooplankton in relation to the discontinuity layer. J. Cons. Cons. int. Explor. Mer 17: 231–241.

Hay, S. J., J. R. G. Hislop & A. M. Shanks, 1990. North Sea scyphomedusae: summer distribution, estimated biomass and significance particularly for 0-group gadoid fish. Neth. J. Sea Res. 25: 113–130.

Hernroth, L. & H. Ackefors, 1979. The zooplankton of the Baltic proper. Report Institute of Marine Research Fishery Board of Sweden 2: 1–60.

Hsieh, Y-H.P., F-M. Leong & J. Rudloe, 2001. Jellyfish as food. Hydrobiologia 451 (Dev. Hydrobiol. 155): 11–17.

Ishii, H. & F. Tanaka, 2001. Food and feeding of *Aurelia aurita* in Tokyo Bay with an analysis of stomach contents and a meas-urement of digestion times. Hydrobiologia 451 (Dev. Hydrobiol. 155): 311–320.

Ishii, H., S. Tadokoro, H. Yamanaka & M. Omori, 1995. Population dynamics of the jellyfish, *Aurelia aurita* in Tokyo Bay in 1993 with determination of ATP-related compounds. Bull. Plankton Soc. Japan 42: 171–176.

Janas, U. & Z. Witek, 1993. The occurrence of medusae in the southern Baltic and their importance in the ecosystem, with special emphasis on *Aurelia aurita*. Oceanologia 34: 69–84.

Justić, D., N. N. Rabalais, R. E. Turner & Q. Dortch, 1995. Changes in nutrient structure of river-dominated coastal waters: Stoichiometric nutrient balance and its consequences. Estuar. coast. shelf Sci. 40: 339–356.

Keister, J. E., E. D. Houde & D. L. Breitburg, 2000. Effects of bottom-layer hypoxia on abundance and depth distributions of organisms in Patuxent River, Chesapeake Bay. Mar. Ecol. Prog. Ser. 205: 43–59.

Kennish, M. J., 1997. Pollution Impacts on Marine Biotic Communities. CRC Press, Boca Raton, Florida: 296 pp.

Kideys, A. E., 1994. Recent dramatic changes in the Black Sea ecosystem: the reason for the sharp decline in Turkish anchovy fisheries. J. mar. Syst. 5: 171–181.

Kideys, A. E., A. V. Kovalev, G. Shulman, A. Gordina & F. Bingel, 2000. A review of zooplankton investigations of the Black Sea over the last decade. J. mar. Syst. 24: 355–371.

Kovalev, A. V. & S. A. Piontkovski, 1998. Interannual changes in the biomass of the Black Sea gelatinous zooplankton. J. Plankton Res. 20: 1377–1385.

Kramp, P. L. & D. Damas, 1925. Les méduses de la Norvège. Introduction et Partie speciale I. Vidensk. Medd. Dan. Naturhist. Foren. 80: 217–323, pl. 35.

Kremer, P., 1994. Patterns of abundance for *Mnemiopsis* in US coastal waters: a comparative overview. ICES J. mar. Sci. 51: 347–354.

Kuhl, H., 1962. Die Hydromedusen der Elbmündung. Abh. Verh. naturwiss. Verh. Hamb. 6: 209–232.

Kuwabara, R., S. Sato & N. Noguchi, 1969. Ecological studies on the medusa, *Aurelia aurita* Lamarck - I. Distribution of *Aurelia* patches in the north-east region of Tokyo Bay in summer 1966 and 1967. Bull. Jpn. Soc. Sci. Fish. 35: 156–162 (in Japanese with English abstract).

Lapointe, B. E. & M. W. Clark, 1992. Nutrient inputs from the watershed and coastal eutrophication in the Florida Keys. Estuaries 15: 465–476.

Lapointe, B. E., D. A. Tomasko & W. R. Matzie, 1994. Eutrophication and trophic state classification of seagrass communities in the Florida Keys. Bull. mar. Sci. 54: 696–717.

Larson, R. J., 1997. Feeding behaviour of Caribbean scyphomedusae: *Cassiopea frondosa* (Pallas) and *Cassiopea xamachana* Bigelow. Uitg. Natuurwet. Studiekring Caraibisch (Stud. Nat. Hist. Carribean Region) 73: 43–54.

Legović, T. & D. Justić, 1997. When do phytoplankton blooms cause the most intense hypoxia in the northern Adriatic Sea? Oceanol. Acta 20: 91–99.

Lin, A. L. & P. L. Zubkoff, 1977. Enyymes associated with carbohydrate metabolism of scyphistomae of *Aurelia aurita* and *Chrysaora quinquecirrha* (Scyphozoa: Semaeostomae). Comp. Biochem. Physiol. 57B: 303–308.

Liu, P., Y. Yu & C. Liu, 1991. Studies on the situation of pollution and countermeasures of control of the oceanic environment in Zhoushan fishing ground – the largest fishing ground in China. Mar. Pollut. Bull. 23: 281–288.

Lucas, C. H., 1996. Population dynamics of *Aurelia aurita* (Scyphozoa) from an isolated brackish lake, with particular reference to sexual reproduction. J. Plankton Res. 18: 987–1007.

Lucas, C. H., A. G. Hirst & J. A. Williams, 1997. Plankton dynamics and *Aurelia aurita* production in two contrasting ecosystems: comparisons and consequences. Estuar. coast. mar. Sci. 45: 209–219.

Maaden, H. van der, 1942. Beobachtungen über Medusen am Strande von Katwijk aan Zee (Holland) in den Jahren 1933–1937. Arch Neerl. Zool. 6: 347–362.

Malakoff, D., 1998. Death by suffocation in the Gulf of Mexico. Science (Wash. D. C.) 281: 190–192.

Malone, T. C., A. Malej, L. W. Harding Jr., N. Smodlaka & R. E. Turner (eds), 1999. Ecosystems at the Land-Sea Margin: Drainage Basin to Coastal Sea. Coastal estuar. Stud. 55: 1–381.

Margonski, P. & K. Horbowa, 1996. Vertical distribution of cod eggs and medusae in the Bornholm Basin. Medd. Havsfiskelab. Lysekil 327: 7–17.

Matsueda, N., 1969. Presentation of *Aurelia aurita* at thermal power station. Bull. mar. Biol. Stn. Asamushi 13: 187–191.

Mayer, A. G., 1910. Medusae of the World Volume III The Scyphomedusae. Carnegie Institution of Washington, Washington: 735 pp.

Merino, M., A. Gonzalez, E. Reyes, M. Gallegos & S. Czitrom, 1992. Eutrophication in the lagoons of Cancun, Mexico. Sci. Total Envir. Suppl.: 861–870.

Mianzan, H. W., N. Mari, B. Prenski & F. Sanchez, 1996. Fish predation on neritic ctenophores from the Argentine continental shelf: a neglected food resource? Fish. Res. 27: 69–79.

Mianzan, H., M. Pájaro, G. Alvarez Colombo & A. Madirolas, 2001. Feeding on survival-food: gelatinous plankton as a source of food for anchovies. Hydrobiologia 451 (Dev. Hydrobiol. 155): 45–53.

Mills, C. E., 2001. Jellyfish blooms: are populations increasing globally in response to changing ocean conditions? Hydrobiologia 451 (Dev. Hydrobiol. 155): 55–68.

Mirza, F. B. & J. S. Gray, 1981. The fauna of benthic sediments from the organically enriched Oslofjord, Norway. J. exp. mar. Biol. Ecol. 54: 181–207.

Möller, H., 1980. A summer survey of large zooplankton, particularly scyphomedusae, in North Sea and Baltic. Meeresforsch. 28: 61–68.

Mutlu, E., 1999. Distribution and abundance of ctenophores and their zooplankton food in the Black Sea. II *Mnemiopsis leidyi*. Mar. Biol. 135: 603–613.

Mutlu, E. & F. Bingel, 1999. Distribution and abundance of ctenophores, and their zooplankton food in the Black sea. I. *Pleurobrachia pileus*. Mar. Biol. 135: 589–601.

Mutlu, E., F. Bingel, A. C. Gucu, V. V. Melnikov, U. Niermann, N. A. Ostr & V. E. Zaika, 1994. Distribution of the new invader *Mnemiopsis* sp. and the resident *Aurelia aurita* and *Pleurobrachia pileus* populations in the Black Sea in the years 1991–1993. ICES J. mar. Sci. 51: 407–421.

Nehring, D., 1992. Eutrophication in the Baltic Sea. Sci. Total Environ. Suppl.: 673–682.

Nicholas, K. R. & C. L. J. Frid, 1999. Occurrence of hydromedusae in the plankton off Northumberland (western central North Sea) and the role of planktonic predators. J. mar. biol. Ass. U. K. 79: 979–992.

Nielsen, A. S., A. W. Pedersen & H. U. Riisgard, 1997. Implications of density driven currents for interaction between jellyfish (*Aurelia aurita*) and zooplankton in a Danish fjord. Sarsia 82: 297–305.

Niermann, U. & W. Greve, 1997. Distribution and fluctuation of dominant zooplankton species in the southern Black Sea in comparison to the North Sea and Baltic Sea. In Özsoy, E. & A. Mikaelyan (eds), Sensitivity to Change: Black Sea, Baltic Sea and North Sea. Kluwer Academic Publishers, Dordrecht, The Netherlands: 65–77.

Niermann, U., F. Bingel, G. Ergun & W. Greve, 1998. Fluctuation of dominant mesozooplankton species in the Black Sea, North Sea and the Baltic Sea: is a general trend recognisable? Tr. J. Zool. 22: 63–81.

Nixon, S. W., 1995. Coastal marine eutrophication: a definition, social causes and future concerns. Ophelia: 199–219.

Nomura, H. & T. Ishimaru, 1998. Monitoring the occurrence of medusae and ctenophores in Tokyo Bay, central Japan, in recent 15 years. Umi no Kenkyu 7: 99–104.

Nomura, H. & M. Murano, 1992. Seasonal variation of meso- and macroplankton in Tokyo Bay, central Japan. La Mer 30: 49–56.

Olesen, N. J., 1995. Clearance potential of jellyfish *Aurelia aurita*, and predation impact on zooplankton in a shallow cove. Mar. Ecol. Prog. Ser. 124: 63–72.

Olesen, N. J., K. Frandsen & H. U. Riisgard, 1994. Population dynamics, growth and energetics of jellyfish *Aurelia aurita* in a shallow fjord. Mar. Ecol. Prog. Ser. 105: 9–18.

Olesen, N. J., J. E. Purcell & D. K. Stoecker, 1996. Feeding and growth of ephyrae of scyphomedusae *Chrysaora quinquecirrha*. Mar. Ecol. Prog. Ser. 137: 149–159.

Omori, M., H. Ishii & A. Fujinaga, 1995. Life history strategy of *Aurelia aurita* (Cnidaria, Scyphomedusae) and its impact on the zooplankton community of Tokyo Bay. ICES J. mar. Sci. 52: 597–603.

Omori, M. & E. Nakano, 2001. Jellyfish fisheries in southeast Asia. Hydrobiologia 451 (Dev. Hydrobiol. 155): 19–26.

Otsubo, K., A. Harashima, T. Miyazaki, Y. Yasuoka & K. Muraoka, 1991. Field survey and hydraulic study of 'Aoshio' in Tokyo Bay. Mar. Pollut. Bull. 23: 51–55.

Paerl, H. W., 1995. Coastal eutrophication in relation to atmospheric nitrogen deposition: Current perspectives. Ophelia 41: 237–259.

Palmén, E., 1953. Seasonal occurrence of ephyrae and subsequent instars of *Aurelia aurita* (L.) in the shallow waters of Tvärminne, S. Finland. Arch. Soc. Zool. - Bot. Fenn. 'Vanamo' 8: 122–131.

Panayotidis, P., G. Anagnostaki & I. Siokou-Frangou, 1986. Variations saisonnieres du diametre et de la biomasse de la scyphoméduse *Aurelia aurita* Lam., dans la Baie d'Elefsis (Saronikos, Mer Egee). Nova Thalassia 8, Suppl. 2: 89–92.

Panayotidis, P., E. Papathanassiou, I. Siokou-Frangou & O. Gotis-Skretas, 1985. Etude de la population de la scyphomeduse *Aurelia aurita* Lam. dans la Baie d'Elefsis (Saronikos, Mer Egee). Rapp. Proces-Verb. Reu. Comm. Int. Explor. Sci. Mer. Medit. 29: 191–193.

Papathanassiou, E., P. Panayotidis & K. Anagnostaki, 1986. Reproduction and growth of *Aurelia aurita* in Elefsis Bay. Nova Thalassia 8, Suppl. 2: 83–88.

Papathanassiou, E., P. Panayotidis & K. Anagnostaki, 1987. Notes on the biology and ecology of the jellyfish *Aurelia aurita* Lam. in Elefsis Bay (Saronikos Gulf, Greece). Mar. Ecol. (Pubbl. Stn Zool. Napoli I) 8: 49–58.

Prodanov, K., K. Mikhailov, G. Daskalov, K. Maxim, A. Chashchin, A. Arkhipov, V. Shlyakhov, & E. Ozdamar, 1997. Environmental impact on fish resources in the Black Sea. In Ozsoy, E. & A. Mikaelyan (eds), Sensitivity to Change: Black Sea, Baltic Sea and North Sea. Kluwer Academic Publishers, Dordrecht, The Netherlands: 163–181.

Purcell, J. E., 1992. Effects of predation by the scyphomedusan *Chrysaora quinquecirrha* on zooplankton populations in Chesapeake Bay, U.S.A. Mar. Ecol. Prog. Ser. 87: 65–76.

Purcell, J. E. & M. N. Arai, 2001. Interactions of pelagic cnidarians and ctenophores with fish: a review. Hydrobiologia 451 (Dev. Hydrobiol. 155): 27–44.

Purcell, J. E. & J. H. Cowan, 1995. Predation by the scyphomedusan *Chrysaora quinquecirrha* on *Mnemiopsis leidyi* ctenophores. Mar. Ecol. Prog. Ser. 129: 63–70.

Purcell, J. E. & D. A. Nemazie, 1992. Quantitative feeding ecology of the hydromedusan *Nemopsis bachei* in Chesapeake Bay. Mar. Biol. 113: 305–311.

Purcell, J. E., A. Malej & A. Benović, 1999a. Potential links of jellyfish to eutrophication and fisheries. Coastal estuar. Stud. 55: 241–263.

Purcell, J. E., T. A. Shiganova, M. B. Decker & E. D. Houde, 2001a. The ctenophore *Mnemiopsis* in native and exotic habitats: U.S. estuaries *versus* the Black Sea basin. Hydrobiologia 451 (Dev. Hydrobiol. 155): 145–175.

Purcell, J. E., J. R. White, D. A. Nemazie & D. A. Wright, 1999b. Temperature, salinity and food effects on asexual reproduction and abundance of the scyphozoan *Chrysaora quinquecirrha*. Mar. Ecol. Prog. Ser. 180: 187–196.

Purcell, J. E., D. L. Breitburg, M. B. Decker, W. M. Graham, M. J. Youngbluth & K. A. Raskoff, 2001b. Pelagic cnidarians and ctenophores in low dissolved oxygen environments: a review. In Rabalais, N. N. & R. E. Turner (eds), Coastal Hypoxia: consequences for living resources and ecosystems. American Geophysical Union. Coastal estuar. Stud. 58: 77–100.

Rabalais, N. N. & R. E. Turner (eds), 2001. Coastal hypoxia: consequences for living resources and ecosystems. American Geophysical Union. Coastal estuar. Stud. 58: 463 pp.

Rabalais, N. N., R. E. Turner, D. Justić, Q. Dortch, W. J. Wiseman & B. K. S. Gupta, 1996. Nutrient changes in the Mississippi River and system responses on the adjacent continental shelf. Estuaries 19: 386–407.

Reid, P. C., M. De Fatima Borges & E. Svendsen, 2001. A regime shift in the North Sea circa 1988 linked to changes in the North Sea horse mackerel fishery. Fish. Res. 50: 163–171.

Reyes, E. & M. Merino, 1991. Diel dissolved oxygen dynamics and eutrophication in a shallow, well-mixed tropical lagoon (Cancun, Mexico). Estuaries 14: 372–381.

Richardson, K. & B. B. Jorgensen, 1996. Eutrophication: definition, history and effects. Coastal estuar. Stud. 52: 1–19.

Riisgard, H. U., C. Jurgensen & F. O. Andersen, 1996. Case study: Kertinge Nor. Coastal estuar. Stud. 52: 205–220.

Riisgard, H. U., P. Bondo Christensen, N. J. Olesen, J. K. Petersen, M. M. Moller & P. Andersen, 1995. Biological structure in a shallow cove (Kertinge Nor, Denmark) – Control by benthic nutrient fluxes and suspension-feeding ascidians and jellyfish. Ophelia 41: 329–344.

Roden, C. L., R. R. Lohoefener, C. M. Rogers, K. D. Mullin & B. W. Hoggard, 1990. Aspects of the ecology of the moon jellyfish, *Aurelia aurita*, in the northern Gulf of Mexico. Northeast Gulf Sci. 11: 63–67.

Roginskaya, I. S., 1988. On possible relationship between Nudibranchia predation on the polyps of *Aurelia aurita* (L) and variations in the abundance of this jellyfish in the Black Sea. Ekol. Morya 30: 58–60 [Russian with English abstract].

Rosenberg, R., J. S. Gray, A. B. Josefson & T. H. Pearson, 1987. Petersen's benthic stations revisited. II. Is the Oslofjord and eastern Skagerrak enriched? J. exp. mar. Biol. Ecol. 105: 219–251.

Rottini-Sandrini, L. & M. Avian, 1986. Workshop on jellyfish in the Mediterranean Sea. Nova Thalassia 8 (Suppl. 2): 1–191.

Runge, J. A., P. Pepin & W. Silvert, 1987. Feeding behavior of the Atlantic mackerel *Scomber scombrus* on the hydromedusa *Aglantha digitale*. Mar. Biol. 94: 329–333.

Russell, F. S., 1939. Hydrographical and biological conditions in the North Sea as indicated by plankton organisms. J. Cons. Cons. int. Explor. Mer 14: 171–192.

Sandström, O. & T. Sörlin, 1981. Production ecology in the Northern Baltic. Hydrobiologia 76: 87–96.

Schulz, S., 1989. Changes in the Baltic pelagic ecosystem. In Klekowski, R. Z., E. Styczyńska & L. Falkowski (eds), Proceedings of the Twenty-first European Marine Biology Symposium. Institute of Oceanology. Polish Academy of Sciences, Warsaw: 463–471.

Schulz, S., G. Ertebjerg, G. Behrends, G. Breuel, P. Ciszewski, U. Horstmann, K. Kononen, E. Kostrichkina, J.-M. Leppanen, F. Mohlenberg, O. Dandstrom, M. Viitasalo & T. Willen, 1992. The present state of the Baltic Sea pelagic ecosystem - an assessment. In Colombo G., I. Ferrari, V. U. Ceccherelli & R. Rossi (eds), Marine Eutrophication and Population Dynamics. Olsen & Olsen, Fredenborg: 35–44.

Scott, A., 1914. The mackerel fishery off Walney in 1913. Proc. Trans. Liverpool Biol. Soc. 28: 109–115.

Scott, A., 1924. Food of the Irish Sea herring in 1923. Proc. Trans. Liverpool Biol. Soc. 38: 115–119.

Segerstråle, S. G., 1951. The recent increase in salinity off the coasts of Finland and its influence upon the fauna. J. Cons. perm. int. Explor. Mer 17: 103–110.

Shick, J. M., 1975. Uptake and utilization of dissolved glycine by *Aurelia aurita* scyphistomae: temperature effects on the uptake process; nutritional role of dissolved amino acids. Biol. Bull.: 117–140.

Shiganova, T. A., 1997. *Mnemiopsis leidyi* abundance in the Black Sea and its impact on the pelagic community. In Ozsoy, E. & A. Mikaelyan (eds), Sensitivity to change: Black Sea, Baltic Sea and North Sea. Kluwer Academic Publishers, Dordrecht, The Netherlands: 117–129.

Shiganova, T. A., 1998. Invasion of the Black Sea by the ctenophore *Mnemiopsis leidyi* and recent changes in pelagic community structure. Fish. Oceanogr. 7: 305–310.

Shiganova, T. A. & Y. V. Bulgakova, 2000. Effects of gelatinous plankton on Black Sea and Sea of Azov fish and their food resources. ICES J. mar. Sci. 57: 641–648.

Shiganova, T. A., Yu. V. Bulgakova, S. P. Volovik, Z. A. Mirzoyan & S. I. Dudkin, 2001. The new invader *Beroe ovata* Mayer, 1912 and its effect on the ecosystem in the northeastern Black Sea. Hydrobiologia 451 (Dev. Hydrobiol. 155): 187–197.

Shiganova, T. A., A. E. Kideys, A. C. Gucu, U. Niermann & V. S. Khoroshilov, 1998. Changes in species diversity and abundance of the main components of the Black Sea community during the last decade. In Ivanov, I. I. & T. Oguz (eds), Ecosystem Modeling as a Management Tool for the Black Sea, Volume 1. Kluwer Academic Publishers, Dordrecht, The Netherlands: 171–188.

Shuntov, V. P., E. P. Dulepova, V. I. Radchenko & V. V. Lapko, 1996. New data about communities of plankton and nekton of the far-eastern seas in connection with climate-oceanological reorganization. Fish. Oceanogr. 5: 38–44.

Shushkina, E. A. & E. I. Musayeva, 1983. The role of jellyfish in the energy system of Black Sea plankton communities. Oceanology 23: 92–96.

Shushkina, E. A. & M. Ye. Vinogradov, 1991. Long-term changes in the biomass of plankton in open areas of the Black Sea. Oceanology 31: 716–721.

Smedstad, O. M., 1972. On the biology of *Aglantha digitale rosea* (Forbes) [Coelenterata: Trachymedusae] in the inner Oslofjord. Norw. J. Zool. 20: 111–135.

Smith, H. G., 1936. Contribution to the anatomy and physiology of *Cassiopea frondosa*. Carnegie Inst. Wash. Publ. 475: 19–52.

Stoecker, D. K., A. E. Michaels & L. H. Davis, 1987a. Grazing by the jellyfish, *Aurelia aurita*, on microzooplankton. J. Plankton Res. 9: 901–915.

Stoecker, D. K., P. G. Verity, A. E. Michaels & L. H. Davis, 1987b. Feeding by larval and post-larval ctenophores on microzooplankton. J. Plankton Res. 9: 667–683.

Sugiura, Y., 1980. On the seasonal appearance of the medusae from Harumi, Tokyo Harbour. Dokkyo Univ. Bull. lib. Arts 15: 10–15. (in Japanese).

Sullivan, B. K., D. Van Keuren & M. Clancy, 2001. Timing and size of blooms of the ctenophore *Mnemiopsis leidyi* in relation to temperature in Narragansett Bay, RI. Hydrobiologia 451 (Dev. Hydrobiol. 155): 113–120.

Sverdrup, A., 1921. Planktonundersokelser fra Kristianiafjorden, Hydromeduser. Videnskapsselskapets Skrifter I Mat. -Naturv. Klasse 1, 1–50, pl. 1–4.

Swanberg, N., 1974. The feeding behavior of *Beroe ovata*. Mar. Biol. 24: 69–76.

Tatara, K., 1991. Utilization of the biological production in eutrophicated sea areas by commercial fisheries and the environmental quality standard for fishing ground. Mar. Pollut. Bull. 23: 315–319.

Thuesen, E. V. & J. J. Childress, 1994. Oxygen consumption rates and metabolic enzyme activities of oceanic California medusae in relation to body size and habitat depth. Biol. Bull. 187: 84–98.

Toyokawa, M., T. Inagaki & M. Terazaki, 1997. Distribution of *Aurelia aurita* (Linnaeus, 1758) in Tokyo Bay; observations with echosounder and plankton net. In Den Hartog, J. C. (ed.), Proceedings of the 6th International Conference on Coelenterate Biology, 1995. National Natuurhistorisch Museum, Leiden: 483–490.

Toyokawa, M. & M. Terazaki, 1994. Seasonal variation of medusae and ctenophores in the innermost part of Tokyo Bay. Bull. Plankton Soc. Jpn. 41: 71–75.

Toyokawa, M., T. Furota & M. Terazaki, 2000. Life history and seasonal abundance of *Aurelia aurita* medusae in Tokyo Bay, Japan. Plankton Biol. Ecol. 47: 48–58.

Turner, R. E., N. Qureshi, N. N. Rabalais, Q. Dortch, D. Justi§, R. F. Shaw & J. Cope, 1998. Fluctuating silicate:nitrate ratios and coastal plankton food webs. Proc. natl. Acad. Sci. 95: 13048–13051.

Tviete, S., 1969. Zooplankton and the discontinuity layer in relation to echo traces in the Oslofjord. Fiskeridir. skr. Ser. Havunders. 15: 25–35.

Uchima, M., 1988. Gut content analysis of neritic copepods *Acartia omorii* and *Oithona davisae* by a new method. Mar. Ecol. Prog. Ser. 48: 93–97.

UNEP, 1984. Workshop on Jellyfish Blooms in the Mediterranean. (Athens, 31 October–4 November 1983). United Nations Environment Programme, Mediterranean Action Plan: 221 pp.

UNEP, 1991. Jellyfish Blooms in the Mediterranean, Proceedings of the II Workshop on Jellyfish Blooms in the Mediterranean Sea (Trieste, 2–5 September 1987). Mediterranean Action Plan Tech. Rep. Ser. 47: 1–320.

Uye, S-I., 1994. Replacement of large copepods by small ones with eutrophication of embayments: cause and consequence. Hydrobiologia 292/293: 513–519.

Uye, S-I. & T. Kasuya, 1999. Functional roles of ctenophores in the marine coastal ecosystem. In Okutani T., S. Ohta & R. Ueshima (eds), Update Progress in Aquatic Invertebrate Zoology. Tokai University Press, Tokyo: 57–76 (Japanese with English abstract).

Uye, S-I. & T. Shimazu, 1997. Geographical and seasonal variations in abundance, biomass and estimated production rates of meso- and macrozooplankton in the Inland Sea of Japan. J. Oceanogr. 53: 529–538.

Uye, S-I, N. Iwamoto, T. Ueda, H. Tamaki & K. Nakahira, 1999. Geographical variations in the trophic structure of the plankton community along a eutrophic-mesotrophic-oligotrophic transect. Fish. Oceanogr. 8: 227–237.

Verwey, J., 1942. Die Periodizitat im Auftreten und die aktiven und passiven Bewegungen der Quallen. Arch. Neerl. Zool. 6: 363–468.

Vinogradov, M. Ye. & E. A. Shushkina, 1982. Estimate of the concentration of Black Sea jellyfish, ctenophores and *Calanus*, based on observations from the Argus submersible. Oceanology 22: 351–355.

Vinogradov, M. E., M. V. Flint & E. A. Shushkina, 1985. Vertical distribution of mesoplankton in the open area of the Black Sea. Mar. Biol. 89: 95–107.

Vinogradov, M. E., E. A. Shushkina & Yu. V. Bulgakova, 1996. Consumption of zooplankton by the comb jelly *Mnemiopsis leidyi* and pelagic fishes in the Black Sea. Oceanology 35: 523–527.

Vinogradov, M. Ye., E. A. Shushkina, E. I. Musayeva & P. Yu. Sorokin, 1989. A newly acclimated species in the Black Sea: the ctenophore *Mnemiopsis leidyi* (Ctenophora: Lobata). Oceanology 29: 220–224.

Vućetić, T., 1991. Hydrobiological variablity in the Middle Adriatic in relation with the unusual distribution or behavior of *Pelagia noctiluca*. Mediterranean Action Plan Tech. Rep. Ser. 47: 188–201.

Vuorinen, I., 1987. Is the ctenophore *Pleurobrachia pileus* important in the ecosystem of the Bothnian Sea? Mem. Soc. Fauna Flora Fenn. 63: 91–96.

Vuorinen, I. & E. Ranta, 1987. Dynamics of marine mesozooplankton at Seili, northern Baltic Sea, in 1967–1975. Ophelia 28: 31–48.

Vuorinen, I. & E. Ranta, 1988. Can signs of eutrophication be found in the mesozooplankton of Seili, Archipelago Sea? Kieler Meeresforsch 6: 126–140.

Vuorinen, I. & S. Vihersaari, 1989. Distribution and abundance of *Pleurobrachia pileus* (Ctenophora) in the Baltic Sea. Mem. Soc. Fauna Flora Fenn. 65: 129–131.

Wagenaar Hummelinck, P., 1968. Caribbean scyphomedusae of the genus *Cassiopea*. Uitg natuurw. Studkring Suriname 25: 1–57.

Watanabe, T. & H. Ishii, 2001. *In situ* estimation of ephyrae liberated from polyps of *Aurelia aurita* using settling plates in Tokyo Bay, Japan. Hydrobiologia 451 (Dev. Hydrobiol. 155): 247–258.

Weisse, T. & M.-T. Gomoiu, 2000. Biomass and size structure of the scyphomedusa *Aurelia aurita* in the northwestern Black Sea during spring and summer. J. Plankton Res. 22: 223–239.

Wilkerson, F. P. & R. C. Dugdale, 1984. Possible connections between sewage effluent, nutrient levels and jellyfish blooms. In Workshop on Jellyfish Blooms in the Mediterranean, UNEP, Athens: 195–201.

Wilkerson, F. P. & P. Kremer, 1992. DIN, DON, and $PO_4$ flux by a medusa with algal symbionts. Mar. Ecol. Prog. Ser. 90: 237–250.

Yasuda, T., 1969. Ecological studies on the jelly-fish, *Aurelia aurita*, in Urazoko Bay, Fukui Prefecture - I Occurrence pattern of the medusa. Bull. Jpn. Soc. Sci. Fish. 35: 1–6.

Yasuda, T., 1970. Ecological studies on the jelly-fish, *Aurelia aurita* (b.), in Urazoko Bay, Fukui Prefecture – V. Vertical distribution of the medusa. Ann. Rep. Noto Mar. Lab. 10: 15–22.

Zaika, V. Ye. & N. G. Sergeyeva, 1990. Morphology and development of *Mnemiopsis mccradyi* (Ctenophora, Lobata) in the Black Sea. Hydrobiol. J. 26: 1–6.

Zaitsev, Yu. P., 1992. Recent changes in the trophic structure of the Black Sea. Fish. Oceanogr. 1: 180–189.

Zaitsev, Yu. P., 1994. Etudes sur les pêches et l'environnement dans le bassin de la mer Noire. Partie 2: Effets de l'eutrophisation sur le faune de la mer Noire. Cons. Gen. Pech. Mediterr. Etud. Rev. 64: 59–88.

Zaitsev, Yu. P. & B. G. Alexandrov, 1997. Recent man-made changes in the Black Sea ecosystem. In Ozsoy, E. & A. Mikaelyan (eds), Sensitivity to Change: Black Sea, Baltic Sea and North Sea. Kluwer Academic Publishers, Dordrecht, The Netherlands: 25–31.

Zaitsev, Yu. P. & V. Mamaev, 1997. Marine biological diversity in the Black Sea: a study of change and decline. United Nations Publications, New York: 208 pp.

Zaitsev, Yu. P. & L. N. Polischuk, 1984. An increase in the number of *Aurelia aurita* (L.) in the Black Sea. Ekol. Morya 17: 35–46.

Zamponi, M. O., E. Suarez & R. Gasca, 1990. Hidromedusas (Coelenterata: Hydrozoa) y Escifomedusas (Coelenterata: Scyphozoa) de La Bahia de la Ascension, Reserva de la Biosfera de Sian Ka'an. In Navarro, D. & J. G. Robinson (eds), Diversidad Biologica en la Reserva de la Biosfera de Sian Ka'an Quintana Roo, Mexico, 8. CIQRO, University of Florida: 101–107.

*Hydrobiologia* **451**: 89–95, 2001.
© 2001 *Kluwer Academic Publishers.*

# Effects of low dissolved oxygen on survival and asexual reproduction of scyphozoan polyps (*Chrysaora quinquecirrha*)

Robert H. Condon, Mary Beth Decker[2] & Jennifer E. Purcell[1],*
*Horn Point Laboratory, University of Maryland Center for Environmental Science, Cambridge, MD 21613-0775, U.S.A.*
[1]*Present address: Shannon Point Marine Center, 1900 Shannon Point Road, Anacortes, WA 98221, U.S.A.*
*E-mail: purcell@hpl.umces.edu (*Author for correspondence).*
[2]*Present address: Dept. of Ecology and Evolutionary Biology, Yale University, New Haven, CT 06520, U.S.A.*

*Key words:* asexual reproduction, Chesapeake Bay, hypoxia, jellyfish, survival, polyp

## Abstract

Hypoxic conditions are common in many coastal environments such as Chesapeake Bay. While medusae appear to be quite tolerant of low dissolved oxygen (DO) concentrations, the effects of hypoxia on the benthic polyp stages are unknown. *Chrysaora quinquecirrha* (DeSor) polyps, and were subjected to 5 DO treatments (air-saturated [control], 3.5, 2.5, 1.5 and 0.5 mg l$^{-1}$) in the laboratory. Polyp survival and development were documented over 24 d. Virtually no mortality occurred in any treatment during the first 5 d. Total polyp mortality after 24 d was 59.3% at the lowest DO concentration, whereas <3% mortality was observed in the air-saturated treatment. Formation of stolons and strobilae occurred in all treatments, however, the proportions of polyps undergoing stolonation and strobilation were significantly greater in all DO concentrations above 0.5 mg l$^{-1}$. Polyp encystment was not observed in any treatment over the course of the 24 d experiment. These results indicate that polyps can survive and asexually propagate even during prolonged exposure to hypoxic conditions.

## Introduction

Scyphozoan medusae, *Chrysaora quinquecirrha* (De-Sor), are dominant predators in the Chesapeake Bay estuary. During the summer months, peak numbers of medusae occur in the mesohaline regions of the bay and its tributaries, although abundance varies spatially and annually (Cargo & King, 1990; Purcell, 1992; Purcell et al., 1999). At these times, high predation rates by *C. quinquecirrha* significantly affect zooplankton and ichthyoplankton stocks and may reduce prey available to zooplanktivorous fish populations (Baird & Ulanowicz, 1989; Purcell et al., 1994a, b).

Like most semaeostome species, the life cycle of *C. quinquecirrha* includes a benthic polyp stage (scyphistoma) and a pelagic stage (medusa). Throughout the cold winter months, scyphistomae persist as dormant cysts, which excyst in the spring to form polyps (Cargo & Schultz, 1966, 1967). Polyps reproduce asexually by stolon formation (at-

tached buds) and strobilation, a process that liberates free-swimming ephyrae (Cargo & Rabenold, 1980). Ephyrae are propagated from polyps during spring and summer, and subsequently develop into sexually mature medusae.

Each summer, the bottom waters of the mainstem Chesapeake Bay and its tributaries experience low dissolved oxygen (DO) concentrations (hypoxia, < 2 mg O$_2$ l$^{-1}$) as a result of density stratification of the water column combined with microbial degradation of organic matter (Taft et al., 1980; Sanford et al., 1990; Jonas, 1992; Malone, 1992). Hypoxic waters < 2 mg O$_2$ l$^{-1}$ are stressful to fish and many other organisms, and can be lethal. Organism abundance, vertical distribution and behavior can be significantly affected (Roman et al., 1993; Breitburg, 1994; Breitburg et al., 1994, 1997). Recent studies on *C. quinquecirrha* medusae have shown that the medusan stage can survive up to 48 h in severely hypoxic conditions (0.5 mg O$_2$ l$^{-1}$), however, longer exposure can be fatal (Houde & Zastrow, unpublished data see Purcell et al., 2001).

Hypoxic conditions also can affect their vertical distributions. In the mainstem Chesapeake Bay, more *C. quinquecirrha* medusae were found above and within the pycnocline than below it in summer (Purcell et al., 1994b). In a tributary of Chesapeake Bay, medusae were absent from bottom waters only when DO was < 1.0 mg $l^{-1}$ (Keister et al., 2000).

Although medusae appear to be tolerant of low DO concentrations at least for short periods, the effect of hypoxia on the polyp stage is unknown. Polyps occur on oyster reefs above 11 m depth (Cargo & Schultz, 1966, 1967), where waters sometimes are exposed to severely low DO. The response of polyps to salinity and temperature stress is to encyst (Cargo & Schultz, 1967), however, the response to oxygen stress has not been tested. The purpose of our study was to examine the effects of long-term exposure to low DO concentrations on the survival and asexual reproduction of *C. quinquecirrha* polyps.

## Materials and methods

### Specimens

*Chrysaora quinquecirrha* polyps were collected on oyster shells dredged from the Choptank River, a sub-estuary of Chesapeake Bay, during August, 1999. The polyps were immediately transported to laboratory facilities and were temporarily maintained in 20 l buckets in flowing water. Encrusting materials and epifauna were removed from shells having polyps by use of jeweler's forceps and a dissecting microscope (×20). Unhealthy polyps (e.g. those that showed evidence of bacterial infection) were also removed. Only early stage scyphistomae were kept for experimentation (i.e. polyps with 4–8 tentacles and no buds or stolons; see Calder, 1982). Each shell was mapped and individual polyp locations recorded prior to experimentation. Shells used in the experiment had approximately similar polyp densities (ca. 3–4 polyps $cm^{-2}$).

### Experimental design and conditions

*Chrysaora quinquecirrha* scyphistomae on oyster shells were subjected to experimental conditions in the laboratory for 24 d in 20 l plastic containers filled with filtered (1 μm) Choptank River water. Shells with polyps were distributed among 5 DO treatments and acclimated for 12 h (Table 1). Two shells were placed in each treatment, and one in the air-saturated control.

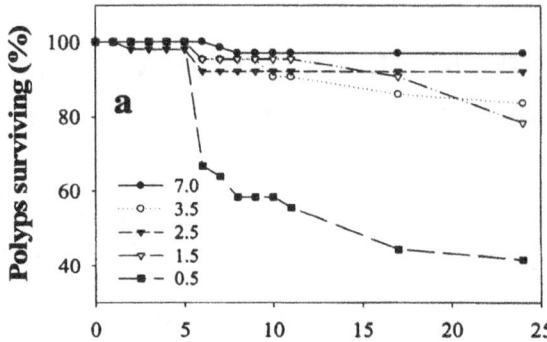
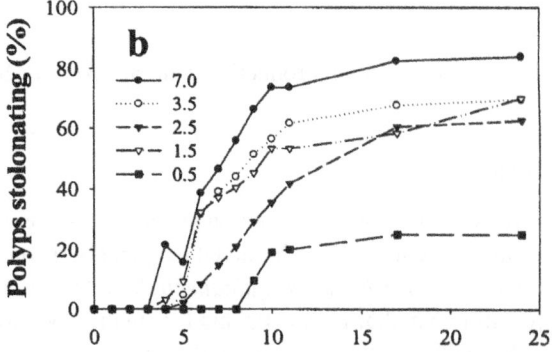
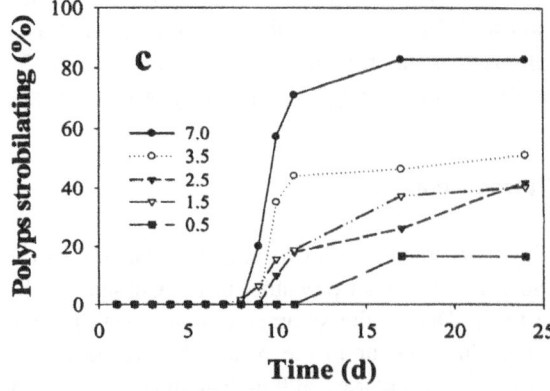

*Figure 1.* Survival and development of *Chrysaora quinquecirrha* polyps at five dissolved oxygen concentrations (mg $l^{-1}$) over 24 d in a laboratory experiment; (a) the proportions of polyps surviving, (b) the proportions of polyps with stolons, and (c) the proportions of polyps undergoing strobilation.

In all treatments except the control, target DO concentrations (3.5, 2.5, 1.5 and 0.5 mg $l^{-1}$) were established and maintained by sparging nitrogen gas into each experimental chamber (see Breitburg et al., 1994). The control container was bubbled with ambient air. All containers were sealed airtight to prevent the diffusion of air into the containers. These con-

*Table 1.* Experimental and physical conditions in each dissolved oxygen (mg l$^{-1}$) treatment throughout the 24 d *Chrysaora quinquecirrha* polyp experiment.

| Treatment | Polyps (No.) | DO (mg l$^{-1}$) | Temp range (°C) | Salinity range |
|-----------|-------------|------------------|------------------|----------------|
| Air sat   | 69          | 7.1±0.5          | 22.0–24.0        | 13.1–15.6      |
| 3.5       | 43          | 3.5±0.2          | 22.0–24.0        | 13.1–15.6      |
| 2.5       | 52          | 2.5±0.2          | 22.0–24.0        | 13.1–15.6      |
| 1.5       | 65          | 1.6±0.2          | 22.0–24.0        | 13.1–15.6      |
| 0.5       | 36          | 0.5±0.1          | 22.0–24.0        | 13.1–15.6      |

tainers maintained stable oxygen concentrations for > 12 h.

A *YSI*$^{TM}$30 Temperature, Conductivity, Salinity meter and a *YSI*$^{TM}$55 Dissolved Oxygen meter were used to monitor temperature, salinity and DO within each chamber 4 times daily (Table 1). Probes were inserted through small holes in the container lid, which were resealed after the measurements and adjustments were made to DO concentrations, as needed. To prevent bacterial infection of the polyps, the water in each container was replaced once every 5 days with filtered Choptank River water previously adjusted to the appropriate DO level.

*Polyp monitoring and maintenance*

All polyps were fed daily on cultured brine shrimp nauplii (*Artemia* sp.) over a 1 h period in separate 4 l plastic containers. Target DO concentrations in the feeding containers were established according to the method described above. The polyps were fed in their respective DO conditions, however, due to high prey mortality in the 0.5 mg l$^{-1}$ DO treatment, those polyps were fed at 1.0 mg O$_2$ l$^{-1}$. Polyps in all treatments readily captured prey items with their tentacles, and in most cases ingested more than one *Artemia* sp. nauplius at each feeding. Temperature, salinity and DO levels within each feeding chamber were recorded before and after feeding and varied little (<1%) from the conditions in the respective 20 l experimental container.

Survival and asexual reproduction of polyps were monitored on days 1 through 11, 17 and 24 of the experiment by placing the oyster shells in small glass dishes containing water from their experimental chamber and observing them with a dissecting microscope (×20). Whitish color, reduced or deformed tentacles and the presence of wounds or bacterial infestations indicated polyp death. Moribund polyps were removed

from the shells and their locations noted. Asexual development was measured in terms of the numbers of buds, stolons and strobilae discs produced per polyp, and the numbers of polyps undergoing stolonation and strobilation (as defined in Cargo & Rabenold, 1980). This process took approximately 15 min to complete for each treatment.

*Data analysis*

Data were analyzed using *Sigma Stat*$^{TM}$ (Version 2.0) software. Chi-square ($\chi^2$) analyses were used to test for independence between measures of polyp survival and asexual development at day 24 and DO concentration. When the $\chi^2$ test detected significant differences across all treatments, pair-wise $\chi^2$ comparisons were made to determine which treatments differed. The null hypothesis ($H_0$) for each $\chi^2$ test was that no relationships existed between the numbers of surviving polyps or the numbers of asexual products and DO concentration. Data are presented as means ± 1 standard error (SE).

**Results**

*Effects of hypoxia on survivorship*

High polyp survival was observed during the first 5 d of the experiment in all treatments (Fig. 1a). Survival at 5 d in air-saturated, 3.5, 1.5 and 0.5 mg O$_2$ l$^{-1}$ was 100% and in 2.5 mg O$_2$ l$^{-1}$ was 98.0%. After 24 d in the treatments, the fewest polyps survived in the 0.5 mg O$_2$ l$^{-1}$ treatment (41.7% survival). Polyp survival in the 0.5 mg l$^{-1}$ treatment was significantly lower than all other DO treatments (Table 2). The most polyps survived in the air-saturated (97.1%) and 2.5 mg l$^{-1}$ (92.2%) treatments (Fig. 1a). Polyp survival in these treatments was significantly greater than in the 3.5 mg l$^{-1}$ (83.7%) and 1.5 mg l$^{-1}$ (78.5%)

treatments (Table 2). $\chi^2$ comparisons failed to show significant differences in polyp survivorship between any other DO treatments. We did not observe polyp encystment in any of the treatments over the course of the 24 d experiment.

### Effects of hypoxia on asexual reproduction

Stolons were first observed on day 4, and the percentages of polyps with stolons increased rapidly between days 5 and 10 of the 24 d experiment (Fig. 1b). The average number of stolons produced per polyp at day 24 was greatest in the air-saturated control ($2.5\pm0.2$ stolons polyp$^{-1}$). By contrast, polyps in the 0.5 O$_2$ mg l$^{-1}$ treatment produced far fewer stolons per polyp ($0.3\pm0.2$) than any other treatment at 24 d (Fig. 2a). A $\chi^2$ test indicated that there were significant relationships between the proportions of polyps producing stolons (stolonation) and DO concentration (Table 2). $\chi^2$ pair-wise comparisons demonstrated that proportion of polyps stolonating was significantly lower at 24 d in the 0.5 O$_2$ mg l$^{-1}$ treatment (25.0%) than in the other DO treatments (62.5–83.8%) (Fig. 2c; Table 2). In addition, stolonation in the 2.5 mg l$^{-1}$ treatment (62.5%) was significantly lower than in the control (83.8%). $\chi^2$ comparisons failed to detect significant differences in polyp stolonation between any other DO treatments (Table 2).

The transverse segmentation characteristic of strobilating polyps (formation of strobilae discs) was first observed on day 8 (Fig. 1c), however, no ephyrae were released in any treatment during the experiment. The proportions of strobilating polyps increased rapidly between days 8 and 11, and more slowly thereafter. Results of a $\chi^2$ analysis indicated significant relationships between the proportions of polyps strobilating at 24 d and DO concentration (Table 2). The greatest proportion of polyps strobilating occurred in the air-saturated control (82.9%; Figs 1c and 2d; Table 2), which was significantly different from all other DO treatments. The proportion of polyps strobilating in the 0.5 mg l$^{-1}$ treatment was significantly lower (16.7%) than in the 2.5 (42.0%) and 3.5 mg O$_2$ l$^{-1}$ (51.2%) treatments, although the $\chi^2$ test did not detect significant differences from the 1.5 mg l$^{-1}$ treatment (Table 2).

The numbers of strobilae discs produced per polyp were highest in the air-saturated treatment after 24 days ($5.5\pm0.4$), although strobilae were observed in all treatments (Fig. 2b). In general, the average number of strobilae produced per polyp decreased linearly with DO concentration (Fig. 2b).

### Discussion

#### Effects of hypoxia on survivorship

Few previous studies have examined the effects of hypoxia on benthic cnidarian polyps. Benović et al. (1987) documented the disappearance of 31 species of meroplanktonic hydromedusae from the northern Adriatic Sea over a 20-year period and concurrent increases in bottom water hypoxic events. They attributed these findings to the inability of benthic hydroids to survive hypoxic conditions.

In Chesapeake Bay, C. quinquecirrha polyps primarily occur at depths above 11 m (Cargo & Schultz, 1966, 1967), where they generally are exposed to favorable DO concentrations (Breitburg, 1990; Sanford et al., 1990). However, polyps distributed within these high productivity zones may also be subjected to short-term fluxes of anoxia (0 mg l$^{-1}$) or severe hypoxia (<0.5 mg l$^{-1}$) through the natural seiching of oxygen-depleted bottom waters onto the flanks of Chesapeake Bay (Breitburg, 1990). Seiching events may persist for as long as 2 days at DO concentrations <2 mg l$^{-1}$, and can occur during approximately 40% of the days in the summer. The overall frequency of events, time scale and severity of hypoxia are dependent upon several physical parameters, including tides and wind strength and direction (Breitburg, 1990; Sanford et al., 1990). DO concentrations also can fluctuate dramatically over relatively short time periods (e.g. decrease of 6 mg l$^{-1}$ in only 4 h). Even at depths as shallow as 4 m, concentrations may drop below 0.5 mg l$^{-1}$ and persist for up to 10 h.

The results of this study suggest that C. quinquecirrha polyps can survive in moderately hypoxic conditions ($\geq1.5$ mg l$^{-1}$) for extended periods of time (24 d; Fig. 1). Furthermore, polyps can persist in severe hypoxic conditions (0.5 mg l$^{-1}$) for short periods (5 d), after which survivorship is detrimentally affected (Fig. 1a). Therefore, our results suggest that C. quinquecirrha polyps would be able to survive hypoxic seiching events in Chesapeake Bay for periods up to 5 d and possibly longer.

High polyp mortality was observed after prolonged exposure to 0.5 and 1.5 mg O$_2$ l$^{-1}$ compared with very low mortality in the air-saturated control (Fig. 1a). In our experiment, polyps submerged in water of the appropriate DO concentration were monitored in open

*Table 2. Chrysaora quinquecirrha.* Results of Chi-square tests of independence between dissolved oxygen concentration (Treatments) and measures of polyp survivorship and asexual propagation after 24 d in a laboratory experiment.

| Treatments (mg l$^{-1}$) | Survival | | | Stolonation | | | Strobilation | | |
|---|---|---|---|---|---|---|---|---|---|
| | $\chi^2$ | df | $p$ | $\chi^2$ | df | $p$ | $\chi^2$ | df | $p$ |
| Overall | 53.7 | 4 | *** | 20.8 | 4 | *** | 52.8 | 4 | *** |
| 7.0 vs. 3.5 | 4.6 | 1 | * | 2.1 | 1 | NS | 10.8 | 1 | *** |
| 7.0 vs. 2.5 | 0.6 | 1 | NS | 5.7 | 1 | * | 19.9 | 1 | *** |
| 7.0 vs. 1.5 | 9.2 | 1 | ** | 2.1 | 1 | NS | 31.2 | 1 | *** |
| 7.0 vs. 0.5 | 38.8 | 1 | *** | 17.8 | 1 | *** | 39.5 | 1 | *** |
| 3.5 vs. 2.5 | 0.9 | 1 | NS | 0.2 | 1 | NS | 0.6 | 1 | NS |
| 3.5 vs. 1.5 | 0.2 | 1 | NS | 0.004 | 1 | NS | 2.4 | 1 | NS |
| 3.5 vs. 0.5 | 13.4 | 1 | *** | 6.3 | 1 | * | 8.7 | 1 | ** |
| 2.5 vs. 1.5 | 3.1 | 1 | NS | 0.5 | 1 | NS | 0.3 | 1 | NS |
| 2.5 vs. 0.5 | 23.9 | 1 | *** | 4.6 | 1 | * | 4.8 | 1 | * |
| 1.5 vs. 0.5 | 12.3 | 1 | *** | 7.9 | 1 | ** | 3.0 | 1 | NS |

\* = $p < 0.05$; \** = $p < 0.01$; \*** = $p < 0.001$; NS = not significant; overall = comparisons among treatments.

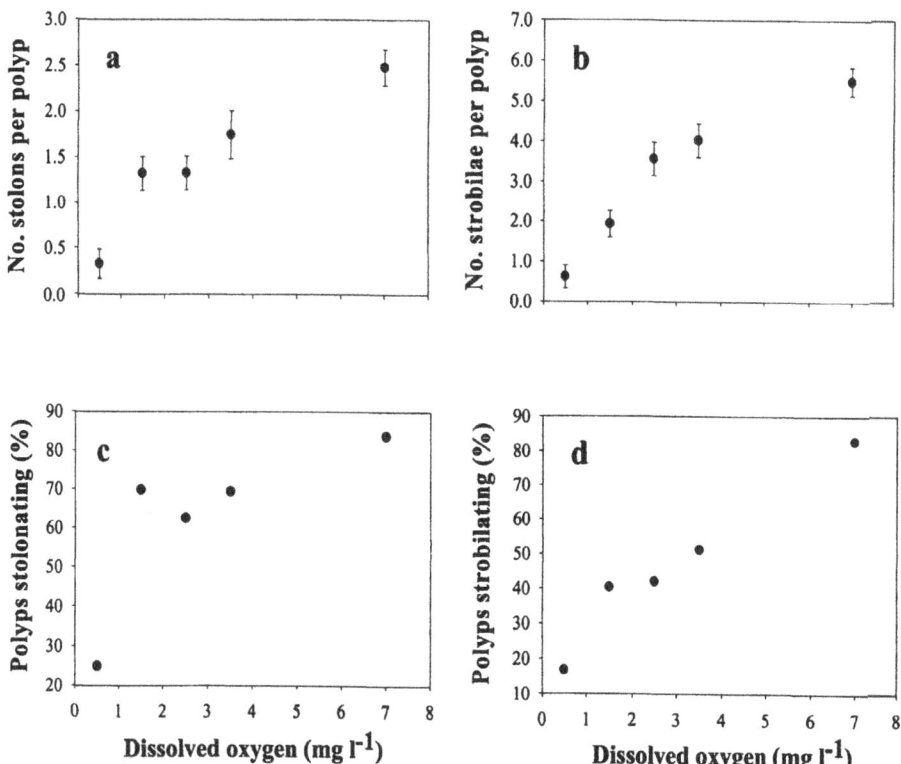

*Figure 2.* Stolonation, strobilation and stolon and strobilae disc formation of *Chrysaora quinquecirrha* polyps at 24 d versus dissolved oxygen concentration; (a) the numbers of stolons produced per polyp (means ± SE), (b) the numbers of strobilae produced per polyp (means ± SE), (c) the proportions of polyps with stolons, and (d) the proportions of polyps undergoing strobilation.

glass dishes for up to 15 min 13 times during the experiment, which may have allowed oxygen concentrations to increase slightly. Also, polyps in the 0.5 mg $O_2$ $l^{-1}$ treatment were fed at 1.0 mg $O_2$ $l^{-1}$ for 1 h daily. It is possible that these brief exposures to slightly higher oxygen conditions allowed some of the polyps in the severely hypoxic treatments to survive the 24 d exposure.

Childress & Seibel (1998) proposed 3 physiological adaptations that animals in oceanic oxygen minimum layers exhibit in order to survive: (1) development of mechanisms for highly effective removal of oxygen from water, (2) reduction in metabolic rates, and (3) use of anaerobic metabolism. The tolerance of C. quinquecirrha polyps to potentially fatal DO concentrations may be due to their ability to use anaerobic metabolism. C. quinquecirrha polyps contain the enzyme phosphoenol pyruvate carboxykinase, which is present in many anaerobic pathways (Lin & Zubkoff, 1977). It is plausible that the polyps can survive short-term hypoxia by utilizing this anaerobic pathway. To date, no studies have examined the metabolic rates and aerobic capabilities of any scyphozoan polyps under oxygen stress. In our study, approximately 40% of the polyps survived severely hypoxic conditions (0.5 mg $O_2$ $l^{-1}$; Fig. 1a). It is doubtful that polyps can sustain adequate energy levels through anaerobic pathways (glycolysis) in order to survive long-term hypoxia (Erik Thuesen, pers. comm.). A more likely explanation is that polyps alternate between aerobic and anaerobic metabolism during prolonged exposure to hypoxia, as do several related medusan species (Thuesen & Childress, 1994; Childress & Seibel, 1998).

The ability of polyps to form dormant cysts when conditions are unfavorable is an important survival strategy of C. quinquecirrha (Cargo & Schultz, 1966, 1967; Cargo & Rabenold, 1980). For example, C. quinquecirrha polyps often form cysts in response to seasonal changes in temperature and salinity. Laboratory experiments showed encystment at extreme temperatures ($\leq 3$ °C and $\geq 34$ °C) and salinities of 5, with incipient encystment at $\geq 30$ (Cargo & Schultz, 1967). It was surprising that we did not observe encystment of polyps in even the most severe treatment (0.5 mg $O_2$ $l^{-1}$). Cargo & Schultz (1966) induced polyp encystment in aquaria "by causing an oxygen-depleted situation associated with some hydrogen sulfide, in which the polyps formed cysts within a few weeks". The authors did not present data relating to temperature, salinity or DO during those observations. Consequently,

it is difficult to compare our results with theirs, however, anoxia and hydrogen sulfide did not occur in our treatments. Encystment probably is a metabolically demanding process, and is not quickly reversible. Therefore, encystment most likely is used to survive long-term (i.e. seasonal) changes, not short-term environmental fluctuations (i.e. periodic exposure to hypoxia). Since polyps in Chesapeake Bay probably only experience short-term reductions in DO on the scale of days (see Breitburg, 1990), low DO may not provide a stimulus for C. quinquecirrha polyps to encyst.

*Effects of hypoxia on asexual reproduction*

It is remarkable that C. quinquecirrha polyps not only survived, but also began asexual reproduction during prolonged exposure to severe hypoxia. Several studies have demonstrated that unfavorable environmental conditions, such as decreases in water temperature, high or low salinity (<7 and >25 ‰), and physical disturbance, can negatively affect asexual propagation of C. quinquecirrha polyps (Cargo & Schultz, 1966, 1967; Calder, 1974; Cargo & Rabenold, 1980; Purcell et al., 1999). Temperature also affects strobilation of other semaeostome species (e.g. Spangenberg, 1968; Gröndahl & Hernroth, 1987; Gröndahl, 1988). Our study showed that prolonged exposure of polyps to low DO concentrations also reduces strobilation and stolon formation, however, this is dependent on the severity of hypoxia and length of exposure (Figs. 1b–c and 2). A general decreasing trend was observed from the high (air-saturated) to low (0.5 mg $l^{-1}$) DO concentrations with respect to stolon and strobilae production per polyp, as well as the proportion of polyps undergoing strobilation. As a result of suppression of polyp asexual reproduction, hypoxia could directly reduce C. quinquecirrha medusae populations within Chesapeake Bay.

Despite their essential role in the life cycle of jellyfish, benthic polyps of scyphozoans have received little attention. In order to gain greater insights into the causes of jellyfish blooms, further research is required on the ecological, behavioral and physiological responses of polyps to environmental factors.

## Acknowledgements

This research was funded by EPA Grant No. R827097-01-0 to D. L. Breitburg, J. E. Purcell, M. B. Decker and K. A. Rose. The authors would like to thank the following people for assistance on this project: J. Matanoski, X. Ma, M. Leonard, D. Meritt, M. Roman and P. Anderson of the Horn Point Laboratory, C. Walter of the Natural History Museum, Smithsonian Institution; J. Davis, K. Richter and M. Norman for reading earlier versions of this manuscript; and the Condon family for support and encouragement. UMCES Contribution No. 3389.

## References

Baird, D. & R. E. Ulanowicz, 1989. The seasonal dynamics of the Chesapeake Bay ecosystem. Ecol. Monogr. 59: 329–364.

Benović, A., D. Justić & A. Bender, 1987. Enigmatic changes in the hydromedusan fauna of the northern Adriatic Sea. Nature 326: 597–600.

Breitburg, D. L., 1990. Near-shore hypoxia in the Chesapeake Bay: patterns and relationships among physical factors. Est. coast. shelf Sci. 30: 593–609.

Breitburg, D. L., 1994. Behavioral responses of fish larvae to low oxygen risk in a stratified water column. Mar. Biol. 120: 615–625.

Breitburg, D. L., T. Loher, C. A. Pacey & A. Gerstein, 1997. Varying effects of low dissolved oxygen on trophic interactions in an estuarine food web. Ecol. Monogr. 67: 489–507.

Breitburg, D. L., N. Steinburg, S. DeBeau, C. Cooksey & E. D. Houde, 1994. Effects of low oxygen on predation on estuarine fish larvae. Mar. Ecol. Prog. Ser. 104: 235–246.

Calder, D. R., 1974. Strobilation of the sea nettle, *Chrysaora quinquecirrha*, under field conditions. Biol. Bull. 146: 326–334.

Calder, D. R., 1982. Life history of the cannonball jellyfish, *Stomolophus meleagris* L. Agassiz, 1860 (Scyphozoa, Rhizostomida). Biol. Bull. 162: 149–162.

Cargo, D. G. & D. R. King, 1990. Forecasting the abundance of the sea nettle, *Chrysaora quinquecirrha*, in the Chesapeake Bay. Estuaries 13: 486–491.

Cargo, D. G. & G. E. Rabenold, 1980. Observations on the sexual reproductive activities of the sessile stages of the sea nettle *Chrysaora quinquecirrha* (Scyphozoa). Estuaries 3: 20–27.

Cargo, D. G. & L. P. Schultz, 1966. Notes on the biology of the sea nettle, *Chrysaora quinquecirrha*, in Chesapeake Bay. Chesapeake Sci. 7: 95–100.

Cargo, D. G. & L. P. Schultz, 1967. Further observations on the biology of the sea nettle and jellyfishes in Chesapeake Bay. Chesapeake Sci. 8: 209–220.

Childress, J. J. & B. A. Seibel, 1998. Life at stable low oxygen levels: adaptations of animals to oceanic oxygen minimum layers. J. exp. Biol. 201: 1223–1232.

Gröndahl, F., 1988. A comparative ecological study on the scyphozoans *Aurelia aurita*, *Cyanea capillata* and *C. lamarckii* in the Gullmar Fjord, western Sweden, 1982 to 1986. Mar. Biol. 97: 541–550.

Gröndahl, F. & L. Hernroth, 1987. Release and growth of *Cyanea capillata* (L.) ephyrae in the Gullmar Fjord, western Sweden. J. exp. mar. Biol. Ecol. 106: 91–101.

Jonas, R., 1992. Microbial processes, organic matter and oxygen demand in the water column. In Smith, D. E., M. Leffler & G. Mackiernan (eds), Oxygen Dynamics in the Chesapeake Bay: a Synthesis of Recent Research. Maryland Sea Grant, College Park: 113–148.

Keister, J. E., E. D. Houde & D. L. Breitburg, 2000. Effects of bottom-layer hypoxia on abundances and depth distributions of organisms in Patuxent River, Chesapeake Bay. Mar. Ecol. Prog. Ser. 205: 43–59

Lin, A. L. & R. L. Zubkoff, 1977. Enzymes associated with carbohydrate metabolism of scyphistomae of *Aurelia aurita* and *Chrysaora quinquecirrha* (Scyphozoa: Semaeostomeae). Comp. Biochem. Physiol. 57B: 303–308.

Malone, T. C., 1992. Effects of water column processes on dissolved oxygen, nutrients, phytoplankton and zooplankton. In Smith, D.E., M. Leffler & G. Mackiernan (eds), Oxygen Dynamics in the Chesapeake Bay: a Synthesis of Recent Research. Maryland Sea Grant, College Park: 61–112.

Purcell, J. E., 1992. Effects of predation by the scyphomedusan *Chrysaora quinquecirrha* on zooplankton populations in Chesapeake Bay. Mar. Ecol. Prog. Ser. 87: 65–76.

Purcell, J. E., J. R. White & M. R. Roman, 1994a. Predation by gelatinous zooplankton and resource limitation as potential controls of *Acartia tonsa* copepod populations in Chesapeake Bay. Limnol. Oceanogr. 39: 263–278.

Purcell, J. E., D. A. Nemazie, S. E. Dorsey, E. D. Houde & J. C. Gamble, 1994b. Predation mortality of bay anchovy *Anchoa mitchilli* eggs and larvae due to scyphomedusae and ctenophores in Chesapeake Bay. Mar. Ecol. Prog. Ser. 114: 47–58.

Purcell, J. E., J. R. White, D. A. Nemazie & D. A. Wright, 1999. Temperature, salinity and food effects on asexual reproduction and abundance of the scyphozoan *Chrysaora quinquecirrha*. Mar. Ecol. Prog. Ser. 180: 187–196.

Purcell, J. E., D. L. Breitburg, M. B. Decker, W. M. Graham, M. J. Youngbluth & K. A. Raskoff, 2001. Pelagic cnidarians and ctenophores in low dissolved oxygen enviroments: a review. In Rabalais, N. N. & R. E. Turner (eds), Coastal Hypoxia: Consequences for Living Resources and Ecosystems American Geophysical Union, Coastal & Estuar. Stud. 58: 77–100.

Roman, M. R., A. L. Gauzens, W. K. Rhinehart & J. R. White, 1993. Effects of low oxygen waters on Chesapeake Bay zooplankton. Limnol. Oceanogr. 38: 1603–1614.

Sanford, L. P., K. G. Sellner & D. L. Breitburg, 1990. Covariability of dissolved oxygen with physical processes in the summertime Chesapeake Bay. J. mar. Res. 48: 567–590.

Spangenberg, D. B., 1968. Recent studies of strobilation in jellyfish. Oceanogr. mar. Biol. Ann. Rev. 6: 231–247.

Taft, J. L., W. R. Taylor, E. O. Hartwig & R. Loftus, 1980. Seasonal oxygen depletion in Chesapeake Bay. Estuaries 4: 242–247.

Thuesen, E. V. & J. J. Childress, 1994. Oxygen consumption rates and metabolic enzyme activities of oceanic Californian medusae in relation to body size and habitat depth. Biol. Bull. 187: 84–98.

*Hydrobiologia* **451**: 97–111, 2001.
© 2001 *Kluwer Academic Publishers.*

# Numerical increases and distributional shifts of *Chrysaora quinquecirrha* (Desor) and *Aurelia aurita* (Linné) (Cnidaria: Scyphozoa) in the northern Gulf of Mexico

W. M. Graham

*Dauphin Island Sea Lab and Department of Marine Science, University of South Alabama,*
*101 Bienville Blvd, Dauphin Island, AL, 36528, U.S.A.*
*E-mail: mgraham@disl.org*

*Key words:* jellyfish, medusae, Mississippi River, SEAMAP, eutrophication, hypoxia

## Abstract

Fisheries resource trawl survey data from the National Marine Fisheries Service from a 11–13-year period to 1997 were examined to quantify numerical and distributional changes of two species of northern Gulf of Mexico scyphomedusae: the Atlantic sea nettle, *Chrysaora quinquecirrha* (Desor), and the moon jelly, *Aurelia aurita* (Linné). Trawl surveys were grouped into 10 statistical regions from Mobile Bay, Alabama to the southern extent of Texas, and extended seaward to the shelf break. Records of summertime *C. quinquecirrha* medusa populations show both an overall numerical increase and a distributional expansion away from shore in the down-stream productivity field of two major river system outflows: Mobile Bay and the Mississippi-Atchafalaya Rivers. In addition, there is a significant overlap between summer *C. quinquecirrha* and lower water column hypoxia on the Louisiana shelf. In trawl surveys from the fall, *A. aurita* medusae showed significant trends of numerical increase in over half of the regions analyzed. For both species, there were statistical regions of no significant change, but there were no regions that showed significant decrease in number or distribution. The relationships between natural and human-induced (e.g. coastal eutrophication, fishing activity and hard substrate supplementation) ecosystem modifications are very complex in the Gulf of Mexico, and the potential impact of increased jellyfish populations in one of North America's most valuable fishing grounds is a most critical issue. Several hypotheses are developed and discussed to guide future research efforts in the Gulf of Mexico.

## Introduction

The role of jellyfish in long-term ecosystem change is receiving increased attention. A number of marine ecosystems, identified by Mills (2001) have either documented or suspected cases of long-term ecological variations that involve jellyfish populations. However, systematically collected, long-term data-sets involving jellyfish numbers or biomass are rare. Among the notable cases where ecological change is best documented with respect to jellyfish are the Black and Azov Seas (Kideys, 1994; Kovalev & Piontkovski, 1998; Shiganova, 1998; Purcell et al., 2001), the Bering Sea (Brodeur et al., 1999) and the Mediterranean Sea (Goy et al., 1989).

Increased jellyfish production in marine ecosystems is perhaps a symptom of larger ecosystem degradation due to coastal eutrophication and overfishing (Caddy, 1993; Mills, 1995). During 'bloom' events, jellyfish are capable of exerting considerable control over the flow of energy and nutrients through the ecosystem due to extremely high consumption rates (Purcell, 1989, 1992, 1997). As such, coastal seas with a high degree of susceptibility to eutrophication and with high fisheries yields should be closely watched for similar ecological change. Yet, again, the availability of existing long-term data-sets involving jellyfish numbers is rare.

The northern Gulf of Mexico continental shelf is among the most productive and highly fished regions of North America. Pulsed delivery of nutrients to the Gulf of Mexico through the Mississippi River Delta and numerous other river-dominated estuaries of the northern Gulf account for a cumulative regional estuarine surface area of 30 000 km$^2$. The drainage area emptying into the Gulf of Mexico is over 4 million km$^2$ or approximately 55% of the conterminous United States with the Mississippi River and Mobile Bay estuary discharging the 1st and 4th largest volumes, respectively. To complicate ecological variations, the Gulf of Mexico yields about 1/3 of the total United States fishery production and supports the largest fishery by volume in North America in the planktivorous Gulf Menhaden, *Brevoortia patronus*.

Fluvial discharge of nutrients directly onto the shelf is responsible for the high production rates (Lohrenz et al., 1997). Rivers and estuaries of the northern Gulf of Mexico, from the Florida panhandle to southern Texas (Fig. 1), discharge the greatest volume of water during winter and spring months. Low estuarine residence times due to high freshwater discharge and typically shallow estuarine geomorphology rapidly displace nutrients and production out onto the shelf (Pennock et al., 1999). The greatest seasonal primary production rates in the northern Gulf of Mexico are associated with the Mississippi-Atchafalya River system and Mobile Bay estuary outflows; productivity rates are depressed accordingly with the lower-discharge estuaries of Texas (reviewed in Pennock et al., 1999). Zooplankton grazing and secondary production are intense at the coastal transition zone (Dagg & Whitledge, 1991; Dagg, 1995; Ortner & Dagg, 1995) and trophic transfer of this energy to fish is highly coupled to estuarine delivery in the Gulf (Deegan et al., 1986). The alternate pathway of energy to gelatinous zooplankton predators has thus far received little, if any, attention.

The historical lack of interest in the ecological role of jellyfish is surprising. The Gulf of Mexico supports among the greatest diversity of pelagic cnidarians in the world. Over 115 epipelagic species were listed by Phillips (1971) including 16 species of Scyphozoa and Cubozoa. Given that Phillips' (1971) synopsis is nearly 30 years old and lacks deep-water information, certainly this number severely under-estimates the real diversity. Given such diversity and potential importance of jellyfish in this very productive system, long-term data on jellyfish variations are crucial for realizing current or future ecological changes as identified by Caddy (1993) and Mills (1995).

The goal of the present study is to analyze more than 10 years of data from large-scale trawling efforts in the northern Gulf of Mexico. These fishery resource survey data will be used to assess long-term variations in two important jellyfish species: the Atlantic sea nettle, *Chrysaora quinquecirrha* (Desor), and the moon jellyfish, *Aurelia aurita* (Linné). Similar analyses of fisheries trawl data have been used previously to identify long-term changes of jellyfish populations in the Bering Sea (Brodeur et al., 1999).

## Methods

### SEAMAP trawl data-set

The data presented in this analysis are from the United States National Marine Fisheries Service (NMFS) Southeast Area Monitoring and Assessment Program (SEAMAP) managed through the Southeast Fisheries Science Center in St. Petersburg, Florida. The specific subset of data used for this study were collected as part of the twice-yearly shrimp/groundfish surveys from 1985 to 1997 in the northern Gulf of Mexico between Mobile Bay, Alabama and the southern border of Texas. Though SEAMAP shrimp/groundfish surveys were initiated in 1982, I have omitted 1982–1984 entirely from analysis because of inconsistent coverage.

The SEAMAP shrimp/groundfish survey protocol is detailed in Stuntz et al. (1985). In summary, surveys over the entire sampling area were divided into 11 statistical regions. Only NMFS designated regions 11 and 13 through 21 were used; regions 1–10 east of this area did not have shrimp/groundfish survey data, and region 12 was only occasionally surveyed. Surveys were conducted twice-annually: once in the summer (May–July) and once in the fall (October–November). Trawl sites (typically 30–50 trawls per statistical region per survey) were randomly located within each region in 5 depth strata. The location of stations within depth strata caused sampling frequency to be higher near the coast but consistent between regions and years. An example of station density and arrangement (for fall 1991) is given in Figure 1. A preliminary analysis of variance was performed as a check on the randomness of distribution within and between statistical regions and depth strata in order to avoid errors in the interpretation of the data. Because of this, I eliminated statistical region 12 entirely from the study, and the years 1985–1986 from analysis of the summer trawls.

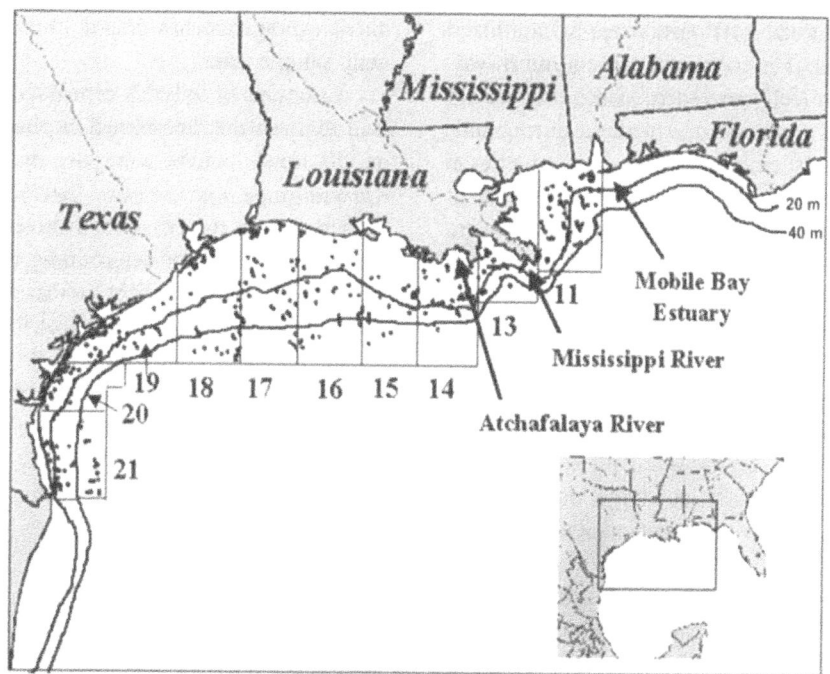

*Figure 1.* Map of study region in the northern Gulf of Mexico. Boxes indicate 10 statistical regions of the SEAMAP sampling program. Points within the boxes are an example of station distribution for a single trawl series in the Fall of 1991 (446 stations). The 20 m and 40 m isobaths are indicated with heavy lines.

*Table 1.* Summary of SEAMAP trawls conducted in the northern Gulf of Mexico from 1985 to 1997. *Chrysaora quinquecirrha* medusae were analyzed only from summer trawls and *Aurelia aurita* medusae were analyzed from fall trawls

| Year | No. Trawls Collected | |
|------|--------|------|
|      | Summer | Fall |
| 1985 | . . . | 404 |
| 1986 | . . . | 236 |
| 1987 | 615 | 365 |
| 1988 | 500 | 701 |
| 1989 | 359 | 631 |
| 1990 | 432 | 451 |
| 1991 | 433 | 446 |
| 1992 | 420 | 364 |
| 1993 | 454 | 402 |
| 1994 | 483 | 380 |
| 1995 | 363 | 337 |
| 1996 | 388 | 438 |
| 1997 | 365 | 377 |
| Total | 4812 | 5532 |

I limited the present study to the two species of scyphomedusae that are most abundant and widely distributed throughout the northern Gulf of Mexico. In addition to their prevalence, both species also reach sizes large enough to be captured in the trawl netting, and they are hardy enough to be retained without so much damage that they are unrecognizable or uncountable. *Chrysaora quinquecirrha* medusae occur during summer months in the Gulf of Mexico with peak abundance usually in June–July (Burke, 1975, 1976). Some reporting of '*Dactylometra quinquecirrha*' from Texas regions has been combined with the *C. quinquecirrha* since these are widely considered to be the same species. *Aurelia aurita* medusae occur during the fall months with peak abundance usually in October–November (Burke, 1975, 1976). Therefore, analysis of *C. quinquecirrha* is limited to the summer surveys (1987–1997) and *A. aurita* is limited to fall surveys (1985–1997). As seen in the trawl summary of Table 1, over 10 000 individual trawls were included in the present analysis.

Numerical trawl data are reported here as a standardized catch. Other studies have used indexed trawl data in this fashion (Brodeur et al., 1999), and a standardized trawl allows for sufficient comparison between years and between regions even if some differences in gear or vessels existed. Most collections were made with a standard 12 m wide shrimp trawl for a maximum 60 min tow. In some nearshore cases, a 5 m wide trawl was used; these data have been standardized to a 12 m swath. All data were standardized to a 60 min trawl. The trawl and cod-end mesh was nominally 2.5 cm (fully stretched). Since, the typical size of *Chrysaora quinquecirrha* medusae during mid-summer is about 10 cm and *Aurelia aurita* medusae is 35–40 cm (Phillips, 1971), it is assumed that at least individuals larger than 5 cm of both species were retained in the trawl.

Numerical recording of all by-catch by taxon has been consistent during the SEAMAP surveys, however, the trawls were conducted for the purpose of collecting shrimp and groundfish on the sea floor. Therefore, collections of jellyfish and other water column fauna should be viewed with some degree of caution, since they were not target species. The expectation here is that the large number of trawls collected over a long period of time will allow a robust view of changes in numbers and distributions of jellyfish. Also, the jellyfish data reflect an integrated sample of the entire water column. As such, they must be interpreted as an integrated or areal concentration of jellyfish rather than as a volumetric concentration. Some biomass and length data are available for jellyfish collected in trawls, but these are not consistent over the same time frame, and I chose to omit them from this analysis.

Dissolved oxygen records from the SEAMAP database also were used in the present analysis. Water samples from the lower water column were collected by NMFS scientists using Niskin-type bottle samplers. Dissolved oxygen was measured either by polarographic oxygen probe or by Winkler titration. Oxygen probes were calibrated by the air saturation technique prior to each cruise. Specific sampling technique depended on participating institutional resources, however, measurements were primarily made with YSI Oxygen Probes.

### Analytical considerations

The standardized catch data used to derive means and standard deviations presented in Figures 2 and 3 were not conducive to traditional parametric analyses despite attempts to normalize the data through various transformations. In order to test whether variations in jellyfish populations increased or decreased with time, an alternative non-parametric Spearman rank-order analysis was employed. This was accomplished by using the regionally averaged standard catch data presented in Figures 2 and 3 and then testing whether

those average catches tended to increase or decrease with sample year.

Variations of jellyfish distribution within a statistical region were determined as changes in frequency in all trawls within a region that recorded either *Chrysaora quinquecirrha* or *Aurelia aurita*. Change in distribution in the cross-shelf direction is an important consideration for determining whether numerical changes of jellyfish reflect increased local abundance or whether there has been a distributional expansion across the shelf. Expansion of the population's distribution from shore was determined by frequency of species occurrence in trawls from stations in less than 20 m total water depth or greater than 20 m water depth.

## Results

### Variations in jellyfish numbers

Time-series of jellyfish catches in summer (*Chrysaora quinquecirrha*) and fall (*Aurelia aurita*) are plotted as regional averages each year (Figs 2 and 3). Qualitatively, and independent of longer trends, the magnitude of year-to-year variations in regional *A. aurita* density was greater than that of *C. quinquecirrha* variations. Log-transformed data in Figures 2 and 3 are used to resolve better the long-term trends. Variations in regional *A. aurita* densities between consecutive years frequently vary by as much as an order of magnitude; similar magnitude variations in *C. quinquecirrha* were less frequent. It is apparent that differences in number and pattern are not consistent across statistical regions within or between species. It is important, therefore, to separate statistically those regions that exhibit consistent long-term changes from those regions that simply have large year-to-year variations.

Only regions 15 and 16 experienced statistically significant long-term (defined as the entire period of study) numerical increase of *Chrysaora quinquecirrha* (rank order test, $p < 0.05$). Proximity of regions 15 and 16 suggests that this increase may have been over a larger, contiguous area of the Louisiana shelf of the northern Gulf of Mexico. In regions 15 and 16, *C. quinquecirrha* medusae were absent prior to 1992, but between 1992 and 1997, they increased numerically by as much as 2 orders of magnitude.

Despite lack of significance in the long-term trend in other regions, a particularly interesting numerical jump in *Chrysaora quinquecirrha* also occurred at

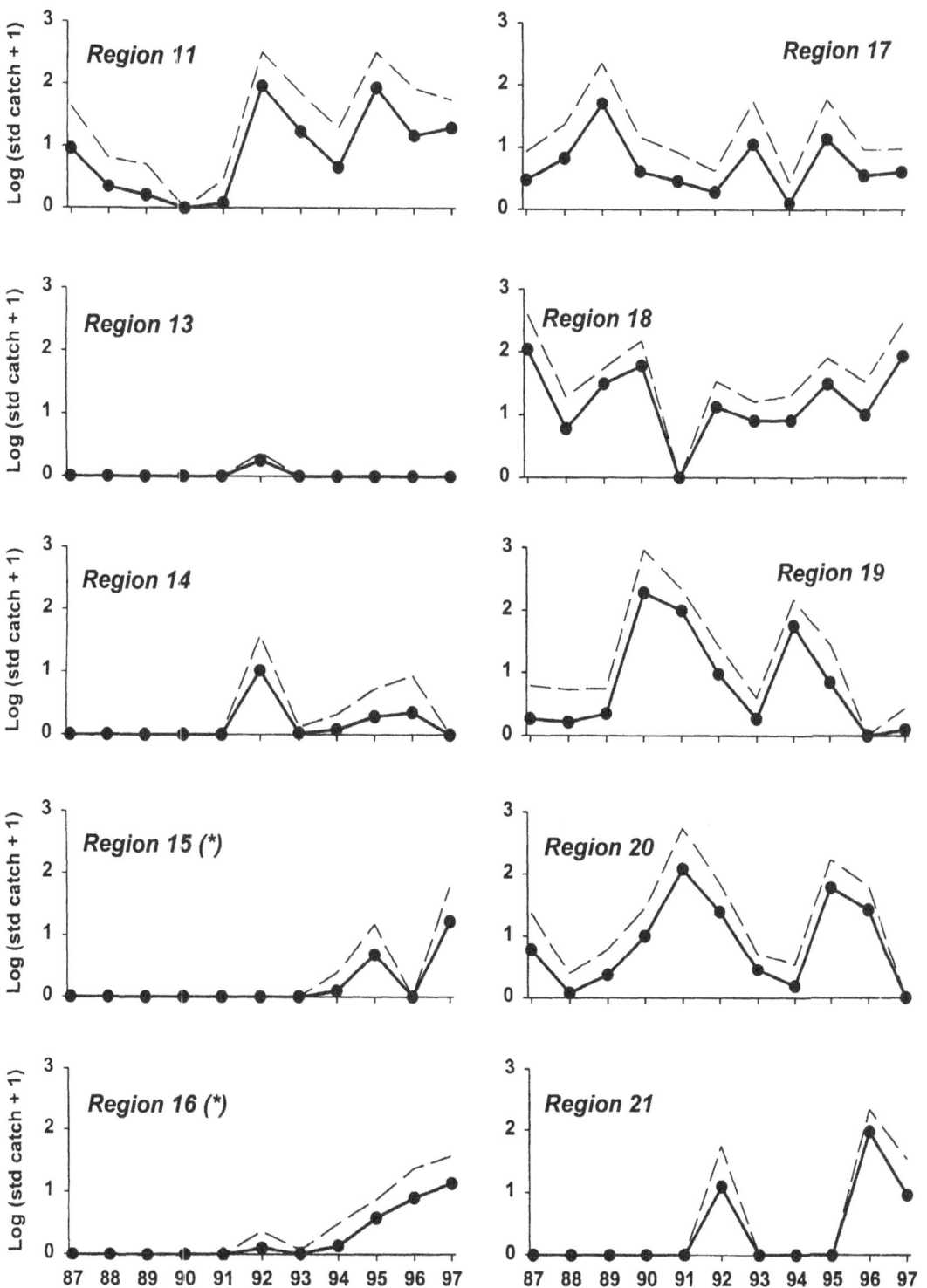

*Figure 2.* Standardized catch of *Chrysaora quinquecirrha* from summer SEAMAP trawl samples, 1987–1997, in each statistical region identified in Figure 1. Solid lines represent the mean catch; the dashed line represents one standard deviation above the mean. An asterisk beside the region number (*) indicates significant increase with time at $p < 0.05$. Standardized catch reflects a 60 minute trawl using a 12 m wide trawl.

102

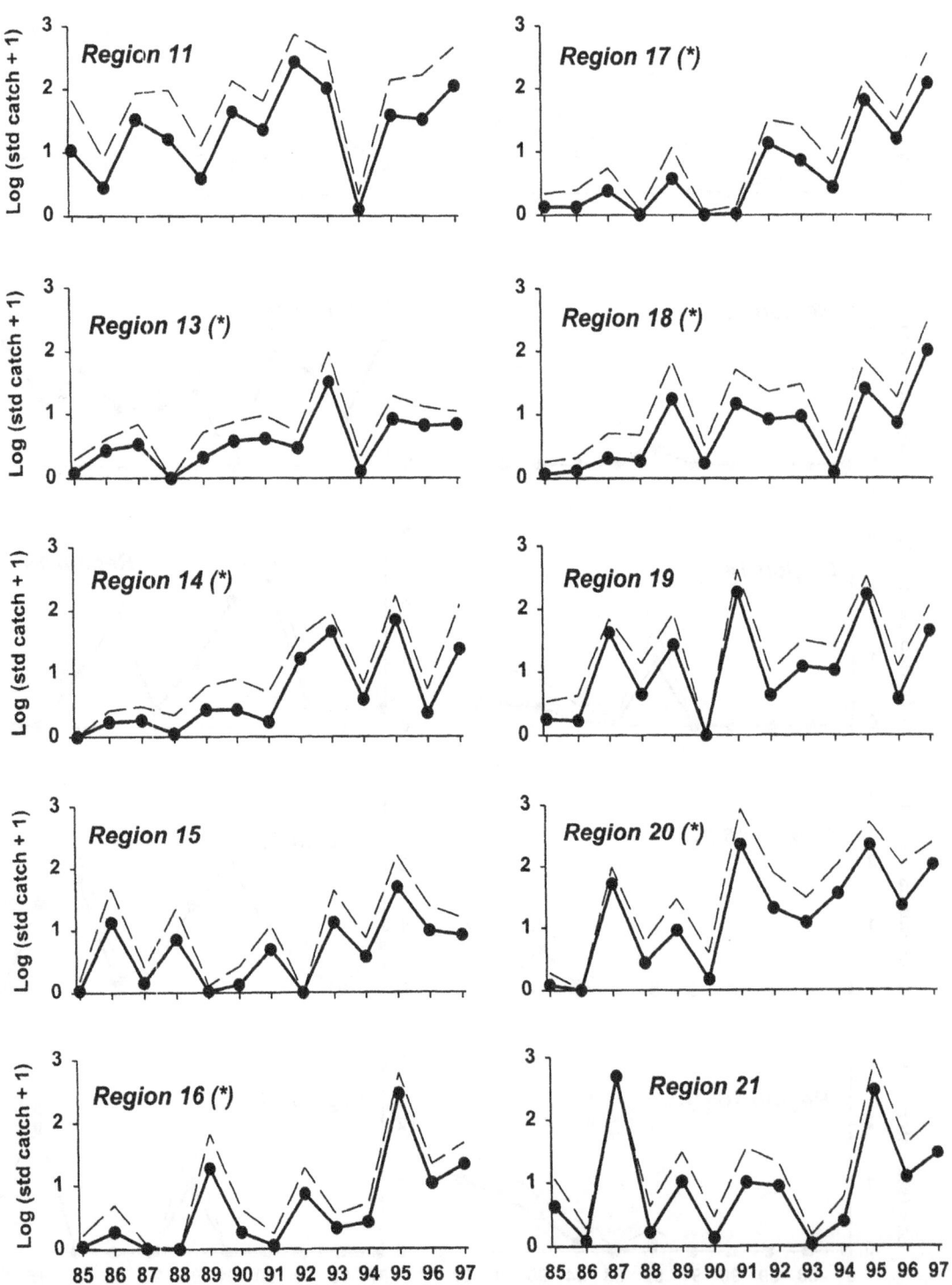

*Figure 3.* Standardized catch of *Aurelia aurita* from fall SEAMAP trawl samples, 1987–1997, in each statistical region identified in Figure 1. All else as in Figure 2.

regions 11, 13 and 14 in the northern Gulf (Fig. 2). Region 11 at the eastern extent of the study area showed a dramatic increase by two orders of magnitude in 1992, yet only a statistically marginal long-term increase in this species with time (rank order analysis, $p = 0.055$). However, in region 11, this transition marked a period when *C. quinquecirrha* densities remained consistently higher than pre-1992 numbers (Fig. 2). Region 13, near the outflow of the Mississippi river, recorded *C. quinquecirrha* only in 1992. This species was not recorded in Region 14 from 1987 to 1991, however, their appearance in 1992 in region 14 marked a 5 year period of presence. Significant trends of *C. quinquecirrha* densities did not occur to the west of region 16 on the Texas shelf. With the exception of region 21 at the southwestern extent of the study area, *C. quinquecirrha* was usually present in varying abundance in all years (Fig. 2). The apparent cycling of medusae of this species in regions 19 and 20 (and possibly 21) is note-worthy and seems consistent among these adjacent regions, but independent of variations in any other region.

Significant trends of numerical *Aurelia aurita* increase occurred in 6 of the 10 statistical regions (Fig. 3). Regions 13, 14 and 16 on the Louisiana shelf showed a 5–10-fold increase *A. aurita* densities between the mid 1980s and the mid 1990s; in fact, the species was virtually absent over much of the study area in the 1980s compared with densities of the mid-1990s. Toward the Texas shelf, regions 17, 18 and 20 each showed nearly a two order of magnitude increase from 1985 to 1986 numbers by 1997. Particularly interesting features across most of the study area were consistent depressions in *A. aurita* densities in 1988 and 1994 across the study area.

*Distribution shifts*

Variations in the distribution of summer *Chrysaora quinquecirrha* and fall *Aurelia aurita* medusae are presented in Figures 4 and 5. The data in these figures are summarized into three trawl-depth categories (0–20 m, 20–40 m and >40 m) in order to resolve the extent that a population occurs offshore in comparison with isobaths (Fig. 1). Table 2 summarizes results of the rank order test of whether frequency of occurrence tends to increase (or decrease) with time either within the entire statistical region or as changes in frequency of occurrence in <20 m or >20 m deep trawl stations. In Table 2, increased frequency of occurrence in the

>20 m trawl stations is assumed to represent offshore expansion of the population.

Overall, *Chrysaora quinquecirrha* medusae exhibited a trend toward increased distribution offshore in 4 of the 10 statistical regions, which is in contrast to the numerical increases in only 2 regions. Between 1987 and 1992 in regions 11–16 (Alabama–Louisiana), *C. quinquecirrha* occurred only rarely and was never recorded at stations seaward of the 20 m isobath (Fig. 4). In 1992, however, frequency of *C. quinquecirrha* occurrence in trawl collections increased markedly at both inshore and offshore stations on the Alabama-Louisiana shelf. For *C. quinquecirrha*, distributional expansion with time was significant throughout regions 11, 15 and 16 (Table 2). By 1997, *C. quinquecirrha* were recovered from 20 to 25% of all trawl stations in regions 11, 15 and 16; 40–75% of these stations were seaward of the 20 m isobath. Interestingly, although few *C. quinquecirrha* were recovered in region 13 during 1992, they occurred in more than 30% of all trawl stations in that region (Fig. 4). To the west of region 16 (i.e. the Texas shelf), the only offshore shift of *C. quinquecirrha* was in region 17, but this did not result in an overall increase in frequency of occurrence (Table 2). *Chrysaora quinquecirrha* was widely distributed among stations within all statistical regions on the Texas shelf during all years with the exception of region 21.

Significant shifts toward offshore distribution of *Aurelia aurita* medusae occurred in only 3 of the 10 statistical regions, which contrasts with numerical increases in 6 of the regions. Since the study began in 1985, *A. aurita* has been widely distributed across the study area (Fig. 5). Significant trends of increased catch frequency of *A. aurita* occurred from the mouth of the Mississippi river to the northeastern Texas coast (regions 13, 15–18). In these five regions, the magnitude of increased frequency of *A. aurita* occurrence ranged from 2-fold to more than 10-fold between 1985 and 1997. The majority of this increase in overall frequency with time can be explained by increased frequency of occurrence at stations inshore of the 20 m isobath (i.e. no offshore expansion of *A. aurita* distribution, Table 2). However, significant expansion of the *A. aurita* distribution away from shore occurred in regions 14, 16 and 17 on the Louisiana shelf.

104

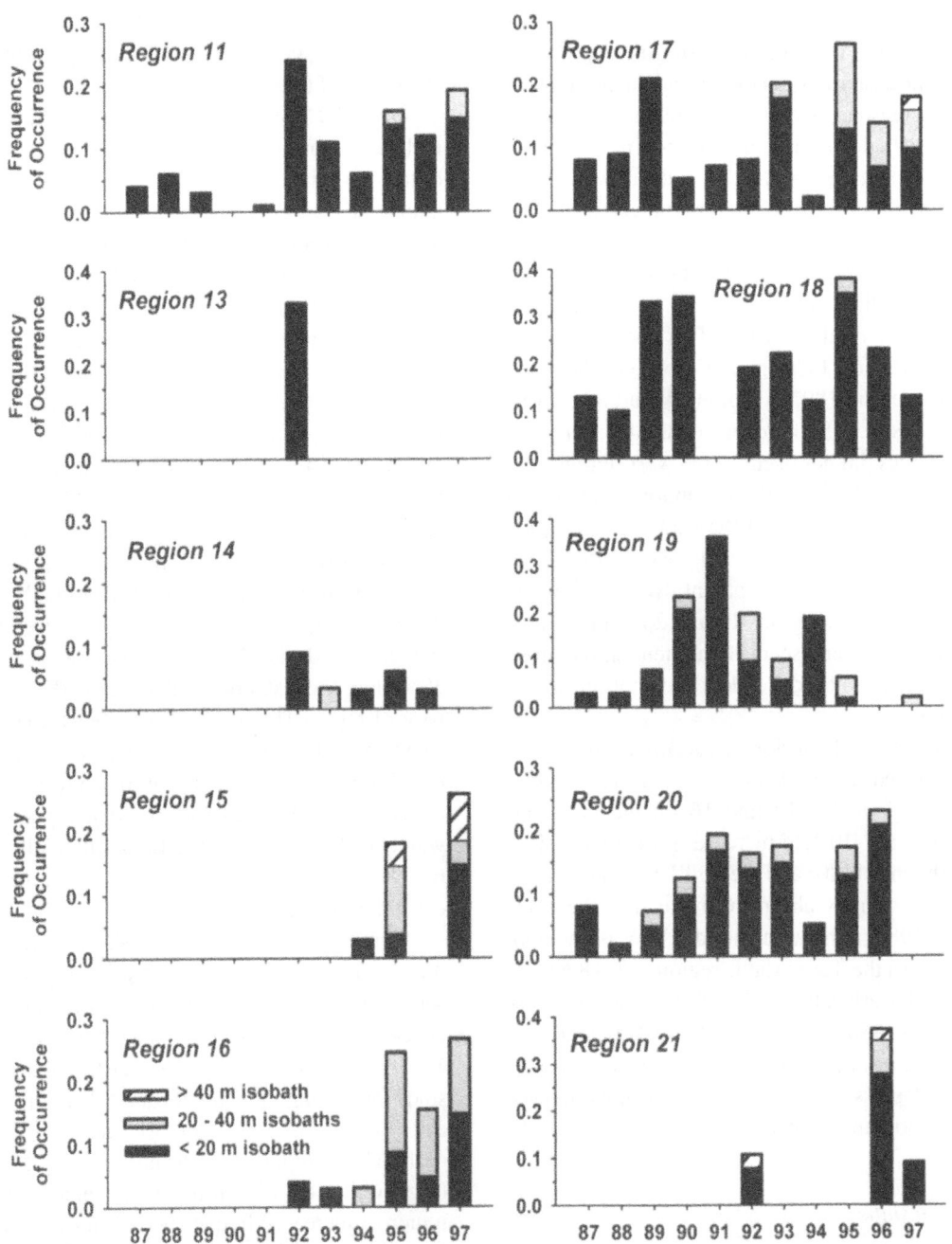

*Figure 4.* Variations in frequency of occurrence of *Chrysaora quinquecirrha* in summer SEAMAP trawl collections, 1987–1997 at stations <20 m, 20–40 m and >40 m depth. Depth bins are used as surrogates of distance from shore; refer to Figure 1 for the location of the 20 m and 40 m isobaths within the study regions.

*Relationship between summer hypoxia and* Chrysaora quinquecirrha *distribution*

The relationship between summer hypoxia and summertime occurrence of *Chrysaora quinquecirrha* medusae was explored in Regions 15 and 16 only since these regions have experienced the most consistent change in both distribution and in numbers of this spe-

cies. The extent of summertime hypoxia (stations with bottom water dissolved oxygen below 2 mg l$^{-1}$) in the northern Gulf of Mexico over the period of study is represented in Figures 6 and 7. Figure 6 presents the spatial contrast in hypoxia distribution between the years 1987 and 1997. By 1997, hypoxic bottom waters had extended to a band across the Louisiana shelf, which coincides with statistical regions 13–17.

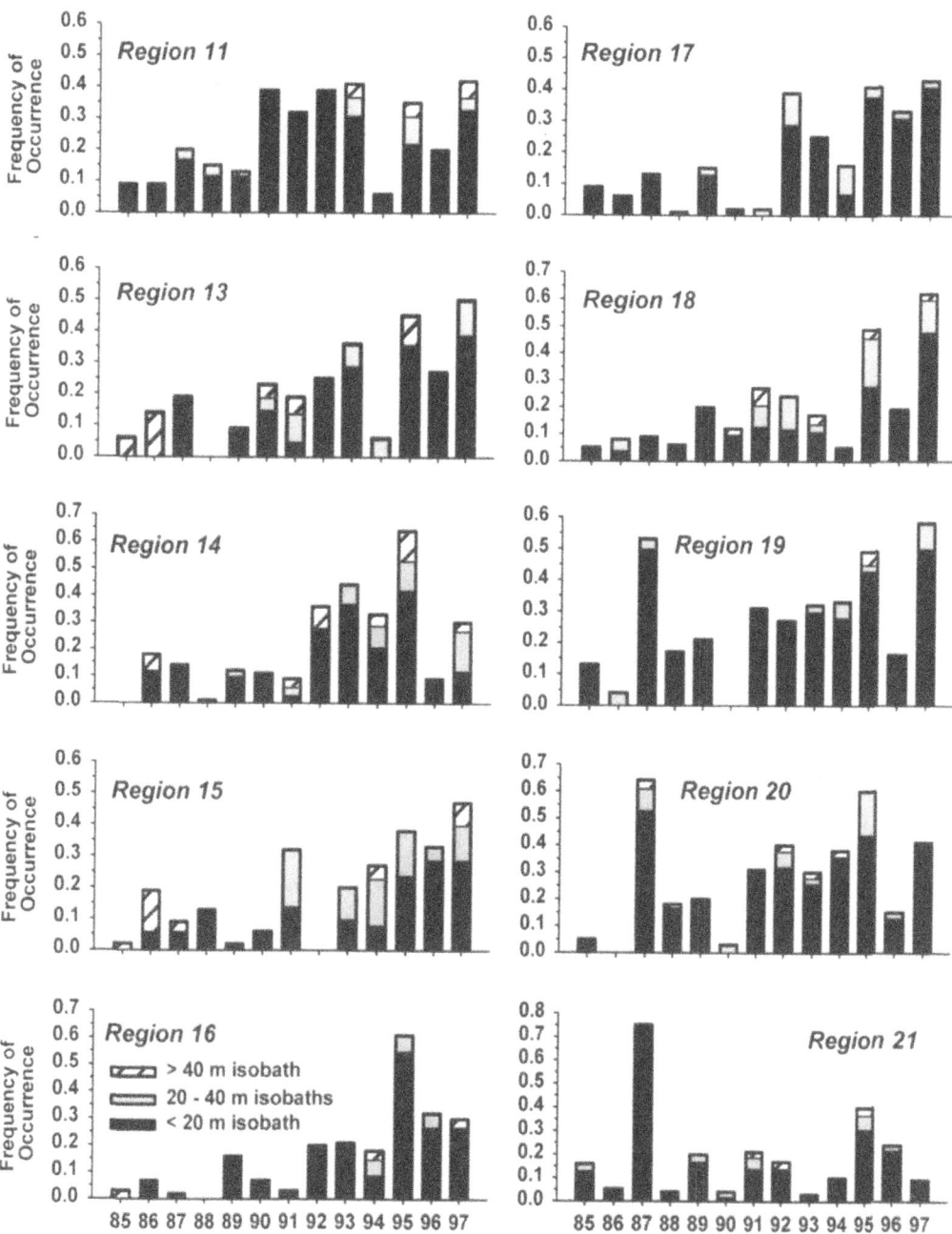

*Figure 5.* Variations in frequency of occurrence of *Aurelia aurita* in summer SEAMAP trawl collections, 1985–1997 at stations <20 m, 20–40 m and >40 m depth. Depth bins are used as surrogates of distance from shore; refer to Figure 1 for the location of the 20 m and 40 m isobaths within the study regions.

Additional patches of hypoxic waters occurred off the Texas and Alabama coasts but were not as extensive as hypoxia on the Louisiana shelf. Though both hypoxia and *C. quinquecirrha* increased in frequency with time during the study period, there is no significant correlation between frequency of occurrence of hypoxia and of this species.

Despite the lack of a temporal relationship, there was substantial spatial overlap between hypoxia and the *Chrysaora quinquecirrha* medusa distribution on the Louisiana shelf. Mean bottom water oxygen concentration was compared between stations that did and stations that did not record *C. quinquecirrha* in trawls. Over all years, stations that recorded this species in regions 15 and 16 had significantly lower dissolved

*Table 2.* Summary of the results of Spearman rank-order analysis of jellyfish distributional shift. Data are presented for the statistical regions identified in Figure 1. Significance within a statistical region indicates that there was a trend toward increased frequency of occurrence with time at stations within or seaward of the 20 m isobath, or over the entire region (overall). 1987–1997 for *Chrysaora quinquecirrha*; 1985–1997 for *Aurelia aurita*. (* = $p<0.05$; ** = $p<0.01$; *** = $p<0.001$; ns = not significant)

| Statistical region | *Chrysaora quinquecirrha* | | | *Aurelia aurita* | | |
|---|---|---|---|---|---|---|
| | <20 m | >20 m | Overall | <20 m | >20 m | Overall |
| 11 | * | * | * | ns | ns | ns |
| 13 | ns | ns | ns | ** | ns | ** |
| 14 | ns | ns | ns | ns | * | ns |
| 15 | * | * | * | ** | ns | ** |
| 16 | *** | *** | *** | *** | * | *** |
| 17 | ns | ** | ns | * | * | ** |
| 18 | ns | ns | ns | ** | ns | * |
| 19 | ns | ns | ns | ns | ns | ns |
| 20 | ns | Ns | ns | ns | ns | ns |
| 21 | * | Ns | ns | ns | ns | ns |

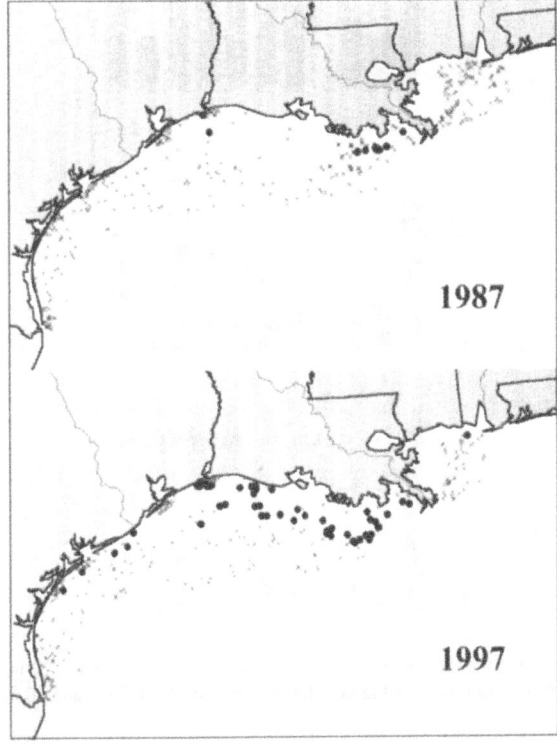

*Figure 6.* Distribution of hypoxic bottom waters in summers of 1987 and in 1997. Solid circles represent stations where dissolved oxygen concentrations were measured below 2 mg l$^{-1}$; 'x' represents stations where oxygen was measured above 2 mg l$^{-1}$.

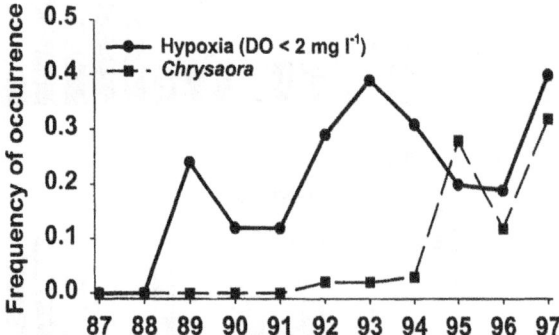

*Figure 7.* Variations in the frequency of occurrence of hypoxia and summer *Chrysaora quinquecirrha* catch in SEAMAP trawls (as percentage of all stations sampled) in Statistical Regions 15 and 16 only.

## Discussion

The fisheries resource survey data used in this study show significant increasing trends in the number and distribution of jellyfish within the northern and north-western Gulf of Mexico over the past 11–13 years. However, these numerical and distributional trends were neither uniform nor consistent between regions and species. Changes in summer *Chrysaora quinque-cirrha* medusae tended to be more in distribution than in abundance, whereas trends involving fall *Aurelia aurita* medusae tended towards increased numbers, but with little distributional change. Both numerical and distributional changes occurred in the northern Gulf of Mexico, within regions 14–17 on the Louisiana shelf. *Chrysaora quinquecirrha* showed the

oxygen in the lower water column (Student's $T$-test, $t = -4.287, p < 0.001$) (Fig. 8).

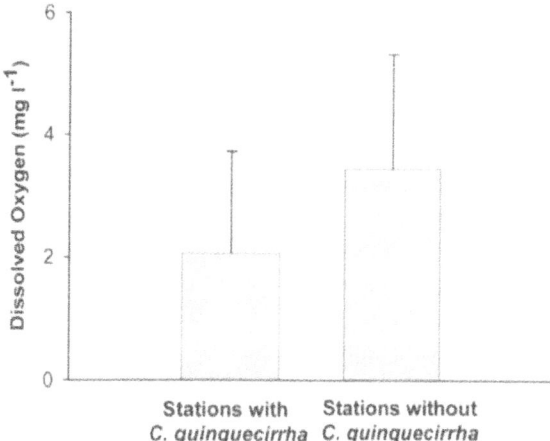

*Figure 8.* Mean and standard deviation of dissolved oxygen concentration in the lower water column at stations (regions 15 and 16) where *Chrysaora quinquecirrha* was collected in trawls and at stations where *C. quinquecirrha* was absent in all years. There is a significant difference between the two means (Student's *t*-test, $p < 0.05$).

greatest change in regions 15 and 16, and *A. aurita* showed the greatest change in regions 14, 16 and 17. As such, the following discussion emphasizes the changes that have occurred on the Louisiana shelf.

Long-term ecosystem change that favors jellyfish has been the focus of recent discussion since increased jellyfish production may be symptomatic of degraded ecosystem health (Caddy, 1993; Mills, 1995, 2001; Pauly et al., 1998). While changes in Gulf of Mexico jellyfish populations over the past 11–13 years are unequivocal, whether these changes reflect a degraded ecosystem remains to be determined through future research. As a foundation for future jellyfish research in the Gulf of Mexico, it is instructive to discuss several key hypotheses that might individually or collectively explain the long-term trends identified in this paper.

*Increased trophic transfer to jellyfish: effects of coastal eutrophication and overfishing*

Numerical increases of jellyfish as a function of increased trophic transfer can result from independent effects of overfishing or eutrophication (Legović, 1987) or from their combined, synergistic effects (Caddy, 1993). This has happened in the Black Sea where combined effects of eutrophication and severe overfishing, working in concert with species invasions and altered hydrology, have led to an increase in the zooplanktivorous ctenophore *Mnemiopsis leidyi* (A. Agassiz) (in Kideys, 1994; Shiganova, 1998). Semi-enclosed seas like the Black Sea are understand-

ably more sensitive to the effects of eutrophication and overfishing (Caddy, 1993). By contrast, the size of the Gulf of Mexico and openness of its northern shelf waters to exchange makes eutrophication, at least superficially, a dubious explanation for increased jellyfish production.

Very large rivers like the Mississippi-Atchafalaya River system, which alone discharges about 90% of the Gulf's freshwater input, can in fact regulate production rates and patterns on the shelf (Lohrenz et al., 1997). The Mississippi River, like most Gulf estuaries, has greatest discharge rates of freshwater and nutrients during winter and spring and lowest discharge by late summer and fall (Lohrenz et al., 1997). Light limitation due to turbidity is responsible for a lag in primary production away from the river mouth (Randall & Day, 1987; Lohrenz et al., 1990). Peak nutrient utilization by phytoplankton (Lohrenz et al., 1990) and subsequent secondary production (Dagg, 1995; Ortner & Dagg, 1995) occurs well downstream of the discharge point. This is typically toward the west along the Louisiana shelf (Wiseman & Garvine, 1995). Although rapid utilization of nutrients probably occurs within 100 km (Nelson & Dortch, 1996) once suspended sediments are removed through settling (Lohrenz et al., 1990), decoupled downstream (i.e. westward) effects will continue due to the high organic load carried by the coastal current.

The combined effects of high nutrient loading and retention of productive water on the shelf to the west of the river mouth makes the Louisiana shelf susceptible to the ecological effects of eutrophication typically experienced by semi-enclosed seas. Moreover, the high volume of freshwater and nutrient discharge coupled with the low residence time of nutrients within shallow Gulf estuaries makes the shelf ecosystem particularly sensitive to long-term land-use changes (Bricker & Stevenson, 1996; Pennock et al., 1999). In the Mississippi River, nutrient concentrations have more than doubled in four decades (Bratkovich et al., 1994; Turner & Rabalais, 1994) corresponding with a long-term production increase on the shelf (Walsh et al., 1989; Müller-Karger et al., 1991; Eadie et al., 1994; Turner & Rabalais, 1994; Rabalais et al., 1996).

An indirect effect of eutrophication on jellyfish ecology is the seasonal development of hypoxia in the northern Gulf of Mexico. Hypoxic waters with dissolved oxygen concentrations $< 2$ mg $l^{-1}$ appear to favour the distribution of jellies through enhanced ecological interactions with other organisms. These interactions can be manifested as increased jellyfish predation rates (Breitburg et al., 1999) and reduced competition or predation with less hypoxia-tolerant species (reviewed in Purcell et al., 2001). Severe summertime hypoxia in the northern Gulf of Mexico may explain, in part, the success of *Chrysaora quinquecirrha* on the Louisiana shelf. Hypoxic bottom waters on the Louisiana shelf are a result of seasonally high organic loading via particle sinking and high stratification suppressing deep-water oxygenation (Justić et al., 1993).

The relationship between *Chrysaora quinquecirrha* and reduced oxygen concentration is not clear (Figs 7 and 8). This species overlaps extensively with hypoxic waters, yet there is no significant relationship between occurrence of hypoxia and its occurrence. It could be concluded, then, that while hypoxia does not appear to cause distribution shifts of *C. quinquecirrha*, hypoxia potentially creates an advantageous environment promoting its success on the Louisiana shelf.

The potential interaction between overfishing and jellyfish populations in the Gulf of Mexico is less clear than for coastal eutrophication, and these two effects may be difficult to separate. The Gulf of Mexico supports among the highest regional fisheries production in North America due to nutrient inputs through Gulf estuaries (Deegan et al., 1986). This includes the largest fishery by volume in Gulf menhaden, which has production cycles mediated by nutrient inputs from the Mississippi River (Warlen, 1988). Despite the heavy fishing pressure placed on Gulf menhaden, which may be a zooplanktivorous competitor of jellyfish in the northern Gulf of Mexico, there has been no formal stock assessment to indicate that Gulf menhaden are overfished (R. Shipp, pers. comm.). Thus, it is not possible in this case to attribute fish removal alone as a cause of enhanced trophic transfer to jellyfish populations as has been suggested by Pauly et al. (1998). Certainly, this is an area of intense interest for future research in the Gulf of Mexico since fishing effects are so pervasive to jellyfish dynamics in other ecosystems (Caddy, 1993; Pauly et al., 1998; Purcell & Arai, 2001) and may explain, in part, increased jellyfish biomass in other 'open' ecosystems like the Bering Sea (Brodeur et al., 1999).

### Expansion of benthic polyp populations

Changes in medusa populations and distribution in the northern Gulf of Mexico may alternatively reflect a change in the distribution or production of benthic polyp stages. While both *Chrysaora quinquecirrha* and *Aurelia aurita* have polyp stages that are dependent on hard substrate, it cannot be concluded that local presence of adult medusae reflects local polyp populations. One enigmatic characteristic of *C. quinquecirrha* in the vicinity of Mobile Bay and the Mississippi Sound is that large, mature medusae arrive in large numbers despite the lack of scyphopolyp populations or the occurrence of ephyrae and young medusae in the region (Burke, 1975; Johnson et al., 2001; Graham, unpublished data).

The location of *Chrysaora quinquecirrha* and *Aurelia aurita* polyp populations remains speculative in the absence of basin-wide benthic surveys; however, a great deal of hard substrate exists on the Gulf of Mexico shelf. This includes both contemporary and ancient natural oyster reef, which Johnson et al. (2001) suspect as the source, and artificially placed structures such as fixed oil and gas production rigs and artificial reefs. While natural hard substrate has likely remained relatively constant over ecological time scales, artificial substrate has increased dramatically in recent decades. On the continental shelf between Texas and Alabama, there are currently about 6000 operating and discontinued oil and gas structures. Most of these structures extend from the seafloor to sea-surface such that polyp populations could thrive in appropriate physical and chemical conditions within the water column (e.g. away from deleterious effects of low dissolved oxygen: Condon et al., 2001). Interestingly, the greatest density of oil and gas structures occurs on the Louisiana shelf between statistical regions 13–17 (Fig. 1), which coincidentally are the regions of consistent numerical and distributional change for *C. quinquecirrha* and *A. aurita* medusae.

### Long-term climate variations

The 11–13-year changes documented in this study might be a reflection of longer period climate variations. Climate oscillations such as the El Niño Southern Oscillation (ENSO) and the North Atlantic Oscillation (NAO) force large-scale variations in the hydrological cycle of North America, and consequently the rate of discharge of freshwater and nutrients into the northern Gulf of Mexico. Large-scale climate fluctuations may serve as a major source of the interannual variability in jellyfish populations of the Gulf of Mexico as has been identified in the Bering Sea (Brodeur et al., 1999) and in Narragansett Bay, Rhode Island (Sullivan et al., 2001).

The years 1992 and 1993 serve as an example of how jellyfish populations respond to hydrological changes. In 1992, an unusually dry winter-spring over the eastern half of the United States dramatically decreased freshwater discharge, and subsequently depressed salinity proximal to the Mississippi River and Mobile Bay estuary (Lohrenz et al., 1997; Pennock et al., 1999). Consequently, very high primary production occurred close to the discharge point of the Mississippi River (Prasad et al., 1995; Lohrenz et al., 1997). Locally increased primary production and salinities likely allowed the incursion of large numbers of *Chrysaora quinquecirrha* into regions 13 and 14, a distribution much closer to the Mississippi River discharge than observed in other non-drought years. In Chesapeake Bay, production of ephyrae from the polyp stage is enhanced by food, and is also very sensitive to salinity regime (Purcell et al., 1999).

By contrast, by mid-late summer 1993, the Mississippi River experienced record flooding as a result of 'El Niño'-induced weather patterns. The excessive input of freshwater and nutrients onto the Louisiana shelf during the summer 1993 led to the most extensive hypoxic zone recorded on the shelf in 1993 with effects that carried into 1994 (Rabalais et al., 1998). Depressions in the number and distribution of summer *Chrysaora quinquecirrha* occurred during the height of the flooding on the Louisiana shelf (Figs 2 and 4). Fall *Aurelia aurita* numbers and distributions were consistently reduced only in 1994, a 1 year lag behind the flood. This might suggest that perhaps the flood impacted polyp habitat more than it impacted existing medusa populations. Purcell et al. (1999) have shown experimentally that success of *C. quinquecirrha* polyps is strongly tied to salinity regime. Since climate variations occur at decadal scales or longer, I cannot at this point resolve their relevance to population changes experienced in the past 11–13 years. Climate variations may also play a key role in Gulf circulation such as spin-off eddies from the tropical Loop Current, which may be critical in advection of medusae from their origin to adult nearshore habitats (e.g. Johnson et al., 2001). Loop Current spin-off eddies have been implicated in the transport of sporadic tropical cubomedusae (Graham, 1998) and large numbers of the rhizostome medusa, *Phyllorhiza punctata* von Lendenfeld in 2000 (Graham, unpublished data).

Another source of long-term variability is hurricane activity. The depression in *Aurelia aurita* numbers and distribution in fall 1988 across the western and northwestern Gulf of Mexico was likely a consequence of flooding due to the passage of two major hurricanes. Hurricane Florence landed on the Mississippi River delta in early September 1988 and Hurricane Gilbert – one of the strongest Atlantic storms in recorded history – made landfall 2 weeks later in northern Mexico and traveled over southern and central Texas causing widespread flooding. The sudden flooding by these two storms most likely resulted in depressed coastal salinities, primary production and trophic transfer to jellyfish.

### Future trends in jellyfish populations of the northern Gulf of Mexico

What might be the long-term prognosis for jellyfish population changes across the northern Gulf of Mexico? Given our current understanding of how jellyfish populations function in coastal ecosystems, this question is probably best addressed in the context of nutrient loading to the shelf from large river systems. Additional nutrient loadings to the Gulf of Mexico shelf could enhance jellyfish populations if loading led to increased secondary production on the shelf and, perhaps, to increased distribution and magnitude of hypoxia. Inputs of new dissolved nutrients into the northern Gulf of Mexico are linked to the hydrological cycle and climate, and if global warming is incorporated into this scenario nutrient loading to the shelf may indeed increase. In a recent study, Justić et al. (1997) modeled a 50% increased in shelf primary production as a function of a 20% increase in Mississippi River discharge – a reasonable expectation under current global warming scenarios for the coming decades. An additional consequence of this climate change scenario is the potential expansion and duration of hypoxic bottom waters in the northern Gulf of Mexico (Justić et al., 1996). Therefore, if shelf production and hypoxia are in fact driving jellyfish populations, a long-term increase in population numbers may be expected.

### Summary and conclusions

*Chrysaora quinquecirrha* medusae have increased both numerically and in breadth of distribution during summers 1987–1997 in the northern Gulf of Mexico. These increases have occurred in the vicinity of two major river systems: Mobile Bay and the Mississippi–Atchafalaya River systems. Most importantly, these total numerical increases are linked to an offshore expansion of the population's distribution. This is especially troublesome off the Louisiana and northeast

Texas shelf where summertime hypoxia significantly overlaps with the distribution of *C. quinquecirrha* medusae, and it is predicted that long-term changes in nutrient loadings may yield greater densities and distributions of *C. quinquecirrha* medusae in the future.

*Aurelia aurita* medusae have increased numerically within fall collections 1985–1997 to a greater extent than summer *Chrysaora quinquecirrha* although these increases typically did not correspond to widespread change in distribution. Numerical increase and distributional expansion away from shore occurred together in three regions located off the mouth of the Mississippi River and the western Louisiana shelf.

This paper presents evidence that two species of Gulf of Mexico jellyfish have increased both in distribution and abundance over a 11–13-year period. However, beyond the distributional overlap between summer *Chrysaora quinquecirrha* and hypoxia, it is not possible to definitively link this ecological change to human activity either through coastal fertilization or through increased habitat availability. Rather, it is suggested that long-term variations in jellyfish abundance and distribution may be influenced by environmental changes. A great deal of additional work is now needed so that we can understand how jellyfish populations respond to both natural and human-induced modifications of the Gulf of Mexico shelf ecosystem.

## Acknowledgements

I thank Mr Kenneth Savastano and his staff at the United States National Marine Fisheries Service for providing access to, and assistance with, the SEAMAP data-set. I also thank my laboratory group at DISL for thoughtful discussions leading to this paper and J. Cowan for statistical discussions. I thank J. Purcell and C. Mills for their thoughtful editorial and scientific comments. Funding in support of these analyses were provided, in part, through National Science Foundation Grant OCE-9733441.

## References

Bratkovich, A., S. P. Dinnel & D. A. Goolsby, 1994. Variability and prediction of freshwater and nitrate fluxes for the Louisiana-Texas shelf: Mississippi and Atchafalaya River source functions. Estuaries 17: 766–778.

Breitburg, D. L., K. A. Rose & J. J. H. Cowan, 1999. Linking water quality to larval survival: predation mortality of fish larvae in an oxygen-stratified water column. Mar. Ecol. Prog. Ser. 178: 39–54.

Bricker, S. B. & J. C. Stevenson, 1996. Nutrients in coastal waters: a chronology and synopsis of research. Estuaries 19: 337–341.

Brodeur, R. D., C. E. Mills, J. E. Overland, G. E. Walters & J. D. Schumacher, 1999. Recent increase in jellyfish biomass in the Bering Sea: possible links to climate change. Fish. Oceanogr. 8: 296–306.

Burke, W. D., 1975. Pelagic Cnidaria of Mississippi Sound and adjacent waters. Gulf Res. Rep. 5: 23–38.

Burke, W. D., 1976. Biology and distribution of the macrocoelenterates of Mississippi Sound and adjacent waters. Gulf Res. Rep. 5: 17–28.

Caddy, J. F., 1993. Toward a comparative evaluation of human impacts on fishery ecosystems of enclosed and semi-enclosed seas. Rev. Fish. Sci. 1: 57–95.

Condon, R. H., M. B. Decker & J. E. Purcell, 2001, Effects of low dissolved oxygen on survival and asexual reproduction of scyphozoan polyps (*Chrysaora quinquecirrha*). Hydrobiologia 451 (Dev. Hydrobiol. 155): 89–95.

Dagg, M. J., 1995. Copepod grazing and the fate of phytoplankton in the northern Gulf of Mexico. Cont. Shelf Res. 15: 1303–1317.

Dagg, M. J. & T. E. Whitledge, 1991. Concentration of copepod nauplii associated with the nutrient-rich plume of the Mississippi River. Cont. Shelf Res. 11: 1409–1423.

Deegan, L. A., J. W. Day, J. G. Gosselink, A. Yanez-Arancibia, G. S. Chavez & P. Sanchez-Gil, 1986. Relationships among physical characteristics, vegetation distribution and fisheries yield in the Gulf of Mexico. In Wolfe, D. A. (ed.), Estuarine Variability, Academic Press New York: 83–100.

Eadie, B. J., B. A. McKee, M. B. Lansing, J. A. Robbins, S. Metz & J. H. Trefry, 1994. Records of nutrient-enhanced coastal ocean productivity in sediments from the Louisiana continental shelf. Estuaries 17: 754–765.

Goy, J., P. Morand & M. Etienne, 1989. Long-term fluctuations of *Pelagia noctiluca* (Cnidaria, Scyphomedusa) in the western Mediterranean Sea. Prediction by climatic variables. Deep-Sea Res. 36: 269–279.

Graham, W. M., 1998. First report of *Carybdea alata* var. *grandis* (Reynaud 1830) (Cnidaria: Cubozoa) from the Gulf of Mexico. Gulf Mex. Sci. 16: 28–30.

Johnson, D. R., H. M. Perry & W. D. Burke, 2001. Developing jellyfish strategy hypotheses using circulation models. Hydrobiologia 451 (Dev. Hydrobiol. 155): 213–221.

Justić, D., N. N. Rabalais & R. E. Turner, 1996. Effects of climate change on hypoxia in coastal waters: a doubled $CO_2$ scenario for the northern Gulf of Mexico. Limnol. Oceanogr. 41: 992–1003.

Justić, D., N. N. Rabalais & R. E. Turner, 1997. Impacts of climate change on net productivity of coastal waters: implications for carbon budgets and hypoxia. Clim. Res. 8: 225–237.

Justić, D., N. N. Rabalais, R. E. Turner & J. W. W. Wiseman, 1993. Seasonal coupling between riverborne nutrients, net productivity and hypoxia. Mar. Pollut. Bull. 26: 184–189.

Kideys, A. E., 1994. Recent dramatic changes in the Black Sea ecosystem: the reason for the sharp decline in Turkish anchovy fisheries. J. mar. Sys. 5: 171–181.

Kovalev, A. V. & S. A. Piontkovski, 1998. Interannual changes in the biomass of the Black Sea gelatinous zooplankton. J. Plankton Res. 20: 1377–1385.

Legović, T., 1987. A recent increase in jellyfish populations: a predator–prey model and its implications. Ecol. Model. 38: 243–256.

Lohrenz, S. E., M. J. Dagg & T. E. Whitledge, 1990. Enhanced primary production at the plume/oceanic interface of the Mississippi River. Cont. Shelf Res. 10: 639–664.

Lohrenz, S. E., G. L. Fahnenstiel, D. G. Redalje, G. A. Lang, X. Chen & M. J. Dagg, 1997. Variations in primary production of northern Gulf of Mexico continental shelf waters linked to nutrient inputs from the Mississippi River. Mar. Ecol. Prog. Ser. 155: 45–54.

Mills, C. E., 1995. Medusae, siphonophores and ctenophores as planktivorous predators in changing global ecosystems. ICES J. mar. Sci. 52: 575–581.

Mills, C. E., 2001. Jellyfish blooms: are populations increasing globally in response to changing ocean conditions? Hydrobiologia 451 (Dev. Hydrobiol. 155): 55–68.

Müller-Karger, F. E., J. J. Walsh, R. H. Evans & M. B. Meyers, 1991. On the seasonal phytoplankton concentrations and sea surface temperature cycles of the Gulf of Mexico as determined by satellites. J. Geophys. Res. 96: 12645–12665.

Nelson, D. M. & Q. Dortch, 1996. Silicic acid depletion and silicon limitations in the plume of the Mississippi River: evidence from kinetics studies in spring and summer. Mar. Ecol. Prog. Ser. 136: 163–178.

Ortner, P. B. & M. J. Dagg, 1995. Nutrient-enhanced coastal ocean productivity explored in the Gulf of Mexico. EOS, Trans., AGU 76: 97–109.

Pauly, D., V. Christensen, J. Dalsgaard, R. Froese & F. Torres, Jr., 1998. Fishing down marine food webs. Science 279: 860–863.

Pennock, J. R., J. N. Boyer, J. A. Herrera-Silveira, R. L. Iverson, T. E. Whitledge, B. Mortazavi & F. A. Comin, 1999. Nutrient behavior and phytoplankton production in Gulf of Mexico estuaries. In Bianchi, T. S., J. R. Pennock & R. R. Twilley (eds), Biogeochemistry of Gulf of Mexico Estuaries. John Wiley & Sons, Inc. New York: 109–162.

Phillips, P. J., 1971. The pelagic Cnidaria of the Gulf of Mexico. Ph. D. Dissertation, Texas A & M University, College Station: 212 pp.

Prasad, K. S., S. E. Lohrenz, D. G. Redalje & G. L. Fahnenstiel, 1995. Primary production in the Gulf of Mexico coastal waters using 'remotely-sensed' trophic category approach. Cont. Shelf Res. 15: 1355–1368.

Purcell, J. E., 1989. Predation on fish larvae and eggs by the hydromedusa Aequorea victoria at a herring spawning ground in British Columbia. Can. J. Fish. aquat. Sci. 46: 1415–1427.

Purcell, J. E., 1992. Effects of predation by the scyphomedusan Chrysaora quinquecirrha on zooplankton populations in Chesapeake Bay, U.S.A. Mar. Ecol. Prog. Ser. 87: 65–76.

Purcell, J. E., 1997. Pelagic cnidarians and ctenophores as predators: selective predation, feeding rates, and effects on prey populations. Ann. Inst. Oceanogr., Paris. 73: 125–137.

Purcell, J. E. & M. N. Arai, 2001. Interactions of pelagic cnidarians and ctenophores with fish: a review. Hydrobiologia 451 (Dev. Hydrobiol. 155): 27–44.

Purcell, J. E., T. A. Shiganova, M. B. Decker & E. D. Houde, 2001. The ctenophore Mnemiopsis in native and exotic habitats: U.S. estuaries versus the Black Sea basin. Hydrobiologia 451 (Dev. Hydrobiol. 155): 145–175.

Purcell, J. E., J. R. White, D. A. Nemazie & D. A. Wright, 1999. Temperature and salinity effects on asexual reproduction and abundance of the scyphozoan Chrysaora quinquecirrha. Mar. Ecol Prog. Ser. 180: 187–196.

Purcell, J. E., D. L. Breitburg, M. B. Decker, W. M. Graham, M. J. Youngbluth & K. A. Raskoff, 2001. Pelagic cnidarians and ctenophores in low dissolved oxygen environments: a review. In Rabalais, N. N. & R. E. Turner (eds), Coastal Hypoxia: Consequences for Living Resources and Ecosystems. American Geophysical Union, Coastal estuar. Stud. 58: 77–100.

Rabalais, N. N., R. E. Turner, W. J. Wiseman, Jr. & Q. Dortch, 1998. Consequences of the 1993 Mississippi River flood in the Gulf of Mexico. Regul. Rivers: Res. Mgmt. 14: 161–177.

Rabalais, N. N., R. E. Turner, Q. Dortch, J. W. J. Wiseman & B. K. S. Gupta, 1996. Nutrient changes in the Mississippi River and system responses on the adjacent continental shelf. Estuaries 19: 386–407.

Randall, J. M. & J. J. W. Day, 1987. Effects of river discharge and vertical circulation on aquatic primary production in a turbid Louisiana (U.S.A.) estuary. Neth. J. Sea Res. 21: 231–242.

Shiganova, T. A., 1998. Invasion of the Black Sea by the ctenophore Mnemiopsis leidyi and recent changes in pelagic community structure. Fish. Oceanogr. 7: 305–310.

Stuntz, W. E., C. E. Bryan, K. Savastano, R. S. Waller & P. A. Thompson, 1985. SEAMAP environmental and biological atlas of the Gulf of Mexico, 1982. Gulf States Fishery Commission Report: 145 pp.

Sullivan, B. K., D. Van Keuren & M. Clancy, 2001. Timing and size of blooms of the ctenophore Mnemiopsis leidyi in relation to temperature in Narragansett Bay, RI. Hydrobiologia 451 (Dev. Hydrobiol. 155): 113–120.

Turner, R. E. & N. N. Rabalais, 1994. Coastal eutrophication near the Mississippi river delta. Nature 368: 619–621.

Walsh, J. J., D. A. Dieterle, M. B. Meyers & F. E. Müller-Karger, 1989. Nitrogen exchange at the continental margin: a numerical study of the Gulf of Mexico. Prog. Oceanogr. 23: 245–301.

Warlen, S. M., 1988. Age and growth of larval Gulf Menhaden, Brevoortia patronus, in the northern Gulf of Mexico. Fish. Bull. 86: 77–90.

Wiseman, W. J., Jr. & R. W. Garvine, 1995. Plumes and coastal currents near large river mouths. Estuaries 18: 509–517.

*Hydrobiologia* **451**: 113–120, 2001.
© 2001 *Kluwer Academic Publishers.*

# Timing and size of blooms of the ctenophore *Mnemiopsis leidyi* in relation to temperature in Narragansett Bay, RI

Barbara K. Sullivan[1], Donna Van Keuren[1] & Michael Clancy[2]
[1]*Graduate School of Oceanography, University of Rhode Island, Narragansett, RI 02882, U.S.A.*
[2]*Department of Biological Sciences, University of Rhode Island, Kingston, RI 02881, U.S.A.*

*Key words:* Ctenophore, predation, climate change, fish eggs, fish larvae

## Abstract

The ctenophore *Mnemiopsis leidyi* is at the northern extreme of its geographic range in Narragansett Bay, an estuary on the northeast coast of the United States. Blooms have typically been observed in late summer and fall according to records from 1950 to 1979. We document an expansion of the seasonal range of this important planktonic predator to include springtime blooms during the 1980s and 1990s. This shift to an earlier seasonal maximum is associated with increasing water temperature in Narragansett Bay. Temperatures in spring have risen, on average, 2 °C from 1950 to 1999 with warm years being associated with the positive phase of the North Atlantic Oscillation. During 1999, *M. leidyi* appeared earlier in spring and was more abundant than during any previous year for which records are available. Changes in the seasonal pattern and abundance of this predator are likely to have important effects on planktonic ecosystem dynamics of Narragansett Bay. These include reduction of zooplankton abundance in spring followed by increases in size and frequency of summer phytoplankton blooms. Earlier blooms of *M. leidyi* may also reduce survival of eggs and larvae of fish because, as in 1999, they coincide with the period of peak spawning.

## Introduction

The ctenophore *Mnemiopsis leidyi* Agassiz inhabits estuaries along the Atlantic coastline of the United States sometimes occurring in dense blooms. Its range extends north to Cape Cod (Mayer, 1912). Thus, in Narragansett Bay, RI, this species is near the northern limit of its distribution. Typically, this species has bloomed later in the summer in estuaries of the northeast compared to spring or early summer blooms typical of warmer waters to the south (Kremer, 1994; Purcell et al., 1994). The earliest reports of *M. leidyi* in Narragansett Bay (Fewkes, 1881) describe blooms in "... the latter part of summer and autumn...".

Ctenophores in Narragansett Bay were the focus of intensive studies in the 1970s which documented the seasonal distribution of *Mnemiopsis leidyi* and established the importance of this predator in regulating both zooplankton and phytoplankton dynamics in Narragansett Bay. Peak abundance of *M. leidyi* was observed in August and September (Kremer, 1975; Kremer, 1979; Deason, 1982). In 4 of 6 years studied,

the pulse of ctenophore numbers was accompanied by a rapid decline in copepod abundance and a summer phytoplankton bloom following relaxation of grazing pressure by copepods (Deason & Smayda, 1982). The timing and magnitude of plankton events and their relationship to the magnitude of the ctenophore pulse provided evidence of top-down control exerted by *M. leidyi*. Despite the key part played by *M. leidyi* in planktonic food web dynamics of Narragansett Bay, there have been no recent published studies describing current patterns of abundance there.

Anecdotal reports of 'unseasonably' early spring appearances of *Mnemiopsis leidyi* (Costello, pers. com.) during the 1990s indicated the possibility of important shifts in the bay ecosystem. The fact that a warming trend in water temperatures of Narragansett Bay has been reported (Hawk, 1998) suggests the hypothesis that these early blooms were related to temperature. Hawk (1998) documented a significant long-term increase of 1.73 °C in Narragansett Bay from 1956 to 1998 and found that winter temperatures correlated with the North Atlantic Oscillation (NAO).

*Table 1.* *Mnemiopsis leidyi* >1 cm. Date of first appearance and bloom peak obtained from published and unpublished records from 1950 to 1999. Bloom peak was determined by the date at which maximum numbers of *M. leidyi* were reported. No data available = n.d.

| Date | First appearance | Bloom peak | Peak Abundance no m$^{-3}$ | Average Temperature in May °C | Reference |
|------|------------------|------------|----------------------------|-------------------------------|-----------|
| 1950 | July 25 | Oct 2 | n.d. | n.d. | Frolander (1955) |
| 1951 | July 25 | Sept 7 | n.d. | n.d. | Frolander (1955) |
| 1971 | July 29 | Aug 20 | 27 | 9.67 | Kremer (1975) |
| 1972 | July 6 | Sept 8 | 18 | 10.61 | Kremer (1975) |
| 1973 | July 15 | July 31 | 70 | 12.11 | Hulsizer (1976) |
| 1974 | July 15 | Aug 20 | 38 | 10.94 | Kremer (1975) |
| 1975 | July 20 | Aug 3 | 27 | 10.89 | Deason & Smayda (1982) |
| 1976 | July 26 | Aug 1 | 5 | 12.11 | Deason & Smayda (1982) |
| 1977 | n.d. | Aug 30 | 7 | 13.22 | Deason & Smayda (1982) |
| 1978 | July 20 | Sept 23 | 8[a] | 11.61 | Deason (1982) |
| 1979 | n.d. | Aug 11 | 70[a] | 13.56 | Deason (1982) |
| 1983 | Aug 15 | Sept 5 | 80 | 12.39 | MERL (unpublished) |
| 1985 | June 10 | June 16 | 250 | 12.50 | MERL (unpublished) |
| 1986 | May 18 | n.d. | 60.5 | 13.00 | MERL (unpublished) |
| 1990 | n.d. | early July | n.a. | 12.06 | Keller et al. (1999) |
| 1999 | May 18 | June 12 | 350 | 14.60 | this study |

[a]Estimated number >1 cm from total biomass and total number, Figure 2, using >1 cm = 10% of total as in Deason & Smayda (1982).

The period 1940–1979 *versus* 1980–1998 exhibited extremes in the NAO index; winter water temperatures in the earlier period were as much as 3.6 °C colder than during the latter period.

Water temperature can be expected to influence metabolic and reproductive rates of pelagic species as well as their latitudinal distribution patterns. In Narragansett Bay, Jeffries & Terciero (1985) found a strong correlation between warmer winters and the collapse of the winter flounder, a species with northern affinities. Several studies report correlation of ecosystem dynamics with temperature increases associated with the NAO (Colebrook, 1986; Dickson et al., 1988). Similarly, abundance of copepods (Fromentin & Planque, 1996) fish, shellfish and lobster stocks (Mann & Drinkwater, 1994) appears to relate to the NAO.

Here, we describe changing patterns in seasonal distribution and increasing abundance of *Mnemiopsis leidyi* associated with warmer spring temperatures in Narragansett Bay. We also discuss the possible trophic consequences of a shift in predation pressure on copepods from fall to spring. Impacts on fisheries could also shift because peak numbers of *M. leidyi* could coincide with the peak spawning period of many species

of fish during warm years. We calculate predation rates on fish eggs in Narragansett Bay using published data on filtering rates of *M. leidyi* on fish eggs together with ctenophore abundance reported in this study. Given that fish eggs and larvae have experienced 2–4-fold reductions in abundance in Narragansett Bay since the 1970s (Keller et al., 1999), we believe further attention to this important predator and competitor of larval fish is indeed warranted by fisheries managers.

## Materials and methods

Collections for ctenophores were made from the dock of the Graduate School of Oceanography (GSO) at the University of Rhode Island and from transects near the southern edge of the Bay using small boats (Fig. 1). Samples from the GSO dock have been acquired by the Marine Ecosystem Research Laboratory (MERL) to assess patterns in Narragansett Bay plankton since 1980. Unpublished MERL data for the years of 1982–1986 are used in this study to compare abundance and distribution with data from published studies 1950–1979 (Table 1).

Samples were taken approximately weekly during MERL experiments and from June through Novem-

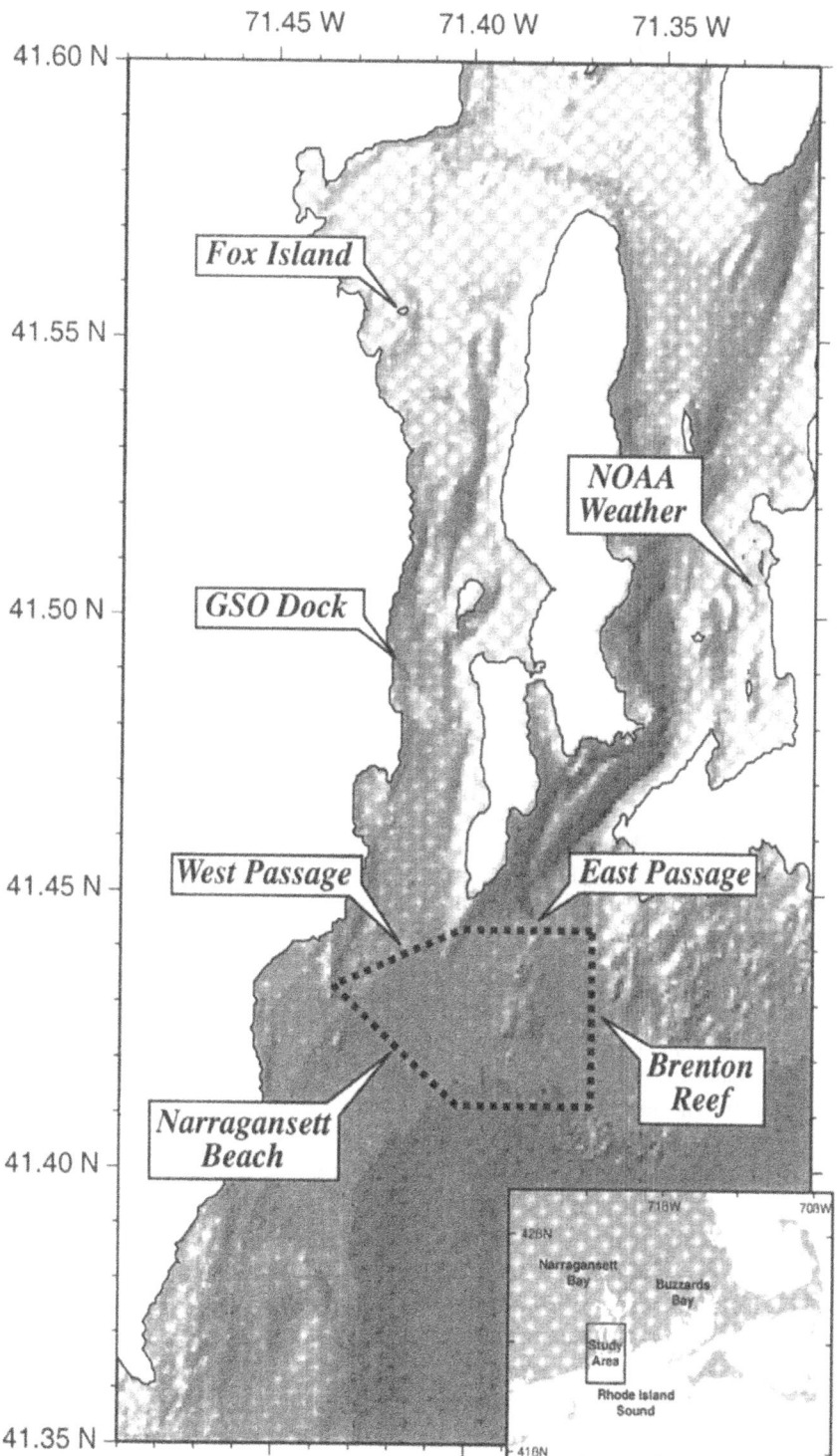

*Figure 1.* Lower Narragansett Bay and locations of stations sampled in 1999 and earlier studies. The Graduate School of Oceanography (GSO) Dock was the site of samples for this study, Frolander (1955), Kremer's (1975) station 3 and unpublished MERL studies. Transects used in 1999 (East and West Passages, Narragansett Beach and Brenton Reef) are indicated by dotted lines. Fox Island indicates the site of studies by Hulsizer (1976), Deason (1982), Deason & Smayda (1982) and temperature measurements for 1994–1999. The NOAA weather station used for temperatures during 1955–1994 is also indicated.

ber 1999 using a plankton net with a 0.25 m diameter opening and 64 $\mu$m mesh. Tows were vertical from top to bottom of the water column (3–6 m depending on tide) and made only at the slack water phase of the tide. Initial tows were made with a flow meter but since the meter produced inaccurate readings for such short tows, volume filtered was calculated from the length of the vertical tow. The tow was sufficiently short to prevent clogging so that this provided an accurate measure of volume filtered.

Samples were counted immediately without preservation under a dissecting scope. During 1999 each individual was measured or 100 live ctenophores were randomly sampled to measure diameter to the nearest 0.1 mm for smaller specimens. Total length was measured to the nearest 1.0 mm for larger ctenophores.

Additional samples for 1999 were collected from transects near the mouth of Narragansett Bay for a survey of larval fish beginning in May (Fig. 1). A 75 cm ring net with 500 $\mu$m mesh was used on the first sampling date. Subsequent samples employed a 1-m$^2$ Tucker Trawl with 1000 $\mu$m mesh. Volume filtered was measured with a flow meter or estimated from metered tows of similar duration and varied from 225 to 550 m$^3$. Except on 4 June, ctenophores were separated from the live sample, drained and total volume of ctenophores measured in ml. Abundance of ctenophores on 4 June was too great to quantify (filling up to 1/3 of the net). Ctenophores were not measured individually.

For the time period prior to 1999, we used 14 years of data from both published and unpublished reports to establish date of first appearance of *M. leidyi* and date of maximum abundance for each of those years (Table 1). Since different mesh nets were used for these collections comparisons were made only for ctenophores >1 cm in diameter. All comparisons using earlier studies include only those stations in lower Narragansett Bay closest to the 1999 sampling locations (Fig. 1). Regression analysis was used to determine the relationship of the date of first appearance of ctenophores to average monthly temperatures in March–May. Timing of the bloom peak was determined simply as the date at which maximum abundance was observed.

Temperature records were obtained from the NOAA weather station in Newport, Rhode Island for 1955–1994 (National Ocean Service, Rockville Maryland). Monthly means were calculated from hourly data. NOAA ceased temperature recording in Narragansett Bay in 1994. Weekly temperatures for 1994–

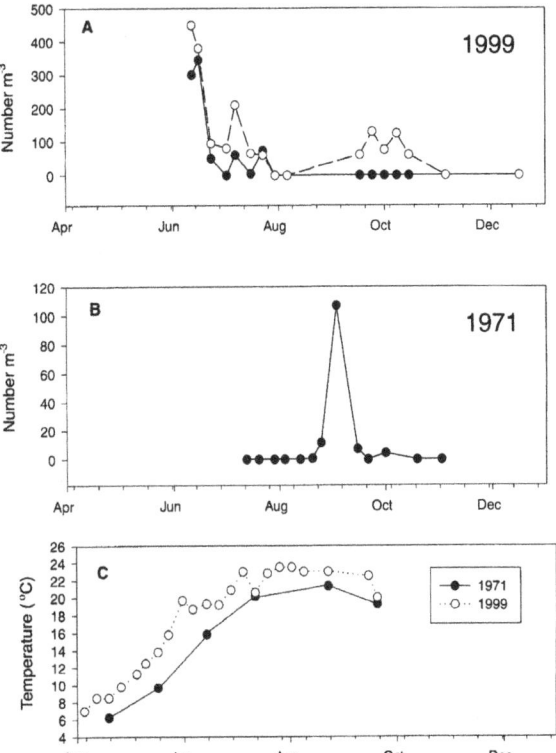

*Figure 2.* Abundance of *Mnemiopsis leidyi* and water temperature in Narragansett Bay. (A) Abundance in 1999 at the Graduate School of Oceanography dock. Open circles are ctenophores <1 cm length; closed circles are >1 cm in length. (B) Abundance of *M. leidyi* >1 cm during 1971 from Kremer (1975), Appendix II, Station 3. (C) Surface temperatures in Narragansett Bay during spring and summer of 1971 and 1999.

1999 were obtained from the University of Rhode Island Fish Trawl Survey at the Fox Island station and averaged to obtain monthly means.

## Results

*Mnemiopsis leidyi* were already very abundant on the first sampling date, 12 June, 1999, at the Graduate School of Oceanography dock station (Fig. 2). The bloom peak at the dock station probably occurred in mid-June or earlier, since ctenophores became less abundant during early July. Ctenophores ranging in size from 0.1 to 5 cm long were collected. Small ctenophores (<1 cm) were present in low numbers all summer and became more abundant again in the fall.

*M. leidyi* were observed during May 1999 at the mouth of Narragansett Bay on the first sampling date for the larval fish study. On 4 June, ctenophores were so abundant in either deep or shallow tows at all 5

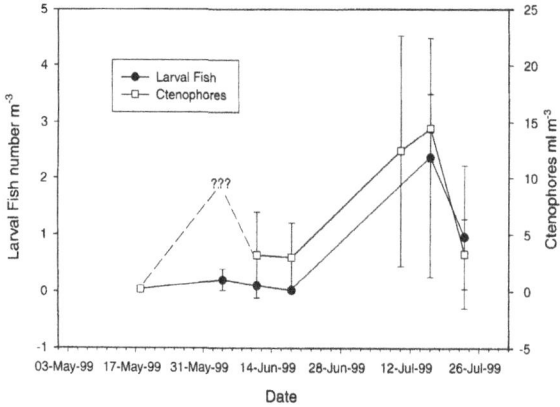

*Figure 3.* Abundance of *Mnemiopsis leidyi* (open squares) and larval fish (closed circles) in 1999 averaged for all stations in lower Narragansett Bay. Error bars indicate the standard deviation of the average values. Question marks denote samples of ctenophores too abundant to quantify.

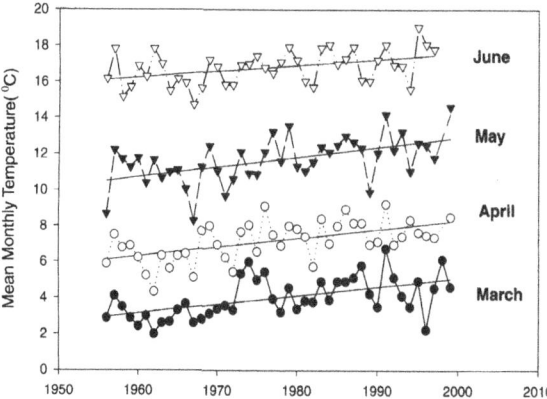

*Figure 4.* Surface water temperature in Narragansett Bay from 1955 to 1999. Mean monthly averages were calculated from weekly data for the months of March through June. Regression lines indicate statistically significant trends of increasing temperature over the time period for all months (see text).

stations that is was not possible to quantify them accurately (Fig. 3). A second peak in abundance at the lower Bay stations was observed in July and coincided with the peak abundance of larval fish collected in the same samples (Fig. 3).

The ctenophore bloom reached peak numbers fully 2 months earlier in 1999 than during 1971, a typical year during the early 1970s (Fig. 2). In addition, ctenophores were 2–3 times more abundant in 1999 than in 1971. Temperatures were warmer throughout the spring of 1999 than in spring 1971 (Fig. 2). An examination of temperature records for the entire period for which records of ctenophores exist, 1950–1999, indicates a warming trend, although there were some years with cold springs even into the 1990s (Fig. 4).

Regression analysis indicated a significant increase in average temperature during the months of March, April, May and June (March: $F = 18.1$, $p < 0.001$, $n = 44$; April: $F = 18.9$, $p < 0.001$, $n = 43$; May: $F = 16.5$, $p < 0.001$, $n = 43$; June: $F = 8.5$, $p = 0.005$, $n = 42$). During this time, ctenophores were observed earlier in the year or reached maximum abundance in spring or early summer in 1985, 1986, 1990 and 1999 (Table 1).

Average temperature during the month of May explained 25.8% of the variation in date on which maximum numbers were observed ($F = 5.172$; $p = 0.044$; $n = 13$). Regression analysis showed no significant relationship between temperatures in March or April and bloom peak, but the average temperature over March – May was significantly related to bloom date ($r^2 = 0.383$; $F = 6.83$; $p = 0.024$; $n = 13$). There was no relationship between temperatures in January or February and bloom date.

Maximum abundance of ctenophores >1 cm (Table 1) also appears to have increased significantly since 1971 (Mann–Whitney Rank Sum Test, $T = 44$; $p=0.017$). Median bloom size of ctenophores >1 cm during 1971–1979 was 46 m$^{-3}$ versus 300 m$^{-3}$ during 1983–1999.

## Conclusions

Prior to 1985, there were no records of spring or early summer blooms of *Mnemiopsis leidyi* in lower Narragansett Bay. We report here 4 recent years in which *M. leidyi* were first observed in May or June and reached peak abundance in June or July. In addition, maximum numbers of ctenophores in these years were higher than had been previously typical and higher than reported in other estuaries on the Atlantic coast of the United States (Kremer & Nixon, 1976; Kremer, 1994; Purcell et al., 1994a). Unfortunately, a lack of monitoring programs to collect the species in the 1990s precludes obtaining a complete picture of how frequent these early blooms have become. Nevertheless, it is clear that in recent years, the seasonal pattern of ctenophore distribution in Narragansett Bay has more nearly resembled that typical of warmer estuaries to the south (Kremer, 1994; Purcell et al., 1994a).

It is tempting to ascribe the change in the seasonal pattern of abundance of *Mnemiopsis leidyi* to the warming of Bay water temperatures during this time period; regression analysis did reveal a significant correlation, but temperature explained only 25–38% of the variation in bloom date. Also consistent with

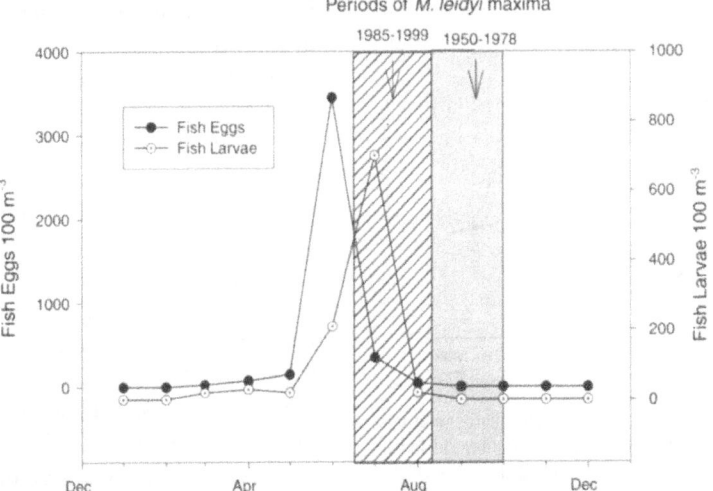

*Figure 5.* Temporal pattern of abundance of fish eggs and fish larvae in Narragansett Bay (redrawn from Keller et al., 1999). The hatched box indicates the possible coincidence in the ctenophore bloom peak with the ichthyoplankton during warm years, indicating potential for predation on and competition with fish eggs and larvae. During 1950–1978, ctenophore blooms peaks typically occurred in months indicated in the gray box.

our hypothesis is the observation of higher densities of *M. leidyi* following warm winters in other estuaries (Purcell et al., 2001). It has been suggested that other factors, primarily zooplankton prey abundance, strongly limit size of ctenophore blooms in Narragansett Bay (Deason & Smayda, 1982); but the effect of zooplankton availability on timing of ctenophore blooms has never been specifically investigated. Hulsizer (1976) did suggest a relationship between early warming in spring of 1973 and increased zooplankton populations followed by a large and relatively early (31 July) ctenophore bloom compared with 1972. We did not have sufficient data to examine influence of zooplankton prey availability for the 1980–1999 period. Examination of available data lead us to hypothesize that both a warm spring and an adequate food supply are necessary for early bloom formation. For example, in 1977, ctenophores were low in number and the bloom was delayed until mid-August despite quite a warm spring, but this was also a year with low numbers of zooplankton preceding the ctenophore bloom (Deason & Smayda, 1982). A resumption of monitoring of both zooplankton and ctenophores in Narragansett Bay and other northeast estuaries could resolve this issue. It is surprising that remarkably few data exist for ctenophores in the northeast coastal waters of the United States despite the capacity for this predator to exert top down control on zooplankton populations (Deason & Smayda, 1982; Feigenbaum & Kelly, 1984). In addition, this species

can be a significant predator of ichthyoplankton and potentially compete for food with larval fish (Monteleone & Duguay, 1988; Cowan & Houde, 1993; Purcell et al., 1994 a, b, 2001; Purcell & Arai, 2001; Purcell et al., 2001).

Although we cannot conclusively determine at this time the cause of earlier blooms of *Mnemiopsis leidyi* in Narragansett Bay, it is important to investigate potential consequences of an increased prevalence of this predator in the Narragansett Bay ecosystem and possibly other estuaries at the northern edge of its range. Increased numbers of ctenophores could lead to decreased zooplankton abundance, increased phytoplankton blooms in summer due to relaxation of grazing pressure by copepods, and decreased survival of fish eggs and larvae resulting from predation as well as competition with larvae for zooplankton food supplies.

*Mnemiopsis leidyi* were estimated to have considerable impact on zooplankton populations in Narragansett Bay even at abundance levels typical in the 1970s (Deason, 1982). Numbers of *M. leidyi* (Table 1) generally exceeded those reported for *M. leidyi* in the Chesapeake Bay (Purcell et al., 1994a) where copepod populations were not found to be limited by gelatinous zooplankton. Deason (1982) estimated that *M. leidyi* removed a bay-wide mean of 20% of the zooplankton standing stock daily in August of 1975 and 1976. A rapid decline in zooplankton abundance followed the summer pulse of *M. leidyi* and was associated with late summer phytoplankton blooms (Deason &

Smayda, 1982). Earlier appearance of abundant *M. leidyi* may result in early declines in zooplankton and early summer phytoplankton blooms.

A change in timing of ctenophore blooms could result in a significant impact on fisheries in Narragansett Bay (Fig. 5). There could be depletion of zooplankton stocks at a critical time for larval feeding. The previous pattern of late summer blooms of ctenophores posed no significant problem because spawning of abundant species occurred in June, prior to the appearance of ctenophores (Bourne & Govoni, 1988). Since the seasonal pattern of spawning does not appeared to have changed with warming of the Bay (Keller et al., 1999), earlier ctenophore blooms are likely to overlap the period of peak abundance of fish eggs and larvae. In fact, peak abundance in both ctenophores and larval fish did coincide in 1999 (Fig. 3).

Early blooms of *Mnemiopsis leidyi* could impact fish recruitment in Narragansett Bay directly by predation on fish eggs. Predation by *M. leidyi* on fish eggs has been confirmed. Purcell et al. (1994b) detected eggs of bay anchovy in guts of *M. leidyi* collected from Chesapeake Bay, reporting average clearance rates of $128 \, l \, d^{-1} predator^{-1}$. Cowan & Houde (1993) reported clearance rates of $3.7–50 \, l \, d^{-1} ctenophore^{-1}$ in large enclosures; rates were independent of ctenophore size. A similar range was reported by Monteleone & Duguay (1988) for 2.0–2.5 cm *M. leidyi* feeding in a variety of container sizes. As a conservative estimate of predation, we applied these lower rates to our June, 1999 data on *M. leidyi* (1.7 cm average length). At the dock station, 300 ctenophores $m^{-3}$ would have cleared $1110–15000 \, l \, m^{-3} \, d^{-1} (= 111–1500\% \, d^{-1})$. Few eggs would be expected to survive in this high abundance of ctenophores. In lower Narragansett Bay, ctenophores were less dense (averaging 15 ml $m^{-3}$ or approximately 7.5 ctenophores $m^{-3}$). Here they would have cleared $26–350 \, l \, m^{-3} \, d^{-1}$ or 2.5–35% of any eggs present per day.

Predation rates by *Mnemiopsis leidyi* on fish larvae are not yet well quantified. Purcell et al. (1994b) did not find larvae in gut contents of field collected ctenophores; Cowan & Houde (1992) reported variable clearance rates on larvae due to presence of alternative prey and container size. Nevertheless, predation by *M. leidyi* frequently results in precipitous declines in zooplankton abundance in Narragansett Bay (Deason & Smayda, 1982); such extensive depletion of the prey of larval fish is likely to reduce larval growth and survival, making *M. leidyi* a potentially important competitor.

The potential for *Mnemiopsis leidyi* to influence ecosystem structure and have significant impacts on fisheries is clearly evident in the Black and Azov Seas where declines in ichthyoplankton, mesozooplankton, and weight and catches of zooplanktivorous fish have been attributed to introduction of *M. leidyi* to these ecosystems (Purcell et al., 2001). Warm winters in the Black Sea also are correlated with high numbers of *M. leidyi* there (Purcell et al., 2001). In areas like Narragansett Bay where the ecological role of *M. leidyi* may be increasing in importance due to expansion of the seasonal range, renewed attention to programs that monitor ctenophores is certainly warranted.

## Acknowledgements

This study was partially funded by Rhode Island Sea Grant # NA86RG0076 to J. Collie and J.S. Cobb. C. A. Oviatt encouraged the 1999 collections of ctenophores from the GSO dock. Tim Feehan, Joe Cofone, David Medici and Dave Cobb assisted in 1999 collections for the lower Bay larval fish study.

## References

Bourne, D. W. & J. J. Govoni, 1988. Distribution of fish eggs and larvae and patterns of water circulation in Narragansett Bay, 1972–1973. Am. Fish. Soc. Symp. 3: 132–148.

Colebrook, J. M., 1986. Environmental influences on long term variability in marine plankton. Hydrobiologia 142: 309–325.

Cowan, J. H. Jr. & E. D. Houde, 1992. Size-dependent predation on marine fish larvae by ctenophores, scyphomedusae and planktivorous fish. Fisheries Oceanography 1: 113–126.

Cowan, J. H. Jr. & E. D. Houde, 1993. Relative predation potentials of scyphomedusae, ctenophores and planktivorous fish on ichthyoplankton in Chesapeake Bay. Mar. Ecol. Prog. Ser. 95: 55–65.

Deason, E. E., 1982. *Mnemiopsis leidyi* (Ctenophora) in Narragansett Bay, 1975–1979: Abundance, size composition and estimation of grazing. Estuar. coast. shelf Sci. 15: 121–134.

Deason, E. E & T. J. Smayda, 1982. Ctenophore-zooplankton-phytoplankton interactions in Narragansett Bay, Rhode Island. J. Plankton. Res. 4: 219–236.

Dickson, R. R., P. M. Kelly, J. M. Colebrook, W. S. Wooster & D. H. Cushing, 1988. North winds and production in the eastern North Atlantic, J. Plankton Res. 10: 151–169.

Fewkes, J. W., 1881. Studies of the Jelly-fishes of Narragansett Bay. Bull. Museum Comp. Zool. Harvard Vol. VIII: 141–182.

Feigenbaum, D. L. & M. Kelly, 1984. Changes in the lower Chesapeake Bay food chain in the presence of the sea nettle. Mar. Ecol. Prog. Ser. 19: 39–47.

Frolander, H., 1955. The biology of the zooplankton of the Narragansett Bay area. Ph.D. Thesis, Brown University: 93 pp.

120

Fromentin, J-M. & B. Planque, 1996. *Calanus* and environment in the eastern North Atlantic.. II. Influence of the North Atlantic Oscillation on *C. finmarchicus* and *C. helgolandicus*. Mar. Ecol. Prog. Ser. 134: 111–118.

Hawk, J. D., 1998. The role of the North Atlantic Oscillation in winter climate variability as it relates to the winter–spring bloom in Narragansett Bay. M.S. Thesis, University of Rhode Island: 148 p.

Hulsizer, E. E., 1976. Zooplankton of lower Narragansett Bay, 1972–1973. Chesapeake Sci. 17: 260–270.

Jeffries, H. P. & M. Terceiro, 1985. Cycle of changing abundances in the fishes of the Narragansett Bay area, Mar. Ecol. Prog. Ser. 25: 239–244.

Keller, A. A., G. Klein-MacPhee & J. St. Onge Burns, 1999. Abundance and distribution of ichthyoplankton in Narragansett Bay, Rhode Island, 1989–1990. Estuaries 22: 149–163.

Kremer, P. M., 1975. The ecology of the ctenophore *Mnemiopsis leidyi* in Narragansett Bay. Ph.D. Thesis, University of Rhode Island: 311 p

Kremer, P., 1979. Predation by the ctenophore *Mnemiopsis leidyi* in Narragansett Bay, Rhode Island. Estuaries 2: 97–105.

Kremer, P., 1994. Patterns of abundance for *Mnemiopsis* in U.S. coastal waters: a comparative overview. ICES J. mar. Sci. 51: 347–354.

Kremer, P. & S. Nixon, 1976. Distribution and abundance if the ctenophore, *Mnemiopsis leidyi* in Narragansett Bay. Estuar. coastal mar. Sci. 4: 627–639.

Mann, K. H. & K. F. Drinkwater, 1994. Environmental influences on fish and shellfish production in the Northwest Atlantic. Envir. Rev. 2: 16–32.

Mayer, A. G., 1912. Ctenophores of the Atlantic coast of North America. Pub. 162. Carnegie Institute of Washington: 58 p.

Monteleone, D. M. & L. E. Duguay, 1988. Laboratory studies of predation by the ctenophore *Mnemiopsis leidyi* on the early stages in the life history of the bay anchovy, *Anchoa mitchilli*. J. Plankton Res. 10: 359–372.

Purcell, J. E. & M. N. Arai, 2001. Interactions of pelagic cnidarians and ctenophores with fish: a review. Hydrobiologia 451 (Dev. Hydrobiol. 155): 27–44.

Purcell, J. E , J. R. White, & M. R. Roman, 1994a. Predation by gelatinous zooplankton and resource limitation on populations in Chesapeake Bay. Limnol. Oceanogr. 39: 263–278.

Purcell, J. E., D. A. Nemazie, S. E. Dorsey, E. D. Houde & J. C. Gamble, 1994b. Predation mortality of bay anchovy *Anchoa mitchilli* eggs and larvae due to scyphomedusae and ctenophores in Chesapeake Bay. Mar. Ecol. Prog. Ser. 114: 47–58.

Purcell, J. E, T.A., Shiganova, M. B. Decker & E. D. Houde, 2001. The ctenophore *Mnemiopsis* in native and exotic habitats: U.S. estuaries *versus* the Black Sea basin. Hydrobiologia 451 (Dev. Hydrobiol. 155): 145–175.

*Hydrobiologia* **451**: 121–129, 2001.

# The impact of El Niño events on populations of mesopelagic hydromedusae

Kevin A. Raskoff

*Monterey Bay Aquarium Research Institute, 7700 Sandholdt Road, Moss Landing, CA 95039, U.S.A.,*
*Department of Biology, Box 951606, University of California, Los Angeles, CA 90095-1606, U.S.A.*
*E-mail: kraskoff@mbari.org*

*Key words:* Cnidaria, blooms, *Mitrocoma*, *Colobonema*, jellyfish, ROV

## Abstract

For over 10 years, the midwater ecology group at MBARI has compiled video and accompanying physical data with the ROV *Ventana* operating in mesopelagic depths of Monterey Bay, CA in order to elucidate patterns in midwater ecology. Two El Niño events have occurred during this time period, in 1991–92 and in 1997–98. The oceanographic metric of spiciness combines temperature and salinity data into one sensitive measurement. Although temperature and salinity measurements alone revealed no clear patterns, clear signals of spiciness were observed that corresponded to water mass intrusions into the deep waters of the bay during the two El Niño events. During these events, some seldom-seen species were observed in high numbers in the midwater, while historically common species became rare. During non-El Niño years, the leptomedusa *Mitrocoma cellularia* (A. Agassiz, 1865) was common in the surface waters (0–50 m) of Monterey Bay, but it was not abundant at depth, while the trachymedusa *Colobonema sericeum* Vanhöffen, 1902 was found in relatively high numbers at mesopelagic depths. During the last two El Niño events, *M. cellularia* was observed in higher numbers at mesopelagic depths, whereas *C. sericeum* was scarce. *M. cellularia* was found in a wider range of temperatures, salinities, and dissolved oxygen values than was *C. sericeum*. Transport and tolerance hypotheses are proposed to explain differences in the presence and numerical density of the medusae.

## Introduction

El Niño events are associated with a variety of physical oceanographic phenomena that affect the distribution, abundance, growth and reproduction of marine organisms throughout the Pacific basin. It is now known that the El Niño-Southern Oscillation is the single most important factor in governing interannual oceanic variability on the sub-decadal to decadal time scale in the Pacific (Chavez, 1996; Chavez et al., 1999). This variability is often expressed off California as unusually warm surface ocean temperatures. The consequences of this mass transport of warm water are primarily a reduction of upwelling and a concomitant reduction in phytoplankton productivity (Glynn, 1988; Chavez, 1996).

During the 1997–98 El Niño, a weakening and reversal of trade winds preceded the generation of high temperature surface waters in the equatorial Pacific

(Chavez et al., 1999). One of the consequences of this warming was the cessation of equatorial upwelling. Normal upwelling did not recur until the trade winds abruptly returned in the eastern Pacific in mid-May of 1998, at which time the cold subsurface waters were upwelled to the surface (Chavez et al., 1999).

Pennington & Chavez (2000) present a long-term (1989–1996) characterization of the physical and biological (primary production) properties of Monterey Bay, California. Primary productivity in Monterey Bay during the 1991–92 El Niño showed a 25% decrease from the 1989 to 91 period (Lenarz et al., 1995). In addition, the beginning of the productive season was delayed by several months (Lenarz et al., 1995; Pennington & Chavez, 2000). This delay and reduction in primary productivity and the elevated temperatures caused substantial changes in local marine populations. Graham (1994) found that the scyphozoan *Chrysaora fuscescens* Brandt, 1835 in

Monterey Bay was several orders of magnitude less abundant during the 1991–92 El Niño event than in the preceding non-El Niño years. He suggested that *C. fuscescens* may have shifted north, although the data to support this contention was inconclusive.

El Niño events have also affected the distribution and abundance of fish throughout the Pacific (Barber & Chavez, 1983, 1986; Glynn, 1988; Bailey et al., 1995; Dorn, 1995; Hammann et al., 1995; Lenarz et al., 1995). The most widely discussed El Niño impact is the crash of the Peruvian anchovy, *Engaulis ringens*, once the world's largest fishery, with the 1972–73 and 1982–83 El Niño events (Barber & Chavez, 1983, 1986; Glynn, 1988). In addition to those on fishes, there are well documented studies of the effects of El Niño on surface-dwelling populations of marine reptiles, birds, and mammals (Barber & Chavez, 1983; Gibbs & Grant, 1987; Glynn, 1988; Laurie, 1990; Ainley et al., 1995; Culik & Luna-Jorquera, 1997).

The effects of El Niño events on mesopelagic communities are not understood, however, recent studies have shown that the surface waters and the mesopelagic community are inextricably linked. Robison et al. (1998) and Silguero & Robison (2000) have shown that the physonect siphonophore, *Nanomia bijuga*, and calycophoran siphonophores have seasonal cycles that are significantly correlated with both the onset of upwelling in the surface waters, and with peaks in primary production. *N. bijuga's* seasonal peaks were found to follow the peak chlorophyll values by a 3–4 month lag, while the calycophoran siphonophores lagged by 1–2 months. The dominant prey of *N. bijuga* were euphausiid krill, which were able to feed directly on the phytoplankton. This short, trophic link between the surface waters and the mesopelagic gelatinous community shows that physical changes in the epipelagic waters can have a direct and dramatic impact on the organisms of the mesopelagic.

Many of the dominant macroscopic organisms in the mesopelagic depths of Monterey Bay are cnidarians (Widder et al., 1989; Raskoff, 1998; Robison et al., 1998; Silguero & Robison, 2000). Hydromedusae are often the most abundant organism in the lower mesopelagic depths, especially in low oxygen regions (Purcell et al., 2001). Medusae typically occupy high trophic positions, which are normally thought of as dominated by fish (Mills, 1995), and thereby play an important role in the overall cycling of nutrients in the deepsea.

*Mitrocoma cellularia* is a hydromedusa (Leptomedusae: Mitrocomidae) found from the Bering Sea to southern California (Kramp, 1968; Arai & Brinkmann-Voss, 1980). It reaches a maximum diameter of 100 mm and has over 300 fine tentacles. In Friday Harbor, Washington it is found seasonally from April-November (Arai & Brinkmann-Voss, 1980; Mills, 1983, 1993). In Monterey Bay, *M. cellularia* is a common member of the epipelagic gelatinous community (surface to 100 m) from fall to earlyspring (Wrobel & Mills, 1998), but it can be found in low numbers at other times of the year (Raskoff, pers. obs.). *Colobonema sericeum* is a hydromedusa (Trachymedusae: Rhopalonematidae) found in the mesopelagic depths of the Pacific, Atlantic and Indian Oceans (Kramp, 1968; Wrobel & Mills, 1998). Its bell diameter reaches 45 mm and it has 32 thick tentacles.

This study focused on two important medusae in the mesopelagic gelatinous community of Monterey Bay, California. Observations were made with the Monterey Bay Aquarium Research Institute's (MBARI) remotely operated vehicle (ROV) over a 9 year time period (1990 through 1998). Temperature, salinity and dissolved oxygen measurements taken in tandem with the video observations have permitted physical characterization of the midwater habitat. The ROV observations overlapped the El Niño events of 1991–92 and 1997–98. The resultant MBARI video database has permitted evaluation of patterns of abundance of midwater medusae as they relate to El Niño events.

## Materials and methods

### Midwater study site

The MBARI midwater time-series site (MWTS) is located at 36° 42′ 00″ N – 122° 02′ 00″ W, at the mouth of Monterey Bay (Robison et. al. 1998). This site lies over the meandering axis of the Monterey Submarine Canyon in 1600 m of water. This location allows access to deepwater, as it enters the bay, yet permits transport of organisms back to the laboratory within 1.5 h. The canyon head is at Moss Landing, and the canyon stretches west southwest through the bay to join the continental rise at 3000 m, roughly 80 km off shore. The deep water currents of Monterey Bay are not well understood. The flow is thought to be primarily up-canyon at the mesopelagic depths, with across canyon currents with differing directions depending on depth (Breaker & Broenkow, 1994). The MWTS site was sampled with a ROV twice a month, on average, from 1989 to the present. Data for this study are from 1990 to 1998.

*Remotely Operated Vehicle (ROV) data and spiciness*

The R/V *Point Lobos* and the ROV *Ventana* have been in operation at MBARI since 1988. Since that time, an active research program on the ecology of the mesopelagic zone (100–1000 m in depth) has been underway at the MWTS site. The *Ventana* has completed over 290 mesopelagic dives at this MBARI midwater site, making it the most visited location in the world for mesopelagic studies. The sub is an ISE Hysub 40 with a variety of sensors and modifications added for midwater research (Robison, 1993). A Falmouth Scientific, Inc. Micro-CTD (conductivity, temperature and depth) sensor and a Sea-Bird Electronics SBE 13 dissolved oxygen (DO) sensor with Beckman oxygen electrode, recorded data every 4 s. The data were then automatically averaged and logged into the database in 15 s intervals. CTD-DO data were collected starting at the end of 1990, therefore only data from 1991 to 1998 are presented. The ROV dove no deeper than 400–500 m until the middle of 1991, at which point it began regular dives to 1000 m. Secondary processing of the CTD data determined that the information from June, 1995 to June, 1996 were erroneous and those data were excluded from analysis.

Individual plots of temperature and salinity data were insufficient to detect differences in water masses brought into Monterey Bay by El Niño events at mesopelagic depths (Raskoff, unpubl. data). It was necessary to look at an oceanographic metric that was more sensitive to changes in water masses than either temperature or salinity individually. The dynamic interaction of temperature and salinity can be described by the calculation of a metric called spiciness (Flament, 1986; Lynn et al., 1995). This metric has been used in the past to identify water mass intrusions and the onset of El Niño events (Lynn et al., 1995). Warm, salty water has a high spiciness, while cold, fresh water has a low spiciness (Jackett & McDougall, 1985; Flament, 1986; Lynn et al., 1995; Ramp et al., 1997; Schlining, 1999; Schmitt, 1999). Spiciness calculations were created from ROV CTD data from the dives at the MWTS site between 1990 and 1998 (excluding June 1995–June 1996). Missing data are shown in Figure 1 as blank areas. Twenty meter depth strata were binned and averaged over weekly intervals. From this, an average CTD profile was created by averaging all

the binned depth strata over the 9 years. From this average profile, anomalies of spiciness, in terms of standard deviations from the mean, were calculated and plotted onto a contour graph. Spiciness anomalies were used to elucidate the very subtle differences found in water masses at these mesopelagic depths. The calculations and graphing of spiciness were performed in Matlab (MathWorks, Inc., Version 5.3) and are based on the algorithms of Flament (1986) as modified and programmed by Schlining (1999). Contours of spiciness were gridded using the Kriging method (Keckler, 1995).

BetaCam video tapes were annotated by the MBARI video lab staff and all resulting records of organisms were placed into a database. Individual annotations were linked to their corresponding CTD-DO data by timecode. The database was queried for organisms of interest and the resulting numbers were normalized to a proxy of abundance, number observed $h^{-1}$. Normalization was accomplished by dividing the number of organisms observed in a particular parcel of water by the amount of time the ROV spent in that same parcel. Data were then binned into 100 m depth strata by month, for the 9 years of the study (1990–1998). Contour plots of medusae vertical distributions (Fig. 2) were constructed for analysis with the inverse distance gridding method (SPSS Inc., 1998).

The CTD-DO data associated with the individual organisms of interest were culled from the data base. The organism-based temperature (°C), salinity, and dissolved oxygen (ml $l^{-1}$) bubble plots (Fig. 3) were created from weekly averaged data over the 9-year time period. This provides a descriptive way of depicting the physical parameters of the water in which the organisms were found. In addition, the values for each species were compared statistically to test for differences in the means with the nonparametric Mann–Whitney $U$ two-sample test. These nonparametric tests were used because of the data's violation of parametric tests requirement of equal variances (Zar, 1996).

## Results

### ROV CTD data

Contours of spiciness anomalies for the MWTS site over the 9-year period showed rapid and marked changes in the waters of the mesopelagic depths when warm, salty waters were brought into the bay during

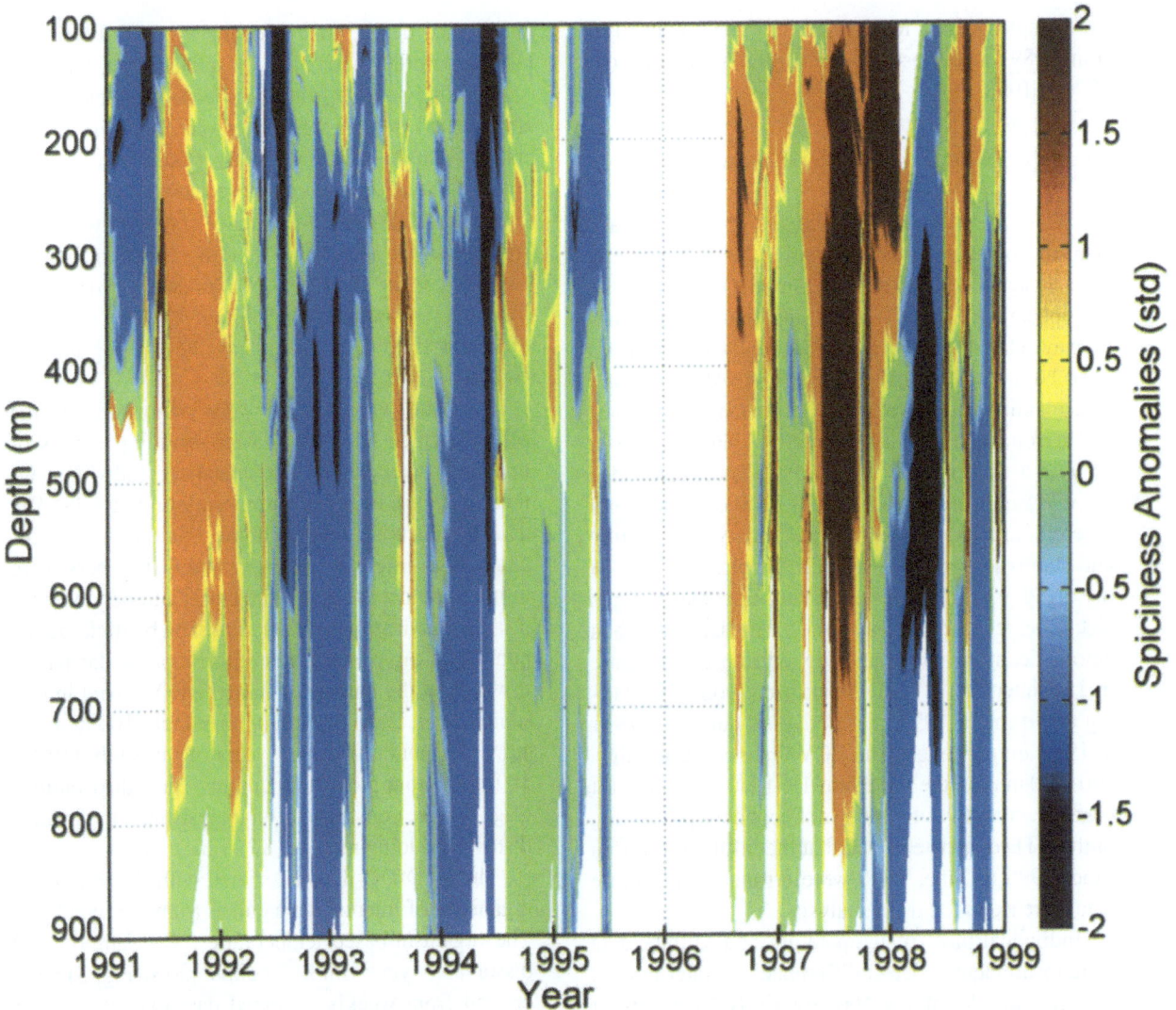

*Figure 1.* Contour of spiciness anomalies for the MWTS site in Monterey Bay, California from 1991–1998. Contours are spiciness anomalies in standard deviations from the mean (see text for details). Red/Orange areas are high in spiciness, blue areas are low.

the two El Niño events, as indicated by the orange/red coloration in Figure 1. Although consistent data collection was not deeper than 400–500 m from January, 1990 to June, 1991, the onset of the 1991–92 El Niño in June/July, 1991 apparently was captured with the ROV's CTD. The near instantaneous onset of the spiciness signal is observed from the surface waters down to 800 m in the middle of 1991. The collapse of the spiciness signal showed a gradual decline in spice in the surface waters down to 300 m over a 3 month period from October to December, 1991. During this time, there were still high spice waters in the 300–600 m depth range, which finally left abruptly in February, 1992. The overall magnitude of the 1991–92 El Niño was moderate as determined by the

spiciness anomalies (averaged 1.0 standard deviation (SD) from the mean). The years following the 1991–92 El Niño were marked with a fairly regular pattern of elevated spiciness from July to October. These results correspond to those found by Schlining (1999), who observed that these elevations were the signature of the California Undercurrent, which seasonally runs up the coast. An unusually strong undercurrent signal was seen at the end of 1996, but missing data in the CTD record for the beginning of that year make interpretation difficult.

The onset of the 1997–98 El Niño was observed in the midwaters in June, 1997. This strongly positive anomaly signal was found from near-surface waters down to 800 m and was nearly twice as strong as the

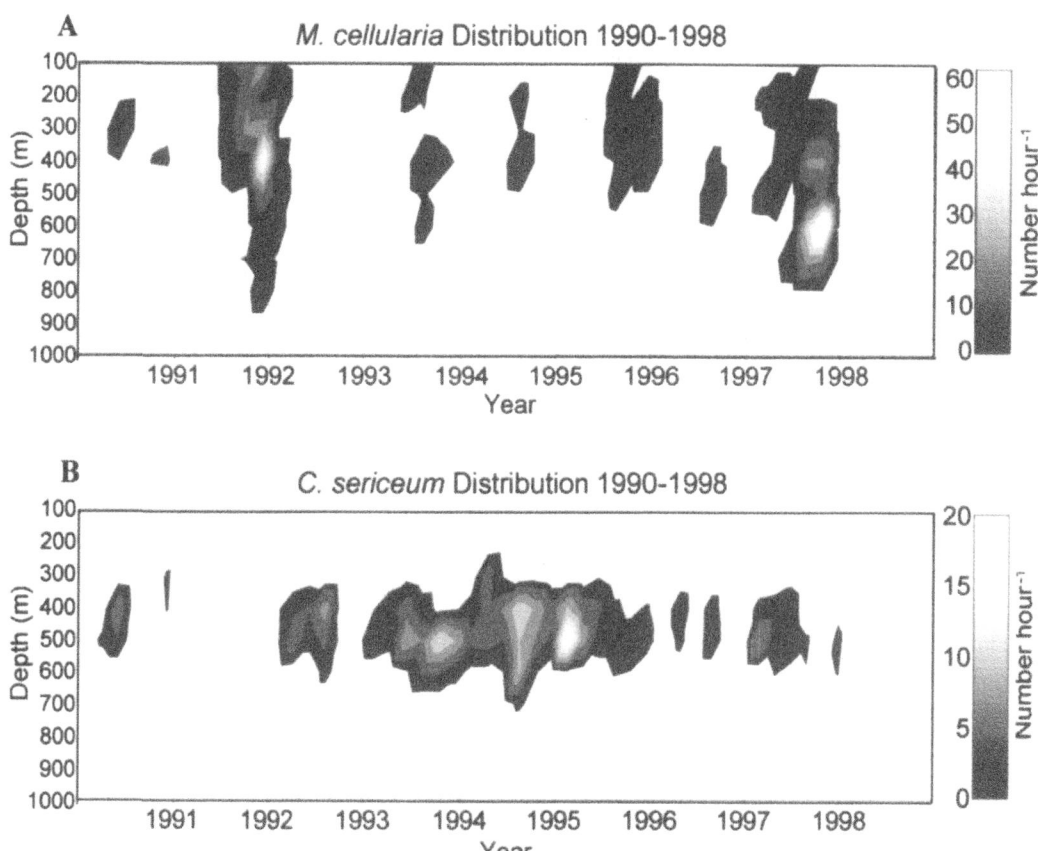

*Figure 2.* Vertical distribution and abundance of *Mitrocoma cellularia* (A) and *Colobonema sericeum* (B) from 1990 to 1998 in Monterey Bay, California. Abundance axes (number h$^{-1}$) are scaled differently for each species. Dots in plot A show sampling opportunities by the ROV.

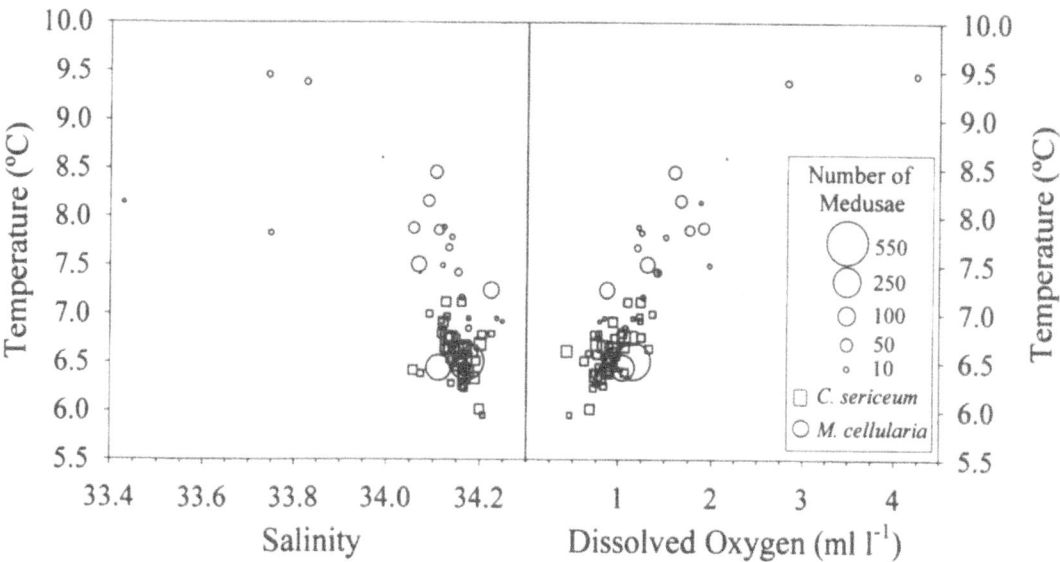

*Figure 3.* Temperature, salinity and dissolved oxygen plots for *Mitrocoma cellularia* and *Colobonema sericeum* in Monterey Bay, California from 1991 to 1998. Bubble plots show the number of medusae observed in waters with listed physical characteristics.

1991–92 El Niño (average of 2.0 SD from the mean). Like the 1991–92 El Niño, the 1997–98 signal had a rapid onset. The decline of the 1997–98 signal was, however, much different than in the previous El Niño. Whereas in 1991–92, the deep mesopelagic waters were the last to recede back to normal spiciness values, in 1997–98 they were the first to fall. There was a gradual decline in spiciness from October, 1997 to January, 1998. Unlike the 1991–92 El Niño, the 1997–98 El Niño was followed not by a period of average spiciness, but by a very strong negative spiciness signal (dark blue), which could be indicative of a major upwelling event.

## Mitrocoma cellularia *distribution*

The population of *Mitrocoma cellularia* experienced a pronounced increase (Fig. 2A) that coincided with the El Niño signatures found in the ROV CTD physical data (Fig. 1). Both the onsets of the population increase and the elevated spiciness signal from the 1991 to 92 El Niño were first observed in June/July, 1991. Medusa abundance decreased in February of 1992. The high spiciness signals also began to decrease at this same time. High medusa abundance in 1992 reached from the surface waters down to >500 m, with a peak abundance of 36 ind. $h^{-1}$, centered at 300–400 m in November, 1991. From 1992 to 97, there were low numbers of *M. cellularia* in the midwaters, with a few medusae appearing in summer/fall. High abundance of medusae recurred in June/July, 1997 (Fig. 2A) at the same time that the spiciness signal from the 1997 to 98 El Niño was also first observed. Medusa abundance declined in November. In 1997–98, the *M. cellularia* population was abundant from the surface waters down to 900 m, with peak abundance in October, 1997 of 52 ind. $h^{-1}$, centered at 600–700 m. The decrease of medusae in 1997 was correlated with the spiciness decrease, as it was in 1991–92. Although the onset of the *M. cellularia* blooms in the bay closely matched the onset of the spiciness signals, the peaks in abundance for both the 1991–92 and 1997–98 blooms did not correspond to the beginning of the spice signal but instead they followed the onset of the spice signal by several months.

## Colobonema sericeum *distribution*

The vertical distribution of *Colobonema sericeum* was very different from that found for *M. cellularia*. *C. sericeum*'s vertical distribution (Fig. 2B) appeared to be inversely related to high spice events (Fig. 1). In years with a high spiciness signal, 1991–92 and 1997–98, *C. sericeum* were scarce, whereas during the years in-between these El Niño events *C. sericeum* were abundant. March/April, 1993 began a 3-year period of relatively high abundance. Peaks in abundance were observed from September to November in both 1993 and in 1994, while 1995 showed an early peak in February, with over 20 medusae observed $h^{-1}$. By July, 1995 the numbers of *C. sericeum* had dropped to more typical levels. With the onset of the 1997–98 El Niño, *C. sericeum* populations dropped to very low levels in the mesopelagic waters of the bay. Over the entire nine years of the study, the depth range of *C. sericeum* varied little, with the population centered around 400–500 m.

## *Organismal T-S-DO plots*

Analysis of temperature, salinity and dissolved oxygen (T-S-DO) plots provide information on the physical parameters of the water mass in which the medusae are associated (Fig. 3). The axis scales represent the minimum and maximum values the medusae are subject to in their natural environment.

The T-S-DO bounding values for *M. cellularia* were: temperature, 4.21–14.09 °C (mean = 7.36, SD = 1.69); salinity, 32.39–34.70 (mean = 34.09, SD = 0.23); dissolved oxygen, 0.30–6.75 ml $l^{-1}$ (mean = 1.55, SD = 1.23). The bounding values for *C. sericeum* were: temperature, 3.78–7.64 °C (mean = 6.49, SD = 0.57); salinity, 34.16–34.78 (mean = 34.16, SD = 0.09); dissolved oxygen, 0.07–3.42 ml $l^{-1}$ (mean = 0.90, SD = 0.43). Thus *M. cellularia* was found to occupy a wide range of temperature and dissolved oxygen values. Conversely, *C. sericeum* was found in a narrow range of temperature and dissolved oxygen values. The temperature, salinity and dissolved oxygen values found with the two medusae were compared with a Mann–Whitney $U$ test to examine the hypothesis that the means were equal for both species. The resulting $P$ values: temperature (°C) was found to be significantly different ($P < 0.001$), salinity was not found to be different ($P = 0.239$), and dissolved oxygen (ml $l^{-1}$) was found to be significantly different ($P < 0.001$). The standard deviations of the three physical parameters were consistently much greater for *M. cellularia* than for *C. sericeum*. In summary, *M. cellularia* could be found in waters almost twice as warm, of lower salinity, and with twice the dissolved oxygen than *C. sericeum*.

## Discussion

CTD data recorded from nine years of ROV dives at the MWTS site were used to characterize the meso-pelagic environment of Monterey Bay. These data, when examined with the sensitive calculation of spiciness, permit visualization of the El Niño water masses as these enter the bay (Fig. 1). The quick onset of both the 1991–92 and 1997–98 El Niños was observed. In addition to the quick onset, spice signals were found down to depths of 800 m and these signals persisted for many months. Further analysis is needed to determine the origin of these intruding water masses, as high-spiciness water masses could have been transported northward from off southern California or could represent an on-shore transport of waters residing initially far off the coast of central California. A similar temperature signal was observed down to 200 m (the limit of the sensors) from mooring data during the 1991–92 El Niño (Pennington & Chavez, 2000). These results show that the spiciness metric can be of great value in finding and visualizing subtle changes in water masses, which can be biologically important.

Patterns of change in the vertical distribution and abundance of both species are associated, both directly and inversely, with the two El Niño events measured (Fig. 2). *Mitrocoma cellularia* vertical distribution and abundance closely mirrored the spiciness signals of the two El Niño events (Fig. 2A). In both 1991–92 and 1997–98 the *M. cellularia* population had a sudden increase in numbers as well as a widening of depth range. During the El Niño events, the depth range of the population extended from the surface waters down to 800 m. Conversely, *Colobonema sericeum* vertical distribution and abundance were inversely related to the El Niño events (Fig. 2B). During the two El Niño events, *C. sericeum* vertical abundance dropped to very low levels. In the years between the two events, *C. sericeum* was found in high numbers.

A comparison of the physical properties of the water in which the two medusae were found was performed to ascertain the possibility of physiological differences which might lead to the differing responses to the El Niño events. The ROV CTD-DO measurements which corresponded with the individual medusae observed were used to construct T-S-DO bubble plots of abundance related to the various physical parameters. *M. cellularia* showed a much wider tolerance for temperature, salinity and dissolved oxygen ranges than did *C. sericeum* (Fig. 3). Although the means for the three parameters were similar between the two

medusae, the amount of variation exhibited by *M. cellularia* proved to be large enough for temperature and dissolved oxygen to be significantly different from each other. *M. cellularia* was found over a much wider range of physical values than *C. sericeum*. A large part of this variation was due to the fact that *M. cellularia* was found over a wider range of depths (surface to 900 m) compared with *C. sericeum* (200–700 m). These results indicate there may be distinct physiological differences in the two medusae with respect to their preferred habitat and their tolerance of changing water properties.

A number of hypotheses for the different distribution patterns of the two medusae can be considered. The first, a tolerance hypothesis, supposes that the medusae are limited by their physiological tolerance to various physical parameters. This premise has some support from the T-S-DO plots. *C. sericeum* occured within a much narrower range of salinity, temperature and dissolved oxygen values than did *M. cellularia*. *M. cellularia* was not found in the midwaters of Monterey Bay all year long, yet its observed T-S-DO ranges should allow it to effectively live in the deep waters of the bay nearly year-round. It is not known what affect changes in T, S and DO would have on the reproduction and survivability of these species. Experimental data are needed to assess the relative importance of each of these factors.

Other researchers have found that variation in physical parameters can have a pronounced effect on the reproductive success of jellyfish. Purcell et al. (1999) found that changes in temperature and salinity had significant affects on ephyra and polyp production in the scyphozoan *Chrysaora quinquecirrha*. Increasing temperature, in combination with high salinity, provided for increased ephyra production, while decreasing temperature delayed the strobilation event. Salinity was also important, with more ephyra and polyps produced at intermediate salinities than at the low or high end. Dawson et al. (2001) found that high temperatures associated with the 1997–98 El Niño caused a prolonged crash of the rhizostome *Mastigias* sp. in the marine "Jellyfish Lake" of Palau. Both reproduction and survivability of the medusae and the scyphistomae decreased dramatically, partly due to bleaching of the zooxanthellae in the polyps. Only after the waters started to cool down did the population start to rebound.

A second possible explanation for the differences in patterns of abundance is related to transport, with medusae carried into or out of the bay with the intruding water masses, as was suggested for the scyphomedusa *Chrysaora fuscescens* during the 1991–92 El Niño (Graham, 1994). In both 1991–92 and 1997–98 El Niños, the *M. cellularia* bloom onset corresponded with the beginning of the spice signal (intrusion event), however peaks of abundance were not observed for several months thereafter. This delay could be explained if the jellies were invected and then reproduce locally in the months following their arrival in the bay. *M. cellularia* is a medusa with a benthic polyp stage which could settle from free-spawned eggs and sperm and reproduce additional medusae asexually within the 3–4 month period. Alternatively, the medusae could have been produced from resident polyps in the bay, which responded to the physical changes in the water by increasing medusa production, as was found for *C. quinquecirrha* (Purcell et al., 1999). Barring a detectable genetic signature which could delineate between *M. cellularia* of southern or local origin, these two possibilities would be very difficult to separate. The transport hypothesis also best explains the *C. sericeum* data reported in this study. These populations declined during the El Niño events, but that may simply represent a northward transport of medusae.

There are also potentially significant life cycle differences between these two medusae. *C. sericeum* exhibits holoplanktonic, direct development, while *M. cellularia* has a benthic polyp form in its life cycle. Little is known about reproduction and development time in *C. sericeum*. Without a resident benthic polyp stage, it may be that it would take *C. sericeum* longer to re-colonize the bay if the population was advected away, regardless of how quickly this holoplanktonic species can reproduce. Whereas the benthic reservoir of *M. cellularia* polyps might enable this species to not only remain in the area, but react quickly to advantageous environmental change. More research is needed to understand the importance of life cycles in relation to population fluctuations.

Although the reasons for these different patterns of distribution are not well understood, it is clear that there are pronounced differences in the gelatinous community with respect to episodic events such as El Niño. More *in situ* research must be done, at even greater depths, if we are to fully understand the link between large scale oceanic and atmospheric events and the ecology of deep sea organisms.

## Acknowledgements

All data are from Bruce Robison's midwater ecology program at MBARI. I would like to thank Brian Schlining for calculating and graphing spiciness. Bruce Robison, Kim Reisenbichler and Rob Sherlock provided assistance with data collection and analysis. N. Jacobsen, K. Rodgers-Walz, K. Schlining and the rest of the past and present video lab staff annotated the ROV dives. K. A. Shirley, S. H. D. Haddock, W. M. Hamner, B. H. Robison, C. E. Mills, J. E. Purcell and an anonymous reviewer provided valuable comments on earlier versions which contributed greatly to this manuscript. Many thanks to the skilled work of the R/V *Point Lobos* crew and the pilots of the ROV *Ventana*. This research was supported by the David and Lucille Packard Foundation through the Midwater Ecology Group at MBARI and by W. M. Hamner at the University of California, Los Angeles.

## References

Ainley, D. G., R. L. Veit, S. G. Allen, L. B. Spear & P. Pyle, 1995. Variations in marine bird communities of the California Current, 1986–1994. CALCOFI Rep. 36: 72–77.

Arai, M. N. & A. Brinckmann-Voss, 1980. Hydromedusae of British Columbia and Puget Sound. Can. Bull. Fish. aquat. Sci. 204: 1–192.

Bailey, K. M., J. F. Piatt, T. C. Royer, S. A. Macklin, R. K. Reed, M. Shima, R. C. Francis, A. B. Hollowed, D. A. Somerton, R. D. Brodeur, W. J. Ingraham, P. J. Anderson & W. S. Wooster, 1995. ENSO events in the northern Gulf of Alaska, and effects on selected marine fisheries. CALCOFI Rep. 36: 78–96.

Barber, R. T. & F. P. Chavez, 1983. Biological consequences of El Niño. Science 222: 1203–1210.

Barber, R. T. & F. P. Chavez, 1986. Ocean variability in relation to living resources during the 1982-83 El Niño. Nature 319: 279–285.

Breaker, L. C. & W. W. Broenkow, 1994. The circulation of Monterey Bay and related processes. Oceanogr. mar. Biol. 32: 1–64.

Chavez, F. P., 1996. Forcing and biological impacts of onset of the 1992 El Niño in central California. Geo. Phys. Res. Let. 23: 265–268.

Chavez, F. P., P. G. Strutton, G. E. Friederich, R. A. Feely, G. C. Feldman, D. G. Foley & M. J. McPhaden, 1999. Biological and chemical response of the equatorial Pacific Ocean to the 1997–98 El Niño. Science 286: 2126–2131.

Culik, B. M. & G. Luna-Jorquera, 1997. Satellite tracking of Humboldt penguins (*Spheniscus humboldti*) in northern Chile. Mar. Biol. 128: 547–556.

Dawson, M. N., L. E. Martin & L. K. Penland, 2001. Jellyfish swarms, tourists and the Christ-child. Hydrobiologia 451 (Dev. Hydrobiol. 155): 131–144.

Dorn, M. W., 1995. The effects of age composition and oceanographic conditions on the annual migration of Pacific whiting, *Merluccius productus*. CALCOFI Rep. 36: 97–105.

Flament, P., 1986. Subduction and fine structure associated with upwelling filaments. Ph.D. dissertation, University of California, San Diego.

Gibbs, H. L. & P. R. Grant, 1987. Ecological consequences of an exceptionally strong El Niño event on Darwin's finches. Ecology 68: 1735–1746.

Glynn, P. W., 1988. El Niño-Southern Oscillation 1982–1983: nearshore population, community and ecosystem responses. Ann. Rev. Ecol. Syst. 19: 309–345.

Graham, W. M., 1994. The physical oceanography and ecology of upwelling shadows. Ph.D. dissertation, University of California, Santa Cruz.

Hammann, M. G., J. S. P. Nayar & O. S. Nishizaki, 1995. The effects of the 1992 El Niño on the fisheries of Baja California, Mexico. CALCOFI Rep. 36: 127–133.

Jackett, D. R. & T. J. McDougall, 1985. An oceanographic variable for the characterization of intrusions and water masses. Deep Sea Res. 32: 1195–1207.

Keckler, D., 1995. Surfer for Windows, Version 6 User's Guide. Golden Software, Inc., Golden, CO, 480 p.

Kramp, P. L., 1968. The hydromedusae of the Pacific and Indian Oceans. Sect. II and III. Dana-Rep. Carlsberg Found. 72: 1–200.

Laurie, W. A., 1990. Population biology of marine iguanas (Amblyrhychus cristatus). I. Changes in fecundity related to a population crash. J. anim. Ecol. 59: 515–528.

Lenarz, W. H., F. B. Schwing, D. A. Ventresca, F. Chavez & W. M. Graham, 1995. Explorations of El Niño events and associated biological population dynamics off central California. CALCOFI Rep. 36: 106–119.

Lynn, R. J., F. B. Schwing & T. L. Hayward, 1995. The effect of the 1991–1993 ENSO on the California current system. CALCOFI Rep. 36: 57–71.

Mills, C. E., 1983. Vertical migration and diel activity patterns of hydromedusae: studies in a large tank. J. Plankton Res. 5: 619–635.

Mills, C. E., 1993. Natural mortality in NE Pacific coastal hydromedusae: Grazing predation, wound healing and senescence. Bull. mar. Sci. 53: 194–203.

Mills, C. E., 1995. Medusae, siphonophores and ctenophores as planktivorous predators in changing global ecosystems. ICES J. mar. Sci 52: 575–581.

Pennington, J. T. & F. P. Chavez, 2000. Seasonal fluctuations of temperature, salinity, nitrate, chlorophyll and primary production at station H3/M1 over 1989–1996 in Monterey Bay, California. Deep Sea Res. II 47: 947–973.

Purcell, J. E., J. R. White, D. A. Nemazie & D. A. Wright, 1999. Temperature, salinity and food effects on asexual reproduction and abundance of the scyphozoan Chrysaora quinquecirrha. Mar. Ecol. Prog. Ser. 180: 187–196.

Purcell, J. E., D. L. Breitburg, M. B. Decker, W. M. Graham, M. J. Youngbluth & K. A. Raskoff, 2001. Pelagic cnidarians and ctenophores in low dissolved oxygen environments. In Rabalais, N. N. & R. E. Turner (eds), Coastal Hypoxia: Consequences for Living Resources and Ecosystems. American Geophysical Union. Coastal and Estuar. Stud. 58: 77–100.

Ramp, S. R., J. L. McClean, C. A. Collins & A. J. Semtner, 1997. Observations and modeling of the 1991–1992 El Niño signal off central California. J. Geophys. Res. 102: 5553–5582.

Raskoff, K. A., 1998. Distributions and trophic interactions of mesopelagic hydromedusae in Monterey Bay, CA: In situ studies with the MBARI ROVs Ventana and Tiburon. Ocean Sciences - San Diego, CA. Eos, Transactions, American Geophysical Union. (abstracts) 79: OS1.

Robison, B. H., 1993. Midwater research methods with MBARI's ROV. Mar. Tech. Soc. Jour. 26: 32–39.

Robison, B. H., K. R. Reisenbichler, R. E. Sherlock, J. M. B. Silguero & F. P. Chavez, 1998. Seasonal abundance of the siphonophore, Nanomia bijuga, in Monterey Bay. Deep Sea Res. II 45: 1741–1751.

Schlining, B., 1999. Seasonal intrusions of equatorial waters in Monterey Bay and their effects on mesopelagic animal distributions. Masters thesis, Moss Landing Marine Laboratories.

Schmitt, R. W., 1999. Spice and the demon. Science 283: 498–499.

Silguero, J. M. B. & B. H. Robison, 2000. Seasonal abundance and vertical distribution of mesopelagic calycophoran siphonophores in Monterey Bay, CA. J. Plankton Res. 22: 1139–1153.

SPSS, Inc., 1998. SigmaPlot 5.0 User's Guide. SPSS Science Marketing Department, Chicago, IL, 448 p.

Widder, E. A., S. A. Bernstein, D. F. Bracher, J. F. Case, K. R. Reisenbichler, J. J. Torres & B. H. Robison, 1989. Bioluminescence in the Monterey Submarine canyon: image analysis of video recordings from a mid-water submersible. Mar. Biol. 100: 541–551.

Wrobel, D. & C. Mills, 1998. Pacific Coast Pelagic Invertebrates: a Guide to the Common Gelatinous Animals. Sea Challengers/Monterey bay Aquarium, Monterey, 108 pp.

Zar, J. H., 1996. Biostatistical Analysis. Simon and Schuster, Upper Saddle River, 662 pp.

*Hydrobiologia* **451**: 131–144, 2001.
© 2001 *Kluwer Academic Publishers.*

# Jellyfish swarms, tourists, and the Christ-child

Mike N Dawson[1,2], Laura E. Martin[1,2] & Lolita K. Penland[2]

[1]*Department of Organismic Biology, Ecology, and Evolution, University of California, P.O. Box 951606, Los Angeles, CA 90095-1606, U.S.A.*
[2]*Coral Reef Research Foundation, Box 1765, Koror, PW 96940, Republic of Palau*
*E-mail: mikend@ucla.edu*

*Key words:* bleaching, El Niño, marine lake, *Mastigias*, Palau, perturbation, tourism

## Abstract

One of the most remarkable sights in the Western Pacific is a perennial swarm of 1.5 million golden medusae (*Mastigias* sp.) crowded into a land-locked marine lake in Palau, Micronesia. This 'Jellyfish Lake' became a popular off-gassing stopover for SCUBA divers and a destination in its own right for non-diving tourists in the mid-1980s. Since then, tourism in Palau has boomed, increasing 500% between 1986 and 1997. However, in December 1998, the golden-medusae disappeared. Apart from patchy occurrences between December 1998 and April 1999, the medusae have since been absent from the lake. Field measurements, including temperature and salinity depth profiles, *Mastigias* medusae population sizes, and the distribution of scyphistomae, in 'Jellyfish Lake' between 1979 and 1999 were integrated with laboratory-based experiments on the effects of salinity, temperature, sunscreen and zooxanthellae enrichment on *Mastigias* scyphistomae or medusae. These studies indicated that the disappearance of medusae was due to physical changes in lake structure, including a substantial increase in temperature, initiated by the 1997–98 El Niño. Here, we describe these studies, the changes in Jellyfish Lake and their probable influence on the *Mastigias*. We further elucidate the changes in Jellyfish Lake by reference to coincident changes in three other 'jellyfish lakes' in Palau: Big Jellyfish Lake, Clear Lake and Goby Lake.

## Introduction

Several tropical marine lakes in Palau harbor immense perennial populations of two scyphozoan jellyfish: the golden-jellyfish, *Mastigias* sp., and the moon-jellyfish, *Aurelia* sp. (Hamner & Hauri, 1981; Hamner, 1982; Hamner et al., 1982). The golden medusae are behaviorally, morphologically and physiologically unique, the manifestations of many thousands of years of evolution in isolated ecosystems (Hamner & Hauri, 1981; Hamner, 1982; Hamner et al., 1982; Muscatine and Marian, 1982; Hamner & Hamner, 1998). A similar situation must be true for the moon-jellyfish although published data are lacking. These unique adaptations, most notably precise diurnal horizontal migrations by *Mastigias*, the high densities and perennial presences of both jellyfish species, and the closed nature of these ecosystems, in which the jellyfish are the top planktivores, make the marine lakes extremely attractive sites for a range of ecological and evolution-

ary investigations. These same features also make the lakes very attractive to tourists and the media.

Since the marine lakes were brought to the attention of the general public (Hamner, 1982), many nature and SCUBA magazines, radio and television shows, and even an IMAX production, 'The Living Sea', have featured the marine lakes in Palau. This publicity has established Jellyfish Lake, formally named 'Ongeim'l Tketau' (roughly 'fifth lake' in English), as one of the most popular snorkeling sites in the tropical Pacific (Hamner & Hamner, 1998; see also PVA, 1999). Between 1986, when Ongeim'l Tketau first was incorporated into dive tours (Etpison, 1997), and 1997, tourism in Palau increased 500% (PCS, 1999a). Almost 75 000 tourists visited Palau in 1997 and most likely between half and three-quarters of them visited Ongeim'l Tketau (Hamner, 1994; PCS, 1999a; pers. obs.). In recent years, tours have been expanded to incorporate other marine lakes that contain similar populations of *Mastigias*. These lakes, which include Goby Lake and Big Jellyfish Lake, have been host to several thousands of visitors (pers. obs.).

Tourists are attracted to the lakes by a strange mix of fact and fiction purveyed by tourist-oriented publications and other media. The most popular story is that *Mastigias* evolved into perennial sting-less farmers of an edible algal crop during millions of years of confinement in the lakes (e.g. Barbour, 1990). In reality, however, these *Mastigias* can sting humans mildly and kill microscopic prey, they are symbiotic with zooxanthellae but their feeding ecology is hardly agricultural (Muscatine et al., 1986), and they have been confined in the lakes probably for less than 20 000 years (Hamner et al., 1982). The perennial abundance of *Mastigias*, however, is a distinguishing feature of marine lake populations and has supported tourism year-round. Measurements made between 1979 and 1997 indicate that the lakes, and perhaps jellyfish populations, have been remarkably stable on time scales of months to millenia (Hamner et al., 1982; Hamner & Hamner, 1998). In the two decades following the first measurements, there has been only one report to the contrary: in 1987, many *Mastigias* medusae disappeared (Etpison, 1997). However, their disappearance was attributed to disturbance of the toxic monimolimnion by divers using SCUBA rather than to natural perturbations (Etpison, 1997).

This view of long-term ecosystem stability was altered catastrophically in fall, 1998, when the population of *Mastigias* medusae went into an unprecedented decline and disappeared. The disappearance was attributed variously to the 1997–98 El Niño (or translated, 'the Christ-child'), to stagnation of the lake because conduits that normally allowed tidal flux were somehow blocked, or to tourists who may have stolen or eaten the jellyfish, poisoned the medusae with sunscreen, or polluted the lake with urine.

During summer in 1999, we investigated a number of the hypotheses. Field observations and laboratory experiments indicated several factors quite likely contributed to the decline and disappearance of golden jellyfish from 'Jellyfish Lake' among which high water temperature may have been predominant. Ultimately, the disappearance probably can be traced to physical perturbations in lake structure that were initiated by the 1997–98 El Niño event.

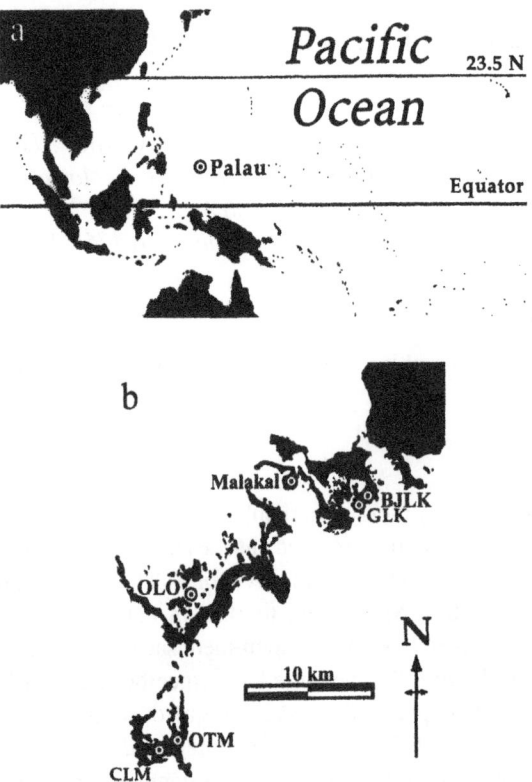

*Figure 1.* (a) Location of Palau in the western Pacific and (b) the marine lakes and other locations mentioned in the text. BJLK – Big Jellyfish Lake, Koror; GLK – Goby Lake, Koror; OLO – Ongael Lake, Ongael; CLM – Clear Lake, Mecherchar; OTM – Ongeim'l Tketau, Mecherchar (better known as "Jellyfish Lake", Hamner et al., 1982).

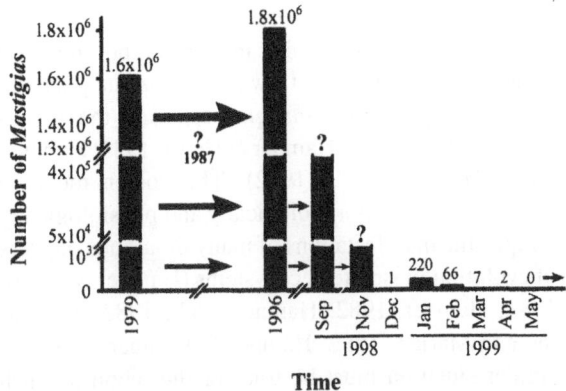

*Figure 2.* Estimates of *Mastigias* medusae population size in Ongeim'l Tketau between 1979 and 1999. Mean estimated population (± standard deviation) are shown for each quantitative sampling event. Qualitative observations were made in September and November, 1998 (L. J. Bell, M.N Dawson, L.E. Martin, B. Yates) for which population size is estimated from the size of aggregations, medusae size, and visual estimates of population size.

## Materials and methods

### Brief description of five marine lakes

The biological, chemical and physical characteristics of Ongeim'l Tketau, Big Jellyfish Lake, Clear Lake and Goby Lake (Figure 1) have been described by Hamner and colleagues (Hamner & Hauri, 1981; Hamner, 1982; Hamner et al., 1982; Hamner & Hamner, 1998). All four lakes are meromictic, or partly mixed, consisting of an upper mixolimnion floating above a lower monimolimnion. The salinity and temperature of the mixolimnion generally increase gradually with depth until a sharp increase in salinity and decrease in temperature establish a robust pycnocline marked by dense purple-sulfur bacteria (Hamner et al., 1982; Hamner & Hamner, 1998). Dissolved oxygen, in contrast, decreases from its maximum near the surface to zero at the pycnocline, the monimolimnetic waters below the pycnocline are highly reducing. However, these four lakes can be differentiated on the basis of surface area, depth of the mixolimnion, maximum depth, shape, orientation, surrounding topography and somewhat different faunas (Hamner & Hamner, 1998; pers. obs.). Ongael Lake (OLO) is a smaller and shallower lake whose holomictic waters are similar physically to those in the adjacent lagoon; its fauna is depauperate, but also lagoon-like. *Mastigias* inhabit all five lakes.

### Field measurements and observations

From December, 1998, to December, 1999, 2 field-trips, approximately two weeks apart, were made to Ongeim'l Tketau (OTM) each month. During one trip, the lake was searched thoroughly, to depths of 6–8 m, for *Mastigias* sp. medusae by 3–4 snorkelers. All medusae found were measured. Medusae >10 mm were measured to the nearest one centimeter by flattening the exumbrella against a rule while in the lake; smaller medusae were measured under a dissecting microscope at the Coral Reef Research Foundation (CRRF), then preserved in DMSO+NaCl (Seutin et al., 1991; Dawson et al., 1998) for subsequent examination. During the second trip, 2 weeks later, the search for *Mastigias* medusae was repeated and vertical temperature, salinity and oxygen (TSO) profiles were measured using a YSI 85 meter (object, not distance). Additional trips to Ongeim'l Tketau using SCUBA were made during July–September, 1999, to search for the benthic scyphistoma stage of *Mastigias*. Scyphistomae were collected and taken to CRRF

where they were examined under a microscope for zooxanthellae, planuloids and apparent health prior to use in experiments (see below). During the final trip on 9 September, 1999, we assessed the tidal range on Ongeim'l Tketau using synchronized half-hourly measurements of tidal height in the lake and the adjacent lagoon for one half tidal cycle (see Hamner & Hamner, 1998).

Monthly field-trips were made to Big Jellyfish Lake (BJLK) and Goby Lake (GLK) during which TSO profiles were measured and the size distributions of *Mastigias* sp. were estimated by visual censuses. Visual censuses were made by 3–4 snorkelers swimming multiple transects across the length and width of each lake. At the entrance into each lake, each snorkeler closed their eyes, finned 5 strokes (along different transects), then opened their eyes and caught the first medusa seen. The bell diameter of that medusa was measured as above. Each snorkeler then repeated the blind swim to find another medusa to measure; efforts were made to minimize the temptation to look for animals nearest the surface by varying the line of sight between cycles. The cycle was repeated until all transects were finished. In addition to the visual survey, in August 1999, the *Mastigias* sp. populations in these lakes also were surveyed using a 0.1095 m², 1 mm² mesh, net, hauled vertically from the lake bottom or below the chemocline, depending which was shallower. These same methods were used to measure jellyfish populations and water column structure in Clear Lake, Mercherchar, during trips in fall 1998 and summer 1999. Also reported here are observations from Clear Lake (CLM) that have particular relevance to this paper

### Laboratory based experiments

*Mastigias* scyphistomae were collected by divers using SCUBA in OTM and GLK and by snorkelers in CLM. Aggregations of scyphistomae on their natural substrates were placed into rigid plastic 250–1000 ml jars full of lake water and transported to CRRF within 2 hours. At CRRF, the scyphistomae, still on their natural substrate, were transferred to 1 L plastic containers (diameter = height) of 1 $\mu$m filtered seawater adjusted to lake salinity and temperature. They were maintained this way for 24 h at which time they were examined under a dissecting microscope (magnification 7×–40×) and scored in five ways: on a four-point scale (0–3) for the zooxanthellae density in their tissues (see below), a Boolean (true/false) scale for strobilation and for the presence of planuloids, and simple counts of the numbers of planuloids and strobilae present.

*Mastigias* scyphistomae collected in OTM rarely contained visible zooxanthellae (cf. scyphistomae populations in other lakes which always had some zooxanthellae). Therefore, prior to experimentation, these scyphistomae were infected with zooxanthellae by feeding to them homogenized tissue from local *Mastigias papua* Lesson and leaving them in running seawater (~29 °C, 34‰) and natural shade for up to several weeks. One batch of scyphistomae apparently was infected by zooxanthellae from anthozoans or *Tridacna* situated upstream.

Zooxanthellae density was assessed by examining scyphistomae under a dissecting microscope at $10\times$-$40\times$ magnification. The area of the calyx and oral disc pigmented by zooxanthellae and the intensity of that pigmentation were used to score all, n, scyphistomae in each replicate on a four-point (0–3, d) relative scale. These scores then were used to calculate an index, ZD, of the change in the mean zooxanthellae density in each replicate at time $t$ relative to the mean density at the start of the experiment.

$$\mathrm{ZD}_t = \left( \sum_{i=1}^{n} d_{it}/n_t \right) \bigg/ \left( \sum_{i=1}^{n} d_{i0}/n_0 \right).$$

In the special case of an initial density equal to zero, i.e. when homogenized *M. papua* tissues were fed to OTM scyphistomae immediately preceding experimental manipulation, the change in mean zooxanthellae density was calculated as

$$\mathrm{ZD}_t = \left( \sum_{i=1}^{n} d_{it}/n_t \right) - \left( \sum_{i=1}^{n} d_{i0}/n_0 \right).$$

Throughout all experiments, scyphistomae were scored by the five methods described above. Scyphistomae were illuminated by 'coral' aquarium lights but were not fed zooplankton because coloured food in the gastric cavity made it difficult to identify zooxanthellae and, therefore, compromised the assessment of zooxanthellae density. No animals were used for more than one experiment.

### Effect of salinity on scyphistomae-zooxanthellae

OTM scyphistomae infected with zooxanthellae were divided equally into 10 separate vessels, duplicates were acclimated to five salinities ranging from 20‰ to 45‰ (original salinity, ~32.5‰) over a period of 1 week and then maintained at target salinities for another 11 days. Seawater was adjusted to the appropriate salinity by dilution with rainwater or addition of aquarium salts in steps of about 1.5‰ per day. Seawater was replaced approximately every fourth day. Scyphistomae in all salinity treatments were kept at the same temperature (means 30.4–31.1 °C; s.d., 1.06–1.32 °C; ANOVA, df = 4,90, $p = 0.7866$). Scyphistomae were scored daily.

### Effect of zooxanthellae enrichment on scyphistomae-zooxanthellae

Scyphistomae freshly collected from OTM were fed homogenized *M. papua* tissues and left in ambient conditions for several days. Control groups were treated similarly, except they were not fed homogenized *M. papua* tissues. One$\mu$m-filtered, 34‰ seawater was replaced every other day. The scyphistomae were scored on days 0 and 4.

### Effect of temperature on scyphistomae-zooxanthellae

Ongeim'l Tketau scyphistomae infected with zooxanthellae were divided equally among 10 one-liter containers and acclimated to different temperature treatments. Eight of the ten containers were transferred from ambient (~29 °C) to a 31 °C waterbath and then, at 2 h intervals, transferred sequentially, in pairs, to approximately 32 °C, 34 °C, and 35 °C waterbaths, until one pair of containers was at each of the five temperatures. The scyphistomae were maintained at these temperatures for 2 weeks. Polyps were scored daily and 34‰ surface seawater at the appropriate temperature was replaced every second day.

The experiment subsequently was repeated using freshly collected, azooxanthellate, scyphistomae which were fed *M. papua* tissues immediately before they were acclimated, stepwise, to the different temperature treatments. The scyphistomae were scored daily and 1 $\mu$m-filtered, 34‰ seawater was replaced every other day for two weeks.

In a third experiment, Goby Lake scyphistomae were divided equally between 10 one-liter containers of filtered, 27‰ seawater and placed into a 28 °C waterbath at CRRF. Filtered seawater prevented infection of scyphistomae with allochthanous sources of zooxanthellae. Two days later, the scyphistomae were acclimated to different temperature treatments and observed for 2 weeks. Scyphistomae were scored daily and seawater was replaced every second day for 2 weeks.

## Effect of temperature on medusae

Medusae, ≤15 mm (bell diameter) were collected from BJLK and GLK in the same manner as polyps. Immediately on arrival at CRRF, medusae were distributed equally among 10 five-liter containers (9 medusae per container) and pairs of containers acclimated incrementally to five temperatures between 28 °C and 36 °C (as described for scyphistomae, above). Medusae were illuminated from 6 a.m. to 6 p.m. using household fluorescent lighting situated several meters from the aquaria. Except for monitoring periods, medusae were not illuminated between 6 p.m. and 6 a.m. Medusae were scored at 6-hour intervals, for up to 7 days, for the rate and strength of bell contractions and for their position in the water column. Medusae were considered terminally damaged when observed on two consecutive occasions pulsing very poorly on the container's bottom; these medusae did not recuperate if transferred to cooler temperatures.

## Effect of sunscreen on medusae

Medusae collected as above were distributed evenly among four 3.5 L containers. Sunscreen (Coppertone 'Water Babies' UVA/UVB sunblock lotion, SPF 45, or BullFrog 'the QuickGel' SPF 36) was added to three containers in concentrations of $10^{-4}$, $10^{-6}$, and $10^{-9}$ (g ml$^{-1}$). Sunscreen was not added to the fourth, control, container. All four containers were kept in the open air at 26–28 °C under a canopy that afforded protection from rain and direct sunlight. The health of the medusae was assessed at 6-hour intervals for 5 days. This experiment was repeated three times, using 32 medusae from BJLK and OLO and 16 medusae from GLK.

## Statistics

Statistical analyses were completed using SYSTAT v.6.0 for Windows 3.1.

# Results

## Field measurements and observations

In September, 1998, a large population of *Mastigias* medusae of all sizes was present in OTM although these medusae possibly were slightly fewer and, on average, larger than usual (Dawson & Martin, pers. obs.). By November, 1998, there were fewer large and many small medusae (L. J. Bell and B. Yates, pers. comm.). On 14th December, 1998, a thorough search

*Table 1.* Zooxanthellae content and reproductive condition of *Mastigias* scyphistomae in three marine lakes, Summer 1999

| Percentage of polyps that were | | Ongeim'l Tketau[1] | Goby Lake[1] | Clear Lake[1] |
|---|---|---|---|---|
| Producing planuloids | % | 7 | 1 | 9 |
| Zooxanthellate | % | ~2 | 89 | 45 |
| | ZD | 0.02 | 1.10 | 0.48 |
| Strobilae | % | 0 | 9 | 0 |

All counts were made on polyps kept in filtered seawater for 24 h after collection. Ongeim'l Tketau scyphistomae were collected on several different occasions between 26 Jul. and 4 Sep., 1999, n = 1,215.
Goby Lake polyps were collected on 12 Jan. 1999, $n = 305$.
Clear Lake polyps were collected on 26 Aug. 1999, $n = 168$.
ZD, zooxanthellae density index.

throughout the lake revealed just one, 20 mm, *Mastigias* (Martin & Dawson, pers. obs.). Subsequently, medusae were encountered in small and declining numbers between January and April 1999 (Fig. 2). These medusae were all ≤1 cm, swimming normally and had a normal complement of zooxanthellae. There apparently was no strobilation of medusae after April 1999. Scyphistomae in OTM in summer 1999, found between approximately 6 and 14 m depth, were not strobilating and appeared azooxanthellate under the microscope (L. J. Bell, pers. comm. and Table 1).

Other organisms in Ongeim'l Tketau also exhibited signs of stress during summer 1999. For example, more than 50% of the endemic anemones, *Entacmaea medusivora* (Fautin & Fitt, 1991), appeared flaccid, their tentacles largely hidden within invaginated oral discs, in contrast to their usual upright posture with tentacles extended. Many moon-jellyfish, *Aurelia* sp., were tattered and torn, inverted, or misshapen, although the population seemed no smaller than usual. Also, for the first time in four years, we saw a dead goby, *Acentrogobius janthinopterus* (Bleeker), floating at the surface of the lake.

The disappearance of *Mastigias* medusae in December, 1998, coincided with the presence of an abnormally hot and salty mixolimnion in Ongeim'l Tketau (Fig. 3a). The temperature of the mixolimnion continued to rise after December, 1998, peaking in April, 1999. It has remained high since. In contrast, the mean salinity of the mixolimnion has declined from its high in December 1998, although it is still above the mean of previous measurements. For most of 1999, a steep pycnocline at about 6 m depth maintained a hot and high salinity water mass at depth despite tidal flux involving shallower waters (Hamner et al., 1982; Hamner & Hamner, 1998). In September, 1999, the tidal pattern in the lake was normal, being damped to approximately 40% (0.69 m) and delayed by about 2.5 h compared with the lagoon (see Hamner & Hamner, 1998).

Other 'jellyfish lakes' also changed. The *Mastigias* population in Goby Lake in December 1998 was unusually small and consisted solely of animals less than 6 cm. However, strobilation by zooxanthellate *Mastigias* scyphistomae, located between approximately 4 and 6 m depth, and growth of medusae restored a normal population of medusae in Goby Lake by mid-1999 (Fig. 4; Table 1). The reduced populations of *Mastigias* medusae in Goby Lake in December 1998 coincided with an unusually hot and salty mixolimnion (Fig. 3b). The subsequent recovery of the populations coincided with the return of the mixolimnion towards its pre-1998 state.

In Clear Lake, the size distribution of *Mastigias* was heavily skewed towards large medusae in August 1999. Thousands of medusae $\geq 12$ cm were present but we found only two medusae that were 2 cm and a few 6–8 cm bell diameter despite searching along transects, totaling approximately 2 km, throughout the lake. Although Clear Lake was not unusually hot, the mixolimnion was approximately 5‰ more saline than ever recorded previously. The shallowest scyphistomae that we found were at approximately 6 meters depth and these contained zooxanthellae but were not strobilating (Table 1).

Clear Lake also contained a large population of golden-*Mastigias* medusae in 1994, but typically (1979, 1995–97) has contained few or none of these medusae (Hamner et al., unpublished data). However, in May 1997, there was a small population (<1000 individuals) of white, apparently azooxanthellate, *Mastigias* medusae (Fig. 5), none larger than ~35 mm and most less than ~25 mm. Not one of ~150 medusae observed *in situ* had any pigmentation and thorough microscopical analyses, including staining with Fluorescent Brightener 28 (Sigma Chemicals), found no remnants of zooxanthellae in 11 preserved medusae (Fig. 5). Three days later, the population was in decline, with a decrease in density of jellyfish and with only animals $\geq 10$ mm showing no physical deterioration. We collected 40 medusae from this apparently senescent population and placed them in an aquarium where, despite the presence of an inoculum of zooxanthellae and successful feeding on *Artemia* nauplii, they died within 2 days.

Unlike the populations of medusae in Ongeim'l Tketau, Goby Lake, and Clear Lake, the population of *Mastigias* medusae in Big Jellyfish Lake was not affected obviously by changes in the mixolimnion during 1998–99 (Fig. 3c), nor were there any other obvious biotic changes. Both temperature and salinity have ameliorated since December 1998 although both still are above the means of pre-1998 measurements.

*Laboratory based experiments*

*Effect of salinity on scyphistomae-zooxanthellae*
A wide range of salinities and salinity changes, greater than any measured in Ongeim'l Tketau, allowed zooxanthellae increases of up to 160% (Fig. 6). Salinities below 25‰ generally caused the zooxanthellae density index to decrease (i.e. bleaching) whereas salinities as high as 45‰ permitted an increase in zooxanthellae density.

There was a weak positive but marginally non-significant correlation between salinity and survival of scyphistomae ($y = 0.0086x + 0.53$, $R^2 = 0.37$, $p = 0.062$). However, there was no correlation between survival and zooxanthellae density, indicating independence of these two effects.

*Effect of zooxanthellae enrichment on scyphistomae-zooxanthellae*
The zooxanthellae densities in OTM scyphistomae in the control and enriched treatments did not differ at the start of the experiment (*t*-test, df = 10, $p = 0.77$). However, 4 days later, zooxanthellae were significantly more dense in the enriched treatment than in the control (*t*-test, df = 10, $p = 0.05$; Fig. 7).

*Effect of temperature on scyphistomae-zooxanthellae*
Although there was some variation between experiments, the mean zooxanthellae density of infected scyphistomae, on average, remained stable in the 28.7 °C and 31.5 °C treatments. In contrast, in all experiments, mean zooxanthellae density decreased with each increment in temperature above 31.5 °C (Fig. 8a). Grouping the results from all experiments into 1 °C classes (beginning at 28.0 °C) and calculating their mean ranks indicated a significant negative effect of increased temperature on zooxanthellae density (Kruskal–Wallis Test, df = 4, $p = 0.04$).

A similar pattern was evident in scyphistomae infected with zooxanthellae immediately prior to temperature treatment. Greater increases in zooxanthellae density occurred at 28.7 °C and 31.5 °C than at 33.3

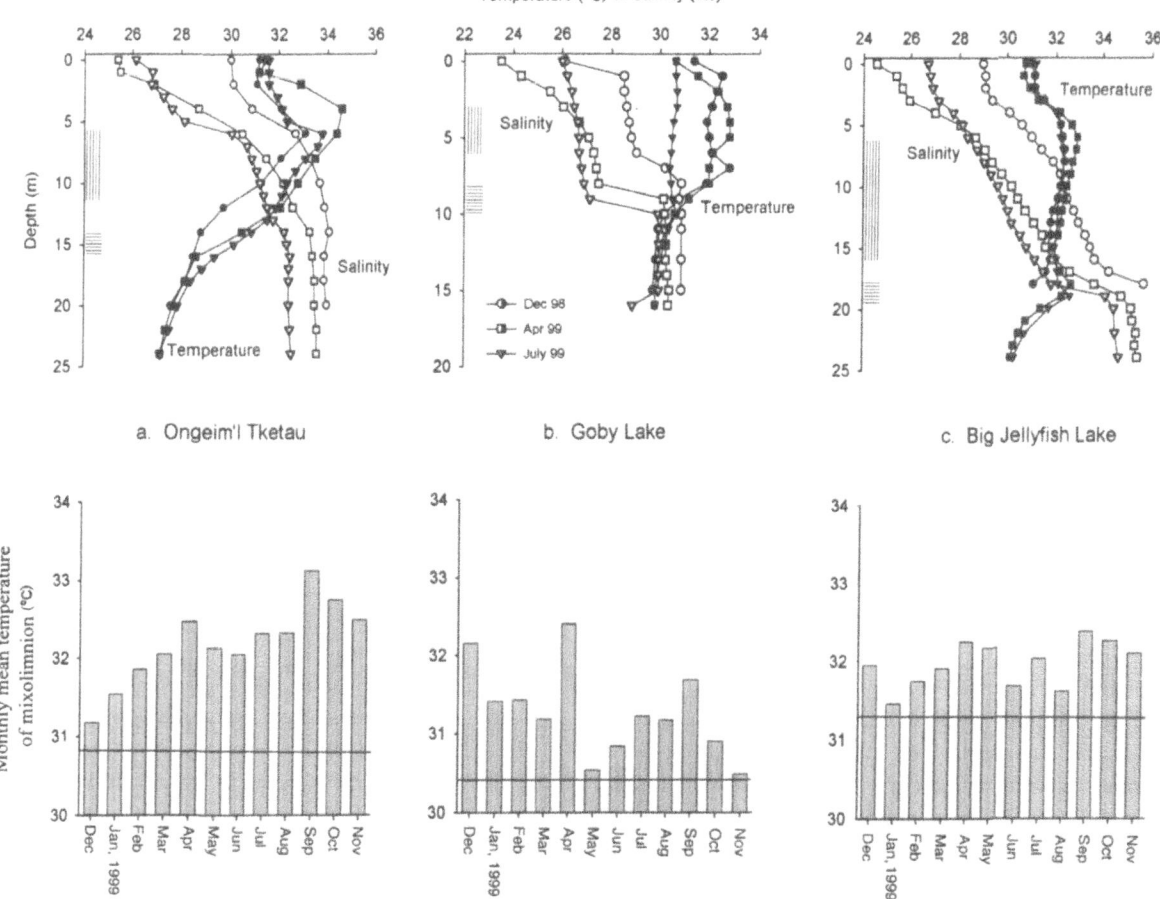

*Figure 3.* Physical structure of three meromictic marine lakes 1998–99. (a) Ongeim'l Tketau. (b) Goby Lake, Koror. (c) Big Jellyfish Lake, Koror. Top row: vertical profiles of the salinity and temperature of the lakes in December 1998 (circles), April 1999 (squares), and July 1999 (triangles). Vertical hatching against the left axis of each graph indicates the depth-range in which *Mastigias* polyps were found. Horizontal hatching against the left axes indicates the approximate depth of the chemocline separating the upper mixolimnion from the lower monimolimnion. Bottom row: monthly mean temperature of the mixolimnion (bars: average of monthly vertical temperature profiles excepting the upper 2 m which vary according to recent rainfall and tidal-cycles), compared to the mean of all such measurements made between 1978 and 1998 (line: Hamner et al., unpubl. data).

°C or 34.4 °C (Fig. 8b). However, this effect was not significant (Kruskal–Wallis Test, df = 3, $p$ = 0.8).

Scyphistomae transferred subsequently from the 33.8 °C and 35.2 °C treatments to 28.7 °C began to show increases in zooxanthellae density after two weeks. Scyphistomae transferred in the opposite direction began bleaching within 3 days.

The survival of scyphistomae generally declined with increasing temperature (Fig. 9). Grouping the results from all experiments into 1 °C classes (beginning at 28.0 °C) and calculating their mean ranks indicated a marginally non-significant negative effect of increased temperature on survival (Kruskal–Wallis Test, df = 4, $p$ = 0.055). There was no relationship between temperature and production of planuloids by

scyphistomae (ANOVA, df = 7, 39, $p$ > 0.5). The frequency of strobilation was greatest at 31.5 °C (12% of scyphopolyps), and less at both cooler (28.7 °C = 4%) and higher (33.3 °C = 5%; 34.4 °C = 4%) temperatures. Strobilation was not observed in either 33.8 °C or 35.2 °C treatments. No scyphopolyps were seen simultaneously strobilating and producing planuloids. Also, the frequency of strobilation and planuloid production in natural populations appeared to be inversely related (Table 1).

*Effect of temperature on medusae*

Increasing temperature significantly decreased the survivorship of small medusae from both Goby Lake and Big Jellyfish Lake (Fig. 10). This trend was exag-

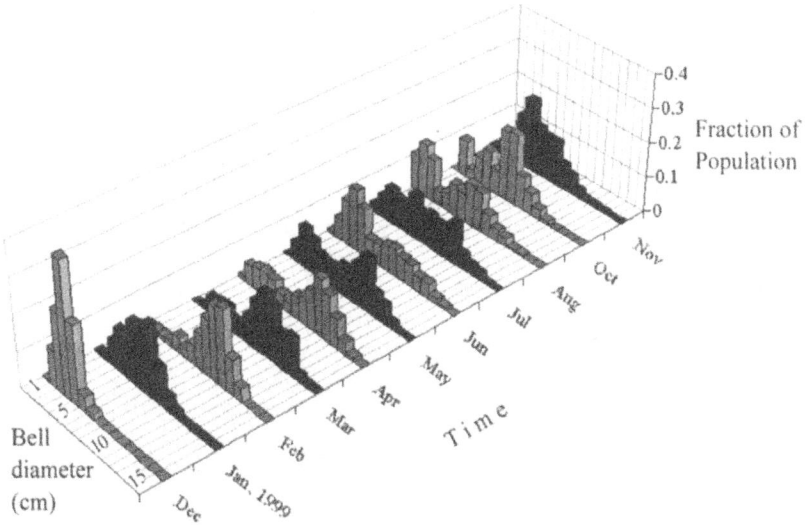

*Figure 4.* Size distribution of *Mastigias* medusae in Goby Lake, 1998–99. Compared to previous years, the mean size of *Mastigias* medusae in Goby Lake was unusually small in December 1998 (Hamner & Hauri, 1981; Dawson, pers. obs.). Strobilation and growth of medusae early in 1999 eradicated this discrepancy. Alternative months are shaded differently to distinguish adjacent months. Data were not collected in September 1999.

gerated at temperatures approaching and greater than 35 °C. Although this pattern is consistent between lakes, the magnitude of the effect is not. Medusae from Goby Lake died significantly more quickly than their Big Jellyfish Lake counterparts at all experimental temperatures (as demonstrated by 95% CI, Fig. 10), consistent with either container effects or other differences between these populations. Terminally damaged medusae did not appear bleached.

*Effect of sunscreen on medusae*

Coppertone sunscreen at concentrations of $\leq 10^{-6}$ had no discernible effect on the survival of medusae from BJLK and GLK within 5 days, but concentrations of $10^{-4}$ caused the death of all medusae within a few minutes (GLK) to several hours (BJLK). BullFrog was lethal to all OLO medusae within 30 min at a concentration of $10^{-4}$, within 6 h at $10^{-5}$, and within 30 h at $10^{-6}$. Only one jellyfish died in one control experiment (BJLK).

# Discussion

*Jellyfish swarms and the Christ-child*

The 1997–98 El Niño was, by some measures, the most severe on record (McPhaden, 1999; Wilkinson et al., 1999). In the western Pacific, dramatic changes in the prevailing winds, depth of the ther-

mocline, drought, unusually high sea-levels, and elevated air and sea-surface temperatures (McPhaden, 1999, NOAA, 1999a) precipitated important biological changes (e.g. Chavez et al., 1999). In Palau, 'the Christ-child' brought a severe drought, which killed many large jungle trees and depleted Palau's only fresh-water lake, and caused exceptionally high tides, high sea surface temperatures, and widespread coral bleaching (PCS, 1999b; CRRF, unpubl. data). Coral bleaching – the perturbation of the coral-zooxanthellae symbiosis (e.g. Hoegh-Guldberg & Smith, 1989; Kleppel, 1989) – is an hallmark of the high sea surface temperatures associated with El Niño events (Brown et al., 1996; Jones et al., 1997; Winter et al., 1998) and, on a global scale, bleaching in 1997–98 was more extensive than any recorded previously (Wilkinson et al., 1999; Aronson et al., 2000). In Palau, at least 30% of all scleractinians and as many as 99% of some species died (PCS, 1999b).

The physical and biotic changes that occurred in the marine lakes in Palau during 1998 were of far greater magnitude than any fluctuations measured previously. Whether the perturbations were truly unprecedented, however, is unclear. Etpison (1997) reported that the medusa population in Ongeim'l Tketau was decimated in 1987 and, although she attributed the loss of medusae to scientific diving and SCUBA generated turbulence that disturbed the toxic monimolimnion, its coincidence with the El Niño of 1986–88 (Brown et al., 1996; NOAA, 1999b) now suggests a close link

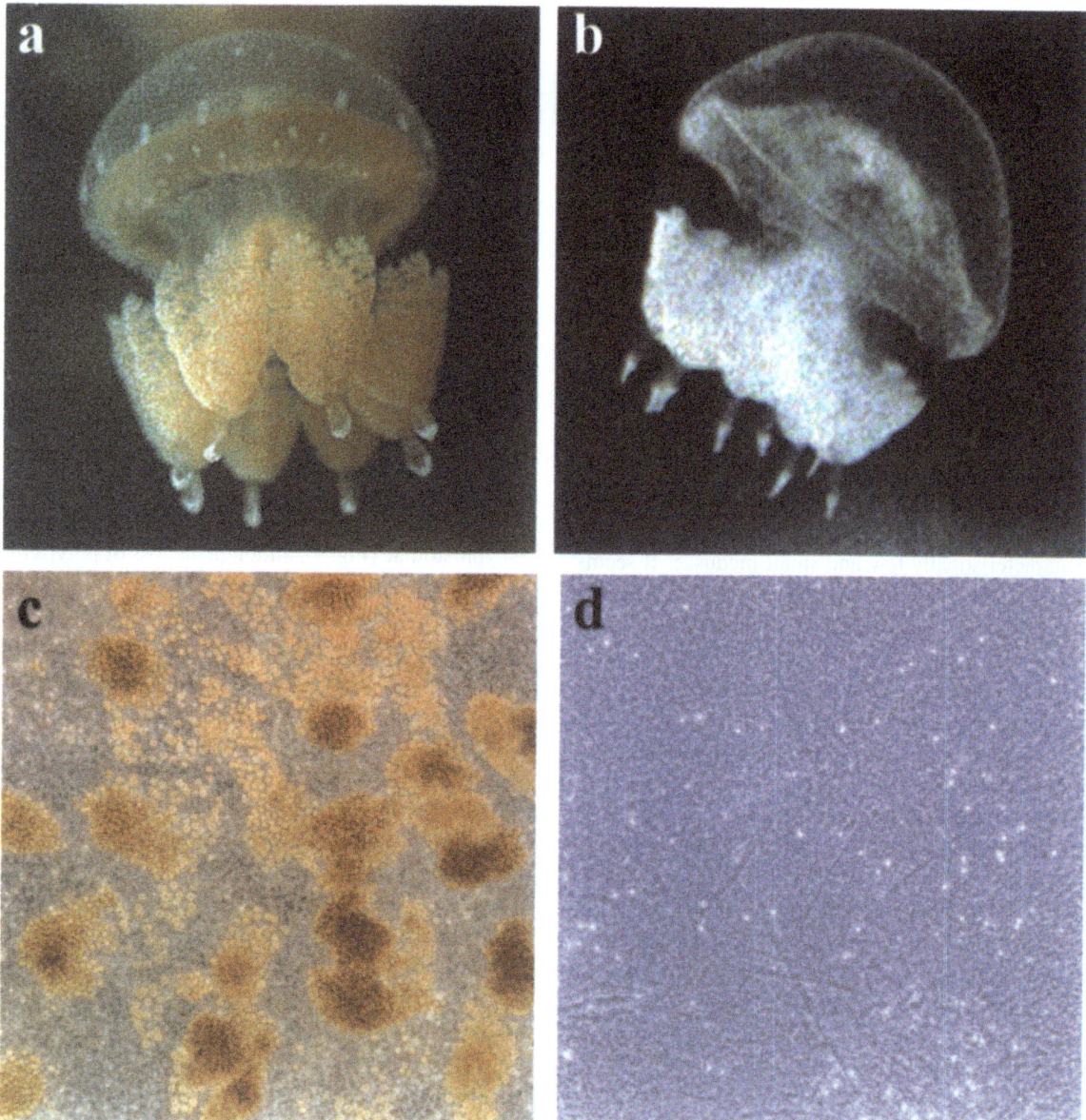

*Figure 5.* (a) A symbiotic golden-*Mastigias* medusa, approximately 30 mm bell diameter and (b) an azooxanthellate white-*Mastigias* of similar size. Under phase-contrast microscopy (magnification ×125), (c) tissue of golden-*Mastigias* is replete with clusters of heavily pigmented zooxanthellae (each zooxanthella is ~9 μm diameter [Muscatine et al., 1986]) while (d) zooxanthellae are conspicuously absent from tissue of white-*Mastigias*.

between climate change and marine lake ecosystem dynamics.

The most striking changes measured in the marine lakes in 1998–99 occurred in Ongeim'l Tketau where extremely high water temperatures and unusual stratification were associated with the disappearance of approximately 1.5 million *Mastigias* medusae. By contrast, *Mastigias* medusae in Goby, Clear, and Big Jellyfish lakes did not suffer the same fate. These lakes, which did not stratify to the same degree, never

became as hot and generally were more variable physically than OTM. Their *Mastigias* populations either were not affected detrimentally or recovered relatively quickly. The coincidence of strong physical and biological changes in the lakes after two decades of apparent stability suggests they were related causally. Laboratory experiments support this interpretation – temperatures as high as those observed in Ongeim'l Tketau can have severe effects on *Mastigias*, including bleaching and increased mortality of scyphistomae

(Figs 8 and 9), inhibition of strobilation, and increased mortality of medusae (Fig. 10).

### Ongeim'l Tketau

The disappearance of *Mastigias* medusae from Ongeim'l Tketau likely resulted from the coincidence of several unusual demographic events. Ongeim'l Tketau normally contains *Mastigias* medusae of all sizes indicating strobilation, growth, and mortality during all, or most, months of each year (Hamner et al., 1982). However, the medusa population in September, 1998, probably had a higher than normal proportion of large medusae because the production or recruitment of small and medium-sized medusae had been less frequent than usual during summer 1998. The large medusae present in September died before the end of November, 1998, leaving in the lake only uncharacteristically few and small medusae. It seems likely that the large, mature, medusae died as a natural consequence of reproductive senescence, as observed in other medusae (Arai, 1997:182–3). Mass mortality of large, presumably post-reproductive, *Mastigias* was observed previously in Ongeim'l Tketau in April, 1997, when hundreds to thousands of motionless medusae carpeted the lake bottom, underneath the previous afternoon's swarm (Dawson, pers. obs.). In contrast, the disappearance of all smaller medusae was anomalous.

The death of small and medium medusae from Ongeim'l Tektau in 1998 appears attributable more to high mortality of small medusae than to lack of strobilation. Between late-1998 and April, 1999, thousands of medusae were strobilated but none survived. Strobilation did not cease until May, 1999. There may be several reasons ephyrae and small medusae did not survive. One possibility is exposure to unusually high temperatures (Fig. 10), which may have occurred nightly when medusae migrated into deeper waters (Hamner et al., 1982). Alternatively, exposure to high temperatures may have been chronic if medusae became trapped behaviorally by and below the thermocline and halocline at 5–6 m (see Lance, 1962; Harder, 1968; Arai, 1973, 1976, 1992; Hamner et al., 1994; Dawson, 2000). Similarly, depending on the relationship between water column structure and the distributions of *Mastigias* medusae and their resources, the physical changes in OTM may have limited the availability of suitable prey (see also Uye, 1994) and access to light above the pycnocline or nutrients below it, at the chemocline. Notably, *Mastigias* medusae <40 mm rely heavily on zooplankton

to meet their metabolic demands (McCloskey et al., 1994). However, the uninterrupted abundance of *Aurelia* in OTM throughout 1998–99 suggests prey were not scarce. Thus, the decline in *Mastigias* seems likely to be attributable to another factor, or factors, perhaps affecting their symbiosis with zooxanthellae.

Irrespective of the causes of their decline, medusae were present until April, 1999, after which strobilation ceased, probably inhibited principally by the unusually high temperature of the mixolimnion (but see also Purcell et al., 1999; Condon et al., 2001; Lucas, 2001; Watanabe & Ishii, 2001). High temperatures likely caused scyphistomae to bleach (Fig. 8), thus removing or reducing the zooxanthellae presumably required for strobilation (Sugiura, 1964, 1965; but see below). In addition, high temperatures may have reduced strobilation by increasing the mortality rate of scyphistomae (Fig. 9), although the loss of polyps may have been mitigated by the production of planuloids (Table 1). Finally, after April, 1999, water temperatures in Ongeim'l Tketau fluctuated little, relative to those in other jellyfish lakes, perhaps depriving scyphistomae of the temperature changes presumably needed to stimulate strobilation (Sugiura, 1964, 1965; see also Figs 3 and 4). Thus, on the basis of laboratory experiments and field measurements, relative to other jellyfish lakes, there should have been few, if any, zooxanthellate or strobilating scyphistomae in Ongeim'l Tketau after April, 1999, and field collections in summer 1999 demonstrated that this was the case (Table 1).

### Goby, Clear and Big Jellyfish lakes

The small body size and number of *Mastigias* medusae in Goby Lake in December, 1998 (Fig. 4), suggests the recent dynamics of the golden-jellyfish population may have been similar to those in Ongeim'l Tketau. However, the cohort of small medusae in Goby Lake in December, 1998, grew successfully. By August 1999, the golden-jellyfish population had returned to approximately its pre-1998 condition and there were many zooxanthellate strobilae in the lake (Table 1).

By contrast, Clear Lake, in August 1999, contained almost exclusively *Mastigias* medusae ≥12 cm and the scyphistomae collected were only moderately infected with zooxanthellae and not strobilating. The dearth of small medusae and strobilation in CLM may have been attributable to low levels of zooxanthellae (Table 1) which are important for strobilation (Sugiura, 1964, 1965; Arai, 1997). However, other factors also are implicated (e.g. Purcell et al., 1999; Con-

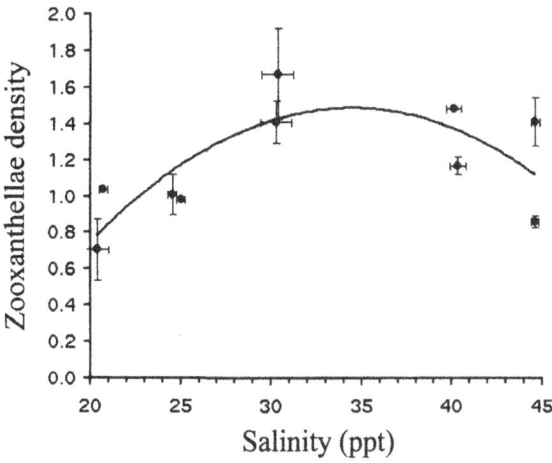

Figure 6. The effect of salinity on the zooxanthellae density index for *Mastigias* scyphistomae collected in Ongeim'l Tketau. Scyphistomae were collected from salinities of 31–32‰ and acclimated to 34‰ during the 3 weeks prior to the experiment. Subsequently, the scyphistomae were acclimated to the experimental salinities at which they were held for 11 days. Fitted curve: $y = -0.0032x^2 + 0.222x - 2.4623$, $R^2 = 0.5488$. 95% Confidence Interval (CI) error bars are shown.

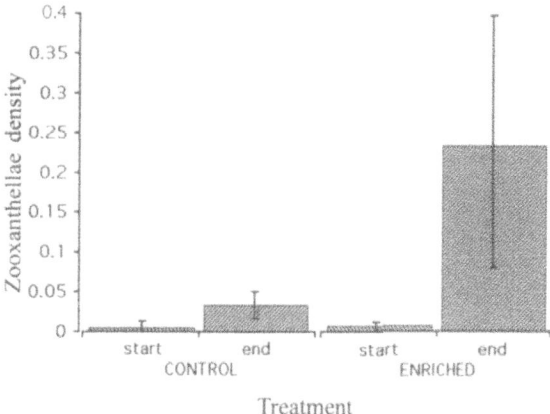

Figure 7. The effect of zooxanthellae enrichment on the zooxanthellae density index for scyphistomae from Ongeim'l Tketau. Homogenized mouth-arm tissues of *Mastigias papua* were added to 'enriched' treatments but not to controls. Student's *t*-test showed a significantly greater ($p = 0.05$) increase in the density of zooxanthellae in 'enriched' over 'control' treatments after 4 days. 95% CI error bars.

don et al., 2001; Lucas, 2001; Watanabe & Ishii, 2001) because zooxanthellae are not absolutely necessary for strobilation – the white *Mastigias* of 1997 more likely were strobilated aposymbiotically (e.g. Rahat & Adar, 1980) than bleached post-strobilation because bleached animals typically retain some remnants of zooxanthellae (Glynn & D'Croz, 1990; Hayes & Bush, 1990; Kuroki & van Woesik, 1999; but see Fig. 5). Irrespective of the reason for the dearth of small and medium *Mastigias* medusae, CLM in August 1999 may have resembled OTM in Fall 1998, just prior to the disappearance of the *Mastigias* population.

Alone among the four lakes reported here, Big Jellyfish Lake experienced no obvious changes in its biota. Possibly this was because temperature extremes in Big Jellyfish Lake were more modest than in other lakes (Fig. 3). However, it also is possible that BJLK *Mastigias* are adapted better to temperature extremes than *Mastigias* in other lakes (Fig. 10). Such local adaptation might explain the demographic differences observed between lake populations in 1998–99 and is consistent with previous evidence of lake-specific adaptations in marine lake *Mastigias* (Hamner & Hauri, 1981; Muscatine & Marian, 1982; McCloskey et al., 1994) and co-adaptation of zooxanthellae and hosts (Rowan & Knowlton, 1995; see also Fig. 8a). Moreover, the events of 1998-99 suggest that natural selection and adaptation may be ongoing in at least some populations of *Mastigias*.

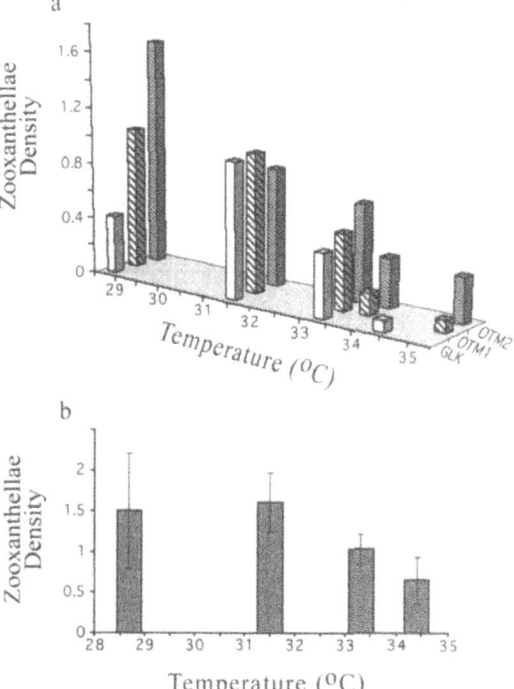

Figure 8. The effect of temperature on the zooxanthellae density index for *Mastigias* scyphistomae. (a) Change in zooxanthellae density in scyphistomae containing established populations of zooxanthellae. GLK, Goby Lake, Koror; OTM1, Ongeim'l Tketau, Mecherchar, infected with non-*Mastigias* zooxanthellae. OTM2, Ongeim'l Tketau, Mecherchar, enriched with zooxanthellae from *Mastigias papua*. (b) The increase in zooxanthellae density (± 95% CI) in scyphistomae that were inoculated with *M. papua* zooxanthellae immediately prior to the experiment. Center of bar indicates temperature treatment.

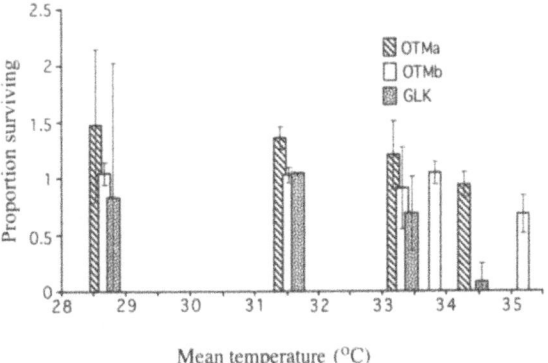

*Figure 9.* The effect of temperature on mean survival (± 95% CI) of *Mastigias* scyphistomae inoculated with *M. papua* zooxanthellae immediately prior to the experiment (OTMa) and those with established populations of zooxanthellae (OTMb, GLK). Center of white bar indicates temperature treatment.

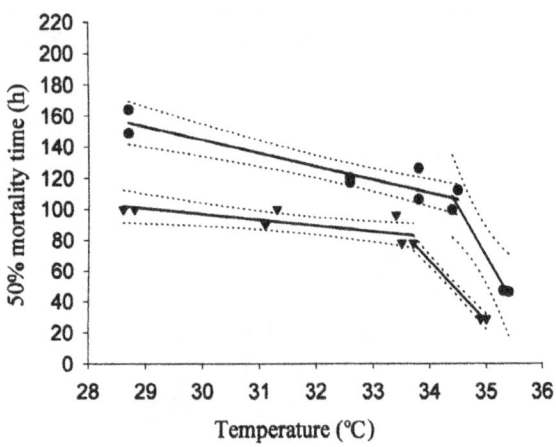

*Figure 10.* Effect of temperature on the 50%-mortality time of *Mastigias* medusae from BJLK (circles) and GLK (triangles). Increasing temperature decreases survivorship (i.e. regressions are significantly different from zero, see below), an effect that is particularly severe at the highest temperatures studied. The linear regressions fitted to BJLK data are, $y = -8.552x + 401.11$, $R^2 = 0.859$, $p = 0.001$, and $y = -64.024x + 2310.45$, $R^2 = 0.943$, $p = 0.029$. The regressions fitted to GLK data are, $y = -3.781x + 210.087$, $R^2 = 0.6075$, $p = 0.023$, and $y = -39.07x + 1394.75$, $R^2 = 0.997$, $p = 0.002$. Dotted lines show 95% confidence intervals on the regressions.

### Tourists and other factors

Tourists probably played little role in the disappearance of *Mastigias* medusae. Populations of *Mastigias* medusae declined almost synchronously in Ongeim'l Tketau and Goby Lake although few tourists visited Goby Lake, and it simply was not logistically feasible that tourists ate, stole or otherwise killed all one-and-a-half million medusae in Ongeim'l Tketau. Moreover, *Mastigias* medusae had not been affected deleteriously by many previous years of sometimes higher tourism. In addition, although sunblock can kill golden-medusae, it is highly unlikely that acutely lethal concentrations, which turn seawater milky, ever have been present in Ongeim'l Tketau. Whether there are detrimental effects due to chronic exposure of *Mastigias* to sunscreen, however, is yet to be investigated experimentally.

The disappearance of *Mastigias* medusae also was not attributable to stagnation of the lake caused by reduced tidal exchange, because the conduits connecting the lake and lagoon were not blocked. The extreme stratification and over-heating of the mixolimnion of Ongeim'l Tketau most likely were induced by El Niño weather patterns. High salinity also is an unlikely explanation for the disappearance because, although salinity was high in December 1998, *Mastigias* did not disappear entirely until April 1999, when salinity profiles already had returned to normal (Fig. 2; Hamner & Hamner, 1998). Moreover, *Mastigias* medusae frequently migrate vertically between waters differing by 3‰ (Hamner et al., 1982) and experimental salinity changes of 10‰ in 7 days generally are somewhat deleterious, but not lethal, to *Mastigias* scyphistomae (Fig. 6; also see Weiler & Black, 1991). Thus, at this time, there is little evidence to implicate factors other than temperature in the disappearance of *Mastigias* medusae from Ongeim'l Tketau.

### Closing remarks

In January, 2000, *Mastigias* medusae were observed in OTM for the first time since April 1999. A small strobilation event occurred after the lake had cooled, by December 1999, to its lowest temperature (max. 32.8 °C) since January 1999. This cohort of medusae matured and grew successfully to as large as 19 cm. Since then, three additional strobilation events have produced mature medusae up to 23 cm. The *Mastigias* population in OTM has increased gradually in size but, as of May 2000, remains probably less than $10^4$ strong.

### Acknowledgements

We thank the Division of Marine Resources and Koror State Government for allowing us to work in Palau. We also are grateful for the continued support of Palau Conservation Society (PCS) and Koror State Rangers. Personal thanks particularly to Tom Graham (PCS) and Yalap P. Yalap (Belau Tourism Authority), for their many efforts on our behalf. We are indebted to Peggy Hamner, Bill Hamner, Lori J. Bell and Pat Colin for their unerring support of and contributions to our research in Palau. Sara Sawyer and Garen Baghdasarian contributed much to our laboratory analyses of white-*Mastigias*. Suggestions made by Lori J. Bell, Pat Colin, Ruth Gates, Bill Hamner , Jenny Purcell and two anonymous reviewers considerably improved the manuscript. This work was funded in part by a grant to William M. Hamner, MND and LEM from the National Sea Grant College Program, National Oceanic and Atmospheric Administration, U.S. Department of Commerce, under grant number NA66RG0477, project number R/C-49PD through the California and Hawaii Sea Grant College Systems. The views expressed herein are those of the authors and do not necessarily reflect the views of NOAA or any of its sub-agencies. The U.S. Government is authorized to reproduce and distribute for governmental purposes. This research also was funded by a PADI Foundation grant to LEM and MND. Equipment used during this work was supported by NSF Grant # DBI-9714179 to the Coral Reef Research Foundation.

# References

Arai, M. N., 1973. Behaviour of the planktonic coelenterates, *Sarsia tubulosa*, *Phialidium gregarium* and *Pleurobrachia pileus* in salinity discontinuity layers. J. Fish. Res. Bd Can. 30: 1105–1110.

Arai, M. N., 1976. Behaviour of planktonic coelenterates in temperature and salinity discontinuity layers. In Mackie, G. O. (ed.), Coelenterate Ecology and Behaviour. Plenum Press, New York: 211–218.

Arai, M. N., 1992. Active and passive factors affecting aggregations of hydromedusae: a review. Sci. mar. 56: 99–108.

Arai, M. N., 1997. A functional biology of Scyphozoa. Chapman & Hall, London: 316 pp.

Aronson, R. B., W. F. Precht, I. G. Macintyre & T. J. T. Murdoch, 2000. Coral bleach-out in Belize. Nature 405: 36.

Barbour, N., 1990. Palau. Full Court Press, San Francisco, 160 pp.

Brown, B. E., R. P. Dunne & H. Chasang, 1996. Coral bleaching relative to elevated seawater temperature in the Andaman Sea (Indian Ocean) over the last 50 years. Coral Reefs 15: 151–152.

Chavez, F. P., P. G. Strutton, G. E. Friederich, R. A. Feely, G. C. Feldman, D. G. Foley & M. J. McPhaden, 1999. Biological and chemical response of the equatorial Pacific Ocean to the 1997–98 El Niño. Science 286: 2126–2131.

Condon, R., M. B. Decker & J. E. Purcell, 2001. Effects of low dissolved oxygen on survival and asexual reproduction of scyphozoan polyps (*Chrysaora quinquecirrha*). Hydrobiologia 451 (Dev. Hydrobiol. 155): 89–95.

Dawson, M. N, 2000. Variegated mesocosms as alternatives to shore-based planktonkreisels: notes on the husbandry of jellyfish from marine lakes. J. Plankton Res. 22: 1673–1682.

Dawson, M. N., K. A. Raskoff & D. K. Jacobs, 1998. Preservation of marine invertebrate tissues for DNA analyses. Molec. Mar. Biol. Biotech. 7: 145–152.

Etpison, M. T., 1997. Palau: portrait of paradise. Neco Marine, Koror: 250 pp.

Fautin, D. G. & W. K. Fitt, 1991. A jellyfish-eating anenome (Cnidaria, Actinaria) from Palau – *Entacmaea medusivora* - sp. nov. Hydrobiologia 216: 453–461.

Glynn, P. W. & L. D'Croz, 1990. Experimental evidence for high temperature stress as the cause of El Niño-coincident coral mortality. Coral Reefs 8: 181–191.

Hamner, W. M., 1982. Strange world of Palau's salt lakes. Nat. Geogr. 161: 264–282.

Hamner, W. M., 1994. Current biological status (1994) of Jellyfish Lake, Macharchar. Report prepared for Division of Marine Resources, Republic of Palau.

Hamner, W. M. & P. P. Hamner, 1998. Stratified marine lakes of Palau (Western Caroline Islands). Phys. Geogr. 19: 175–220.

Hamner, W. M. & I. R. Hauri, 1981. Long-distance horizontal migrations of zooplankton (Scyphomedusae: *Mastigias*). Limnol. Oceanogr. 26: 414–423.

Hamner, W. M., R. W. Gilmer & P. P. Hamner, 1982. The physical, chemical and biological characteristics of a stratified, saline, sulfide lake in Palau. Limnol. Oceanogr. 27: 896–909.

Hamner, W. M., P. P. Hamner & S. W. Strand, 1994. Sun-compass migration by *Aurelia aurita* (Scyphozoa): population retention and reproduction in Saanich Inlet, British Columbia. Mar. Biol. 119: 347–356.

Harder, W., 1968. Reactions of planktonic organisms to water stratification. Limnol. Oceanogr. 13: 156–168.

Hayes, R. L. & P. G. Bush, 1990. Microscopic observations of recovery in the reef-building scleractinian coral, *Montastrea annularis*, after bleaching on a Cayman reef. Coral Reefs 8: 203–209.

Hoegh-Guldberg, O. & J. G. Smith, 1989. The effect of sudden changes in temperature, light and salinity on the population density and export of zooxanthellae from the reef corals *Stylophora pistillata* Esper and *Seriatopora hystrix* Dana. J. exp. mar. Biol. Ecol. 129: 279–303.

Jones, R. J., R. Berkelmans & J. K. Oliver, 1997. Recurrent bleaching of corals at Magnetic Island (Australia) relative to air and seawater temperature. Mar. Ecol. Prog. Ser. 158: 289–292.

Kleppel, G. S., R. E. Dodge & C. J. Reese, 1989. Changes in pigmentation associated with the bleaching of stony corals. Limnol. Oceanogr. 34: 1331–1335.

Kuroki, T. & R. Van Woesik, 1999. Changes in zooxanthellae characteristics in the hermatypic coral *Stylophora pistillata* during a 'bleaching event'. Galaxea 1: 97–101.

Lance, J., 1962. Effects of water of reduced salinity on the vertical migration of zooplankton. J. mar. biol. Ass. U.K. 42: 131–154.

Lucas, C. H., 2001. Reproduction and life history strategies of the common jellyfish, *Aurelia aurita*, in relation to its ambient environment. Hydrobiologia 451 (Dev. Hydrobiol. 155): 229–246.

McCloskey, L. R., L. Muscatine & F. P. Wilkerson, 1994. Daily photosynthesis, respiration and carbon budgets in a tropical marine jellyfish (*Mastigias* sp.). Mar. Biol. 119: 13–22.

McPhaden, M. J., 1999. Genesis and evolution of the 1997–98 El Niño. Science 283: 950–954.

Muscatine, L. & R. E. Marian, 1982. Dissolved inorganic nitrogen flux in symbiotic and nonsymbiotic medusae. Limnol. Oceanogr. 27: 910–917.

Muscatine, L., F. P. Wilkerson & L. R. McCloskey, 1986. Regulation of population density of symbiotic algae in a tropical marine jellyfish (*Mastigias* sp.). Mar. Ecol. Prog. Ser. 32: 279–290.

NOAA, 1999a. TAO/TRITON times series plot. http://www.pmel.noaa.gov/toga-tao/realtime.html

144

NOAA, 1999b. El Niño theme page. http://www.pmel.noaa.gov/toga-tao/el-n ino/nino-home.html

PCS, 1999a. Palau's Taiwanese tourism industry: assessment of issues and suggestions for the future. PCS report 99-04. Palau Conservation Society, Koror: 42 pp.

PCS, 1999b. Current status of Palau's corals. PCS newsletter, No.11, July 1999: 4.

PVA, 1999. Palau recognized as a top world destination. Palau Visitors Authority press release. Tia Belau, Koror, 5th December 1999.

Purcell, J. E., J. R. White, D. A. Nemazie & D. A. Wright, 1999. Temperature, salinity and food effects on asexual reproduction and abundance of the scyphozoan *Chrysaora quinquecirrha*. Mar. Ecol. Prog. Ser. 180: 187–196.

Rahat, M. & O. Adar, 1980. Effect of symbiotic zooxanthellae and temperature on budding and strobilation in *Cassiopeia andromeda* (Eschscholz). Biol. Bull. 159: 394–401.

Rowan, R. & N. Knowlton, 1995. Intraspecific diversity and ecological zonation in coral-algal symbiosis. Proc. natl. Acad.Sci. U.S.A. 92: 2850–2853.

Seutin, G., B. N. White & P. T. Boag, 1991. Preservation of avian blood and tissue samples for DNA analyses. Can. J. Zool. 69: 82–90.

Sugiura, Y., 1964. On the life-history of rhizostome medusae. II. Indispensability of zooxanthellae for strobilation in *Mastigias papua*. Embryologia 8: 223–233.

Sugiura, Y., 1965. On the life-history of rhizostome medusae. III. On the effects of temperature on the strobilation of *Mastigias papua*. Biol. Bull. 128: 493–496.

Uye, S., 1994. Replacement of large copepods by small ones with eutrophication of embayments: causes and consequences. Hydrobiologia 292/293: 513–519.

Watanabe, T. & H. Ishii, 2001. *In situ* estimation of ephyrae liberated from polyps of *Aurelia aurita* using settling plates in Tokyo Bay, Japan. Hydrobiologia 451 (Dev. Hydrobiol. 155): 247–258.

Weiler, K. S. & R. E. Black, 1991. Immediate decrease in apparent translation rate caused by salinity change in scyphistomae of *Aurelia aurita* (Linnaeus). Comp. Biochem. Physiol. 99A: 199–202.

Wilkinson, C., O. Linden, H. Cesar, G. Hodgson, J. Rubens & A. E. Strong, 1999. Ecological and socioeconomic impacts of 1998 coral mortality in the Indian Ocean: an ENSO impact and a warning of future change? Ambio 28: 188–196.

Winter, A., R. S. Appeldoorn, A. Bruckner, E. H. Williams Jr. & C. Goenaga, 1998. Sea surface temperatures and coral reef bleaching off La Parguera, Puerto Rico (northeastern Caribbean Sea). Coral Reefs 17: 377–382.

*Hydrobiologia* **451**: 145–176, 2001.
© 2001 *Kluwer Academic Publishers.*

# The ctenophore *Mnemiopsis* in native and exotic habitats: U.S. estuaries *versus* the Black Sea basin

Jennifer E. Purcell[1,2], Tamara A. Shiganova[3], Mary Beth Decker[1,4] & Edward D. Houde[5]

[1] *University of Maryland Center for Environmental Science, Horn Point Laboratory, P.O. Box 775, Cambridge, MD 21613, U.S.A.*
[2] *Present address: Shannon Point Marine Center, 1900 Shannon Point Rd., Anacortes, WA 98221, U.S.A.*
*E-mail: purcell@hpl.umces.edu*
[3] *P.P. Shirshov Institute of Oceanology, Russian Academy of Sciences, 36 Nakhimovskiy Pr., Moscow 117851, Russia*
[4] *Present address: Dept. of Ecology and Evolutionary Biology, Yale University, New Haven, CT 06520, U.S.A.*
[5] *University of Maryland Center for Environmental Science, Chesapeake Biological Laboratory, P.O. Box 38, Solomons, MD 20688-0038, U.S.A.*

*Key words: Beroe*, Chesapeake Bay, fish, jellyfish, predation, zooplankton

## Abstract

The native habitats of the ctenophore, *Mnemiopsis*, are temperate to subtropical estuaries along the Atlantic coast of North and South America, where it is found in an extremely wide range of environmental conditions (winter low and summer high temperatures of 2 and 32 °C, respectively, and salinities of <2–38). In the early 1980s, it was accidentally introduced to the Black Sea, where it flourished and expanded into the Azov, Marmara, Mediterranean and Caspian Seas. We compile data showing that *Mnemiopsis* has high potentials of growth, reproduction and feeding that enable this species to be a predominant zooplanktivore in a wide variety of habitats; review the population distributions and dynamics of *Mnemiopsis* in U.S. waters and in the Black Sea region; and examine the effects of temperature and salinity, zooplankton availability and predator abundance on *Mnemiopsis* population size in both regions, and the effects of *Mnemiopsis* on zooplankton, ichthyoplankton and fish populations, focusing on Chesapeake Bay and the Black Sea. In both regions, *Mnemiopsis* populations are restricted by low winter temperatures (<2 °C). In native habitats, predators of *Mnemiopsis* often limit their populations, and zooplanktivorous fish are abundant and may compete with the ctenophores for food. By contrast, in the Black Sea region, no obvious predators of *Mnemiopsis* were present during the decade following introduction when the ctenophore populations flourished. Additionally, zooplanktivorous fish populations had been severely reduced by over fishing prior to the ctenophore outbreak. Thus, small populations of potential predators and competitors for food enabled *Mnemiopsis* populations to swell in the new habitats. In Chesapeake Bay, *Mnemiopsis* consumes substantial proportions of zooplankton daily, but may only noticeably reduce zooplankton populations when predators of *Mnemiopsis* are uncommon. *Mnemiopsis* also is an important predator of fish eggs in both locations. In the Black Sea, reductions in zooplankton, ichthyoplankton and zooplanktivorous fish populations have been attributed to *Mnemiopsis*. We conclude that the enormous impact of *Mnemiopsis* on the Black Sea ecosystem occurred because of the shortage of predators and competitors in the late 1980s and early 1990s. The appearance of the ctenophore, *Beroe ovata*, may promote the recovery of the Black Sea ecosystem from the effects of the *Mnemiopsis* invasion.

## Introduction

Comb jellies comprise a diverse phylum (Ctenophora) of delicate, gelatinous species living throughout the world's oceans. In estuarine and coastal waters, ctenophores can reach great abundances. They feed at high rates on zooplankton and ichthyoplankton, and thereby may be detrimental to fish populations. They are well adapted for rapid population growth, having high rates of feeding, growth and reproduction.

The native habitats of ctenophores in the genus *Mnemiopsis* are estuaries along the eastern coastline of North and South America. *Mnemiopsis* is probably the most-studied ctenophore genus in the world because of its great abundance in estuaries in heavily populated areas of the U.S., and because of its explosive population growth after accidental introduction into the Black Sea in the early 1980s. It has spread from the Black Sea to the Seas of Azov, Marmara and eastern Mediterranean, and in 1999, it appeared for the first time in the Caspian Sea. At present, its impact is a serious problem for the ecosystems of these basins.

The identities of species within the genus *Mnemiopsis* are uncertain. Two species names have been commonly used, *M. leidyi* A. Agassiz, which has been used for areas above Cape Hatteras in the U.S., and *M. mccradyi* Mayer, 1900, which has been used below Cape Hatteras (Larson, 1988), and both species are listed from South America (Mianzan, 1999). Mayer (1912) considered the high body density and the warts on the body surface to be specific features of *M. mccradyi*, however, these features were formerly noted in *M. leidyi* (in A. Agassiz, 1865; Fewkes, 1881). Recent investigations reveal indistinguishable specimens from North and South America (K. Bayha, pers. comm.). Experimental results on respiration, excretion, egg production, feeding and growth on *Mnemiopsis* from northern and southern U.S. are indistinguishable (Kremer, 1994).

Identification of the ctenophore that appeared in the Black Sea was confused by inconsistencies in description of morphological features of the genus *Mnemiopsis* by its first researchers. The introduced ctenophore was called both *M. leidyi* and *M. mccradyi* (e.g. Vinogradov et al., 1989; Zaika & Sergeeva, 1990). Seravin (1994) revised the genus *Mnemiopsis*, concluding that it includes only one polymorphic species, *M. leidyi*. Due to the uncertainty of clear species, we will refer to the ctenophores by the genus name.

In this review, we explore the effects of the environment on populations of *Mnemiopsis*. First, we consider how temperature, salinity and prey densities affect the ctenophore's composition, metabolism, individual growth and reproduction. Then, we examine the conditions found in the habitats of *Mnemiopsis* and how their populations vary with physical and biological factors, first considering native habitats and then exotic habitats. We have chosen to concentrate on comparisons between the Chesapeake Bay and the Black Sea, which are at the same latitude, have similar ranges of environmental conditions, and similar zooplankton and zooplanktivorous fish assemblages.

## Body composition, metabolism, growth and reproduction of *Mnemiopsis*

### Body composition

The relationships among various measures of body size of *Mnemiopsis* may vary with ctenophore size, salinity and feeding history (Table 1). Oral-aboral length measurements are somewhat subjective due to the flaccid body, as illustrated by the variation in length-to-wet weight conversions (Table 1). Biovolume (V in ml) typically is measured in field studies and is roughly equivalent to wet weight (WW in g). Live WW can be calculated from tentacle bulb length in formalin-preserved samples (Purcell, 1988). Dry weight (DW) commonly is measured, but results differ depending on the salt content of the ctenophore, which reflects ambient salinity. Carbon (C) and nitrogen (N) contents as percent of DW vary, generally being inversely related to ctenophore size and directly related to prey density (Reeve et al., 1989; Borodkin & Korzhikova, 1991). C%DW ranged from 13.1% in 1.1 mm ctenophores to 1.5% in 20 mm specimens. C%DW ranged from 1% when fed at $\leq 20$ prey $l^{-1}$ to 2% at 200 prey $l^{-1}$. Chemical composition (glycogen, mono- and polysaccharides) of *Mnemiopsis* has been further studied by Anninsky et al. (1998).

### Metabolism

More extensive research has been conducted on the respiration and excretion of *Mnemiopsis* than on any other gelatinous species. The average respiratory quotient (molecular ratio) of $CO_2:O_2$ of 0.74 and $O_2:NH_4$ of 6.7 (O:N atomic ratio = 13.3) are indicative of protein and lipid metabolism, and suggest that *Mnemiopsis* are strict carnivores (Kremer, 1977). Weight-specific metabolism of *Mnemiopsis* and other ctenophores has been reported to depend on body size (Miller, 1970; Nemazie et al., 1993) (Table 2), however, Kremer (1975a, 1977, 1982) found no weight-specific effects. Total nitrogen excretion is comprised of 66% $NH_4^+$ and 34% dissolved organic nitrogen (Kremer, 1977, 1982). Ammonium excretion rates per g DW reported from Chesapeake Bay are 3-fold higher than for Narragansett Bay, but are similar when

Table 1. Biometric conversions for *Mnemiopsis*. V = live volume in ml, L = total length in mm, WW = wet weight in g, DW = dry weight in g, C = carbon in mg, N = nitrogen in mg

| Conversion | Salinity | Source |
|---|---|---|
| WW = 1.017V - 0.122 | 31 | Kremer & Nixon (1976) |
| WW = $0.009L^{1.872}$ | 31 | Kremer & Nixon (1976) |
| WW=$0.00236L^{2.35}$ | 18 | Vinogradov et al. (1989) |
| WW=$0.00079 L^{2.51}$ | 18 | Pavlova & Minkina (1993) |
| WW= $0.001074L^{2.76}$ | — | Finenko et al. (1995) |
| WW=$0.0023L^{2.05}$, R=0.94 (Spring) | 18 | Shiganova (unpublished) |
| WW=$0.0061L^{1.81}$, R=0.95 (Summer) | 18 | Shiganova (unpublished) |
| DW = 0.034WW | 31 | Kremer & Nixon (1976) |
| DW=$0.0331WW^{0.939}$ | 18 | Tzikhon-Lukanina & Reznichenko (1991) |
| DW = 0.0095WW - 0.0014 | 6–12 | Nemazie et al. (1993) |
| C = 0.0005V | 6–12 | calculated from Nemazie et al. (1993) |
| C%WW=0.00012% | 18 | Vinogradov et al. (1989) |
| C%DW = 1.7% | 31 | Kremer & Nixon (1976) |
| C% DW = 4.2% | 18 | Finenko et al. (1995) |
| C%DW = 5.1% | 6–12 | Nemazie et al. (1993) |
| N% DW = 0.5% | 31 | Kremer (1976) |
| N%DW = 1.3% | 6–12 | Nemazie et al. (1993) |

Table 2. Metabolic rates of *Mnemiopsis*. Excretion (E) in $\mu$g atoms $NH_4^+$-N g $DW^{-1}$ $h^{-1}$. Respiration (R) in $\mu$g atoms $O_2$ g $DW^{-1}$ $h^{-1}$, $R_1$ in ml $O_2$ $ind^{-1}$ $h^{-1}$, $R_2$ in $\mu$l $O_2$ g $DW^{-1}$ $h^{-1}$, $R_3$ in $\mu$l $O_2$ $ind^{-1}$ $h^{-1}$. T = temperature in °C, DW = g dry weight. * = calculated from the equations for ctenophores 0.4 g DW at 22 °C, and using the conversions 1 ml $O_2$ = 1.42 mg $O_2$ = 44.88 $\mu$mol $O_2$ = 89.76 $\mu$g atoms $O_2$

| Rate equation | Temperature (°C) | Salinity | R or E at 22 °C, 0.4 g DW ($\mu$g atoms g $DW^{-1}$ $h^{-1}$)* | Source |
|---|---|---|---|---|
| E = 0.059 $e^{0.13T}$ | 10–24 | 31 | 1.03 | Kremer (1975b) |
| E = 1.952 $DW^{0.742}$ | 18–28 | 6–12 | 1.17 | Nemazie et al. (1993) |
| E = 1.118+0.605 log DW + 0.053 T | 18–28 | 6–12 | 0.98 | Nemazie et al. (1993) |
| R = 0.79 $e^{0.13T}$ | 10–24 | 31 | 13.8 | Kremer (1977) |
| $R_1$ = 0.045 $DW^{0.584}$ | 20–21 | 18 | 5.8 | Finenko et al. (1995) |
| $R_2$ = 3.2 $DW^{0.89}$ | 23 | 18 | 0.127 | Pavlova & Minkina (1993) |
| $R_3$ = 2.50 $WW^{0.83}$ | 12–14 | – | 5.1 | Anninsky et al. (1995) |

standardized per g N (Nemazie et al., 1993). The authors conclude that the 3-fold salinity differences in the two locations explain the differences in excretion rates standardized by DW.

*Mnemiopsis* are very sensitive to temperature in both U. S. waters and the Black Sea, showing the metabolism increased by a factor of 3.4–3.7 for a 10 °C increase in temperature ($Q_{10}$) (Kremer, 1977; Nemazie et al., 1993; Finenko et al., 1995). Kremer (1977) compares results from Narragansett Bay with earlier results from Chesapeake Bay and Biscayne

Bay, where summer temperatures were higher, and concludes that ctenophores from the warmer locations were less sensitive to high temperatures. Metabolic rates also are sensitive to feeding history (Kremer, 1982; Kremer & Reeve, 1989; Finenko et al., 1995), with the highest rates measured for *Mnemiopsis* at high prey densities (Table 3).

Metabolic rates can be used to estimate minimum ingestion rates. *Mnemiopsis* turns over 6–18% $d^{-1}$ of its body carbon and nitrogen (Kremer, 1975, 1982; Nemazie et al., 1993). Those estimates increase

*Table 3.* Metabolic rates of *Mnemiopsis* at 22 °C relative to feeding history from Kremer (1982). Rates are for ctenophores 0.01–0.4 g DW. Respiration (R) in $\mu$g atoms $O_2$ g $DW^{-1}$ $h^{-1}$. Excretion (E) in $\mu$g atoms $NH_4^+$-N g $DW^{-1}$ $h^{-1}$

| Conditions | Respiration | Excretion |
|---|---|---|
| Freshly-collected <13 h | 8.7 | 0.70 |
| Starved 40 h | 6.0 | 0.42 |
| 5 prey $l^{-1}$ | 11.9 | 0.82 |
| 50 prey $l^{-1}$ | 16.6 | 1.41 |
| 200 prey $l^{-1}$ | 26.4 | 1.80 |

by 34% if organic nitrogen is included. Therefore, weight-specific ingestion would need to be 8–24% $d^{-1}$ to balance metabolic costs at 26 °C, but less at 21 °C.

Nitrogen excretion has been used to evaluate the importance of *Mnemiopsis* to nutrient recycling. At high ctenophore abundances, the contribution of *Mnemiopsis* excretion to the ammonium pool was 3–15% $d^{-1}$ (Kremer, 1975b; Nemazie et al., 1993), but was small when compared with demands of the whole microplankton community in Chesapeake Bay (<0.6% $d^{-1}$; Nemazie et al., 1993). The contribution towards ammonium recycling by *Mnemiopsis* is similar to that by mesozooplankton in Narragansett and Chesapeake Bays (Kremer, 1975b; Nemazie et al., 1993).

*Growth and reproduction*

*Mnemiopsis* is a simultaneous hermaphrodite with direct development. The spherical cydippid larvae hatch in 20–24 h, feed with two tentacles, and develop lobes at about 10 mm length. The combination of high feeding, growth and reproduction rates enable *Mnemiopsis* populations to increase rapidly in favourable conditions.

Individual growth rates of *Mnemiopsis* are related to prey densities. Young animals (about 1 mg DW) die after 3 weeks without food, losing 90% of their body mass (Finenko et al., 1995). Specific growth (carbon) of small ctenophores during 4 d without food ranged from −0.02 to −0.18 $d^{-1}$ (Reeve et al., 1989). Ctenophores <7 d old can double their weight daily with abundant food (Stanlaw et al., 1981). Specific growth rates (carbon) of small ctenophores (6 mm initial size) increased rapidly over 4 d to 13–17 mm at prey densities of 20–200 $l^{-1}$, achieving a maximum of over 0.8 $d^{-1}$ (more than a doubling in biomass daily) at the highest prey level (Table 4, Reeve et al., 1989). Specific growth rates of larger ctenophores (17 mm initial

size) were lower overall with a maximum of 0.34 $d^{-1}$ at 200 prey $l^{-1}$ in a 16-d experiment with final sizes of 52–84 mm at 20–200 prey $l^{-1}$. The gross growth efficiencies were generally in the range of 30–50% at 20–200 prey $l^{-1}$, with lower efficiencies at 200 prey $l^{-1}$ due to larger size and lower assimilation efficiency. More carbon was allocated to somatic growth than reproduction, and egg production changed the gross growth efficiencies little. Individual growth models have been constructed by Kremer (1976) and Kremer & Reeve (1989). Mean specific growth rates of Black Sea *Mnemiopsis* at prey densities of 60–100 copepods $l^{-1}$ ranged from 0.08 to 0.24% DW $d^{-1}$, increasing with food concentration (Finenko et al., 1995). Those rates are much lower than reported from the U. S., possibly due to use of small containers (1 l) and dry weight instead of carbon.

Egg production in *Mnemiopsis* varies with ctenophore size, food availability and temperature (Kremer, 1976; Reeve et al., 1989; Zaika & Revkov, 1994) (Table 4). In U. S. waters, appreciable egg production generally begins at about 30 mm length (1.2 g WW, 1 mg C), and fecundity increases linearly with ctenophore size, but is variable *in situ* (Kremer, 1976). The maximum egg production observed in a field-collected specimen (27 g WW) was 14 000 eggs $d^{-1}$ (Kremer, 1976). In the laboratory, well-fed ctenophores (80 mm and 25 g WW) produced 1000–3000 eggs $d^{-1}$. Each egg was calculated to contain about 0.1 $\mu$g C. Egg production increases with prey density, but represents a small fraction (<0.03% $d^{-1}$) of the body carbon (Reeve et al., 1989). Egg production ceases after only 2–4 d without food, and resumes in a similar time frame (U. Båmstedt, unpublished data). Thus, egg production is very sensitive to food conditions.

Egg production occurred at all temperatures (11.8–29 °C) and salinities (5.6–29.8) sampled in mid-April to early May and in mid-June to mid-August, 1997 and 1998 in Chesapeake Bay (Brink & Purcell, unpublished data). Weight-specific egg production was greater in summer (Fig. 1a), when temperatures, salinities, and prey densities all were higher, than in spring. Maximum spawning occurs between 02:00 and 04:00 h (P. Kremer, U. Båmstadt, L. Brink, unpublished data).

Black Sea ctenophores begin sexual reproduction at 10 mm oral-aboral length (13.5 mg DW) with optimal food (Finenko et al., 1995). During the next 10 d, they produce up to 12 000 eggs, however, fecundity depends upon prey densities and environmental conditions (Zaika & Revkov, 1994). Egg size is 0.25–0.3

*Table 4.* Specific growth and gross growth efficiencies (GGE) calculated from carbon biomasses, and egg production of *Mnemiopsis* relative to prey concentration at 26 °C. In Experiment 1, 15 ctenophores (6 mm initial length) were maintained for 4 d. In Experiment 2, 8 ctenophores (17 mm initial length) were maintained for 16 d (from Reeve et al., 1989)

| Experiment 1 | | | Experiment 2 | | | |
|---|---|---|---|---|---|---|
| Prey (No. $l^{-1}$) | Specific growth ($d^{-1}$) | GGE (%) | Specific growth ($d^{-1}$) | GGE (%) | Total eggs (No. ind.$^{-1}$) | GGE with eggs (%) |
| 20 | 0.43 | 32.3 | 0.23 | 44.2 | 42 | 44.5 |
| 50 | 0.67 | 40.6 | 0.29 | 45.3 | 196 | 46.0 |
| 100 | 0.83 | 39.1 | 0.31 | 36.1 | 531 | 36.9 |
| 200 | 0.87 | 28.1 | 0.34 | 28.2 | 1431 | 29.5 |

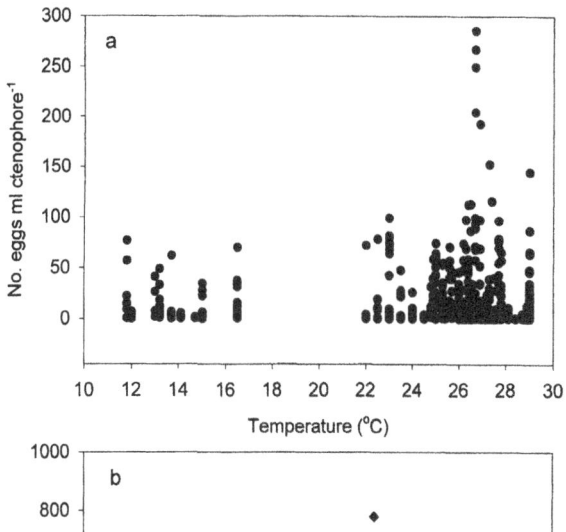

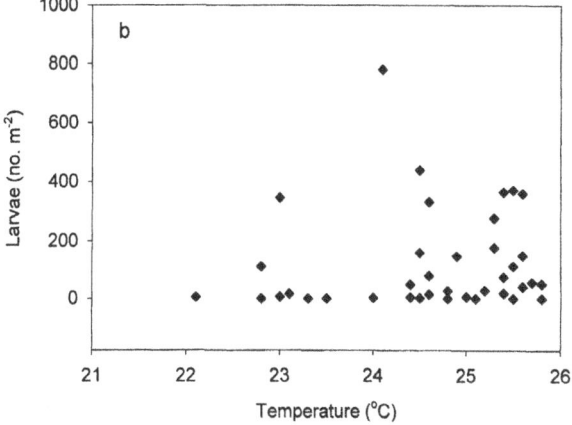

*Figure 1.* (a) Size-specific egg production of *Mnemiopsis* from ship board incubations *versus* surface water temperature in Chesapeake Bay during April and August, 1997 and 1998. Each dot represents one ctenophore (Brink & Purcell, unpublished data). (b) numbers of *Mnemiopsis* larvae <1 mm diameter *versus* temperature in the Black Sea during August, 1994. Vertical tows were made from 150 to 0 m, or from the bottom to surface inshore using a Bogorov-Rass (BR) net with 500 $\mu$m mesh (Shiganova, unpublished data).

mm. The mean egg production rate in 1 l containers is low, only 3 eggs $d^{-1}$, and is not representative of natural conditions. Spawning in the Black Sea begins in late evening, and peaks at midnight to 02:00 h (Zaika & Revkov, 1994). Field observations and experiments showed that no spawning occurs in spring at 8–16 °C (Shiganova, unpublished results), and that intensive spawning occurs in late summer–autumn when temperatures reach 23 °C (Shiganova, 1998) (Fig. 1b).

## *Mnemiopsis* in native habitats

### *Environmental conditions and regional patterns*

Kremer (1994) reviews habitat characteristics of four locations from the northern to southern United States where *Mnemiopsis* has been studied (Table 5). The environments are shallow estuaries and bays that range in average depth from 2 to 30 m. *Mnemiopsis* also occurs in coastal waters (Mayer, 1912; Purcell et al., 2001). Temperatures range from 0 °C in northern locations in the winter, to 32 °C in the southern estuaries during the summer, and salinities range from ≤2 to 38. These environments also have high zooplankton biomasses, ranging from 11 to 200 mg C m$^{-3}$, where the copepod *Acartia tonsa* Dana predominates. Kremer (1994) concludes that three factors act in a hierarchy to determine the abundance of *Mnemiopsis*, with temperature being the most important, food availability second and mortality (predation) third.

The southern locations (Biscayne Bay, Florida, and Nueces Estuary, Texas) experience higher temperatures, 7–18 °C minima and 31–32 °C maxima than in the north (Kremer, 1994). The *Mnemiopsis* population in Biscayne Bay peaks in December to February, when temperatures are lower (~20 °C) than the rest of the year (Fig. 2). *Mnemiopsis* in Nueces Estuary experiences several small population peaks throughout the year (Fig. 2). The biomasses in both southern locations

*Table 5.* Comparison of systems over the range of *Mnemiopsis* in the U.S. Modified from Kremer (1994)

| Location | Av. depth (m) | Temp. (°C) | Salinity | Zooplankton biomass (mg C m$^{-3}$) | | | Ctenophore biomass (ml m$^{-3}$) | | | |
|---|---|---|---|---|---|---|---|---|---|---|
| | | | | Season | Peak | Peak range | Season | Peak | Peak range | No. years |
| Narragansett Bay, RI | 9 | 1–25 | 25–32 | June–July | 70[a] | 30–110 | Aug–Sept | 50 | 6–100 | >8 |
| Mid Chesapeake Bay, MD | 5–10 | 2–30 | 5–16 | Summer | 90[b] | 30–180 | June–Sept | 60 | 10–100 | 16 |
| Biscayne Bay, FL | 2 | 18–32 | 14–45 | Fall to winter | 11[c] | nd | Fall | 30 | nd | 1+ |
| Nueces Estuary, TX | 2.4 | 7–31 | 20–38 | Variable | 50 | nd | Summer | 15 | 8–20 | 1 |

[a] >153 $\mu$m fraction, assuming C = 35%DW, or 1 ml displacement volume = 60 mg C.

[b] Converted from counts assuming 3 $\mu$g C per copepodite or adult.

[c] >202 $\mu$m fraction, assuming C = 35% DW.

nd = no data.

are lower than in the north, with maxima around 20–30 ml m$^{-3}$ (Table 5).

The northern locations (Narragansett Bay, Rhode Island, and Chesapeake Bay, Maryland) are typical of temperate environments, with summertime (July–August) temperature maxima of 25–26 °C, and wintertime (February) minima of 1–2 °C (Table 5). The *Mnemiopsis* populations typically exhibit one annual peak in biomass, in August–September for Chesapeake Bay (Fig. 2), and in September–October for Narragansett Bay, although this is occurring earlier with rising temperatures in recent years (Sullivan et al., 2001). Peak biomasses in Narragansett Bay are up to about 100 ml m$^{-3}$ (50 mg C m$^{-3}$) (Kremer, 1994). A simulation model was constructed from experimental results on fecundity, feeding and metabolism to simulate the seasonal dynamics of ctenophore population biomass in Narragansett Bay (Kremer, 1976). In Chesapeake Bay, *Mnemiopsis* biomasses were <50 ml m$^{-3}$ in summers (1987, 1988, 1991) with abundant scyphomedusan predators, *Chrysaora quinquecirrha* (Desor) (Purcell et al., 1994a, b), however, peak biomasses sampled in the summers of 1995–1998 ranged from 200 to 600 ml m$^{-3}$ when scyphomedusae were less abundant with a peak baywide average of 60 ml m$^{-3}$ (Purcell et al., unpublished data). *Mnemiopsis* also occurs during the winter in Chesapeake Bay, generally in low numbers and biomass (Miller, 1974; Burrell & Van Engel, 1976; Purcell, unpublished data). Subsequently, we focus on *Mnemiopsis* in Chesapeake Bay, where conditions are similar to the Black Sea, and where the most extensive recent research has been conducted.

The Chesapeake Bay is a partially-mixed temperate estuary with 70% of the freshwater entering from the Susquehanna River at the northern end. There also is an extensive system of smaller tributaries that, to-

gether with the mainstem bay, total 11 500 km$^2$ in surface area and 74 km$^3$ in volume. Chesapeake Bay is shallow, with the greatest depth of 30 m and average depth of the mainstem and its tributaries of 6.5 m. Water flow and salinities vary seasonally, with strong spring runoff from rain and snowmelt, and low freshwater inputs during summer and fall. Substantial variation in biological processes are driven by interannual differences in precipitation (e.g. Malone et al., 1999). Saline water from the ocean enters the bay mouth and flows north in the bottom layer. The system is strongly stratified from May through October, and hypoxic waters are persistent below the pycnocline during that period. Circulation is driven by the two-layer structure of the estuary, and is also affected by winds, tides, and gravitational forces. The two-layer circulation pattern and long (320 km), narrow (mean 20 km) shape of Chesapeake Bay contribute to retention of nutrients and organisms within the bay. More complete discussions of the characteristics of Chesapeake Bay are in Malone et al. (1999).

Chesapeake Bay has been strongly affected by human populations since European settlement in the 1600s (Malone et al., 1999). Eutrophication accelerated in the mid-1800s when use of inorganic fertilizers began. Today, 29% of the drainage basin is in agricultural use and 11% is urban. The total annual loadings of nitrogen and phosphorous are estimated to be 6.2 and 16.5 times greater than before European settlement. The Chesapeake Bay now exhibits many symptoms of extensive human utilization, including nutrient enrichment, bottom-water anoxia, harmful algal blooms, fish kills, habitat loss, over-exploitation of fish and invertebrate fisheries, and establishment of exotic species (in Malone et al., 1999).

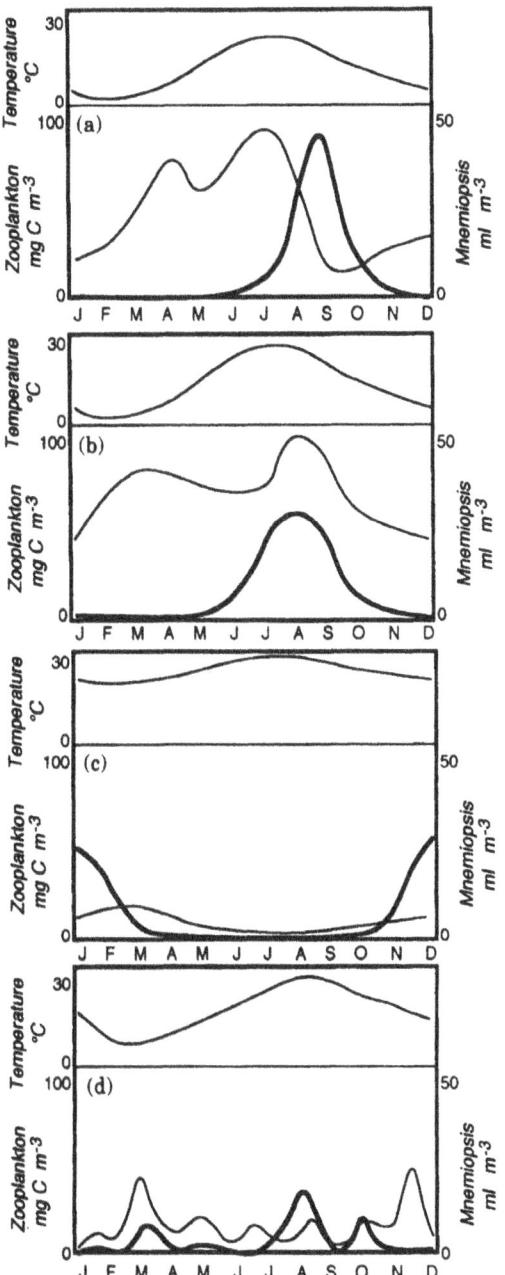

*Figure 2.* Schematic representations of annual cycles for temperature (upper portion), zooplankton biomass (light line), and *Mnemiopsis* biomass (heavy line) for four regions of the United States. (a) Northern (Narragansett Bay); (b) mid-Atlantic (Chesapeake Bay); (c) South Atlantic (Biscayne Bay). (d) Texas Bays (Nueces Estuary) (From Kremer, 1994).

## Distribution and abundance of Mnemiopsis

### Vertical distribution

Studies of the vertical distribution of *Mnemiopsis* have given different results, perhaps depending on wa-

ter depth and degree of vertical stratification. Miller (1974) found most ctenophores to be near the surface in the daytime, and evenly distributed at night on one date in November in a ≤6 m water column that probably was well mixed (Pamlico Sound, North Carolina). He also reported that the ctenophores leave the surface under choppy conditions. Kremer & Nixon (1976) also found homogeneous vertical distributions in nighttime sampling in Narragansett Bay at ≤8 m depth. In a deeper, stratified water column (bottom depth 16–27 m) in Chesapeake Bay, *Mnemiopsis* had greater biomasses above the pycnocline (about 11 m depth) during the day and night than below it in day or night when two depth intervals (surface-pycnocline and pycnocline-bottom) were sampled with an opening-closing net (Purcell et al., 1994b). Fine-scale observations by SCUBA divers and video showed no ctenophores below the sharp pycnocline at dissolved oxygen concentrations <0.2 mg l$^{-1}$ (Purcell, M. B. Decker & D. L. Breitburg, unpublished results).

Discrete depth sampling with a Tucker trawl in the Patuxent River tributary of Chesapeake Bay indicated that *Mnemiopsis* occurred in bottom waters with low dissolved oxygen concentrations (Keister et al., 2000). *Mnemiopsis* was rare or absent in samples from bottom waters with 0.2 mg l$^{-1}$ dissolved oxygen, but was found at higher densities in the bottom layer than in surface or pycnocline waters when bottom dissolved oxygen concentrations were 1.3 and 2.3 mg l$^{-1}$. Laboratory experiments also show great tolerance of *Mnemiopsis* to low dissolved oxygen concentrations. In 96-h duration experiments, survival of ctenophores (mean 2 cm length, 0.7 ml volume) was 100% at dissolved oxygen concentrations ≥0.5 mg l$^{-1}$ (Purcell et al., 2001).

### Seasonal dynamics

*Mnemiopsis* population sizes in temperate locations are small during cold winter temperatures, and increase with reproduction in the spring (Kremer, 1994). The greatest numbers of ctenophores in Chesapeake Bay occur in the spring, when numerous, small (<1 g WW) individuals occur (Purcell, 1988; Purcell et al., 1994a). Densities of larval ctenophores were measured by use of a camera net system up to 76 m$^{-3}$ in June 1992 (Purcell et al., unpublished data).

*Mnemiopsis* is found over a wide range of salinity, from ≤2 to 38 (reviewed in Kremer, 1994; Purcell et al., 1999a). At low winter temperatures in Chesapeake Bay waters, ctenophores were found only in higher salinities (>6) than in the summer (Miller, 1974; Burrell

152

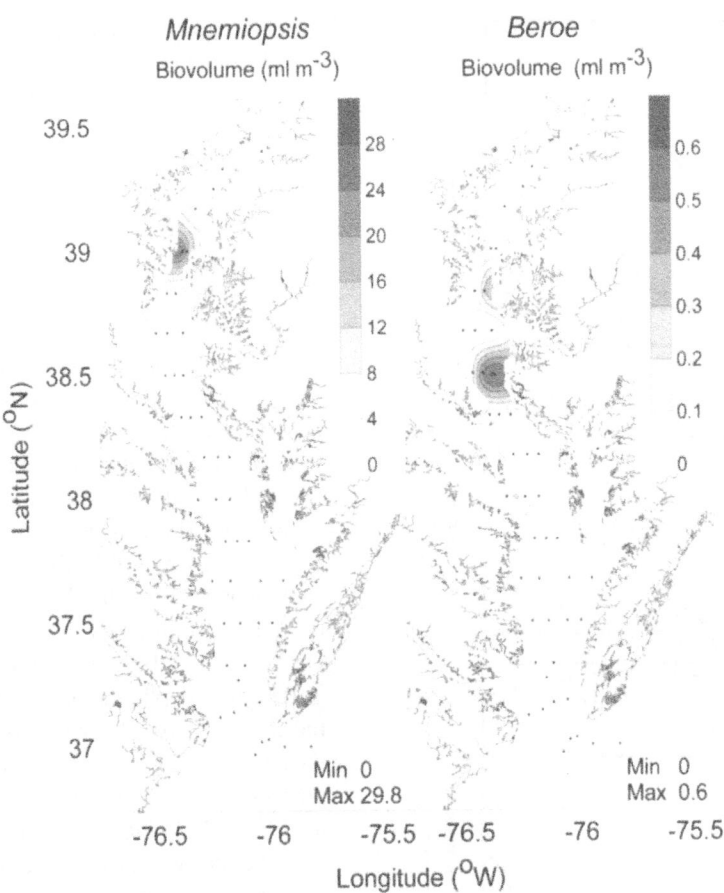

*Figure 3.* Biovolumes (ml m$^{-3}$) of *Mnemiopsis* and their predator, the ctenophore, *Beroe ovata*, in October, 1995 throughout the water column in Chesapeake Bay. Ctenophores were collected in a 1 m$^2$ Tucker trawl with 280 $\mu$m mesh (Purcell et al., unpublished data).

& Van Engel, 1976). In the summer, low and high salinity waters may serve as a refuge from predation by the scyphomedusan, *Chrysaora quinquecirrha*, which prefers salinities of 10–25 (Cargo & Schultz, 1966, 1967; Purcell et al., 1999b).

The greatest biomass of *Mnemiopsis* in Chesapeake Bay occurs in July and August, when individuals are large (1–50 g WW) and abundant (means 30–60 ml m$^{-3}$; Table 6). Ctenophore distribution and abundance do not appear to be reduced by physical parameters or prey availability in summer, however, predation by *Chrysaora quinquecirrha* medusae can have a great effect (see predators section). Ctenophore biovolumes decline in the autumn when temperatures decrease, and are low in October ($\leq$1 ml m$^{-3}$; Fig. 3).

*Factors affecting interannual variation in population size*

In areas where *Mnemiopsis* populations have been sampled for several years, they show marked variation in abundances and biomasses. We show previously unpublished data from Chesapeake Bay collected in late April, mid-July–early August, and late October, 1995–1998, when the entire water column was sampled using a Tucker Trawl with a 1 m$^2$ opening and 280 $\mu$m mesh nets (Tables 6 and 7, Figs 3–6). Those surveys suggest that interannual variation in springtime temperatures affects ctenophore size. Low spring (April) temperatures (11–13.5 °C) are significantly related to smaller ctenophore sizes than when temperatures were higher (14–15.5 °C) (Fig. 4), however, there is no significant relationship between salinity

*Table 6.* Densities and biomass estimates for *Mnemiopsis* ctenophores and their predators, *Chrysaora quinquecirrha* medusae, sampled from the whole water column of Chesapeake Bay, and surface temperature and salinity during July in four years. Organisms were collected in a 1 m² Tucker trawl with 280 μm mesh. Numbers are means (±1 standard error) (Purcell et al., unpublished data)

|  | 23–28 Jul 1995 | 17–25 Jul 1996 | 11–23 Jul 1997 | 5–12 Aug 1998 |
|---|---|---|---|---|
| No. of stations | 47 | 38 | 50 | 49 |
| *Mnemiopsis* density (no. m⁻³) | 2.8 (±1.2) | 12.7 (±1.9) | 11.9 (±2.6) | 5.7 (±1.1) |
| *Mnemiopsis* biovolume (ml m⁻³) | 31.9 (±13.2) | 60.0 (±9.2) | 40.5 (±7.3) | 38.9 (±6.8) |
| *Mnemiopsis* length (mm) | 55.5 (±1.8) | 33.6 (±1.0) | 37.4 (± 2.5) | 41.1 (±1.4) |
| *Chrysaora* density (no. m⁻³) | 0.02 (±0.0) | 0.001 (±0.0) | 0.004 (±0.0) | 0.002 (±0.0) |
| *Chrysaora* biovolume (ml m⁻³) | 0.8 (±0.2) | 0.3 (±0.2) | 0.4 (±0.1) | 0.3 (±0.1) |
| Surface temperature (°C) | 29.2 (±0.1) | 25.6 (±0.2) | 26.8 (± 0.1) | 26.7 (±0.1) |
| Surface salinity | 15.4 (±0.7) | 8.2 (±1.1) | 13.2 (±0.9) | 14.8 (±0.7) |

*Table 7.* Estimated potential clearance of copepods, fish eggs and fish larvae by *Mnemiopsis leidyi,* and densities fish eggs and fish larvae in the whole water column, and surface zooplankton densities in Chesapeake Bay during summer in four years. Clearance rates ($C$ in l ind⁻¹ d⁻¹) were calculated according to the following equations, and multiplied by ctenophore densities at each station: zooplankton ($C = 11.22$ $WW^{0.5413}$; Purcell, unpublished data), fish eggs from field data ($C = 128$ l ind⁻¹ d⁻¹; Purcell et al., 1994a), fish larvae with alternative prey (0.7 l d⁻¹ ml ctenophore⁻¹; Cowan & Houde, 1993a, b). Zooplankton sampling was by a 1-m Tucker trawl, 280 μm mesh in 1995, and by pump, 200 μm mesh in 1996–1998. Ichthyoplankton densities are from the Tucker trawl in all years. Numbers are means (±1 standard error). Abundance data (unpublished) are from Houde (ichthyoplankton) and M. R. Roman (zooplankton). Temperatures, and salinities as in Table 6

|  | 23–28 Jul 1995 | 17–25 Jul 1996 | 11–23 Jul 1997 | 5–12 Aug 1998 |
|---|---|---|---|---|
| Number of stations | 47 | 38 | 50 | 49 |
| Clearance of copepods (l m⁻³ d⁻¹) | 109.3 (±45.7) | 321.9 (±47.4) | 233.9(±43.3) | 174.1 (±30.8) |
| Clearance of fish eggs (l m⁻³ d⁻¹) | 358.3 (±157.3) | 1623.9 (±236.4) | 1526.8(±337.5) | 728.2 (±133.9) |
| Clearance of fish larvae (l m⁻³ d⁻¹) | 22.4 (±9.2) | 42.0 (±6.4) | 28.4 (±5.1) | 27.2 (±4.8) |
| Zooplankton (no. m⁻³) | 4463.1 (±760.7) | 787.1 (±135.0) | 7478.3 (±1176.9) | 7503.8 (±1621.4) |
| Fish eggs (no. m⁻³) | 5.0 (±1.5) | 1.2 (±0.4) | 28.8 (±4.7) | 23.9 (±11.5) |
| Fish larvae (no. m⁻³) | 1.4 (±0.4) | 0.3 (±0.1) | 3.2 (±0.7) | 3.3 (±0.7) |

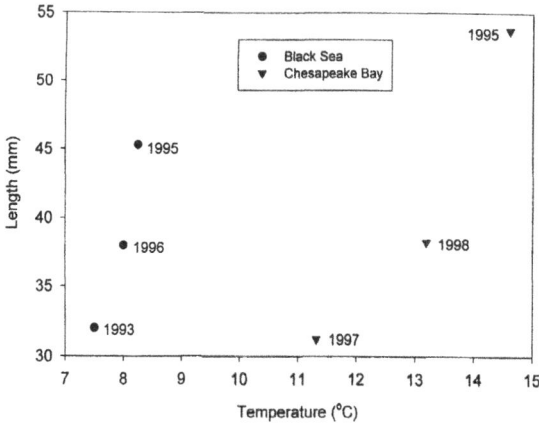

*Figure 4. Mnemiopsis* average length relative to temperature in April of 3 years in Chesapeake Bay (triangles), and in March of 3 years in the Black Sea (circles). Ctenophores in Chesapeake Bay were collected in a 1 m² Tucker trawl with 280 μm mesh (Purcell et al., unpublished data).

(7–16) and ctenophore body size (data not shown). Large body size in the spring could lead to large ctenophore populations by increasing reproduction, and because large ctenophores are less susceptible to predation by *Chrysaora quinquecirrha* medusae, which generally appear after June (Purcell & Cowan, 1995; Kreps et al., 1997). Earlier multiple-year studies have not differed much in springtime temperatures or did not evaluate the possible effects of temperature on subsequent ctenophore population size (Miller, 1974; Kremer, 1976; Deason & Smayda, 1982), however, Sullivan et al. (2001) suggests that increasing temperatures from 1950 to 1999 have resulted in earlier seasonal maxima of *Mnemiopsis* in Narragansett Bay.

Zooplankton abundance may directly affect ctenophore abundance. *Mnemiopsis* is found in waters with high zooplankton stocks, and were missing where zooplankton biomass was <3 mg C m⁻³ (reviewed by

154

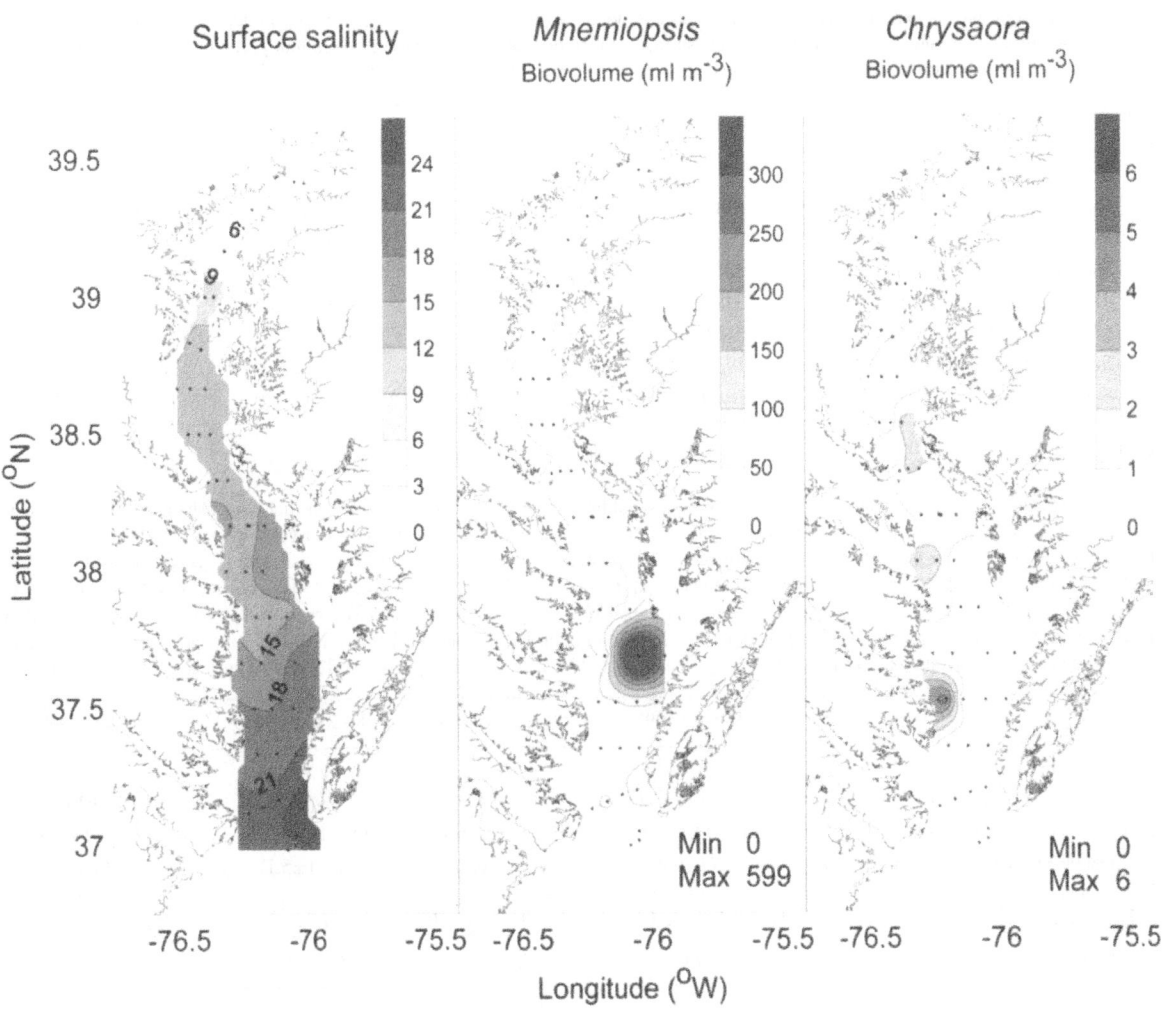

*Figure 5.* Biovolumes (ml m$^{-3}$) of *Mnemiopsis* and their predator, the scyphomedusan, *Chrysaora quinquecirrha*, throughout the water column, and surface layer salinity in July, 1995. Organisms were collected in a 1 m$^2$ Tucker trawl with 280 $\mu$m mesh (Purcell et al., unpublished data). Stations, dates, temperatures, and salinities are the same as in Tables 6, 7 and 9 (summer). The winter–spring of 1995 had low levels of precipitation.

Kremer, 1994). In the northern U. S. locations, peak zooplankton biomass averaged about 100 mg C m$^{-3}$, and in the southern locations, peak zooplankton biomass averaged 25–50 mg C m$^{-3}$ (reviewed by Kremer, 1994). In Narragansett Bay, densities of *Mnemiopsis* integrated over July and August ranged from 1 to 11 ctenophores m$^{-3}$ and were strongly correlated ($r$ = 0.92) with zooplankton biomass (60–190 mg DW m$^{-3}$) at the beginning of that period in 6 years (Deason & Smayda, 1982). Therefore, zooplankton can have a great influence on *Mnemiopsis* population abundance.

Interannual variation in abundance of *Mnemiopsis* is strongly related to predator abundance in U.S. waters. Ctenophores in the genus *Beroe* are known predators of other ctenophore species. The popula-

tion abundance of *Mnemiopsis* in Narragansett Bay decreased dramatically in September, 1974 with increasing numbers of *Beroe ovata* Brugiére (in Kremer & Nixon, 1976). This pattern contrasted with other years sampled (1971, 1972), when *B. ovata* did not occur, and *Mnemiopsis* populations decreased later in the autumn.

*Beroe ovata* is found in Chesapeake Bay only at relatively high salinities and, therefore, they do not reduce *Mnemiopsis* during most of its season through much of the bay. *B. ovata* appears in mid summer (July) in the lower bay, where they seemed to eliminate *Mnemiopsis* from southern stations in the York River tributary (Burrell & Van Engel, 1976). *B. ovata* is first found in late summer (August–September) in mid-bay.

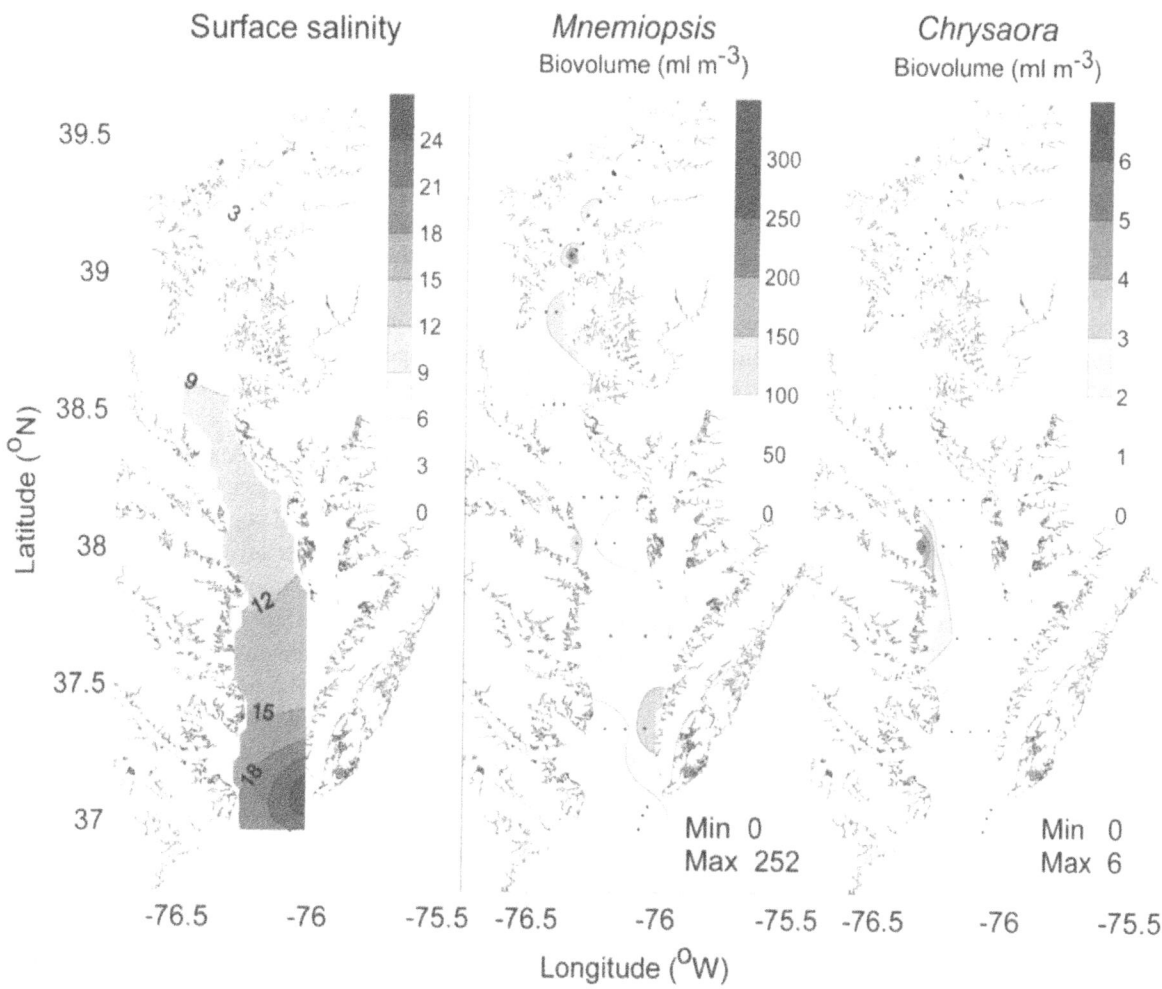

*Figure 6.* Biovolumes (ml m$^{-3}$) of *Mnemiopsis* and their predators, *Chrysaora quinquecirrha* medusae, throughout the water column, and surface layer salinity in July, 1996 (Purcell et al., unpublished data). Methods, stations, dates, temperatures and salinities are the same as in Figure 5. The winter–spring of 1996 had record high levels of precipitation.

During October 1995–1998, baywide average densities of *B. ovata* were ≤0.1 m$^{-3}$ (0.3–0.8 ml m$^{-3}$), while *Mnemiopsis* densities were 0.1–0.4 m$^{-3}$ (0.7–1.2 ml m$^{-3}$). In October 1995, when *B. ovata* was abundant, the population centers of the two ctenophore species did not overlap, and may suggest a predation effect (Fig. 3).

*Cyanea capillata* (Linnaeus) scyphomedusae generally occur in low numbers (<0.2 m$^{-3}$) in spring in the mesohaline mainstem bay (Purcell et al., unpublished data), and usually disappear by May. *C. capillata* is known to eat ctenophores (Båmstedt et al., 1997), however, the effects of *C. capillata* on *Mnemiopsis* populations are unknown.

*Chrysaora quinquecirrha* scyphomedusae first appear in May or June in the tributaries of the mesohaline region of Chesapeake Bay when temperatures exceed

17 °C, and about a month later in the mainstem bay. The medusae are more abundant in the tributaries, where densities were measured at 16 medusae m$^{-3}$ (Purcell, 1992), than in the main channel, where densities typically are <0.2 medusae m$^{-3}$ (Purcell et al., 1994a, 1999b; Table 6). Polyps and medusae of *C. quinquecirrha* are found in Chesapeake Bay at salinities above 5 and below 25 (Cargo & Schultz, 1966, 1967), and budding of ephyrae from the polyps is reduced at salinities <10 and >25 (Purcell et al., 1999b).

Abundances of *Mnemiopsis* and *Chrysaora quinquecirrha* medusae have been shown to vary inversely in tributaries of Chesapeake Bay (Miller, 1974; Feigenbaum & Kelly, 1984; Purcell & Cowan, 1995). Ctenophores in the tributaries are numerous in spring before the medusae are large or abundant, then decrease or disappear when medusae become numerous, and rebound when the medusae die in the autumn (op. cit.). Predation rates on ctenophores by medusae were sufficient to eliminate ctenophores from the tributaries, where medusae were abundant, but not in the main bay, where medusae were less abundant (Purcell & Cowan, 1995). Densities and biomasses of medusae and ctenophores also vary inversely in the mainstem Chesapeake Bay (Table 6).

The outcome of interactions between *Mnemiopsis* and *Chrysaora quinquecirrha* depend on the relative sizes of the predator and prey. Predation begins early in the life histories, when ephyrae consume larval ctenophores at higher rates than protozoan or crustacean zooplankton (Olesen et al., 1996). Ephyrae and small medusae (1–23 mm diameter) consumed whole ctenophores that were less or equal in length to the diameter of the medusa, and ate parts of ctenophores that were larger (Purcell & Cowan, 1995). Larger ctenophores (25–85 mm long) escaped from 97% of free-swimming contacts with medusae 30–150 mm in diameter (Kreps et al., 1997).

The apparent refuge of large size has important implications for ctenophore populations. If medusae are early and numerous, the ctenophore population can be eliminated from the tributaries and limited in the main bay, as in 1995 (Fig. 5). If, by contrast, ctenophores are large or numerous before medusae appear, or the medusa population is suppressed because of low temperatures or salinities, then medusae do not restrict the ctenophore population in Chesapeake Bay, as in 1996 (Fig. 6).

A variety of fishes are known to consume gelatinous species, including ctenophores (reviewed by Purcell & Arai, 2001). Harvestfish, *Peprilus alepidotus* (Linnaeus), and butterfish *P. triacanthus* (Peck), are known predators of *Mnemiopsis* (in Harbison, 1993; GESAMP, 1997), but because of the low salinities in much of Chesapeake Bay, they are found mostly in the southern bay. Oviatt & Kremer (1977) estimated that butterfish could eat 4–184 ml of ctenophore g fish $DW^{-1} h^{-1}$, and that this predation could account for the autumn decline of the *Mnemiopsis* population in Narragansett Bay. No estimates of fish predation on *Mnemiopsis* exist elsewhere.

### Trophic ecology of Mnemiopsis

#### Prey consumed

The gut contents of *Mnemiopsis* in U.S. estuaries include a wide variety of prey taxa, with 75–93% of the items being copepods, copepod nauplii, barnacle nauplii and bivalve veligers (Nelson, 1925; Burrell & Van Engel, 1976; Larson, 1987). Other common prey items include polychaete larvae, cladocerans, crab and shrimp larvae, epibenthic crustaceans and ichthyoplankton (op. cit.). In Chesapeake Bay in August, 50 ml specimens contained an average of 374 prey each, including copepod nauplii (59%), copepods (38%), bivalve veligers (1.6%), cladocerans (0.8%) and barnacle nauplii (0.5%) (Purcell, unpublished data). In October, *Mnemiopsis* larvae $\leq 5$ mm diameter from Chesapeake Bay contained diatoms (52% of the prey items), dinoflagellates (34%) and ciliates (14%), averaging 15 prey $larva^{-1}$; larvae 6–10 mm contained diatoms (55%), dinoflagellates (25%), ciliates (15%), copepod nauplii (3%) and copepods (2%), averaging 31 prey $larva^{-1}$ (Purcell et al., unpublished data). Larval and lobate *Mnemiopsis* consumed ciliates in laboratory experiments (Stoecker et al., 1987).

The time required for digestion of prey items by *Mnemiopsis* increases with prey size and number. Gut clearance (digestion) times for 1–10 prey items averaged 0.5 h for prey <1 mm (copepod nauplii, *Oithona* copepods) and 1 h for prey 1–2 mm (*Acartia* copepods) at 25–27 °C (Larson, 1987). Digestion times at 21 °C by larval ctenophores were approximately twice those estimates for copepod nauplii (1.1 h) and copepods (2.2 h; Stanlaw et al., 1981).

Prey selection does not appear to be strong in *Mnemiopsis*, since many authors have concluded there is no selection. Deason (1982) reports selection of copepod nauplii by *Mnemiopsis* larvae, but Stanlaw et al. (1981) infers selection for copepodites and adults. Nevertheless, selection by lobate specimens was positive in 6 of 6 samples for the calanoid copepod, *Acartia tonsa*, harpacticoid copepods, barnacle nauplii and bivalve veligers, and negative for the cyclopoid copepod, *Oithona* and copepod nauplii (4 of 6 samples; Larson, 1987). Clearance rates by *Mnemiopsis* differ dramatically among various prey types, from 2.4 l d^{-1} for *Oithona* copepods and 17–53 l d^{-1} for *Acartia* copepods to 13–366 l d^{-1} for bay anchovy eggs (Table 8). Kremer (1979) found that feeding rates on cyclopoids, and bivalve veligers were 25% of the feeding rates on calanoids and cladocerans. Such differences may reflect the catchability of the prey (Purcell, 1997). Costello et al. (1999) examined individual copepod–ctenophore interactions and found that capture success of *A. tonsa* and *Oithona colcarva* Bowman copepods was similarly high (74%) after contact with the interior surfaces of the ctenophore's oral lobes. Larson (1987)

*Table 8.* Clearance rates (liters cleared ctenophore$^{-1}$ d$^{-1}$) by *Mnemiopsis* approximately 50 mm in length (14 ml volume and 14 g WW). Clearance rates were calculated from data in the sources

| Prey type | Container size | Temperature (°C) | Clearance rates | Source |
|---|---|---|---|---|
| *Acartia* copepods | 4 l | 22–24 | 32.2 | Quaglietta (1987) |
| *Acartia* copepods | 20 l | 20–25 | 16.8 | Kremer (1979) |
| *Acartia* copepods | 1 m$^3$ | 25 | 46.0 | Purcell (unpublished data) |
| *Acartia* copepods | 68 m$^3$ | 27–31 | 52.8 | Walter (1976) |
| *Acartia* copepods | field | 27–31 | 21.6 | Larson (1987) |
| *Oithona* copepods | field | 27–31 | 2.4 | Larson (1987) |
| Copepod nauplii | field | 27–31 | 4.8 | Larson (1987) |
| Barnacle nauplii | field | 27–31 | 31.2 | Larson (1987) |
| Bivalve veligers | field | 27–31 | 14.4 | Larson (1987) |
| Bivalve veligers | 4 l | 22–24 | 20.7 | Quaglietta (1987) |
| Bay anchovy eggs | 15 l | 21–24 | 13–42[a] | Monteleone & Duguay (1988) |
| Bay anchovy eggs | 200 l | 21–24 | 60–170[a] | Monteleone & Duguay (1988) |
| Bay anchovy eggs | 750 l | 21–27.6 | 128[a], 82 | Cowan & Houde (1993b) |
| Bay anchovy eggs | 3 m$^3$ | 21–27.6 | 366±58[a] | Cowan & Houde (1993a) |
| Bay anchovy eggs | field | 26 | 128±58[b] | Purcell et al. (1994a) |
| Bay anchovy larvae | 3 m$^3$ | 21–27.6 | 172[a], 15 | Cowan & Houde (1993a, b) |

[a] Without alternative prey.

[b] Mean ctenophores volume = 41 ml.

and Costello et al. (1999) speculate that the less active swimming of *Oithona* and copepod nauplii results in fewer encounters with ctenophore feeding surfaces than the active swimming of *Acartia* and barnacle nauplii.

*Feeding rates*

The feeding rates of *Mnemiopsis* have been measured in several studies. Typically, laboratory experiments measure the changes in prey density over time, and clearance rates are calculated from the following equation:

$$C = [V/(n \times t)]\ln(C_0/C_t), \qquad (1)$$

where $V$ = volume of the experimental container (l); $n$ = number of ctenophores; $t$ = incubation time (h); and $C_0$ and $C_t$ = number of prey organisms at times 0 and $t$, respectively.

Several important insights into the characteristics of feeding by gelatinous species have emerged. First, feeding by lobate *Mnemiopsis* is directly proportional to prey concentration over an extremely wide range of prey density ($\leq$3600 copepods l$^{-1}$) (Bishop, 1967; Miller, 1970; Reeve et al., 1978; Kremer, 1979), except at densities <4 prey l$^{-1}$ (Reeve et al., 1989). Therefore, lobate *Mnemiopsis* does not satiate at natural prey concentrations. Feeding behavior is not necessarily invariable, as starved animals

have higher ingestion rates than fed ones (Reeve et al., 1989), and swimming activity differs in the presence of prey (Reeve & Walter, 1978; Reeve et al., 1989). Feeding by tentaculate larvae <10 mm appears to satiate above 300 prey l$^{-1}$ (Reeve & Walter, 1978; Deason, 1982). Second, although large ctenophores clear greater volumes of water than small ones, the weight specific clearance rates decrease with increasing ctenophore size (Kremer, 1979). Third, clearance rates were lower at low temperatures (10–15 °C) than at high temperatures (20–25 °C; Kremer, 1979). Fourth, clearance rates increase with increasing container size (Table 8). Fifth, clearance rates differ among various kinds of prey (Table 8).

Measurements of clearance rates by *Mnemiopsis* feeding on *Acartia tonsa* copepods vary among studies, at least in part due to differences in container size. We compare clearance rates for ctenophores 50 mm in length (about 14 ml V and 14 g WW) (Table 8). From Kremer's (1979) experiments in 20 l containers, a ctenophore of that size would clear 16.8 l d$^{-1}$, but in 1 m$^3$ mesocosms, it would clear 46.0 l d$^{-1}$ (Purcell, unpublished data), and in 68 m$^3$ containers, it would clear 52.8 l d$^{-1}$ (Walter, 1976). This indicates that feeding by *Mnemiopsis* increases with increasing container size, especially at container volumes <1 m$^3$. From field gut contents and digestion rates, Larson

(1987) estimated clearance rates on *A. tonsa* of 21.6 l ctenophore$^{-1}$ d$^{-1}$, which is surprisingly low, because feeding rates of other species determined from gut contents generally are higher than from containers (reviewed in Purcell, 1997).

*Effects on zooplankton populations*

All areas of Chesapeake Bay have abundant mesozooplankton populations, reflecting the high productivity of the bay. Species indigenous to both fresh and estuarine waters are found in the oligohaline regions (Brownlee & Jacobs, 1987). Cladocerans and cyclopoid copepods can be abundant, but calanoid copepods predominate, with *Acartia tonsa* in summer (May–November) and *Eurytemora affinis* (Poppe) in winter (December through April). In the mesohaline regions, *E. affinis* and *A. tonsa* predominate, as above, with some cyclopoid copepods (*Oithona colcarva*), and larvae of benthic invertebrates (polychaetes, barnacles, bivalves) sometimes being extremely abundant. Species diversity is greatest in the polyhaline southern bay, where both estuarine and marine species occur. There, *A. tonsa* is the dominant copepod from June through December, and *A. hudsonica* Pinhey is most abundant from January through May, and meroplankton also are abundant and diverse during summer. Total zooplankton densities and biomasses in the bay are high from May through September (Brownlee & Jacobs, 1987; Purcell et al., 1994a). Production of *A. tonsa* in the mesohaline region is high during that period, ranging from 0.3 to 1.2 $\mu$g C ($\mu$g C)$^{-1}$ d$^{-1}$, equivalent to 30–120% of female C daily, with highest rates in summer–early fall (Purcell et al., 1994a). In summer, densities of copepods can reach 73 000 m$^{-3}$, and nauplii can exceed 100 000 m$^{-3}$ in the mesohaline region (64 $\mu$m mesh; Roman et al., 1993). Zooplankton (280 $\mu$m mesh) in July varied among years (1995–1998) from baywide averages of 800–8230 m$^{-3}$ (Table 7).

Several studies report inverse correlations of *Mnemiopsis* and zooplankton abundances (summarized in Purcell, 1988), however, some estimates of zooplankton consumption do not support the inference that ctenophore predation reduced zooplankton populations. *Mnemiopsis* were estimated to consume <7% d$^{-1}$ of the copepod standing stock in the mainstem mesohaline portion of Chesapeake Bay in 1987 and 1988, when *Chrysaora quinquecirrha* medusae were abundant, with clearance of zooplankton being greatest during July and August (Purcell et al., 1994a). *Mnemiopsis* did not appear to control copepod populations during those years in the mainstem bay, where copepod production averaged 87% d$^{-1}$ (Purcell et al., 1994a). Potential clearance of zooplankton by *Mnemiopsis* in July 1995–1998 was estimated by use of the following experimentally-measured clearance rate equation, the numbers of ctenophores m$^{-3}$ and mean ctenophore size at each station (Table 7). Clearance rates were calculated from the following equation, which was determined from 48 h experiments in 1 m$^3$ mesocosms on 99 ctenophores 3–25 g WW: $C = 11.22WW^{0.5413}$, where $C$ = clearance in 1 ind.$^{-1}$ d$^{-1}$, and WW = grams wet weight ($r^2 = 0.65$, $p < 2 \times 10^{-12}$; Purcell, unpublished data). In years with smaller *Mnemiopsis* populations (1995, 1998), baywide average clearance by ctenophores was estimated to remove 11–17% d$^{-1}$ of the zooplankton, but when ctenophore populations were greater (1996, 1997), estimated average clearance was 23–32% d$^{-1}$ (Tables 6 and 7). Zooplankton densities were markedly lower in 1996 when *Mnemiopsis* predation was greatest.

*Effects on ichthyoplankton*

Eggs and larvae of fishes are abundant in Chesapeake Bay and are dominated by bay anchovy, *Anchoa mitchilli* (Valenciennes). Bay anchovy eggs and larvae constituted, on average, 84.6 and 84.2%, respectively, of the combined-species ichthyoplankton abundances during July cruises in 1995–1998. Peak spawning by bay anchovy and great abundances of *Mnemiopsis* coincide in July (Cowan & Houde, 1993a; Rilling & Houde, 1999a, b). Anchovy eggs and larvae are eaten by *Mnemiopsis* (Table 8), and anchovy larvae and adults potentially compete with ctenophores for zooplankton prey. Negative correlations between bay anchovy larval abundances and gelatinous zooplankton biovolumes are usually observed in Chesapeake Bay (MacGregor & Houde, 1996; Rilling & Houde, 1999a), suggesting significant control via competition or predator-prey interactions. Densities of bay anchovy ichthyoplankton vary by more than an order of magnitude from year to year; during July, when anchovy eggs and larvae and ctenophores are abundant, mean densities of eggs and larvae range from 1.2 to 28.8 and 0.3 to 3.3 m$^{-3}$, respectively (Table 7). Thus, their potential role as prey and as competitors with ctenophores must vary substantially interannually.

Clearance rates of *Mnemiopsis* feeding on ichthyoplankton have been measured in containers ranging in size from 15 l to 3.2 m$^3$. Clearance rates of bay anchovy eggs without alternative prey increased with container size from ≤42 l ctenophore$^{-1}$ d$^{-1}$ in 15 l containers to 366 l ctenophore$^{-1}$ d$^{-1}$ in 3 m$^3$ meso-cosms, however, the presence of zooplankton reduced predation on eggs by 36% (Table 7; Cowan & Houde, 1993b). Ctenophores ($n = 75$) in the field contained 0–3 bay anchovy eggs, and digested them in 0.6–1 h (Purcell et al., 1994b). Clearance rates on bay anchovy eggs determined from gut contents *in situ* and diges-tion rates for ctenophores averaging 40 g WW were 128±58 l ctenophore$^{-1}$ d$^{-1}$ (Table 7).

Clearance rates of *Mnemiopsis* feeding on bay an-chovy larvae were 172 l ctenophore$^{-1}$ d$^{-1}$ in 3 m$^3$ mesocosms without alternative prey, however, the presence of zooplankton reduced predation on lar-vae by 91% (Table 8). *Mnemiopsis* ate all sizes of fish larvae available (3.0–9.5 mm), but based on size, swimming speed and behavior, small larvae are more susceptible to predation by the ctenophores (Cowan & Houde, 1992). Gut contents of ctenophores ($n = 75$) from mid Chesapeake Bay contained no larvae even though larvae were abundant (Purcell et al., 1994b), perhaps because digestion of small bay anchovy larvae is very rapid.

Consumption of ichthyoplankton by *Mnemiopsis* can be very important in Chesapeake Bay (Cowan & Houde, 1993a, b; Purcell et al., 1994b). *Mnemi-opsis* was estimated to consume 20–40% d$^{-1}$ of bay anchovy eggs and larvae (Cowan & Houde, 1993a), and 0–41% d$^{-1}$ of eggs during 8 d in July, 1991 (Purcell et al., 1994b). The potential clearance of ich-thyoplankton by *Mnemiopsis* in July 1995–1998 was estimated using *in situ* clearance rates for eggs (Purcell et al., 1994b), and clearance rates from 3 m$^3$ enclos-ures with alternative prey for larvae (Cowan & Houde, 1993a, b). The estimated potential clearance of eggs was 36 to >100% d$^{-1}$, however, the estimated poten-tial clearance of fish larvae was only 2–4% d$^{-1}$ (Table 7). These estimates suggest that *Mnemiopsis* can have great effects on ichythyoplankton, particularly at the egg and yolksac larva stages.

*Relationships of* Mnemiopsis *with other zooplanktivores*

*Mnemiopsis* populations potentially could affect, or be affected by, other zooplanktivorous species through competition for zooplankton prey. Four species of scyphomedusae occur in Chesapeake Bay. *Aurelia aurita* (Linnaeus) medusae are found in low numbers in the lower and middle bay in mid to late sum-mer. *Rhopilema verrilli* (Fewkes) medusae are seen infrequently in summer. *Cyanea capillata* medusae

generally appear in January or February, occur in low numbers (<0.2 m$^{-3}$) in April in the mesohaline mainstem bay (Purcell et al., unpublished data), and usually disappear in May. *C. capillata* medusae eat zooplankton as well as ctenophores (Båmstedt et al., 1997; Purcell & Sturdevant, 2001), and are both po-tential competitors and predators of *Mnemiopsis* in the spring, however, the effects of *C. capillata* on zo-oplankton and ctenophore populations are unknown. During the summer, *Chrysaora quinquecirrha* medu-sae attain high densities in the mesohaline portions of tributaries of Chesapeake Bay (as many as 16 me-dusae m$^{-3}$, Purcell, 1992), but in the mainstem bay, densities are usually <2 m$^{-3}$ (Purcell et al., 1994a; also Table 6). *C. quinquecirrha* consumes the same prey as *Mnemiopsis*, and can have greater effects than the ctenophores on zooplankton at the flanks of the bay, however, both species together did not reduce zo-oplankton populations in 1987 or 1988 (Purcell et al., 1994a). *C. quinquecirrha* also is an important predator of *Mnemiopsis* (see predators section).

Twenty-three species of hydromedusae were found in southern Chesapeake Bay (Calder, 1971), however, only a few species are abundant (Purcell et al., 1999a). Recent bay-wide surveys in April, July and Octo-ber, 1995–1998 show that hydromedusan abundances were highest in the southern bay, where *Mnemiopsis* is not abundant (Hood et al., 1999; Purcell et al., 1999a, unpublished data). Combined species densities usually are <7 m$^{-3}$, however, occasionally, *Liriope tetraphylla* (Chamisso and Eysenhardt) can be very abundant in the southern bay (up to 111 m$^{-3}$ in Oc-tober, 1995) and *Nemopsis bachei* L. Agassiz reaches densities >100 m$^{-3}$ during spring in the mid bay. Hydromedusa abundances are low (<1 m$^{-3}$) in the summer when *Mnemiopsis* biomass is greatest (Purcell et al., 1994a, b, 1999a, unpublished data), possibly due to reduction of their prey by *Mnemiopsis*.

The spatial and temporal distributions of medusae do not overlap much with the distribution of *Mnemi-opsis*, except for *Chrysaora quinquecirrha*. Therefore, although they all consume zooplankton foods, it is un-likely that they compete for food with *Mnemiopsis*. The diets and distributions of *Mnemiopsis* and *C. quinquecirrha* overlap (Purcell et al., 1994a, b), and their populations vary inversely (Feigenbaum & Kelly, 1984; Purcell & Cowan, 1995), however, predation by *C. quinquecirrha* on *Mnemiopsis* is probably more important than competition for food.

The dominant planktivorous fish in Chesapeake Bay is the bay anchovy, *Anchoa mitchilli*, which

Table 9. Mean relative abundances (catch per 20-min tow of an 18-m$^2$ mouth-opening midwater trawl) of bay anchovy $\geq 30$ mm total length from seasonal collections in Chesapeake Bay, 1995–1999 (Houde, unpublished data). Means ($\pm 1$ standard error). Dates, temperatures, and salinities in summer 1995–1998 are the same as in Tables 6 and 8

| Months | 1995 | 1996 | 1997 | 1998 | 1999 |
|--------|------|------|------|------|------|
| Apr–May | 43 | 75 | 236 | 81 | 129 |
| | ($\pm 12$) | ($\pm 27$) | ($\pm 63$) | ($\pm 12$) | ($\pm 23$) |
| Jun–Jul | 843 | 94 | 195 | 268 | 97 |
| | ($\pm 381$) | ($\pm 27$) | ($\pm 45$) | ($\pm 75$) | ($\pm 18$) |
| Oct–Nov | 901 | 581 | 1508 | 4724 | 2172 |
| | ($\pm 309$) | ($\pm 117$) | ($\pm 246$) | ($\pm 665$) | ($\pm 455$) |

is a consumer of zooplankton, principally copepods (Houde & Zastrow, 1991). In baywide midwater trawl surveys from 1995 to 1999, relative abundances of bay anchovy, a potential competitor of *Mnemiopsis* for zooplankton prey, have varied 8-fold in October when recruitment is complete, as well as in July during the season of peak overlap between bay anchovy and *Mnemiopsis* (Table 9). The anchovy and ctenophore broadly overlap in distribution and both are potentially major zooplankton consumers in Chesapeake Bay. Vasquez (1989) and Klebasko (1991) found that bay anchovy can consume >15% of their body weight daily in summer, and Luo & Brandt (1993) believe that zooplankton abundances limited bay anchovy production in the bay. Although bay anchovy and *Mnemiopsis* overlap spatially and temporally, there has been no careful analysis to determine the potential for competitive interactions. Bay anchovy populations apparently declined markedly in the early to mid-1990s (Virginia Marine Resources Commission, Maryland Dept of Natural Resources, unpublished data), however, there was no corresponding increase in ctenophore populations that would suggest that the anchovy decline was due to competition with or predation by ctenophores.

## *Mnemiopsis* in exotic habitats

### Environmental conditions

Since its accidental introduction to the Black Sea, *Mnemiopsis* has spread to adjacent bodies of water, inhabiting waters of salinities ranging from 3 in the Sea of Azov to 39 in the eastern Mediterranean, and temperatures ranging from 4 °C in winter to 31 °C in summer. The Mediterranean basin is a system of semi-closed seas connected by straits (Fig. 7). The eastern Mediterranean Sea has high salinities ranging from 38.7 to 39.1, and temperatures from 13.3 to 14.1 °C in winter, and 24–28 °C in summer (Arkhipkin & Dobrolubov, 1999). The other seas in the basin are estuarine, with lower salinities and colder winter temperatures (Table 10).

The Black Sea is the largest semi-closed basin in the World, connected via the Bosporous Strait with the Sea of Marmara and via the Kerch Strait with the brackish Sea of Azov. Because of the restricted water exchange, waters below the permanent halocline at 60–200 m depth are anoxic (87% of the sea volume). The Cold Intermediate layer with temperatures of 6–8 °C occurs between the permanent halocline and the seasonal thermocline in warm seasons. In this layer, the oxycline is the main factor determining the lower boundary of the planktonic community.

The surface waters of the Black Sea are typical of temperate estuaries. This surface aerated zone, which includes the upper mixed layer, is above the seasonal thermocline at 15–25 m depth. Salinity is low at the surface (average 18) due to discharge of great rivers in the northwest, and 21.9–22.3 at depth. The surface waters are warmed to 24–27 °C in summer, and cooled to 2–8 °C and sometimes to negative temperatures in the northwest in winter (Table 10). In winter, the seasonal thermocline breaks down, and the isothermal layer extends from the surface to 70–80 m (Ovchinnikov & Titov, 1990).

The Black Sea began to change in the 1960s because of several anthropogenic factors, most importantly, decrease of fresh-water runoff, eutrophication, selective and overfishing, and alien species introduction (Ivanov & Beverton, 1985; Caddy & Griffiths, 1990). The greatest effects have been in the northern part of the Black Sea, where the Danube, Dnepr and Dnestr rivers determine the hydrological and hydrochemical regime. The northern Black Sea was the most important spawning area for all commercial fish species until the middle of the 1970s (Ivanov & Beverton, 1985). Stocks of piscivorous species of fish, which migrate to the Black Sea from the Mediterranean for spawning and feeding in spring, decreased greatly during the 1980s due to overfishing and reduced numbers of migrating fish (Ivanov & Beverton, 1985; Caddy & Griffiths, 1990). Nitrogen and phosphorous inputs dramatically increased in the 1960s, causing eutrophication in the coastal ecosystem of the northwestern Black Sea and the Sea of Azov. Eutrophication led to

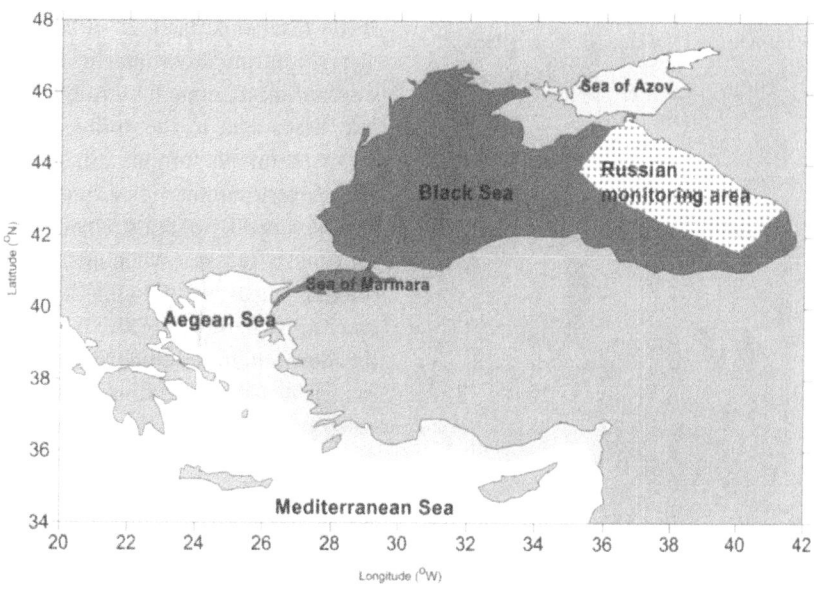

*Figure 7.* Patterns of occurrence of *Mnemiopsis* in the Mediterranean basin as of summer, 1999. Dark areas = present all year, lighter shading = seasonal occurrence. Russian monitoring area is marked by a grid.

*Table 10.* Geomorphological and hydrological characteristics of the seas of the Mediterranean basin

| Location | Depth (m) | Winter temp. (°C) | Summer temp. (°C) | Salinity |
|---|---|---|---|---|
| Black Sea | oxic layer 60–200 | 0–8 | 24–27 | 18–22.3 |
| Sea of Azov | 4–14.5 | −0.8 to +1.2 | 24–30 | 0–14 |
| Sea of Marmara | 10–1335 | 8–15 | 24–29 | 18–29 |
| Caspian Sea | 5–788 | 0–11 | 24–28 | 0.1–11 |
| Aegean Sea | 20–500 | 13.3–14.1 | 24–29 | 38.7–39.1 |

changes in zooplankton structure and blooms of the heterotrophic dinoflagellate, *Noctiluca scintillans* Kofoid & Swezy, in the Black Sea (Aleksandrova et al., 1996; Zaitsev & Alexandrov, 1997).

The Sea of Azov is a shallow (4–14.5 m depth), brackish basin. In the 1970s, decreases in discharge from the Don and the Cuban rivers caused increases in salinity (Kuropatkin, 1998). Salinities now range from < 1 to 16. Temperature varies from −2.4 to 1.2 °C in the winter and 24–32.8 °C in the summer (Kuropatkin, 1998). Prior to the 1980s, the Black Sea and the Sea of Azov had high zooplankton biomass and were productive regions for fisheries (Rass, 1992; Kovalev et al., 1998).

### Distribution and abundance of Mnemiopsis

#### Introduction and population expansion

*Mnemiopsis* was first found in Sudak Bay of the Black Sea in November, 1982 (Pereladov, 1983). By summer–autumn, 1988 it had spread throughout the Black Sea, with average biomasses of up to 1 kg WW m$^{-2}$ (40 g WW m$^{-3}$) and average numbers of up to 310 ctenophores m$^{-2}$ (12.4 m$^{-3}$) (Vinogradov et al., 1989). *Mnemiopsis* biomass in inshore and offshore waters of the Black Sea from 1988 to 1999 is shown in Figure 8. In the autumn of 1989, the greatest mean biomass of 4.6 kg WW m$^{-2}$ (184 g WW m$^{-3}$) and the greatest average numerical density of 7600 ctenophores m$^{-2}$ (304 m$^{-3}$) were measured in the open sea (Vinogradov et al., 1989). In spring, 1990, the abundance of *Mnemiopsis* was still very high, but by summer, ctenophores began to decrease in numbers and biomass (Vinogradov et al., 1992). This decreasing trend continued until summer, 1993. A second peak in ctenophore biomass occurred in September, 1994, with an average biomass of 2.7 kg WW m$^{-2}$ (108 g WW m$^{-3}$) in the open sea, and much greater values in the inshore waters (maximal 9.7 kg WW

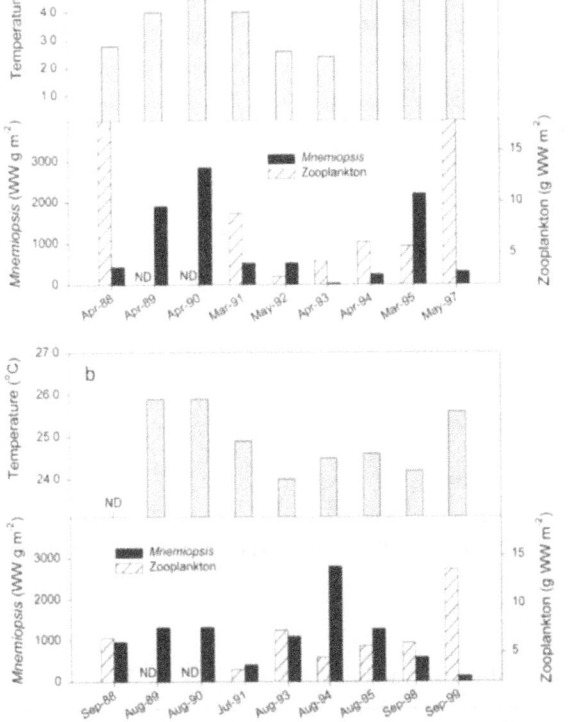

*Figure 8.* Long term variations in *Mnemiopsis* biomass (g WW m$^{-2}$), zooplankton biomass (g WW m$^{-2}$), and temperature in inshore and offshore waters of the Black Sea since 1988. (a) Average winter air temperature, and spring biomasses, (b) Average surface water temperature and biomasses in August–September. ND = no data. Ctenophores were sampled with vertical tows (0–150 m) of a BR net, 200 $\mu$m mesh, and the data multiplied by coefficients for insignificant catchability (<45 mm, = 2; >45 mm = 2.3). Zooplankton were sampled at the same depths by water bottle in 1988–1992, and with a Juday net, 500 $\mu$m mesh in 1993–1999. Data from 1988 to 1992 (Vinogradov et al., 1992; Khoroshilov, 1993). Data from 1993 to 1999 (Shiganova, 1998, unpublished data).

m$^{-2}$, 176 g WW m$^{-3}$; average 4.5 kg WW m$^{-2}$, 110 g WW m$^{-3}$). Ctenophore abundances during this second outbreak were not as great in the open sea as in 1989, however, greater biomasses were measured in inshore waters than previously. After 1995, the *Mnemiopsis* population decreased until 1998, when it increased in the offshore waters to an average biomass of 876 g WW m$^{-2}$ (35 g WW m$^{-3}$) and an average density of 463 ctenophores m$^{-2}$ (18 m$^{-3}$). These data demonstrate that *Mnemiopsis* populations show marked interannual variation in the Black Sea, probably resulting from environmental interactions and food availability (Tzikhon-Lukanina et al., 1992; Shiganova, 1998).

After its introduction to the Black Sea, *Mnemiopsis* has moved through the straits to adjacent basins (Fig. 7). It was first observed in the Sea of Azov in August,

1988 (Studenikina et al., 1991). *Mnemiopsis* cannot survive during the winter in the Sea of Azov, and is re-introduced annually through the Kerch Strait from the Black Sea in the spring or summer, depending on wind-driven currents. Since 1989, it has bloomed there every summer to early autumn, with the highest biomass in July–August when inoculation is early, or in August–October when inoculation is late. The peak biomass measured, 936 g WWm$^{-2}$ (103 g WW m$^{-3}$), was in September, 1989. In August, 1991, biomass again was high, reaching 812 g WW m$^{-2}$ (91.2 g WW m$^{-3}$), possibly due to early re-introduction and unusually high average water temperatures in July and August (26 °C). *Mnemiopsis* biomass decreased in 1992–1994 with the least biomass in 1994, when the maximum measured was 362 g WW m$^{-2}$ (53.9 g WW m$^{-3}$) in August. From 1995, *Mnemiopsis* biomass has increased in the Sea of Azov, with a high biomass, 610 g WW m$^{-2}$ (74.5 g WW m$^{-3}$), in August, 1995, and in 1999, 1075 g WW m$^{-2}$ (131.5 g WW m$^{-3}$) (reviewed by Shiganova et al., 2001b). Although it can attain great population biomass, *Mnemiopsis* remains relatively small (1.5–2.5 cm) in the Sea of Azov.

*Mnemiopsis* first penetrated to the Sea of Marmara from the Black Sea with the upper Bosporus current. It occurs year-round in the upper water layer of the Sea of Marmara. In early October, 1992, the average biomass was 152 g WW m$^{-3}$ and numbers were 27 ctenophores m$^{-3}$ (Shiganova, 1993), however, in July, 1993, the numbers of *Mnemiopsis* were only 2.5% of that in the previous year (Kideys & Niermann, 1994).

*Mnemiopsis* also has spread into the Mediterranean Sea. It was first recorded during late spring-summer, 1990 in the Aegean Sea (Saronikos Gulf, 45–75 ind. m$^{-2}$), and in the following years until 1996, it was found mainly in spring in low density. A large swarm (up to 10 ind. m$^{-3}$ or 150 ind. m$^{-2}$) was observed in January, 1998. In 1996–1998, *Mnemiopsis* was observed in low density in several coastal areas of the Aegean Sea islands in spring, summer and autumn. In June, 1998, it was found near the Dardanelles Strait. The abundance of *Mnemiopsis* in the northern Aegean Sea is consistently lower than in the Black Sea and Sea of Azov. Adult specimens in the Aegean Sea (6–7 cm) also are smaller than in the Black Sea (12–18 cm). *Mnemiopsis* may have been transported in the ballast waters of ships (Shiganova et al., 2001b) to new parts of the eastern Mediterranean (Levantine Sea), where it was found in Mersin Bay in spring, 1992 (Kideys & Niermann, 1994), and in Syrian coastal waters in October, 1993 (Shiganova, 1997).

## Vertical distribution

*Mnemiopsis* inhabits mainly the surface layer from 0 to 15–25 m, above seasonal thermocline, during the warm season. Some large specimens penetrate below the thermocline but remain above the pycnocline (60–80 m). In winter, *Mnemiopsis* is found throughout the isothermal layer above the pycnocline, but most of the population occurs above 50 m. In March during the R/V 'Bilim' survey, small *Mnemiopsis* were found at depths of 105–150 m in low oxygen concentrations (Mutlu, 1999). *Mnemiopsis* generally occurs above the hypoxic/anoxic waters in the Black Sea, nevertheless, 2.4–4.7 mm long specimens survived in hypoxic water (0.5 mg $O_2$ $l^{-1}$) in 1 l containers for 96 h at 21 °C; 60–80 mm specimens did not survive, however (Shiganova, unpublished results).

Adult *Mnemiopsis* fed and spawned above the thermocline (15–25 m) at night in August (Shiganova, unpublished data). Large specimens of *Mnemiopsis* occur at the surface at night, and small ones occur there in daytime (Zaika & Ivanova, 1992). In winter, most ctenophores were concentrated in the upper layer (0–50 m) during the day (10:00 h), and only adults were observed at greater depths at 18:00 h, however, this pattern was reversed after midnight (Mutlu, 1999). In the shallow Sea of Azov, *Mnemiopsis* occurs throughout the water column (4–11 m). In the deep Sea of Marmara, it occurs only above thermocline (15–30 m) in the surface water mass that originates from the Black Sea (Shiganova, 1993).

## Seasonal dynamics and factors controlling population size

The most critical time for *Mnemiopsis* survival is winter. The ctenophores cannot survive in the low salinity waters of the Black Sea and Sea of Azov during cold winters when air and water temperatures are < 2 °C and <4 °C, respectively (Volovik et al., 1993; Shiganova, 1998). Ctenophore body size in the Black Sea in spring increases with temperature (Fig. 4). Similarly, spring ctenophore population size is low after cold winters (Fig. 9a). The relative size of the population in spring is reflected later in the year, as shown by the high abundances of *Mnemiopsis* in both March and August, 1995 following a warm winter (Fig. 9b, c). At the end of winter, moderate sized individuals predominate (Fig. 4).

Between February and June, somatic growth of the overwintering population increases. In June and July, the population is composed mostly of adult individu-

als. The mean individual weight is greatest in July or August (27 g WW in 1993). While the biomass of *Mnemiopsis* normally peaks in August–September, numbers peak in September–November when reproduction is greatest and the population contains many larvae and small individuals. Spawning is possible episodically in some locations with high zooplankton concentrations and temperature higher than 21 °C in early summer, but intensive reproduction begins in the inshore waters in late July or early to mid August, and continues until October–November. Intensive reproduction begins when water temperatures reach 23.5–24 °C, and peaks at 24.5–25.5 °C (Fig. 1). Food supply also controls the initiation and intensity of reproduction (Tzikhon-Lukanina et al., 1993a). The main areas of reproduction are the inshore waters, which have more abundant mesozoo- and meroplankton, however, reproducing ctenophores also spread to the open sea.

During October–November, the population biomass of small ctenophores <3 g WW is greater than that of large individuals. The density depends on the success of spawning in the autumn and surface water temperatures. In late October to early November, reproduction decreases and then stops, first in the inshore waters where temperatures drop earlier than in the open sea.

In the Sea of Azov, the timing of reproduction by *Mnemiopsis* depends on when it penetrates into the Sea of Azov from the Black Sea, and on prey availability (Volovik et al., 1993). Intensive reproduction occurs in July–October with peak in July–September when its arrival is early, and in August–October when its arrival is later. The intensity of reproduction seems greater at higher temperatures. Reproduction occurs throughout the Sea since it is shallow and has similar conditions to coastal areas of the Black Sea (Shiganova et al., 2001b).

For more than a decade after the introduction of *Mnemiopsis*, few predators of ctenophores were present in the Black Sea. The only potential fish predators of ctenophores there are mackerels (*Scomber scombrus* Linnaeus and *S. japonicus* Houttuyn) and the Black Sea whiting, *Merlangus merlangus euxinus* (Linnaeus), which are known to eat gelatinous species elsewhere (GESAMP, 1997; Purcell & Arai, 2001). Populations of scombrid species had been greatly reduced before 1980 by overfishing (Caddy & Griffiths, 1990; Rass, 1992).

There were no gelatinous predators of *Mnemiopsis* in the Black Sea until 1997, when the ctenophore

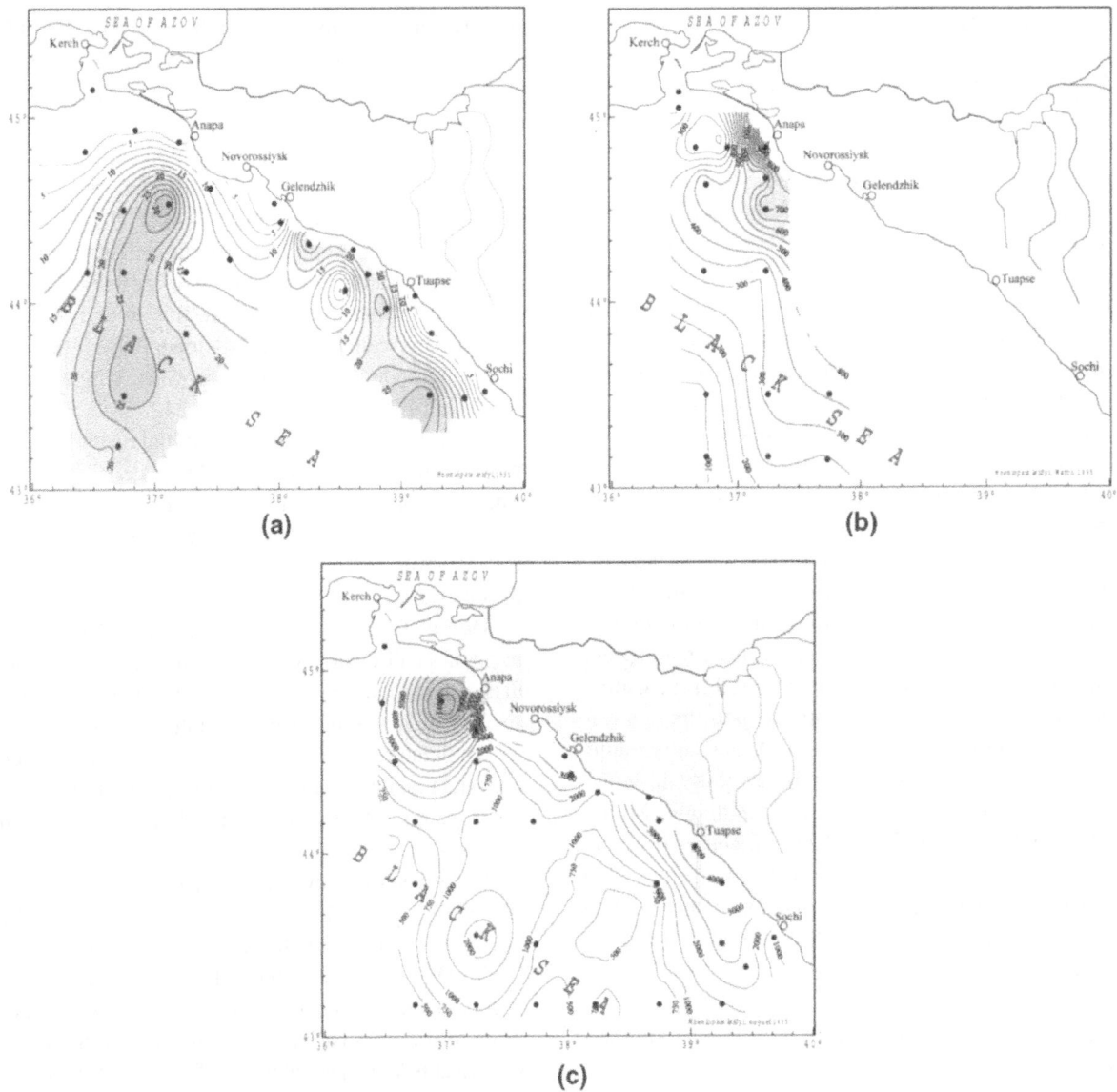

*Figure 9.* Abundance of *Mnemiopsis* (no. m$^{-2}$) (a) in spring after a cold winter (0–30 ind. m$^{-2}$ in March, 1993), and (b, c) after a warm winter (100–900 ind. m$^{-2}$ in March, and 500–10 000 ind. m$^{-2}$ in August, 1995) in the Russian monitoring area of the Black Sea (Shiganova, 1998, unpublished data).

*Beroe ovata* first appeared, probably arriving through the Sea of Marmara from the Mediterranean Sea. During 1997 and 1998, it occurred in some coastal areas in spring (Nastenko & Polishchuk, 1999) and mainly in late summer/autumn (Konsulov & Kamburska, 1998; Shiganova et al., 2001a,b; Finenko et al., 2001). In August, 1999, the first bloom of *B. ovata* was recorded, with average numbers and biomass of 1.1 ind. m$^{-2}$ and 31 g WW m$^{-2}$, respectively. The abundance of *Mnemiopsis* was greatly reduced at that time (17.3 ind. m$^{-2}$ and 155 g WW m$^{-2}$) (Fig. 8b).

*Trophic ecology of* Mnemiopsis

*Prey consumed*

The prey of *Mnemiopsis* in the Black Sea varies depending on season, place and time of the day. The gut contents of *Mnemiopsis* from the Black Sea include a wide variety of prey taxa, depending mainly on the ambient zooplankton species composition. In the open northern sea in summer, 1989, 17 prey taxa occurred in the gut contents, with 45.5–75% being copepods (mainly *Acartia clausi* Giesbrecht and *Calanus euxinus* Hulsemann), 24–25.9% cladocerans (mainly *Penilia avirostris* Linnaeus), 8% copepod nauplii; barnacle cyprid larvae, bivalve veligers and fish larvae also were eaten (Tzikhon-Lukanina et al. 1991, 1993a, b). Small ctenophores ate more copepods (50%) than cladocerans (30%), while large ones ate more cladocerans (53%) than copepods (27%) (Tsikhon-Lukanina & Reznichenko, 1991; Tsikhon-Lukanina et al., 1991, 1992). Mean prey size is 0.75–1 mm. Ctenophores in coastal waters also eat meroplankton, especially mussel veligers (Sergeeva et al., 1990; Tzikhon-Lukanina et al., 1991). Ctenophore feeding *in situ* was maximal at midnight (Sergeeva et al., 1990). In July, 1992 in the northern region, gut contents of *Mnemiopsis* contained 47% cladocerans (*Pleopis polyphemoides* (Leuckart)), 26% bivalve veligers, 15% copepods, as well as copepod ovae, appendicularians, tintinnids, cyprid, gastropod and polychaete larvae, and fish eggs (Zaika & Revkov, 1998). During 1991–1995 in the southern region, *Acartia clausi* was the most numerous prey item (37%), followed by *Calanus euxinus* (35%), and *Pseudocalanus elongatus* (Boeck), *Oithona similis* Claus and *Paracalanus parvus* (Claus). Bivalve larvae were the predominant prey item (84%) in February–May, 1994 in the southern Black Sea (Multu, 1999).

*Feeding rates*

Feeding by *Mnemiopsis* is directly proportional to prey concentration. Prey consumption by *Mnemiopsis* increases continuously from 1.87 to 15.25 mg g DW$^{-1}$ d$^{-1}$ at *Acartia clausi* densities of 20–200 copepods l$^{-1}$, and from 0.70 to 12 mg g DW$^{-1}$ d$^{-1}$ at barnacle nauplii densities of 20–180 ind. l$^{-1}$ (Finenko et al., 1995). Daily rations of *Mnemiopsis* range from 0.2 to 1.5% DW, depending on prey concentration in small (1 l) containers, and increase with container size (Finenko et al., 1995). Feeding by *Mnemiopsis* was not detectable at prey densities of 3 copepods l$^{-1}$ (0.6 mg l$^{-1}$) (Tzikhon-Lukanina et al., 1991).

Digestion times depend on ctenophore size, the numbers of food items in the stomodeum and prey type. Digestion of zooplankton requires 2–3 h (average 2.4 h) at 20–23 °C, increasing with more food items in gut contents (Sergeeva et al., 1990). Digestion times at 18–20 °C of individual copepods increase with increasing copepod size and decreasing ctenophore size from 0.8–2.3 h for small *Acartia clausi*, to 1.1–4.5 h for *Calanus euxinus*, and 2.4–7 h for large *Pontella mediterranea* (Claus), with the low rates in these ranges being from 67 mm ctenophores and the high rates from 6 mm ctenophores (Tzikhon-Lukanina

et al., 1991). Digestion of fish larvae at 18–20 °C ranges from 1 to 4.5 h (mean 2.8±1.2 h), increasing with larva size (6–32 mm, mean 14.4±8.5 mm) (Tzikhon-Lukanina et al., 1993b).

*Effects on zooplankton*

Mesozooplankton communities in the Black Sea are diverse historically (about 150 holomesozooplankton species and about 50 meroplankton species), due to their varied origins from subtropical species of the surface layer and moderate cold-water species of the cold intermediate layer to brackish and fresh-water species. During recent decades, the species composition of zooplankton has changed due to a variety of factors, such as eutrophication, pollution, reduced fresh water flow and the invasion of *Mnemiopsis*. Some species practically disappeared, while some new species arrived from the Mediterranean Sea (Kovalev et al., 1998).

In the late 1960s, the structure of the zooplanktonic communities began to change due to eutrophication, which indirectly affected zooplankton species diversity through its impact on the phytoplankton. The abundance of detritivorous and herbivorous zooplankton species increased. For example, the biomass of *Acartia clausi* and *Pleopis polyphemoides* increased 5 times, while *Centropages ponticus* Karavaev, *Paracalanus parvus*, *Oithona nana* Giesbrecht biomass declined in the 1980s. This was particularly noticeable in the northwestern Black Sea (Kovalev, 1993; Petranu, 1997).

Zooplankton species structure, abundance and biomass in the Black Sea differ greatly among areas, seasons and years. The community of the open sea differs from the coastal areas. In winter, the biomass of zooplankton decreases in the open sea, but not as much as in coastal areas. In warm winters when water temperatures remain above 8 °C, the abundance of eurythermal *Acartia clausi, Oithona similis, O. nana* and cold water *Calanus euxinus* and *Pseudocalanus elongatus* is high in both the open sea and inshore areas (Vinogradov et al., 1992).

In March, before the seasonal thermocline develops, *Acartia clausi, Oithona similis, Paracalanus parvus* and *O. nana* reproduce at 15–30 m in the open sea. In late March–April when the seasonal thermocline forms, *P. parvus* occurs in small numbers in upper layer, *A. clausi* occurs just above the thermocline at 21–82 ind. m$^{-3}$ (18–50% of total zooplankton abundance), *C. euxinus* and *P. elongatus* occur below the thermocline at 35–48 ind. m$^{-3}$ (24–43%) and 42–64 ind. m$^{-3}$ (26–50%), respectively. The first peak

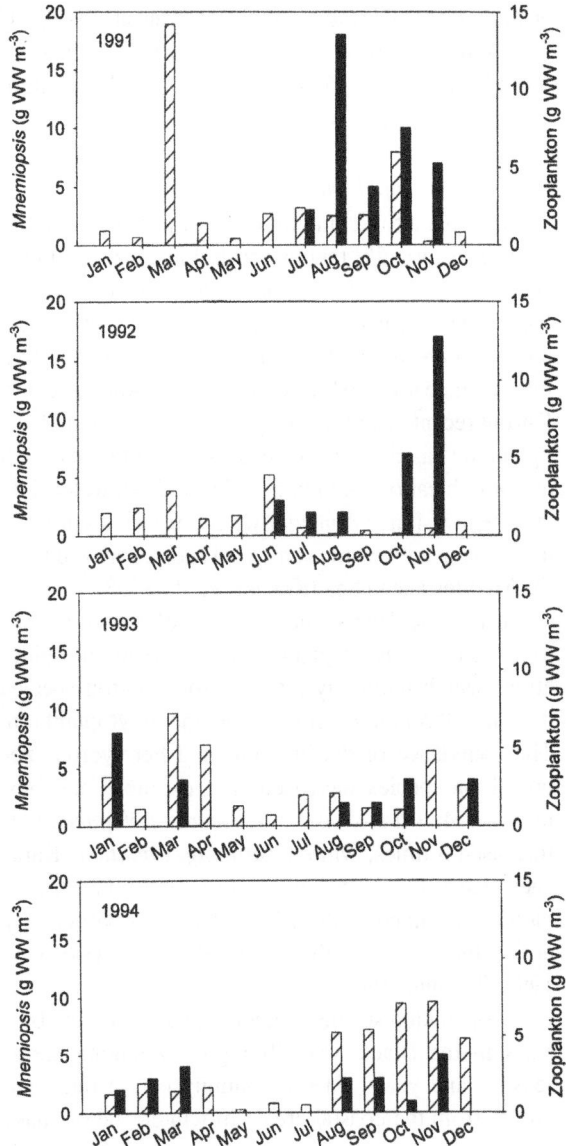

*Figure 10.* Annual cycles of zooplankton biomass (g WW m$^{-3}$, hatched bars) and *Mnemiopsis* biomass (g WW m$^{-3}$, black bars) during four years in Gelendzhik (Blue) Bay in the northeastern Black Sea. Vertical tows (15–0 m) were made with a Juday net, 170 μm mesh (From Khoroshilov & Lukasheva, 1999).

in zooplankton densities generally occurs in May or June, but can be earlier (March) in some years (Fig. 10).

In summer, warm-water species occur in the upper layer in both the open sea and inshore waters of the Black Sea. In recent years, the most abundant species has been the cladoceran *Penilia avirostris*, particularly at the end of summer. In coastal areas, other cladocerans, warm-water *Pleopis polyphemoides* and

less-abundant, *Evadne spinifera* P. E. Muller, begin to develop in June–July, particularly in the northwestern area. Reproduction of eurythermal copepod species continues in summer, and currently, *Acartia clausi* is the most abundant at 7.8–14.2 ind. m$^{-3}$. Larvae of benthic gastropods, bivalves and polychaetes also appear in the plankton in the summer, mainly in the inshore waters. In late summer or early autumn, the second peak of zooplankton abundance occurs due to increases of *P. avirostris, A. clausi, Oithona similis* and meroplanktonic larvae (Petranu, 1997; Shiganova, unpublished data).

After the invasion of *Mnemiopsis*, a precipitous decline occurred in the numbers of mesozooplankton in the Black Sea (Fig. 8). The changes were greatest in the northern region, which already was severely damaged by eutrophication and predation by *Aurelia aurita* medusae on zooplankton (Shiganova et al., 1998). Since the summer of 1989, the abundance of *Paracalanus parvus, Oithona similis, Acartia clausi*, all species of cladocerans, appendicularians and polychaete and gastropod larvae have decreased, particularly in the upper layer and coastal areas. Species such as *O. nana* and representatives of the family Pontellidae completely disappeared from samples (Kovalev et al., 1998). By autumn of 1989, zooplankton biomass in the open sea was only 23% that in the summer of 1988, and since 1990, the abundance of *Calanus euxinus* also decreased (Vinogradov et al., 1992).

Reciprocal oscillations of ctenophores and zooplankton reflect the effects of *Mnemiopsis* predation (Fig. 8). During years of low ctenophore abundance, such as 1992–1993, zooplankton populations, including *Calanus euxinus* and *Pseudocalanus elongatus*, began to recover in the Black Sea. Zooplankton biomass dropped again in the autumn of 1994, when *Mnemiopsis* biomass again was high. In 1996, when *Mnemiopsis* abundance decreased, significant increases in zooplankton biomass occurred, particularly of *C. euxinus*, as well as increased diversity of other copepod species, including *Paracalanus parvus*. Speciles that had disappeared, such as *Pontella mediterranea* and *Centropages ponticus*, reappeared in small numbers.

Great abundance of recently-introduced *Beroe ovata* in 1999 caused a marked decrease in *Mnemiopsis* abundance, and consequently, an increase in zooplankton numbers and biomass in the Black Sea (Fig. 8). The biomass of mesozooplankton increased up to about 11 g WW m$^{-2}$ in the open northeastern area and to 13 g WW m$^{-2}$ in the inshore areas, which is much higher than during the 10 years following the *Mnemiopsis* invasion. The biomass of *Calanus euxinus* and *Pseudocalanus elongatus* stayed mostly the same, while the biomass of other copepods, primarily in the surface layer, increased 3-fold, reaching 1.4 g WW m$^{-2}$. Cladocerans increased up to 150–300 thousand ind. m$^{-2}$, with *Penillia avirostris* being most abundant, and meroplankton densities also increased greatly. After the disappearance of *Pontella mediterranea* and *Centropages ponticus* in the early 1990s, they were recorded again in samples in 1999 (Shiganova et al., 2001a). Mediterranean species appeared in the Black Sea during the 1990s mainly in the southern and northwestern areas (Kovalev et al., 1998).

*Effects on ichthyoplankton*

Dekhnik (1973) lists 56 species and subspecies of the Black Sea marine fish that have pelagic development, 28 species with both pelagic eggs and larvae and 28 with only pelagic larvae. The diverse fauna includes summer-spawning warm-water species of Mediterranean origin and winter-spawning moderate cold-water boreal species. The highest density of *Mnemiopsis* coincides with spawning of the warm-water fish species, which begins in late spring and lasts until July–August or August–September.

The abundance and species diversity of summer ichthyoplankton was already greatly reduced in the 1980s before the first outbreak of *Mnemiopsis* in the Black Sea (Fig. 11). The eggs and larvae of mainly zooplanktivorous species, *Trachurus mediterraneus ponticus* Aleev (Mediterranean horse mackerel) and particularly, *Engraulis encrasicolus ponticus* Aleksandrov (Black Sea anchovy), had become the most abundant. After the *Mnemiopsis* outbreak in 1989, the numbers of anchovy eggs further decreased and remained low through 1991, even though ctenophore abundance decreased (Fig. 11). Decreased densities of anchovy eggs and larvae again were observed in 1994 and 1995 with increased *Mnemiopsis* abundance. In 1993 and 1996, abundances of anchovy eggs and larvae were somewhat greater when *Mnemiopsis* abundances were low (Fig. 11). Anchovy egg and larval densities have remained low in the north, while being substantially greater in the south and southwest (Niermann et al., 1994; Gordina & Klimova, 1995; Shiganova et al., 1998). The greatest abundances of anchovy eggs since the invasion of *Mnemiopsis* were measured in 1996 in the south (90 eggs m$^{-2}$) and in the north (12 eggs m$^{-2}$) (Kideys et al., 1998). After the appearance of *Beroe ovata* in 1999, the numbers of fish eggs increased greatly (Fig. 11), particularly

those of anchovy, Mediterranean horse mackerel and *Diplodus annularis* (Linnaeus). We speculate that this resulted from the arrival of *B. ovata* and consequent decrease in *Mnemiopsis* density because we observed increased egg numbers only on 17–20 August when *B. ovata* already was abundant in the Black Sea. Those eggs were spawned one day before sampling, however, larval numbers remained low, which we suspect was a result of *Mnemiopsis* eating eggs and larvae during the summer when *B. ovata* was not abundant.

Potential predation on ichthyoplankton by *Mnemiopsis* varies greatly with ctenophore and fish egg and larval abundances. Tzikhon-Lukanina et al. (1993b) estimated the consumption of fish larvae during an outbreak of *Mnemiopsis* in the northeastern coastal Black Sea in May–June, 1990), using the experimentally-determined consumption rate of 5 larvae d$^{-1}$ ctenophore$^{-1}$. During the outbreak, *Mnemiopsis* abundance averaged 286 ind. m$^{-2}$ (Shushkina & Vinogradov, 1991), and fish larvae in this area occurred at 47 ind. m$^{-2}$ (Oven et al., 1991). Tzikhon-Lukanina et al. (1993b) assumed that 2.6% of the ctenophores ate larvae, which equaled 35 fish larvae d$^{-1}$ m$^{-2}$ or 74% of larval abundance during the ctenophore outbreak. Consumption of larvae was estimated to be only 7% d$^{-1}$ when the *Mnemiopsis* population is relatively low (10%) between outbreaks (Tzikhon-Lukanina et al., 1993b). *In situ* observations showed that 1% of *Mnemiopsis* had larvae in the stomodeum, and 2–10% had 1–8 eggs (Shiganova, unpublished data).

*Relationships of* Mnemiopsis *with other zooplanktivores*

Three indigenous gelatinous species occur in the Black Sea: two scyphozoan medusae (*Rhizostoma pulmo* (Macri) and *Aurelia aurita)* and the ctenophore, *Pleurobrachia pileus* (O. F. Müller). While all three species are zooplanktivorous, only *A. aurita* overlaps spatially and temporally with *Mnemiopsis*. *R. pulmo* mainly inhabits contaminated coastal areas of the Black Sea. *P. pileus* inhabits the interzonal layer (15–150 m) and in most cases, no correlation between abundances of *Mnemiopsis* and *P. pileus* has been found (Shiganova et al., 1998).

*Aurelia aurita* medusae occur throughout the Black Sea, but in greatest abundance in the inshore waters. Beginning in the 1970s, the population of *A. aurita* grew explosively, reaching its peak in the early 1980s when average medusa biomass was 0.6–1.0 kg WW

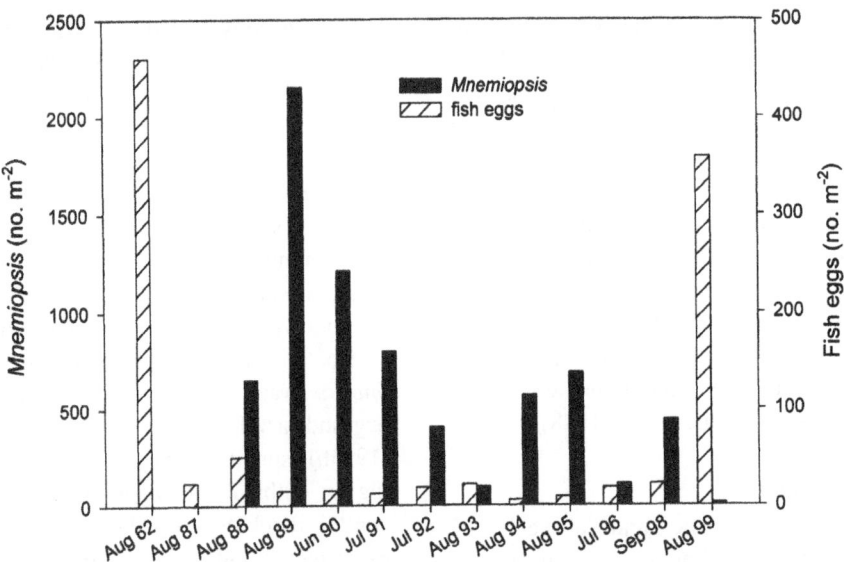

*Figure 11.* Average abundances (no. m$^{-2}$) of fish eggs and *Mnemiopsis* in June–August in the northern Black Sea. The first bar shows numbers in 1962. The last date is after the arrival of the ctenophore, *Beroe* ovata, which eats *Mnemiopsis* (Shiganova et al., 2001).

m$^{-2}$ and total biomass for the sea was estimated at 3–5 × 10$^8$ tonnes (Lebedeva & Shushkina, 1991). After the outbreak of *Mnemiopsis* in 1989, the abundance of *A. aurita* medusae decreased. There is a significant negative correlation between the number of *Mnemiopsis* and biomass of *A. aurita* in subsequent years ($n = 14, r = -0.80, p = 0.005$). This suggests competition for zooplankton prey among these species. *A. aurita* penetrated into the Sea of Azov in 1972 when salinity increased as a result of decreased River Don discharge (Zakhutsky et al., 1983), however, since the invasion of *Mnemiopsis, A. aurita* medusae are seldom found there (Z. A. Mirsoyan, pers. comm.). In the Aegean Sea, the abundance of *A. aurita* is much lower when *Mnemiopsis* is present (Shiganova et al., 2001b).

Another zooplanktivore, the chaetognath, *Sagitta setosa* Muller, virtually disappeared from the Black Sea after *Mnemiopsis* arrived, possibly due to reduction of zooplankton prey and direct predation by ctenophores on chaetognaths. By the autumn of 1989, zooplankton biomass in the open sea was 23% of that in the summer of 1988, and *S. setosa* had decreased to 3% of earlier numbers (Vinogradov et al., 1992). During years of low *Mnemiopsis* biomass, zooplankton populations and *S. setosa* began to recover. With the outbreak of *Beroe ovata* in 1999, the numbers of *S. setosa* greatly increased up to 6–15 thousand ind. m$^{-2}$. Chaetognaths were found in low numbers in the gut contents (1% of prey items) of *Mnemiopsis* from the open sea in summer, 1989 (Tzikhon-Lukanina et

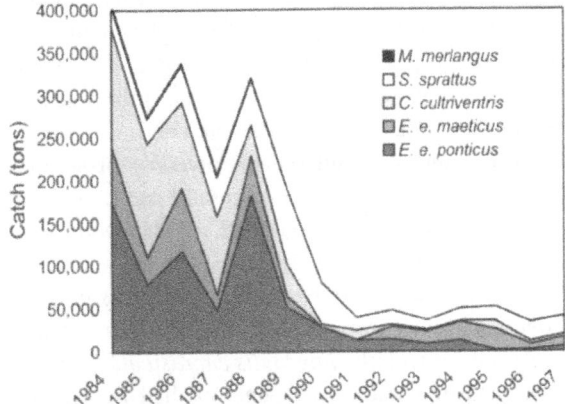

*Figure 12.* Annual catches of zooplanktivorous fish in the Black Sea and Sea of Azov in 1984–1997 (Russia, Ukraine, Georgia) (FAO data 1984–1994; AzNIIRCH data 1994–1997, in Shiganova & Bulgakova, 2000). *Merlangus merlangus euxinus* (Black Sea); *Sprattus sprattus phalericus* (Black Sea); *Clupeonella cultriventris* (Sea of Azov); *Engraulis encrasicolus maeticus* (Sea of Azov); *Engraulis encrasicolus ponticus* (Black Sea).

al., 1993a), and in summer, 1992 (0.03%) (Zaika & Revkov, 1998), therefore, direct predation probably is not the main negative effect of *Mnemiopsis* on *S. setosa*.

There are several zooplanktivorous fish species in the Black Sea, the Black Sea anchovy, Mediterranean horse mackerel, and sprat (*Sprattus sprattus phalericus* Risso). They became the main commercial species during the 1980s after great decreases of piscivorous fish and dolphins (Caddy & Griffiths, 1990). But their stocks and catches declined dramatically during the first and most intensive blooms of *Mnemiopsis* in the Black Sea and Sea of Azov (Fig. 12). The most severe decline was recorded for warm-water species spawning during the summer, the Black Sea anchovy and Mediterranean horse mackerel. The diet composition and rations of these species deteriorated due to reduced zooplankton abundance and species composition. Anchovy foods changed to low-calorie barnacle and bivalve larvae, ostracods, and fish, including their own larvae (Shiganova & Bulgakova, 2000). As a result, the growth and weight of anchovy decreased, as well as the frequency of spawning and fecundity (Lisovenko et al., 1997). The Mediterranean horse mackerel completely disappeared from Russian commercial catches a few years after *Mnemiopsis* was introduced. The Black Sea anchovy occurs mainly in Turkish catches now.

The abundance and diet of sprat, a moderate cold-water species, also have been greatly changed by *Mnemiopsis*. Biomass of sprat declined sharply during the *Mnemiopsis* outbreak in 1989–1991, but began to increase in 1992–1993 after first decrease in ctenophore abundance (Prodanov et al., 1997). The diet composition of sprat began to change in the 1970s in the northwestern Black Sea, where eutrophication caused great decreases of warm-water copepods, which were its main prey. After the *Mnemiopsis* invasion, sprat became almost monophagous, consuming only *Calanus euxinus*. In the northeastern region, the effect of eutrophication on sprat feeding was not noticeable, but after the *Mnemiopsis* invasion, the warm-water copepods disappeared from its diet and only *C. euxinus* was eaten (Shiganova & Bulgakova, 2000).

The situation is much worse in the Azov Sea, where pelagic fish stocks gradually have declined since the 1970s (Volovik & Chikharev, 1998). When *Mnemiopsis* is re-introduced in the spring, the Azov anchovy, *E. encrasicolus maeticus*, and Azov kilka, *Clupeonella cultriventris* (Nordmann), have insufficient food to support spawning, and their larvae have insufficient food for survival because *Mnemiopsis* depletes mesozooplankton during early summer in the Sea of Azov (Budnichenko et al., 1999). The spawning stocks of the fish are very poor under these circumstances. When *Mnemiopsis* is re-introduced in June or July, fish spawning stocks are larger (Volovik & Chikharev, 1998). During the last several decades, the diet of the Azov anchovy has changed, with decreasing proportions of copepods and polychaetes, and increasing proportions of low-calorie meroplankton.

Consequently, the average length, weight, and fat content of the Azov anchovy have decreased, overwintering mortality has increased and fishery catches have decreased (Fig. 12) (Shiganova & Bulgakova, 2000).

*Ecosystem models of* Mnemiopsis *in the Black Sea*

Several papers feature model simulations of the effects of *Mnemiopsis* on the Black Sea ecosystem. In the first two-layer model, Lebedeva & Shushkina (1994) attempt to estimate and compare functioning of the plankton community before and after the introduction of *Mnemiopsis*. Included in the model are dissolved organic matter and phosphate, detritus, primary production, phytoplankton, microplankton (bacteria, Protozoa), mesozooplankton, *Aurelia aurita* and *Pleurobrachia pileus* taken together (although they inhabit different layers and areas), and *Mnemiopsis*. Inputs to the model were incomplete in not including nitrogen and the upper layer mesozooplankton. They assumed that reproduction occurred all year and a quasi-constant interannual level of *Mnemiopsis* biomass, and they did not include interannual variability of environmental conditions. Those assumptions and the lack of long term data (only 1988–1992 were used) lead to errors about the seasonal dynamics of *Mnemiopsis* biomass and individual weight. Their simulations showed the great effect of *Mnemiopsis* on mesozooplankton, which lead to increasing phytoplankton.

The next model of Lebedeva (1998) is devoted to *Mnemiopsis* spatial distribution, growth and temporal variability. She considers that transport with currents, reproduction, growth and mortality of *Mnemiopsis* affect its distribution. Lebedeva uses a two-dimensional model that contains both hydrophysical and biological components. Seasonal and interannual variability of the densities and biomasses of *Mnemiopsis* in the Black Sea and the total *Mnemiopsis* biomass in the sea for different seasons also are included. The water masses were subdivided into three types – shelf waters where *Mnemiopsis* is most abundant, rich areas of open waters where *Mnemiopsis* also is abundant, and low productivity centers of eastern and western gyres where *Mnemiopsis* abundance is very low. She concludes that the biomass of *Mnemiopsis* is higher where the shelf is wider, which is true for summer and early autumn when *Mnemiopsis* reproduces in the inshore waters, and that *Mnemiopsis* biomass increases in the years when zooplankton biomass is higher. The simulated seasonal dynamics of *Mnemiopsis* and change

*Table 11.* Abundant congeneric zooplankton species in brackish waters of Chesapeake Bay and the surface layer of the Black Sea (* indicates predominant species in spring, summer and fall)

| Chesapeake Bay | Black Sea |
|---|---|
| *Centropages* spp. | *Centropages ponticus* |
| *Paracalanus* spp. | *Paracalanus parvus* |
| *Acartia tonsa**, *A. hudsonica* | *Acartia clausi** |
| *Eurytemora affinis* | *Eurytemora affinis* |
| *Oithona colcarva* | *Oithona similis** |
| *Penilia avirostris* | *Penilia avirostris** |
| *Pleopis polyphemoides* | *Pleopis polyphemoides* |
| *Podon leuckarti* (G. O. Sars) | *Podon leuckarti* |
| *Evadne tergestina* | *Evadne tergestina, E. spinifera* |

of individual weight are similar to findings of empirical studies during that period (Shiganova, 1997, 1998), however, we now know that high spring ctenophore biomass predicted by the model occurs only after warm winters.

The model of Van Eeckhout & Lancelot (1997) describes changes in the ecosystem of the northwestern part of the Black Sea, which is highly sensitive to eutrophication because it receives the discharge from the Danube, Dnestr and Dnepr rivers. This model derives from the assemblage of five different modules describing the dynamics of phytoplankton, micro- and mesozooplankton, gelatinous organisms (*Noctilluca scintillans, Aurelia aurita, Mnemiopsis*), microbial degradation in the water column, and early diagenetic processes in the sediments. The authors consider that the complex food web is composed of three branches, specifically, linear, microbial and gelatinous pathways. Cycling of carbon, nitrogen, phosphorus and silicon is simulated through aggregated chemical and biological components, which include the planktonic and benthic components of the coastal area. This model includes the hypothesis of Shiganova (1997) on the influence of diminished fish stocks on population growth of *Mnemiopsis*. The model also allows the testing of different hypotheses regarding the structure of the ecosystem and the forcing functions on it, for example, the impact of fishing pressure on the diet composition of *Mnemiopsis* and the effects of nutrient load reduction.

A one-dimensional, vertically resolved, coupled physical-biochemical model by Oguz et al. (2001) builds on the results of earlier models that did not include *Mnemiopsis* (Oguz et al., 1996, 1998, 1999).

Included in this most recent model are dissolved and particulate organic and inorganic forms of nitrogen, diatoms, dinoflagellates, bacterioplankton, microzooplankton, omnivorous mesozooplankton, *Noctiluca scintillans, Aurelia aurita* and *Mnemiopsis*. Simulations using parameters from the late 1970s and early 1980s reproduce the observed plankton food web in the Gelendezhik area of the northeastern Black Sea before the introduction of *Mnemiopsis* when *A. aurita* had substantial effects. Simulations from the 1980s and early 1990s show the great effects of *Mnemiopsis* in reducing mesozooplankton grazing, which lead to increased phytoplankton blooms, as observed in the Black Sea.

## Summary: similarities and contrasts of *Mnemiopsis* in native and exotic habitats

The native habitats of the ctenophore, *Mnemiopsis*, are temperate to subtropical estuaries along the Atlantic coasts of North and South America. In the early 1980s, it was introduced to the Black Sea, where it has flourished. Characteristics of *Mnemiopsis* that enable it to predominate as a zooplanktivore in its native waters also have enabled it to be an extremely successful invader. *Mnemiopsis* is euryhaline and eurythermal, and has high reproductive and growth potentials that permit rapid population increases in favourable conditions. Growth and reproduction rates increase with increasing ctenophore size, temperature and prey densities. Reproduction occurs in spring through autumn at temperatures $\geq 12$ °C, but may require higher temperatures in the Black Sea. Reproduction peaks at temperatures of 24–28 °C.

In its native estuaries, *Mnemiopsis* encounters a wide range of conditions – temperatures from 0 °C in winter to 32 °C in summer and salinities of <2–38. *Mnemiopsis* was introduced to the Black Sea where the surface waters are typical of a temperate estuary. From there, *Mnemiopsis* has expanded its range to the Azov, Marmara, Mediterranean and Caspian Seas, which encompass the range of environmental conditions found in the native habitats. *Mnemiopsis* does not seem to be limited by salinities >2, however, low winter temperatures, particularly in combination with low salinities, prevent survival through winter in some locations (e.g. the Sea of Azov). In addition, the size of surviving ctenophores in spring and resulting population size increase with warmer water temperatures. The peak of the *Mnemiopsis* popula-

*Table 12.* Mean biomass (mg C m$^{-3}$) and production (mg C m$^{-3}$) of *Acartia clausi* (>125 $\mu$m) and *A. tonsa* (>200 $\mu$m) in the Black Sea and the mesohaline Chesapeake Bay, respectively. Data are from the upper 40 m in summer in Kamish Bay before *Mnemiopsis* (1960s) and primarily from the northwestern and southern Black Sea with *Mnemiopsis* (from Ostrovskaya et al., 1998). Data from the oxygenated surface layer (11 m) of Chesapeake Bay are averages of June and July (from Purcell et al., 1994b)

| Location | Biomass (mg C m$^{-3}$) | Production (mg C m$^{-3}$) |
| --- | --- | --- |
| Black Sea (before *Mnemiopsis*) | 33 | 3.1 |
| Black Sea (with *Mnemiopsis*) | 13 | 1.4 |
| Chesapeake Bay | 6 | 1.3 |

tion occurs in July–August in Chesapeake Bay, but in August–September or October in the Black Sea, where temperatures are lower all year. Average ctenophore population biomasses are similar in both locations (roughly 50 g WW m$^{-3}$), however, peak average biomass in the Black Sea (184 g WW m$^{-3}$) may be greater than measured in Chesapeake Bay (60 g WW m$^{-3}$). Maximum body size also may be greater in the Black Sea (12–18 cm) than in Chesapeake Bay (<12 cm), but this may reflect different methods of measurement.

Biotic controls on *Mnemiopsis* populations that are important in native locations were not conspicuous during the last decade in the Black Sea. Although *Mnemiopsis* must share zooplankton resources with zooplanktivorous jellyfish and fish species, the ctenophores seem to out-compete jellyfish and chaetognaths for food. Zooplanktivorous fish such as anchovies, which could be important competitors for food in Chesapeake Bay, had been greatly reduced by commercial fishing in the Black Sea prior to the outbreak of *Mnemiopsis*. Similarly, fish predators of *Mnemiopsis* are not fished heavily in Chesapeake Bay, but populations of potential fish predators in the Black Sea had been reduced by over-fishing (Purcell & Arai, 2001). Additionally, gelatinous predators of *Mnemiopsis* that limit *Mnemiopsis* populations in native habitats were absent from the Black Sea until 1999, when *Beroe ovata* arrived. Small populations of potential competitors and predators enabled the *Mnemiopsis* population to bloom unchecked in the Black Sea from 1989–1999.

Zooplankton population size can determine the size of the *Mnemiopsis* population, and *Mnemiopsis* populations also can greatly affect zooplankton. Zooplankton diversity is high in the Black Sea due to several sources of origin of the fauna. Similar species of subtropical warm-water or eurythermal zooplankton inhabit Chesapeake Bay and the upper layers of the Black Sea (Table 11). The most abundant species are *Acartia tonsa* in Chesapeake Bay and *A. clausi* in the Black Sea. *A. tonsa* was recently introduced into the Black Sea, probably from the Mediterranean Sea, and sometimes reaches high densities in the northwestern Black Sea (Ostrovkaya et al., 1998). Biomass and production of *A. clausi* in the Black Sea before the introduction of *Mnemiopsis* was about twice as great as afterwards (Table 12). After the introduction of *Mnemiopsis*, *A. clausi* and *A. tonsa* biomass and production are comparable in the Black Sea and Chesapeake Bay, respectively (Table 12).

The diets of *Mnemiopsis* in both regions include the range of available zooplankton, with copepods being the predominant prey. Feeding by lobate *Mnemiopsis* does not satiate at natural zooplankton densities. Thus, their general diet and voracious feeding contribute to their success in diverse environments. During high abundances of *Mnemiopsis*, zooplankton populations are low in Chesapeake Bay and in the Black Sea region, however, reduction of zooplankton stocks may only occur when predators of *Mnemiopsis* are not abundant.

Several different ecological groups of fish occur in both Chesapeake Bay and the Black Sea; however, the species diversity of fish is much higher in the Black Sea. The most abundant fish species in both regions is the zooplanktivorous anchovy, represented by the family Engraulidae (*Engraulis encrasicolus* in the Black Sea and *Anchoa mitchilli* in Chesapeake Bay). These species are potentially the main fish competitors of *Mnemiopsis* in both regions. Piscivorous bluefish (*Pomatomus saltator* Linnaeus) and anadromous shads (genus *Alosa*) also occur in both areas.

*Mnemiopsis* also is an important predator of ichthyoplankton, particularly of fish eggs. *Mnemiopsis* populations are capable of clearing 30–100% d$^{-1}$ of the fish eggs in Chesapeake Bay, but <5% of the fish larvae. In the Black Sea, reductions of fish eggs coincide with high ctenophore abundances. *Mnemiopsis* adversely affects zooplanktivorous fish populations by direct predation on the early life stages as well as consuming the same zooplankton foods. Fish populations sharply declined and remained low during the decade of high ctenophore abundances in the Black Sea.

In the Black Sea, reductions in zooplankton, ichthyoplankton and zooplanktivorous fish populations have been attributed to *Mnemiopsis*, but similar reductions have not been seen in Chesapeake Bay. We conclude that the enormous impact of *Mnemiopsis* on the Black Sea ecosystem occurred because of the shortage of predators and competitors there in the late 1980s and 1990s. The appearance of the ctenophore, *Beroe ovata*, may promote the recovery of the Black Sea ecosystem from the effects of the *Mnemiopsis* invasion.

## Acknowledgements

Previously unpublished results from field studies in Chesapeake Bay during 1995–1998 were funded by NSF grant OCE-9633607, and during 1992, by Maryland Sea Grant College grant R/P-35 to J. E. P. Logistical support, environmental, ichthyoplankton and fish data from 1995 to 1999 were funded by NSF grant DEB-9412113. We especially thank K. B. Heidelberg and D. A. Nemazie, and E. D. H.'s group for assistance in 1992 and 1995–1999, respectively, and M. R. Roman for use of 1996–1998 zooplankton data. The previously unpublished mesocosm study was funded by EPA grant R819640. Russian investigations were funded by RBFI-00-05-64827. We also thank P. Kremer and G. Finenko for constructive comments on the manuscript. UMCES Contribution No. 3388.

## References

Agassiz, A., 1865. North American Acalephae. Illustr. catalog Mus. Compar. Zool. 2: 1–234.

Aleksandrova, Z. V., T. E. Baskakova & E. V. Makarov, 1996. Assessment of the parameters that describe the eutrophication of the Sea of Azov with usage of multiple analyses. In Makarov, E. A. (ed.), The Main Problems of Fisheries in the Azov Sea Basin. AzNIIRHK, Rostov-on-Don: 20–28.

Anninsky, V. E. & A. D. Gubanova, 1998. Chemical composition and state of ctenophore *Mnemiopsis mccradyi* in the Black Sea. Gydrobiologichesky Zhurnal 34:36–43.

Anninsky, B. E., Z. A. Romanova, G. I. Abolmasova, A. C. Gucu & A. E. Kideys, 1998. The ecological and physiological state of the ctenophore *Mnemiopsis leidyi* (Agassiz) in the Black Sea in autumn 1996. In Ivanov, L. I. & T. Oguz (eds), Ecosystem Modeling as a Management Tool for the Black Sea. Kluwer Academic Publishers, Dordecht, The Netherlands: 249–262.

Arkhipkin, V. S. & S. A. Dobrolubov, 1999. Steric variation of the Mediterranean and Black Sea level. In Balopoulos, E. Th. & A. Iona (eds), Oceanography of the Eastern Mediterranean and Black Sea. Sakellariou, Athens: 5–6.

Båmstedt, U., H. Ishii & M. B. Martinussen, 1997. Is the scyphomedusa *Cyanea capillata* (L.) dependent on gelatinous prey for its early development? Sarsia 82: 269–273.

Bishop, J. W., 1967. Feeding rates of the ctenophore *Mnemiopsis leidyi*. Chesapeake Sci. 8: 259–264.

Borodkin, S. O. & L. I. Korzhikova, 1991. Chemical composition of the ctenophore *Mnemiopsis leidyi* and evaluation of its role in transformation of biogenic elements in the Black Sea. Oceanology 31: 555–558.

Brownlee, D. C. & F. Jacobs, 1987. Mesozooplankton and microzooplankton in the Chesapeake Bay. In Majumdar, S. K., L. W. Hall, Jr. & H. M. Austin (eds), Contaminant Problems and Management of Living Chesapeake Bay Resources. The Pennsylvania Academy of Science: 217–269.

Budnichenko, E. V., A. V. Firulina & Yu. V. Bulgakova, 1999. Feeding conditions of Azov anchovy in summer and autumn of 1995–1996. J. Ichthyol. 39: 233–240.

Burrell, V. G. Jr. & W. A. Van Engel, 1976. Predation by and distribution of a ctenophore, *Mnemiopsis leidyi* A. Agassiz, in the York River estuary. Estuar. coast. mar. Sci. 4: 235–242.

Caddy, J. F. & R. C. Griffiths, 1990. A perspective on recent fishery-related events in the Black Sea. Studies and Reviews. GFCM, 63: 43–71.

Calder, D. R., 1971. Hydroids and hydromedusae of southern Chesapeake Bay, Virginia Inst. Mar. Sci. Spec. Papers in Mar. Sci. 1: 1–125.

Cargo, D. G. & L. P. Schultz, 1966. Notes on the biology of the sea nettle, *Chrysaora quinquecirrha*, in Chesapeake Bay. Chesapeake Sci. 7: 95–100.

Cargo, D. G. & L. P. Schultz, 1967. Further observations on the biology of the sea nettle and jellyfishes in the Chesapeake Bay. Chesapeake Sci. 8: 209–220.

Costello, J. H., R. Loftus & R. Waggett, 1999. Influence of prey detection on capture success for the ctenophore *Mnemiopsis leidyi* feeding upon adult *Acartia tonsa* and *Oithona colcarva*. Mar. Biol. 191: 207–216.

Cowan, J. H., Jr. & E. D. Houde, 1993a. Relative predation potentials of scyphomedusae, ctenophores and planktivorous fish on ichthyoplankton in Chesapeake Bay. Mar. Ecol. Prog. Ser. 95: 55–65.

Cowan, J. H., Jr. & E. D. Houde, 1993b. Size-dependent predation on marine fish larvae by ctenophores, scyphomedusae and planktivorous fish. Fish. Oceanogr. 2: 113–126.

Deason, E. E., 1982. *Mnemiopsis leidyi* (Ctenophora) in Narragansett Bay, 1975–79: abundance, size composition and estimation of grazing. Estuar. coast. shelf Sci. 15: 121–134.

Deason, E. E. & T. J. Smayda, 1982. Ctenophore–zooplankton–phytoplankton interactions in Narragansett Bay, Rhode Island, U.S.A., during 1972–1977. J. Plankton Res. 4: 203–217.

Dekhnik, T. V., 1973. Ichthyoplankton of the Black Sea. Naukova Dumka, Kiev: 234 pp.

Feigenbaum, D. & M. Kelly, 1984. Changes in the lower Chesapeake Bay food chain in presence of the sea nettle *Chrysaora quinquecirrha* (Scyphomedusa). Mar. Ecol. Prog. Ser. 19: 39–47.

Fewkes, J. W., 1881. Studies of the jelly-fish of Narragansett Bay. Bull. Mus. Compar. Zool. 8: 141–182.

Finenko, G. A., G. I. Abolmasova & Z. A. Romanova, 1995. Intensity of the nutrition, respiration and growth of *Mnemiopsis mccradyi* in relation to grazing conditions. Biologia Morya 21: 315–320 (in Russian).

Finenko, G. A., Z. A. Romanova & G. I. Abolmasova, 2000. The ctenophore *Beroe ovata* is a recent invader to the Black Sea. Ecologiya Morya 50: 21–25 (in Russian).

Finenko, G. A., B. E. Anninsky, Z. A. Romanova, G. I. Abolmasova & A. E. Kideys, 2001. Chemical composition, respiration and feeding rates of the new alien ctenophore, *Beroe ovata*, in the Black Sea. Hydrobiologia 451 (Dev. Hydrobiol. 155): 177–186.

GESAMP (IMO/FAO/UNESCO-IOC/WMO/WHO/IAEA/UN/UNEP Joint Group of Experts on the Scientific Aspects of Marine Environmental Protection), 1997. Opportunistic settlers and the problem of the ctenophore *Mnemiopsis leidyi* invasion in the Black Sea. Rept Stud. GESAMP 58: 84 pp.

Gordina, A. D. & T. N. Klimova, 1995. Dynanics of species composition and number of ichthyoplankton in coastal and open waters. In Konovalov, S. M. (ed.), Modern State of the Ichthyofauna of the Black Sea. Naukova Dumka, Kiev: 74–92.

Hood, R. R., H. V. Wang, J. E. Purcell, E. D. Houde & L. W. Harding, Jr., 1999. Modeling particles and pelagic organisms in Chesapeake Bay: convergent features control plankton distributions. J. Geophys. Res. 104: 1223–1243 and 3289–3290.

Harbison, G. R., 1993. The potential of fishes for the control of gelatinous zooplankton, ICES C.M. 1993/L: 74: 1–10.

Houde, E. D. & C. E. Zastrow, 1991. Bay anchovy. In Funderburk, S. L., J. A. Mihursky, S. J. Jordan & D. Riley (eds), Habitat Requirements for Chesapeake Bay Living Resources. 2nd edn. Living Resources Subcommittee, Chesapeake Bay Program, Annapolis: 8–1 to 8–14.

Houde, E. D., S. Juki -Peladi, S. Brandt & S. D. Leach, 1999. Fisheries: trends in catches, abundance and management. In Malone, T. C., A. Malej, L. W. Harding, Jr., N. Smodlaka & R. E. Turner (eds), Ecosystems at the Land-Sea Margin: Drainage Basin to Coastal Sea. American Geophysical Union, Coastal and Estuarine Studies 55: 341–366.

Ivanov, L. & R. J. H. Beverton, 1985. The fisheries resources of the Mediterranean. Part two. Black Sea. GFCM FAO, Rome: 70 pp.

Keister, J. E., E. D. Houde & D. L Breitburg, 2000. Effects of bottom-layer hypoxia on abundances and depth distributions of organisms in Patuxent River, Chesapeake Bay, Mar. Ecol. Prog. Ser. 204: 43–59.

Khoroshilov, V. S., 1993. Seasonal dynamics of the Black Sea population of ctenophore *Mnemiopsis leidyi*. Oceanology 33: 558–562 (in Russian).

Khoroshilov, V. S. & T. A. Lukasheva, 1999. Changes of zooplankton community of the Blue Bay after introduction of *Mnemiopsis* in the Black Sea. Oceanology 39: 1–6.

Kideys, A. E. & U. Niermann, 1994. Occurrence of *Mnemiopsis* along the Turkish coast ICES J. mar. Sci. 51: 423–427.

Kideys, A. E., A. O. Gordina, U. Nierman, Z. Usal, T. A. Shiganova & F. Bingel, 1998. Distribution of eggs and larvae of anchovy with respect to ambient conditions in the southern Black Sea during 1993 and 1996. In Ivanov, L. & T. Oguz (eds), Ecosystem Modeling as a Management Tool for the Black Sea. Kluwer Academic Publishers, Dordrecht, The Netherlands: 47: 189–198.

Klebasko, M. J., 1991. Feeding ecology and daily ration of bay anchovy (*Anchoa mitchilli*) in the mid-Chesapeake Bay. M. S. Thesis, University of Maryland, College Park: 103 pp.

Konsulov, A. S. & L. T. Kamburska, 1998. Ecological determination of the new Ctenophora – *Beroe ovata* invasion in the Black Sea. Oceanology. Proc. Inst. Oceanol. Varna 2: 195–198.

Kovalev, A. V., 1993. Mesozooplankton. In Kovalev, A. V. & Z. Z. Finenko (eds), Plankton. Naukova Dumka, Kiev: 144–165.

Kovalev, A. V., S. Besiktepe, J. Zagorodnyaya & A. Kideys, 1998. Mediterraneanization of the Black Sea zooplankton is continuing. In Ivanov, L. & T. Oguz (eds), Ecosystem Modeling as a Management Tool for the Black Sea. Kluwer Academic Publishers, Dordrecht, The Netherlands: 47: 199–207.

Kremer, P., 1975a. Excretion and body composition of the ctenophore *Mnemiopsis leidyi* (A. Agassiz): comparisons and consequences. In Persoone, G. & E. Jaspers (eds), 10th European Symposium on Marine Biology. Universa Press, Weeteren 2: 351–362.

Kremer, P., 1975b. Nitrogen regeneration by the ctenophore *Mnemiopsis leidyi*. In Howell, F. G., J. B. Howell & M. H. Smith (eds), Mineral Cycling in Southeastern Ecosystems. ERDA Symposium Series (Cong-740513), Kingston, Rhode Island: 279–290.

Kremer, P., 1976. Population dynamics and ecological energetics of a pulsed zooplankton predator, the ctenophore *Mnemiopsis leidyi*. In Wiley, M. L. (ed.), Estuarine Processes. Academic Press, New York 1: 197–215.

Kremer, P., 1977. Respiration and excretion by the ctenophore *Mnemiopsis leidyi*. Mar. Biol. 44: 43–50.

Kremer, P., 1979. Predation by the ctenophore *Mnemiopsis leidyi* in Narragansett Bay, Rhode Island. Estuaries 2: 97–105.

Kremer, P., 1982. Effect of food availability on the metabolism of the ctenophore *Mnemiopsis mccradyi*. Mar. Biol. 71: 149–156.

Kremer, P., 1994. Patterns of abundance for *Mnemiopsis* in U.S. coastal waters: a comparative overview. ICES J. mar. Sci. 51: 347–354.

Kremer, P. & S. Nixon, 1976. Distribution and abundance of the ctenophore *Mnemiopsis leidyi* in Narragansett Bay. Estuar. coast. mar. Sci. 4: 627–639.

Kremer, P. & M. R. Reeve, 1989. Growth dynamics of a ctenophore (*Mnemiopsis*) in relation to variable food supply. II. Carbon budgets and growth model. J. Plankton Res. 11: 553–574.

Kreps, T. A., J. E. Purcell & K. B. Heidelberg, 1997. Escape of the ctenophore, *Mnemiopsis leidyi* from the scyphomedusa predator, *Chrysaora quinquecirrha*. Mar. Biol. 128: 441–446.

Kuropatkin, A. P., 1998. Changes in salinity and in the vertical stability of the Azov Sea water in the present day conditions. In Makarov, E. V. (ed.), The Main Problems of Fisheries in the Azov Sea Basin. AzNIIRHK Rostov-on-Don: 30–33.

Larson, R. J., 1987. Feeding and functional morphology of the lobate ctenophore *Mnemiopsis mccradyi*. Estuar. coastal shelf Sci. 27: 495–502.

Lebedeva, L. P., 1998. Variability of the number and biomass of Ctenophora *Mnemiopsis* in the Black Sea (a Model Research). Oceanology 38: 727–733.

Lebedeva, L. P. & E. A. Shushkina, 1991. The estimation of population characteristics of *Aurelia aurita* in the Black Sea. Oceanology 31: 434–441.

Lebedeva, L. P. & E. A. Shushkina, 1994. The model investigation of the Black Sea plankton community changes caused by *Mnemiopsis*. Oceanology 34:79–87.

Lisovenko, L. A., D. P. Andrianov & Yu. V. Bulgakova, 1997. Reproductive ecology of the Black Sea anchovy *Engraulis encrasicolus ponticus*. II. Quantitative parameters of spawning. J. Ichthyol. 37: 639–646.

Luo, J. & S. B. Brandt, 1993. Bay anchovy, *Anchoa mitchilli*, production and consumption in mid-Chesapeake Bay based on a bioenergetics model and acoustic measures of fish abundance. Mar. Ecol. Prog. Ser. 140: 271–283.

174

MacGregor, J. M. & E. D. Houde, 1996. Onshore-offshore pattern and variability in distribution and abundance of bay anchovy *Anchoa mitchilli* eggs and larvae in Chesapeake Bay. Mar. Ecol. Prog. Ser. 138: 15–25.

Malone, T. C., A. Malej, L. W. Harding, Jr., N. Smodlaka & R. E. Turner (eds), 1999. Ecosystems at the Land–Sea Margin: Drainage Basin to Coastal Sea. American Geophysical Union, Coastal & Estuarine Studies 55: 1–381.

Mayer, A. G., 1912. Ctenophores of the Atlantic coast of North America. Publ. Carnegie Inst., Washington, D.C. 162: 1–58.

Mianzan, H., 1999. Ctenophora. In Boltovskoy, D. (ed.), South Atlantic Zooplankton. Backhuys Publishers, Leiden: 561–573.

Miller, R. J., 1970. Distribution and energetics of an estuarine population of the ctenophore, *Mnemiopsis leidyi*. Ph.D. Thesis, North Carolina State Univ., Raleigh, N. C.: 78 pp.

Miller, R. J., 1974. Distribution and biomass of an estuarine ctenophore population, *Mnemiopsis leidyi* (A. Agassiz). Chesapeake Sci. 15: 1–8.

Monteleone, D. M. & L. E. Duguay, 1988. Laboratory studies of predation by the ctenophore *Mnemiopsis leidyi* on the early stages in the life history of the bay anchovy, *Anchoa mitchilli*. J. Plankton Res. 10: 359–372.

Mutlu, E., 1999. Distribution and abundance of ctenophores and their zooplankton food in the Black Sea. II. *Mnemiopsis leidyi*. Mar. Biol.135: 603–614.

Nastenko, E. V. & L. M. Polishchuk, 1999. The comb jelly *Beroe* (Ctenophora: Beroida) in the Black Sea. Dopovidi Nazionalnoi Akademii Nauk, Ukraine 11: 159–161.

Nelson, T. C., 1925. On the occurrence and food habits of ctenophores in New Jersey inland coastal waters. Biol. Bull. 48: 92–111.

Nemazie, D. A., J. E. Purcell & P. M. Glibert, 1993. Ammonium excretion by gelatinous zooplankton and their contribution to the ammonium requirements of microplankton in Chesapeake Bay. Mar. Biol. 116: 451–458.

Niermann, U., F. Bingel, A. Gorban, A. D. Gordina, A. C. Gugu, A. E. Kideys, A. Konsulov, G. Radu, A. A. Subbotin & V. E. Zaika, 1994. Distribution of anchovy eggs and larvae (*Engraulis engrasicolus* Cuv.) in the Black Sea in 1991–1992. ICES J. mar. Sci. 51: 395–406.

Oguz, T., H. W. Ducklow, J. E. Purcell & P. Malanotte-Rizzoli, 2001. Modeling the response of top-down control exerted by gelatinous carnivores on the Black Sea pelagic food web. J. Geophys. Res. 106: 4543–4564.

Oguz, T., H. Ducklow, P. Malanotte-Rizzoli, J. W. Murray, V. I. Vedernikov & U. Unluata, 1999. A physical–biochemical model of plankton productivity and nitrogen cycling in the Black Sea. Deep-Sea Res. 46: 598–635.

Oguz, T., H. Ducklow, P. Malanotte-Rizzoli, S.Tugrul, N. Nezlin & U. Unluata, 1996. Simulation of annual plankton productivity cycle in the Black Sea by a one-dimensional physical–biological model. J. Geophys. Res. 101: 16 585–16 599.

Oguz, T., H. Ducklow, E. A. Shushkina, P. Malanotte-Rizzoli, S. Tugrul & L. P. Lebedeva, 1998. Simulation of upper layer biochemical structure in the Black Sea. In Ivanov, L. & T. Oguz (eds), Ecosystem Modeling as a Management Tool for the Black Sea. Kluwer Academic Publishers, Dordrecht, The Netherlands: 257–300.

Olesen, N. J., J. E. Purcell & D. K. Stoecker, 1996. Feeding and growth by ephyrae of scyphomedusae *Chrysaora quinquecirrha*. Mar. Ecol. Prog. Ser. 137: 149–159.

Ovchinnikov, I. M. & V. B. Titov, 1990. Antycyclonic eddies of currents in the coastal area of the Black Sea. DAN USSR 314: 739–746.

Oven, L. S., A. D. Gordina & V. Ye. Giragosov, 1991. Current state of stocks of certain exploited populations of fish in the Black Sea. In Vinogradov, M. W. (ed.), Izmeneniya Ekosistemy Chernogo Morya: Yestestvennye I Antropogennye Factory (Variability of the Black Sea Ecosystem: Natural and Anthropogenic Factors). Nauka, Moscow: 241–247.

Oviatt, C. A. & P. M. Kremer, 1977. Predation on the ctenophore, *Mnemiopsis leidyi*, by butterfish, *Peprilus triacanthus*, in Narragansett Bay, Rhode Island. Chesapeake Sci. 18: 236–240.

Pavlova, E. V. & N. I. Minkina, 1993. The respiration rate of the Black Sea invader ctenophore (Ctenophora. Lobata: *Mnemiopsis*) Dokl. RAS 333 (5): 682–683.

Pereladov, M. V., 1988. Some observations for biota of Sudak Bay of the Black Sea. III All-Russian conference of marine biology. Naukova Dumka, Kiev 1: 237–238 (in Russian).

Petranu, A., 1997. Black Sea Biological Diversity. Romania. BSEP, 4. UN Publications. New York: 315 pp.

Prodanov, K., K. Mikhailov, G. Daskalov, K. Maxim, A. Chashchin, A. Arkhipov, V. Shlyakhov & E. Ozdamar, 1997.Environmental impact on fish resources in the Black Sea. In Ozsoy, E. & A. Mikaelyan (eds), Sensivity of North Sea, Baltic Sea and Black Sea to Anthropogenic and Climatic Changes. Kluwer Academic Publishers, Dordrecht, The Netherlands: 163–181.

Purcell, J. E., 1988. Quantification of *Mnemiopsis leidyi* (Ctenophora, Lobata) from formalin-preserved plankton samples. Mar. Ecol. Prog. Ser. 45: 197–200.

Purcell, J. E., 1992. Effects of predation by the scyphomedusan *Chrysaora quinquecirrha* on zooplankton populations in Chesapeake Bay. Mar. Ecol. Prog. Ser. 87: 65–76.

Purcell, J. E., 1997. Pelagic cnidarians and ctenophores as predators: Selective predation, feeding rates and effects on prey populations. Ann. Inst. Oceanogr., Paris 73: 125–137.

Purcell, J. E. & M. N. Arai, 2001. Interactions of pelagic cnidarians and ctenophores with fishes: a review. Hydrobiologia 451 (Dev. Hydrobiol. 155): 27–44.

Purcell, J. E. & J. H. Cowan, Jr., 1995. Predation by the scyphomedusan *Chrysaora quinquecirrha* on *Mnemiopsis leidyi* ctenophores. Mar. Ecol. Prog. Ser. 128: 63–70.

Purcell, J. E. & M. V. Sturdevant, 2001. Prey selection and dietary overlap among zooplanktivorous jellyfish and juvenile fishes in Prince William Sound, Alaska. Mar. Ecol. Prog. Ser. 210: 67–83.

Purcell, J. E., A. Malej & A. Benović, 1999a. Potential links of jellyfish to eutrophication and fisheries. In Malone, T. C., A. Malej, L. W. Harding, Jr., N. Smodlaka & R. E. Turner (eds), Ecosystems at the Land–Sea Margin: Drainage Basin to Coastal Sea. American Geophysical Union, Coastal and Estuarine Studies 55: 241–263.

Purcell, J. E., J. R. White & M. R. Roman, 1994a. Predation by gelatinous zooplankton and resource limitation as potential controls of *Acartia tonsa* copepod populations in Chesapeake Bay. Limnol. Oceanogr. 39: 263–278.

Purcell, J. E., J. R. White, D. A. Nemazie & D. A. Wright, 1999b. Temperature, salinity and food effects on asexual reproduction and abundance of the scyphozoan *Chrysaora quinquecirrha*. Mar. Ecol. Prog. Ser. 180: 187–196.

Purcell, J. E., D. A. Nemazie, S. E. Dorsey, E. D. Houde & J. C. Gamble, 1994b. Predation mortality of bay anchovy (*Anchoa mitchilli*) eggs and larvae due to scyphomedusae and ctenophores in Chesapeake Bay. Mar. Ecol. Prog. Ser. 114: 47–58.

Purcell, J. E., D. L. Breitburg, M. B. Decker, W. M. Graham, M. J. Youngbluth & K. A. Raskoff, 2001. Pelagic cnidarians and ctenophores in low dissolved oxygen environments: a review. In Rabalais, N. N. & R. E. Turner (eds), Coastal Hypoxia:

Consequences for Living Resources and Ecosystems, American Geophysical Union, Coastal and Estuar. Stud. 58: 77–100.

Quaglietta, C. E., 1987. Predation by *Mnemiopsis leidyi* on hard clam larvae and other natural zooplankton in Great South Bay, N.Y. M. S. thesis, Marine Sciences Research Center, State University of New York, Stony Brook: 66 pp.

Rass, T. S., 1992. Changes in the fish resources of the Black Sea. Oceanology 32: 197–203.

Reeve, M. R. & M. A. Walter, 1978. Nutritional ecology of ctenophores – a review of recent research. Adv. mar. Biol. 15: 249–287.

Reeve, M. R., M. A. Syms & P. Kremer, 1989. Growth dynamics of a ctenophore (*Mnemiopsis*) in relation to variable food supply. I. Carbon biomass, feeding, egg production, growth and assimilation efficiency. J. Plankton Res. 11: 535–552.

Rilling, G. C. & E. D. Houde, 1999a. Regional and temporal variability in distribution and abundance of bay anchovy (*Anchoa mitchilli*) eggs, larvae and adult biomass in the Chesapeake Bay. Estuaries 22: 1096–1109.

Rilling, G. C. & E. D. Houde, 1999b. Regional and temporal variability in growth and mortality of bay anchovy, *Anchoa mitchilli*, larvae in Chesapeake Bay. Fish. Bull. U. S. 97: 555–569.

Roman, M. R., A. L. Gauzens, W. K. Rhinehart & J. R. White, 1993. Effects of low oxygen waters on Chesapeake Bay zooplankton. Limnol. Oceanogr. 38: 1603–1614.

Seravin, L. N., 1994. The systematic revision of the genus *Mnemiopsis* (Ctenophora, Lobata). Zool. Zhurnal 73: 9–18 (in Russian).

Sergeeva, N. G., V. E. Zaika & T. V. Mikhailova, 1990. Nutrition of ctenophore *Mnemiopsis mccradyi* (Ctenophora, Lobata) in the Black Sea. Zool. J. Ecologia Morya 35: 18–22 (in Russian).

Shiganova, T. A., 1993. Ctenophore *Mnemiopsis leidyi* and ichthyoplankton in the Sea of Marmara in October of 1992. Oceanology 33: 900–903.

Shiganova, T. A., 1997. *Mnemiopsis leidyi* abundance in the Black Sea and its impact on the pelagic community. In Ozsoy, E. & A. Mikaelyan (eds), Sensitivity of North Sea, Baltic Sea and Black Sea to Anthropogenic and Climatic Changes. Kluwer Academic Publishers, Dordrecht, The Netherlands: 117–130.

Shiganova, T. A., 1998. Invasion of the Black Sea by the ctenophore *Mnemiopsis leidyi* and recent changes in pelagic community structure. Fish. Oceanogr. 7: 305–310.

Shiganova, T. A. & Y. V. Bulgakova, 2000. Effect of gelatinous plankton on the Black and Azov Sea fish and their food resources. ICES J. mar. Sci. 57: 641–648.

Shiganova, T. A, J. V. Bulgakova, P. Yu. Sorokin & Yu. F. Lukashev, 2000. Investigation of new invader *Beroe ovata* in the Black Sea. Russian Biology (as is confused with U.S.) Biol. Bull. 27: 247–255.

Shiganova, T. A., Yu. V. Bulgakova, S. P. Volovik, Z. A. Mirzoyan & S. I. Dudkin, 2001a. The new invader *Beroe ovata* Mayer, 1912 and its effect on the ecosystem in the northeastern Black Sea. Hydrobiologia 451 (Dev. Hydrobiol. 155): 187–197.

Shiganova, T. A., A. E. Kideys, A. S. Gucu, U. Niermann & V. S. Khoroshilov, 1998. Changes of species diversity and their abundance in the main components of pelagic community during last decades. In Ivanov, L. I. & T. Oguz (eds), Ecosystem Modeling as a Management Tool for the Black Sea. Kluwer Academic Publishers, Dordrecht, The Netherlands: 171–188.

Shiganova, T. A., I. A. Mirzoyan, E. A. Studenikina, S. P. Volovik, I. Siokoi-Frangou, S. Zervoudaki, E. D. Christou, A. Yu. Skirta & H. J. Dumont, 2001b. Comparison of spatial and temporal distribution of the invader ctenophore *Mnemiopsis leidyi* in the Black Sea and adjacent seas of the Mediterranean basin. Mar. Biol. In press.

Shushkina, E. A. & M. E. Vinogradov, 1991. Long-term changes in the biomass of plankton in open areas of the Black Sea. Oceanology 31: 716–721.

Stanlaw, K. A., M. R. Reeve & M. A. Walter, 1981. Growth, food and vulnerability to damage of the ctenophore *Mnemiopsis maccradyi* in its early life history stages. Limnol. Oceanogr. 26: 224–234.

Stoecker, D. K., P. G. Verity, A. E. Michaels & L. H. Davis, 1987. Feeding by larval and post-larval ctenophores on microzooplankton. J. Plankton Res. 9: 667–683.

Studenikina, E. I., S. P. Volovik, I. A. Miryozan & G. I. Luts, 1991. The ctenophore *Mnemiopsis leidyi* in the Sea of Azov. Oceanology 3: 722–725.

Sullivan B. K., D. Van Keuren & M. Clancy, 2001. Timing and size of blooms of the ctenophore *Mnemiopsis leidyi* in relation to temperature in Narragansett Bay, RI. Hydrobiologia 451 (Dev. Hydrobiol. 155): 113–120.

Tzikhon-Lukanina, E. A. & O. G. Reznichenko, 1991. Feeding peculiarities of different size specimens of ctenophore *Mnemiopsis* in the Black Sea. Oceanology 31: 442–446.

Tzikhon-Lukanina, E. A., O. G. Reznichenko & T. A. Lukasheva, 1991. Quantitative aspects of feeding in the Black Sea ctenophore *Mnemiopsis leidyi*. Oceanology 31: 272–276.

Tzikhon-Lukanina, E. A., O. G. Reznichenko & T. A. Lukasheva, 1992. What ctenophore *Mnemiopsis* eats in the Black Sea inshore waters? Oceanology 32: 724–729.

Tzikhon-Lukanina, E. A., O. G. Reznichenko & T. A. Lukasheva, 1993a. Predation rates on fish larvae by the ctenophore *Mnemiopsis leidyi* in the Black Sea inshore waters. Oceanology 33: 895–899.

Tzikhon-Lukanina, E. A., O. G. Reznichenko & T. A. Lukasheva, 1993b. Ecological variation of comb-jelly *Mnemiopsis leidyi* (Ctenophora) in the Black Sea. Zhurnal obszhei Biologii 54: 713–724 (in Russian).

Van Eeckhout, D. & C. Lancelot, 1997. Modeling the functioning of the northwestern Black Sea ecosystem from 1960 to present. In Ozsoy, E. & A. Mikaelyan (eds), Sensitivity to Change: Black Sea, Baltic Sea and North Sea. Kluwer Academic Publishers, Dordrecht, The Netherlands: 455–469.

Vasquez, A. V., 1989. Energetics, trophic relationships and chemical composition of bay anchovy, *Anchoa mitchilli*, in the Chesapeake Bay. M. S. Thesis, University of Maryland, College Park: 166 pp.

Vinogradov, M. E., V. V. Sapozhnikov & E. A. Shushkina, 1992. The Black Sea Ecosystem. Nauka, Moscow: 112 pp. (in Russian).

Vinogradov, M. E., E. A. Shushkina, E. I. Musaeva & P. Yu. Sorokin, 1989. Ctenophore *Mnemiopsis leidyi* (A. Agassiz) (Ctenophora: Lobata) – new settlers in the Black Sea. Oceanology 29: 293–298.

Volovik, S. P. & A. S. Chikhachev, 1998. Man-provoked transformations of the ichthyofauna in the Azov Sea basin. In Makarov, E. B. (ed.), The Main Problems of Fisheries in the Azov Sea Basin. AzNIIRHK, Rostov-on-Don: 7–22.

Volovik, S. P., I. A. Mirzoyan & G. S. Volovik, 1993. *Mnemiopsis leidyi*: biology, population dynamics, impact to the ecosystem and fisheries. ICES (Biol. Oceanogr. Committee) 69: 1–11.

Walter, M. A., 1976. Quantitative observations on the nutritional ecology of ctenophores with special reference to *Mnemiopsis mccradyi*. M. S. Thesis, University of Miami.

Zaika, V. Ye. & N. I. Ivanova, 1992. Ctenophore *Mnemiopsis mccradyi* in autumn hyponeuston of the Black Sea. Ecol. Morya 42: 6–10.

176

Zaika, V. E. & N. K. Revkov, 1994. Anatomy of gonads and regime of spawning of ctenophore *Mnemiopsis* sp. in the Black Sea. Zool. Zhurnal 73: 5–10 (in Russian).

Zaika, V. E. & N. K. Revkov, 1998. Nutrition of the Black Sea ctenophore in dependence on zooplankton abundance. Gydrobiologichesky Zhurnal. 34: 29–35 (in Russian).

Zaika, V. Ye. & N. G. Sergeeva, 1990. Morphology and development of *Mnemiopsis mccradyi* (Ctenophora, Lobata) in the Black Sea. Zool. Zhurnal 69: 5–11 (in Russian).

Zaitsev, Yu. P. & B. G. Aleksandrov, 1997. Recent man-made changes in the Black Sea ecosystem. In Ozsoy, E. & A. Mikaelyan (eds), Sensitivity of North Sea, Baltic Sea and Black Sea to Anthropogenic and Climatic Changes. Kluwer Academic Publishers, Dordrecht, The Netherlands: 25–32.

Zakhutsky, V. P., G. I. Lutz & V. M. Shishkin, 1983. The numbers and biomass of jelly fish in the Sea of Azov. Rybnoe Khozyaistvo 8: 33–34 (in Russian).

*Hydrobiologia* **451**: 177–186, 2001.
© 2001 *Kluwer Academic Publishers.*

# Chemical composition, respiration and feeding rates of the new alien ctenophore, *Beroe ovata*, in the Black Sea

G. A. Finenko, B. E. Anninsky, Z. A. Romanova, G. I. Abolmasova & A. E. Kideys[1,*]
*Institute of Biology of the Southern Seas, Nakhimov Av. 2, Sevastopol 335011, Ukraine*
[1]*Institute of Marine Sciences, Middle East Technical University, P.O. Box 28, Erdemli 33731, Turkey*
*E-mail: kideys@ims.metu.edu.tr*
(*Author for correspondence)

*Key words:* Black Sea, ctenophore, *Beroe*, *Mnemiopsis*, feeding, respiration

## Abstract

Maximum daily rations of the ctenophore *Beroe ovata* Brugière and predatory impacts on the *Mnemiopsis leidyi* A. Agassiz population were estimated via digestion time, prey biomass and predator and prey density in Sevastopol Bay and adjacent water regions. Digestion times ranged from 0.5 to 5.2 h and depended on the prey/predator weight ratio. Overall, the mean daily ration was 45% of *B. ovata* wet weight. Preliminary conclusions are given on the *B. ovata* population as an effective control of the *M. leidyi* population and on the dynamics and structure of the planktonic community as a whole.

## Introduction

The massive population explosion of the invading ctenophore, *Mnemiopsis leidyi* at the end of the 1980s led to tremendous changes in the Black Sea ecosystem, which was already suffering due to eutrophication. Feeding voraciously on zooplankton as well as on fish eggs and larvae, *Mnemiopsis* was one of the most important reasons for the adverse changes in the planktonic community structure (Vinogradov et al., 1989; Gordina & Klimova, 1996; Zaitsev, 1998; Kovalev et al., 1998; Shiganova et al., 1998; Kideys et al., 2000; Purcell et al., 2001) and pelagic fish stocks of the Black Sea and adjacent seas (Volovik et al., 1991; Kideys, 1994; Niermann et al., 1994; Kideys et al., 1999). The lack of natural predators feeding on *M. leidyi* resulted in temperature and food conditions being the only factors apparently controlling its population in the Black Sea (Purcell et al., 2001). Such an unprecedented impact caused this alien species to attract great attention from the scientific community, so much so that UNEP intervened to develop a strategy and to recommend measures to overcome the ctenophore and prevent similar invasions in other parts of the world, using the Black Sea region as a key example (GESAMP, 1997). One of the strategies recommen-

ded was the introduction of a predator to the Black Sea; along with the American butterfish, *Peprilus triacanthus*, ctenophores in the genus *Beroe* emerged as the best candidates for reducing the high biomass levels of *Mnemiopsis*.

Interestingly, by October 1997, the ctenophore, *Beroe ovata*, had already appeared in shallow waters of the Black Sea (Konsulov & Kamburska, 1998; Zaitsev, 1998), in September 1998 in the Sea of Marmara (A. E. Kideys, unpublished data) and in August–September 1999 in Sevastopol Bay and adjacent water regions as well as in the northeastern Black Sea (Finenko et al., 2000, Shiganova et al., 2000, Vostokov et al., 2000, Vinogradov et al., 2000). As in the genus *Mnemiopsis* (Order Lobata), species of *Beroe* spp. (Order Beroida) are difficult to differentiate (see Mills et al., 1996 for review), and *Beroe* in the Black Sea was identified either as *B. ovata* (in Konsulov & Kamburska, 1998; Shiganova et al., 2000, 2001), or as *B. cucumis* Fabricius (in Zaitsev, 1998). Although here it is cited as *B. ovata*, it could just as well be *B. cucumis*. The main criterion for distinguishing the two species is the presence of connection (i.e. anastomosing) of the meridional channels; whilst they are not joined in *B. cucumis*, *B. ovata* have anastomosing meridional channels (Gosner, 1971; Mianzan, 1999).

Apparently both structures are observed in the Black Sea. It is worth noting that *Beroe ovata* already had been reported from the Sea of Marmara by October, 1992 (Shiganova et al. 1994).

All previous studies suggest that species of *Beroe* almost exclusively feed on other ctenophores (Kamshilov, 1960; Greve, 1970; Swanberg, 1974; Harbison et al., 1978; Matsumoto & Harbison, 1993; Mianzan, 1999). In studies of predator–prey relationships of beroid and lobate ctenophore species, both Greve (1970) for *B. cucumis* vs *Bolinopsis infundibulum* (O. F. Müller, 1776) and Swanberg (1974) for *Beroe ovata* vs *Bolinopsis vitrea* (L. Agassiz, 1860) claimed that the feeding interactions among ctenophores form an ecological feed-back system that also affects other compartments of the planktonic community. Within each feed-back system, quantitative predator–prey relations are direct; both members of the system affect each other immediately. While these primary effects are more obvious and immediate, evaluation of secondary, tertiary etc. effects on other compartments of the ecosystem will take a longer time and effort to assess. By analyzing the long-term distribution of *M. leidyi*, it has already been observed that the biomass of this ctenophore has been decreasing since the appearance of its predator *B. ovata* in the southern Black Sea (Kideys & Romanova, 2001).

Here, we attempt to quantify the predation impact of the new alien ctenophore *B. ovata* on the earlier invading *M. leidyi* population by studying digestion and respiration rates of *B. ovata* in the laboratory along with abundance, biomass and population structure of both ctenophores from the shallow waters of the Black Sea.

## Methods

### Abundance, biomass and population structure

Ctenophores (*M. leidyi* and *B. ovata*) were collected by vertical tows (10–0 m) with a Bogorov-Rass net (500 $\mu$m mesh size and 80 cm diameter) at 11 stations in Sevastopol Bay and adjacent water regions during September–November 1999 (Fig. 1). Simultaneously, with the same net, horizontal tows were performed from the surface (0–0.5 m). The net filtered 77.2 m$^3$ of seawater in each 5 min horizontal tow. Immediately upon retrieval, samples were examined, ctenophores were counted and their lengths (total length of *B. ovata* and oral-aboral length of *M. leidyi*) were measured to

the nearest 1 mm without removing them from the water-filled jars. For the estimation of length-weight equations, 22 *B. ovata* and 230 *M. leidyi* were weighed individually on a balance to the nearest 0.01 g and simultaneously length measurements were taken.

### Chemical analysis

For the analysis of chemical composition as well as respiration and feeding experiments, ctenophores were collected gently with hand-held jars from the bay shore. Only undamaged and active animals were used for the experiments. In order to determine the dry weight of *B. ovata* 17 newly caught ctenophores were individually weighed to the nearest 0.01 g and then dried at 60 °C to a constant weight. Individually homogenized dry tissue of nine *B. ovata* were stored at −20 °C for further analysis of organic matter, protein, lipid and carbohydrate contents to calculate calorific value of *B. ovata*. The body length and wet weight of these specimens ranged from 35 to 103 mm, and from 5.9 to 65.3 g, respectively.

The biochemical assays, which are routine colourimetric techniques (Anninsky, 1994), were made within 10 days following the sample collection. Protein was measured by the Lowry method with an HSA (human serum albumin) standard; free amino acids were measured by the Pochinok method with a D,L-$\alpha$ - alanine standard; carbohydrate was measured by the Dubois method with a D-glucose standard. Lipid was determined according to the Amenta method, following the Folch chloroform/methanol (2:1) extraction. The standard was triolein/cholesterol (1:1). The energy content of *B. ovata* specimens was calculated by dry weight and the calorific value of each main biochemical compound determined (i.e. 5.65 cal mg$^{-1}$ for protein, 9.45 cal mg$^{-1}$ for lipid and 4.10 cal mg$^{-1}$ for carbohydrate; Winberg, 1971).

### Respiration

Sixteen *B. ovata* specimens were incubated at 21±1 °C after minimal delay (<2 h) to determine the freshly collected metabolic rate. The individual ctenophores were kept in the dark in respiration chambers (0.25–3.27 l capacity according to specimen size) filled with 120 $\mu$m filtered sea water for 17–19 h. Calculations of the metabolic rates were made from the measured difference in oxygen concentration between bottles with and without ctenophores. At the end of the incubation period, oxygen concentrations were measured in subsamples of seawater transferred into 60 ml

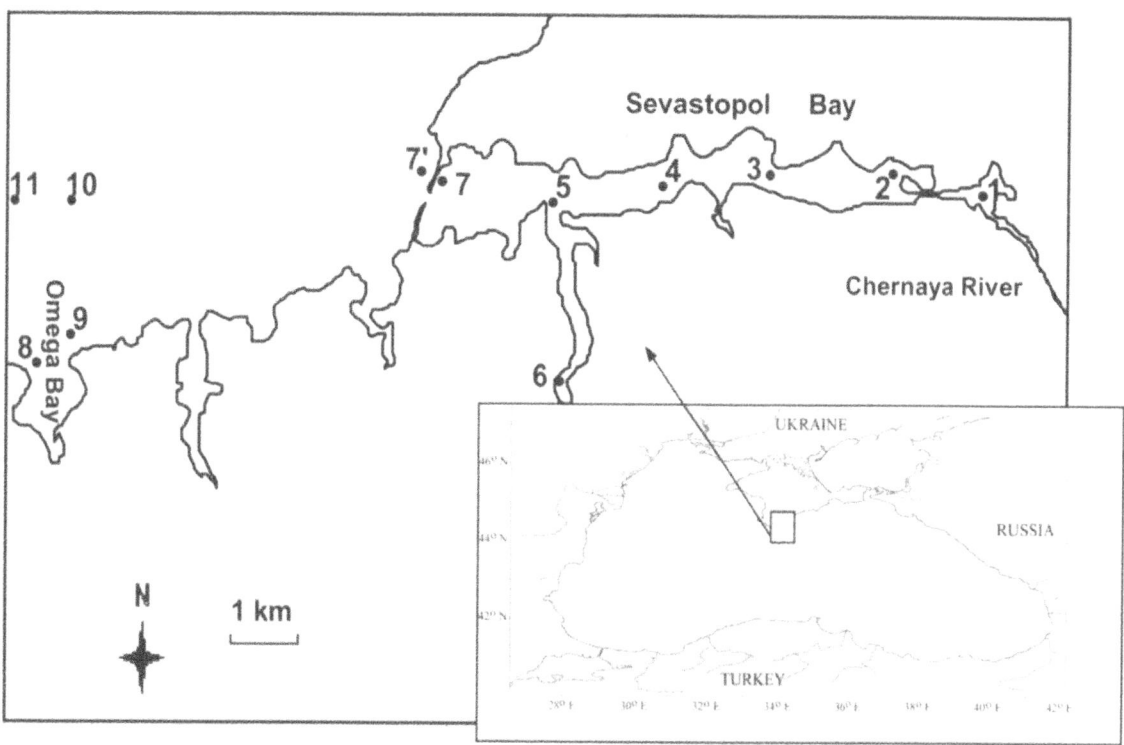

*Figure 1.* Sampling stations in Sevastopol Bay and adjacent regions in the Black Sea, during September–November 1999.

biochemical-oxygen demand bottles. Oxygen concentrations were determined by titration (Omori & Ikeda, 1984). For the conversion of respiration rate into energy values, the coefficient value of 4.86 cal ml $O_2^{-1}$ was used (Omori & Ikeda, 1984).

*Estimation of digestion time and rate of consumption*

Feeding experiments were conducted during 23 September–18 October using two series of experiments. All experiments were carried out in the dark at $21\pm1$ °C (close to the ambient temperature for Sevastopol Bay for this time) using containers of 5 l capacity.

Two independent methods were used to determine the consumption rate of *B. ovata*. In the first series of experiments, maximum feeding rate was measured under experimental conditions. These experiments were run over a relatively longer period (duration varied from 2 to 4 days) with a single *B. ovata* and 3–5 *M. leidyi* being held in the same container. In this condition, *B. ovata* had many prey available and could feed ad lib over the whole period. The number of *M. leidyi* in the containers was counted daily and new prey animals were added to ensure that prey items were never exhausted. The wet weight of *B. ovata* in these

experiments (total 19) ranged from 4.85 to 21.31 g and *M. leidyi* from 0.93 to 25.0 g. The consumption rates were calculated from the measured differences in concentration and total weight of the prey at the beginning and at the end of observations.

The second series of feeding experiments were designed to determine the digestion time with respect to the size ratio of prey and predator. Then, knowing the size distribution of ctenophores from field sampling, this relationship could be used to calculate the maximum potential consumption rate in field. In these experiments, one specimen each of *B. ovata* and *M. leidyi* were placed in each container after starving the predator for 24 h. There were 4 containers present in each observation during 8–10 h. The wet weight of *B. ovata* ranged from 1.4 to 58.6 g and *M. leidyi* ranged from 0.25 to 13.69 g during 19 estimations of digestion time. The duration between the start of the experiment and commencement of ingestion was also noted. Once ingestion occured, the *B. ovata* specimens were monitored every 15 min until defecation was complete and the gut was empty. Daily consumption rates were calculated from such digestion time and wet weight of the prey.

In both series of experiments, B. ovata mainly preferred prey smaller than itself. Only once was a larger prey swallowed whole. Although there were no partial feeding during the experiments, in aquariums we observed partial predation of B. ovata on M. leidyi as reported by Swanberg (1974). On some occasions, B. ovata refused to feed during the entire experimental period. Unfed ctenophores were discounted during the calculation of miscellaneous experimental values, however, an allowance was made for unfed animals in the calculation of field consumption rates. The potential impact of predation by B. ovata was estimated from ctenophore densities in the field and the daily ration.

## Results

### Abundance, biomass and population structure of ctenophores

Biomasses of two ctenophore species were estimated from their numerical abundance and mean wet weight. The wet weight was calculated from the following relationships between wet weight (WW, mg) and body length ($L$, mm) for three groups of ctenophores:

B. ovata (14–120 mm):   $WW = 1.77L^{2.23}$

($n = 22; r^2 = 0.982$)

M. leidyi (2–10 mm):   $WW = 1.074L^{2.76}$

($n = 135; r^2 = 0.986$)

M. leidyi (11–65 mm):   $WW = 1.31L^{2.49}$

($n = 95; r^2 = 0.956$)

The mean values of abundance and biomass of B. ovata and M. leidyi for all 11 stations are shown in Table 1. In September–October, B. ovata was present in 10–25% of vertical tows and 60–70% of horizontal tows. In November, it was observed in 2% of horizontal tows only. B. ovata length ranged from 14 to 120 mm, weight was 0.5–70.0 g. Ctenophores of between 20 and 40 mm long were the most abundant (51%) in September, and this range was 30–60 mm in October (57%). In November some large B. ovata (about 100 mm) were found in the horizontal tows. There were eggs and juvenile ctenophores (0.4–1.5 mm length) found in several stations of the bay in September–October.

M. leidyi was more abundant (0.2 m$^{-3}$) than B. ovata (0.022–0.075 m$^{-3}$). The lengths for M. leidyi

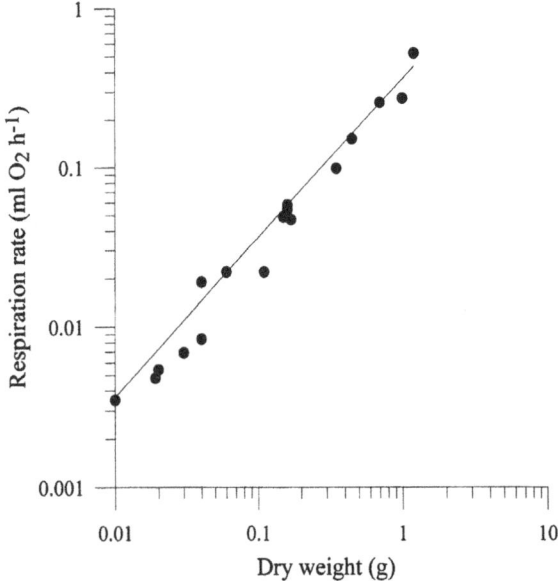

*Figure 2.* Relationship of *Beroe ovata* body weight and its oxygen consumption rate at 21 °C.

were 0.5–65.0 mm, and weights were 0.16–43.0 g, respectively. Possible reproduction of M. leidyi occurred during the study period, because the majority of the population consisted of juvenile ctenophores (with lengths of 10–20 mm) in September (52%) and 5–15 mm in October (72%).

### Chemical composition

Data on the chemical composition of B. ovata are shown in Table 2. Dry weight was equal to 2.40 ± 0.13% of wet weight. The main component of organic matter was protein (80%). The mean calorific value of B. ovata tissue was determined from the known energetic standards of protein, lipid and carbohydrate as 16.0±2.4 cal g$^{-1}$ WW, or 640±96 cal g$^{-1}$ DW.

### Respiration rate

The relationship between oxygen consumption rate ($Q$) and dry weight (DW) of B. ovata was expressed by the allometric equation (Fig. 2):

$$Q = 0.341DW^{1.04},$$

$$(r^2 = 0.98; n = 17; \text{ at } 21 \pm 1\ °C);$$

where $Q$ is respiration rate (ml $O_2$ g$^{-1}$ h$^{-1}$), and DW is dry weight (g). The slope of 1.04 indicated that the weight-specific respiration rate was independent of weight over the measured weight range (0.01–0.9 g

*Table 1.* Abundance, biomass and average individual wet weight (WW) of ctenophores *Beroe ovata* and *Mnemiopsis leidyi* at 11 stations in Sevastopol Bay and adjacent water regions in 1999 estimated from vertical sampling

| Species | Date | Temperature ($^\circ$C) | Abundance (ind $m^{-3}$) | Biomass (g WW $m^{-3}$) | Mean WW* (g) |
|---------|------|-------------|-----------|---------|---------|
| *B. ovata* | 23 Sep. | 22.5±1.2 | 0.02±0.07 | 0.11±0.33 | 14.87±14.76 |
| *B. ovata* | 25 Oct. | 16.2±0.6 | 0.08±0.14 | 1.74±3.66 | 23.24±25.09 |
| *B. ovata* | 10 Nov. | 13.8±1.2 | 0 | 0 | 51.50±0 |
| *M. leidyi* | 23 Sep. | 22.5±1.2 | 0.21±0.20 | 0.53±0.60 | 2.79±4.87 |
| *M. leidyi* | 25 Oct. | 16.2±0.6 | 0.20±0.19 | 0.43±0.68 | 2.85±6.47 |
| *M. leidyi* | 10 Nov. | 13.8±1.2 | 0.34±0.60 | 1.55±3.79 | 3.66±4.384 |

*From horizontal tows.

*Table 2.* The proximate chemical composition of the ctenophore *Beroe ovata* in the Black Sea (mean ± 1 SD from 9 individuals)

| Chemical components | Concentration (mg $g^{-1}$ WW) | As % of organic matter |
|---------------------|-----------------|----------------|
| Organic matter | 2.72±0.43 | 100.0 |
| Protein | 2.17±0.36 | 79.8 |
| Lipid | 0.26±0.04 | 9.6 |
| Free amino acids | 0.15±0.06 | 5.5 |
| Carbohydrate | 0.14±0.04 | 5.2 |

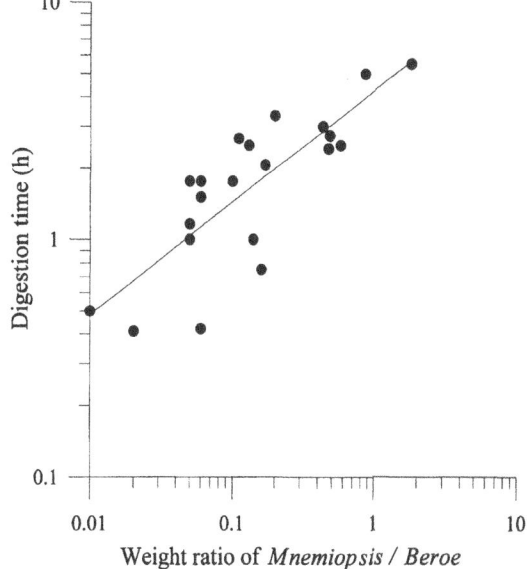

*Figure 3.* Relationship of the prey/predator weight ratio and the time required for *Beroe ovata* to digest *Mnemiopsis leidyi* ctenophores at 21 $^\circ$C.

DW ). The cost of respiration was calculated to be equal to 10% $d^{-1}$ of *B. ovata* DW.

*Digestion time and daily ration*

In the laboratory, *B. ovata* was not very active. It either remained motionless or swam slowly. Individuals did not immediately locate prey that had been newly placed in the tank. In our experiments, the duration between the start of the experiment and commencement of ingestion was on average 2.3±1.2 h and did not depend on prey or predator size. Similarly, the ctenophores did not engulf new prey just after recent digestion. *M. leidyi* was generally completely engulfed, except on a few occasions when the prey was larger than *B. ovata* and was partly eaten. There was no cannibalism behaviour by *B. ovata*, however, once one individual engulfed another *B. ovata*, but after 30-40 min, it was regurgitated alive and healthy.

In long-term experiments, the daily rations of *B. ovata* specimens of 4.85–21.31 g WW varied from 3.4 to 27.4 g WW $ind^{-1}$ $day^{-1}$ or from 60 to 128% of its own WW (Table 3) and were close to the rations calculated from digestion time in the first series of

experiments (see below). *B. ovata* fed in only 8 out of 19 long-term experiments; therefore, if we assume that half of the ctenophores in population feeds daily, the average ration of *B. ovata* in the field amounts to about 45% of its wet weight daily. Because the dry: wet weight ratios and calorific values of *B. ovata* and *M. leidyi* are similar (Finenko & Romanova, 2000), the daily ration in terms of energy content would be 45% of body energy content.

Digestion time of *B. ovata* feeding on *M. leidyi* at 21±1 $^\circ$C varied from 0.5 to 5.5 h in the studied weight range of both ctenophores. *B. ovata* of every size consumed both small and large *M. leidyi*, and the ratio between prey and predator weight ($P'$, range 0.01–

Table 3. The daily ration of *Beroe ovata* in long-term laboratory experiments with unlimited prey (*Mnemiopsis leidyi*)

| *B. ovata* weight (g) | Duration of feeding (d) | Total consumed food (g WW ind$^{-1}$) | Ration (g WW ind$^{-1}$ d$^{-1}$) | Ration as % *B. ovata* WW |
|---|---|---|---|---|
| 4.85 | 3.0 | 12.51 | 4.17 | 85.9 |
| 5.57 | 4.0 | 13.40 | 3.35 | 60.1 |
| 16.00 | 3.0 | 40.20 | 13.40 | 83.7 |
| 21.31 | 2.0 | 54.80 | 27.40 | 128.6 |
| | | | Mean | 89.6 |

Table 4. Estimated maximum daily predation rate of the *Beroe ovata* population on *Mnemiopsis leidyi* biomass for Sevastopol Bay in autumn 1999. Biomasses of ctenophores are in Table 1. The digestion time (*D*) and ration (*R*) of *B. ovata* were estimated from the regression of digestion time and wet weight ratio of ctenophores from the second series of feeding experiments

| Date | WW ratios of ctenophores $W_M/W_B$ (*P*) | Digestion time (h) (*D*) | Ration (g WW ind$^{-1}$ d$^{-1}$) (*R*) | Ration as % *B. ovata* WW | Ration in field (g m$^{-3}$ d$^{-1}$) | Daily predation as % of *M. leidyi* biomass |
|---|---|---|---|---|---|---|
| 23 Sep. | 0.188 | 1.92 | 15.87 | 106.7 | 0.05 | 9.3 |
| 29 Sep. | 0.125 | 1.57 | 11.35 | 77.5 | – | – |
| 12 Oct. | 0.024 | 0.72 | 2.15 | 19.5 | – | – |
| 25 Oct. | 0.123 | 1.56 | 17.72 | 76.4 | 0.57[a] | 132.1 |

[a] With temperature correction Q$_{10}$= 2.2.

<2.0) affected digestion time (*D*). The relationship between these values was expressed by the allometric equation (Fig. 3):

$$D = 4.26P^{0.478}, (n = 19; r^2 = 0.65).$$

The maximum potential daily ration of *B. ovata* for the field in September–October in Sevastopol Bay was calculated using digestion time from this equation and the *M. leidyi/B. ovata* weight ratio and the mean weight of *M. leidyi* from these data. Since the duration between the completion of digestion and commencement of new feeding was on average 2.3 h in the experiments, the number of meals would be 24/(2.3 + *D*) in a day. In our experiments, *B. ovata* was never observed to consume more than one prey at a time and the period between the completion of digestion and commencement of new feeding did not depend on how many prey were available. We suggest that the same situation exists in the sea. Then daily rations (*C* in g WW ind$^{-1}$ d$^{-1}$) of *B. ovata* calculated as *C* = 24*W*/(2.3 + *D*), where *W* and *D* represent the most weight (g) of *M. leidyi* consumed and digestion time (h), respectively, ranged from 2.2 to 17.7 g ind$^{-1}$

d$^{-1}$ or from 20 to 107% of its own WW (Table 4). The wide range in daily ration was due to wide range of *M. leidyi/B. ovata* wet weight ratio as well as *M. leidyi* weight that varied over about one order of magnitude. The maximum daily ration occurred at maximum P (the $W_M/W_B$ ratio) and the minimum ration when this ratio was low.

*The potential predatory impact of* B. ovata *on the* M. leidyi *population*

On the basis of data on abundance and maximum daily ration, we estimated the predatory impact of *B. ovata* on the *M. leidyi* population in Sevastopol Bay. Data on the abundance of *B. ovata* and *M. leidyi* were taken from Table 1. The daily ration of *B. ovata* at 22 °C amounted to 45% of its biomass if we assume only half of the *B. ovata* population feeds daily (Table 3).

We estimate that in September, the *B. ovata* population consumed 9.3% and in October 132% of the *M. leidyi* biomass ( Table 4). It should be pointed out that in October, *B. ovata* biomass was one order of magnitude higher than that in September. The grazing values determined here should be considered as maximum ones with continuous feeding of *B. ovata* when food is constantly available. In the field, the proportion of *B. ovata* with *M. leidyi* in the stomadeum could be more or less than a half as we suggested (12–20%, Shiganova et al., 2000; Vostokov et al., 2000). In these cases, our values calculated would decrease to 2–53%.

A comparison of mean daily specific growth rate of the *M. leidyi* population for the year in Sevastopol Bay (3.9% of population biomass; Finenko & Romanova, 2000) with predatory impact of *B. ovata* population on *M. leidyi* (2–53%) showed that in inshore waters of the Black Sea, despite their low numbers, the *B. ovata* population may control abundances of their prey population in certain periods.

## Discussion

### Abundance, biomass and population structure of ctenophores

*B. ovata* was found in the Black Sea in October 1997 for the first time in the inshore waters of Bulgaria (Konsulov & Kamburska, 1998). Later it appeared during May–August 1998 in a coastal zone of the northwestern Black Sea stretching from Odessa to the mouth of the Danube (Nastenko & Polishchuk, 1999). Numerical abundance of *B.ovata* determined from zooplankton samples in this region varied from 0.2 to 0.4 ind m$^{-3}$ in May and from 35 to 392 ind m$^{-3}$ in August. The population was comprised of young ctenophores ranging from 0.2 to 9 mm long, and as a result the population biomass was low (0.00001–0.097 g m$^{-3}$). In August–October 1999, *B. ovata* occurred in the northeastern Black Sea near Gelendzhik (Anokhina et al., 2000; Shiganova et al., 2000, 2001), where numbers were 0.06–0.13 ind m$^{-3}$ and biomass was from 18.6 to 42.8 g m$^{-3}$. The ctenophores ranged in size between 20 and 120 mm in length in the population.

This is the first report about the occurrence of *B. ovata* in the inshore waters near Sevastopol based on samples obtained in September–October 1999. Here, the abundance in Sevastopol Bay and adjacent waters ranged from 0.02 to 0.07 ind m$^{-3}$ and biomass from 0.11 to 1.74 g m$^{-3}$. The ctenophores measuring 20–60 mm in length were most numerous. *B. ovata* biomass in Sevastopol Bay was an order of magnitude lower than in coastal waters of the northeastern Black Sea near Gelendzhik.

The numerical abundance of *M. leidyi* in September–October 1999 was greatly reduced as compared with that in previous years. In September–October 1995, abundance and biomass of *M. leidyi* were 125 ind m$^{-3}$ and 33 g WW m$^{-3}$, respectively (Finenko & Romanova, 2000) in September–October 1999, *M. leidyi* abundance and biomass were only 0.2 ind m$^{-3}$ and 0.5 g WW m$^{-3}$. This marked decrease was probably caused by grazing of *M. leidyi* by *B. ovata*. Similarly, since the summer of 1997 when *B. ovata* was first noted, a clear decreasing trend in *M. leidyi* biomass from the southern Black Sea was also noticed (Kideys & Romanova, 2001).

### Respiration rate

To compare the respiration rate of *B. ovata* with the data of other authors under similar temperature conditions, we have calculated oxygen consumption in ml O$_2$ g$^{-1}$h$^{-1}$ and recalculated for standard temperature 20 °C by introducing the temperature correction $Q_{10}$ = 2.2 (Winberg, 1983). In our case, it corresponded to 0.34 ml O$_2$ g$^{-1}$ h$^{-1}$, whereas in Kremer et al. (1986) and Shiganova et al. (2000), these estimates were 0.14 and 0.35 ml O$_2$ g$^{-1}$ h$^{-1}$, respectively. Our value is quite similar compared with that of Shiganova et al. (2000a), but higher than is obtained by Kremer et al. (1986).

### Chemical composition

Our data on the chemical composition of *B. ovata* are in good agreement with the literature for many other species of epipelagic ctenophores. This is true not only for the total organic content, which usually amounts in these animals to 1–8 mg g$^{-1}$ WW (Kremer et al., 1986; Clarke et al., 1992; Bailey et al., 1994, 1995), but also for relative values, where the organic matter typically contains 60–80% protein (Hoeger, 1983; Schneider, 1989; Anninsky, 1994). By contrast, some Arctic or deep sea specimens sometimes accumulate lipids in their bodies (Larson & Harbison, 1989).

At the same time, with respect to the amount of organic matter, a wide range of values have been reported within the species of Beroida. According to Kremer et al. (1986), the organic content of *B. ovata* calculated by carbon (C × 1.9) usually varied from 1.33 to 3.99 mg g$^{-1}$ WW, while the values obtained for *B. cucumis* (Hoeger, 1983) and *B. gracilis* Künne (Bailey et al., 1994) amounted to 6.56 and 4.99 mg g$^{-1}$ WW, respectively. Besides, there are other values, being equal to 5.8–7.3 mg g$^{-1}$ WW of organic matter for *Beroe* sp. (Clarke et al., 1992; Bailey et al., 1994).

Evidently, the organic content in *B. ovata* is the lowest among the Beroida.

On the whole, *B. ovata* ctenophores have a greater organic content (2.4–3.1 mg g$^{-1}$ WW) and caloric values (14.0–18.7 cal g$^{-1}$ of wet weight), compared with other ctenophore species of the Black Sea. For instance, the organic content of *Pleurobrachia rhodopis* Chun, 1880 (synonym *P. pileus* (O. F. Müller, 1776)) is 2.0–2.9 mg g$^{-1}$ WW, and of *M. leidyi* is 1.0–1.5 mg g$^{-1}$ (Anninsky, 1994). Energetic value of their tissues ranged from 12.6 to 17.6 cal g$^{-1}$ WW and from 7.4 to 13.4 cal g$^{-1}$ WW, respectively. The highest values for *B. ovata* could be caused by the specific carnivorous life style of this species.

*Digestion time, consumption rate and the potential predatory impact of* B. ovata *on the* M. leidyi *population*

Digestion times and rations of *B. ovata* in the Black Sea determined in our experiments and by other authors, are very close. According to Shiganova et al. (2000) digestion time varied from 3 to 5 h at 24–26 °C and the daily minimum ration calculated from respiratory demands was 30% of wet weight. In September 1999 in the northeastern Black Sea, they estimated that the *B. ovata* population grazed 0.7–5.7% of *M. leidyi* biomass daily.

The food ingestion rate in our experiments far exceeded respiratory demands (average 5.72±1.35% of *B. ovata* energy content) of the ctenophore for the studied weight range. The food rations of *B. ovata* observed in the experiments would be considered as maximal under abundant food conditions. It was shown that in a homogenous environment, *B. ovata* could find 1 prey item approximately every 17 h (Swanberg, 1974). Kamshilov (1960) claimed that a *B. cucumis* requires a 10 mm *Bolinopsis infundibulum* to support a 1.6% growth increment every 18.3 h. In nature, *B. ovata* probably are able to realize their high feeding activity only rarely and over short periods. Moreover, in nature, partial predation, lower prey densities, and escape of *M. leidyi* (in Kreps et al., 1997) would result in lower consumption rates. Probably the ingestion rate of ctenophores in the Black Sea was limited mainly by the rate at which the predators captured prey rather than by the rate at which they digested it. The large capacity of the predator's stomach enables them to consume the large prey items and thus to receive at once sufficient food to meet their respiratory demands for a long time. This ability is an adaptation tool to take advantage of episodic periods of high prey density.

It is obvious that ctenophores of the genus *Beroe* form an important link in pelagic food chains, however, little is known of their general feeding biology. The species of order Beroida feed mainly on other ctenophores (Kamshilov, 1960; Swanberg, 1974; Matsumoto and Harbison, 1993; Shiganova et al., 2000). In the Black Sea, *B. ovata* seems to be restricted to the surface waters and therefore their food would mainly consist of *M. leidyi* because the other ctenophore species, *Pleurobrachia pileus*, is mainly distributed in deeper waters with low oxygen concentrations (Mutlu, 1999; Kideys & Romanova, 2001). The high concentrations of *B. ovata* were found in areas of high *M. leidyi* abundance (Shiganova et al., 2000). *Beroe* spp. can have a substantial impact on the population structure of planktivorous ctenophore communities. *B. ovata* reduced *M. leidyi* populations in the autumn in Narragansett Bay (Kremer et al. 1986). Greve & Reiners (1980) showed that predation by the ctenophore *B. gracilis* may be a factor in controlling the *P. pileus* population in the southeastern North Sea and Van Der Veer & Sadee (1984) reported similarly for the Dutch Wadden Sea.

Kamshilov (1960) claimed that *B. cucumis* indirectly modified the composition of other zooplankton by feeding on the zooplanktivorous *Bolinopsis infundibulum*. Purcell & Cowan (1995) also suggested that feeding by the cnidarian *Chrysaora quinquecirrha* (Desor), on *M. leidyi* populations may reduce mortality of zooplankton and ichthyoplankton. We suppose that due to its intensive feeding on *M. leidyi*, *B. ovata* could cause an increase in the concentration of edible zooplankton in the Black Sea. If this holds true, schooling zooplanktivorous fishes, which are the main food competitors of *M. leidyi*, will greatly benefit from the consequences of this new arrival to the Black Sea.

## References

Anninsky, B. E., 1994. Organic matter composition of the jelly-fish *Aurelia aurita* and two species of ctenophores from the Black Sea. Biolgya Morya 20 (4): 291–295 (in Russian).

Anokhina, L. L., E. I. Musaeva, L. I. Loginova & E. A. Shushkina, 2000. The concentration of the ctenophore *Beroe* and other zooplanktonic invaders in the northeastern Black Sea. In Matishov, G. G. (ed.), Vidy- Vselenzy v Evropeiskikh Moryakh Rossii. Tezisy dokladov nauchnogo seminara (Murmansk, 27–28 yanvar 2000). Murmansk. (Species-invaders in the European seas in Russia. Abstracts of the presentations of the scientific seminar Murmansk, January 27–28, 2000): 14–15 (in Russian).

Bailey, T. G., M. J Youngbluth & G. P. Owen, 1994. Chemical composition and oxygen consumption rates of the ctenophore *Bolinopsis infundibulum* from the Gulf of Maine. J. Plankton Res. 16: 673–679.

Bailey, T. G., M. J. Youngbluth & G. P. Owen, 1995. Chemical composition and metabolic rates of gelatinous zooplankton from midwater and benthic boundary layer environments off Cape Hatteras, North Carolina, U.S.A. Mar. Ecol. Prog. Ser. 122: 121–134.

Clarke, A., L. J. Holmes & D. J. Gore, 1992. Proximate and elemental composition of gelatinous zooplankton from the Southern Ocean. J. exp. mar. Biol. Ecol. 155: 55–68.

Finenko F. A., Z. A. Romanova & G. I. Abolmasova, 2000. The ctenophore *Beroe ovata* is a recent invader to the Black Sea. Ecologiya morya 50: 21–25 (in Russian).

Finenko, G. A. & Z. A. Romanova, 2000. Population dynamics and energetics of the ctenophore *Mnemiopsis leidyi* in Sevastopol Bay. Oceanology 40: (in Russian) 677–685.

GESAMP    (IMO/FAO/UNESCO-IOC/WMO/WHO/IAEA/UN/ UNEP Joint Group of Experts on the Scientific Aspects of Marine Environmental Protection), 1997. Opportunistic settlers and the problem of the ctenophore *Mnemiopsis leidyi* invasion in the Black sea. Rep. Stud. GESAMP, 58: 84 pp.

Gordina, A. D. & T. N. Klimova, 1996. Species composition and ichtyoplankton abundance dynamics in offshore and pelagic waters of the Black Sea. In Konovalov, S. M. (ed.), Sovremennoe Sostoyanie Ichtioplanktona Chernogo Morya. Morskoi Hydrophysicheskii Institut, Sevastopol: 74–94 (in Russian).

Gosner, L. K., 1971. Guide to identification of marine and estuarine invertebrates – Cape Hatteras to the Bay of Fundy. Wiley Interscience, New York.

Greve, W., 1970. Cultivation experiments on North Sea ctenophores. Helgolander wiss. Meeresunters. 20: 304–317.

Greve, W. & F. Reiners, 1980. The impact of prey–predator waves from estuaries on the planktonic marine ecosystem. In: Estuarine Perspectives. Proceedings of the Fifth Biennial International Estuarine Research Conference. Academic Press, New York: 405–421.

Harbison, G. R., L. P. Madin & N. R. Swanberg, 1978. On the natural history and distribution of oceanic ctenophores. Deep-Sea Res. 25: 233–256.

Hoeger, U., 1983. Biochemical composition of ctenophores. J. exp. mar. Biol. Ecol. 72: 251–261.

Kamshilov, M. M., 1960. Feeding of ctenophore *Beroe cucumis* Fabr. Dokl. Acad. Nauk, S.S.S.R., 130 (5): 1138–1140 (in Russian).

Kideys, A. E., 1994. Recent dramatic changes in the Black Sea ecosystem: the reason for the sharp decline in Turkish anchovy fisheries. J. mar. Syst. 5: 171–181.

Kideys, A. E., A. V. Kovalev, G. Shulman, A. D. Gordina & F. Bingel, 2000. A review of zooplankton investigations of the Black Sea over the last decade. J. mar. Syst. 24: 355–371.

Kideys, A. E., A. D. Gordina, F. Bingel & U. Niermann, 1999. The effect of environmental conditions on the distribution of eggs and larvae of anchovy (*Engraulis encrasicolus* L.) in the Black Sea. ICES J. mar. Sci. 56: 58–64.

Konsulov, A. S. & L. T. Kamburska, 1998. Ecological determination of the new Ctenophora – *Beroe ovata* invasion in the Black Sea . Oceanology (Bulgaria) 2: 195–198.

Kovalev, A. V., A. D Gubanova, A. E. Kideys, V. V. Melnikov, U. Niermann, N. A. Ostrovskaya, I. Yu. Prusova, V. A. Skryabin, Z. Uysal & Ju. A. Zagarodnyaya, 1998. Long-term changes in the biomass and composition of fodder zooplankton in a coastal regions of the Black Sea during the period 1957–1996. In Ivanov,

L. & T. Oguz (eds), Ecosystem Modelling as a Management Tool for the Black Sea. Kluwer Academic Publishers, Dordrecht, The Netherlands: 209–220.

Kremer, P., M. F. Canino & R. F. Gilmer, 1986. Metabolism of epipelagic tropical ctenophores. Mar. Biol. 90: 403–412.

Kreps, T. A., J. E. Purcell & K. B. Heidelberg, 1997. Escape of the ctenophore *Mnemiopsis leidyi* from the scyphomedusa predator *Chrysaora quinquecirrha*. Mar. Biol. 128: 441–446.

Larson, R. J. & G. R. Harbison, 1989. Source and fate of lipids in polar gelatinous zooplankton. Arctic 42 (4): 339–346.

Matsumoto, G. I. & G. R. Harbison, 1993. In situ observations of foraging, feeding, and escape behaviour in three orders of oceanic ctenophores: Lobata, Cestida and Beroida. Mar. Biol. 117: 279–287.

Mianzan, H. W., 1999. Ctenophora. In Boltovskoy, D. (ed.), South Atlantic Zooplankton. Backhuys Publ., Leiden: 561–573.

Mills, C. E., P. R. Pugh, G. R. Harbison & S. H. D. Haddock, 1996. Medusae, siphonophores and ctenophores of the Alboran Sea, south western Mediterranean. Sci. mar. 60: 145–163.

Mutlu, E. & F. Bingel, 1999. Distribution and abundance of ctenophores, and their zooplankton food in the Black Sea. I. *Pleurobrachia pileus*. Mar. Biol. 135: 589–601.

Nastenko, E. V. & L. M. Polishchuk, 1999. The comb jelly *Beroe* (Ctenophora: Beroida) in the Black Sea. Dopovidi Nazionalnoi Akademii Nauk, Ukraine 11: 159–161.

Niermann, U., F. Bingel, A. Gorban, A. D. Gordina, A. C. Gücü, A. E. Kideys, A. Konsulov, G. Radu, A. A. Subbotin & V. E. Zaika, 1994. Distribution of anchovy eggs and larvae (*Engraulis encrasicolus* Cuv.) in the Black Sea in 1991–1992. ICES J. mar. Sci. 51: 395–406.

Omori, M. & T. Ikeda, 1984. Methods in Marine Zooplankton Ecology. Wiley & Sons, New York. 232 pp.

Purcell, J. E. & J. H. Cowan Jr, 1995. Predation by the scyphomedusan *Chrysaora quinquecirrha* on *Mnemiopsis leidyi* ctenophores. Mar. Ecol. Prog. Ser. 129: 63–70.

Schneider, G., 1989. Zur chemischen Zusammensetzung der Ctenophore *Pleurobrachia pileus* in der Kieler Bucht. Helgolanders Meeresunter. 43 (1): 67–76.

Shiganova, T. A., B. Ozturk & A. Dede, 1994. Distribution of the ichtlyo, jelly-and zooplankton in the Sea of Marmara. FAO Fisheries Report 495: 141–145.

Shiganova, T. A., A. E. Kideys, A. C. Gucu, U. Niermann & V. S. Khoroshilov, 1998. Changes in species diversity and abundance of the main components of the Black Sea pelagic community during the last decade. In Ivanov, L. & T. Oguz (eds), NATO TU-Black Sea Project: Ecosystem Modeling as a Management Tool for the Black Sea, Symposium on Scientific Results. Kluwer Academic Publishers, Dordrecht, The Netherlands: 171–188.

Shiganova, T. A., Yu. V. Bulgakova, P. Yu. Sorokin & Yu. F. Lukashev, 2000. Investigation of a new settler *Beroe ovata* in the Black Sea. Biology Bull. 27(2): 202–209.

Shiganova, T. A., Yu. V. Bulgakova, S. P. Volovik, Z. A. Mirzoyan & S. I. Dudkin, 2001. The new invader *Beroe ovata* Mayer, 1912 and its effect on the ecosystem in the northeastern Black Sea. Hydrobiologia 451 (Dev. Hydrobiol. 155): 187–197.

Swanberg, N., 1974. The feeding behavior of *Beroe ovata*. Mar. Biol. 24: 69–76.

Van Der Veer, H. W. & C. F. M. Sadee, 1984. Seasonal occurrence of the ctenophore *Pleurobrachia pileus* in the western Dutch Wadden Sea. Mar. Biol. 79: 219–227.

Vinogradov, M. E., E. A. Shushkina, E. I. Musaeva & P. Yu. Sorokin, 1989. Ctenophore *Mnemiopsis leidyi* (A. Agassiz) (Ctenophora: Lobata) – a new settler in the Black Sea. Oceanology 29: 293–298 (in Russian).

Vinogradov, M. E., E. A Shushkina, L. L. Anokhina, S. V. Vostokov, N. V. Kucheruk & T. A. Lukashova, 2000. Dense aggregations of the ctenophore *Beroe ovata* (Eschscholtz) near the north-east shore of the Black Sea. Oceanology 40: 52–55 (in Russian).

Volovik, S. P., G. I. Lutz, Z. A. Mirzoyan, Yu. V. Pryakhin, S. F. Rogov, E. I. Studenikina & N. I. Revina, 1991. Introduction of the ctenophore *Mnemiopsis* to the Azov Sea: Preliminary assessment of the effect. Rynb. Khoz. 1: 47–49.

Vostokov, S. V., E. G. Arashkevich, A. V. Drits, T. A. Lukasheva & A. N. Tolomeev, 2000. The investigations of the peculiarities of biology of the ctenophores *Beroe ovata* and *Mnemiopsis leidyi* invaders into the Black Sea. In Matishov, G. G. (ed.), Vidy- vselenzy v Evropeiskikh moryakh Rossii. Tezisy dokladov nauchnogo seminara (Murmansk, 27–28 yanvar 2000). Murmansk. (Species-invaders in the European seas in Russia. Abstracts of the presentations of the scientific seminar (Murmansk, January 27–28, 2000) Murmansk: 28–29 (in Russian).

Winberg, G. G., 1971. Methods for the Estimation of Production of Aquatic Animals. Academic Press, London: 175 pp.

Winberg, G. G., 1983. The Vant-Goff temperature factor and the Arrhenius equation in biology. Zhurnal Obshchei Biologii. 44 (1): 31–42 (in Russian).

Zaitsev, Yu. P., 1998. Marine hydrobiological investigations of National Academy of Science of Ukraine during the 1990s in XX century: Shelf and coastal water bodies of the Black Sea. Hydrobiol. J. 6: 3–21 (in Russian).

*Hydrobiologia* **451**: 187–197, 2001.
© 2001 *Kluwer Academic Publishers.*

# The new invader *Beroe ovata* Mayer 1912 and its effect on the ecosystem in the northeastern Black Sea

Tamara A. Shiganova[1], Yulia V. Bulgakova[1], Stanislav P. Volovik[2], Zinaida A. Mirzoyan[2] & Sergey I. Dudkin[2]

[1]*P. P. Shirshov Institute of Oceanology, Russian Academy of Sciences, 36 Nakhimovskiy Pr.,
Moscow, 117851, Russia
E-mail: shiganov@ecoys.sio.rssi.ru*
[2]*State Unitary Enterprise Research, Institute of the Azov Sea Fishery Problems 21/2 Beregovaya,
Rostov-on-Don, 344007, Russia*

*Key words:* Azov Sea, *Mnemiopsis leidyi*, feeding, respiration, ichthyoplankton, zooplankton

## Abstract

The abundance, biomass and distribution of the ctenophore, *Beroe ovata* Mayer 1912 were assessed along with several parameters associated with composition, respiration and feeding. Digestion time of *B. ovata* feeding on other ctenophores ranged from 4–5.5 h for *Mnemiopsis leidyi* A. Agassiz to 7–8 h for *Pleurobrachia pileus* (O. F. Müller). Daily ration was estimated as 20–80% of wet weight based on field observations of feeding frequency coupled with digestion time. Calculations indicate that the measured population of *B. ovata* ingested up to 10% of the *M. leidyi* population daily. A marked decrease in *M. leidyi* density was recorded. The abundance of zooplankton increased about 5-fold and ichthyoplankton about 20-fold compared with the same season in previous years following the *M. leidyi* invasion.

## Introduction

The invasion of alien species has become a great problem for many seas of the world due to increasing international commerce combined with the use of ballast water on ships. The discharge of ballast water has resulted in the transfer of species to regions where they are not indigenous.

As a rule, the invasion of alien species has a damaging effect on ecosystems. For example, the introduction of the ctenophore, *Mnemiopsis leidyi*, with ballast water from ships at the beginning of the 1980s was a real catastrophe for the Black and Azov Seas (Vinogradov et al., 1989; Volovik et al., 1993). These ecosystems were already damaged due to hydrological and hydrochemical changes resulting from decreased river discharge, eutrophication and overfishing (Ivanov & Beverton, 1985; Caddy & Griffiths, 1990; Zaitzev & Aleksandrov, 1997). The invasive *M. leidyi* demonstrated explosive development in 1989, when its total biomass reached 1 billion tons for the entire Black Sea. Co-incident with high ctenophore

abundance, the biomass of trophic mesozooplankton, its main food, sharply declined (Vinogradov et al., 1989). The stocks of zooplanktivorous fish (anchovy, Mediterranean horse mackerel and sprat) dropped, presumably due to competition with *M. leidyi* for food and predation by *M. leidyi* on fish eggs and larvae (Tzikhon-Lukanina et al., 1993; Shiganova et al., 1998; Shiganova & Bulgakova, 2000).

In response to this situation, a group of experts from the international commission of the Joint Group of Experts on the Scientific Aspects of Marine Pollution (GESAMP) (IMO/FAO/UNESCO-IOC/WMO/WHO/IAEA/UN/UNEP) (GESAMP, 1997) proposed the introduction of potential predators for *Mnemiopsis leidyi*. Among suggested species was another ctenophore, *Beroe ovata*. Although this suggestion was not intentionally implemented, *Beroe* sp. nevertheless appeared in the Black Sea in 1997 for the first time. During 1997–1998, *Beroe* sp. occurred in some coastal areas in the northern part of the Black Sea (Konsulov & Kamburska, 1998; Nastenko & Polishchuk, 1999; N. Lupova, (Saint Petersburg University, Russia) pers. com.; Prof. M. Gomoiu, (Institute of Geoecology, Constanta, Romania) pers. com.). By late August 1999, *Beroe* sp. spread throughout the northeastern Black Sea (Shiganova et al., 2000a,b) and was also found in the northwestern and southern regions (Finenko et al., 2000a,b). In October 1999, *Beroe* sp. was observed in the Sea of Azov for the first time (Shiganova et al., 2000b).

## Material and methods

Investigations of the ctenophore, *Beroe ovata,* were conducted from 27 August to 15 September 1999. They included experimental studies in both the laboratory of the Southern Branch of the P. P. Shirshov Institute of Oceanology, Russian Academy of Sciences and the laboratory of the Novorossiisk research station, and also field surveys in the northeastern Black Sea on 17–24 August, 7–15 September 1999 and 12–19 April 2000. In addition, long-term field data of the authors were used for comparison.

The dry weight (DW) was determined for 31 specimens (length range 40–112 mm), of them, 6 specimens were weighed individually and additional weights were determined for groups of ctenophores with mean lengths of 50, 75 and 105 mm. All animals had empty guts. Length was measured for each animal, then excess moisture was removed by blotting before individuals were weighed (wet weight, WW). Ctenophores were dried at 60 °C constant to obtain a dry weight. Twenty five ctenophores (lengths 40–110 mm) were analyzed for lipids by the Swahn method (Swahn, 1953), for protein by the Lowry method (Pushkina, 1963), and for carbohydrates by the Orcin method.

Tolerance to different salinities of *Beroe ovata* was also studied using 3 l containers, with a single ctenophore in each container. Water of different salinity was prepared by diluting Black Sea water with distilled water. Experiments were carried out with gradual salinity changes from 18 to 15, 18 to 13.5, 18 to 10.8, 18 to 7.2, 18 to 4.5. Each experiment was conducted for 1 h. Freshly collected ctenophores were used for every experiment after a 1 h acclimation period. After the experiment testing *B. ovata* tolerance to different salinities, ctenophores were moved to a salinity of 18 for 1 h rehabilitation. The experimental temperature was kept at 25–26 °C.

Respiration rate was determined in closed containers (volume 1.5 l) using 1 freshly collected large ctenophore or 2–3 small ctenophores per container. The duration of the experiments was 4 h; temperatures ranged from 24.6 to 26.1°C. Metabolic rates were estimated as differences between oxygen concentration in the container without (control) and with ctenophores. Oxygen concentrations were determined by the Winkler method (Romanenko & Kuznetzov, 1974). The oxygen concentration in the experimental bottles decreased to 78–89% of the initial oxygen concentration during the incubation. Twenty-one measurements were made at a salinity of 18 and another 3 measurements at a salinity of 10.8 after the 1 h acclimation period. The experiments for tolerance to low oxygen were carried out at 21 °C. Ctenophores were kept in closed containers until they died.

*Beroe ovata* was studied in laboratory and by *in situ* observations. Feeding rate, digestion time, prey selection and behavior were conducted in the laboratory from 27 August to 6 September and on 7–13 September on board R/V 'Akvanavt'. The freshly caught specimens of *B. ovata* and their potential prey – *Mnemiopsis leidyi, Aurelia aurita* (L), *Pleurobrachia pileus*, fish larvae and *Acartia clausi* Giesbrecht (Copepoda) were placed in aquaria. To avoid damaging the ctenophores being used in experiments, weights were estimated using a length-weight relationship based on 43 specimens (Fig. 1a). Weights of *Mnemiopsis leidyi* used in the feeding experiments were estimated using the equation

$$y = 0.0056x^{1.85},$$

where $x$ is the length in mm and $y$ is the wet weight in grams.

The study of the spatial and vertical distribution of *Beroe ovata*, other gelatinous plankton, zoo- and ichthyoplankton was conducted aboard the R/V 'Akvanavt' by Institute of Oceanology, Russian Academy of Sciences. An additional ichthyological survey was carried out on 17–24 August by the Institute of the Azov Sea Fishery Problems, and those ichthyoplankton data were included in our study. Gelatinous plankton and ichthyoplankton were sampled with a Bogorov-Rass net (square net opening of 1 m$^2$, mesh size 500 $\mu$m) in vertical hauls from 150 m to the surface in deep water and from the bottom to the surface in shallow areas. A Juday net (square net opening of 0.1 m$^2$, mesh size 200 $\mu$m) with a closing device was used for depth-stratified hauls from the thermocline to the surface (15–25 to 0 m), from the pycnocline to the thermocline (60–80 m to 15–25 m) and from the anoxic layer to the pycnocline (120–200 m to 60–80 m).

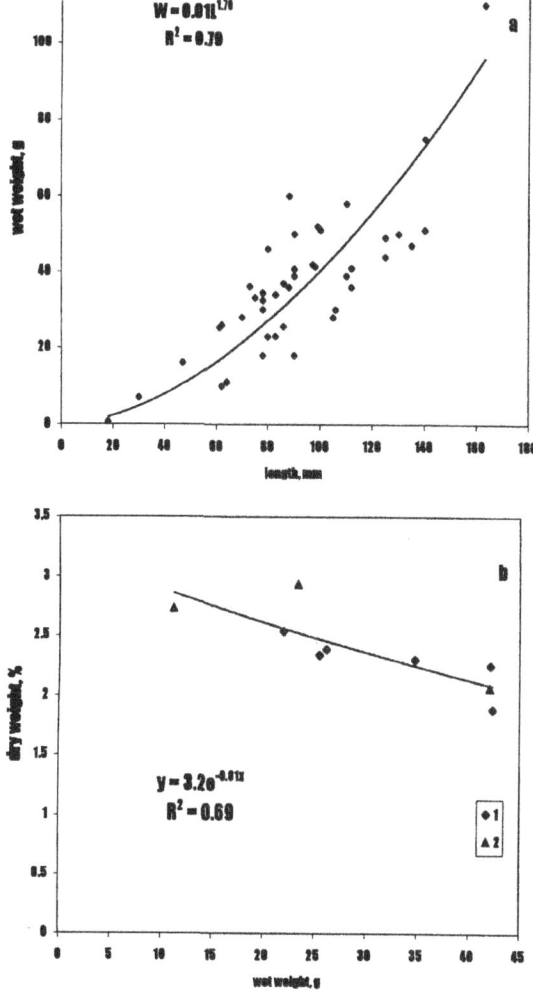

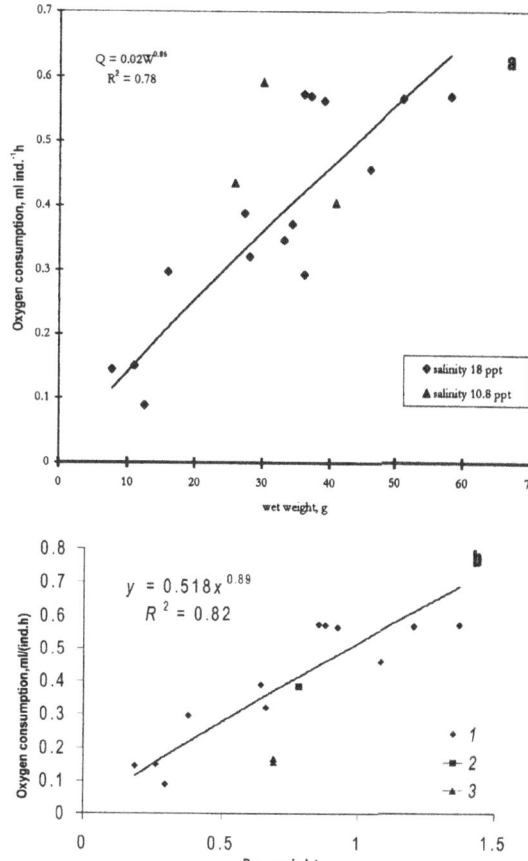

*Figure 2.* Oxygen consumption of *Beroe ovata*. (a) – dependent on wet weight; (b) – dependent on the dry weight.

*Figure 1.* (a) Dependence of wet weight on length of *Beroe ovata*, (b) – dry weight as percentage of wet weight: 1-individual measurements; 2- group measurements.

During bad weather, only two hauls were made per station. A total of 50 stations were occupied and 148 hauls were made in September 1999 and 42 stations, 84 hauls were made in April 2000.

## Results

### Taxonomy, body structure, size and chemical composition of Beroe ovata

At least two species of beroid ctenophores inhabit the Mediterranean Sea – *Beroe ovata* Esch and *Beroe forskalii* Chun (in Tregouboff & Rose, 1957). Of these species, only *Beroe ovata* has been identified in the Sea of Marmara (Shiganova et al., 1994). With the

appearance of a beroid in the Black Sea, the question of species identification was relevant. Some scientists of the Black Sea countries identified this invasive species as *Beroe cucumis* (Zaitzev, 1998), introduced with ballast waters from the North Atlantic, others identified the ctenophore as *Beroe ovata* (in Konsulov & Kamburska, 1998; Shiganova et al., 2000a, b). Analyses of morphology, conducted by Prof. L. N. Seravin (Saint Petersburg University, Russia) determined it to be *Beroe ovata* Mayer, 1912 (Seravin et al., 2001) which was transferred with ballast waters from northern atlantic area.

The ctenophores generally have a pink colour, and the largest were coloured with a brown tint. The body of this ctenophore is oval, wider at the oral end and tapered but not pointed at the aboral end. The meridian canals have anostomoses between them, which are characteristic for *Beroe ovata* (in Seravin, 1998). The young specimens are wider in both the oral and aboral ends of the body.

The relationship between length (L) and wet weight (WW) for *Beroe ovata* was expressed by a power regression: WW = 0.01 L $^{1.78}$ ($R^2$= 0.79, $n$ = 43, $p < 0.001$, std. error = 0.0092) (Fig. 1a). We collected mostly moderate to large *B. ovata* (60–162 mm) in offshore waters. Small specimens (16–18 mm) were found only in coastal areas with shallow depths of 2–4 m. The most common sizes of *B. ovata* collected were 70–90 mm for specimens from Blue Bay, and 80–100 mm for specimens from the open sea.

In the coastal area, the examined adult specimens were mature with both male and female gonads. Egg production was observed in a single field-collected specimen. Eggs have a round shape. Hatched larvae were beroid type larvae without tentacles. The occurrence of small individuals (16–18 mm) in the coastal area along with larger sizes supports the conjecture that *Beroe ovata* can both grow and reproduce in the Black Sea.

Dependence of dry weight (DW) on the length was expressed by the equation: DW = $0.0016L^{1.36}$ ($R^2 = 0.93$, $n = 9$, $p < 0.01$). The proportion of DW to wet weight (WW) decreased with an increase in *Beroe ovata* size (Fig. 1b). Mean ratio of DW to WW for 31 specimens (40–112 mm) was $2.4 \pm 0.3\%$. Mean protein content was 2.47 mg g$^{-1}$, lipid content was 0.28 mg g$^{-1}$, carbohydrates were 0.54 mg g$^{-1}$ of WW. The mean calorific value of *B. ovata* tissue was determined from the known energetic standards of protein (5.7 cal mg$^{-1}$), lipid (9.45 cal mg$^{-1}$) and carbohydrate (4.10 cal mg$^{-1}$) as 18.8 cal g$^{-1}$ of WW. The carbon composition of *B. ovata* specimens from the Black Sea was 7.8% of DW, higher than in ocean waters (2–6%) (Kremer et al., 1986). This relatively high percentage of carbon per unit dry weight and the relatively low ratio of dry to wet weight can be explained by the lower salinity of Black Sea water compared with other regions where *B. ovata* has been studied.

*Tolerance to salinity*

Results of experiments showed that behaviour of ctenophores did not change when the salinity was decreased from18 to13.5. When ctenophores were moved from a salinity of 18 to a salinity of 15, they first stopped beating their cilia. However, in approximately 15 min, the cilia beating recovered and the ctenophores were again very active. When *Beroe ovata* was moved from the Black Sea water (salinity 18) to water of salinity 13.5, it sank down to the bottom,

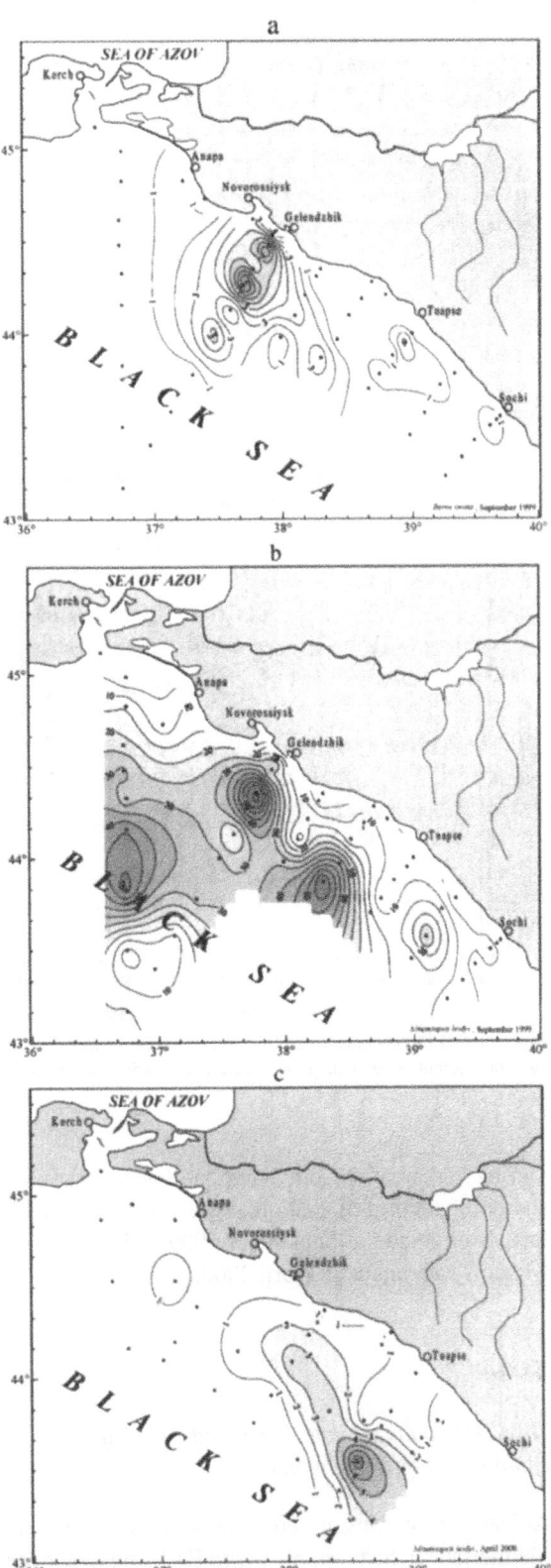

*Figure 3.* Spatial distribution of ctenophores in the northeastern Black Sea in September 1999: (a) – *Beroe ovata*; (b) – *Mnemiopsis leidyi* in September 1999; (c) – *Mnemiopsis leidyi* in April 2000.

made some movements there with actively beating cilia, but did not go to the surface. Then the ctenophore began to produce mucus, but later it recovered active movements. In the water with salinity of 10.8, *B. ovata* sank down to the bottom, cilia beating ceased for 15 min but then the ctenophore completely recovered active movements and natural behaviour.

With the treatment at salinity of 7.2, *Beroe ovata* laid on the bottom almost without any movements. Its body grew turbid, but in 20–30 min. the ctenophore began to move and its behaviour was the same as in previous experiments. In a salinity of 4.5 the ctenophore sank to the bottom without movement, body density decreased and its tissue turned a whitish color and began to disintegrate. Later some recovery of movement was observed. When *B. ovata* was moved from Black Sea water (salinity 18) to the water with salinity 3 it immediately sank to the bottom and laid there without any movements and no recovery. These data are preliminary but showed that *B. ovata* can tolerate short-term salinity shocks.

*Respiration rate*

Oxygen consumption by *Beroe ovata* was measured at 2 salinities, 18 and 10.8. In the first case, the respiration rate $Q$ (ml ind$^{-1}$h$^{-1}$) varied nearly linearly with wet weight (WW) where $Q = 0.02 WW^{0.86}$ ($R^2 = 0.78$, $n = 21$, $p < 0.001$). Two of three measurements in water with salinity 10.8 gave higher estimates than in experiments with salinity of 18 (Table 1, Fig. 2a). When the respiration rate was expressed in terms of DW (Fig. 2b), no regression was significant. The oxygen consumption ranged from 0.25 to 0.71, with a mean of $0.54 \pm 0.131$ ml g WW$^{-1}$ h$^{-1}$.

Our results of metabolic rate, re-calculated for standard temperature (20 °C) according to the normal Krogh's curve, were close to results of Finenko et al. (2000 a, b; 2001), but were two times higher than results obtained for *Beroe ovata* in the Bahamas (Kremer et al., 1986) (Table 1). This difference probably was due to salinity differences in the habitats, and therefore on the dry weight, of *B. ovata*. If the results are expressed in terms of body C, we obtain 6.9 ml $O_2$g C$^{-1}$ h$^{-1}$, which is similar the value re-calculated from Kremer et al. (1986) of 7.3 ml $O_2$g C$^{-1}$ h$^{-1}$.

*Feeding*

*In situ* observations in the coastal area of the Blue Bay showed *Beroe ovata* swam near the surface when not feeding. Most of the time it was oriented vertically,

with the mouth closed 5–10 cm from the surface. Less often *B. ovata* was 30–50 cm from the surface with its body oriented horizontally.

Of the specimens found near the shore, only 2.5% of them had *Mnemiopsis leidyi* in the gut contents. Further from shore in depths of 2–4 m, approximately 20% of *B. ovata* had *M. leidyi* in the gastrovascular cavity. About 2% of *B. ovata* had fish larvae in the stomodeum or a larva was attached on the body outside close to the mouth.

In initial observations, *Beroe ovata* with newly ingested *Mnemiopsis leidyi, Aurelia aurita*, or Mediterranean horse mackerel (*Trachurus mediterraneus ponticus* Aleev) and mullet (*Mugil saliens* Risso) fry were placed individually in aquaria (20 l). On one occasion, the mullet larva, which was on the external side of *B. ovata* escaped and survived in the aquarium. In a few other occasions, larvae attached on the external part of the body were observed already dead with damaged skin. One ingested *A. aurita* (diameter 32 mm) was alive in the *B. ovata* (size 80 mm) stomodeum, and continued to make swimming movements. The mouth of *B. ovata* was closed and it kept the medusa in its stomodeum for 8 h before *A. aurita* was ejected and survived. Once a specimen of *B. ovata* tried to envelop another one in the aquarium, but the swallowed *B. ovata* was finally egested. *B. ovata* also did not consume the copepod *A. clausi* and if swallowed, it egested *A. clausi*. In aquaria, non-feeding *B. ovata* oriented vertically near the surface with the closed mouth upwards or near the bottom of the tank with the mouth turned towards the bottom.

During digestion observations on the R/V 'Akvanavt', newly collected *Beroe ovata* and their prey, *Mnemiopsis leidyi, Pleurobrachia pileus* and *Aurelia aurita*, were measured and placed in an aquarium (250 l). *B. ovata* were very active and swallowed *M. leidyi* in two ways – either enveloping them gradually or opening the mouth widely, bending their body and rapidly sucking in the entire prey. Biting off parts of the *M. leidyi* body with macrocillia was not observed. During digestion, the prey was gradually macerated and advanced with stomodeum cilia to the aboral pole where it accumulated in whitish clumps near the preinfundibular complex. Then whitish material entered the preinfundibulum in fractions and passed in meridional canals and excreted with moisture though the pores. In addition to *M. leidyi, B. ovata* ingested *P. pileus*. One *B. ovata* caught as many as three *P. pileus* or small *M. leidyi* and *P. pileus* at the same time. *B. ovata* did not consume *A. aurita* in these experiments. Once *B. ovata*

*Table 1.* Beroe ovata oxygen consumption at experimental temperatures, and recalculated to standard temperature (20 °C)

| Author | DW (g) | Duration (h) | Temperature (°C) | Salinity | Oxygen consumption (ml g DW$^{-1}$ h$^{-1}$) | Oxygen consumption at 20°C (ml g DW$^{-1}$ h$^{-1}$) |
|---|---|---|---|---|---|---|
| Kremer et al. (1986) | 0.01–0.56 | 3–5 | 25 | 35 | 0.27 | 0.19 |
| Finenko et al. (2000 a, b) | 0.01–0.9 | 17–18 | 21 | 18 | 0.35–0.44 | 0.34 |
| Shiganova et al. (2000a, b) | 0.31–1.04 | 4 | 24.6–26 | 18 | 0.25–0.71 (0.54±0.131) | 0.37 |
|  |  |  |  | 10.8 | 0.47–0.83 (0.66) | 0.46 |

swallowed two *P. pileus* which had *Calanus euxinus* Hulsemann in their gut contents. *P. pileus* was digested by *B. ovata*, but *C. euxinus* was egested through the mouth. Digestion times for *M. leidyi* prey were 4–5.5 h at 21–26° C, for *P. pileus* were 7–8 hours at 25 ° C (Table 2).

We estimated daily ration by two methods. First, the cost of respiration at 24–26 °C equalled 2.4 cal g DW$^{-1}$ h$^{-1}$. Consequently the minimum daily ration of *B. ovata* (assimilation rate = 0.7) would be about 80 cal g DW$^{-1}$ d$^{-1}$ or about 2 cal g WW$^{-1}$ d$^{-1}$. Feeding on *Mnemiopsis leidyi* (caloricity = 10 cal g WW$^{-1}$ (Vinogradov et al., 1989)), the minimum daily ration of *B. ovata* should be about 20% of its wet weight. Second, if *B. ovata* were assumed to feed continuously on *M. leidyi*, one at a time, then the digestion time measurements suggest that the daily rations would range from 80 to 400% measured as wet weight of both predator *B. ovata* and prey *M. leidyi*. Such rates are unlikely *in situ*. Our observations *in situ* during August-September 1999 indicated that approximately 20% of *B. ovata* contained *M. leidyi*. Therefore, we estimate that during this period, daily ration of *B. ovata* was in the range of 16–80% of its wet weight.

*Distribution and abundance of the main components of the Black Sea ecosystem*

In late August–early September 1999, *Beroe ovata* occurred throughout the northeastern part of the Black Sea above the seasonal thermocline (15–25 m). The average abundance was 0.62 ind. m$^{-2}$ in the inshore waters and 1.27 ind. m$^{-2}$ in the open sea, with an average abundance in the whole investigated area of 1.1 ind. m$^{-2}$, and an average WW biomass of 31 g m$^{-2}$ (Fig. 3a). The most abundant aggregations of *B. ovata* generally coincided with areas where *Mnemiopsis leidyi* was also abundant (Fig. 3a,b). In the open sea, there were only moderate to large sized *B. ovata*, while small specimens were found only in the coastal

areas and only in small numbers in late August–early September.

After the arrival of *Beroe ovata*, the sharpest drop among other gelatinous species was observed for both biomass and density of the ctenophore *Mnemiopsis leidyi* in 1999 compared with late summer data during the previous ten years. In September 1999, the average abundance was only 17.3 ind. m$^{-2}$, and the biomass was 155 g m$^{-2}$ (Fig. 4). The proportions of size groups of *M. leidyi* greatly changed, too. Previously, there had been intensive reproduction in August–September resulting in an abundance of small specimens of *M. leidyi* (<10 mm) (Shiganova, 1997). By contrast, in September 1999, small *M. leidyi* made up only 11.7% of the population inshore and 16.4% in the open sea. The most intensive reproduction usually occurs in the inshore waters, but during August–September 1999, the small *M. leidyi* were proportionally less abundant in the inshore waters than in the open sea. During our survey in April 2000, *M. leidyi* was found only in the warmest locations and its density was 1.2 ind. m$^{-2}$ and biomass was 28 g WW m$^{-2}$, but there were no individuals of *B. ovata* at all (Fig. 3c).

By contrast, populations of gelatinous species not eaten much by *Beroe ovata* were unaffected. The biomass of the medusa *Aurelia aurita* was comparable in both September 1998 and 1999 (Fig. 4): 283.6 g m$^{-2}$ in 1999 and 305 g m$^{-2}$ in 1998. The biomass and density of the ctenophore *Pleurobrachia pileus* remained practically unaltered relative to the previous year, the biomass was 83 g m$^{-2}$ in 1999 and 86.2 g m$^{-2}$ in 1998, density comprised 176.6 ind. m$^{-2}$ in 1999, and 183.3 ind. m$^{-2}$ in 1998 (Fig. 4). The abundance of the dinoflagellate, *Noctiluca scintillans* Kofoid & Swezy, decreased markedly from 30 to 37 × 10$^3$ ind. m$^{-2}$ in September 1998 to 9–10 × 10$^3$ ind. m$^{-2}$ in 1999 (Fig. 5).

The biomass of zooplankton was several times higher than observed in the same season during all years since the *Mnemiopsis leidyi* invasion. The mean

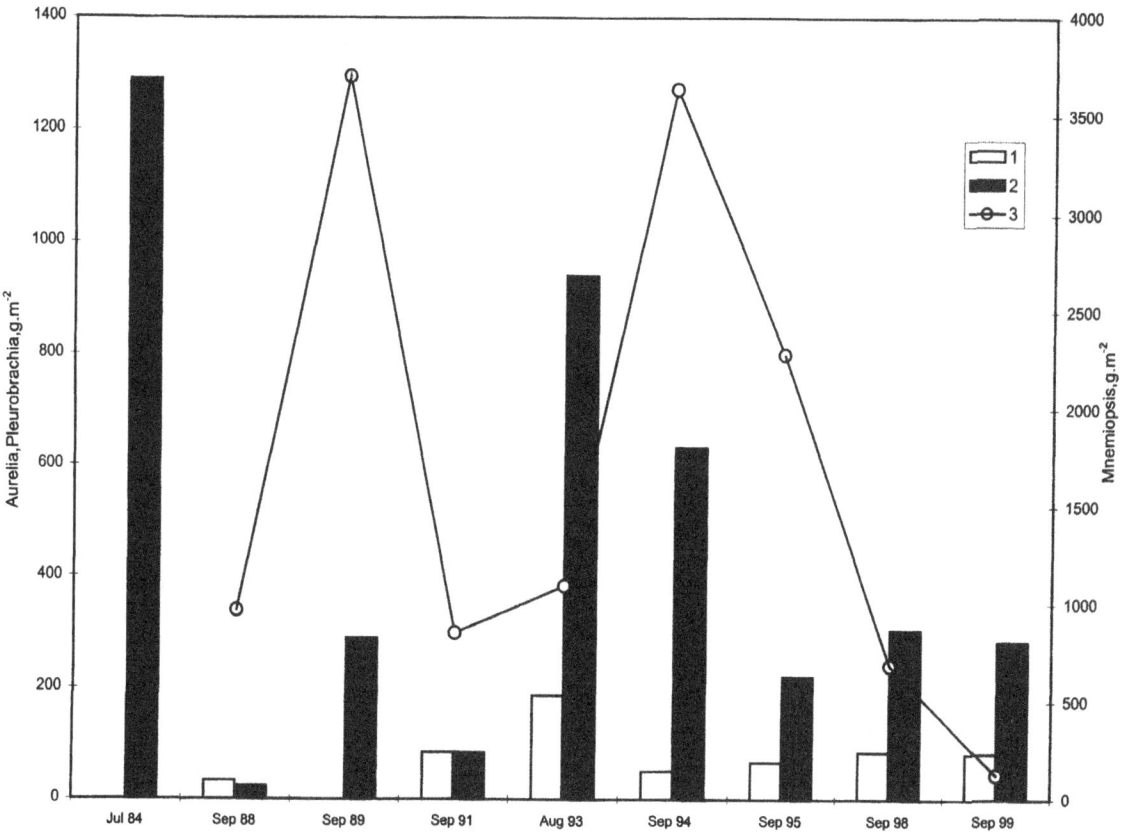

*Figure 4.* Biomass of gelatinous animals in August–September in the Black Sea 1 – *Pleurobrachia pileus*; 2 – *Aurelia aurita*; 3 – *Mnemiopsis leidyi* (Data from 1989–1992 in Vinogradov et al. (1989, 1992, 1993) and Khoroshilov (1993).

biomass of zooplankton was about 11g m$^{-2}$ in the open sea and about 13 g m$^{-2}$ in the inshore waters in early September 1999. Compared with1998, biomasses of *Calanus euxinus* and *Pseudocalanus elongatus* Boeck remained approximately constant in 1999, but biomass of other copepods increased more than three fold and reached 1.4 g m$^{-2}$ and >4 × 10$^4$ ind m$^{-2}$ between these periods (Fig. 5). The density of *Sagitta setosa* Muller also increased substantially to 6–15 × 10$^3$ ind. m$^{-2}$ and the density of meroplankton also went up. The most conspicuous increase in mesozooplankton was for Cladocera, which increased up to 150–300 × 10$^3$ ind. m$^{-2}$. The most abundant cladoceran species was *Penielia avirostris* Dana. After many years of absence from the Black Sea zooplankton samples species such as *Pontella mediterranea* Claus appeared and *Centropages ponticus* Karaw was found in the 1999 samples in large numbers.

In 1999, twenty-four species of fish eggs and larvae were selected as part of this study (Fig. 6a, b). The eggs of the anchovy *Engraulis encrasicolus ponticus* Aleksandrov predominated (average 323 eggs m$^{-2}$),

followed by the eggs of *Trachurus mediterraneus ponticus* (average 11.1 eggs m$^{-2}$), and the eggs of *Mugil saliens* and *Diplodus annularis* (Linne) (average 1.2 eggs m$^{-2}$). Until these collections such high densities of *E. encrasicolus* and *T. mediterraneus* eggs had not recorded since the *Mnemiopsis leidyi* invasion (Fig. 6a).

## Discussion

In late August–early September 1999, *Beroe ovata* occurred everywhere in the northeastern Black Sea, both in the deep sea and in the inshore waters. Our observations demonstrated that *B. ovata* consumes both other species of ctenophores in Black Sea, *Mnemiopsis leidyi* and *Pleurobrachia pileus*, primarily feeding on the more available *M. leidyi*. Our observations indicated that *B. ovata* did not digest crustacean zooplankton, the medusa *A. aurita* or fish larvae, although they were caught occasionally. Our results for digestion times were similar with other researchers (Table 2). *B. ovata* digested large *Bolinopsis* in 4–5 h at 24–25° C

*Table 2.* Digestion times of *Beroe* spp. feeding on lobate ctenophores. Numbers in parentheses are means

| Author | Temp. (°C) | Predator | Prey predator | Length (mm) prey | Length (mm) predator | Weight (g) prey | Weight (g) time (h) | Digestion |
|---|---|---|---|---|---|---|---|---|
| Kamshilov (1960) | | *B.cucumis* | *Bolinopsis* | 30–40 | 15–20 | | | 3 |
| Swanberg (1974) | 24–25 | *B. ovata* | *Bolinopsis* | 12–40 (27) | 12–90 (50) | | | 4.5 |
| Finenko et al. (2000b) | 21 | *B. ovata* | *Mnemiopsis* | | | 1.4–59 | 0.25–14 | 0.5–5.5 |
| Shiganova et al. (2000a, b) | 21–26 | *B. ovata* | *Mnemiopsis* | 50–98 | 30–88 | 11–37 | 3–22 | 4–5.5 |

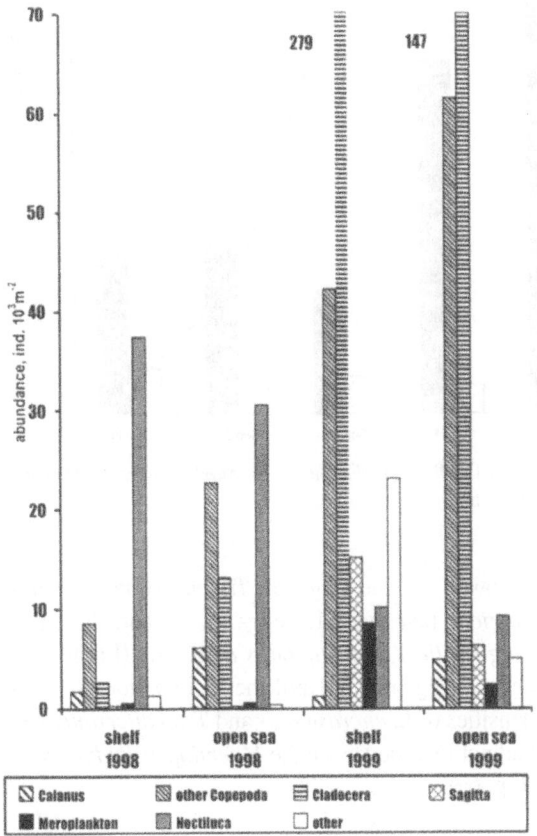

*Figure 5.* Density of zooplankton in the northeastern Black Sea in September 1998–1999.

(Swanberg, 1974). *B. cucumis* Fabricius from the Barents Sea digested small *Bolinopsis* in 3 h (Kamshilov, 1960 a, b). In our experiments the larger lobates, *M. leidyi*, were digested considerably faster than smaller cydippid *P. pileus*.

Maximum size of *Beroe ovata* collected in the Black Sea was larger than for collections it in their native regions. *B. ovata* reaches 115 mm near the Atlantic coast of North America (Mayer, 1912) as well as in the Mediterranean Sea (Tregouboff & Rose, 1957).

In Great Bahama Bank, the specimens were observed from 12 to 40 mm ($x = 27$ mm) (Swanberg, 1974). Specimens of *B. ovata* within the size range 10–50 mm were found in the Bahamas (Kremer et al., 1986). It is not clear the reasons for the size difference between locations. We suppose that *B. ovata* in the Black Sea reach a larger size due to the abundance of available food, *Mnemiopsis leidyi*.

The metabolic rate estimated in our experiments was very close to the value of Finenko et al. (2000a,b, 2001) for the northwestern Black Sea in 1999 (Table 1), but significantly higher on a dry weight basis than for the same species from the northern Atlantic (Kremer et al., 1986). Most of this difference can be explained by salinity differences if we assume *Beroe ovata* composition is in isotonic equilibrium with its surrounding water, where the Black Sea is approximately half the salinity of the water of the Bahamas.

The minimum daily ration for *Beroe ovata* estimated from our metabolic data was about 20% of its wet weight. The observed digestion rate suggests that if *B. ovata* fed continuously, they could consume and digest 5–8 specimens of *Mnemiopsis leidyi* daily. In the field in August–September 1999, we observed about 20% *B. ovata* with *M. leidyi* in their guts, equivalent to an ingestion rate of 1–1.6 *M. leidyi* per day. During this period, mean *B. ovata* biomass was 31 g WW m$^{-2}$ (1.1 ind. m$^{-2}$) and *M. leidyi* biomass was 155 g WW m$^{-2}$ (17.3 ind. m$^{-2}$). Consumption by *B. ovata* that we calculated from field and experimental feeding rates amounted to a daily removal of 6.5–10% of the *M. leidyi* population. About 4% of the *M. leidyi* population would need to be ingested per day to meet the metabolic requirement of the *B. ovata* population as estimated from measured respiration rates of *B. ovata*.

A sharp decrease in the density and biomass of *Mnemiopsis leidyi* in the northeastern Black Sea in September 1999, as compared to September of previous years (Fig. 4), implicates *Beroe ovata* as a po-

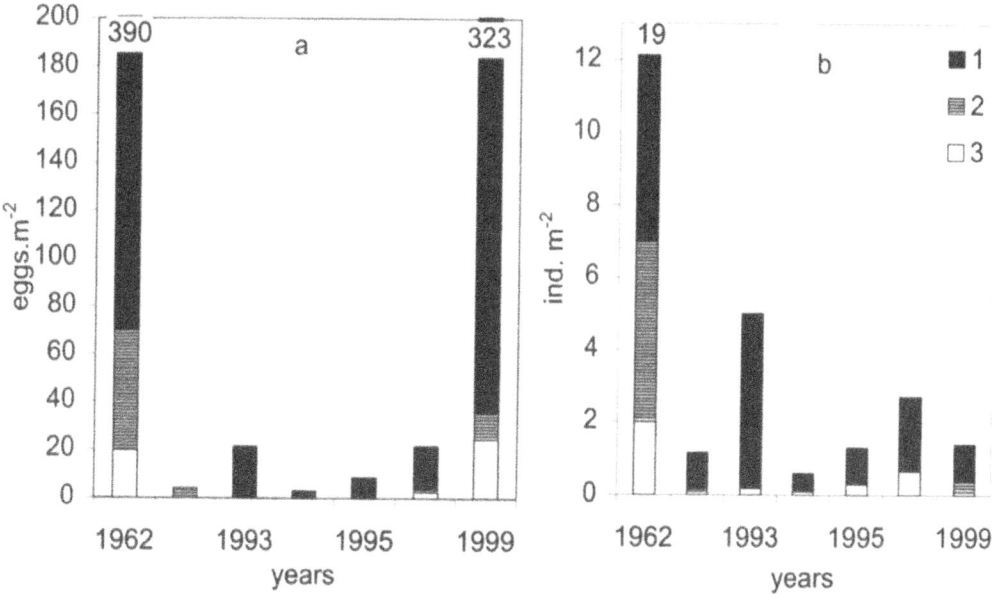

*Figure 6.* Long-term dynamics of ichthyoplankton density in the northeastern Black Sea in July–August. (a) – eggs; (b) – larvae. Data from 1962 in Dekhnik (1973). 1 – anchovy; 2 – Mediterranean horse mackerel, 3 – others.

tentially important source of mortality. The *M. leidyi* population usually increases in spring after a warm winter (Shiganova, 1998); considering that the 1998–1999 winter was warm and that water temperature increased by 1.5–2 °C in June–July 1999 (data of Hydrophysical laboratory of the Southern Branch of the P. P. Shirshov Institute of Oceanology, Russian Academy of Sciences), a high density of *M. leidyi* would have been expected in August–September, as observed in previous years (Purcell et al., 2001). In addition, *M. leidyi* usually actively reproduces at that time (Shiganova, 1997, 1998). By contrast, in September 1999, we observed an unusually low abundance of *M. leidyi* and low numbers of small specimens relative to the previous September (Shiganova et al., 2000 b). Probably *B. ovata* initially grazed small specimens in the inshore waters, due to the availability and accessibility of young *M. leidyi* as prey.

Observations of previous years showed that seasonal population dynamics of *Mnemiopsis leidyi* were similar in the Black and Azov Seas (Shiganova et al., 2001). In September 1999, however, biomass of *M. leidyi* was very high in the Sea of Azov (Fig. 7), while the population in the Black Sea was the lowest in 10 years (Fig. 6). This circumstantial evidence implicates *B. ovata* as a major predator of *M. leidyi* in the Black Sea.

The abundance of the ctenophore *Pleurobrachia pileus* in late summer 1999 was essentially unchanged from the previous year. This may be because *Beroe ovata* inhabits the surface layer while *P. pileus* predominantly inhabits the cold intermediate layer and below. Possibly, *B. ovata* ingests some *P. pileus* at night, however, only a small part of the *P. pileus* population ascends into the surface layer (Vinogradov et al., 1987). The population of the medusa *Aurelia aurita* was also similar in 1999 to the 1998 level (Fig. 4), and slightly higher than average levels for the past 5 years (Shiganova et al., 1998), suggesting that its population also was not affected by *B. ovata*.

The density and biomass of mesozooplankton and its species diversity greatly increased in September 1999, after the appearance of *Beroe ovata,* and were similar to values observed prior to the *Mnemiopsis leidyi* invasion (Shiganova, 1998). The species inhabiting cold intermediate layer (*Calanus euxinus, Pseudocalanus elongatus*) did not change in density and biomass, while the density of copepods inhabiting surface layer greatly increased. The density of Cladocera, mainly *Penilia avirostris,* inhabiting the surface layer also greatly increased. In addition, several Mediterranean species (Copepoda: *Euchaeta marina* Pustandrea, *Rhyncalanus nasutus* Giesbrecht, *Pleuromamma gracilis* Claus, and Ostracoda *Philomedes globosa*) were found in the northeastern region for the first time (E. I. Musaeva, P. P. Shirshov Institute of Oceanology, Russian Academy of Sciences, pers. com.).

196

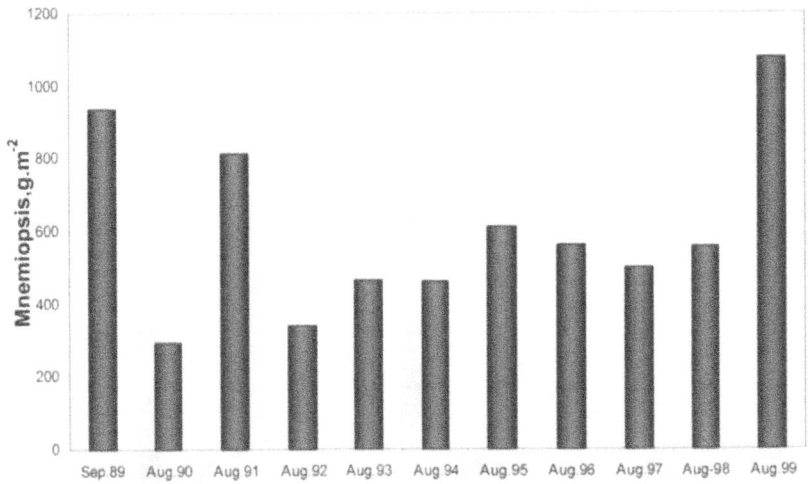

*Figure 7.* Long-term dynamics of *Mnemiopsis leidyi* density in Sea of Azov in August.

Species diversity and total numerical abundance of ichthyoplankton were unusually high for the northeastern region in late August 1999. The density of eggs of both *Engraulis encrasicolus ponticus* and *Trachurus mediterraneus ponticus* increased to levels comparable to those before the *Mnemiopsis leidyi* invasion (Fig. 6a). The density of larvae was still low in August after the introduction of *Beroe ovata*. Unlike the collected eggs, which were spawned only one day before sampling, when *B. ovata* was present and *M. leidyi* density was low, the larvae possibly were from spawnings before *B. ovata* was present. Thus, the 1999 data suggest the appearance of *B. ovata* in the Black Sea decreased the abundance of *M. leidyi* sufficiently to modify populations of other zooplanktivores. All effects of *B. ovata* may not be positive, however. Preliminary observations indicate that fish larvae, although not ingested by *B. ovata*, may be killed though occasional direct contact with *B. ovata*.

Certainly, all the above improvements were not merely the results of the invasion by *Beroe ovata*. Since the first decrease of *Mnemiopsis leidyi* density in 1993, some improvements in parameters of the ecosystem and recovery of biodiversity were recorded (Shiganova et al., 1998). But it was not until 1999, with the advent of *B. ovata,* that there was a sharp increase in the density and diversity of mesozooplankton and ichthyoplankton.

Our data demonstrate that *Beroe ovata* can reproduce, feed and grow in the northeastern Black Sea. The same conclusions were made for the northwestern area (Finenko et al., 2000 a, b, 2001). If *B. ovata* continues to be found in the Black Sea,

neither mesozooplankton nor endemic gelatinous species (*Aurelia aurita, Pleurobrachia pileus)* should be affected directly. Due to its feeding preference and vertical distribution, *B. ovata* should primarily impact the population of *Mnemiopsis leidyi* and indirectly affect mesozooplankton.

We demonstrated that *Beroe ovata* can live in water with rather low salinity, and therefore may be able to penetrate into the Sea of Azov, however, it is unknown if *B. ovata* could live and reproduce in the Sea of Azov, if it would be a constant or temporary inhabitant there, or if it could influence the recovery of its ecosystem.

## References

Caddy, J. F. & R. C. Griffiths, 1990. A perspective on recent fishery-related events in the Black Sea. Studies and Reviews. GFCM 63: 43–71.

Dekhnik, T. V., 1973. Ichthyoplankton of the Black Sea. Naukova Dumka: 234 pp

Finenko, G. A., Z. A. Romanova & G. I. Abolmasova, 2000 a. New settler into the Black Sea – ctenophore *Beroe ovata*. In Matishov, G. G. (ed.), Thesis of the Report at Scientific Seminar "Species - Invaders in the European Seas of Russia", Murmansk, 27–28 January 2000: 95–96 (in Russian).

Finenko, G. A., Z. A. Romanova & G. I. Abolmasova, 2000 b. The ctenophore *Beroe ovata* is a recent invader to the Black Sea. Ecologiya morya 50: 21–25 (in Russian).

Finenko, G. A., B. E. Anninsky, Z. A. Romanova, G. I. Abolmasova & A. E. Kideys, 2001. Chemical composition, respiration and feeding rates of the new alien ctenophore, *Beroe ovata*, in the Black Sea. Hydrobiologia 451 (Dev. Hydrobiol. 155): 177–186.

GESAMP (IMO/FAO/UNESCO-IOC/WMO/WHO/IAEA/UN/UNEP Joint Group of Experts on the Scientific Aspects of Marine Environmental Protection), 1997. Opportunistic settlers and the problem of the ctenophore *Mnemiopsis leidyi* invasion in the Black Sea. Rep. Stud. GESAMP 58: 84 pp.

Ivanov, L. & R. J. H. Beverton, 1985. The fisheries resources of the Mediterranean. Part two. Black Sea. GFCM FAO, Rome: 70 pp.

Kamshilov, M. M., 1960a. Feeding of ctenophore *Beroe cucumis* Fab. Dokl. Acad. Nauk SSSR 130: 1138–1140 (in Russian).

Kamshilov, M. M., 1960b. The dependence of ctenophore *Beroe cucumis* Fab sizes from feeding. Dokl. Acad. Nauk SSSR 131: 957–960 (in Russian).

Khoroshilov, V. S., 1993. Sesonal dynamics of the Black Sea population of ctenophore *Mnemiopsis leidyi*. Oceanology 33: 558–562.

Konsulov, A. & L. Kamburska, 1998. Ecological determination of the new Ctenophora - *Beroe ovata* invasion in the Black Sea. Tr. Ins. Oceanology, Varna: 195–197.

Kovalev, A. V., S. Besiktepe, J. Zagorodnyaya & A. Kideys, 1998. Mediterraneanization of the Black Sea zooplankton is continuing. In Ivanov, L. & T. Oguz (eds), Ecosystem Modeling as a Management Tool for the Black Sea. Kluwer Academic Publishers, Dordrecht 47: 199–207.

Kremer, P., M. F. Canino & R. W. Gilmer, 1986. Metabolism of epipelagic tropical ctenophores. Mar. Biol.: 403–412.

Mayer, A. G., 1912. Ctenophores of the Atlantic coast of North America. Publs. Carnegie Inst. 162: 1–58.

Nastenko, E. V. & L. M. Polishchuk, 1999. The comb jelly *Beroe* (Ctenophora: Beroida) in the Black Sea. Dopovidi Nazionalnoi Akademii Nauk, Ukraine 11: 159–161.

Pushkina, N. I., 1963. Biokhimicheskie metody issledovanii. Biochemical methods of investigations. Moscow, MedGIZ: 63 pp.

Purcell, J. E., T. A. Shiganova, M. B. Decker & E. D. Houde, 2001. The ctenophore *Mnemiopsis* in native and exotic habitats: U. S. estuaries versus the Black Sea basin. Hydrobiologia, this volume.

Romanenko, V. I. & S. I. Kuznetzov, 1974. Opredelenie rastvorennogo v vode kisloroda po metody Winklera. Determination of dissolved oxygen according to Winkler method. Ekologia mikroorganizmov presnikh vodoemov, Leningrad, Nauka: 176–177.

Seravin, L. N., 1998. Ctenophora. Dobrovol'sky, A. (ed.), Omsk Biological Scientific Institute, St.-Petersburg: 1–84.

Seravin, L. N., Shiganova, T. A., Lupova, N. E., 2001 The history of study of ctenophore *Beroe ovata* (ctenophora, atentaculata, beroida) and some peculiarities of morphology of the representative from the Black Sea. Zoologichesky jour. in press.

Shiganova, T. A., 1997. *Mnemiopsis leidyi* abundance in the Black Sea and its impact on the pelagic community. Sensitivity of the North, Baltic Sea and Black Sea to Antropogenic and Climatic Changes. In Ozsoy, E. & A. Mikaelyan (eds), Kluwer Academic Publishers, Dordrecht, The Netherlands: 117–130.

Shiganova, T. A., 1998. Invasion of the Black Sea by the ctenophore *Mnemiopsis leidyi* and recent changes in pelagic community structure. Fish. Oceanogr. GLOBEC Special Issue: 305–310.

Shiganova, T. A. & Y. V. Bulgakova, 2000. Effect of gelatinous plankton on the Black and Azov Sea fish and their food resources. ICES J. mar. Sci. 57: 641–648.

Shiganova, T. A., B. Ozturk & A. Dede, 1994. Distribution of the ichthyo-, jelly- and zooplankton in the Sea of Marmara. FAO Fisheries report 495: 141–145.

Shiganova, T. A., Yu. V. Bulgakova, P. Yu. Sorokin & Yu. F. Lukashev, 2000a. Preliminary results of investigations of *Beroe ovata*, a new invader into the Black Sea, and its effect on the pelagic ecosystem. In Matishev, G.G. (ed.), Thesis of the Report at Scientific Seminar "Species - Invaders in the European Seas of Russia", Murmansk: 105–108 (In Russian).

Shiganova, T. A., Y. V. Bulgakova, P. Yu. Sorokin & Yu. F. Lukashev, 2000b. Investigations of new settler *Beroe ovata* in the Black Sea. Biol. Bull. 2: 247–255.

Shiganova, T. A., U. Niermann, A. C. Gucu, A. Kideys & V. S. Khoroshilov, 1998. Changes of species diversity and their abundance in the main components of pelagic community after *Mnemiopsis leidyi* invasion. In Ivanov, L. & T. T. Oguz (eds), Kluwer Academic Publishers, Dordrecht, The Netherlands: 171–188.

Shiganova, T. A., I. A. Mirzoyan, E. A. Studenikina, S. P. Volovik, I. Siokoi-Frangou, S. Zervoudaki, E. D. Christou, A. Yu. Skirta, & H. J. Dumont, 2001. Development of the Ctenophore *Mnemiopsis leidyi* (A. Agassiz) in the Black Sea and in the other seas of the Mediterranean basin. Mar. Biol. In press.

Swahn, B., 1953. Method identification of lipids. Scand. J. Clin. Lab. Inv. 5: 9.

Swanberg, N., 1974. The feeding behavior of *Beroe ovata*. Mar. Biol. 24: 69–76.

Tregouboff, G. & M. Rose, 1957. Manuel de Planctonologie Mediterraneenne. Centre Nat. de la Rech. Sci. Paris 1: 1–587.

Tzikhon-Lukanina, E. A., O. G. Reznichenko & T. A. Lukasheva, 1993. Ecological variation of combjelly *Mnemiopsis leidyi* (Ctenophora) in the Black Sea. J. obschei biologii 6: 713–721 (in Russian).

Vinogradov, M. E., M. V. Flint & G. G. Nikolaeva, 1987. Vertical distribution of mesoplankton in the open Black Sea in spring. In Musaeva, E. I. (ed), Sovremennoe sostoyanie ecosystemy Chernogo morja. Nauka, Moscow: 144–186 (in Russian).

Vinogradov, M. E., V. V. Sapozhnikov & E. A. Shushkina, 1992. The Black Sea Ecosystem. Nauka, Moscow: 112 pp.

Vinogradov, M. E., E. A. Shushkina & G. G. Nikolaeva, 1993. The state of zoocenosis in the open Black Sea in the end of summer 1992. Oceanology 33: 382–387.

Vinogradov, M. E., E. A. Shushkina, E. I. Musaeva & P. Yu. Sorokin, 1989. Ctenophore *Mnemiopsis leidyi* (A. Agassiz) (Ctenophora: Lobata) – new settlers in the Black Sea. Oceanology 29: 293–298.

Volovik, S. P., I. A. Mirzoyan & G. S. Volovik, 1993. *Mnemiopsis leidyi*: biology, population dynamics, impact to the ecosystem and fisheries. ICES. Statutory meeting C. M. 1993/L:69: 1–12.

Zaitsev, Yu. P., 1998. Marine hydrobiological investigations of National Academy of Science of Ukraine during the 1990s in XX century: shelf and coastal water bodies of the Black Sea. Hydrobiol. Zhurnal 6: 3–21 (in Russian).

Zaitsev, Yu. P. & B. G. Aleksandrov, 1997. Recent man-made changes in the Black Sea ecosystem. In Ozsoy, E. & A. Mikaelyan (eds), Sensitivity of North Sea, Baltic Sea and Black Sea to Anthropogenic and Climatic Changes. Kluwer Academic Publishers, Dordrecht, The Netherlands: 25–32.

*Hydrobiologia* **451**: 199–212, 2001.
© 2001 *Kluwer Academic Publishers.*

# A physical context for gelatinous zooplankton aggregations: a review

William M. Graham[1], Fransesc Pagès[2] & William M. Hamner[3]
[1]*Dauphin Island Sea Lab and Department of Marine Sciences, University of South Alabama,*
*101 Bienville Boulevard, Dauphin Island, AL, 36528, U.S.A.*
*E-mail: mgraham@disl.org*
[2]*Institut de Ciències del Mar (CSIC), Plaça del Mar s/n, 08039 Barcelona, Catalonia, Spain*
[3]*University of California, Los Angeles, Department of Organismal Biology, Ecology and Evolution, Box 951606,*
*Los Angeles, CA, 90095-1606, U.S.A.*

*Key words:* jellyfish blooms, fronts, behavior, thermoclines, haloclines

**Abstract**

The magnitude and extent of jellyfish blooms are influenced not only by the biology and behavior of the animal, but also by the geographic setting and physical environment. Hydrography alone is often thought to cause or favor gelatinous zooplankton aggregations, however, it is clear that interactions between biology of the animal and physics of the water are very important sources of population variations, especially at local scales. We summarize the role of physical processes and phenomena that promote aggregations of gelatinous zooplankton. We have identified and discussed a suite of physical gradients that can be perceived by gelatinous zooplankton. These include light, gravity, temperature, salinity, pressure and turbulence. A recurring theme is accumulation of jellyfish around physical discontinuities such as fronts (shelf-break, upwelling, tidal and estuarine) and pycnoclines (thermoclines and haloclines). Interestingly, there are few data to suggest that large-scale, quasi-stationary features, such as the largest oceanic fronts, serve to physically aggregate gelatinous animals at a similar scale. Rather, examples of local aggregations appear to dominate the literature. We also discuss various jellyfish behaviors that are theorized to promote aggregation, feeding and reproduction in relation to physical discontinuities.

## Introduction

The sudden appearance and disappearance of jellyfish aggregations or swarms is a common yet enigmatic characteristic of jellyfish populations. Most gelatinous zooplankters have life-histories with asexual reproduction and extraordinary growth rates that allow them to undergo rapid population increases. Moreover, their gelatinous structure, absence of complex neural framework and apparent weak swimming suggest that they passively drift with the water current. To the naive observer, therefore, the sudden appearance of large numbers of jellyfish implies that a bloom has somehow occurred when no one was watching. Indeed, blooms, by definition, appear or occur unexpectedly or in surprising quantity, almost over night. We now know, however, that the apparently simple body plan of jellyfish often disguises sophisticated interactions with their biological, chemical and physical environ-

ment that contribute to patchy distribution of jellyfish throughout the marine environment. Advances in the use of *in situ* techniques over the past 30 years (Hamner, 1985) have enormously increased our knowledge about the behavioral complexity of gelatinous animals.

Rapid changes in jellyfish concentrations can be due either to rapid population growth (a true bloom) or to a re-distribution or re-dispersion of a stable population (an apparent bloom). Since population changes are addressed elsewhere in this volume, we will address only physical factors implicated in apparent blooms. We distinguish between advective and concentrating factors. Advection means that a patch or aggregation of animals is translocated to a previously uninhabited location, whereas concentration means that the density of a population has changed. Advection of planktonic populations occurs passively whenever a water mass moves, whereas concentration changes, in the absence of population growth or mortality, require an element of active behavior or physiological alteration by the jellyfish that promotes changes in local density.

Animal aggregations constitute a fascinating and complex research topic that requires theory and modeling studies linked to empirical work (Parrish & Hamner, 1997; Parrish & Edelstein-Keshet, 1999). Nonetheless, *in situ* observations using SCUBA (Hamner et al., 1975), research submersibles and remotely operated vehicles (Larson et al., 1992; Mills et al., 1996; Brodeur, 1998; Raskoff, 2001) and profiling video (Paffenhöffer et al., 1991; Davis et al., 1992), have been crucial for investigating the relationship between biology, behavior and the physical environment of marine organisms (Hamner, 1985). In this paper, we review those physical factors that contribute to aggregation of gelatinous zooplankton. As the literature base is comparatively new and rarely pre-dates the 1970s, we cast a broad net to include mostly medusae, but also siphonophores, ctenophores and pelagic tunicates. The ultimate goal of the paper is to provide the underlying physical context of aggregations as they occur in conjunction with, or the absence of, biology and behavior of the animal, and when possible we discuss adaptive implications of these interactions. While we recognize the importance of physical transport of a bloom as an important source of apparent population variability (e.g., Nielsen et al., 1997), we emphasize here only physical attributes that lead to local aggregation of gelatinous animals.

This review is organized to provide the reader first with an understanding of how gelatinous animals perceive, and respond to, specific physical characteristics and physical gradients. Biological and chemical gradients may also mediate gelatinous zooplankton distribution (e.g. Arai, 1992; Purcell et al., 1994; Purcell et al., 2001), but we are limiting this discussion to physical-biological interactions. We then move to specific examples of aggregations that have been described in both the horizontal and vertical dimensions. And finally, we discuss future directions that scientists might follow to advance our understanding of gelatinous zooplankton aggregations.

## Passive and active responses to environmental physical gradients

Changes in physical properties of the environment serve as important cues for initiating biological and ecological processes that result in gelatinous zo-

oplankton aggregation. Physical cues provide the primary, proximal information for movements of individuals within an aggregation, as well as for movement of the entire aggregation. Active responses to physical cues can also facilitate aggregation through the interplay between local circulation and the directed swimming movements of the animals. Another response, but not described in further detail, is seasonal changes of physical environmental variables that can herald the onset of ultimately favorable conditions for reproduction, resulting in population changes that also culminate in locally high densities of plankton, as in blooms of red tide. In the following section, we discuss the role of physical properties and gradients as cues mediating gelatinous zooplankton migrations and aggregations.

Aggregations of gelatinous zooplankton are three-dimensional entities, and therefore both vertical and horizontal physical gradients facilitating aggregation will be considered. Gradients and variations of physical properties are far more pronounced in the vertical dimension than in the horizontal, therefore physical–biological interactions tend to be more easily observed and documented in the vertical axis. Jellyfish aggregations in the vertical dimension can be envisioned as 'layers' in three dimensions. The vertical extent of these layers may be from a few centimeters to hundreds of meters thick. By contrast, horizontal dimensions of aggregations can be on the order of tens of meters to hundreds of kilometers. Physical gradients and discontinuities implicated in gelatinous zooplankton aggregations include light, pressure, turbulence, currents, temperature and salinity (thus density). While active responses to both horizontal and vertical gradients or discontinuities are often mediated by behavior, we also discuss evidence for passive accumulation along physical discontinuities.

### Light-mediated migrations

Jellyfish are generally considered to be planktonic, i.e., not capable of extensive movements against currents. Consequently, their vertical migration patterns have been documented more thoroughly than have their horizontal migrations. Vertical migrations for zooplankton have been described repeatedly in the literature, with induction of migration implicitly linked to the diel cycle of light. 'Diurnal Vertical Migration', or DVM, usually refers to a pattern of diel behavior in the water column wherein individuals swim upward toward the surface during the night, and swim downward toward the bottom during the day. The behavioral swimming response seen during DVM can result in either the entire population moving *en masse* as a migrating layer, or a vertically dispersed population may become locally concentrated at the sea surface during a specific time of day (usually at night). In fact, daily nocturnal ascent and diurnal descent is by far the dominant type of vertical migratory behavior in nature (Bayly, 1986). Perhaps the most studied migrating layer of gelatinous zooplankton is the 'Deep Scattering Layer' (DSL) discovered in the late 1940s from the acoustic properties of physonect siphonophores - primarily *Nanomia bijuga* delle Chiaje (Barham, 1963; see Pugh, 1977 for review).

Light-mediated DVMs range from a few to hundreds of meters, and they are widespread among hydromedusae (Russell, 1927; Arai, 1973, 1992; Mills, 1981; Arkett, 1985; Mills & Goy, 1988), siphonophores (Barham, 1963; Pagès & Gili, 1991) and scyphomedusae (Yasuda, 1973; Mackie et al., 1981; Hamner, 1995; Schuyler & Sullivan, 1997). There is little evidence that ctenophores exhibit DVMs mediated by light (Frank & Widder, 1997). This is perhaps due to the lack of light sensing machinery within the phylum. Vertically migrating animals must have biological equipment that is sensitive to directional environmental information. Many gelatinous zooplankton have sensory apparati that can detect gravity or light, and these two types of sense organs clearly are sufficient to provide the information needed to discriminate 'up' from 'down'. While ocelli are typically present in light-mediated migrators, ocelli are not specifically required for perception of light by medusae, although medusae seem to require illumination to stimulate migrations (Mackie et al., 1981). Direct photo-stimulation of neurons has been described for jellyfish (Anderson & Mackie, 1977), and a number of cases of DVM have been described in species believed to lack ocelli: *Solmissus albescens* Gegenbaur (in Mills & Goy, 1988), *Chrysaora quinquecirrha* (Desor) (in Schuyler & Sullivan, 1997) and *Pelagia noctiluca* (WMG, unpublished observations). Moreover, direction and magnitude of illumination or the rate of change of illumination driving migrations seems to be species dependent (e.g. Mackie et al., 1981; Arkett, 1985; Schuyler & Sullivan, 1997).

Almost all experiments and observations regarding DVM, for all oceanic animals, have involved only light and gravity (Bayly, 1986). DVMs invariably correspond with diel changes in ambient illumination within the water column. It is tempting, therefore, to conclude that animals migrate up or down in simple, direct response to changes in the intensity of light during a 24-h day, by maintaining a species-specific, constant level of illumination as long as possible during twilight, the so-called 'constant isolume hypothesis' (Bayly, 1986). Nonetheless, there is surprisingly little evidence in the literature to suggest that the coincidence of swimming depth and penetration depth of light during twilight is anything more than a necessary consequence of the irrevocable rise and set of the sun [but see Backus et al. (1965) and Enright & Hamner (1967)].

Various alternative hypotheses regarding the cues that trigger DVM are also tenable. For example, changes in light intensity at dawn may provide the cue to initiate downward swimming, whereas the depth to which the population descends during the day might be determined by a secondary cue such as pressure, temperature, salinity or light intensity. At dusk, alternatively, upward swimming might well be triggered by decreasing light intensity but the upward or downward extent of the migration may be constrained simply by the physical presence of the sea surface, by pressure, or by temperature or salinity discontinuities (e.g. Purcell & Madin, 1991). Another hypothesis invokes the presence of an internal circadian rhythm with which the animals tell time of day and thereby anticipate dawn and dusk (Enright & Hamner, 1967). An internal clock could trigger directed swimming behavior, with final position in the water column adjusted according to information provided by pressure, temperature, salinity, light intensity or the presence of the sea surface (Enright & Hamner, 1967). Unfortunately, there is no evidence in the literature, to our knowledge, about how any gelatinous zooplankter determines when to migrate vertically or about what cues are used to determine a suitable depth to terminate vertical migrations.

Although the physical cues required to trigger DVM are poorly understood, there is no question that many species of medusae regularly engage in diurnal vertical migratory behavior. For example, in several of the marine lakes in Palau, two species of scyphomedusae exhibit quite different DVM patterns. Hamner et al. (1982) found that the moon jelly, *Aurelia aurita* (L.), engaged in typical DVM, swimming to the surface at night and down during the day. However, unlike many diurnal vertical migrators that descend during the day presumably to avoid visual predators, *A. aurita* has no pelagic predators in these lakes. *Aurelia aurita* may simply be tracking their copepod prey, which migrate down to darker waters in daytime, probably to avoid visual predators such as silverside fishes. Thus, the vertical diurnal migration patterns of *A. aurita* in Jellyfish Lake may be the result of optimal foraging behavior rather than a response to daily changes in ambient illumination. *Aurelia aurita* does engage in active foraging behavior in the laboratory. Bailey & Batty (1983) demonstrated that starved *A. aurita* remained quiescent at the top of a laboratory aquarium, but immediately after the introduction and capture of one fish larva the jellyfish began vigorously swimming throughout the tank, clearly engaged in active foraging behavior.

Variations in migratory behavior within the genus *Aurelia* (and quite possibly within the species *A. aurita*) are pronounced. During the summer in Saanich Inlet, Vancouver Island, Canada, a large population of presumably *A. aurita* (possibly *A. labiata* Chamisso & Eysenhardt) does not engage in DVM but remains at the surface both day and night (Hamner et al., 1994). By contrast, *A. aurita* in Eil Malk Jellyfish Lake (now called Tourist Lake), Palau, engaged in normal vertical diurnal migration behavior until 1998. At that time an extreme 'La Niña' occurred in the western Pacific and the marine lakes in Palau became exceptionally warm. The high lake temperature did not affect the size of the *A. aurita* medusa population, but all 1.6 million of the *Mastigias* sp. medusae disappeared (Dawson et al., 2001). For unknown reasons, the remaining population of *A. aurita* ceased its daily vertical diurnal migration and these medusae now remain at the surface both day and night. While other environmental changes cannot be discounted, this behavioral shift suggests that *A. aurita* can exhibit considerable behavioral plasticity. Prior to their disappearance, the *Mastigias* sp. exhibited an unusual example of reverse vertical migrations, wherein the animals remained at the surface throughout the day and migrated to depth during the night. The *Mastigias* sp. made individual, repeated vertical excursions at night to the permanent chemocline of this meromictic marine lake, presumably to bathe the symbiotic zooxanthellae in nutrients, but the medusae remained at the sunlit surface waters during the day presumably to provide the zooxanthellae with light for photosynthesis (Hamner et al., 1982).

DVM is also reported in several salp species, including *Salpa aspera* Chamisso (Wiebe et al., 1979), *Salpa fusiformis* Cuvier (Franqueville, 1971), and *Cyclosalpa bakeri* (in Purcell & Madin, 1991). Wiebe et al. (1979) described long-distance vertical migra-

tions of at least 800 m for the large salp, *S. aspera*, in northwestern Atlantic slope water. These long-distance vertical migrations were attributed to nighttime feeding in surface waters. The similar, but shallower (<30 m), migration pattern of *Cyclosalpa bakeri* in the subarctic Pacific was attributed instead to near-surface spawning (Purcell & Madin, 1991). In fact, as discussed by Purcell & Madin (1991), diel periodicity of light seems to be inherently linked to both swimming activity and gamete release in many gelatinous zooplankton.

Horizontal movements of jellyfish were overlooked for the most part until the 1980's when *in situ* ethological descriptions of planktonic organisms began to appear in the literature (reviewed in Hamner, 1995). Published accounts of horizontal migrations are still somewhat rare, yet in existing examples some generalizations have emerged about the physical cues used during horizontal migration and, in at least one instance, the adaptive advantages of horizontal migration seem clear. Horizontal migrations, like vertical migrations, are possible only if the animals can perceive, and respond to, horizontal environmental cues. Furthermore, horizontal cues must be scaled to jellyfish swimming to elicit an effective response. As such, only small-scale horizontal movements have been documented for gelatinous zooplankton. Light appears to be the dominant cue used by gelatinous zooplankton for directed horizontal migrations that lead to aggregation.

The first documented case of horizontal migration in a jellyfish was for *Mastigias* sp. in the marine lakes of Palau (Hamner & Hauri, 1981). *Mastigias* has symbiotic zooxanthellae in its tissues and obtains much of its daily nutrition from these algae (Muscatine & Marian, 1982; Muscatine et al., 1986; McCloskey et al., 1994). Sunlight governs the life of *Mastigias* and 70% of the population is concentrated in the top 2.5 m of the water column during daytime. In Eil Malk Jellyfish Lake the entire population of *Mastigias* migrates horizontally each morning towards the east end, a maximum distance of about 0.5 km. The jellyfish stop swimming eastward when they reach the shadow line beneath the mangroves at the end of the lake. When the population exceeded one million medusae, they were compressed into an intense, milling aggregation at the eastern end by mid-day, with densities often exceeding 1000 medusae $m^{-3}$. In the present smaller population, the accumulation at the eastern shadow line is less dense, but still the medusae all reorient to the west by the early afternoon. By late afternoon all of the medusae have migrated back to the western basin, where they remain until dawn and engage in nocturnal vertical migration, as described above.

Since *Mastigias* sp. are nutritionally dependent on photosynthesis by their symbiotic algae, it is tempting to postulate that their migrations are simple eastward phototactic responses toward the sun in the morning and westward toward the sun in the afternoon. But on partially overcast days *Mastigias* sp. often begins its westward migration as early as 10:00 h two hours before the sun has passed its zenith, before there is clear phototactic directional information from the sun (L. Colin, pers. com.). Moreover, in two other jellyfish lakes in Palau, *Mastigias* sp. migrates in quite different compass directions during the day, west to east in one lake and north to south in the second (Hamner & Hauri, 1981). Consequently, we do not yet fully understand the environmental cues used by *Mastigias* sp. for horizontal migration.

The 'sun-compass' hypothesis for oriented swimming is supported by a large population of *Aurelia aurita* (or *A. labiata*) in Saanich Inlet, British Columbia (Hamner et al., 1994). When the sun is shining, *Aurelia* sp. exhibits southeasterly migration during the morning, however, when the sun is obscured either by clouds or by shadows cast by the eastern ridge of the fjord, the population becomes randomly oriented. Likewise, the population is also randomly oriented at night. In the absence of solar cues, the population is dispersed by both randomly oriented swimming and tidal currents, but in the presence of direct sun, concentrated aggregations of medusae form along the southeastern shore of the fjord. This type of navigation also requires time-compensated celestial navigation ability because the medusae swim toward the southeast all day long, irrespective of the position of the sun in the sky. Directed swimming of the population toward the southeast during the day increases population retention in the fjord and locally also greatly increases the population density along the south east side of the fjord. This results in a high density of sexually mature medusae, facilitating fertilization by minimizing gamete dilution. Two of us (WMG and WMH) have observed similar patterns for *A. aurita* along the eastern Monterey Bay during a recent investigation. While medusae can clearly navigate, the neurological mechanisms mediating the behavior are unknown. More specifically, no one yet understands how a rotating, radially symmetrical medusa can orient to a point source of light.

*Vertical layers associated with temperature, salinity and density discontinuities*

Zooplankton aggregations are quite often most concentrated around sharp density discontinuities (Owen, 1989). Earliest evidence of this distributional pattern are from SCUBA observations. Gradients of temperature and salinity (and hence density) are especially pronounced in fjords, estuaries, and the coastal transition zone. Therefore it is not surprising that gelatinous zooplankton layers are most pronounced in these physical regimes as well. Layering of jellies along gradients and discontinuities can be attributed to both active behavioral responses and passive accumulation.

Most of the experimental work to date has been on the interaction of gelatinous zooplankton with haloclines since, logistically at least, artificially constructed layers of temperature are far more difficult to construct than salinity layers (Harder, 1968). However, Arai (1976) did make a direct comparison between temperature and salinity stratification. With few exceptions, for instance some upwelling centers, vertical structure of coastal water columns is often established by salt content rather than temperature. While temperature contributes to water column stabilization during periods of surface warming, rapid cooling during the night-time or during cold front passage also leads to water column destabilization.

As jellyfish approach sharp haloclines, two situations develop that can independently lead to localized accumulation along the discontinuity. The first is passive accumulation. Since jellies are mostly water of the same ionic concentration of the surrounding seawater, they tend to remain along isohalines. The medusae also may exhibit active behavioral or physiological responses to the sudden osmotic stresses encountered at sharp haloclines, which may cause jellyfish to either slow their swimming while salts are being adjusted in tissues, or perhaps the animals simply turn around to remain in a specific range of physiological tolerance.

Much of our understanding of jellyfish around haloclines derives from a series of simple artificial water column experiments conducted in the 1960s and 1970s. The first experiments by Harder (1968) tested the general tendency of zooplankton to accumulate at discontinuities using artificially created haloclines and thermoclines in 2 l graduated cylinders. The breadth of taxa used by Harder (1968) included non-gelatinous zooplankton (e.g. copepods, mysids, barnacle larvae, veliger larvae and fish eggs), as well as the ctenophore, *Pleurobrachia pileus* O. F. Müller. Most test animals, including *P. pileus,* accumulated at an artificial halocline that was only a few millimeters thick and represented a <3 ppt gradient. Harder concluded that accumulation of zooplankton at the artificial halocline was entirely passive.

Arai (1973) expanded on Harder's (1968) single observation of *P. pileus* by stepping the 2–3 ppt salinity gradient over a broader range of absolute salinities. She found that swimming behaviors accounted for changes in position within the cylinders with respect to salinity discontinuity. In higher salinity layers (>25 ppt), animals tended to accumulate higher in the column, while in lower salinity layers (~20 ppt), animals accumulated at the bottom. A similar reaction was observed for the hydromedusa *Sarsia tubulosa* M. Sars but was less clear for the hydromedusa *Phialidium gregarium* L. Agassiz (= *Clytia gregaria*).

Jellyfish are osmoconformers, and they maintain an ionic concentration of tissue and mesoglea that reflects the surrounding seawater medium (reviewed by Arai, 1997). Gelatinous zooplankton may actively pump sulfate ions to modify density as a form of buoyancy regulation in stratified water columns (Bidigare & Biggs, 1980). However, a simple experiment by Mills & Vogt (1984) showed sulfate ion regulation was not responsible for short-term density changes and that vertical migration in medusae and ctenophores was most likely accomplished by swimming. As such, small jellyfish such as hydromedusae and ctenophores attempting to cross a sharp halocline will be constrained within the halocline until the relatively slow process of osmosis across cell membranes has occurred (Mills, 1984). Recent experimental work by Wright & Purcell (1997) indicates that regulation of specific cations such as potassium, magnesium and sulfate is important in the estuarine jellyfish *Chrysaora quinquecirrha*, but that sulfate equilibration time for *C. quinquecirrha* was on the order of 40 h, too long to be of value for osmotic adjustments when medusae cross sharp haloclines.

All of the experimental evidence for passive accumulation of jellyfish along haloclines has been for small hydromedusae and ctenophores. Similar experimental studies have not been performed on the larger, faster swimming scyphomedusae. However, *in situ* observations from profiling video of large scyphomedusae and from SCUBA shows that larger medusae exhibit little difficulty in transiting sharp haloclines or thermoclines (WMG unpublished observations). Observations made on large (35–50 cm bell diameter) *Phyllorhiza punctata* von Lendenfeld in the north-

ern Gulf of Mexico showed that these scyphomedusae made extensive vertical excursions across steep haloclines (gradients of >10 ppt) without slowing as they approached the salinity discontinuity. Therefore, larger medusae that accumulate along a salinity discontinuity may be behaviorally motivated.

Interactions of jellyfish with thermoclines are less clear because an experimental approach with artificial water columns is difficult (e.g. Harder, 1968) and because the *in situ* effect of thermocline and halocline cannot be easily separated because the density effects of thermoclines and haloclines invariably interact. Arai (1976) showed that *Pleurobrachia pileus* and *Sarsia tubulosa* had a modest affinity to the thermal discontinuity that was similar to the effect to salinity. Her results indicated that a combined effect of thermal and salinity stratification led to greater concentration than experienced by either factor alone.

A drawback of the experimental studies described here is that they were all performed under artificial laboratory circumstances of reduced scale and an absence of natural halocline-related turbulence. While salt gradients are often very steep in estuaries and fjords, natural settings will always have a higher degree of shear creating turbulent mixing at discontinuities. Thus unmixed laboratory water columns tend to create gradients over millimeters to a few centimeters, whereas in nature these gradients tend to exist from centimeters to meters. Scaling is very important since the animals in question range from millimeters to almost meters in length. Future experiments in the laboratory will need to recreate realistic scales of salinity and temperature stratification.

Field evidence of thermocline effects is difficult to evaluate since there are few places where thermoclines exist in the absence of haloclines. Graham (1994) showed layering of a large scyphomedusa, *Chrysaora fuscescens* Brandt, along the thermocline in the Monterey Bay upwelling system. Solar heating controls structure of the water column in this and other upwelling regions, and the absence of a sharp halocline suggests that the thermocline is important in determining layering of this species. By contrast, *Chrysaora hysoscella* (L.) swims unimpeded through steep thermal gradients (Pagès & Gili, 1992). Pagès & Gili (1991) reported that upward migrations of some siphonophore species were restricted by a thermocline that created a density barrier. They also showed evidence that *Chelophyes appendiculata* Eschscholtz can pass easily through steep thermal gradients. In the subarctic Pacific, the 30 m thermocline (and halocline) serves as a lower boundary to the daytime depth distribution of the salp, *Cyclosalpa bakeri* (Purcell & Madin, 1991). However, long-distance migrations of *Salpa aspera* are unaffected by the presence of a thermocline (Wiebe et al., 1979).

One physical characteristic that is difficult to evaluate experimentally is current shear. Current shear occurs at the air–sea interface, the benthic-boundary layer and within strong density discontinuities. Though there are a number of examples where gelatinous animals populate boundary-layers, mechanisms that function to perceive shear are very limited. In the western Mediterranean, the population of the mesopelagic appendicularian, *Oikopleura villafrancae* Fenaux, is associated with a persistent convergent flow that produces localized concentrations of food particles along the strongly stratified isopycnals (Gorsky et al., 1991). The ctenophore *Pleurobrachia pileus* also has been observed to vertically migrate in the Seine estuary in response to currents (Wang et al., 1995). A time-series of *P. pileus* in relation to the semi-diurnal tidal cycle showed that ctenophores generally had a hyperbenthic distribution, but the distribution extended to the surface in phase with tidal cycle during flood tides. This migration pattern is believed to contribute to estuarine retention of the *P. pileus* population (Wang et al., 1995). Strong horizontal shear within a surface front (Graham & Largier, 1997) may be important in maintaining aggregations of *Chrysaora fuscescens* in Monterey Bay (Graham, 1994).

*Other cues: hydrostatic pressure and turbulence*

Hydrostatic pressure, more than any other physical characteristic of the sea including light attenuation, potentially provides the best indication of depth to planktonic organisms. In a survey of hydrostatic pressure effects on swimming orientation, Rice (1964) illustrates that gelatinous zooplankton tend to respond to pressure variations using gravity as the proximal directional cue (with the exception of anthomedusae which lack a statocyst). A variety of threshold behavioral responses to pressure (termed barokinesis) for gelatinous zooplankton are discussed by Knight-Jones & Morgan (1966). Some of these response thresholds are remarkably small. For instance, the hydromedusae, *Gossea corynetes* Gosse and *Clytia hemispherica,* increase pulse rates following pressure increases of only 3–5 decibars. The physonect siphonophre, *Nanomia bijuga,* a major component of the

DVM community, is similarly responsive to pressure change (Jacobs, 1937). Among the Ctenophora, *Pleurobrachia pileus* and *Mnemiopsis* sp. are responsive over this same range of pressure change. Knight-Jones & Morgan (1966) argue that this consistent range of response threshold may play an important role in the feeding process as many gelatinous animals rely on activity (swimming) followed by inactivity while performing tentaculate fishing. Sensitivity in this manner to pressure changes would allow for maintenance of a layered aggregation of animals that still require a degree of vertical movement to feed. Unfortunately, pressure receptors for most of these taxa have not been identified.

Turbulence might be sensed as pressure waves detected across the body of an animal (Knight-Jones & Morgan, 1966). Many gelatinous animals are susceptible to tissue damage from excessive turbulence near the sea surface or close to shore where waves and bottom currents create intense shear stresses that could easily damage them. The abilities of gelatinous animals to sense and, accordingly, move away from these areas can contribute to observed layering and migrations of gelatinous zooplankton (e.g. Kopacz, 1994). In the Pamlico River estuary, wind and turbulence strongly affected the distribution of the ctenophore *Mnemiopsis leidyi* A. Agassiz (in Miller, 1974). Ctenophores remained at the surface until encountering turbulence that caused them to sound. They were then transported toward the shore in a bottom countercurrent until turbulence diminished and they returned to the surface. This behavior tended to keep the ctenophores off the beach and out of rough water, unfavorable conditions for weakly swimming, soft bodied organisms. Shanks & Graham (1987) described a similar turbulence avoidance behavior for the rhizostome medusa, *Stomolophus meleagris* L. Agassiz, in coastal waters of North Carolina.

**Accumulation at the shoreline**

The formation and maintenance of gelatinous zooplankton aggregations often depends on impedance of swimming by physical barriers. In the next two sections, we review the literature that invokes physical barriers as mechanisms of gelatinous zooplankton aggregation. For this purpose, physical barriers include both geological (i.e. shoreline and seafloor) and hydrographic (i.e. horizontal fronts and vertical discontinuities) and therefore do not imply that the barrier is impermeable to exchange across the barrier. In the following discussion, it is important for the reader to draw upon the earlier examples of behaviorally mediated swimming because it is the inherent interaction of behavior with physical barriers that creates aggregation of gelatinous zooplankton.

Dense aggregations of jellyfish frequently occur in the surface waters along coastal margins. Accumulations appear to be strongly influenced by the direction and energy of prevailing winds and surface currents such that populations of jellyfish are compressed along the shoreline (e.g. Shenker, 1984; Larson, 1990). Pleustonic species, such as *Physalia physalis* (L.) and *Velella velella* (L.), are routinely blown onto shore by sustained onshore winds as 'mass strandings'(Kennedy, 1972; Evans, 1986). Mass strandings of pleustonic jellies are typically seasonal and dependent on wind direction and duration. Interestingly, the origin of these populations is often unknown (Bieri, 1977; Shannon & Chapman, 1983).

Wind alone often does not explain formation and maintenance of jellyfish aggregations along shorelines. In many cases, local hydrology interacts with topography or coastal prominences to produce retentive features that support entrainment of aggregations. It is often the specialized circulation along the coastline that interacts with swimming behavior to create or maintain aggregations. One such example is the entrainment of upwelled water by a coastal prominences in northern Monterey Bay, California. This feature, described by Graham et al. (1992) as 'upwelling shadows' is a site of a semi-persistent aggregation of *Chrysaora fuscescens* (Graham, 1994; Lenarz et al., 1995).

Another example of shoreline interaction on jellyfish blooms is in the Mediterranean Sea where blooms of the scyphomedusa *Pelagia noctiluca* (Forskål) occur over approximately decadal cycles (Goy et al., 1989). These blooms have been particularly problematic during the past twenty years (Malej, 1989). Dense subsurface swarms of *P. noctiluca* up to 20 m thick and extending for several kilometers along the shore (Malej, 1989) show marked behavioral patterns that facilitate aggregation under varying wind and current regimes. In calm waters, individuals within a subsurface aggregation swim actively but are oriented randomly (Zavodnik, 1987). However, when a coastal geostrophic current exists, the swimming direction of individuals within the aggregation becomes uniform and directed with current. As we have already indicated, we know little about the mechanistic nature of

how medusae can 'sense' a current. Perhaps medusae sense current shear and not the current itself. After several days of sustained wind, large aggregations of jellyfish form along the mainland and island shores. Tidal currents are also important in concentrating *P. noctiluca* along the shore. During the flood tide, hundreds of medusae $m^{-2}$ accumulate near the sea surface along the coast. On the ebb tide, these medusae leave the surface layer, and they are then driven into deeper layers away from shore (Zavodnik, 1987).

## Accumulations at surface convergences

### Fronts

Ocean fronts are a class of circulation that develops along the interface between water bodies of different origin. As such, fronts are ubiquitous in the World Ocean and exist across a number of scales from the Antarctic Polar Frontal Zone at tens of thousands of kilometers to Langmuir circulation at tens of meters. Because of the breadth in scale, fronts are considered one of the principle mechanisms for the re-distribution of biological patterns and processes in the sea (Owen, 1981; Mackas et al., 1985). Fronts are also quasi-ordered phenomena and tend to follow a set relationship between spatial and temporal scale (Mackas et al., 1985), i.e. larger features tend to be temporally persistent (but see Graham, 1993). Characteristic circulation of ocean fronts, in both the horizontal and vertical dimensions, manifest as convergence, divergence and shear. While divergent flow tends to disperse gelatinous zooplankton, convergent flow at fronts is often implicated in the formation and maintenance of gelatinous zooplankton aggregations. Current shear is a potential mechanism for orientation and thus may be an important, albeit poorly understood, mechanism for gelatinous zooplankton aggregation as well. Since the research on current shear perception by jellies is still in its infancy, our overview of surface fronts emphasizes convergent features.

Accumulation of gelatinous zooplankton within surface fronts is probably the most commonly reported type of jellyfish 'patchiness' in the sea. Yet, there are surprisingly few published examples describing the fine-scale mechanistic relationships between gelatinous zooplankton and fronts. Moreover, there is the common misperception that accumulation of gelatinous animals, especially large medusae, is entirely passive. This misperception is almost certainly based on the high water content of gelatinous animals, which would appear to make them either weak swimmers or passive tracers. In fact, many large medusae are quite capable of swimming speeds that approach or exceed vertical (plunging) flow velocities within convergent fronts.

While the term 'front' is often applied generically to describe the interface between water bodies, this is a gross over-simplification for contemporary research on the dynamics of frontal systems (Federov, 1983). Conceptually, fronts have two characteristics relevant to biological systems. These are maximal horizontal gradients (Federov, 1983) and laterally convergent flow at the surface with induced vertical flow below (Owen, 1981). In this sense, we discuss the role of fronts in forming and maintaining gelatinous zooplankton aggregations without describing the abundance and complexities of the various types of fronts (e.g. upwelling, estuarine, tidal, shelf-break, etc.) since all of these fronts share characteristics of horizontal gradients and three-dimensional flow. However, we do differentiate fronts by scale as suggested by Federov (1983) into (i) large-scale, quasi-stationary, (ii) meso-scale, and (iii) small-scale, local origin. We will also discuss a fourth category of surface feature that includes even smaller convergent structures such as Langmuir cells and internal waves.

Ocean fronts serve as regions of intense trophic activity. Phytoplankton and small zooplankton grazers usually have generation times appropriately scaled to express increased production in many stable frontal regions (e.g. Pingree et al., 1977; Yamamoto & Nishizawa, 1986; Wolanksi & Hamner, 1988; Franks, 1992). Within the gelatinous zooplankton, perhaps only pelagic tunicates (Alldredge, 1982; Deibel, 1985; Purcell & Madin, 1991), ctenophores (Kremer, 1994; Sullivan et al., 2001) and siphonophores (Pagès et al., 2001) also have reproductive life-history characteristics that allow population responses to increased production within fronts. However, both hydromedusae and scyphomedusae have relatively long generation times when accounting for the benthic polyp stages. It is unlikely that aggregation of large medusae within fronts reflects locally increased reproduction, but rather an accumulation of animals by an interaction between the front's circulation and the animal's swimming behavior (or relative density).

There are relatively few data suggesting that the large, quasi-stationary oceanic frontal regions physically aggregate gelatinous zooplankton. Perhaps this is due to a paucity of large-scale cross-frontal surveys, or

perhaps it indicates that smaller scale, local circulation is more coherent with jellyfish aggregations. Mackas et al. (1985) suggested that ecological processes like behavioral aggregation are matched to physical processes that occur at scales of a few kilometers and less. Therefore, we would expect the largest scale fronts to reflect population and community level increases in abundance due to increased reproduction and population growth. Pagès et al. (1996) and Pagès & Schnack-Schiel (1996) have conducted large-scale frontal surveys in the vicinity of the Antarctic Polar Frontal Zone and Antarctic Slope Front, respectively. They indicate that these zones serve as boundaries for entire communities of gelatinous animals, but the evidence for physical aggregation in this large front is weak. Young et al. (1996) found no increase in biomass of gelatinous zooplankton in the sub-tropical convergence between East Australian Current water and sub-Antarctic water.

Meso-scale frontal regions such as shelf/slope fronts and meso-scale eddies are associated more with population-level changes of organisms than with physical-behavioral aggregation (Mackas et al., 1985). However, secondary circulation that develops along the boundaries of these systems may be quite important in creating local conditions that facilitate aggregation of gelatinous animals. A divergent shelf/slope front along the Catalan coast is a region of increased secondary production (Sabatés et al., 1989), however, only salps appear to be concentrated in this particular front. This indicates that salps are accumulating through population increases (i.e. blooms) rather than physical accumulation, which is unexpected in a divergent flow. Pagès & Gili (1992) reported substantially increased abundance of siphonophores and medusae in an upwelling front within the Agulhas Current system.

The literature is replete with examples of small fronts with local origin that serve as aggregation centers for gelatinous zooplankton. Coyle & Cooney (1993) conducted an acoustic study of zooplankton abundances around hydrographic fronts in the vicinity of the Pribilof Islands, Bering Sea. Fronts in this region exhibit both strong salinity and temperature gradients and contribute greatly to the overall variability of biological production in the southeast Bering Sea (Coachman, 1986). In the study by Coyle & Cooney (1993), sound-scattering around surface frontal features was dominated by cnidarian assemblages. However, the authors attributed this increased abundance to the elevated chlorophyll, trophic transfer in the region and numerical increase due to reproduction. We suggest that medusae were probably physically accumulated in fronts, but other gelatinous (and non-crustacean) animal populations, such as chaetognaths and larvaceans, may have experienced population growth. Brodeur et al. (1997) also found that populations of large medusae are delineated by numerous tidal fronts in the eastern Bering Sea.

Local circulation patterns found within sounds, bays and estuaries also contribute to physical aggregation of gelatinous animals. In Prince William Sound, Alaska, Purcell et al. (2000) observed aggregations of up to millions of *Aurelia labiata*. Vertical swimming behaviors, up and down, likely promoted concentration within numerous convergences in this dynamic coastal region. Small swarms (20 m long, 5 m deep) of *Liriope tetraphylla* Chamisso & Eysenhardt in concentrations up to 3000 m$^{-2}$ were reported in Hiroshima Bay within an estuarine front (Ueno & Mitsutani, 1994). Although the front's role could not be ascertained, they concluded that individuals within the swarm were reproducing. A similar swarm was described in the Río de la Plata estuary, Argentina, by Mianzan et al. (In press). In Tokyo Bay, *Aurelia aurita* aggregate at the innermost part near the estuarine frontal region, especially at the edge of the low salinity water mass (Toyokawa et al., 1997). Aggregations of gelatinous predators have been described from river-plume fronts in the Chesapeake Bay (MacGregor & Houde, 1996). In Monterey Bay, California, a frontal region between warm nearshore water and colder offshore water is a consistent location for large aggregations of *Chrysaora fuscescens* (Graham, 1994). Oriented swimming in Monterey Bay populations of *C. fuscescens* is described by Graham (1994), and this behavior likely contributes to their nearshore concentration. However, the cues for *C. fuscescens* swimming are not completely understood.

Temporary aggregations of gelatinous zooplankton often develop in small linear surface convergences during Langmuir circulation. Langmuir circulation cells typically develop under sustained winds that exceed about 2 m s$^{-1}$ (but break down at wind speeds exceeding about 10 m s$^{-1}$). Under these conditions, surface convection cells form along the axis of the wind, creating alternating patterns of convergence and divergence. Within convergences, trapping of buoyant material is nearly 100% (Owen, 1981). Pleustonic animals like the Portuguese Man-O'-War, (*Physalia physalis*) are effectively entrained by Langmuir circulation cells (Woodcock, 1944). Upward-swimming epiplanktonic animals are also entrained within Lang-

muir cells. Alldredge (1982) observed long, parallel rows of appendicularians, *Oikopleura longicauda* Vogt, that were created by Langmuir circulation. High concentrations of appendicularians (up to 3565 ind l$^{-1}$) in these aggregations were believed to be spawning. A separate mechanism was proposed to explain the formation of rows of *O. longicauda* in surface waters by Owen (1966) who observed dense reddish concentrations of this appendicularian in calm weather. He suggested that under calm wind, small thermohaline circulation cells called Bénard cells (Owen, 1981) develop surface slicks that can also effectively trap organisms. Owen (1981) suggests that under a freshening wind, Langmuir circulation cells may develop from Bénard circulation cells.

There are also a number of examples of medusae being concentrated within Langmuir cells. In the Bering Sea, both hydromedusae and scyphomedusae form dense aggregations at the surface in summer (Hamner & Schneider, 1986). On windy nights, medusae reached concentrations of 1000 ind m$^{-3}$ in regularly spaced, linear rows that ran parallel to the wind. In the Caribbean Sea, the small scyphomedusan, *Linuche unguiculata* Schwartz, formed dense, elongated patches with the elongated axis being parallel to the wind (Larson, 1992). These medusae were apparently maintained in these convergences by upward swimming. Patch-maintenance behavior of *L. unguiculata* appears to be a form of reproductive swarming (Larson, 1992). Shanks & Graham (1987) also observed dense aggregations of the scyphomedusa, *Stomolophus meleagris*, in Langmuir circulation cells. Development of *S. meleagris* aggregations was apparently facilitated by oriented swimming of medusae into or against wind and /or wave direction. Kingsford et al. (1991) reported concentrations of *Aurelia aurita* up to 29 ind m$^{-3}$ in Langmuir slicks in an Australian coral reef lagoon.

In addition to the often-cited benefit of spawning within locally concentrated aggregations, there are a number of other additional advantages for gelatinous zooplankton. Purcell et al. (2000) suggest that, in addition to increased fertilization, accumulation of medusae within nearshore fronts may aid in the retention of aggregations close to shore and near hard substrate required by benthic polyps. In addition to retention, locally increased zooplankton prey populations are also distributed within these features (e.g. Graham, 1994; Purcell et al., 2000). Purcell et al. (2000) also noted that large aggregations of jellies might be an effective defense against predation, especially other gelatinous predators. We offer a final advantage that highly concentrated aggregations of jellyfish may serve to exclude other potential competitors such as zooplanktivorous fishes and possibly other gelatinous species. Such exclusion of competitors might be facilitated by exudation of chemical 'scents' or unfired nematocysts (Shanks & Graham, 1988).

## Conclusions and future directions for research

In this review, we have suggested that the highly aggregated dispersion of gelatinous zooplankton populations may, in part, be attributed to the distribution of physical processes and gradients in the sea. In many instances, local perception of large numbers of jellyfish may be misinterpreted as a real population increase, when in fact it is the local re-distribution of a population by some physical and/or behavioral mechanism. Moreover, long-term population changes may be important on an ecosystem scale, but local aggregations mediated by physical processes may be vastly more important on short-term ecological scales.

In many instances, locally enhanced concentrations of ctenophores, siphonophores and pelagic tunicates within fronts may be due to a rapid population increase. However, the presence of large medusae at physical discontinuities such as fronts and thermo- and haloclines indicates that physical accumulation has occurred because population increases of these large animals is almost always decoupled from water column processes. The physical accumulation of medusae may be linked to a variety of behavioral swimming responses to environmental cues, many of which are physical cues such as light, gravity, temperature, salinity (and density), current shear, pressure and turbulence. In some cases, weaker swimming hydromedusae and ctenophores may be passively accumulated in haloclines due to lagged adjustment of tissue salts.

Oceanographers have understood the distribution and dynamics of physical gradients and discontinuities for at least a hundred years longer than they have understood how these gradients influence marine organisms. Yet, most of the research we have reviewed here is largely descriptive and qualitative, and the few experimental studies that have been conducted were performed at unrealistically small scales (i.e., in small cylinders and aquaria). We suspect that inherent difficulties of working with gelatinous animals contribute to the paucity of information. With the continued de-

velopment of novel *in situ* optical and acoustical techniques (e.g., Davis et al., 1992; Monger et al., 1998), we will gain a much more thorough understanding of the relationship between individuals, their behaviors and their physical environment. These developing systems are directly applicable to fragile gelatinous zooplankton distributions and can be integrated with a suite of sensors to characterize fully the physical environment. However, these tools will continue to provide mostly qualitative descriptions unless biological oceanographers work more closely with physical oceanographers and hydrodynamicists. The future of research in this area depends on cross-disciplinary co-operation in order to apply proper theoretical aspects of the physical environment to the biology of these important animals.

## Acknowledgements

We thank J. Purcell and two anonymous reviewers for their helpful comments on an earlier draft of this paper.

## References

Alldredge, A. L., 1982. Aggregation of spawning appendicularians in surface windrows. Bull. mar. Sci. 32: 250–254.

Anderson, P. A. V. & G. O. Mackie, 1977. Electrically coupled, photosensitive neurons control swimming in a jellyfish. Science 197: 186–188.

Arai, M. N., 1973. Behavior of planktonic coelenterates, *Sarsia tubulosa*, *Philiadium gregarium* and *Pleurobrachia pileus* in salinity discontinuity layers. J. Fish. Res. Bd Can. 30: 1105–1110.

Arai, M. N., 1976. Behavior of planktonic coelenterates in temperature and salinity discontinuity layers. In Mackie, G.O. (ed.), Coelenterate Ecology and Behavior, Plenum Press, New York: 211–217.

Arai, M. N., 1992. Active and passive factors affecting aggregations of hydromedusae: a review. In Bouillon, J., F. Boero, F. Cicogna, J. M. Gili & R. G. Hughes (eds), Aspects of Hydrozoan Biology, Sci. Mar. 56: 99–108.

Arai, M. N., 1997. A Functional Biology of Scyphozoa. Chapman & Hall, New York: 316 pp.

Arkett, S. A., 1985. The shadow response of a hydromedusan (*Polyorchis penicillatus*): behavioral mechanisms controlling diel and ontogenetic vertical migration. Biol. Bull. 169: 297–312.

Backus, R. H., R. C. Clark & A. S. Wing, 1965. Behaviour of certain marine organisms during the solar eclipse of July 20, 1963. Nature 205: 989–991.

Bailey, K. M. & R. S. Batty, 1983. A laboratory study of predation by *Aurelia aurita* on larval herring (*Clupea harengus*): experimental observations compared with model predictions. Mar. Biol. 72: 295–301.

Barham, E. G., 1963. Siphonophores and the deep scattering layer. Science 140: 826–828.

Bayly, I. A. E., 1986. Aspects of diel vertical migration in zooplankton, and its enigma variations. In Deckker, P. D. & W.

D. Williams (eds), Limnology in Australia, CSIRO, Melbourne: 349–368.

Bidigare, R. R. & D. C. Biggs, 1980. The role of sulfate exclusion in buoyancy maintenance by siphonophores and other oceanic gelatinous zooplankton. Comp. Biochem. Physiol. 66A: 467–471.

Bieri, R., 1977. The ecological significance of seasonal occurrence and growth rate of *Velella* (Hydrozoa). Publ. Seto Mar. Biol. Lab. 24: 63–76.

Brodeur, R. D., 1998. *In situ* observations of the association between juvenile fishes and scyphomedusae in the Bering Sea. Mar. Ecol. Prog. Ser. 163: 11–20.

Brodeur, R. D., M. T. Wilson & J. M. Napp, 1997. Distribution of juvenile Pollock relative to frontal structure near the Pribilof Islands, Bering Sea. In Forage Fishes in Marine Ecosystems, American Fisheries Society, Lowell Wakefield Fisheries Symposium Series, no. 14: 573–589.

Coachman, L. K., 1986. Circulation, water mass, and fluxes on the southeastern Bering Sea shelf. Cont. Shelf Res. 5: 23–108.

Coyle, K. O. & R. T. Cooney, 1993. Water column sound scattering and hydrography around the Pribilof Islands, Bering Sea. Cont. Shelf Res. 13: 803–827.

Davis, C. S., S. M. Gallager & A. R. Solow, 1992. Microaggregations of oceanic plankton observed by towed video microscopy. Science 257: 230–232.

Dawson, M. N, L. E. Martin & L. K. Penland, 2001. Jellyfish swarms, tourists and the Christ-child. Hydrobiologia 451 (Dev. Hydrobiol. 155): 131–144.

Deibel, D., 1985. Blooms of the pelagic tunicate *Dolioletta gegenbauri*: are they associated with Gulf Stream frontal eddies? J. mar Res. 43: 211–26.

Enright, J. T. & W. M. Hamner. 1967. Vertical diurnal migration and endogenous rhythmicity. Science 157: 137–141.

Evans, F., 1986. *Velella velella* (L), the 'by-the-wind-sailor' in the North Pacific Ocean in 1985. Mar. Obs. 56: 196–200.

Federov, K. N., 1983. The physical nature and structure of oceanic fronts. Lecture Notes on Coastal and Estuarine Studies 19: 333 pp.

Frank, T. M. & E. A. Widder, 1997. The correlation of downwelling irradiance and staggered vertical migration patterns of zooplankton in Wilkinson Basin, Gulf of Maine. J. Plankton Res. 19: 1975–1991.

Franks, P. J. S., 1992. Phytoplankton blooms at fronts: patterns, scales and physical forcing mechanisms. Rev. aquat. Sci. 6: 121–137.

Franqueville, C., 1971. Macroplancton profound (invertébrés) de la Méditerranée Nord-occidentale. Tethys 3: 11–56.

Gorsky, G., N. Lins da Silva, S. Dallot, Ph. Laval, J. C. Braconnot & L. Prieur, 1991. Midwater tunicates: are they related to the permanent front of the Ligurian Sea (NW Mediterranean)? Mar. Ecol. Prog. Ser. 74: 195–204.

Goy, J., P. Morand & M. Etienne, 1989. Long-term fluctuations of *Pelagia noctiluca* (Cnidaria, Scyphomedusa) in the western Mediterranean Sea. Prediction by climatic variables. Deep-Sea Res. 36: 269–279.

Graham, W. M., 1993. Spatio-temporal scale assessment of an 'upwelling shadow' in northern Monterey Bay, California. Estuaries 16: 83–91.

Graham, W. M., 1994. The physical oceanography and ecology of upwelling shadows. Doctoral dissertation, Department of Biology, University of California, Santa Cruz, 204 pp.

Graham, W. M., J. G. Field & D. C. Potts, 1992. Persistent "upwelling shadows" and their influence on zooplankton distributions. Mar. Biol. 114: 561–570.

Graham, W. M. & J. L. Largier, 1997. Upwelling shadows as nearshore retention sites: the example of northern Monterey Bay. Cont. Shelf Res. 17: 509–532.

Hamner, W. M., 1985. The importance of ethology for investigations of marine zooplankton. Bull. mar. Sci. 37: 414–424.

Hamner, W. M., 1995. Sensory ecology of scyphomedusae. Mar. Fresh. Behav. Physiol. 26: 101–118.

Hamner, W. M., L. P. Madin, A. L. Alldredge, R. W. Gilmer & P. P. Hamner, 1975. Underwater observations of gelatinous zooplankton: sampling problems, feeding biology and behavior. Limnol. Oceanogr. 20: 907–917.

Hamner, W. M. & I. R. Hauri, 1981. Long-distance horizontal migrations of zooplankton (scyphomedusae: *Mastigias*). Limnol. Oceanogr. 26: 414–423.

Hamner, W. M., R. W. Gilmer & P. P. Hamner, 1982. The physical, chemical, and biological characteristics of a stratified, saline, sulfide lake in Palau. Limnol. Oceanogr. 27: 896–909.

Hamner, W. M. & D. Schneider, 1986. Regularly spaced rows of medusae in the Bering Sea: role of Langmuir circulation. Limnol. Oceanogr. 31: 171–177.

Hamner, W. M., P. P. Hamner & S. W. Strand, 1994. Sun-compass migration by *Aurelia aurita* (Scyphozoa): population retention and reproduction in Saanich Inlet, British Columbia. Mar. Biol. 119: 347–356.

Harder, W., 1968. Reactions of plankton organisms to water stratification. Limnol. Oceanogr. 13: 156–168.

Jacobs, W., 1937. Beobachtungen über das Schwebender der Siphonophoren. Z. vergleich. Physiol. 24: 583–601.

Kennedy, F. S. Jr., 1972. Distribution and abundance of *Physalia* in Florida waters. Florida Department of Natural Resources, Mar. Res. Lab., Prof. Pap. Ser. 18: 1–38.

Kingsford, M. J., E. Wolanski & J. H. Choat, 1991. Influence of tidally induced fronts and Langmuir circulations on distribution and movements of presettlement fishes around a coral reef. Mar. Biol. 109: 167–180.

Knight-Jones, E. W. & E. Morgan, 1966. Responses of marine animals to changes in hydrostatic pressure. Oceanogr. mar. Biol. ann. Rev. 4: 267–299.

Kopacz, U., 1994. Evidence for tidally-induced vertical migration of some gelatinous zooplankton in the Wadden Sea area near Split. Helgol. Wiss Meeresunters. 48: 333–342.

Kremer, P., 1994. Patterns of abundance for *Mnemiopsis* in U.S. coastal waters: a comparative overview. ICES J. mar. Sci. 51: 347–354.

Larson, R. J., 1990. Scyphomedusae and cubomedusae from the eastern Pacific. Bull. mar. Sci. 47: 546–556.

Larson, R. J., 1992. Riding Langmuir circulations and swimming in circles: a novel form of clustering behavior by the scyphomedusa *Linuche unguiculata*. mar. Biol. 112: 229–235.

Larson, R. J., G. I. Matsumoto, L. P. Madin & L. M. Lewis, 1992. Deep-sea benthic and benthopelagic medusae, recent observations from submersibles and a remotely operated vehicle. Bull. mar. Sci. 51: 277–286.

Lenarz, W. H., D. A. VenTresca, W. M. Graham, F. B. Schwing & F. Chavez, 1995. Explorations of El Niño events and associated biological population dynamics off central California. CalCOFI Rep. 36: 106–119.

MacGregor, J. M. & E. D. Houde, 1996. Onshore-offshore pattern and variability in distribution and abundance of bay anchovy *Anchoa mitchilli* eggs and larvae in Chesapeake Bay. mar. Ecol. Prog. Ser. 138: 15–25.

Mackas, D. L., K. L. Denman & M. R. Abbot, 1985. Plankton patchiness: biology in the physical vernacular. Bull. mar. Sci. 37: 652–674.

Mackie, G. O., R. J. Larson, K. S. Larson & L. M. Passano, 1981. Swimming and vertical migration of *Aurelia aurita* (L.) in a deep tank. Mar. Behav. Physiol. 7: 321–329.

Malej, A., 1989. Behavior and trophic ecology of the jellyfish *Pelagia noctiluca* (Forsskål, 1775). J. exp. mar. Biol. Ecol. 126: 259–270.

McCloskey, L. R., L. Muscatine & F. P. Wilkerson, 1994. Daily photosynthesis, respiration, and carbon budgets in a tropical marine jellyfish (*Mastigias* sp.). Mar. Biol. 19: 13–22.

Mianzan, H., D. Sorarraín, J. Burnett & L. Lutz, In press. Mucocutaneous junctional and lexural paresthesias caused by the holoplanktonic trachymedusae *Liriope tetraphylla*. Dermatology.

Miller, R. J., 1974. Distribution and biomass of an estuarine ctenophore population, *Mnemiopsis leidyi* (A. Agassiz). Chesapeake Sci. 15: 1–8.

Mills, C. E., 1981. Diversity of swimming behaviors in hydromedusae as related to feeding and utilization of space. Mar. Biol. 64: 185–189.

Mills, C. E., 1984. Density is altered in hydromedusae and ctenophores in response to changes in salinity. Biol. Bull. 166: 206–215.

Mills, C. E. & R. G. Vogt, 1984. Evidence that ion regulation in hydromedusae and ctenophores does not facilitate vertical migration. Biol. Bull. 166: 216–227.

Mills, C. E. & J. Goy, 1988. In situ observations of the behavior of mesopelagic *Solmissus* narcomedusae (Cnidaria, Hydrozoa). Bull. mar. Sci. 43: 739–751.

Mills, C. E., P. R. Pugh, G. R. Harbison & S. H. D. Haddock, 1996. Medusae, siphonophores and ctenophores of the Alboran Sea, southwestern Mediterranean. Sci. Mar. 60: 145–163.

Monger, B. C., S. Chinniah-Chandy, E. Meir, S. Billings, C. H. Greene & P. H. Wiebe, 1998. Sound scattering by the gelatinous zooplankters *Aequorea victoria* and *Pleurobrachia bachei*. Deep-Sea Res. 45: 1255–1271.

Muscatine, L. & R. E. Marian, 1982. Dissolved inorganic nitrogen flux in symbiotic and nonsymbiotic medusae. Limnol. Oceanogr. 27: 910–917.

Muscatine, L., F. P. Wilkerson & L. R. McCloskey, 1986. Regulation of population density of symbiotic algae in a tropical marine jellyfish (*Mastigias* sp.). Mar. Ecol. Prog. Ser. 32: 279–290.

Nielsen, A. S., A. M. Pedersen & H. U. Rissgård, 1997. Implications of density driven currents for interaction between jellyfish (*Aurelia aurita*) and zooplankton in a Danish fjord. Sarsia 82: 297–305.

Owen, R. W. Jr., 1966. Small-scale, horizontal vortices in the surface layer of the sea. J. mar. Res. 24: 56–66.

Owen, R. W., 1981. Fronts and eddies in the sea: mechanisms, interactions and biological effects. In Longhurst, A. R. (ed.), Analysis of marine ecosystems. Academic Press, New York: 197–233.

Owen, R. W., 1989. Microscale and finescale variations of small plankton in coastal and pelagic environments. J. mar. Res. 47: 197–240.

Paffenhöffer, G.-A., T. B. Stewart, M. J. Youngbluth & T. G. Bailey, 1991. High-resolution vertical profiles of pelagic tunicates. J. Plankton Res. 13: 971–981.

Pagès, F. & J-M. Gili, 1991. Vertical distribution of epipelagic siphonophores at the confluence between Benguela waters and the Angola Current over 48 hours. Hydrobiologia 216/217: 355–362.

Pagès, F. & J-M. Gili, 1992. Influence of the thermocline on the vertical migration of medusae during a 48 hr sampling period. S. Afr. J. Zool. 27: 50–59.

Pagès, F. & S. B. Schnack-Schiel, 1996. Distribution patterns of the mesoplankton, principally siphonophores and medusae, in the vicinity of the Antarctic Slope Front (eastern Weddell Sea). J. mar. Syst. 9: 231–248.

Pagès, F., M. G. White & P. G. Rodhouse, 1996. Abundance of gelatinous carnivores in the nekton community of the Antarctic Polar Frontal Zone in summer 1994. Mar. Ecol. Prog. Ser. 141: 139–147.

Pagès, F., H. E. González, M. Ramón, M. Sobarzo & J. M. Gili, 2001. Gelatinous zooplankton assemblages associated with water masses in the Humboldt Current System and potential predatory impact by *Bassia bassensis* (Siphonophora: Calycophorae). Mar. Ecol. Prog. Ser.

Parrish, J. K. & W. M. Hamner, 1997. Animal Groups in Three Dimensions. Cambridge Univ. Press, Cambridge: 336 pp.

Parrish, J. K. & L. Edelstein-Keshet, 1999. Complexity, pattern and evolutionary trade-offs in animal aggregation. Science 284: 99–101.

Pingree, R. D., P. M. Holligan & R. N. Head, 1977. Survival of dinoflagellate blooms in the western English Channel. Nature 265: 266–269.

Pugh P. R., 1977. Some observations on the vertical migrations and geographical distribution of the siphonophores in the warm waters of the North Atlantic Ocean. In Proccedings of the Symposium on Warm Water Zooplankton. Natl. Inst. Oceanogr. Goa, India: 362–378.

Purcell, J. E. & L. P. Madin, 1991. Diel patterns of migration, feeding, and spawning by salps in the subarctic Pacific. Mar. Ecol. Prog. Ser. 73: 211–217.

Purcell, J. E., D. A. Nemazie, S. E. Dorsey, E. D. Houde & J. C. Gamble, 1994. Predation mortality of bay anchovy *Anchoa mitchilli* and larvae due to scyphomedusae and ctenophores in Chesapeake Bay. Mar. Ecol. Prog. Ser. 114: 47–58.

Purcell. J. E., E. D. Brown, K. D. E. Stokesbury, L. H. Haldorson & T. C. Shirley, 2000. Aggregations of the jellyfish *Aurelia labiata*: abundance, distribution, association with age-0 walleye pollock, and behaviors promoting aggregation in Prince William Sound, Alaska, U.S.A. Mar. Ecol. Prog. Ser. 195: 145–158.

Purcell J. E., D. L. Breitbug, M. D. Decker, W. M. Graham, M. J. Youngbluth & K. A. Raskoff, 2001. Pelagic cnidarians and ctenophores in low dissolved oxygen environments: a review. In Rabalais, N. N. & R. E. Turner (eds), Effects of Hypoxia on Living Resources and Ecosystems. American Geophysical Union, Coastal and Estuar. Stud. 58: 77–100.

Raskoff, K. A., 2001. The impact of El Niño events on blooms of mesopelagic hydromedusae. Hydrobiologia 451 (Dev. Hydrobiol. 155): 121–129.

Rice, A. L., 1964. Observations on the effects of changes of hydrostatic pressure on the behaviour of some marine animals. J. mar. biol. Assoc. U. K. 44: 163–175.

Russell, F. S., 1927. The vertical distribution of marine macroplankton. V. The distribution of animals caught in the ring-trawl in the daytime in the Plymouth area. J. mar. biol. Assoc. U. K. 14: 557–608.

Sabatès, A., J. M. Gili & F. Pagès, 1989. Relationship between zooplankton distribution, geographic characteristics and hydrographic patterns off the Catalan coast (western Mediterranean). Mar. Biol. 103: 153–159.

Schuyler, Q., & B. K. Sullivan, 1997. Light responses and diel migration of the scyphomedusa *Chrysaora quinquecirrha* in mesocosms. J. Plankton Res. 19: 1417–1428.

Shanks, A. L. & W. M. Graham, 1987. Orientated swimming in the jellyfish *Stomolophus meleagris* L. Agassiz (Scyphozoan: Rhizostomida). J. exp. mar. Biol. Ecol. 108: 159–169.

Shanks, A. L. & W. M. Graham, 1988. Chemical defense in a scyphomedusa. Mar. Ecol. Prog. Ser. 45: 81–86.

Shannon, L. V. & P. Chapman, 1983. Incidence of *Physalia* on beaches in the South Western Cape Province during January 1983. S. Afr. J. Sci. 79: 454–458.

Shenker, J. M., 1984. Scyphomedusae in surface waters near the Oregon coast, May-August, 1981. Estuar. coast. Shelf Sci. 19: 619–632.

Sullivan, B. K., D. Van Keuren & M. Clancy, 2001. Timing and size of blooms of the ctenophore *Mnemiopsis leidyi* in relation to temperature in Narragansett Bay, RI. Hydrobiologia 451 (Dev. Hydrobiol. 155): 113–120.

Toyokawa, M., T. Inagaki & M. Terazaki, 1997. Distribution of *Aurelia aurita* (Linnaeus, 1758) in Tokyo Bay; observations with echosounder and plankton net. Proc. 6th Int. Conf. Coel. Biol. 1995: 483–490.

Ueno, S. & A. Mitsutani, 1994. Small-scale swarm of a hydrozoan medusa *Liriope tetraphylla* in Hiroshima Bay, the Inland Sea of Japan. Bull. Plankton Soc. Japan 41: 93–104.

Wang, Z., E. Thiébaut & J. C. Dauvin, 1995. Spring abundance and distribution of the ctenophore *Pleurobrachia pileus* in the Seine estuary: advective transport and diel vertical migration. Mar. Biol. 124: 313–324.

Wiebe, P. H., L. P. Madin, L. R. Haury, G. R. Harbison & L. M. Philbin, 1979. Diel vertical migration by *Salpa aspera* and its potential for large-scale particulate organic matter transport to the deep-sea. Mar. Biol. 53: 249–255.

Wolanski, E. & W. M. Hamner, 1988. Topographically controlled fronts in the ocean and their biological influence. Science 241: 177–181.

Woodcock, A. H., 1944. A theory of surface water motion deduced from the wind-induced motion of the *Physalia*. J. mar. Res. 5: 196–205.

Wright, D. A. & J. E. Purcell, 1997. Effect of salinity on ionic shifts in mesohaline scyphomedusae, *Chrysaora quinquecirrha*. Biol. Bull. 192: 332–339.

Yamamoto, T. & S. Nishizawa, 1986. Small-scale zooplankton aggregations at the front of a Kuroshio warm-core ring. Deep-Sea Res. 33: 1729–1740.

Yasuda, T., 1973. Ecological studies on the jellyfish, *Aurelia aurita* (Linne), in Urazoko Bay, Fukui Prefecture-VIII. Diel vertical migration of the medusa in early fall, 1969. Publ. Seto Mar. Biol. Lab. 20: 491–500.

Young, J. W., R. W. Bradford, T. D. Lamb & V. D. Lyne, 1996. Biomass of zooplankton and micronekton in the southern bluefin tuna fishing grounds off Tasmania, Australia. Mar. Ecol. Prog. Ser. 138: 1–14.

Zavodnik, D., 1987. Spatial aggregations of the swarming jellyfish *Pelagia noctiluca* (Scyphozoa). Mar. Biol. 94: 265–269.

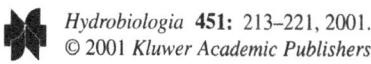

*Hydrobiologia* **451**: 213–221, 2001.
© 2001 *Kluwer Academic Publishers.*

# Developing jellyfish strategy hypotheses using circulation models *

Donald R. Johnson[1], Harriet M. Perry[2] & W. David Burke[2]
[1]*Naval Research Laboratory, Oceanography Division, Stennis Space Center, MS 39529, U.S.A.*
*E-mail: djohnson@drj.nrlssc.navy.mil*
[2]*University of Southern Mississippi, Institute of Marine Sciences, Gulf Coast Research Laboratory,*
*Ocean Springs, MS 39564, U.S.A.*

*Key words:* Scyphozoa, *Chrysaora quinquecirrha*, ocean circulation, numerical modeling, distribution

## Abstract

Little information exists relating life histories of jellyfish species to ocean currents. Successful cycling from sessile polyp to mature jellyfish and back must doubtlessly rely on circulation patterns that serve to retain the species in an optimum environment or disperse the species for other adaptive advantages. In this study, current vectors from a high resolution numerical model of the Gulf of Mexico are applied to a simple advection scheme to develop estimates of time and distance scales from probable polyp habitats to areas in which mature scyphomedusae are observed in the northern Gulf of Mexico. Although seasonal patterns of wind stress form the basis for circulation processes that favour shoreward distribution of medusae of oceanic origin, this dynamic may be altered by deep basin events that occur during critical life history stages. Inter-annual differences in distributional patterns of the sea nettle, *Chrysaora quinquecirrha* (Desor 1848), in Mississippi coastal waters could be explained by Loop Current processes that alter shelf circulation in the Mississippi Bight.

## Introduction

There is little information demonstrating the dependence of jellyfish species on ocean currents to carry them from the sessile polyp to the mature jellyfish and to subsequently return planulae to substrates appropriate for new polyp formation. Both adults and larvae rely on repetitive patterns of currents for distribution and recruitment. Current patterns typically vary inter-annually according to shifts of global oceanic and atmospheric conditions and can produce remarkable fluctuations in local jellyfish population densities. Understanding the mechanism of adult jellyfish distribution and the subsequent replenishment of the over-wintering polyp-cyst stages are germane to understanding population fluctuations.

The present study examines the distribution of *Chrysaora quinquecirrha* (Desor 1848), a prominent scyphomedusa of the northern Gulf of Mexico (GOM). Except for coloration differences along their radial lines, the GOM population is similar to the sea

nettle population of upper Chesapeake Bay (*fide*, J. Purcell). Due to significant impacts on tourism and commercial fisheries, the Chesapeake Bay population has been the focus of a number of studies (Cargo & Schultz, 1966; Schultz & Cargo, 1969; Feigenbaum & Kelly, 1984; Cargo & King, 1990; Purcell et al., 1994a,b). Life cycles of that population (Cargo & Schultz, 1966) include two distinct stages, an estuarine sessile polyp form requiring a hard substrate such as oyster shells, and pelagic ephyra and medusa stages, generated by seasonal strobilation of the polyp. The medusa stage cannot survive the winter, but the polyp can encyst into a highly resistant form which may overwinter, subsequently excysting in the spring and strobilating one or more times (Purcell et al., 1999). It would be reasonable to expect that the GOM sea nettle population would follow a similar life cycle; however, it is not demonstrable that the Gulf race of *C. quinquecirrha* occurs in estuaries in the polyp stage (Burke, 1975a; Graham, 2001). Rather, they appear to be advected into shallow coastal waters from the deeper GOM. Numerical circulation models were used to back trace from observed distributions of adults

---

* The U.S. Government right to retain a non-exclusive, royalty-free licence in and to any copyright is acknowledged.

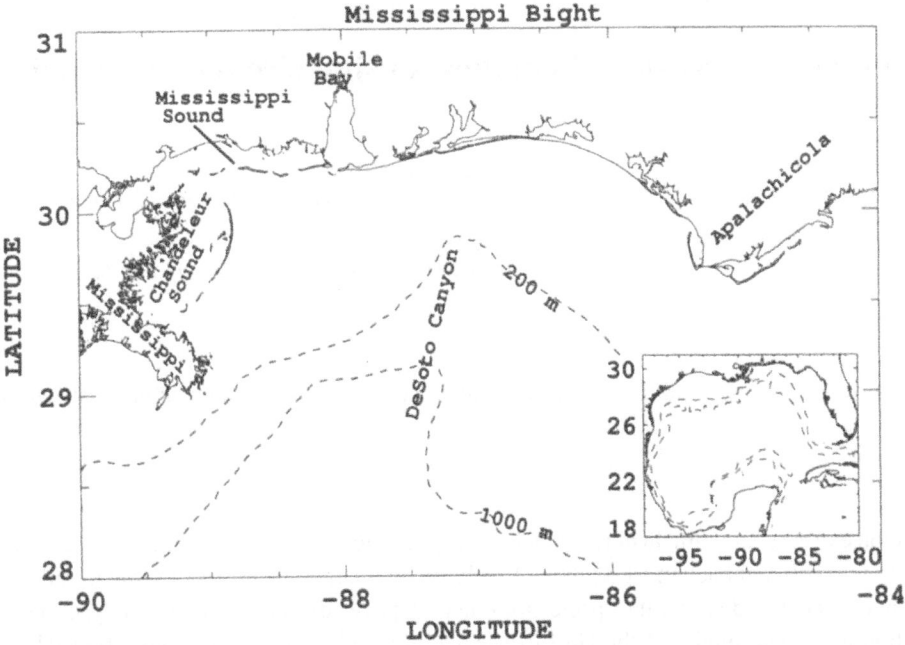

*Figure 1.* Mississippi Bight study location. Inset = Gulf of Mexico. Dashed lines are the 200 and 1000 m isobaths.

nearshore in the Mississippi Bight to potential offshore substrates for polyps. Forward traces from presumed polyp sites were then examined to determine if annual fluctuations in observed distribution patterns can be explained by variations in current patterns. Ostensibly, such analyses could be used to distinguish enhanced production from apparent increases resulting from the vagaries of currents.

The Mississippi Bight is an elongated area of the continental shelf in the northern GOM lying between the Mississippi River Delta in Louisiana and Apalachicola, Florida (Fig. 1). It is divided into two parts by the DeSoto Canyon, which intrudes shoreward near the center of the Bight and provides a mechanism for interaction of deep basin waters with the shallow interior of the shelf. The western half of the Bight is characterized by a series of low, sand barrier islands that separate the inner shelf from the estuarine-like environments of Mississippi and Chandeleur sounds. Fresh water drains into these sounds through complex waterways involving a number of rivers and bays. In contrast, the eastern half of the Bight is characterized by sandy beaches with significantly less fresh water drainage and less turbid waters.

Few studies have been conducted on the distribution and biology of large medusae in the northern GOM. Occurrence and seasonality of planktonic cnidarians in Mississippi Sound was investigated by Burke

(1975b). Burke (1975a) surveyed the distribution and abundance of adult medusae over a 27-month long sampling period along both the interior and oceanic sides of the barrier islands of Mississippi Sound (Fig. 2). These data were analyzed for seasonal and annual variations in abundance and were used to make some suggestion as to possible polyp source areas. Although Burke's surveys used trawls, plankton tows and visual transects along the beaches, all observations were combined into a single data set. Seasonal, geographic and annual distributions of adult *C. quinquecirrha* along the Mississippi barrier islands from these data are given in Figure 3. From the seasonal distribution patterns, peak adult medusae abundance occurred in the shallow waters of the inner shelf in August and September. A minor peak also occurred in May/June. The minor peak may have been due to more distant populations that strobilated earlier in the season (*fide*, M. Graham). Typically by November, all adult medusae were gone and did not re-appear until the following May. From the distribution patterns within locales, it is clear that the overwhelming majority of observations were from outside Mississippi Sound, hence it is reasonable to conclude that the polyp-strobilation source region is offshore, rather than in the estuaries or the sounds. The inference here is that the northern GOM population of nettles is most

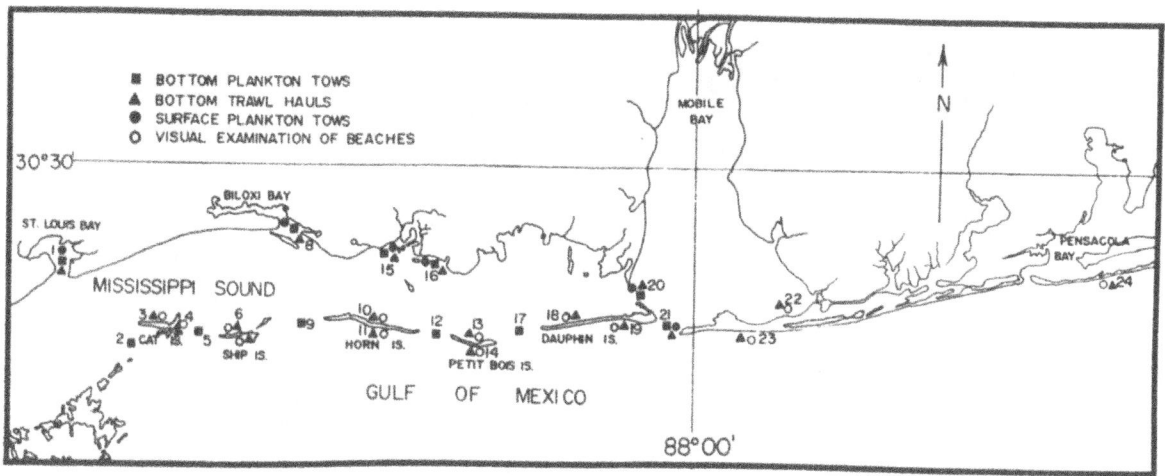

*Figure 2.* Sampling sites in the northern Gulf of Mexico, April 1971–June 1973 (from Burke, 1975a).

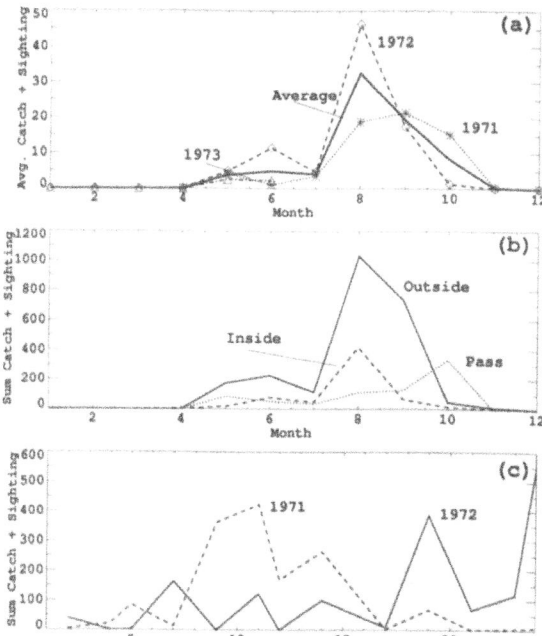

*Figure 3.* Seasonal and areal distribution of *Chrysaora quinquecirrha* in Mississippi coastal waters: (a) total monthly observations by year and average over the three years (note that the 1973 collection extends January through June), (b) total monthly observations separated into: inside the barrier islands, outside the barrier islands, in the barrier island passes, (c) distribution of observations by year and by station number with only those stations located outside the barrier islands. Data from Burke (1975a).

probably oceanic in origin rather than estuarine in contra-distinction to the Chesapeake Bay population.

Wind stress is a significant forcing mechanism for continental shelf environments. Due to the shallow water depths, there is limited capacity in the interior of the water column for storage of energy flux through the surface. This excess energy concentration is quickly converted to fluid momentum. Thus, seasonal winds play a strong role in driving the advective patterns responsible for the appearance of mature *C. quinquecirrha* in the near shore regions of the Mississippi Bight. Deep basin events may also influence shelf circulation dynamics where there is a narrow continental shelf or where canyons intrude into near shore regions. In the Mississippi Bight, there is a severely narrowed shelf off its western end (Mississippi Delta) and an intrusive canyon in the center (Fig. 1). Exceptionally vigorous deep basin events occur in the area through the periodic intrusion of the Loop Current and its spin-off eddies.

The GOM Loop Current (Fig. 4) is part of the North Atlantic western boundary current system, which flows through the Yucatan Straits into the GOM. It forms a loop in the Gulf as it turns anti-cyclonically back southward and then exits eastward into the Straits of Florida before passing into the North Atlantic as the Gulf Stream. The Loop Current is dynamically unstable (Hurlbert & Thompson, 1980), at times intruding far into the Gulf and then pinching off into an eddy of 200 km diameter scale which drifts westward at about 3–4 cm s$^{-1}$ with circulation speeds above 100 cm s$^{-1}$ (Elliot, 1982; Johnson et al., 1992). One way this spin-off eddy decays is by generating cyclonic eddies around its periphery. Interactions of the Loop Current eddy and its associated cyclonic eddies with the outer continental slope in the northern and western GOM and in the De Soto Canyon are common (Huh et al., 1981; Johnson et al., 1992; Oey, 1995; Perry et al., 1998; Johnson & Perry, 1999). If a Loop Current intrusion occurred and a spin-off eddy was

216

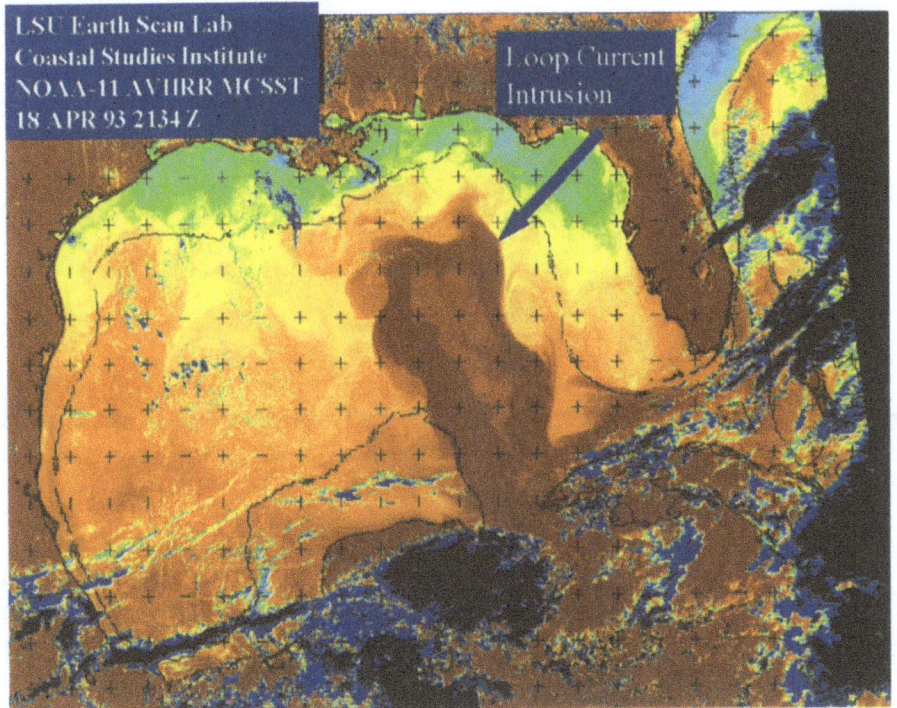

*Figure 4a.* AVHRR images for the Gulf of Mexico (a) during a Loop Current intrusion on 18 April 1993.

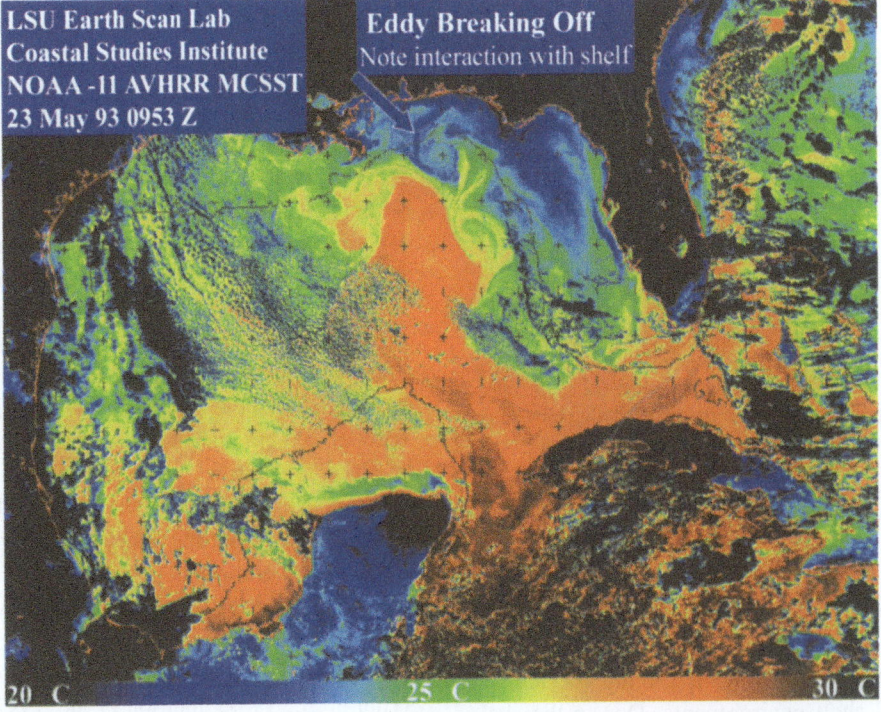

*Figure 4b.* (b) the subsequent breaking off of a large anticyclonic eddy.

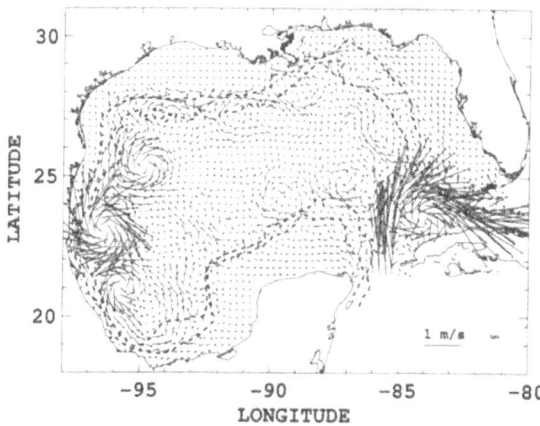

*Figure 5.* Example of surface currents calculated by the GOM circulation model. For clarity in presentation the gridding was reduced by a factor of 3. The Loop Current is in a minimal intrusion state. The dashed contours are the 200 m and 1000 m isobaths.

generated during the mid-summer when jellyfish were under the influence of seasonal shelf circulation patterns, it would be expected that normal distributions would be modified.

In the present study, current vectors from a high resolution numerical model of the GOM were applied to a simple advection scheme to develop estimates of time and distance scales from probable polyp habitats to areas in the northern GOM in which mature jellyfish, *C. quinquecirrha*, occur. Although *C. quinquecirrha* are known to migrate vertically in the water column, only surface currents were used in the tracking process. Lack of information on swimming patterns in GOM populations of *C. quinquecirrha* limited our ability to use vertical structure of currents to model trajectories.

## Materials and methods

The GOM circulation model was developed and run by Choi & Kantha (1997) and the results stored at 10-day intervals. The model is a three-dimensional, primitive equation, sigma-coordinate formulation of the Princeton Ocean Model (Blumberg & Mellor, 1983). The model has 21 levels in the vertical and a horizontal grid resolution of 1/12th degree in longitude and latitude (8–9 km). Fig. 5 shows an example of the surface currents calculated by the model. The model is described more fully and validated in Johnson & Perry (1999), where it was applied in a study of blue crab larval advection.

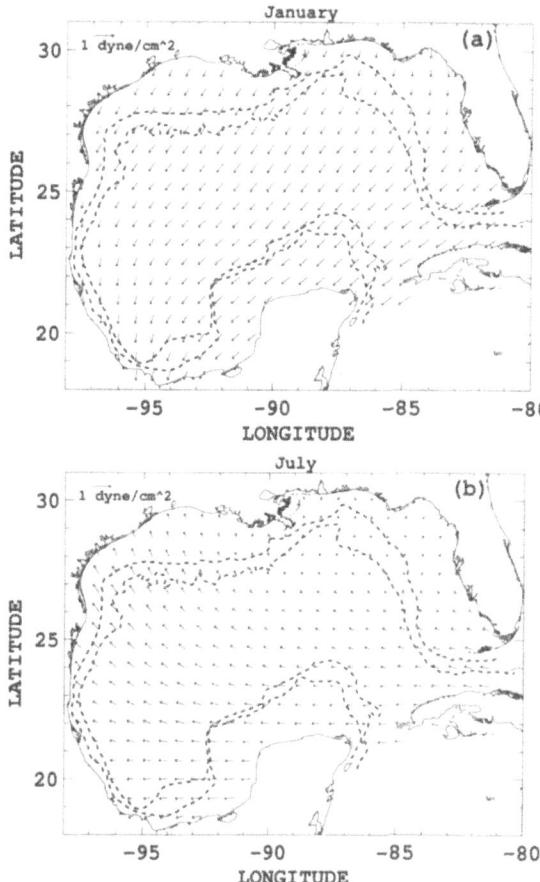

*Figure 6.* Hellerman climatological wind stress patterns for January (a) and July (b).

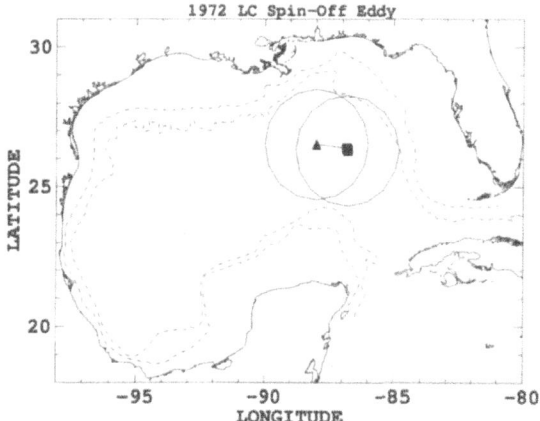

*Figure 7.* Location and movement of newly formed anticyclonic spin-off eddy during 1972. The center of the eddy during two cruises in May, 1972, are located by squares. The center in June, 1972, is located by the triangle, showing westward motion. The larger circles are estimated diameters of the eddies (Elliot, 1982).

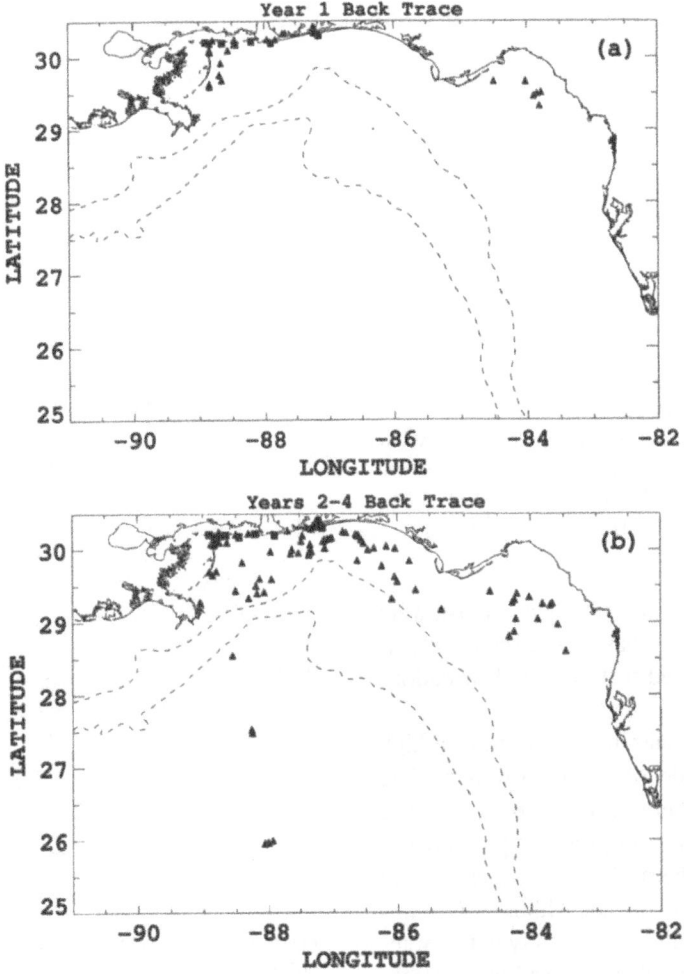

*Figure 8.* Back tracing from the barrier island observation stations (Fig. 2) for 90 days from September and August. Squares represent observation stations where the back trace began and triangles represent ending locations: (a) model year 1 with Loop Current spin-off eddy forming during July, (b) model years 2–4. Trajectories are omitted for clarity of presentation.

It is important to note that the model was forced by climatological monthly averaged winds and damped to climatological sea surface temperature and salinity (Fig. 6). Hellerman monthly averaged climatological winds were converted to surface wind stress and used to force the numerical circulation model. Forcing was ramped between months to produce a smooth transition. Although the model current vector components were only stored at 10 day intervals, the monthly forcing scale and the smooth transition means that the data can be used for advective scale motions (integrations on the order of a month or longer) without significant aliasing. The model was run for 4 model years with the same annual forcing cycle. This means that any inter-annual variations would come from processes other than by winds, such as Loop Current interactions with the continental shelf. Damping to

climatological sea surface temperature and salinity produces a highly smoothed surrogate for seasonal heat flux and seasonal stream inflow.

In this study, only the surface currents are used in the trajectory modeling. To run advective trajectories, the model surface current components, $U_{i,j,k}$, $V_{i,j,k}$, are first retrieved from storage, where $U$ and $V$ represent the positive east and positive north components, respectively; the indices $i$ and $j$ represent longitude and latitude, respectively, and $k$ represents time. A location and time is selected to start the process and the correct starting indices are calculated. From that point, currents are integrated backward (or forward) in time at $\frac{1}{2}$-day intervals, with both space and time linearly interpolated from surrounding index points, and the trajectory is plotted. The medusae are treated as Lagrangian particles in an Eulerian velocity field.

No account is taken of patchiness or dispersion from the mean by turbulence (Hood et al., 1999).

Strobilation of *C. quinquecirrha* in upper Chesapeake bay is triggered most prominently between mid-May and mid-June with adults appearing in quantity in July and August (Cargo & King, 1990). This gives a 60–90 day window for growth to maturity. Assuming the same time scale for the northern GOM population, backward integration for 90 days determines potential sites, and forward integration for 90 days determines distributions from selected sites and assesses inter-annual influences.

## Results

Climatological wind stress patterns for the northern GOM are presented for January and July (Fig. 6). The January averaged wind stress (Fig. 6a) represents the typical wintertime pattern (October–March). Forcing is strongly offshore toward the southwest during this period. In April, the wind stress turns more shoreward toward the northwest. By mid-summer (Fig. 6b), the wind stress is light and directly shoreward in the Mississippi Bight. This pattern of wind stress is consistent with late summer arrival of adult *C. quinquecirrha* in the near shore regions of the Mississippi Bight if strobilation takes place offshore in early- or mid-summer. This does not imply that wind direction triggers strobilation, but advances the idea that seasonal changes that trigger strobilation also create the advective forcing conditions responsible for transport shoreward (Cargo & King, 1990).

Elliot (1982) lists a number of Loop Current spin-off eddies which were documented in the GOM between 1965 and 1972. In 1972, a strong intrusion toward the northern Gulf occurred with a spin-off eddy being created between May and June of that year. The given dimensions of the eddy (Fig. 7) would suggest that a strong interaction with the Mississippi Bight was possible through the De Soto Canyon, and hence could have been responsible for the differences in jellyfish distributions between 1971 and 1972 (Fig. 3c). In 1971, the peak abundance was found toward the west (Horn and Petit Bois Islands). In 1972, the peak abundance was shifted eastward toward the center of the Mississippi Bight (Pensacola, Florida). To explain this distribution and to determine potential polyp sites, model currents were used to back trace from all of the observation stations of Burke (1975a). This was done for the model year (1), in which a break-

off eddy occurred, and for the three model years (2–4), in which no break-off eddy occurred during the summer months. Potential polyp sites associated with high counts of medusae at the observation stations were then forward traced and the resulting model distribution compared to observed distributions.

Figure 8a illustrates back trace results for model year 1. A Loop Current intrusion began in the spring and a spin-off eddy was formed in mid-summer. Although intrusions and spin-off eddies occurred in other model years, they did not occur during the important summer months. Model year 1 provides a reasonable surrogate for conditions during 1972 when an eddy was formed in early summer (Fig. 7), and when the distribution of *C. quinquecirrha* was biased toward the center of Mississippi Bight and away from its western side (Fig. 3c). From the model, three potential sites were identified: the estuaries of the western Mississippi Bight, along the Chandeleur Islands, and in the Apalachee Bay area of northern Florida. Based on Burke's (1975a) observation that *C. quinquecirrha* in the GOM was oceanic in habitat, it would seem probable that juvenile sources lay outside the barrier islands of Mississippi Sound (Fig. 3b).

Back trace results in non-intrusion-summer years show a great deal more variability (Fig. 8b). In addition to the three sites noted in model year 1, there were sites on the mid- and outer-shelf of the Mississippi Bight and in the central deep basin of the GOM. The latter site may represent Caribbean medusae advected northward by the flow through the Yucatan Straits. The currents associated with the Loop Current are sufficiently strong that medusae could be advected to the central Gulf in relatively short time so that an error in distance and time here may not be highly significant.

Since polyps need a hard bottom for attachment, we examined the Mississippi Bight for hard bottom areas, recognizing that much of the Bight is sand and mud. Thompson et al. (1999) presented a synthesis of hard bottom areas from a variety of studies including lease blocks with documented hard bottom (unfortunately this study did not extend into Apalachee Bay). These hard bottom sites are shown in Figure 9 (represented by square symbols).

Starting in the model month of early June, drifters were integrated forward in time from each of the hard bottom sites for 90 days for each of the four model years. In addition, potential sites in Apalachee Bay and along the Chandeleur Islands were examined. As seen with the backward integration (Fig. 8), the differences between model year 1 and model years 2–4

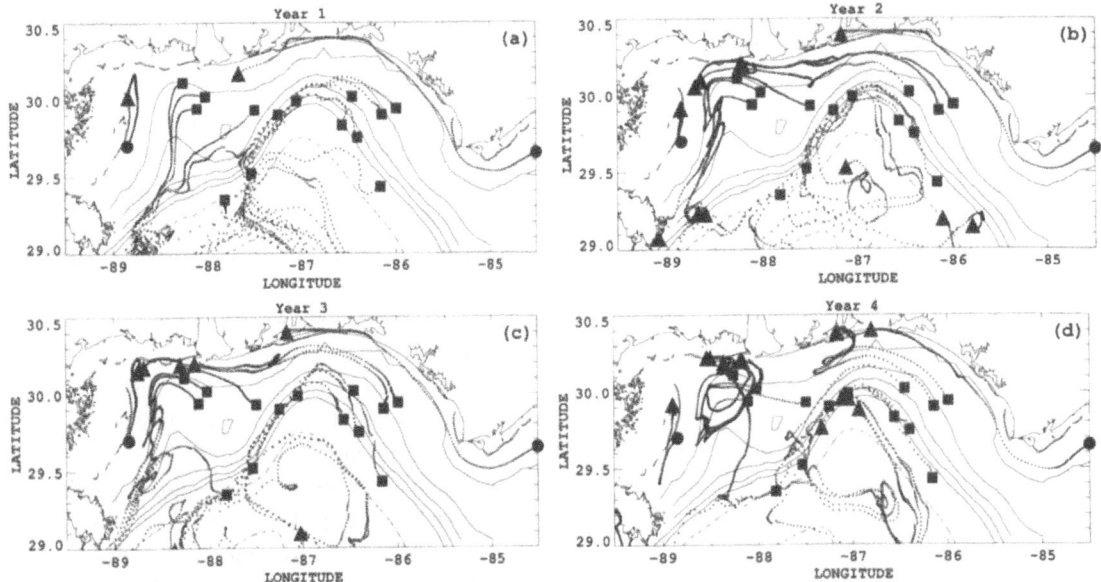

*Figure 9.* Forward tracing from selected hard bottom locations in the Mississippi Bight (square symbols) for 90 days from June through August. The two circle symbols represent possible locations on the northeast Florida Shelf and along the Chandeleur Islands. Small crosses show track locations every 0.5 days and triangles represent ending locations after 90 days. Ending locations outside the Bight are not shown. Isobaths are shown at 20 m intervals from 20 m to 100 m (solid lines of model year 1). The 200 m isobath is shown by a dot-dash line and the 1000 m isobath by the dashed line. (a) model year 1, (b) model year 2, (c) model year 3, and (d) model year 4.

are striking. In model year 1 (Fig. 9a), trajectories for all of the sites within Mississippi Bight go offshore, hence, medusae from these sites would not show up in the observations along the barrier islands of Mississippi Sound. The only successful landing was from the site in Apalachee Bay. This conforms well with the observation that, during a year of summer Loop Current intrusion and eddy formation (1972 observations and model year 1), the higher number of observations occurred at the easternmost stations in the central Bight. In model years 2–4 (Figs 9b–d), successful landing occurred in the western part of the Bight from shallow water hard bottom sites near the DeSoto Canyon in water depths from 20 to 60 m. This conforms well with the data from 1971 (Fig. 3c) that showed more landings in the western part of the Bight.

## Discussion

Observational data to date implies that *C. quinquecirrha* in the northern GOM is oceanic. Identification of *C. quinquecirrha* polyps on shelf molluscs and the absence of polyps in inshore waters suggests that potential polyp sites are located offshore (Burke, 1975a). In this study, this hypothesis was examined using advective currents from a numerical circulation model.

Stored arrays of currents from the model were used to back trace from adult observation sites to areas of potential polyp sites. When the general area was located, hard bottom sites in these regions were identified from the literature. Forward integration from these sites was used to test inter-annual variability.

Wind stress forcing plays a dominant role in driving continental shelf circulation and in establishing annual distribution patterns of scyphomedusae on the shelf. The intrusive deep basin Loop Current and spin-off eddies provide an added mechanism for changing distribution patterns when the intrusion occurs during the important summer months. Model year 1 and survey year 1972 were years with such summer intrusions and eddy formations in the northern Gulf. The inter-annual differences in distribution patterns from both model and survey could be explained by Loop Current processes if the polyp strobilation areas were located in deeper water (20–60 m) near the DeSoto Canyon and on the northern West Florida Shelf. There are potential hard bottom sites near the DeSoto Canyon. Knowledge of such sites for the northern West Florida Shelf does not exist.

Although there is, in general, little exchange between shelves and deep basins, it is demonstrable that exchange occurs through the DeSoto Canyon and along the western edge of the Bight where the shelf is narrow. Approximately 6% of the back calculated trajectories terminate in the central Gulf where it is easy to imagine that they could have been brought to that area by strong currents coming from the Caribbean. Larger medusae, associated with the small peak in observations in May (Fig. 3a), could easily be of Caribbean origin where the long advection time allows greater growth and an early stobilation date is possible (*fide*, M. Graham).

It should be recognized that this study is a presentation of methodology as well as a search for reasonable hypotheses for potential polyp sites. The role that variations in circulation patterns can play in creating fluctuations in abundance and distribution of medusae is promulgated and it is not difficult to see that some years and some areas could experience a 'bloom' that is more related to circulation patterns than to enhanced productivity.

## Acknowledgments

DRJ was funded through the Naval Research Laboratory's Accelerated Research Initiative titled "Coupled bio-physical dynamics across the littoral transition (COBALT)" under program element 61153N sponsored by the Office of Naval Research. HMP and WDB were supported by the University of Southern Mississippi, Institute of Marine Science, Gulf Coast Research Laboratory. We thank Jenny Purcell and Monty Graham for sharing their knowledge of cnidarians and Kirsten Larsen for her technical help in preparation of this manuscript. Model data were provided by L. Kantha and J. Choi. AVHRR imagery was obtained from the LSU Earth Scan Laboratory.

## References

Blumberg, A. F. & G. L. Mellor, 1983. Diagnostic and prognostic numerical circulation studies of the South Atlantic Bight. J. Geophys. Res. 88: 4579–4592.

Burke, W. D., 1975a. Biology and distribution of the macrocoelenterates of Mississippi Sound and adjacent waters. Gulf Res. Repts. 5: 17–28.

Burke, W. D., 1975b. Pelagic Cnidaria of Mississippi Sound and adjacent waters. Gulf Res. Repts. 5: 23–38.

Cargo, D. G. & D. R. King, 1990. Forecasting the abundance of the sea nettle, *Chrysaora quinquecirrha*, in the Chesapeake Bay. Estuaries 13: 486–491.

Cargo, D. G. & L. P. Schultz, 1966. Notes on the biology of the sea nettle, *Chrysaora quinquecirrha*, in the Chesapeake Bay. Ches. Sci. 7: 95–100.

Choi, J. K. & L. H. Kantha, 1997. Refinement and verification of a climatological and forecast model of the Loop Current and associated eddies. Colorado Center for Astrodynamics Research, Univ. Colorado, Report for EJIP CASE.

Elliot, B. A., 1982. Anticyclonic rings in the Gulf of Mexico. J. Phys. Oceanogr. 12: 1292–1309.

Feigenbaum, D. & M. Kelly, 1984. Changes in the lower Chesapeake Bay food chain in the presence of the sea nettle *Chrysaora quinquecirrha* (Scyphomedusa). Mar. Ecol. Prog. Ser. 19: 39–47.

Graham, W. M., 2001. Numerical increases and distributional shifts of *Chrysaora quinquecirrha* (Desor) and *Aurelia aurita* (Linné) (Cnidaria: Scyphozoa) in the northern Gulf of Mexico. Hydrobiologia 451 (Dev. Hydrobiol. 155): 97–111.

Hood, R. R., H. V. Wang, J. E. Purcell, E. D. Houde & L. W. Harding, Jr., 1999. Modeling particles and pelagic organisms in Chesapeake Bay: convergent features control plankton distributions. J. Geophys. Res. 104: 1223–1243.

Huh, O. K., W. J. Wiseman, Jr. & L. R. Rouse, 1981. Intrusion of Loop Current waters onto the west Florida continental shelf. J. Geophys. Res. 86: 4186–4192.

Hurlburt, H. E. & J. D. Thompson, 1980. A numerical study of Loop Current intrusions and eddy shedding. J. Phys. Oceanogr. 10: 1611–1651.

Johnson, D. R. & H. M. Perry, 1999. Blue crab larval dispersion and retention in the Mississippi Bight. Bull. mar. Sci. 65: 129–149.

Johnson, D. R., J. D. Thompson & J. D. Hawkins, 1992. Circulation in the Gulf of Mexico from Geosat altimetry during 1985–1986. J. Geophys. Res. 97: 2201–2214.

Oey, L-Y., 1995. Eddy- and wind-forced shelf circulation. J. Geophys. Res. 100: 8621–8637.

Perry, H. M., D. Johnson, C. Trigg, C. Eleuterius & J. Warren, 1998. Application of remote sensing to settlement of *Callinectes sapidus* megalopae in the Mississippi Bight. J. Shell. Res. 17(5): 1439–1442.

Purcell, J. E., J. R. White & M. R. Roman, 1994a. Predation by gelatinous zooplankton and resource limitation as potential controls of *Acartia tonsa* copepod populations in Chesapeake Bay. Limnol. Oceanogr. 39: 263–278.

Purcell, J. E., D. A. Nemazie, S. E. Dorsey, E. D. Houde & J. C. Gamble, 1994b. Predation mortality of bay anchovy *Anchoa mitchilli* eggs and larvae due to scyphomedusae and ctenophores in Chesapeake Bay. Mar. Ecol. Prog. Ser. 114: 47–58.

Purcell, J. E., J. R. White, D. A. Nemazie & D. A. Wright, 1999. Temperature, salinity and food effects on asexual reproduction and abundance of the scyphozoan, *Chrysaora quinquecirrha*. Mar. Ecol. Prog. Ser. 180: 187–196.

Schultz, L. P. & D. G. Cargo, 1969. Sea nettle barriers for bathing beaches in upper Chesapeake Bay. Natural Resources Institute, University of Maryland, College Park, Maryland. Ref. No. 69–58: 10 pp.

Thompson, M. J., W. W. Schroeder & N. W. Phillips, 1999. Ecology of live bottom habitats of the northeastern Gulf of Mexico: a community profile. U.S. Dept. Interior, Minerals Management Service, OCS Study mmS 99-0004: 1–74.

*Hydrobiologia* **451**: 223–227, 2001.
© 2001 *Kluwer Academic Publishers.*

# Flow and prey capture by the scyphomedusa *Phyllorhiza punctata* von Lendenfeld, 1884

Isabella D'Ambra[1], John H. Costello[2] & Flegra Bentivegna[1]
[1]*Stazione Zoologica 'Anton Dohrn', Villa Comunale 1, 80121 Naples, Italy*
*E-mail: mikidambra@hotmail.com*
[2]*Biology Department, Providence College, Providence, RI 02819, U.S.A.*

*Key words:* jellyfish, feeding, swimming, behaviour

## Abstract

The mechanical basis of prey capture and behaviour of *Phyllorhiza punctata* von Lendenfeld, 1884, as with most members of the Order Rhizostomeae, has not been described. Free-swimming medusae were videotaped in order to quantitatively describe the feeding process of *P. punctata*. Kinematic data demonstrated that adult medusae were surrounded by relatively high Re ($10^2$–$10^3$) flows while swimming. Therefore, momentum dominated these flows and the motions of particles entrained in the fluid surrounding swimming *P. punctata*. *Artemia salina* nauplii entrained within these flows contacted two principle capture surfaces: the oral arm cylinder and the underside of the subumbrellar surface. Prey were ingested by small polyp-like mouthlets located on these surfaces. Ingestion followed capture at these sites. *P. punctata*'s body morphology is highly modified to channel flows into these capture surfaces and feeding is dependent upon this pattern. Swimming activity, and hence the creation of flows used for prey capture, is continuous, as is feeding, and plays a central role in this medusa's foraging behaviour.

## Introduction

Medusae of the scyphomedusan order Rhizostomeae are an important group of planktonic predators and are part of growing 'jellyfish' fisheries in tropical Pacific waters. Despite their increasing economic importance, the feeding biology of rhizostome medusae has remained sparsely studied. Several genera of this group are actively caught and exported to a variety of southeast Asian countries (Hsieh et al., 2001; Omori & Nakano, 2001). Annual catches of these medusae are highly variable (Omori & Nakano, 2001) and little is known of the factors determining annual abundance or distribution of these species. Likewise, the feeding habits of these medusae are not well documented. Although the diets of species such as *Pseudorhiza haeckeli* Haacke (in Fancett, 1988), *Stomolophus meleagris* L. Agassiz (in Larson, 1991) and *Phyllorhiza punctata* (in García & Durbin, 1993) have been quantified, the mechanisms underlying selection and capture of their zooplankton prey are less well known. One rhizostome species, *Stomolophus meleagris*, has been found to entrain zooplankton in currents created by bell pulsation and to subsequently sieve these prey through intricate feeding structures (Larson, 1991; Costello & Colin, 1995). However, comparative data on feeding by other rhizostome medusae has been lacking and so the applicability of the results found for *S. meleagris* for other rhizostome medusae has remained unresolved.

*Phyllorhiza punctata* is a tropical rhizostome scyphomedusa that has been successfully reared through its full life cycle in the laboratory (Lange & Kaiser, 1995). The ability to maintain *P. punctata* in a controlled environment provided an opportunity to quantitatively describe functional aspects of prey capture and broaden understanding of feeding by rhizostome medusae.

## Methods

### Experimental organisms

*Phyllorhiza punctata* medusae of different sizes (1.4–7.4 cm diameter) were received from the Berlin Zoo-Aquarium (Germany) and maintained in a cylindrical kriesel design tank (Tide-Pool, Inc., Japan) at Naples Aquarium (Italy). The tank used a sand bed filter to maintain water quality and four 50 W lamps to supply light for *P. punctata*'s zooxanthellae.

### Microvideography

A backlit optical system similar to that described by Costello & Colin (1994) was used to detail movements of medusae, prey (newly hatched *Artemia salina* nauplii) and their surrounding fluid. Medusae were videotaped while swimming freely within rectangular vessels ranging in dimensions from $30 \times 17 \times 25$ cm to $23 \times 7 \times 23$ cm (height × depth × width). Vessel size was chosen in order to maximize the medusa's opportunity to swim freely and avoid contact with the vessel walls while minimizing the depth of the viewing field. These trade-offs optimized image clarity and our ability to track individual particle motions.

A field counter labelled each sequential S-VHS video frame (1/60 per field) in order to provide temporal information. Spatial characteristics of the optical field were determined from scale bars periodically included in the recordings.

### Kinematics

Bell pulsation provides thrust during medusan swimming and the changes in bell shape during pulsation were measured by the bell fineness ratio. The fineness ratio ($F$) was defined as the proportion of bell height ($h$) to diameter ($d$):

$$F = \frac{h}{d}.$$

Medusa motion was measured from sequential changes in position ($x$) of the tip of the exumbrellar surface at 0.05 s (3 fields) time intervals ($t$). Motion only within a two-dimensional viewing field was ensured by using a sequence in which bell orientation was level as the medusa swam through the viewing field.

The medusa's velocity ($u$) for a specific time interval ($i$) was calculated according to the formula:

$$u_i = \frac{x_i - x_i - l}{t}.$$

Reynolds number (Re) is used to estimate the relative importance of viscous and inertial forces in a fluid (Vogel, 1994). Re-calculations for the flow around swimming *P. punctata*:

$$Re = \frac{du}{\nu}$$

were based upon bell diameter ($d$), medusan velocity ($u$) and the kinematic viscosity of seawater ($\nu$) at 20 °C ($1.047 \times 10^{-6}$ m$^2$ s$^{-1}$).

### Flow fields

Prey trajectories were tracked in the fluid surrounding the bell during the initial stages of the relaxation phase because prior observations indicated that this phase of the bell pulsation cycle was critical for prey entrainment. The pathways of entrained *Artemia salina* nauplii were used to track fluid motions around swimming *P. punctata*. Although *Artemia* nauplii do swim, their swimming velocities are negligible compared to flow velocities created by most scyphomedusae (Costello & Colin, 1994, 1995). Flow field images were taken while the camera was stationary and the medusa swam through the field of view. The flow field diagram was constructed from several pulsation cycles, because no single cycle contained enough appropriately located and focused prey to describe the entire flow field. All the prey paths were measured by superimposing an x–y grid on a video sequence of a free-swimming medusa in order to measure particle motions in relation to the medusa's bell position. All measurements were made at the same phase of the relaxation phase- 0.08 s (5 fields) after the stroke began. Particle velocity was determined from its change in position during the subsequent 0.08 s interval.

## Results

Cyclic bell pulsation resulted in regular variations in medusan swimming velocities. During bell contraction (increasing bell fineness, Fig. 1a) velocities increased (Fig. 1b) as did Re values of the fluid moving past the medusa's bell (Fig. 1c). Fluid flows around a swimming medusa were characterized by peak Re values greater than $10^3$ during bell contraction. Therefore, these flows, and the prey entrained within them, were dominated by inertial forces (Vogel, 1994).

Prey entrained in the high Re flow created by bell contraction were transported past the bell margin

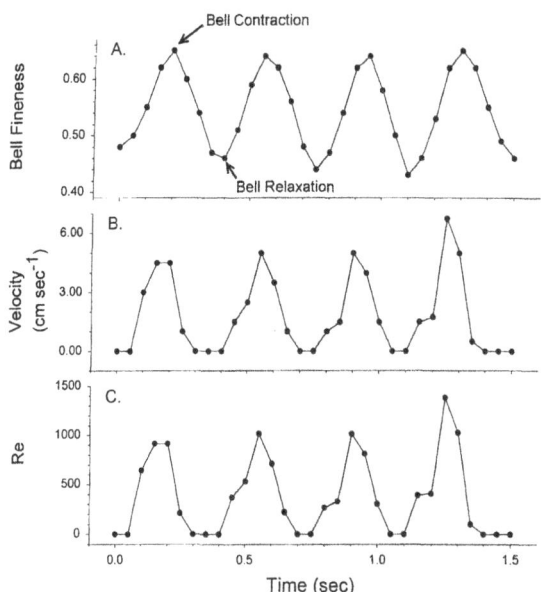

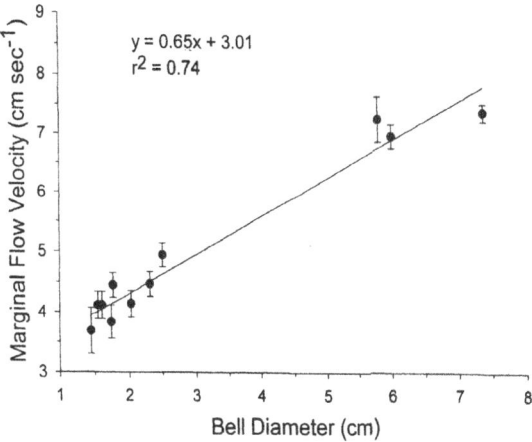

*Figure 3.* Marginal flow velocities of *Phyllorhiza punctata* medusae of different bell diameters. Each data point represents a different medusa. Ten particle velocities were determined for each medusa; error bars represent ± 1 standard deviation from the mean velocity for each medusa.

*Figure 1. Phyllorhiza punctata.* (A) Bell fineness, (B) swimming velocity and (C) Re (bell diameter was used as the reference length scale) determined from a swimming medusa, 2.8 cm diameter.

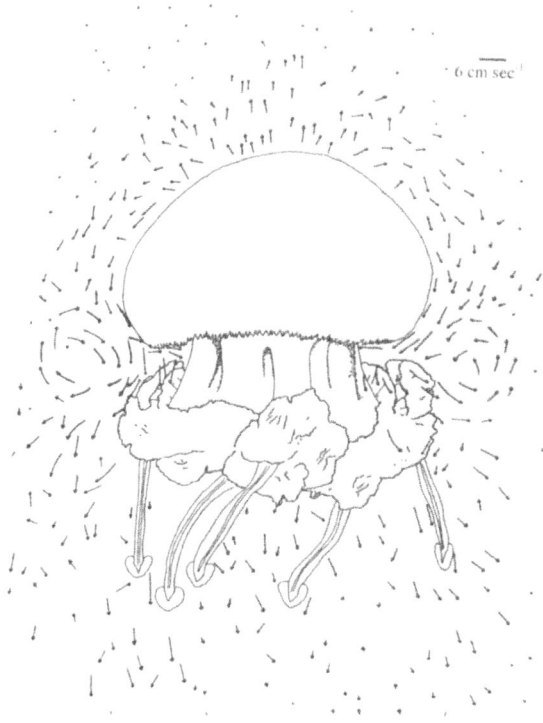

*Figure 2.* Flow patterns surrounding a 6.0 cm *Phyllorhiza punctata* medusa. Arrows represent *Artemia salina* paths at the beginning of the recovery phase of bell pulsation. Particles were tracked for 0.08 s beginning 0.08 s after the start of bell relaxation. The scale bar represents a particle velocity of 6 cm s$^{-1}$.

(Fig. 2) where the flow branched towards either of two principle prey capture areas. These two capture regions were located at the distal end of the oral arm cylinder and at the oral arm convergence near the central region of the bell's subumbrellar surface (Fig. 2). Flow velocities at the bell margin increased as bell diameter increased (Fig. 3). Each of the capture surfaces, and, to a lesser extent, the interior surfaces of the oral arms between the two principle capture areas, was covered with small mouthlets (term from Mianzan & Cornelius, 1999) – pinnately branched polyp-like oral structures that have been described in more detail by Mayer (1910). Prey were observed to be trapped on the tentacle-like extensions of these mouthlets and ingestion followed prey entrapment.

## Discussion

Prey capture via flow entrainment resembles that described for another rhizostome medusa, *S. meleagris* (in Larson, 1991; Costello & Colin, 1995) in several important ways. Both medusae entrain prey in fluid motions created by bell pulsation and these flows transport prey to capture surfaces covered by mouthlets. The position and morphological details of the capture surfaces differ between the two species of medusae, but share essential similarities. For both species, the major concentrations of mouthlets were located at the distal end of the oral cylinder. Bell contraction by either species forces fluid, and entrained prey, along and through these distal feeding structures.

226

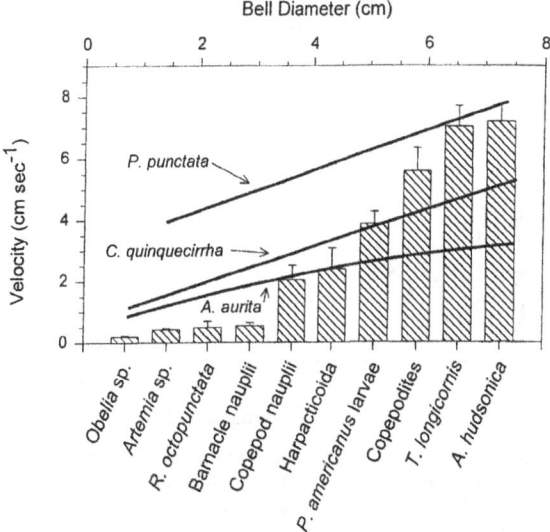

*Figure 4.* Comparison of medusae marginal flow velocity regressions for *Phyllorhiza punctata*, *Chrysaora quinquecirrha* and *Aurelia aurita* with prey escape velocities. *P. punctata* regression from Figure 3, *C. quinquecirrha* regression from Ford et al. (1997), *A. aurita* regression and prey escape velocities from Costello & Colin (1994).

As the bell begins the relaxation phase, a portion of the fluid flowing past the bell margin is drawn into the bell cavity and passes through a second major capture surface lining the oral arm surface nearest the bell. The scapulets of *S. meleagris* are covered with mouthlets which contact prey during the relaxation phase. *P. punctata* does not possess scapulets, but fluid entering the subumbrellar space during the relaxation phase flows over clusters of mouthlets near the base of the oral arm disk and in the center of the fused oral arm cylinder. Thus, both species use the contraction and relaxation phases of bell pulsation to transport prey to different capture surfaces. Based on evidence from these two species and the morphology of other rhizostome medusae (illustrated in Mayer, 1910), we suspect that this is a widespread trait within the Rhizostomeae. In this case, different rhizostome genera utilize variations of this flow-based feeding architecture.

*P. punctata*'s marginal flow velocities are high relative to previously studied scyphomedusae (Fig. 4). For comparable bell diameters, the marginal flow velocities of *P. punctata* exceed those of semaeostome scyphomedusae such as *Aurelia aurita* (Linnaeus) or *Chrysaora quinquecirrha* (Desor). Although not described here, *Cyanea* sp. marginal flow velocities approach, but are still less than, those of *P. punctata* (Costello, unpublished data). These relatively high

values reflect the strong bell contraction pattern of *P. punctata*. They are ecologically important because higher marginal flow velocities allow entrainment of more rapidly escaping prey (Costello & Colin, 1994) and can thereby affect *in situ* prey selection (Sullivan et al., 1994). Several rhizostome species are strong swimmers (Hamner & Hauri, 1981; Shanks & Graham, 1987) and probably share high marginal flow velocities with *P. punctata*. If so, the evolution among rhizostome medusae of morphologies that create, channel and sieve flow may result in different prey selection patterns than co-occurring semaeostome medusae. There are some indications that these patterns do occur in nature. For example, in a coastal Australian community, the rhizostome *Pseudorhiza haeckeli* consumed more rapidly swimming calanoid copepods and fed at significantly higher rates than did the co-occurring semaeostome medusae *Cyanea capillata* (Linnaeus) (in Fancett, 1988). However, the reliability of inter-order comparisons of scyphomedusan feeding impacts is presently limited because relatively few species have been studied in sufficient detail. This level of prey selection detail will need to be determined if the impacts of scyphomedusan population outbreaks are to be accurately evaluated.

## Acknowledgements

The authors thank J. Lange, Curator of the Berlin Aquarium, and his assistant, R. Kaiser, for providing the medusae used in this study. J. Higgins III provided drafting assistance. Travel support for JHC was provided by the National Science Foundation (OCE 9103309).

## References

Costello, J. H. & S. P. Colin, 1994. Morphology, fluid motion and predation by the scyphomedusa *Aurelia aurita*. Mar. Biol. 121: 327–334.

Costello, J. H. & S. P. Colin, 1995. Flow and feeding by swimming scyphomedusae. Mar. Biol. 124: 399–406.

Fancett, M. S., 1988. Diet and selectivity of scyphomedusae from Port Phillip Bay, Australia. Mar. Biol. 98: 503–509.

García, J. R. & E. Durbin, 1993. Zooplanktivorous predation by large scyphomedusae *Phyllorhiza punctata* (Cnidaria: Scyphozoa) in Laguna Joyuda. J. exp. mar. Biol. Ecol. 173: 71–93.

Hamner, W. M. & I. R. Hauri, 1981. Long-distance horizontal migrations of zooplankton (Scyphomedusae: *Mastigias*). Limnol. Oceanogr. 26: 414–423.

Hsieh, Y-H. P., P-M. Leong & J. Rudloe, 2001. Jellyfish as food. Hydrobiologia, this volume.

Lange, J. & R. Kaiser, 1995. The maintenance of pelagic jellyfish in the Zoo-Aquarium Berlin. Int. Zoo Yb. 34: 59–64.

Larson, R. J., 1991. Diet, prey selection and daily ration of *Stomolophus meleagris*, a filter-feeding scyphomedusa from the NE Gulf of Mexico. Estuar. coast. shelf Sci. 32: 511–525.

Mayer, A. G., 1910. Medusae of the World III. The Scyphomedusae. Carnegie Inst. of Washington, Washington, D.C.: 733 pp.

Mianzan, H. W. & P. S. Cornelius, 1999. Cubomedusae and scyphomedusae. In Boltovskoy, D. (ed.), South Atlantic Zooplankton. Backhuys Publishers, Leiden: 513–559.

Omori, M. & E. Nakano, 2001. Jellyfish fishery in southeast Asia. Hydrobiologia 451 (Dev. Hydrobiol. 155): 19–26.

Shanks, A. L & W. M. Graham, 1987. Oriented swimming in the jellyfish *Stomolophus meleagris* L. Agassiz (Scyphozoan: Rhizostomida). J. exp. mar. Biol. Ecol. 108: 159–169.

Sullivan, B. K., J. R. Garcia & G. Klein-MacPhee, 1994. Prey selection by the scyphomedusan predator *Aurelia aurita*. Mar. Biol. 121: 335–341.

Vogel, S., 1994. Life in moving fluids: the physical biology of flow. Princeton Univ. Press, Princeton: 467 pp.

Von Ledenfeld, R., 1884. The scyphomedusae of the Southern Hemisphere. Proc. linn. Soc. New South Wales 9: 259–306.

*Hydrobiologia* **451**: 229–246, 2001.
© 2001 *Kluwer Academic Publishers.*

# Reproduction and life history strategies of the common jellyfish, *Aurelia aurita*, in relation to its ambient environment

Cathy H. Lucas

*School of Ocean & Earth Science, University of Southampton, Southampton Oceanography Centre, European Way, Southampton, SO14 3ZH, U.K.*
*Tel: +44 (0)23 8059 6270. Fax: +44 (0)23 8059 3059. E-mail: Catherine.H.Lucas@soc.soton.ac.uk*

*Key words:* medusa, polyp, reproduction, recruitment, temperature, food

## Abstract

The scyphozoan *Aurelia aurita* (Linnaeus) is a cosmopolitan species, having been reported from a variety of coastal and shelf sea environments around the world. It has been extensively studied over the last 100 years or so, and examination of the literature reveals three striking features: (1) the presence of populations in a wide range of environmental conditions; (2) large inter-population differences in abundance and life history patterns over large and small spatial scales; and (3) inter-annual variability in various aspects of its population dynamics. *A. aurita* is clearly a highly flexible species that can adapt to a wide range of environmental conditions. While various physiological and behavioural characteristics explain how *A. aurita* populations can take advantage of their surrounding environment, they do not explain what governs the observed temporal and spatial patterns of abundance, and the longevity or lifespan of populations. Understanding these features is necessary to predict how bloom populations might form. In a given habitat, the distribution and abundance of benthic marine invertebrates have been found to be maintained by four factors: larval recruitment (sexual reproduction), migration, mortality and asexual reproduction. The aims of this review are to determine the role of reproduction and life history strategies of the benthic and pelagic phases of *A. aurita* in governing populations of medusae, with special attention given to the dynamic interaction between *A. aurita* and its surrounding physical and biological environment.

## Aurelia aurita populations

### General observations

The common moon jellyfish, *Aurelia aurita* (Linnaeus), is a cosmopolitan species with a worldwide distribution in neritic waters between 70° N and 40° S (Kramp, 1961; Russell, 1970). It has been reported from a variety of coastal and shelf sea marine environments particularly in northwestern Europe (Palmén, 1954; Möller, 1980; Hernroth & Gröndahl, 1983; Van Der Veer & Oorthuysen, 1985; Lucas & Williams, 1994; Olesen et al., 1994; Schneider & Behrends, 1994), the Black Sea (Shushkina & Arnotauv, 1985; Lebedeva & Shushkina, 1991; Mutlu et al., 1994), Japan (Yasuda, 1969; Ishii et al., 1995; Omori et al., 1995), and parts of North America (Hamner et al., 1994; Greenberg et al., 1996). *A. aurita* has a complex life cycle, comprising an alteration of generations between an asexual benthic polyp, or scyphistoma, and a sexual pelagic medusae. Examination of the wealth of literature on *A. aurita* populations reveals three striking features, summarised as follows: (1) the presence of populations in a wide range of environmental conditions; (2) large inter-population differences in abundance and life history patterns, and (3) inter-annual variability in abundance, growth, and timing of reproduction within single populations.

### Habitats

*A. aurita* populations are found predominantly in coastal embayments, fjords, and estuaries where there are suitable substrata for the benthic scyphistoma polyp. Within these habitats, the degree of containment, tidal flow, water depth, temperature and salinity, and trophic condition can vary quite considerably (Table 1).

*Table 1.* Summary of the documented habitats of *Aurelia aurita* medusae, together with their maximum abundance and biomass. Location Types: s-e est = semi-enclosed estuary; s-e l = semi-enclosed lagoon; s-e fjd = semi-enclosed fjord; o-fjd = open fjord; o-b = open bay; s-e b = semi-enclosed bay; o = open water. nd = no data. [a]Converted from published data using WW:DW:C relationships reviewed by Hirst & Lucas (1998). References: 1. Lucas & Williams (1994); 2. Lucas (1996); 3. Olesen et al. (1994); 4. Riisgård et al. (1995); 5. Hernroth & Gröndahl (1983); 6. Gröndahl (1988a); 7. Omori et al. (1995); 8. Panayotidis et al. (1988); 9. Hamner et al. (1982); 10. Ishii & Båmstedt (1998); 11. Schneider (1989b); 12. Möller (1980); 13. Schneider & Behrends (1994); 14. Mutlu et al. (1994); 15. Van Der Veer & Oorthuysen (1985); 16. Yasuda (1969); 17. Yasuda (1971); 18. Rasmussen (1973)

| Location | Type | Depth (m) | Notes | Productivity | Max. Abundance (No. m$^{-3}$) | Max. Biomass (mg C m$^{-3}$) | Ref |
|---|---|---|---|---|---|---|---|
| Southampton Water, U.K. | s-e est | 15 | industrial | High (12.8 mg C m$^{-3}$ mesozooplankton) | 8.71 | 30.2 | 1,2 |
| Horsea Lake, U.K. | s-e-l | 6–7 | no river | Low (5.2 mg C m$^{-3}$ mesozooplankton) | 24.9 | 43.2 | 2 |
| Kertinge Nor, Denmark | s-e fjd | 2–3 | | Eutrophic + low mesozooplankton | 300 | 350–450 | 3,4 |
| Gullmarfjord, Sweden | 0-fjd | 120 | winter ice | nd | 14.96 | nd | 5,6 |
| Tokyo Bay, Japan | o-b | nd | | Eutrophic (400 mg DW m$^{-3}$ zooplankton) | 1.53 | 125[a] | 7 |
| Elefsis Bay, Greece | s-e b | 33 | anoxia | Eutrophic (471 mg DW m$^{-3}$ zooplankton) | 44 | 95[a] | 7 |
| Jellyfish Lake, Palau | s-e l | 30 | sulphide | High | 0.30 | nd | 9 |
| Vågsbøpollen, Norway | s-e b | 12 | | Very low zooplankton | 22.3 | 710 | 10 |
| Eckernförde Bay, Kiel Bight | o-fjd | 20 | | 26.7 mg C m$^{-3}$ zooplankton | 0.03–0.23 | 54.8 | 11 |
| Kiel fjord, Kiel Bight | o-fjd | 8 | | nd | 0.33 | ~60 | 12,13 |
| Kiel Bight | o | nd | | nd | ~0.14 | ~35 | 13 |
| Black Sea | o | >200 | *Mneiopsis* | Eutrophic | ? | 0.21 g C m$^{-2}$ | 14 |
| Dutch Wadden Sea | o | 5–10 | North Sea | High ? | 0.49 | 175 | 15 |
| Urazoko Bay, Japan | s-e b | nd | | nd | 71 | nd | 16,17 |
| Isefjord, Denmark | o-fjd | 9 | winter ice | Eutrophic | nd | nd | 18 |

Medusa abundance is generally higher in small, shallow, semi-enclosed or enclosed systems with limited tidal exchange (Olesen et al., 1994; Lucas, 1996; Ishii & Båmstedt, 1998), than in open water systems or where depths exceed several hundreds of metres (Van Der Veer & Oorthuysen, 1985; Mutlu et al., 1994; Schneider & Behrends, 1994). On a seasonal and small-scale spatial basis, temperate estuaries often have widely fluctuating temperature and salinity regimes, and *A. aurita* can be considered both eurythermal and euryhaline in its distribution (Fig. 1). Populations occur in fjords that experience winter ice cover (Rasmussen, 1973; Hernroth & Gröndahl, 1983, 1985a), and in tropical, aseasonal environments such as Jellyfish Lake, Palau, where surface water temperature rarely deviates from 31 to 32 °C (Hamner et al., 1982; Dawson & Martin, 2001). Similarly, *A. aurita* are found in salinities ranging from 14 to 22 in Kertinge Nor, Denmark (Olesen et al., 1994) to 38 in Elefsis Bay, Greece (Papathanassiou et al., 1987), although there are also reports of populations living in salinities of <10 (see Russell, 1970).

As well as the physical parameters described above, one of the most conspicuous aspects of *A.*

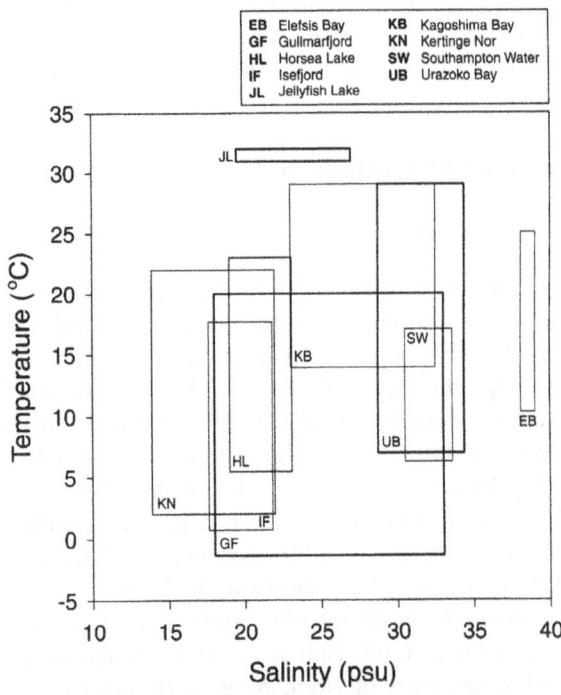

*Figure 1.* Temperature and salinity ranges of *Aurelia aurita*.

*aurita* distribution is its presence in systems with highly contrasting nutrient inputs, productivity and food availability (Table 1). For example, Horsea Lake in southern England has no riverine input to supply nutrients, and a numerically- and species-poor zooplankton community that is controlled by both bottom-up (chlorophyll *a* <2 mg m$^{-3}$) and top-down (*A. aurita* 43 mg C m$^{-3}$) pressure (Lucas et al., 1997). On the other hand, *A. aurita* populations may also be abundant in polluted, eutrophic systems that may have anoxic bottom waters. Examples include Elefsis Bay (Papathanassiou et al., 1987; Panayotidis et al., 1988), Kertinge Nor (Olesen et al., 1994; Riisgård et al., 1995) and the Black Sea (Mutlu et al., 1994).

## Variations among populations

*A. aurita* populations around the world are characterised by a great diversity in population and life history characteristics, such as abundance, growth, timing and periodicity of strobilation, timing of and size at sexual maturation and longevity of the medusa (Fig. 2). Significant differences in population characteristics can occur over small (10s of kilometres) and large (100s of kilometres) spatial scales, with the role of environmental conditions such as temperature and food availability considered to play a key role.

In many of the populations studied, the main period of strobilation, resulting in the liberation of ephyrae, starts in the late winter/early spring (Thiel, 1962; Möller, 1980; Van Der Veer & Oorthuysen, 1985; Lucas & Williams, 1994; Omori et al., 1995). Autumn and spring periods of strobilation have also been reported in parts of northern Europe (Rasmussen, 1973; Hernroth & Gröndahl, 1983), with the autumn generation of ephyrae undergoing a winter diapause in deeper water (Hernroth & Gröndahl, 1983, 1985a). For the most part, the winter–spring recruitment of ephyrae develop into medusae in early spring, which then remain in the water column until the summer or early autumn (Möller, 1980; Papathanassiou et al., 1987; Lucas & Williams, 1994), thus forming distinct seasonal patterns of medusa abundance and biomass. Prolonged or even semi-continuous periods of strobilation have also been reported in some areas, resulting in the presence of ephyrae in the water column for much of the year (Yasuda, 1968; Hamner et al., 1982; Schneider, 1989b; Lucas, 1996). Year-round populations of ephyrae and medusae occur in Urazoko and Tokyo Bays, Japan (Yasuda, 1971; Omori et al., 1995),

Jellyfish Lake (Hamner et al., 1982) and Horsea Lake (Lucas, 1996).

Patterns of growth and the maximum bell diameter attained by medusae are highly variable among populations. Three growth phases are typical of most populations: slow growth during the winter and early spring; exponential growth once temperature and food availability have increased in mid-spring; and finally, shrinkage in the summer and autumn during gamete release (Möller, 1980; Schneider, 1989b; Lucas & Williams, 1994). Medusae typically reach 200–300 mm maximum bell diameter, although they can be as large as 450 mm (M. Omori, pers. comm.). Comparison among populations reveals an inverse relationship between medusa abundance and maximum bell diameter (Fig. 3), suggesting that there is a density-dependent mechanism governing *A. aurita* populations, as was first suggested by Schneider & Behrends (1994) for the Kiel Bight population. In severely food-limited populations such as in Kertinge Nor (Olesen et al., 1994) and Horsea Lake (Lucas, 1996), instantaneous growth rates are <0.1 d$^{-1}$, and mean bell diameter of adult medusae is typically <50 mm.

Medusae typically reach sexual maturation at the upper end of the size range of the population (Lucas & Lawes, 1998). In most situations adult medusae shrink and die following spawning. It is thought that this is caused by extrusion of gastric filaments during gamete release, resulting in morphological degradation and susceptibility to parasitic invasion (Spangenberg, 1965). Longevity of medusae is impossible to quantify if there is semi-continuous recruitment. Even in univoltine populations, the age of an individual is extremely difficult to determine. Nevertheless, it seems that in most environments, medusae live for 4–8 months. Individuals living for >1 year have been reported by Yasuda (1971) and Hamner & Jenssen (1974),whilst Miyake et al. (1997) consider that not only do medusae from Kagoshima Bay, Japan, live for 2 years, but that individuals spawn twice in successive summers. Recently, attempts have been made to use morphological and chemical characteristics to determine age of medusae. Kakinuma et al. (1993) and Miyake et al. (1997) found that the number of branching points on the radial canals of the water vascular system increased with time, independently of environmental factors, and therefore could be used as an age index for *A. aurita*. Ishii et al. (1995) found that *A. aurita* medusae in the shrinkage phase following gamete release had lower concentrations of

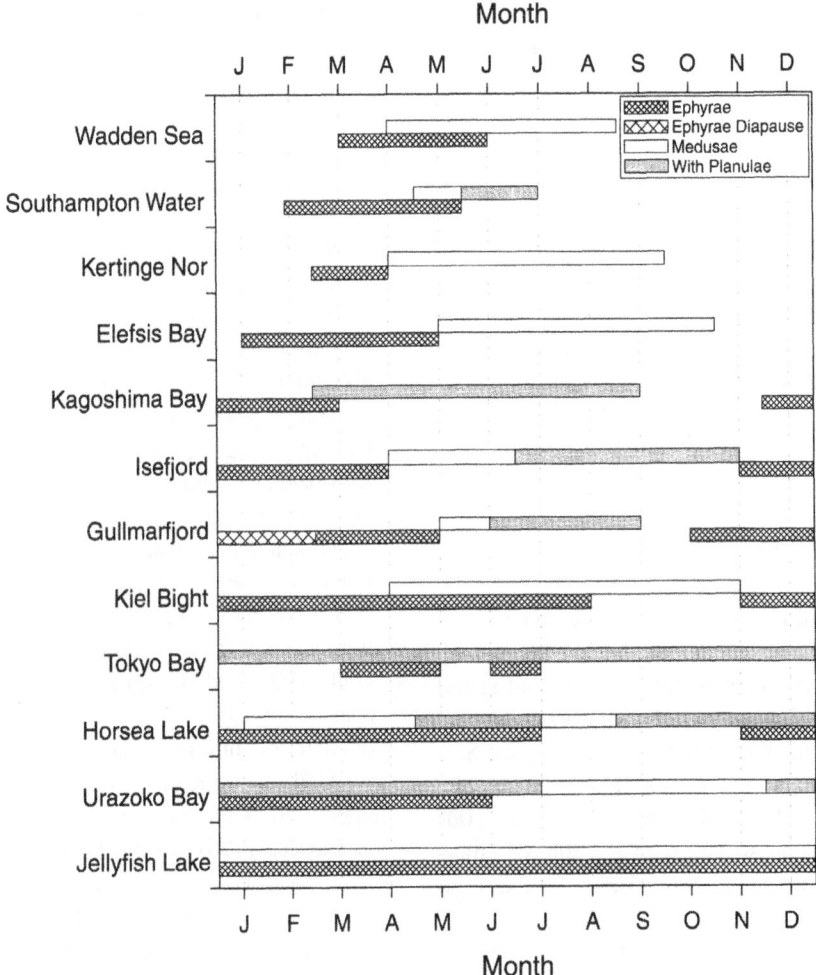

*Figure 2.* Timing of the appearance of ephyrae, medusae and planula-bearing medusae (where determined) in a variety of *Aurelia aurita* populations.

ATP-related compounds compared with those in their growth phase.

*Inter-annual variability*

Although long-term datasets are rather scarce in jelly-fish research, a few regions have been studied over periods of 5 years or more. In addition to the inter-population differences described above, these studies have highlighted that there can be considerable intra-population variability in timing of ephyrae release, abundance, growth, bell diameter and reproductive output, both in the shorter term (i.e. between consec-utive years), and over decadal scales. Over the shorter time-scale, one of the most comprehensive datasets for *A. aurita* is in the western Baltic Sea, with studies by Möller (1980), Schneider (1989b) and Schneider & Behrends (1994, 1998). Years of high numbers of

smaller, lighter medusae were followed by years of low numbers of larger, heavier medusae, indicating a density-dependent mechanism regulating adult size. In addition, the dense populations of medusae pro-duced fewer numbers of high C:N ratio planula larvae, while the lower abundance medusae produced very large numbers of planulae with low amounts of storage compounds (Schneider, 1988).

Inter-annual variability in the timing of ephyrae appearance, maximum abundance and bell diameter has also been observed in Southampton Water (Re-ubold, 1988; Zinger, 1989; Lucas & Williams, 1994, Lucas, unpublished data). Although there has been a considerable decline in medusa abundance between the mid-1980s and early 1990s, the density-dependent relationship between abundance and bell diameter (see Schneider & Behrends, 1994) is not readily apparent

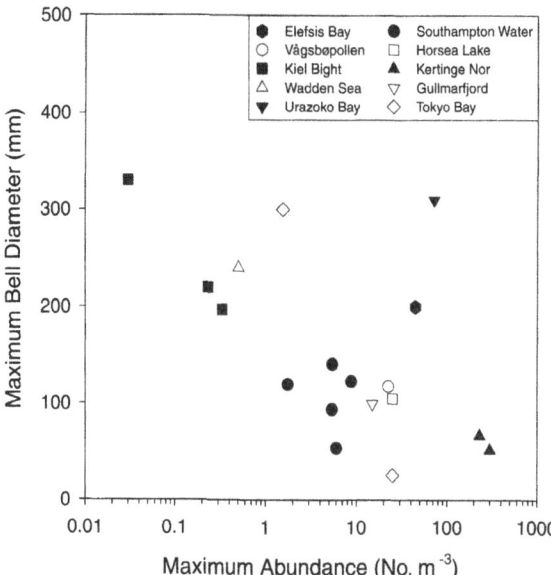

*Figure 3.* Multiple-population relationship between maximum bell diameter and medusa abundance of *Aurelia aurita*.

*Table 2.* Inter-annual variation in the appearance, maximum abundance and bell diameter of the *Aurelia aurita* population in Southampton Water, U.K. Data from Reubold (1988); Zinger (1989); Lucas & Williams (1994); Lucas (unpubl. data); Hirst (1996). No sampling in 1989 and 1995; nd = no data

| Year | Appearance (Julian Days) | | Max. Abundance (No. m$^{-3}$) | Max. Bell Diameter (mm) |
|------|------|------|------|------|
| | From | To | | |
| 1985 | nd | 177 | 29.96 | nd |
| 1986 | 59 | 176 | 1.16 | nd |
| 1987 | nd | nd | 21.0 | 145 |
| 1988 | 42 | 135 | nd | nd |
| 1990 | 61 | 159 | 5.39 | 141 |
| 1991 | 77 | 179 | 8.71 | 123 |
| 1992 | 20 | 133 | 5.98 | nd |
| 1993 | 59 | 173 | 5.35 | 94 |
| 1994 | 42 | 174 | 1.75 | 135 |
| 1996 | nd | 182 | 2.48 | 130 |

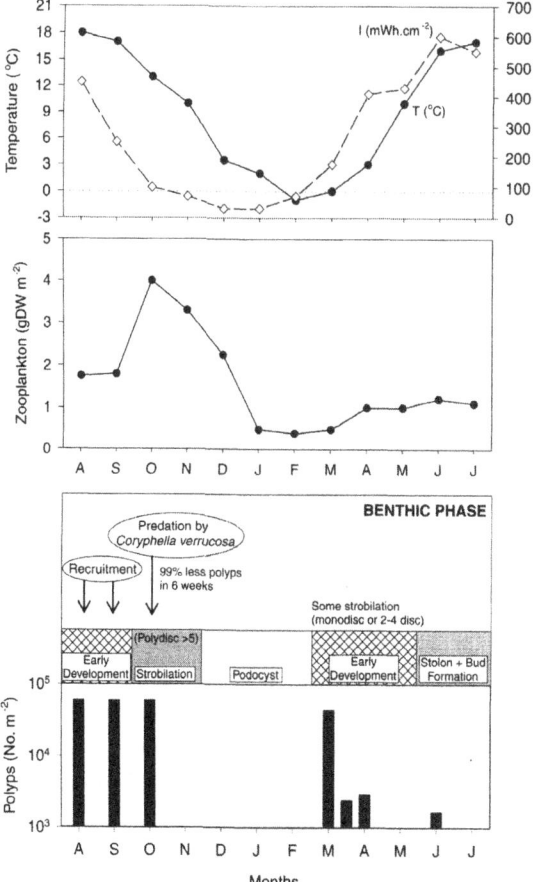

*Figure 4.* Activity of the benthic phase of the life cycle of *Aurelia aurita* in the Gullmarfjord, Sweden, with relevant environmental variables (re-drawn from Gröndahl, 1988a).

in Southampton Water. This implies that factors in addition to food availability regulate size and density of the Southampton Water population. The very small medusae in 1992, for example, may have been due to the temporal shift in their presence to the early part of the year, when low water temperature and food availability would have resulted in low growth rates (Hansson, 1997). Regarding the timing of ephyrae appearance, this has been rather irregular over the 11 years of study, varying by 58 days (Table 2). Nevertheless, longevity of the population has been consistent at 90–115 d, except in 1988. In the Isefjord, longevity of the medusa population has been related to summer temperatures (Rasmussen, 1973), and this may be the case in Southampton Water.

*Impacts of* Aurelia aurita

When abundant, *A. aurita* medusae can have a significant impact on coastal plankton communities. In several temperate coastal systems, it has been reported that the spring-summer reduction in zooplankton biomass is caused by *A. aurita* predation (Schneider & Behrends, 1994, 1998; Lucas et al., 1997). When gelatinous predatory pressure on herbivores is sufficiently intense, and nutrients and light are not limiting, top-down regulation of plankton communities may also occur, resulting in increased phytoplankton blooms and alteration of species composition (Lindahl & Hernroth, 1983; Olsson et al., 1992; Smayda,

1993). Predators such as *A. aurita* may also affect fish standing stocks, either directly by predation on fish larvae, or indirectly by competing with fish larvae for available food resources (Purcell & Arai, 2001).

The role of jellyfish and other gelatinous plankton in marine ecosystems is of interest not only biologically but also socio-economically. Bloom populations of *A. aurita* can have detrimental effects on fisheries, tourism and power stations, and mass occurrences of *A. aurita* have been reported from many regions. The most notable cases include, Elefsis Bay (Papathanassiou et al., 1987; Panayotidis et al., 1988), Kertinge Nor (Olesen et al., 1994), Horsea Lake (Lucas, 1996), Vågsbøpollen, Norway (Ishii & Båmstedt, 1998), and the Black Sea (Shushkina & Musayeva, 1983; Shushkina & Arnautov, 1985; Mutlu et al., 1994). While some of these blooms are considered to have been brought about by restricted water exchange, eutrophication and overfishing (Zaitsev, 1992; Kideys, 1994; Ishii & Båmstedt, 1998), it is clear that *A. aurita* is a highly flexible species that can successfully adapt to a wide range of environmental conditions.

## Causes of population variation

Several aspects of the physiology and behaviour of *A. aurita* enable it to utilise the available food to its maximum advantage. The mucous and ciliary current prey capturing mechanism means that *A. aurita* can be considered a 'generalist feeder', with a wide range of mesozooplankton (Matsakis & Conover, 1991; Sullivan et al., 1994), some microzooplankters (Stoecker et al., 1987; Båmstedt, 1990), and fish larvae (Möller, 1980) observed in the diets of individuals. There is evidence of prey selectivity (Sullivan et al., 1994, 1997), although reports are mixed. Nevertheless, as with most gelatinous predators, they are capable of very high clearance rates, even at prey densities in excess of those found *in situ* (Båmstedt, 1990). Thus, they are able to utilise dense zooplankton populations and prey patches effectively.

When food is sufficient, *A. aurita* is capable of extremely rapid growth because of its low carbon density, and this results in a large surface area and further increase in prey capture capacity. In Kiel Bight, for example, the maximum increase in mean bell diameter was in the order of 5.7 mm d$^{-1}$ (Möller, 1980), while in Southampton Water it was 4.9 mm d$^{-1}$ (Lucas & Williams, 1994). When food is scarce, medusae are also capable of reversible shrinkage, which allows them to survive considerable periods of starvation. There do not appear to be any long-term side-effects, as the medusae grow to their original size, and the gonads, which may have been resorbed, develop and become fertile when food becomes available again (Hamner & Jenssen, 1974).

While these characteristics explain how *A. aurita* populations can take full advantage of their surrounding environment, they do not explain what governs the observed temporal and spatial patterns of abundance and distribution, or the longevity of populations. The most likely explanation for high medusa abundance in shallow, enclosed systems is that there is high recruitment at both the planula and ephyra stages because of the presence of suitable substrata in shallow water for the scyphistomae, and containment of the adult planula-bearing population (Ishii & Båmstedt, 1998). Nonetheless, remarkably little is known about the role of reproduction, recruitment and life history strategies in governing population size of scyphomedusae, and how these might be affected by environmental factors.

## Reproduction and life-history strategies

Scyphozoans have a life cycle comprising a planktonic sexually-reproducing medusa and a benthic asexually-reproducing polyp. Selection occurs at all stages of a life cycle, but mortality during the juvenile stages of planktonic-benthic life cycles can have a major effect on the abundance of the adult population. Thus, mortality of the planula, polyps and ephyrae may be important in the development of large inter-annual variations in the abundance of mature medusae. Gröndahl (1988b) considered mortality of planulae and polyps to be more important than mortality of ephyrae and medusae for the abundance of the medusa population.

In contrast to the medusa, we know very little about the scyphistoma and its variants. The few studies that have examined the benthic phase of the *A. aurita* life cycle include those of Hernroth & Gröndahl (1985a, b) and Gröndahl (1988a, b, 1989) in the Gullmarfjord, Sweden, Brewer (1978), Keen (1987) and Keen & Gong (1989) in North America, and Miyake et al. (1997) and Watanabe & Ishii (2001) in Japan. Production of polyps has also been studied in *Cyanea capillata* (Linnaeus) (Gröndahl, 1988a; Brewer & Feingold, 1991), and *Chrysaora quinquecirrha* (Desor) (Purcell et al., 1999; Condon et al., 2001).

Most of the research on reproduction and life history strategies in the marine environment has been carried out on benthic invertebrates. In a given habitat, the distribution and abundance of benthic marine invertebrates is maintained by 4 factors: recruitment via sexual reproduction, migration, mortality and asexual reproduction (Chia, 1989), with each of these affected by the ambient environment. The roles of food, temperature, salinity and photoperiod have long been considered important in synchronising annual cycles of reproduction and influencing reproductive output in marine invertebrates, with much of the evidence based on correlations between gonad maturation, spawning, and environmental variables (Olive, 1985). Investigations into how the reproductive biology of jellyfish is affected by environmental conditions are, however, extremely rare.

The life cycle and developmental biology of *A. aurita* has been reviewed comprehensively by Arai (1997). Here, I will consider the dynamic interactions between *A. aurita* and its physical and biological environment, and how these can lead to blooms of medusae. In particular, I will draw upon regions that have long-term data sets or have been fairly comprehensively studied: Kiel Bight in the Baltic Sea, the Gullmarfjord in western Sweden, Southampton Water and Horsea Lake in the U. K., and Tokyo Bay and Urazoko Bay in Japan. These studies have been invaluable in providing information on complete life cycles and inter-annual variability in population dynamics. In addition to the influence of environmental variables on *A. aurita*, another possibility is that the recognised species, *A. aurita*, is in fact made up of more than one species, and that genotype causes divergence in physiological and reproductive characteristics (Dawson & Martin, 2001). Brief reference to this topic will be made later.

## Life history stages of *Aurelia aurita* and its environment

### Asexual benthic phase

Brewer & Feingold (1991) recognised that "the temporal distribution of the conspicuous medusa is regulated by factors affecting the occurrence and activity of the inconspicuous polyp", and yet we know very little about what influences the benthic phase of the life cycle of scyphozoans. As with the pelagic medusa, the benthic scyphistoma of *A. aurita* is highly adaptable to a wide range of environmental conditions, and

has similar feeding and growth characteristics. The scyphistoma displays low feeding selectivity (Tsikon-Lukanina et al., 1995), and can also survive several months of starvation (Hiromi et al., 1995). The distribution and abundance of the benthic population is determined by: (1) recruitment of planulae to the seabed, (2) vegetative budding of the scyphistoma; and (3) inter- and intra-specific predation. Recruitment to the pelagic phase usually occurs via strobilation of the scyphistoma, although direct development of ephyrae from planulae has also been reported (Kakinuma, 1975). Each of these activities may be stimulated or influenced in some way by environmental variables.

The timing and periodicity of recruitment of planulae to the seabed obviously depends on sexual maturation of the adult population, which can either be continuous or occur over a 2–3-month period in summer or autumn. Planulae of *A. aurita* are lecithotrophic, spending between 12 h and 1 week in the water column prior to settling. The reason for this time range is unclear, but is probably related to the effect of water temperature on metabolism and survivorship, and on turbulence transporting the larvae to suitable substrata in shallow water (Schneider & Weisse, 1985). Although there is no evidence in the literature, mortality of planulae in the water column is likely to be high. Successful recruitment of planulae depends partly on the reproductive strategy of the medusa. *A. aurita* is considered to be an r-strategist, so that high mortality is offset by the production of large numbers of larvae. The low numbers of larvae that are produced in years of high medusa abundance have high organic content, which would presumably promote increased survivorship to settlement (Schneider, 1988).

Factors affecting the settlement and subsequent metamorphosis of marine invertebrate larvae include: (1) predation by benthic fauna, (2) physical characteristics of the substratum, including boundary layer properties; (3) contact with biofilms; (4) gregarious settlement with conspecifics; and (5) presence of compounds produced by adults. Often, a hierarchical combination of factors is required (see Keen, 1987). Polyps of *A. aurita* are found on the undersides of virtually any hard substrata – bare rock, shells, amphipod and polychaete tubes, ascidians and macroalgae (Russell, 1970; Rasmussen, 1973; Brewer, 1978; Miyake et al., 1997). They also readily colonise glass, ceramic or plastic settling plates. *In situ* experiments have been carried out at a variety of depths between 0.3 and 25 m (Keen, 1987; Hernroth & Gröndahl, 1985b; Brewer &

Feingold, 1991), although Russell (1970) reports that polyps are not found deeper than 20 m.

Brewer (1978) and Gröndahl (1989) found evidence of gregarious settlement and metamorphosis of *A. aurita* planulae, whereas Keen (1987) observed that recruitment of planulae was dependent on areas of low shear stress and relatively thick boundary layer, and independent of conspecific density. Based on the tidal regimes of the study sites of Keen (1987) and Gröndahl (1989), the latter considered that in areas with strong water movement (i.e. Eel Pond, Woods Hole), hydrodynamic processes were more important than conspecific density. Although Gröndahl (1989) found evidence of gregarious settlement and metamorphosis, predation experiments by the same author (Gröndahl, 1988b) also revealed that polyps of *A. aurita* preyed upon planulae of *C. capillata* and, to a lesser degree on planulae of conspecifics. The difference between the two studies was that 4 d old polyps (0.28 mm) were used by Gröndahl (1989), while 10 d old polyps (0.62 mm), which had better-developed feeding tentacles, were used by Gröndahl (1988b).

Dense populations of established *A. aurita* polyps may affect the recruitment of planulae to the seabed because they are competing for space and food. Gröndahl (1989) considers that planulae develop behavioural patterns to settle in areas with conspecific polyps, while polyps develop behavioural patterns to avoid overcrowding. The relative effects of pulsed or continuous inputs of planulae on recruitment success are unknown.

In a major study in the Gullmarfjord (Hernroth & Gröndahl, 1983, 1985a, b; Gröndahl, 1988a, b), settling plates were used to evaluate the annual cycle of *A. aurita* polyp development in relation to environmental variables (Fig. 4). Peak polyp densities of between 60 000 and 400 000 m$^{-2}$ occurred between July and October following recruitment of planulae. Dense aggregations in the field may arise from high planulae recruitment as well as from asexual budding of new polyps. In the Gullmarfjord, this type of cloning occurred in the spring, and so did not contribute to the observed peak densities. During the autumn, the developing scyphistomae became highly polydisc and strobilated between October and mid-November. This is the period of peak ephyrae abundance. Occasionally, monodisc or moderately polydisc strobilation occurred in spring. A dramatic decline in polyp density during October–November resulted from predation by the nudibranch *Coryphella verrucosa* (M. Sars). Because *C. verrucosa* is highly stenophagous, a tight

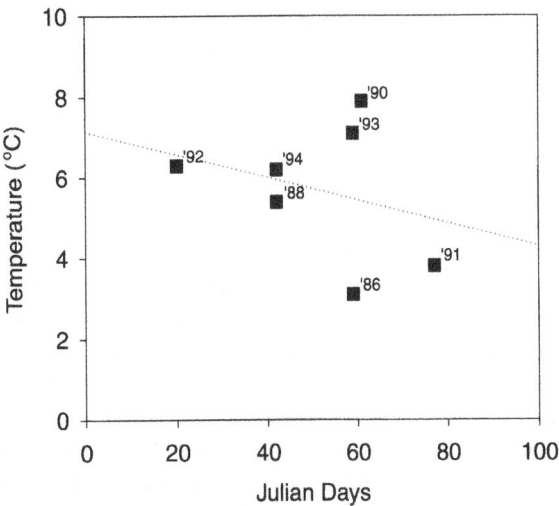

*Figure 5.* Relationship between the initial timing of *Aurelia aurita* ephyrae appearance and minimum winter temperature in Southampton Water (Lucas, unpubl. data.).

coupling between predator and prey abundance was observed, so that nudibranch density decreased rapidly once the polyps had been consumed. Gröndahl (1988a) considered interannual variation in ephyrae abundance to be controlled by predation of the polyps by *C. verrucosa*. Following the outbreak of *C. verrucosa*, the few surviving polyps developed into podocysts, and remained as such throughout the winter and early spring. The nudibranch *Facelina bostoniensis* (Couthouy) (Thiel, 1962) has also been found to prey on *A. aurita* polyps.

Little is known about the factors that stimulate cyst formation, although scyphistomae seem to encyst during unfavourable conditions as a mechanism of ensuring the survival of a cohort through to the next period of strobilation. In the Gullmarfjord, podocysts occur during the winter when both water temperature and zooplankton biomass are at their minimum. By contrast, Brewer & Feingold (1991) observed that *Cyanea* sp. encysted during the summer, but suggested that this was for protection against predators and competitors.

Polyps of scyphozoans are perennials, surviving for up to several years in the laboratory. Reports on *in situ* longevity are scarce, although Brewer & Feingold (1991) considered that polyps in the field could survive for prolonged periods of many months in the encysted state. Vegetative budding (cloning) of daughter polyps helps to counterbalance mortality and increases the survivorship of the polyp population for future strobilation. Budding in *A. aurita* is varied (Berrill, 1949).

Kakinuma (1975) identified four main types of asexual reproduction: (1) budding from the base of the parent; (2) stolon formation; (3) longitudinal fission; and (4) the formation of new polyps from the pedal disc (pedal laceration). The controls over asexual proliferation are very poorly understood. In Kakinuma's laboratory study, elevated temperatures (25 °C) were optimal for all types of bud formation. Keen & Gong (1989) found that budding rate and bud size increased with feeding frequency. This would suggest that at low food levels, the strategy of increasing the surface area to volume ratio and maximising the per-polyp rate of food capture is preferred over the strategy of utilising the limited food resources for producing a few large individuals with potentially larger reproductive output (strobilation of ephyrae). However, the total volume of tissue produced displayed plasticity, depending on genotype, food supply and the interaction between the two. In the field, budding and stolon formation tend to be observed during the summer months, when water temperatures and zooplankton biomass are elevated (see Berrill, 1949; Gröndahl, 1988a).

Initiation of the process of strobilation in *A. aurita* has most frequently been attributed to, or correlated with, changing temperature, levels of irradiance and food supply, although no variable has been singled out as the major regulator (Verwey, 1942; Thiel, 1962; Spangenberg, 1968). Nevertheless, most authors agree that the main period of strobilation occurs following a reduction in temperature. Purcell et al. (1999) also observed that in *C. quinquecirrha*, each 5 °C reduction in temperature delayed peak production of ephyrae by 1 week. The trend of later appearance of *A. aurita* ephyrae at more northerly latitudes in Europe (and North America) has been attributed to the lower temperatures at these locations (Verwey, 1942; Thiel, 1962). However, there are many exceptions to this. Within populations, year-to-year variations in timing of strobilation are more readily explained by differences in winter temperature (Palmén, 1954; Rasmussen, 1973). A similar trend of delayed ephyrae appearance in colder winters is also observed in Southampton Water (Table 2, Fig. 5), although the relationship is not significant. Other factors such as food availability during the winter period may also be important. In contrast to the above observations, some *A. aurita* populations (e.g. Gullmarfjord) consistently release their first ephyrae within a 10 d time-period each year (Gröndahl, 1988a). It is not clear which factor or combination of factors, synchronise strobilation to occur in such a small time frame.

Food supply also plays a role in the strobilation process. Many authors agree that a relatively long period of high food density for the build up of storage products is necessary for strobilation (Thiel, 1962). In addition, the number of segments produced is strongly dependent on food availability (Thiel, 1962; Spangenberg, 1968), presumably because well-fed polyps grow larger and have greater reproductive capacity (Keen & Gong, 1989). Up to 20–30 ephyrae can be produced per scyphistoma (Berrill, 1949). In the Gullmarfjord, polydisc strobilation coincided with maximum zooplankton biomass, while in the spring, when zooplankton biomass was low, strobilation either did not occur or it was monodisc or moderately polydisc. In several locations, such as Kiel Bight and Horsea Lake, there is a secondary peak of ephyra abundance during mid-late spring, which could be a response to elevated food levels. Individual polyps can strobilate several times a year in response to high food levels (Thiel, 1962), and those that are strobilating may become polydisc, so produce more ephyrae per polyp.

Although some broad generalisations have been made regarding strobilation, caution should be applied when directly comparing different populations, and there are many more questions than answers. If strobilation occurs only a few weeks after a drop in temperature (Brewer & Feingold, 1991), one would expect winter to be the main period of ephyrae release. In many populations, however, peak ephyrae release occurs during the spring when water temperatures are rising (Table 3, Fig. 6). In high latitudes where there is winter ice cover, the very low temperatures cause strobilation to stop. Therefore, ephyrae release can not take place until the late spring (Rasmussen, 1973; Kakinuma, 1975). In the Gullmarfjord, however, peak strobilation occurs in the autumn before the ice cover forms. In this case, strobilation is most likely stimulated by a combination of decreasing temperature and maximum zooplankton biomass. The reasons for the spring occurrence of ephyrae in some warmer environments, such as Tokyo and Urazoko Bays in Japan, and Elefsis Bay in Greece are unclear as these locations do not experience particularly low winter temperatures.

What governs the longevity of ephyrae production? Ephyrae may occur for only a couple of months (e.g. Southampton Water), several times a year (e.g. Horsea Lake, Kiel Bight), or year round, as in Jellyfish Lake, where temperatures rarely deviate from 31 to 32 °C. Strobilation can be stimulated throughout the year in the laboratory, but conditions are not usually representative of the ambient environment. If food is a

238

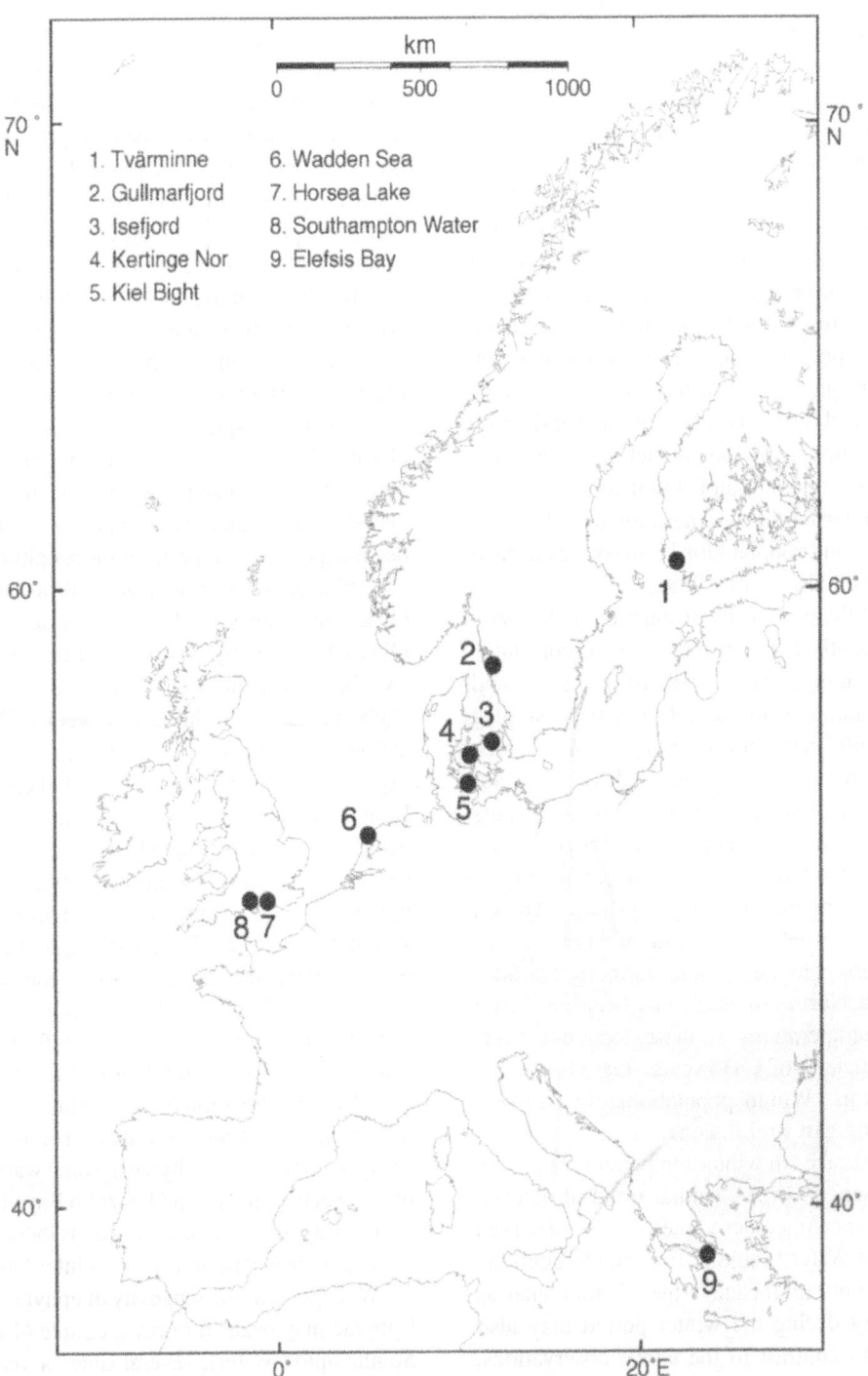

*Figure 6.* Map of Western Europe, showing the locations of *Aurelia aurita* sites summarised in Table 3.

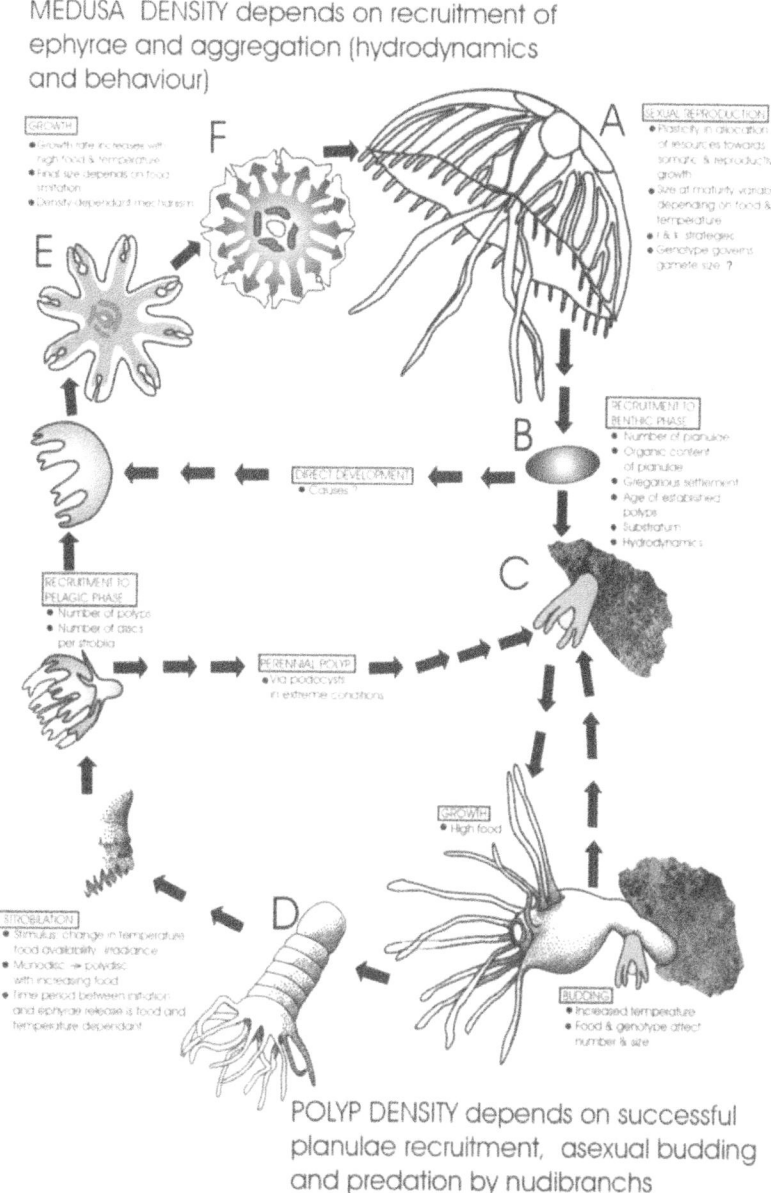

MEDUSA DENSITY depends on recruitment of
ephyrae and aggregation (hydrodynamics
and behaviour)

POLYP DENSITY depends on successful
planulae recruitment, asexual budding
and predation by nudibranchs

*Figure 7.* Summary of the role of environment variables in synchronising and influencing the benthic and pelagic stages of the life cycle of *Aurelia aurita* (re-drawn from Gröndahl, 1988b).

stimulus for multiple strobilation events, as suggested by Thiel (1962), what causes ephyrae production to occur over a 7-month period in Horsea Lake, where there is an extremely numerically- and species-poor zooplankton community? Nearby, in the productive estuary, Southampton Water, ephyrae are present for only a 2–3 month period.

Our understanding of strobilation and other asexual reproductive activities has barely advanced since the early 1980s, and there clearly exists a great need to examine these further. It is most likely that

there exists a dynamic balance between environmental variables, in particular food availability and temperature, synchronising strobilation. The time-period between the initiation of strobilation and actual release of ephyrae probably depends on the winter minimum temperature and food availability affecting the strobilation process. Kakinuma (1975) reported that decreasing temperature in advance-stage strobilating polyps resulted in no separation, while early-stage strobilating polyps reverted to non-strobilating polyps. Nevertheless, we know very little about the combined

*Table 3.* Timing of the appearance of ephyrae across the latitudes in Western Europe, with corresponding minimum winter temperature and temperature at the time of ephyrae release, together with zooplankton abundance or biomass. References: 1. Palmén (1954); 2. Hernroth & Gröndahl (1983, 1985a); 3. Rasmussen (1973); 4. Olesen et al. (1994); Riisgård et al. (1995); 5. Thiel (1962), Schneider (1989b); 6. Van Der Veer & Oorthuysen (1985); 7. Lucas (1996); 8. Lucas & Williams (1994); 9. Papathanassiou et al. (1987), Panayotidis et al. (1988). nd = no data

| Location | Ephyrae appearance | Temperature (°C) | | Zooplankton |
|---|---|---|---|---|
| | | winter minimum | at ephyrae release | |
| 1. Tvärminne, Finland 60° 27′N, 22° 02′ E | May–Aug | 0 (Mar–Apr) | 5–6 | nd |
| 2. Gullmarfjord, Sweden 58° 19N 11° 32′ E | Nov–Dec (+ Feb–Apr) | −1.4–1.0 (Feb–Mar) | 10–12 (Nov) | Max: Sep–Dec (3–6.5 g DW m$^{-2}$) Min: Jan–Mar |
| 3. Isefjord, Denmark 55° 45′ N 11° 50′E | Nov–Dec ? (Feb–Apr) | 0 (Feb 1963) 1 (Feb 1959) | 5 (Mar–Apr 1963) 3–4 (Feb 1959) | Max: spring |
| 4. Kertinge Nor, Denmark 55° 35′N 10° 23′E | Feb–Apr | 3.5 (Dec) | 3.5–6.4 ? | <5 mg C m$^{-3}$ Max: Mar (30 mg C m$^{-3}$), Jul (55 mg C m$^{-3}$) |
| 5. Kiel Bight, W. Baltic Sea 54° 21′N 10° 10′E | Dec–Apr (+ May–Jun) | 1 (Jan–Feb) | 5 (Dec) | Max: May and autumn Min: early spring |
| 6. Wadden Sea, Holland 52° 58′N 04° 46′E | Feb/Mar ?–May | nd | nd | Range: 40–60 mg C m$^{-3}$ in May–Jun |
| 7. Horsea Lake, U.K. 50° 52′N 01° 06′W | Dec–Jun | 5.8 (Jan) | 9.4 (Dec) | Max: Jul (4000 m$^{-3}$) Min: winter and mid-summer |
| 8. Southampton Water, U.K. 50° 54′N 01° 25′W | Jan/Feb–Mar | 2.6–7.9 (Jan) | 2.6–8.7 | Max: May and Oct–Nov (8000 m$^{-3}$) Min: winter and mid-summer |
| 9. Elefsis Bay, Greece 38° 02′N 23° 33′E | Jan–Feb (+Apr–May) | 10.1–12.1 (Jan–Feb) | 10.1–12.1 | Max: Jan–Apr (100–471 mg DW m$^{-3}$) Min: May–Oct (<50 mg DW m$^{-3}$) |

effects of environmental variables. It is also possible that a combination of critical conditions is specific to individual populations, and that the response of the benthic scyphistoma is partly governed by genotype, as is cloning (Keen & Gong, 1989).

## Sexual pelagic phase

It has been demonstrated above that the balance between mortality resulting from inter- and intra-specific predation and the rate of asexual budding and strobilation during the benthic phase of the life cycle can regulate year-to-year fluctuations in medusa abundance. The successful recruitment of planulae to the benthic population is also important, and this depends, in part, on factors that influence sexual reproduction during the pelagic phase of the life cycle.

As with the polyp generation, the distribution and biomass of the medusa population depends on larval recruitment (i.e. ephyrae production), mortality and growth. Mortality of ephyrae and medusae prior to maturation is generally considered to be low (Schneider, 1989a). Predators of *A. aurita* include other pelagic cnidarians such as *C. capillata*, *Aequorea victoria* (Murbach & Shearer) and *Phacellophora camtschatica* (Brandt) (Arai & Jacobs, 1980; Strand & Hamner, 1988; Purcell, 1991; Båmstedt et al., 1994), turtles and various fish species (Arai, 1988; Purcell & Arai, 2001). These may be potentially important regulators of medusa abundance where predator and prey species co-exist, but it is difficult to assess accurately in the field because jelly-fish are not easily identified in gut contents. There are also several parasites in *A. aurita* medusae, with larval trematodes and cestodes (Arai, 1988; Purcell & Arai, 2001), and hyperiid amphipods having received most attention. Infestations of medusae by the amphipod *Hyperia galba* (Montagu) increase rapidly following the maturation of the gonads. It is not clear to what extent this might decrease fecundity or mortality of the host, although Rasmussen (1973) reported ~100% infection of *A. aurita* medusae every summer in the Isefjord could account for their "early" disappearance.

A characteristic feature of all scyphomedusae is their ability to utilise the available food resources for growth and reproduction, quickly and effectively. Although high food availability can result in medusae reaching bell diameters of >400 mm, there appears to be a density-dependent mechanism governing max-imum bell diameter, whereby dense populations of medusae are characterised by small average size. The regulation of medusa size has important implications for the subsequent maturation and eventual reproduct-ive output of the population. In general, the initial ap-pearance of sexually mature individuals occurs when the maximum average size of the population has been reached (Rasmussen, 1973; Lucas, 1996; Lucas & Lawes, 1998). Considering the wide range in max-imum bell diameter among populations, one would also expect considerable variation in size at matur-ity, assuming that it is not size- or age-dependent (Table 4). There is good evidence that the size of sexually mature medusae is influenced by the effect of food availability on growth. In laboratory exper-iments, Ishii & Båmstedt (1998) demonstrated that when food is scarce growth rate slows down, and there is a change in energy allocation towards repro-duction. Maturation therefore occurs at a smaller size than in well-fed individuals. When food is abund-ant, medusae grow to a large size at the expense of gonad development. The *A. aurita* population in Horsea Lake has also been shown to partition the avail-able food resources toward somatic growth when food was abundant, and reproductive effort when food was scarce (Lucas, 1996).

Within individual populations, although the on-set of maturation occurs first in the largest medusae, eventually all medusae mature, even those that are re-latively small. In Southampton Water, a reduction in minimum size at maturity from 118 mm in late May to 64 mm by mid-June was strongly correlated with increasing water temperature (Lucas & Lawes, 1998).

*Table 4.* Variation in the size of sexually mature female medusae of *Aurelia aurita*, both within and among populations. Refer-ences: 1. Lucas (1996); 2. Lucas & Lawes (1998); 3. Ishii & Båmstedt (1998); 4. Spangenberg (1965); 5. Yasuda (1971); 6. Omori et al. (1995). nd = no data

| Location | Size at maturity (mm) | | Ref |
|---|---|---|---|
| | min | max | |
| Horsea Lake, U.K. | 19–63 | 98 | 1,2 |
| Southampton Water, U.K. | 45–118 | 135 | 2 |
| Vågsbøpollen, Norway (*in situ*) | 50 | nd | 3 |
| Vågsbøpollen, Norway (well-fed) | ?(>156) | nd | 4 |
| Laboratory-reared | 55 | nd | 4 |
| Urazoko Bay, Japan | 70 | 310 | 5 |
| Tokyo Bay, Japan | 85–90 | nd | 6 |

Increased temperature may act directly on the meta-bolic rate of *A. aurita*, or indirectly on its food supply through increased primary and secondary production (Lucas & Lawes, 1998). Miyake et al. (1997) also reported that, as a medusa aged, the more acceler-ated was its gonadal development. In Horsea Lake, the relationship between sexual maturation and ambi-ent temperature and food availability is more complex. During the spring when temperatures increased, size at maturity decreased, as described above. However, fol-lowing an isolated zooplankton peak, both overall size and minimum size at maturity increased to 63 mm. Thereafter, it decreased to 19–20 mm in December as zooplankton biomass and temperature declined. It was thought that the extreme food-limitation during the autumn and early winter led to either the pre-cocious maturation of young individuals (Hamner & Jenssen, 1974), or shrinkage of existing adult medu-sae (Lucas & Lawes, 1998). Differences in the size of sexually-mature medusae are likely to influence the reproductive output of individuals, which may sub-sequently affect recruitment success of planulae to the seabed. The number of planulae produced per female increases linearly with body weight (Schneider, 1988; Lucas, 1996), although the number of planulae pro-duced at a particular size may vary, depending on food availability (Lucas, 1996).

Somatic and reproductive growth are clearly in competition for the assimilated food, so that different food conditions will favour either growth or repro-ductive development. The Demographic Theory of life history evolution (reviewed by Olive, 1985) states that when food is limiting, reproductive effort increases, leading to increased fecundity and/or increased survivorship of the offspring. Conversely, when food is abundant, somatic investment increases, leading to increased future output. From a bioenergetic point of view, the amount of assimilated food directed into growth and reproduction when *A. aurita* medusae have abundant food is 35% and 4%, respectively, whilst in food-poor years, these values drop to 18% and 2%, respectively (Schneider, 1989a). Although food limitation may favour an increase in reproductive effort over growth, the overall amount of energy allocated to reproduction is higher when medusae are well fed. Gonad indices (i.e. gonad weight as a percentage of total weight), used as an indicator of reproductive condition, also indicate that the amount of energy invested into reproduction is greater in well-fed populations compared with severely food-limited populations (Lucas & Lawes, 1998).

Not only does the amount of energy invested into reproduction depend on food availability, but also the nature of partitioning of the assimilated energy once metabolic demands have been met. Variation in reproductive output may manifest itself through changes in fecundity and/or organic content of the gametes produced. Most life-history models predict that females produce many small eggs in favourable conditions, and fewer larger eggs in unfavourable conditions (see Olive, 1985). This can be seen in *A. aurita* when comparing populations in Kiel Bight, Southampton Water, and Horsea Lake (Schneider, 1988; Lucas, 1996; Lucas & Lawes, 1998). Populations living in food-rich environments produce large numbers (~65,000) of planulae of small size (average 34.6 $\mu$m dia.) and low organic content (0.28 $\mu$g C). Conversely, food-limited medusae produce fewer planulae (<5000) that are significantly larger (average 40.4 $\mu$m dia.) and have higher organic content (0.68 $\mu$g C); in other words *r* and *k* selection. Eckelbarger & Larson (1988) provided evidence of three types of storage product in vitellogenic oocytes: membrane-bound yolk bodies, glycogen, and lipid droplets. Planulae from food-limited medusae have higher lipid content than those from well-fed medusae (Schneider, 1988). The enhanced investment in each gamete presumably increases the survivorship of offspring, thus improving the chances of successful recruitment of planulae to the benthic generation.

As well as physiological adaptations, behavioural mechanisms may also enhance successful recruitment. *A. aurita* medusae are dioecious. In order for high fertilisation success to occur, spawning must occur in close proximity with individuals of the opposite sex.

This may take place over a considerable period of time as gametogenesis is asynchronous and age structure of the population may be heterogenous, depending on the periodicity of ephyrae release. Active migration of *A. aurita* medusae during the daytime was observed by Hamner et al. (1994) in Saanich Inlet, British Columbia, and this was interpreted as an aggregation mechanism to facilitate reproductive success. The vast majority of evidence indicates that medusae are annuals, and that once spawning has occurred, medusae deteriorate and die, although Miyake et al. (1997) suggested that medusae could reproduce more than once.

*A. aurita* medusae clearly display great plasticity in their population characteristics in response to environmental conditions. Growth, maturation and fecundity can even change within a season (Hamner & Jenssen, 1974; Lucas, 1996). Schneider (1988) suggested, rather speculatively, that fluctuations in the production of planula larvae by *A. aurita* depend on the breeding strategy of the female medusae, and that breeding strategy is regulated by medusa density. However, to what degree is reproductive strategy and output governed by phenotypic plasticity or by genotype? The study of Lucas & Lawes (1998) revealed that although there were significant differences in planula size between two populations with contrasting food supplies, planula size did not vary within a population (Horsea Lake), in spite of differences in food supply.

## The role of genetics

At the beginning of the 20th century, 12 *Aurelia* species were recognised by Mayer (1910). Since then, the number of species has been reduced by synonymy, so that now three species, *A. aurita* (Linné), *A. limbata* (Brandt) and most recently *A. labiata* (Chamisso & Eysenhardt) are recognised (Greenberg et al., 1996; Wrobel & Mills, 1998; Purcell et al., 2000; Dawson & Martin, 2001). In recent years, there has also been much debate as to whether *A. aurita* is a truly cosmopolitan species, or whether there are in fact distinct species, which may have different population characteristics. Evidence from ongoing allozyme studies on *A. aurita* (Dawson & Martin, 2001) suggest that there are six divergent groups that probably represent different 'species', and that at least five additional 'races' are spread within three of the six species. Dawson & Martin (2001) suggest that each species is adapted to its own habitat, so that over a wide range of latitudes and temperatures, physiological (and possibly reproductive) rates are essentially the same.

Regarding the Horsea Lake and Southampton Water *A. aurita* populations, it is possible that the contrasting food and salinity regimes have caused selection of certain alleles, so that each population has adapted to its own particular ambient environment (e.g. Powers et al., 1983). In dynamic models of energy allocation (reviewed by Perrin & Sibly, 1993), assimilated energy is allocated to maintenance, somatic growth, storage or reproduction and the nature of allocation can change with age. The expected evolutionary outcome in different populations is an optimal allocation pattern, which depends on the ambient environment experienced during the evolutionary process. Thus, certain reproductive characteristics, such as planulae size may be genetically pre-determined within a population, regardless of short-term changes in food supply. This is rather speculative, however, as the Horsea Lake population has been isolated for only 15–20 years. Nevertheless, selection within the limited gene pool may have yielded gene combinations different from those of the ancestral population – the so called 'founder effect'.

## Summary and suggested future research directions

A summary of our current understanding of the interactive relationships between the benthic and pelagic phases of *A. aurita* and the ambient environment is illustrated in Figure 7. The balance between growth (planulae recruitment, asexual budding, strobilation) and mortality (recruitment failure, predation by conspecific polyps and nudibranchs, extreme conditions) in benthic population is clearly a major factor in governing recruitment to the pelagic phase and thus medusa abundance. However, the benthic phase of the life cycle, from planulae recruitment to ephyrae production, is very complex, and our knowledge of it is still very limited. This is in spite of the fact that *A. aurita* is probably the most studied jellyfish. Knowledge of other jellyfish is extremely limited.

Both the benthic and pelagic forms of *A. aurita* display great adaptability to their ambient environment, making it possible for this species to exploit a wide range of environments and form mass occurrences. It can do this because of plasticity in reproductive and life history characteristics. An individual polyp can undergo asexual budding, encystment or strobilation, depending on ambient conditions of temperature,

food, irradiance and possibly predation, to ensure survival to the next generation. Similarly, switches in medusa growth and reproductive strategy in response to variable food supply results in the production of viable larvae that will successfully settle and metamorphose. Although it is clear that reproductive strategy is governed by the ambient conditions, it is possible that to a certain extent, the strategies are influenced by the medusae themselves (i.e. density-dependent regulation and genotype).

The vast majority of research carried out on *A. aurita* and other jellyfish species over the years has been concerned with the impact of populations on the zooplankton community. Little is known about life history strategies, and factors governing them. This is somewhat surprising because in order to understand which factors control population size and therefore predator and prey interactions, a basic knowledge of reproduction and recruitment is required. In order to improve our understanding of the factors affecting medusa abundance, particularly in light of the apparent increase in bloom events, I suggest that the following areas need to be addressed in future research:

(1) Greater understanding of the benthic phase of life cycle; in particular factors affecting recruitment to the medusa population. Simultaneous studies of the benthic and pelagic phases of the life cycle should be carried out, in order to determine the relative roles of the two phases of the life cycle in governing population fluctuations.
(2) The relative roles of phenotypic plasticity and genotype in governing population response to the ambient environment.
(3) Further information on medusa reproductive strategy in response to environmental conditions, and therefore environment change.
(4) Greater understanding of the nature and importance of density-dependent mechanisms controlling populations.

## Acknowledgements

The motivation behind this review was provided by the International Conference on Jellyfish Blooms, Alabama, organised by Dr Monty Graham and Dr Jenny Purcell. In particular, I would like to thank Jenny Purcell for her encouragement to write this article and her valuable comments in revising this manuscript. Prof. Paul Tyler and Dr Phil Smith kindly read earlier drafts of this manuscript, and provided useful comments. Anthea Dunkley and Kate Davis helped with the figures. Finally, this review would not have been possible without the dedicated research carried out by the small band of 'jellies' researchers over the years. I thank them for their results and ideas.

# References

Arai, M. N., 1988. Interactions of fish and pelagic coelenterates. Can. J. Zool. 66: 1913–1927.

Arai, M. N., 1997. A functional biology of Scyphozoa. Chapman & Hall, London: 316 pp.

Arai, M. N. & J. R. Jacobs, 1980. Interspecific predation of common Strait of Georgia plankton coelenterates: laboratory evidence. Can. J. Fish. aquat. Sci. 37: 120–123.

Båmstedt, U., 1990. Trophodynamics of the scyphomedusa *Aurelia aurita*. Predation rate in relation to abundance, size and type of organism. J. Plankton Res. 12: 215–229.

Båmstedt, U., M. B. Martinussen & S. Matsakis, 1994. Trophodynamics of the two jellyfishes, *Aurelia aurita* and *Cyanea capillata*, in western Norway. ICES J. mar. Sci. 51: 369–382.

Berrill, N. J., 1949. Developmental analysis of scyphomedusae. Biol. Rev. 24: 393–410.

Brewer, R. H., 1978. Larval settlement behaviour in the jellyfish *Aurelia aurita* (Linnaeus) (Scyphozoa: Semaeostomae). Estuaries 1: 121–122.

Brewer, R. H. & J. S. Feingold, 1991. The effect of temperature on the benthic stages of *Cyanea* (Cnidaria: Scyphozoa), and their seasonal distribution in the Niantic River estuary, Connecticut. J. exp. mar. Biol. Ecol. 152: 49–60.

Chia, F-S., 1989. Differential larval settlement of benthic marine invertebrates. In Ryland, J. S. & P. A. Tyler (eds), Reproduction, Genetics and Distributions of Marine Organisms. Olsen & Olsen, Fredensborg: 3–12.

Condon, P., M. B. Decker & J. E. Purcell, 2001. Effects of low dissolved oxygen on survival and asexual reproduction of scyphozoan polyps (*Chrysaora quinquecirrha*). Hydrobiologia 451 (Dev. Hydrobiol. 155): 89–95.

Dawson, M. N & L. E. Martin, 2001. Geographic variation and ecological adaptation in *Aurelia* (Scyphozoa, Semaeostomeae): some implications from molecular phylogenetics. Hydrobiologia 451 (Dev. Hydrobiol. 155): 259–273.

Eckelbarger, K. J. & R. L. Larson, 1988. Ovarian morphology and oogenesis in *Aurelia aurita* (Scyphozoa: Semaeostomae): ultrastructural evidence of heterosynthetic yolk formation in a primitive metazoan. Mar. Biol. 100: 103–115.

Greenberg, N., R. L. Garthwaite & D. C. Potts, 1996. Allozyme and morphological evidence for a newly introduced species of *Aurelia aurita* in San Francisco Bay, California. Mar. Biol. 125: 401–410.

Gröndahl, F., 1988a. A comparative ecological study on the scyphozoans *Aurelia aurita*, *Cyanea capillata*, and *C. lamarckii* in the Gullmar Fjord, western Sweden, 1982 to 1986. Mar. Biol. 97: 541–550.

Gröndahl, F., 1988b. Interactions between polyps of *Aurelia aurita* and planktonic larvae of scyphozoans: an experimental study. Mar. Ecol. Prog. Ser. 45: 87–93.

Gröndahl, F., 1989. Evidence of gregarious settlement of planula larvae of the scyphozoan *Aurelia aurita*: an experimental study. Mar. Ecol. Prog. Ser. 56: 119–125.

Hamner, W. M. & R. M. Jenssen, 1974. Growth, degrowth, and irreversible cell differentiation in *Aurelia aurita*. Am. Zool. 14: 833–849.

Hamner, W. M., R. W. Gilmer & P. P. Hamner, 1982. The physical, chemical, and biological characteristics of a stratified, saline, sulfide lake in Palau. Limnol. Oceanogr. 27: 896–909.

Hamner, W. M., P. P. Hamner & S. W. Strand, 1994. Sun-compass migration by *Aurelia aurita* (Scyphozoa): population retention and reproduction in Saanich Inlet, British Columbia. Mar. Biol. 119: 347–356.

Hansson, L. J., 1997. Effect of temperature on growth rate of *Aurelia aurita* (Cnidaria, Scyphozoa) from Gullmarsfjorden, Sweden. Mar. Ecol. Prog. Ser. 161: 145–153.

Hernroth, L. & F. Gröndahl, 1983. On the biology of *Aurelia aurita* (L.): 1. Release and growth of *Aurelia aurita* (L.) ephyrae in the Gullmarfjorden, western Sweden. Ophelia 22: 189–199.

Hernroth, L. & F. Gröndahl, 1985a. On the biology of *Aurelia aurita* (L.): 2. Major factors regulating the occurrence of ephyrae and young medusae in the Gullmarfjorden, western Sweden. Bull. mar. Sci. 37: 567–576.

Hernroth, L. & F. Gröndahl, 1985b. On the biology of *Aurelia aurita* (L.): 3. Predation by *Coryphella verrucosa* (Gastropoda, Opisthosobranchia), a major factor regulating the development of *Aurelia aurita* populations in the Gullmarfjord, western Sweden. Ophelia 24: 37–45.

Hiromi, J., T. Yamomoto, Y. Koyama & S. Kadota, 1995. Experimental study on predation of scyphopolyp *Aurelia aurita*. Bull. Coll. Agric. Vet. Med. Nihon Univ. 52: 126–130 (in Japanese; English abstract).

Hirst, A. G., 1996. Zooplankton production and energy flow – towards a biological model of Southampton Water. PhD Thesis, University of Southampton: 445 pp.

Hirst, A. G. & C. H. Lucas, 1998. Salinity influences body weight quantification in the scyphomedusa *Aurelia aurita*: important implications for body weight determination in gelatinous zooplankton. Mar. Ecol. Prog. Ser. 165: 259–269.

Ishii, H. & U. Båmstedt, 1998. Food regulation of growth and maturation in a natural population of *Aurelia aurita* (L.). J. Plankton Res. 20: 805–816.

Ishii, H., S. Tadokoro, H. Yamanaka & M. Omori, 1995. Population dynamics of the jellyfish *Aurelia aurita*, in Tokyo Bay in 1993 with determination of ATP-related compounds. Bull. Plankton. Soc. Jpn. 42: 171–176.

Kakinuma, Y., 1975. An experimental study of the life cycle and organ differentiation of *Aurelia aurita* Lamarck. Bull. mar. Biol. Stat. Asamushi 15: 101–113.

Kakinuma, Y., K. Takada & H. Miyake, 1993. Environmental influence on medusa's size of *Aurelia aurita* and age indicator. Zool. Sci. 10 (suppl.) 163.

Keen, S. L., 1987. Recruitment of *Aurelia aurita* (Cnidaria: Scyphozoa) larvae is position-dependent, and independent of conspecific density, within a settling surface. mar. Ecol. Prog. Ser. 38: 151–160.

Keen, S. L. & J. Gong, 1989. Genotype and feeding frequency affect clone formation in a marine cnidarian (*Aurelia aurita* Lamarck 1816). Funct. Ecol. 3: 735–745.

Kideys, A. E., 1994. Recent dramatic changes in the Black Sea ecosystem – the reason for the sharp decline in Turkish anchovy fisheries. J. mar. Res. 5: 171–181.

Kramp, P. L., 1961 Synopsis of the Medusae of the World. J. mar. biol. Assoc. U. K. 40: 1–469.

Lebedeva, L. P. & E. A. Shushkina, 1991. Evaluation of the population characteristics of the medusa *Aurelia aurita* in the Black Sea. Oceanology (Wash.) 31: 314–319.

Lindahl, O. & L. Hernroth, 1983. Phyto-zooplankton community in coastal waters of western Sweden – an ecosystem off balance. Mar. Ecol. Prog. Ser. 10: 119–126.

Lucas, C. H., 1996. Population dynamics of the scyphomedusa *Aurelia aurita* (L.) from an 'isolated', brackish lake, with particular reference to sexual reproduction. J. Plankton Res. 18: 987–1007.

Lucas, C. H. & S. Lawes, 1998. Sexual reproduction of the scyphomedusa *Aurelia aurita* in relation to temperature and variable food supply. Mar. Biol. 131: 629–638.

Lucas, C. H. & J. A. Williams, 1994. Population dynamics of the scyphomedusa *Aurelia aurita* in Southampton Water. J. Plankton Res. 16: 879–895.

Lucas, C. H., A. G. Hirst & J. A. Williams, 1997. Plankton dynamics and *Aurelia aurita* production from two contrasting ecosystems: causes and consequences. Estuar. coast. shelf Sci. 45: 209–219.

Matsakis, S. & R. J. Conover, 1991. Abundance and feeding of medusae and their potential impact as predators on other zooplankton in Bedford Basin (Nova Scotia, Canada) during spring. Can. J. Fish. aquat. Sci. 48: 1419–1430.

Mayer, A. G., 1910. Medusae of the World. III. The Scyphomedusae. Carnegie Institute of Washington, Washington: 499–735.

Miyake, H., K. Iwao & Y. Kakinuma, 1997. Life history and environment of *Aurelia aurita*. S. Pacific Stud. 17: 273–285.

Möller, H., 1980. Population dynamics of *Aurelia aurita* medusae in Kiel Bight, Germany (FRG). Mar. Biol. 60: 123–128.

Mutlu, E., F. Bingel, A. C. Gücü, V. V. Melnikov, U. Niermann, N. A. Ostr & V. E. Zaika, 1994. Distribution of the new invader *Mnemiopsis* sp. and the resident *Aurelia aurita* and *Pleurobrachia pileus* populations in the Black Sea in the years 1991–1993. ICES J. mar. Sci. 51: 407–421.

Olesen, N. J., K. Frandsen & H. U. Riisgård, 1994. Population dynamics, growth and energetics of jellyfish *Aurelia aurita* in a shallow fjord. Mar. Ecol. Prog. Ser. 105: 9–18.

Olive, P. J. W., 1985. Physiological adaptation and the concepts of optimal reproductive strategy and physiological constraint in marine invertebrates. In Laverack, M. S. (ed.) Physiological Adaptations of Marine Animals. Symposia of the Society for Experimental Biology, No. 39. The company of Biologists Ltd., Cambridge, England: 267–300.

Olsson, P., E. Granéli, P. Carlsson & P. Abreu, 1992. Structuring of a postspring phytoplankton community by manipulation of trophic interactions. J. exp. mar. Biol. Ecol. 158: 249–266.

Omori, M., H. Ishii & A. Fujinaga, 1995. Life history strategy of *Aurelia aurita* (Cnidaria, Scyphomedusae) and its impact on the zooplankton community of Tokyo Bay. ICES J. mar. Sci. 52: 597–603.

Palmén, E., 1954. Seasonal occurrence of ephyrae ad subsequent instars of *Aurelia aurita* (L.) in the shallow waters of Tvarminne, S.Finland. Arch. Soc. Vanamo 8: 122–131.

Panayotidis, P., E. Papathanassiou, I. Siokou-Frangou, K. Anagnostaki & O. Gotsis-Skretas, 1988. Relationship between the medusae *Aurelia aurita* Lam. and zooplancton in Elefsis Bay (Saronikos Gulf, Greece). Thalassiografica 11: 7–17.

Papathanassiou, E., P. Panayotidis & K. Anagnostaki, 1987. Notes on the biology and ecology of the jellyfish *Aurelia aurita* L. in Elefsis Bay (Saronikos Gulf, Greece). Mar. Biol. 8: 49–58.

Perrin, N. & R. M. Sibly, 1993. Dynamic models of energy allocation and investment. Ann. Rev. Ecol. Syst. 24: 379–410.

Powers, D. A., L. DiMichele & A. R. Place, 1983. The use of enzymatic kinetics to predict differences in cellular metabolism, development rate and swimming performance between LDH-B genotypes of the fish *Fundulus heteroclitus*. Isozymes–Curr. T. Biol. 10: 147–170.

Purcell, J. E., 1991. A review of cnidarians and ctenophores feeding on competitors in the plankton. Hydrobiologia 216: 335–342.

Purcell, J. E. & M. N. Arai, 2001. Interactions of pelagic cnidarians and ctenophores with fish: a review. Hydrobiologia 451 (Dev. Hydrobiol. 155): 27–44.

Purcell, J. E., J. R. White, D. A. Nemazie & D. A. Wright, 1999. Temperature, salinity and food effects on asexual reproduction and abundance of the scyphozoan *Chrysaora quinquecirrha*. Mar. Ecol. Prog. Ser. 180: 187–196.

Purcell, J. E., E. D. Brown, K. D. E. Stokesbury, L. H. Haldorson & T. C. Shirley, 2000. Aggregations of the jellyfish *Aurelia labiata*: abundance, distribution, associations with age-0 walleye pollock, and behaviors promoting aggregation in Prince William Sound, Alaska, U.S.A. Mar. Ecol. Prog. Ser. 195: 145–158.

Rasmussen, E., 1973. Systematics and ecology of the Isefjord marine fauna (Denmark). Ophelia 11: 41–46.

Reubold, J., 1988. The biology and feeding strategy of *Aurelia aurita* in Southampton Water: problems associated with laboratory experiments. MSc Thesis, University of Southampton: 84 pp.

Riisgård, H. U., P. Bondo Christensen, N. J. Olesen, J. K. Petersen, M. M. Möller & P. Andersen, 1995. Biological structure in a shallow cove (Kertinge Nor, Denmark) - control by benthic nutrient fluxes and suspension-feeding ascidians and jellyfish. Ophelia 41: 329–344.

Russell, F. S., 1970. The Medusae of the British Isles. II Pelagic Scyphozoa with a Supplement to the First Volume on Hydromedusae. Cambridge University Press, London: 284 pp.

Schneider, G., 1988. Larvae production of the common jellyfish *Aurelia aurita* in the western Baltic 1982–1984. Kieler Meeresforsch. 6: 295–300.

Schneider, G., 1989a. Estimation of food demands of *Aurelia aurita* medusae populations in the Kiel Bight / western Baltic. Ophelia 31: 17–27.

Schneider, G., 1989b. The common jellyfish *Aurelia aurita*: standing stock, excretion, and nutrient regeneration in the Kiel Bight, Western Baltic. Mar. Biol. 100: 507–514.

Schneider, G. & G. Behrends, 1994. Population dynamics and the trophic role of *Aurelia aurita* in the Kiel Bight and western Baltic. ICES J. mar. Sci. 51: 359–367.

Scheider, G. & G. Behrends, 1998. Top-down control in a neritic plankton system by *Aurelia aurita* medusae – a summary. Ophelia 48: 71–82.

Schneider, G. & T. Weisse, 1985. Metabolism measurements of *Aurelia aurita* planulae larvae and calculation of maximal survival period of the free swimming stage. Helgoländer Meeresunters. 39: 43–47.

Shushkina, E. A. & G. N. Arnotauv, 1985. Quantitative distribution of the medusa *Aurelia* and its role in the Black Sea ecosystem. Oceanology 25: 102–106.

Shushkina, E. A. & E. I. Musayeva, 1983. The role of jellyfish in the energy system of Black Sea plankton communities. Oceanology 23: 92–96.

Smayda, T., 1993. Experimental manipulations of phytoplankton + zooplankton + ctenophore communities, and foodweb roles of the ctenophore *Mnemiopsis leidyi*. ICES cm 1993/L:68: 13 pp.

Spangenberg, D. B., 1965. Cultivation of the life stages of *Aurelia aurita* under controlled conditions. J. exp. Zool. 159: 303–318.

Spangenberg, D. B., 1968. Recent studies of strobilation in jellyfish. Oceanogr. mar. biol. Ann. Rev. 6: 231–247.

Stoecker, D. K., A. E. Michaels & L. H. Davis, 1987. Grazing by the jellyfish, *Aurelia aurita*, on microzooplankton. J. Plankton Res. 9: 901–910.

Strand, S. W. & W. M. Hamner, 1988. Predatory behavior of *Phacellophora camtschatica* and size-selective predation upon *Aurelia aurita* (Scyphozoa: Cnidaria) in Saanich Inlet, British Columbia. Mar. Biol. 99: 409–414.

Sullivan, B. K., J. R. Garcia & G. Klein-MacPhee, 1994. Prey selection by the scyphomedusan predator *Aurelia aurita*. Mar. Biol. 121: 335–341.

Sullivan, B. K., C. L. Suchman & J. H. Costello, 1997. Mechanics of prey selection by ephyrae of the scyphomedusa *Aurelia aurita*. Mar. Biol. 130: 213–222.

Thiel, H., 1962. Untersuchungen über die Strobilisation von *Aurelia aurita* Lam. an einer Population der Kieler Förde. Kiel. Meeresforsch. 18: 198–230.

Tsikon-Lukanina, E. A., O. G. Reznichenko & T. A. Lukasheva, 1995. Food intake by scyphistomae of the medusa *Aurelia aurita* in the Black Sea. Okeanologiya 35: 895–899 (in Russian; English abstract).

Van Der Veer, H. W. & W. Oorthuysen, 1985. Abundance, growth and food demand of the scyphomedusan *Aurelia aurita* in the western Wadden Sea. Neth. J. Sea Res. 19: 38–44.

Verwey, J., 1942. Die Periodizität im Auftreten und die aktiven und passiven Bewegungen der Quallen. Arch. Neerl. Zool. 6: 363–468.

Watanabe, T. & H. Ishii, 2001. *In situ* estimation of the number of ephyrae liberated from polyps of *Aurelia aurita* on settling plates in Tokyo Bay, Japan. Hydrobiologia 451 (Dev. Hydrobiol. 155): 247–258.

Wrobel, D. & C. Mills, 1998. Pacific coast pelagic invertebrates: a guide to the common gelatinous animals. Monterey Bay Aquarium, Monterey, CA: 108 pp.

Yasuda, T., 1968. Ecological studies on the jellyfish *Aurelia aurita* in Urazoko Bay, Fukui Prefecture II Occurrence pattern of the ephyrae. Bull. Jap. Soc. Sci. Fish. 34: 983–987 (in Japanese; English abstract).

Yasuda, T., 1969. Ecological studies on the jellyfish *Aurelia aurita* in Urazoko Bay, Fukui Prefecture I Occurrence of the medusa. Bull. Jap. Soc. Sci. 35: 1–6 (in Japanese; English abstract).

Yasuda, T., 1971. Ecological studies on the jellyfish *Aurelia aurita* in Urazoko Bay, Fukui Prefecture IV Monthly change in bell-length composition and breeding season. Bull. Jap. Soc. Sci. Fish. 37: 364–370 (in Japanese; English abstract).

Zaitsev, Y. P., 1992. Recent changes in the structure of the Black Sea. Fish. Oceanogr. 1: 180–189.

Zinger, I., 1989. Zooplankton community structure in Southampton Water and it's potential response to estuary chronic pollution. PhD Thesis, University of Southampton: 377 pp.

*Hydrobiologia* **451**: 247–258, 2001.
© 2001 *Kluwer Academic Publishers.*

# *In situ* estimation of ephyrae liberated from polyps of *Aurelia aurita* using settling plates in Tokyo Bay, Japan

Tomoko Watanabe & Haruto Ishii
*Tokyo University of Fisheries, 4-5-7 Konan Minato-ku, Tokyo, 108-8477, Japan*
*E-mail: ad98204@cc.tokyo-u-fish.ac.jp*

*Key words: Aurelia aurita*, polyp, survivorship, ephyra, reproduction, jellyfish

## Abstract

The continuous changes in the number of newly established polyps of *Aurelia aurita* (L.) on settling plates under natural conditions were observed from August 1998 to September 1999 in Tokyo Bay, Japan. A sharp decline in survivorship of newly settled polyps was observed within the first few days, however, survivorship of polyps settled in October increased by budding up to 399% after two months. The number of discs in each strobila varied from 1 to 6, however, most of the strobilae formed single discs. The percentage ratios of the total number of ephyrae to the initial number of polyps on settling plates were generally lower than 10%, but the highest ratio of 594.4% was estimated for the polyps settled in October. It is considered that most of the liberated ephyrae originate from the polyps settled in October in Tokyo Bay. This study suggests that the occurrence of ripe medusae with planula larvae throughout the year contributes to the success of settlement and growth of the polyp stage in Tokyo Bay.

## Introduction

The scyphomedusan *Aurelia aurita* (L.) is a cosmopolitan jellyfish observed in many coastal waters (e.g. Yasuda, 1971; Van Der Veer & Oorthuysen, 1985; Matsakis & Conover, 1991; Lucas & Williams, 1994; Schneider & Behrends, 1994; Miyake et al., 1997; Ishii & Båmstedt, 1998; Dawson & Martin, 2001; Lucas, 2001) and some previous studies described their large predation impact on pelagic zooplankton communities (Möller, 1984; Van Der Veer & Oorthuysen, 1985; Behrends & Schneider, 1995; Schneider & Behrends, 1998). In Tokyo Bay, mass occurrences of medusae are frequently observed in summer and autumn (Ishii et al., 1995; Omori et al., 1995; Toyokawa et al., 2000) and a decisive role of medusae in the food web is suggested (Ishii & Tanaka, 2001).

The life cycle of *A. aurita* includes an alternation between the benthic polyp and the pelagic medusa stages. The polyp stage reproduces asexually by budding and strobilation (Lucas, 2001), and each disc of a strobila liberates an ephyra. The number of ephyrae produced is equal to the number of discs in each strobila. Therefore, it is important for the prospect of mass occurrence of the medusae to know the *in situ*

changes in the numbers of polyps and discs in each strobila.

Most of the previous studies on polyps were performed by laboratory experiments, and *in situ* observation of polyps under natural conditions is rare (Thiel, 1962; Hernroth & Gröndahl, 1983, 1985a, b; Gröndahl & Hernroth, 1987; Gröndahl, 1988a, 1989). All previous studies described the reduction of polyps under natural conditions, and some studies implicated the predation impact on polyp population (Thiel, 1962; Hernroth & Gröndahl, 1985a, b). Thiel (1962) found that the nudibranch, *Facelina bostoniensis* (Couthouy), was an important predater in Kiel Fjord, Germany. Hernroth & Gröndahl (1985a, b) carried out *in situ* observations and laboratory experiments and stated that predation on *A. aurita* polyps, especially on older polyps by the nudibranch, *Coryphella verrucosa* (M. Sars), was one of the significant factors regulating the size of *A. aurita* medusae populations. However, the degree and seasonal trends of predation effects have not been elucidated.

Strobilation in *A. aurita* is usually induced by decreasing water temperature (Kakinuma, 1962; Custance, 1964; Kato et al., 1980). This phenomenon coincided with *in situ* observations. Thiel (1962) found that most polyps (ca. 100%) strobilated between January and March and the maximum number of discs in each strobila was 3–10 in Kiel Fjord, Germany. *In situ* observations of *A. aurita* polyps on settling plates were also made in Gullmar Fjord, Sweden (Hernroth & Gröndahl, 1983, 1985a, b, Gröndahl & Hernroth, 1987, Gröndahl, 1988a, 1989). Hernroth & Gröndahl (1983, 1985a) reported that strobilating polyps of *A. aurita* formed plural discs during autumn and a single disc during winter. In spite of observations of various numbers of discs in each strobila, the crucial factor affecting the number of discs is unknown. These *in situ* investigations gave information about predation and strobilation of *A. aurita* polyps, however, seasonal changes in the number of newly settled polyps and total number of ephyrae liberated in each polyp are unknown.

In Tokyo Bay, the female medusae with planula larvae occur throughout the year (Miyake et al., 1997). The planula larvae are always present in the water and always potentially can settle on suitable substrates. In this study, we observe the changes in the number of newly settled polyps of *A. aurita* on settling plates under natural conditions and estimate the percentage ratio of the number of ephyrae liberated from polyps to the initial number of newly settled polyps. This is the first study on polyps in Tokyo Bay and the first *in situ* observation of polyps on settling plates continuously throughout one year.

## Materials and methods

### Sampling and in situ experiments

Sampling of *Aurelia aurita* medusae was conducted in daytime once a month with the T.S. "Hiyodori" of Tokyo University of Fisheries in Tokyo Bay, Japan from August 1998 to June 1999 (Fig. 1). Female medusae with planula larvae were scooped from surface aggregations with a hand net (10 mm mesh size) and kept in buckets with ambient seawater. Planula larvae were collected by a pipette from the brood sacs of the oral arms of ripe female medusae and were immediately transferred to glass bottles filled with ambient seawater.

In the laboratory, the samples of planula larvae were cleaned by carefully pouring the planulae through a 330 $\mu$m mesh net into a 1200 ml bowl filled with 1 $\mu$m filtered seawater. Planula larvae were immediately transferred to glass bowls containing 200 ml

of 1 $\mu$m filtered seawater, and a settling plate made of acrylic plastic (5×6 cm) was floated in each glass bowl for the settlement of the planula larvae. After 1 week, the settling plates with newly settled polyps were transferred to petri dishes with 1 $\mu$m filtered seawater and the number of polyps on each settling plate counted using a dissecting microscope. The median and estimated standard error of the number of polyps on each plate was 318±99.8. These settling plates were kept in a plastic container filled with 1 $\mu$m filtered seawater and were carried to Stations A and B within 3 h (Fig. 1).

At the stations, 3–12 settling plates were attached on the upper and lower sides of the base plates (25×25 cm) in a large container (50×67×23 cm) filled with ambient seawater (Fig. 2). Each base plate with the settling plates was horizontally moored from the piers at 0.8 and 2.3 m depths at Station A, and at 0.8 m depth at Station B. We conducted 11 different series with different starting dates. We identified the series by the starting dates of each observation.

The observations of polyps on the settling plates at both stations were made on day 1, 2, 3, 5 and 7 during the first week, and once a week thereafter. For these observations, the base plates were lifted from the sea and placed into large containers (50×67×23 cm) filled with ambient seawater, and the settling plates were detached from the base plates. The settling plates were immediately moved to a transparent plastic box (13.5×13.5×7 cm) filled with ambient seawater. Polyps on the settling plates were observed with a dissecting microscope during the daytime, and the number and diameter of polyps, the number of strobilae and the number of discs in each strobila were determined. Strobilae were defined as polyps with distinctly formed discs. Other organisms that settled on the settling plates were also identified and counted. After the observations, the settling plates were reattached to the base plates and were moored again. All observations were carried out until 27 September 1999.

Surface temperature and salinity were also measured simultaneously with the polyp observations. Vertical profiles of temperature and salinity were measured once a month using a CTD (AST-1000S, Alec Co., Japan). Water sampling was carried out once a month using a Van-Dorn Bottle (3 l) at 1.5 m and 0.8 m depths at Stations A and B, respectively. The water sample was concentrated through a 20 $\mu$m mesh net and subsequently filtered through 1 $\mu$m Whatman GF/C filters. The salt contained in any water remaining on the filters was eliminated with isotonic ammonium formate. The filters were immediately frozen and transferred to the laboratory. For the determination of the total dry weight of micro- and small zooplankton, the filters were dried at 60 °C for 24 h and weighed.

*Calculations*

The mean survivorship of polyps on the settling plates on $t$ day (MS, %) was calculated from the following equation;

$$\text{MS} = \left[ \sum_{i=1}^{n} (P_{it}/P_{i0}) \cdot 100 \right] / n$$

where $P_{it}$ is the number of surviving polyps including strobilae on each settling plate on $t$ day, $P_{io}$ is the initial number of polyps on each settling plate and $n$ is the total number of settling plates.

The cumulative strobilation ratio ($R$, %) on the settling plates on $t$ day was calculated from the following equation;

$$R = \left[ \sum_{i=1}^{n} S_{it}' \Big/ \left( \sum_{i=1}^{n} S_i + \sum_{i=1}^{n} P_i' \right) \right] \cdot 100,$$

where $S_{it}'$ is the cumulative number of strobilae on each settling plate on $t$ day, $S_i$ is the total number of strobilae on each settling plate during the investigated period and $P_i'$ is the number of surviving polyps on each settling plate on the day the last strobila was found.

The ratio of the total number of strobilae to the initial number of polyps ($N$, %) was calculated from the following equation;

$$N = \left( \sum_{i=1}^{n} S_i \Big/ \sum_{i=1}^{n} P_{io} \right) \cdot 100.$$

The ratios of the total number of ephyrae liberated from polyps to the initial number of polyps ($E$, %) were estimated from the following equation;

$$E = \left( \sum_{i=1}^{n} D_i \Big/ \sum_{i=1}^{n} P_{io} \right) \cdot 100,$$

where $D_i$ is the total number of discs on each settling plate.

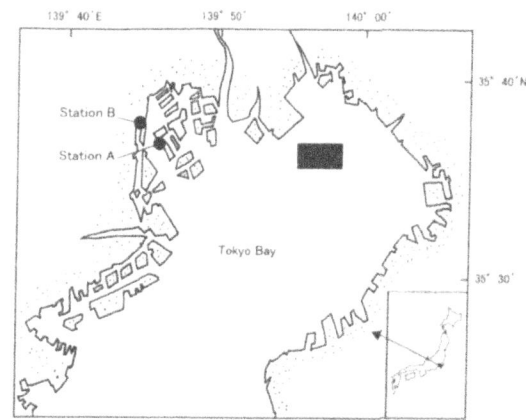

*Figure 1.* Medusa sampling area (solid rectangle) and mooring stations for the observations of polyps (solid circles) in Tokyo Bay.

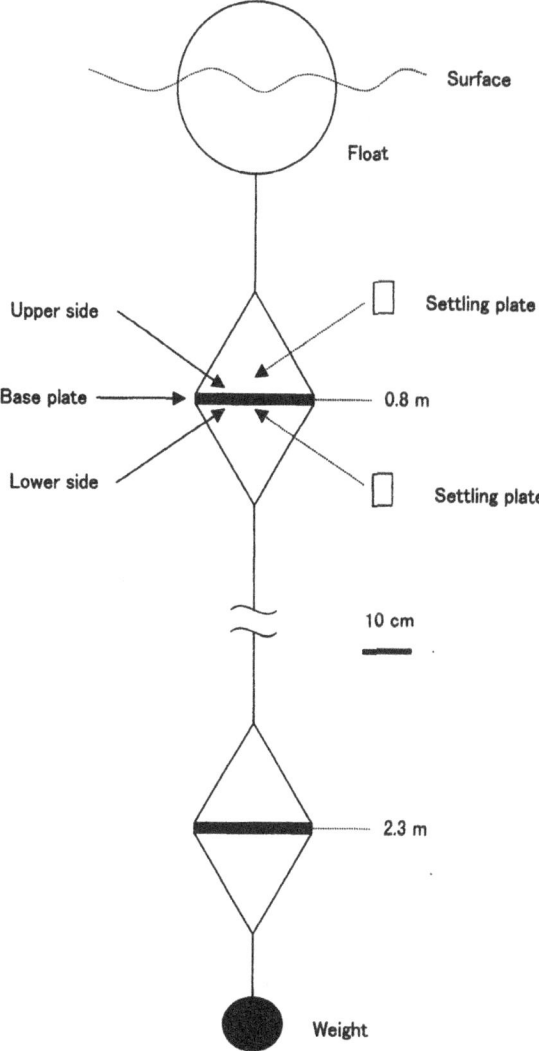

*Figure 2.* Base plates for settling plates. The base plates were horizontally moored from the piers at 0.8 and 2.3 m depths at Station A, and at 0.8 m depth at Station B in Tokyo Bay.

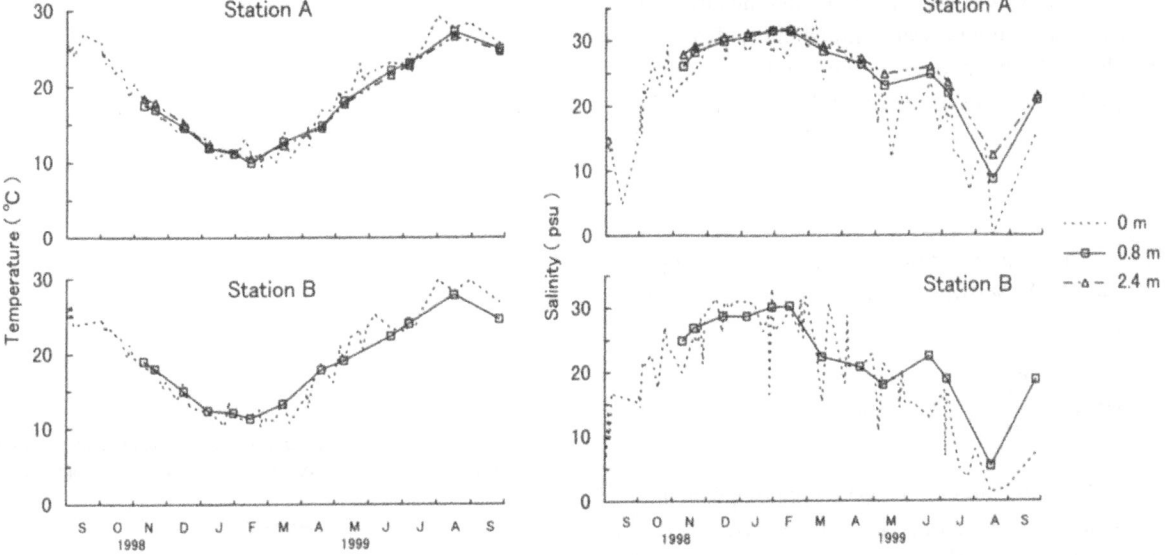

*Figure 3.* Monthly changes of water temperature and salinity at Stations A and B in Tokyo Bay from September 1998 to September 1999.

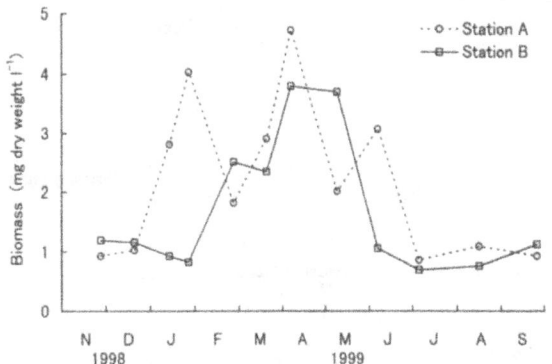

*Figure 4.* Monthly changes in biomass as mg dry weight $l^{-1}$ of zooplankton (> 20 $\mu$m) at Stations A and B in Tokyo Bay from November 1998 to September 1999.

## Results

### Temperature and salinity

Monthly changes of water temperature and salinity at Stations A and B from September 1998 to September 1999 are shown in Figure 3. The maximum surface temperatures were 29.2 °C and 29.8 °C in August and the minimum surface temperatures were 9.4 °C and 10.4 °C in February, at Stations A and B, respectively. Variation of salinity among observation dates was higher in surface waters. The maximum surface salinity was 32.9 psu in March and January, and the minimum surface salinity was <0.1 psu and 1.3 psu in August, at Stations A and B, respectively.

### Zooplankton biomass

Monthly changes in biomass as mg dry weight $l^{-1}$ of zooplankton at Stations A and B from November 1998 to September 1999 are shown in Figure 4. The biomass increased from winter to spring and it decreased during summer and autumn at both stations. Maximum biomass was 4.7 and 3.8 mg dry weight $l^{-1}$ on 8 April, 1999 at Stations A and B, respectively. Minimum biomass was 0.86 and 0.69 mg dry weight $l^{-1}$ on 7 July, 1999 at Stations A and B, respectively.

### Survivorship

Figure 5 shows the changes in mean survivorship of polyps on the upper side of the settling plates at Stations A and B. The numbers of polyps on the settling plates rapidly decreased during the first few days after the start of observations. Polyps on each plate with different starting dates disappeared within 4 months at both stations, except for polyps on the settling plates of 2 October at Station B. A different pattern was found on the settling plates of 2 October; mean survivorship reached 159.0% on 20 November 1998. Significant differences in the mean survivorship of polyps between 0.8 m and 2.3 m depths at Station A and between Stations A and B were not observed except two and four cases, respectively (Table 1).

Figure 6 shows changes in mean survivorship of polyps on the lower side of the settling plates at Stations A and B. The numbers of polyps on the settling plates decreased rapidly with variable patterns dur-

*Table 1.* Results of the statistical analyses by Two-way ANOVA without replication for survivorship of *Aurelia aurita* polyps on the settling plates on each starting date. The results of the analyses among sampling dates were omitted from the table. *F* and *p* indicate *F*-ratios and probability values, respectively (*: $p < 0.05$, **: $p < 0.01$). Degrees of freedom of each result was 1 on each analysis.

| Starting date | Upper side | | | | Lower side | | | | Upper Side – Lower Side | | | | | |
|---|---|---|---|---|---|---|---|---|---|---|---|---|---|---|
| | 0.8 m–2.3 m | | Sta. A – Sta. B | | 0.8 m – 2.3 m | | Sta. A – Sta. B | | Sta. A 0.8 m | | Sta. A 2.3 m | | Sta. B 0.8 m | |
| | F | p | F | p | F | p | F | p | F | p | F | p | F | p |
| 31 Aug. 1998 | 2.422 | 0.140 | 16.595 | 0.004 ** | 0.067 | 0.800 | 9.046 | 0.017 * | 0.002 | 0.968 | 2.077 | 0.170 | 1.071 | 0.359 |
| 2 Oct. 1998 | 5.710 | 0.044 * | 35.877 | < 0.001 ** | 9.526 | 0.005 ** | 24.315 | < 0.001 ** | 25.852 | < 0.001 ** | 33.177 | < 0.001 ** | 36.527 | < 0.001 ** |
| 26 Oct. 1998 | 2.883 | 0.133 | 0.001 | 0.980 | 2.676 | 0.128 | 26.760 | < 0.001 ** | 19.917 | < 0.001 ** | 9.490 | 0.010 ** | 40.591 | < 0.001 ** |
| 24 Nov. 1998 | 2.422 | 0.140 | 1.498 | 0.236 | 16.211 | < 0.001 ** | 29.969 | < 0.001 ** | 0.489 | 0.497 | 10.685 | 0.003 ** | 21.593 | < 0.001 ** |
| 15 Dec. 1998 | 10.779 | 0.006 ** | 12.833 | 0.002 ** | 3.940 | 0.060 | 26.194 | < 0.001 ** | 21.300 | < 0.001 ** | 4.092 | 0.056 | 13.928 | < 0.001 ** |
| 26 Jan. 1999 | 1.444 | 0.253 | 12.364 | 0.005 ** | 12.860 | 0.003 ** | 20.692 | < 0.001 ** | 4.813 | 0.051 | 6.783 | 0.019 * | 14.913 | 0.001 ** |
| 23 Feb. 1999 | 0.006 | 0.941 | 1.128 | 0.309 | 0.845 | 0.373 | 7.135 | 0.017 * | 1.982 | 0.187 | 0.836 | 0.375 | 16.173 | 0.001 ** |
| 6 Apr. 1999 | 1.446 | 0.268 | 1.550 | 0.253 | 0.492 | 0.498 | 11.114 | 0.008 ** | 0.058 | 0.816 | 5.505 | 0.039 * | 2.331 | 0.158 |
| 5 May 1999 | 1.263 | 0.290 | 3.710 | 0.090 | 7.477 | 0.017 * | 2.232 | 0.186 | 2.641 | 0.143 | 13.557 | 0.003 ** | 1.181 | 0.319 |
| 25 May 1999 | 0.923 | 0.381 | 0.043 | 0.845 | 9.293 | 0.023 * | 1.360 | 0.308 | 2.350 | 0.186 | 9.906 | 0.020 * | 1.229 | 0.349 |
| 6 Jul. 1999 | 4.552 | 0.077 | 5.419 | 0.059 | 8.193 | 0.019 * | 0.479 | 0.511 | 7.287 | 0.031 * | 7.424 | 0.023 * | 0.468 | 0.524 |

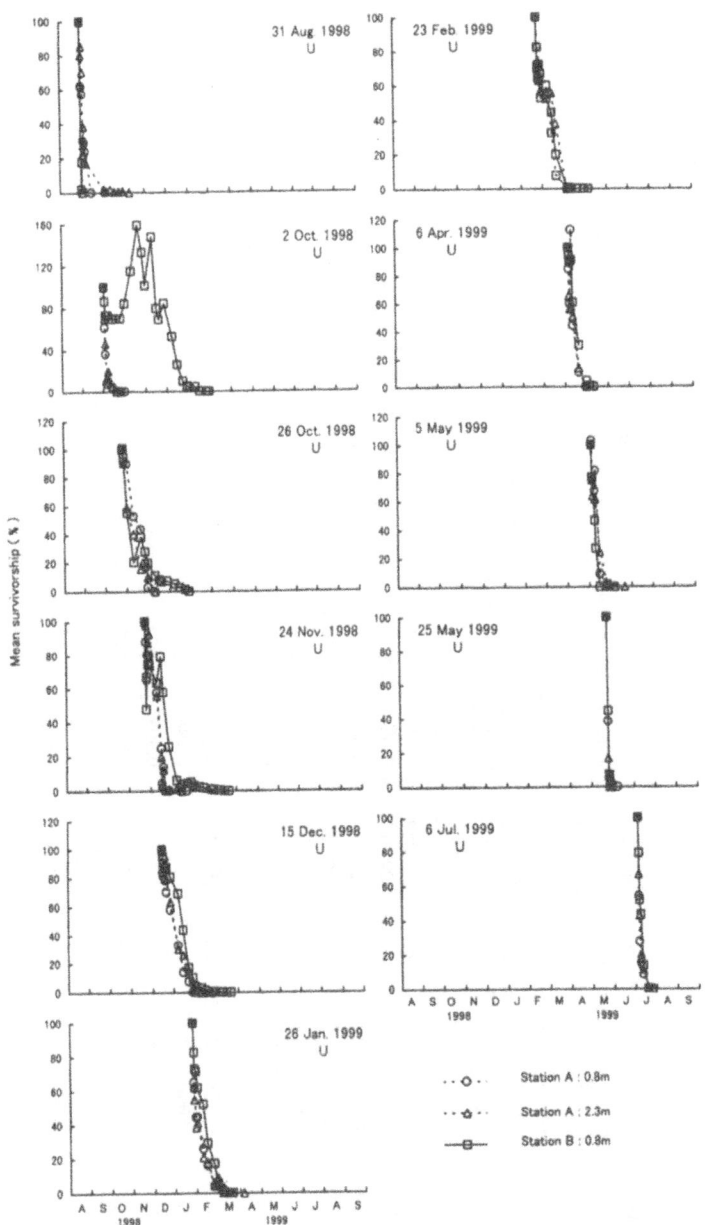

*Figure 5.* Changes in mean survivorship of *Aurelia aurita* polyps on the upper side of the settling plates at Stations A and B in Tokyo Bay. Starting date is also expressed.

ing the first few days after the start of observations. Polyps on settling plates of 31 August, 26 January, 23 February, 6 April, 5 May, 25 May and 6 June disappeared within 4 months at both stations. But polyps on settling plates of 2 October, 26 October, 24 November and 15 December survived more than 4 months. The mean survivorships of polyps of 2 and 26 October reached 320.9% and 393.3% on 10 December and 16 December 1998 at Station B, respectively. Significant differences of the mean survivorship of polyps

between 0.8 m and 2.3 m depths at Station A and between Stations A and B were observed for six and eight cases, respectively (Table 1).

Polyp survivorship was usually higher on the lower side. Statistical analyses of the mean survivorship of polyps between upper and lower sides at each station also showed that there were significant differences on each settling plate, especially for the settling plates of October (Table 1). No significant differences of the survivorship between upper and lower sides of the set-

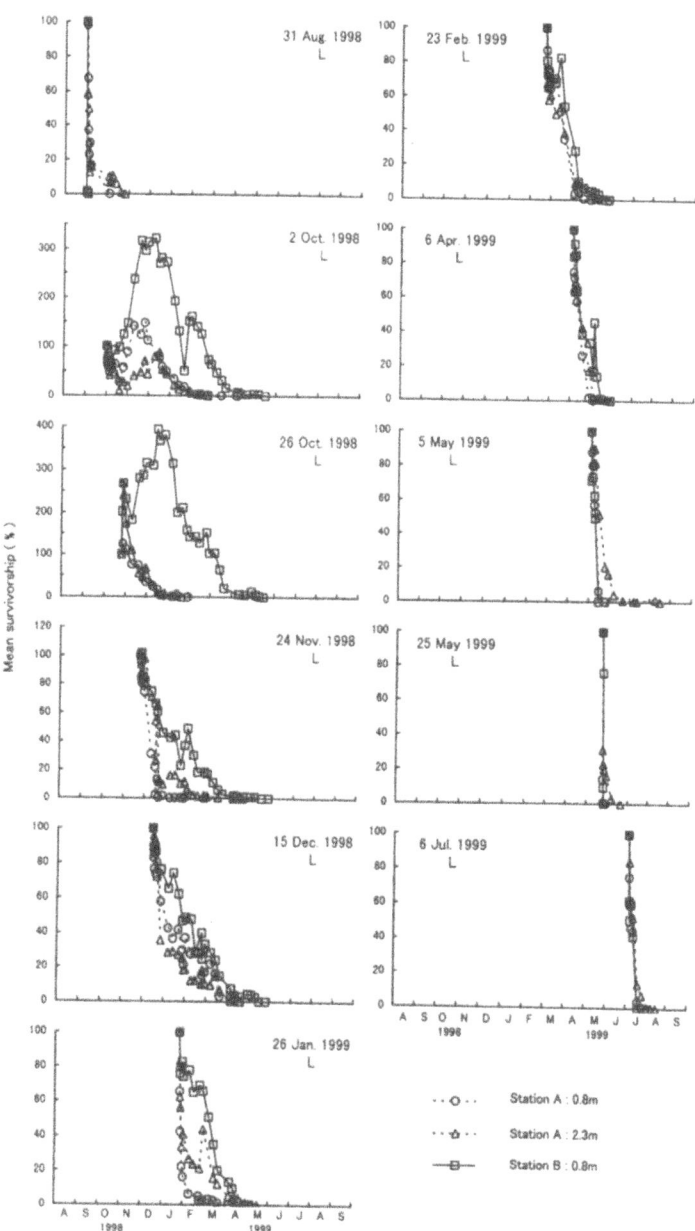

*Figure 6.* Changes in mean survivorship of *Aurelia aurita* polyps on the lower side of the settling plates at Stations A and B in Tokyo Bay. Starting date is also expressed.

tling plates were found for the plates of 31 August at either station (Table 1).

The settling plates were usually invaded by the barnacles, *Balanus eburneus* Gould, *B. improvinsus* Darwin and *B. amphitrite* Darwin, the clam worm, *Polydora ligni* Webster, the mussel, *Mytilus gallo-provincialis* Lamarck, and the amphipod, *Corophium* sp, which grew rapidly on the both sides of the settling plates at both stations. All plates were frequently covered on both sides with dense aggregations of

these organisms, such as *M. galloprovincialis*, which completely covered the settling plates after May 1999.

## Strobilation

Figure 7 shows the changes in cumulative strobilation ratios on the upper and lower sides of the settling plates at Stations A and B. Strobilae were first observed on the lower side of settling plates of 2 October at Station B on 10 December 1998. At Station A,

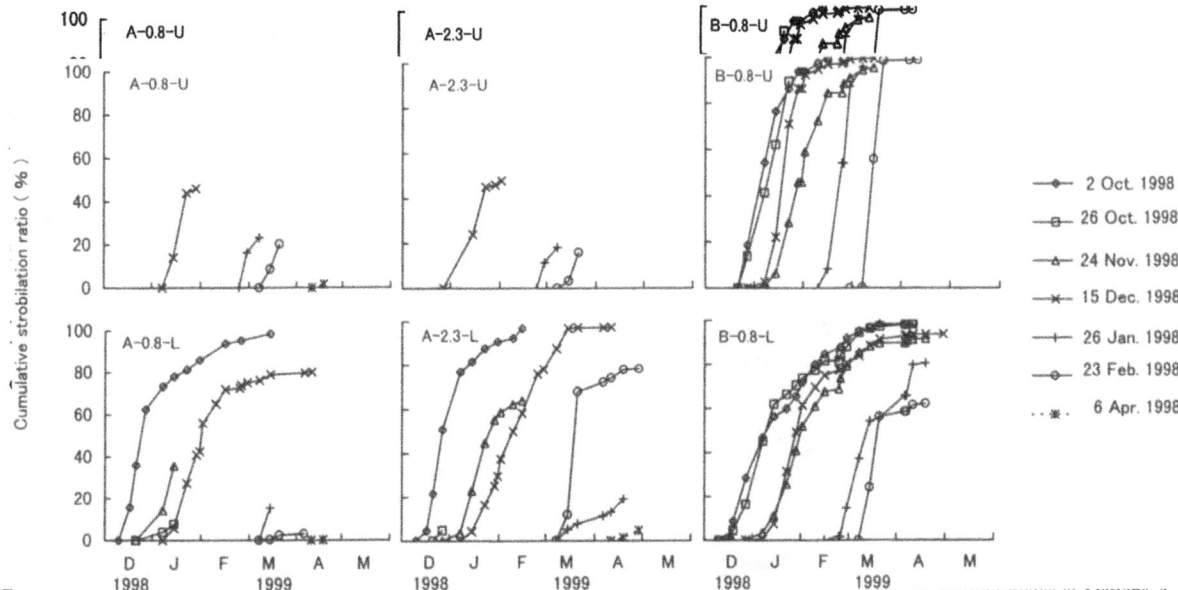

*Figure 7.* Changes in cumulative strobilation ratios of *Aurelia aurita* polyps on the upper and lower sides of the settling plates at Stations A and B in Tokyo Bay. A and B = Station; 0.8 and 2.3 = depth of the settling plates; U and L mean the upper and lower side, respectively. Starting dates of the observations are also shown.

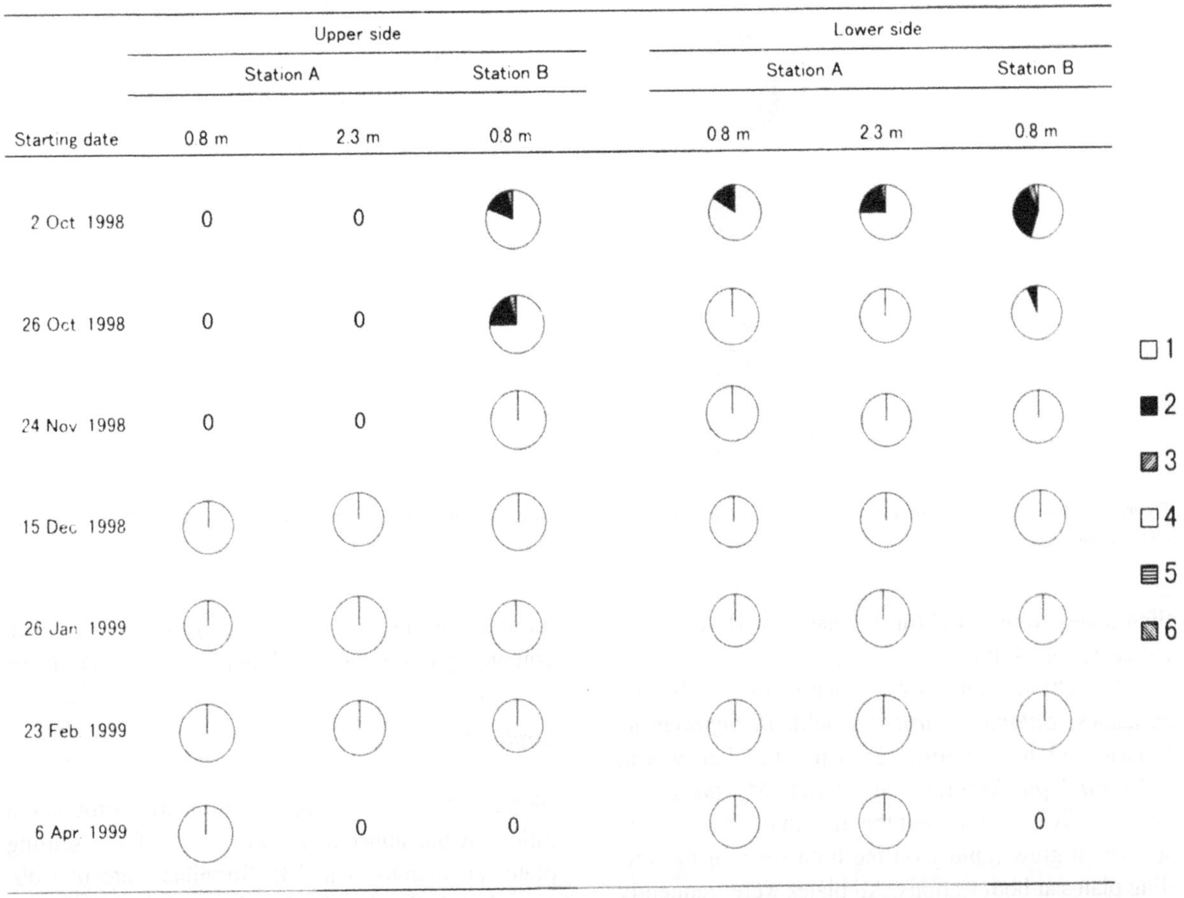

| | Upper side | | | Lower side | | |
|---|---|---|---|---|---|---|
| | Station A | | Station B | Station A | | Station B |
| Starting date | 0.8 m | 2.3 m | 0.8 m | 0.8 m | 2.3 m | 0.8 m |
| 2 Oct 1998 | 0 | 0 | ◕ | ◕ | ◔ | ◕ |
| 26 Oct 1998 | 0 | 0 | ◔ | ○ | ○ | ◔ |
| 24 Nov 1998 | 0 | 0 | ○ | ○ | ○ | ○ |
| 15 Dec 1998 | ○ | ○ | ○ | ○ | ○ | ○ |
| 26 Jan 1999 | ○ | ○ | ○ | ○ | ○ | ○ |
| 23 Feb 1999 | ○ | ○ | ○ | ○ | ○ | ○ |
| 6 Apr 1999 | ○ | 0 | 0 | ○ | ○ | 0 |

□ 1
■ 2
▨ 3
□ 4
▤ 5
▧ 6

*Figure 8.* The frequency of *Aurelia aurita* strobilae having various number discs in each strobilae. 0 denotes no strobilae observed.

the cumulative strobilation ratios on the upper sides were less than 50%, however, on the lower side, 96.9–98.4% of polyps strobilated on the settling plates of 2 October. By contrast, the cumulative strobilation ratios at Station B were nearly 100% on both sides and for all settling plates. Strobilation continued for more than one month. On 1 May 1999, the last strobila was observed on the lower side at Station B, and all polyps disappeared by 31 May 1999.

*Number of discs in each strobila*

Figure 8 shows the frequency of strobilae having various numbers of discs on each strobila on the settling plates. The number of discs varied from 1 to 6 in each strobila, however, 54.9–100% of the strobilae had single disc. Strobilae having plural discs were only observed on the settling plates of 2 and 26 October. Six discs in a strobila was observed on the lower side of the settling plate of 2 October at Station B.

*Number of ephyrae*

We estimated the percentage ratios of the total number of strobilae and anticipated ephyrae to the initial number of polyps (Table 2). The percentage ratios were generally lower than 10%, however, greater than 200% was observed on the lower side of the settling plates of 2 and 26 October at Station B. The highest percentage ratio of 594.4% was observed on the settling plates of 26 October at Station B. Comparisons between the upper and lower sides showed that the estimated ratios of ephyrae liberated from polyps were significantly higher on the lower side of the settling plates at 2.3 m depth at Station A (Wilcoxon's signed-ranks test, $P < 0.01$) and at Station B ($P < 0.025$).

## Discussion

*Survivorship*

The higher survivorship of polyps on the lower side of the settling plates suggests that naturally settled polyps will also be observed on the lower side of substrata *in situ*. This phenomenon is frequently observed in other coastal waters (e.g. Hernroth & Gröndahl, 1983; Gröndahl & Hernroth, 1987; Gröndahl, 1988a; Miyake et al., 1997). In Japanese waters, SCUBA divers observed that polyps settled only on the lower side of floating piers (Miyake et al., 1997). In laboratory experiments, more than 90% of the planula

larvae recruited on the lower side of plastic coverslips (Brewer, 1978). These studies suggest that cnidarian planulae preferentially settle on the lower side of substrata, perhaps enabling polyps to avoid sedimentation. In coral reefs, sedimentation is a serious problem for the survivorship of polyps in many habitats (e.g. Babcock & Davies, 1991; Babcock & Mundy, 1996; Gilmour, 1999). In the present study, sedimentation was observed only on the upper side, where polyp survivorship was lower.

A rapid decrease in the survivorship of newly recruited larvae is also observed in other benthic organisms (Gosselin & Qian, 1997; Hunt & Scheibling, 1997), indicating that one of the significant causes of their mortality is predation by other benthic organisms. Gröndahl (1988b) reported that the mortality rate of newly settled polyps in Gullmar Fjord was 4.5–28% within 10 days after placement *in situ*.

In our present study, invasion and growth of other organisms were observed on all settling plates. From February to September, other organisms such as *M. galloprovincialis* Lamarck completely covered the surface of the settling plates on both sides as a mussel bed. The invasion and growth of other organisms on the plates decreased from October to December, and the abundance of *A. aurita* polyps increased by budding during this period. Of course in nature, some polyps can establish colonies which is different from our investigations; the importance of the colony effect, which could eliminate open space for other organisms, remains to be elucidated. In the present study, some polyps (i.e. plates of October on lower side in Station B) established dense colonies on the settling plates and the lower mortality was recorded on this plate. If polyps grow continuously without initial reduction by competition, they would survive for a long period. These findings suggest that competition for space with other organisms influence to the survivorship of *A. aurita* polyps.

*Strobilation*

A high cumulative strobilation ratio in this study means that most of the polyps surviving throughout the strobilation period will strobilate and liberate ephyrae. Thiel (1962) also reported that ca. 100% of the *A. aurita* polyps in Kiel Fjord, Germany strobilated. Previous laboratory experiments indicated that strobilation by *A. aurita* is initiated by reduction in water temperature (Kakinuma, 1962; Custance, 1964; Kato et al., 1980). Because of water temperature de-

*Table 2.* Percentage ratios of the total number of strobilae and ephyrae liberated from polyps to the initial number of polyps. 0 denotes no strobilae observed.

| | Started date | Station A | | | | Station B | |
| | | 0.8 m | | 2.3 m | | 0.8 m | |
| | | Strobilae | Ephyrae | Strobilae | Ephyrae | Strobilae | Ephyrae |
|---|---|---|---|---|---|---|---|
| **Upper side** | 2 Oct. 1998 | 0 | 0 | 0 | 0 | 29.7 | 46.2 |
| | 26 Oct. 1998 | 0 | 0 | 0 | 0 | 8.6 | 9.2 |
| | 24 Nov. 1998 | 0 | 0 | 0 | 0 | 2.2 | 2.2 |
| | 15 Dec. 1998 | 0.1 | 0.1 | 0.3 | 0.3 | 5.5 | 5.5 |
| | 26 Jan. 1999 | 0.7 | 0.7 | 0.6 | 0.6 | 1.6 | 1.6 |
| | 23 Feb. 1999 | 1.8 | 1.8 | 6.5 | 6.5 | 16.6 | 16.6 |
| | 6 Apr. 1999 | 0.2 | 0.2 | 0 | 0 | 0 | 0 |
| **Lower side** | 2 Oct. 1998 | 12.2 | 14.9 | 19.5 | 25.1 | 237.2 | 292.0 |
| | 26 Oct. 1998 | 0.2 | 0.2 | 0.3 | 0.3 | 455.6 | 594.4 |
| | 24 Nov. 1998 | 0.1 | 0.1 | 1.9 | 1.9 | 9.0 | 9.0 |
| | 15 Dec. 1998 | 3.3 | 3.3 | 2.7 | 2.7 | 27.1 | 27.2 |
| | 26 Jan. 1999 | 0.7 | 0.7 | 1.2 | 1.2 | 6.8 | 6.8 |
| | 23 Feb. 1999 | 2.0 | 2.0 | 9.0 | 9.0 | 11.3 | 11.3 |
| | 6 Apr. 1999 | 0.1 | 0.1 | 0.8 | 0.8 | 0 | 0 |

creases from autumn to winter, the autumn recruitment and growth of polyps observed in this study should reduce the dangerous period when polyps are exposed to competition for space with other benthic organisms. On the other hand, the reduction and loss of the number of polyps after strobilation suggests that polyp lifetime is less than 1 year or until late spring. As mentioned above, the new invasion of other benthic organisms, such as *M. galloprovincialis,* may cause the reduction of the post-strobilated polyps of *A. aurita.* The *in situ* observation of natural polyps of Gröndahl (1988b) and Hernroth & Gröndahl (1985a, b) revealed that the predation impact by other organisms is greater on older polyps of *A. aurita.* We could not observe actual predation by other organisms on the settling plates, however, from the observation that all polyps disappeared by August, we assume that the *in situ* lifetime of polyps is less than 1 year in Tokyo Bay.

Cessation of strobilation and loss of polyps may also be explained by lowering salinity (Purcell et al., 1999). In the present study, water salinity decreased from winter to spring with a minimum of <0.1 psu on 17 August. Purcell et al. (1999) also stated that the low salinity (<11 psu) inhibits strobilation and the production of *Chrysaora quinquecirrha* (DeSor) polyps in Chesapeake Bay, U.S.A.

*Number of discs in each strobila*

Most of the strobilae in the present study formed single discs. In Gullmar Fjord, Sweden, Hernroth & Gröndahl (1983) found plural discs in each strobila during autumn and spring with *in situ* observations and a single disc in each strobila only in midwinter. Hernroth & Gröndahl (1983) and Gröndahl & Hernroth (1987) suggested that the zooplankton biomass in ambient seawater correlated with the occurrence of plural discs. An increase in the numbers of discs in each strobila was reported with an increase in food availability in laboratory experiments by Spangenberg (1967, 1968). Polyps of *A. aurita* usually formed a single disc after starvation (Spangenberg, 1967). Spangenberg (1968) also reported that the sizes of *A. aurita* polyps apparently influenced the number of discs of strobilae and that very small polyps formed a single disc. By contrast, Watanabe (unpublished) found that each strobila produced 22–30 discs when supplied with excess *Artemia* nauplii as food in laboratory experiments. All results suggest that the food availability was not enough to produce plural discs in each strobila during the strobilation period in Tokyo Bay. However, further experiments are needed to know the relationship with food availability and the number of discs produced in each strobila.

*Number of ephyrae*

The number of ephyrae liberated from polyps reflects the survivorship at the polyp stage. The present study revealed that most of the liberated ephyrae would originate from polyps recruited in October. On the lower side in Station B, it was surprising that 2–5 times the initial numbers of settled polyps were produced after several months as polyps. If the ephyrae were produced continuously from December to April, as observed in this study, the cumulative numbers of ephyrae collected by net would be highest from March to April. This suggestion agrees with previous results of the occurrence of ephyrae by net sampling in Tokyo Bay (Sugiura, 1980; Toyokawa & Terazaki, 1994; Omori et al., 1995). Sugiura (1980) reported that ephyrae occurred from December to April. Toyokawa & Terazaki (1994) reported that ephyrae were abundant from January to March, and peaked in March. Omori et al. (1995) also reported that ephyrae appeared in March. This suggests that the number of polyps newly settled in autumn should contribute to the mass occurrence of medusae in next year in Tokyo Bay.

Most of the settled substrate in Tokyo Bay is occupied by the other benthic organisms such as *M. galloprovincialis* (see Furuse & Furota, 1985). This observation means that *A. aurita* polyps are exposed to keen competition for space with other organisms, especially during spring and summer as observed in this study. If the recruitment of planula larvae is restricted to summer as observed in many waters (Van Der Veer & Oorthuysen, 1985; Matsakis & Conover, 1991; Lucas & Williams, 1994; Schneider & Behrends, 1994; Miyake et al., 1997; Ishii & Båmstedt, 1998), the consequent production of ephyrae could be low in the following spring. In Tokyo Bay, ripe medusae with planula larvae are frequently observed throughout the year (Miyake et al., 1997). It is believed that it depends on the abundance of zooplankton, such as *Oithona davisae* Ferrari & Orsi, as food (Ishii & Tanaka, 2001) or on rather warm water temperatures during the winter (Hamner et al., 1982). The presence of ripe medusae with planula larvae, even in autumn and winter, would contribute to increasing settlement and survival of polyps during the period of low recruitment and growth of other benthic organisms, resulting an abundant settlement and high survival during the polyp stage. This flexibility will explain their success in many coastal waters.

## Acknowledgements

We are grateful to Prof. M. Omori, the staff of Museum of Maritime Science, and the Tokyo Box Barge Transportation Cooperative. We also thank the crews of T.S. "Hiyodori" and the members of the Ecology Laboratory, Tokyo University of Fisheries, for their comments, assistance and co-operation. Comments by Mr S. Tadokoro and critical reading of the manuscript by Mr R. Vassallo-Agius are gratefully acknowledged. We would like to thank two anonymous referees for their constructive comments.

## References

Babcock, R. & P. Davies, 1991. Effects of sedimentation on settlement of *Acropora millepora*. Coral Reefs 9: 205–208.

Babcock, R. & C. Mundy, 1996. Coral recruitment: consequences of settlement choice for early growth and survivorship in two scleractinians. J. exp. mar. Biol. Ecol. 206: 179–201.

Behrends G. & G. Schneider, 1995. Impact of *Aurelia aurita* medusae (Cnidaria, Scyphozoa) on the standing stock and community composition of mesozooplankton in the Kiel Bight (western Baltic Sea). Mar. Ecol. Prog. Ser. 127: 39–45.

Brewer, R. H., 1978. Larval settlement behavior in the jelly-fish *Aurelia aurita* (Linnaeus) (Scyphozoa: Semaeostomeae). Estuaries 1: 120–122.

Custance, D. R. N., 1964. Light as inhibitor of strobilation in *Aurelia aurita*. Nature 204: 1219–1220.

Dawson, M. N & L. E. Martin, 2001. Geographic variation and ecological adaptation in *Aurelia* (Scyphozoa, Semaeostomeae): some implications from molecular phylogenetics. Hydrobiologia 451 (Dev. Hydrobiol. 155): 259–273.

Furuse, K. & T. Furota, 1985. An ecological study on the distribution of intertidal sessile animals in inner Tokyo Bay. Marine Fouling 5: 1–6.

Gilmour, J., 1999. Experimental investigation into the effects of suspended sediment on fertilisation, larval survival and settlement in a scleractinian coral. Mar. Biol. 135: 451–462.

Gosselin, L. A. & P. Y. Qian, 1997. Juvenile mortality in benthic marine invertebrates. Mar. Ecol. Prog. Ser. 146: 265–282.

Gröndahl, F., 1988a. A comparative ecological study on the scyphozoans *Aurelia aurita*, *Cyanea capillata* and *C. lamarckii* in the Gullmar Fjord, western Sweden, 1982–1986. Mar. Biol. 97: 541–550.

Gröndahl, F., 1988b. Interactions between polyps of *Aurelia aurita* and planktonic larvae of scyphozoans: an experimental study. Mar. Ecol. Prog. Ser. 45: 87–93.

Gröndahl, F., 1989. Evidence of gregarious settlement of planula larvae of the scyphozoan *Aurelia aurita*: an experimental study. Mar. Ecol. Prog. Ser. 56: 119–125.

Gröndahl, F. & L. Hernroth, 1987. Release and growth of *Cyanea capillata* (L.) ephyrae in the Gullmar Fjord, western Sweden. J. exp. mar. Biol. Ecol. 106: 91–101.

Hamner, W. M., R. W. Gilmer & P. P. Hamner, 1982. The physical, chemical and biological characteristics of a stratified, saline, sulfide lake in Palau. Limnol. Oceanogr. 27: 896–909.

Hernroth, L. & F. Gröndahl, 1983. On the biology of *Aurelia aurita* (L.) 1. Release and growth of *Aurelia aurita* (L.) ephyrae in the Gullmar Fjord, western Sweden, 1982–83. Ophelia 22: 189–199.

258

Hernroth, L. & F. Gröndahl, 1985 a. On the biology of *Aurelia aurita* (L.): 2. Major factors regulating the occurrence of ephyrae and young medusae in the Gullmar Fjord, western Sweden. Bull. mar. Sci. 37: 567–576.

Hernroth, L. & F. Gröndahl, 1985 b. On the biology of *Aurelia aurita* (L.) 3. Predation by *Coryphella verrucosa* (Gastropoda, Opisthobranchia), a major factor regulating the development of *Aurelia* populations in the Gullmar Fjord, western Sweden. Ophelia 24: 37–45.

Hunt, H. L. & R. E. Scheibling, 1997. Role of early post-settlement mortality in recruitment of benthic marine invertebrates. Mar. Ecol. Prog. Ser. 155: 269–301.

Ishii, H. & U. Båmstedt, 1998. Food regulation of growth and maturation in a natural population of *Aurelia aurita* (L.). J. Plankton Res. 20: 805–816.

Ishii, H., S. Tadokoro, H. Yamanaka & M. Omori, 1995. Population dynamics of the jellyfish, *Aurelia aurita*, in Tokyo Bay in 1993 with determination of ATP-related compounds. Bull. Plankton Soc. Japan. 42: 171–176.

Ishii, H. & F. Tanaka, 2001. Food and feeding of *Aurelia aurita* in Tokyo Bay with a analysis of stomach contents and a measurement of digestion times. Hydrobiologia 451 (Dev. Hydrobiol. 155): 311–320.

Kakinuma, Y., 1962. On some factors for the differentiations of *Cladonema uchidai* and of *Aurelia aurita*. Bull. mar. Biol. Stat. Asamushi. 11: 81–85.

Kato, K., T. Tomioka & K. Sakagami, 1980. Morphogenetic patterns of scyphozoan strobilation. In Tardent, P. & R. Tardent (eds), Developmental and Cellular Biology of Coelenterates Elsevier, Amsterdam: 245–250.

Lucas, C. H., 2001. Reproduction and life history strategies of the common jellyfish, *Aurelia aurita*, in relation to its ambient environment. Hydrobiologia 451 (Dev. Hydrobiol. 155): 229–246.

Lucas, C. H. & J. A. Williams, 1994. Population dynamics of the scyphomedusa *Aurelia aurita* in Southampton Water. J. Plankton Res. 16: 879–895.

Matsakis, S. & R. J. Conover, 1991. Abundance and feeding of medusae and their potential impact as predators on other zooplankton in Bedford Basin (Nova Scotia, Canada) during spring. Can. J. Fish. aquat. Sci. 48: 1419–1430.

Miyake, H., K. Iwao & Y. Kakinuma, 1997. Life history and environment of *Aurelia aurita*. South Pacific Study 17: 273–285.

Möller, H., 1984. Reduction of a larval herring population by jellyfish predator. Science 224: 621–622.

Omori, M., H. Ishii & A. Fujinaga, 1995. Life history strategy of *Aurelia aurita* (Cnidaria, Scyphomedusae) and its impact on the zooplankton community of Tokyo Bay. ICES J. mar. Sci. 52: 597–603.

Purcell, J. E., J. R. White, D. A. Nemazie & D. A. Wright, 1999. Temperature, salinity and food effects on asexual reproduction and abundance of the scyphozoan *Chrysaora quinquecirrha*. Mar. Ecol. Prog. Ser. 180: 187–196.

Schneider, G. & G. Behrends, 1994. Population dynamics and the trophic role of *Aurelia aurita* medusae in the Kiel Bight and western Baltic. ICES J. mar. Sci. 51: 359–367.

Schneider, G. & G. Behrends, 1998. Top - down control in a neritic plankton system by *Aurelia aurita* medusae – a summary. Ophelia 48: 71–82.

Spangenberg, D. B., 1967. Iodine induction of metamorphosis in *Aurelia*. J. exp. Zool. 165: 441–450.

Spangenberg, D. B., 1968. Recent studies of strobilation in jellyfish. Oceanogr. mar. Biol. ann. Rev. 6: 231–247.

Sugiura, Y., 1980. On the seasonal appearance of the medusae from Harumi, Tokyo Harbour. Dokkyo Univ. Bull. lib. Arts. 15: 10–15.

Thiel, H., 1962. Untersuchungen über die Strobilisation von *Aurelia aurita*. Lam. an einer population der Kieler Förde. Kieler Meeresforsch. 18: 198–230.

Toyokawa, M. & M. Terazaki, 1994. Seasonal variation of medusae and ctenophores in the innermost part of Tokyo Bay. Bull. Plankton Soc. Japan. 41: 71–75.

Toyokawa, M., T. Furota & M. Terazaki, 2000. Life history and seasonal abundance of *Aurelia aurita* medusae in Tokyo Bay, Japan. Plankton Biol. Ecol. 47: 48–58.

Van Der Veer, H. W. & W. Oorthuysen, 1985. Abundance, growth and food demand of the scyphomedusa *Aurelia aurita* in the western Wadden Sea. Neth. J. Sea Res. 19: 38–44.

Yasuda, T., 1971. Ecological studies on the jelly-fish, *Aurelia aurita* in Urazoko Bay, Fukui Prefecture. Monthly change in the bell-length composition and breeding-season. Bull. Japan Soc. Sci. Fish. 37: 364–370.

*Hydrobiologia* **451**: 259–273, 2001.
© 2001 *Kluwer Academic Publishers.*

# Geographic variation and ecological adaptation in *Aurelia* (Scyphozoa, Semaeostomeae): some implications from molecular phylogenetics

Mike N Dawson & Laura E. Martin

*Department of Organismic Biology, Ecology, and Evolution, University of California,*
*Box 951606, Los Angeles, CA 90095-1606, U.S.A. and Coral Reef Research Foundation, Box 1765,*
*Koror, PW 96940, Republic of Palau*
*E-mail: dawsonmartin@palaunet.com*

*Key words:* counter gradient variation, feeding, growth, introduced species, jellyfish, phylogeography, population dynamics

## Abstract

Mitochondrial and nuclear DNA sequence data indicate considerable phylogeographic structure and at least five sibling species of *Aurelia* in the Pacific Ocean. At least a sixth sibling species can be found in the northwest Atlantic Ocean. These data suggest long histories of geographic and ecological sub-division and divergence of populations, which are inconsistent with current descriptions of *Aurelia* as a tri-typic genus in which most populations belong to one almost ubiquitous ecological generalist, *A. aurita* Linnaeus. Existing ecological and systematic descriptions of *Aurelia*, therefore, should be re-evaluated in light of these molecular data. Reciprocally, such re-evaluations should facilitate interpretation of the molecular data. Here, we introduce new DNA sequence data from Pacific and Black Sea *Aurelia* and novel ecological data describing tropical *Aurelia* inhabiting a marine lake in Palau, Micronesia. Despite large genetic distances between temperate and tropical *Aurelia* and the different environments inhabited by these populations, their rates of feeding, growth, respiration and swimming are similar. We discuss this result in terms of geographic variation and ecological adaptation in *Aurelia* and also comment on population dynamics, blooms, exotic species and the systematics of *Aurelia*. Finally, we consider briefly the implications of these findings for other scyphozoan species.

## Introduction

The moon jellyfish, *Aurelia*, is among the most widely distributed of all scyphozoans (Mayer, 1910; Arai, 1997), ranging circumglobally between 70°N and 55 °S (Fig. 1; Möller, 1980a; Mianzan & Cornelius, 1999). It also is the best studied of all scyphozoans (see references in Arai, 1997). *In situ* observations and experimental field and laboratory studies, ranging in scale from manipulations of groups of organisms to biochemical experiments, describe *Aurelia*, most often *Aurelia aurita*, from Japan eastward and northward to Scandinavia. Consequently, although these data originate from only a fraction of its range, *Aurelia* has provided information sufficient to explore the nature and consequences of geographic variation in scyphozoans.

Ecological studies refer to *A. aurita* as a widespread, ecological generalist (e.g. Kerstan, 1977, cited in Behrends & Schneider, 1995; Olesen et al., 1994; Miyake et al., 1997; Ishii & Båmstedt, 1998), a description supported by its successful invasion of foreign habitats around the world (Greenberg et al., 1996; Wrobel & Mills, 1998). Its aptitude for anthropogenic introduction may reflect a flexible life-history and ecology that result from adaptation to variable environments, which would be encountered by a widely dispersing planktonic species over evolutionary time. For example, *Aurelia* medusae feed on many different prey items, ranging from ciliates (Stoecker et al., 1987) to zooplankton (e.g. Behrends & Schneider, 1995; Ishii & Tanaka, 2001) and fish larvae (e.g. Möller, 1980b), and may absorb dissolved organic matter (Shick, 1973, 1975). *A. aurita* may even sort its food,

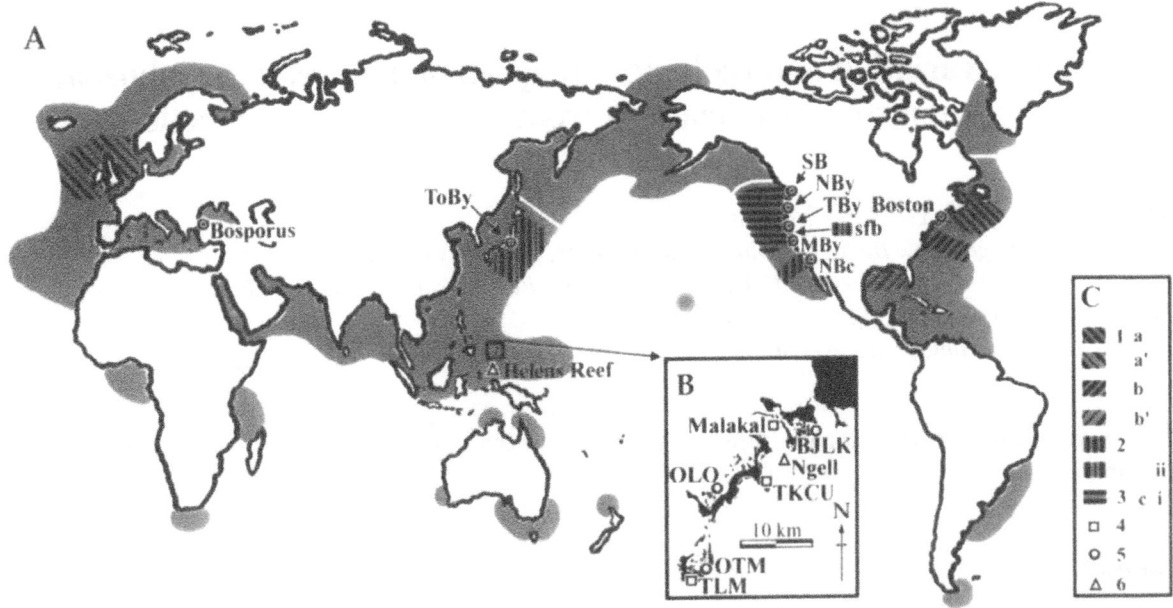

*Figure 1.* (A) The known distribution of the genus *Aurelia* (grey shading; Mayer, 1910; Kramp, 1961, 1965, 1968; Naumov, 1961; Devaney & Eldredge, 1977; Möller, 1980a; Kingsford et al., 1991; Colin & Arneson, 1995; Mianzan & Cornelius, 1999; Dawson & Jacobs, submitted; northwest Australia – P. Alderslade, pers. comm.; southwest Australia – K. Pitt, pers. comm.; New Zealand – J. Starmer, pers. comm.). Black hatching illustrates the geographic locations of populations used in molecular analyses. White bars in the northern Pacific Ocean and northwestern Atlantic Ocean indicate approximately the southern limits of *Aurelia limbata* (Kramp, 1961). (B) The locations of *Aurelia* sampled in Palau. (C) Key to geographic locations of *Aurelia* recognized as genetically distinct by molecular studies. Numbers, 1–6, indicate six putative sibling-species of *Aurelia* based on phylogenetic analyses of mitochondrial and nuclear DNA sequence data (Dawson & Jacobs, 2001; this study). Letters, a–c, indicate up to five groups of *Aurelia* distinguishable by protein electrophoresis; these data indicated close affinities between a and a' and between b and b' (Zubkoff & Linn, 1975). Roman numerals, i–ii, indicate two groups of *Aurelia* distinguishable by protein electrophoresis; these data suggest the introduction of *Aurelia* from Japan into San Francisco Bay (*sfb*; Greenberg et al., 1996). *SB* – Sooke Basin, Vancouver Island, Canada; *NBy* – Newport Bay, Oregon, U.S.A.; *TBy* – Tomales Bay, California, U.S.A.; *MBy* – Monterey Bay, California, U.S.A.; *NBc* – Newport Beach, California, U.S.A.; *ToBy* – Tokyo Bay, Japan; *BJLK* – Big Jellyfish Lake, Koror, Palau; *Ngell* – Ngell Channel, Palau; *TKCU* – Tab Kukau Cove, Urukthapel, Palau; *OLO* – Ongael Lake, Ongael, Palau; *OTM* – Ongeim'l Tketau, Mecherchar, Palau; *TLM* – Tketau Lake, Mecherchar, Palau.

depending upon what is available, presumably to serve best its nutritional or energetic needs (Stoecker et al., 1987; Martin, 1999). Yet, if starved, *Aurelia* can degrow (Hamner & Jenssen, 1974), potentially eluding death by starvation until conditions again are favorable for growth. Moreover, the benthic scyphistomae can reproduce asexually in more than half-a-dozen ways (Arai, 1997; Lucas, 2001; Watanabe & Ishii, 2001), allowing a population to persist and propagate even if environmental conditions do not foster sexual reproduction (via medusae) for one or more years. If conditions are favorable, however, such attributes might also promote the rapid growth of medusae and populations resulting in 'blooms' of *Aurelia*. Large, often dense, swarms of *Aurelia* are characteristic of protected coastal waters from sub-polar (Purcell et al., 2000) and cold-temperate regions (e.g. Hamner et al., 1994; Olesen et al., 1994) to the tropics (Hamner et al., 1982; Martin, 1999), although aggregations of jelly-

fish typically have been related to behavioral (Hamner & Hauri, 1981; Larson, 1992; Hamner et al., 1994; Purcell et al., 2000) and hydrographic effects (Hamner & Schneider, 1986; Larson, 1992; Purcell et al., 2000). Thus, *A. aurita* may be a quintessential generalist with a suite of attributes that can be induced by and befit best the particular local environment.

However, recent molecular data suggest there are at least four cryptic species of *Aurelia* in the North Pacific alone (Dawson & Jacobs, 2001). Thus, variations on the life-cycle and ecology of *Aurelia* may not be adaptation of a single species to a wide variety of habitats but adaptation of many sibling species to local conditions (e.g. Berstadt et al., 1995). The generalist actually may be a composite of specialists (Fig. 1). The molecular data are more akin to the traditional taxonomy of more than 35 years ago, when as many as 20 species or varieties of *Aurelia* were recognized (e.g. Mayer, 1910; see Kramp, 1968), than more recent descriptions of *Aurelia* as a bipartite (Kramp, 1968; Russell, 1970) or tri-partite (Wrobel & Mills, 1998; L. Gershwin, pers. comm.) genus. Unfortunately, while the flux of species names is grist to the mills of taxonomists, it confounds the interpretation of, among others, systematic, evolutionary, conservation and ecological research.

Here, we present new and summarize briefly the existing molecular data that describe *Aurelia*. We also introduce novel ecological data describing a tropical population of *Aurelia* that inhabits a marine lake in Palau, Micronesia. We then compare these novel data with similar data from more northerly populations and discuss their implications for geographical variation and ecological adaptation in *Aurelia*. We also refer briefly to introduced species and the systematics of *Aurelia*. Finally, we discuss the implications of these data for other scyphozoans.

## Methods

### *Phylogenetic analyses*

Mitochondrial cytochrome oxidase *c* subunit I (COI) and nuclear internal transcribed spacer region one (ITS1) sequences were concatenated for 19 specimens for which complete data are available (see Dawson & Jacobs, 2001; Fig. 1). Additional COI and ITS1 sequences were obtained from *Aurelia* medusae collected (during summer and fall 1999) in Tokyo Bay (Japan), Monterey Bay (California), and the Bosporus (Turkey) using the method of Dawson & Jacobs (2001; modified from Dawson et al., 1998), concatenated, and appended to the existing database. These combined sequences (456 bp COI and 397 bp ITS1) were used in 'total evidence' (Kluge, 1998) maximum parsimony analyses. The default options of the branch-and-bound search option in PAUP4.0 b3a (Swofford, 2000) were used in all reconstructions. Both accelerated and delayed transformation were used. Analyses explored the effects of a range of weighting schemes, varying from (1) 'equal weighting' of all characters to (2) weighting 1st:2nd:3rd positions 4:18:1, transitions:transversions 1:3, and COI:ITS1 sequences 1:1.2, respectively. These schemes reflected inversely the number of variable characters in 1st, 2nd and 3rd positions, the frequency of transitions and transversions, and the length discrepancy between COI and ITS1 sequences. Bootstrap analyses and Bremer indices were calculated for unweighted data using

PAUP4.0 b3a, PAUP 3.1.1 (Swofford, 1993) and TreeRot (Sorenson, 1996). Weighted and unweighted data also were analyzed using distance (Neighbor-Joining; Kimura 2-parameter substitution model) and maximum likelihood (quartet-puzzling; default options) criteria in PAUP4.0 b3a.

### *Ecological studies in Palau, Micronesia*

All research was completed with *Aurelia* from Big Jellyfish Lake, Koror (BJLK), either *in situ* or at the nearby Coral Reef Research Foundation (CRRF), Malakal, between September 1994 and December 1998. Many of the physical, chemical and some biological characteristics of BJLK are described by Hamner & Hamner (1998; see also Hamner & Hauri, 1981 and Hamner et al., 1982). Of the many attributes of the *Aurelia* inhabiting BJLK described by Martin (1999), we present here a subset that are methodologically independent such that each attribute, although linked irrevocably via the functional biology of the medusa, is a discrete estimate of the biology of these *Aurelia*.

### *Respiration rates*

Estimates of daily respiration rates were determined at BJLK by enclosing groups of several medusae of similar size in sealed containers (0.87 or 5.4 l depending upon the size of the medusae; mean sizes = 4–26 cm bell diameter) for 4 h. A YSI 85, digital, field oxygen meter with a polarographic Clarke type sensor was used to measure the initial and final dissolved oxygen concentrations and temperatures in the containers. Control vessels without medusae were used to distinguish the respiration of incidentally enclosed plankton from that of *Aurelia*. The diameter and wet weight (WW, c.f. dry weight, DW) of animals were measured at the end of the experiment. The wet weights were used to estimate the displacement volume of animals (assuming neutral buoyancy) and so to calculate the true volume of water used and oxygen consumed in each experimental chamber. Respiration rates were calculated as $\mu l\ O_2\ h^{-1}\ mg\ WW^{-1}$ and converted to $\mu l\ O_2\ h^{-1}\ mg\ DW^{-1}$ using the equation DW = 0.028 WW (Martin, 1999).

### *Daily ration*

Gut contents were collected from *Aurelia* on May 7th, 1996, between 0845 and 0930 hours, on May 9th and 23rd, 1997, at 0800, 1200, and 1600 hours, and on May 24th–25th, 1997, at 2000, 0000, and 0400 hours. *Aurelia* of all sizes were collected in re-sealable bags from all depths by freedivers or divers using SCUBA. *Aurelia* were taken as quickly as possible to the surface where the diameter of each animal was recorded and the gut contents evacuated completely from all 4 gastric cavities using a clean 60 ml syringe. The syringe's contents were expelled immediately into a pre-labeled, microcentrofuge tube containing an aliquot of formalin. The gut contents of each individual were removed and preserved generally within 5–10 min of collection. Gut contents from a total of 197 individuals, ranging from 3 to 29 cm, were collected.

All gut contents were enumerated under a dissecting microscope and counts were used to estimate individual daily feeding rates (FR = ingestion rates), according to the equation:

$$FR = \text{number of prey in guts} \times$$

$$[24/\text{digestion time}] = \text{number prey d}^{-1}$$

The mean digestion times of copepods ($n = 42$) and bivalve veligers ($n = 21$) were determined experimentally at 30°C by feeding freshly caught zooplankton to 21 freshly captured BJLK *Aurelia* of 4–8 cm bell diameter and using a dissection microscope to document *in vivo* digestion of individual plankters or food boluses at quarter-hour intervals. Feeding rates were converted, using values from the literature, to equivalent carbon ingestion rates to estimate specific daily ration (SDR), i.e. the percentage of the individual's carbon content ingested per day (Arai, 1997). The average carbon content of *Acrocalanus inermis* nauplii (0.05 $\mu$g C) and post-naupliar individuals (0.54 $\mu$g C) were taken from Kimmerer (1983) for the same species found in Kaneohe Bay, Hawaii. No values were available specifically for *Oithona oculata* so the mean carbon content for post-naupliar individuals was assumed to equal 0.2 $\mu$g C per individual (equivalent to *Oithona davisae* from Tokyo Bay, Japan; Uye, 1994), and 0.02 $\mu$g C for nauplii (*Oithona similis* nauplii from the Kattegat; Sabitini & Kiorbe,1994). Estimates of the carbon content of bivalve and snail veligers (0.2 $\mu$g C), and fish larvae (6 $\mu$g C) and eggs (10 $\mu$g C), tunicate larvae (1 $\mu$g C), harpacticoid copepods (2 $\mu$g C), and shrimp megalopa (10 $\mu$g C) were taken from Martinussen & Båmstedt (1995). BJLK *Aurelia* carbon content (CC) was calculated using the conversion CC = 0.04DW (see Arai, 1997).

*Population dynamics and growth rates*

During field trips between September 1994 and December 1998, the *Aurelia* population in BJLK was sampled at weekly to 3-monthly intervals using a 1 m$^2$ mouth, 1 mm mesh, net hauled vertically through the entire habitable mixolimnion (18 m to surface). Sequential lifts were made at 10 or 20 m intervals along the lengths of two perpendicular transects that crossed at the center of the lake. Sampling typically was completed during one or two consecutive days. The bell diameters of all *Aurelia* in each haul were measured to the nearest 0.5 cm while animals were lying exumbrella-down on a flat tray. These diameters were used to construct size frequency distributions for each sampling event. Growth rates between 11/20/98 and 1/30/97 were estimated as the change-in-mode divided by time between sampling events. However, the mode was not always a robust measure of the growth of cohorts. Growth rates of *Aurelia* for 3/97, 9/98 and 10/98 were estimated using an horizontal cohort method based on median values that accounts for mortality (Aksnes et al., 1993). This method entails dividing the difference between the median values of consecutive stages, here defined as consecutive sampling events, by the time between those stages. This method was preferred because the median is the most informative 'average' value of skewed distributions (Steel & Torrie, 1980; Wilkinson et al., 1992), always approximated the mean (within 0.8 cm), and was more robust than the mode to sampling error.

*Pulse rates*

In December, 1996, and January, 1997, medusae of all sizes were selected at random and observed for one minute during which time the number of bell contractions (pulses) was counted. A time-budget for swimming was estimated by documenting the behaviour of medusae for up to half-an-hour. Observations made at night used red light. Water temperatures were measured contemporaneously throughout the water column of BJLK. This protocol was repeated at a second marine lake, Ongeim'l Tketau (OTM, 'Jellyfish Lake').

**Results**

*Phylogenetic analyses*

All analyses but one recovered topologically the same tree (Fig. 2) with only minor variations within each of the seven numbered clades. The single exception, maximum likelihood analysis of unweighted data which placed clade 2 outside a group containing clades 3, 4, 5 and 6, likely was attributable to third positions of COI because this topology was not recovered

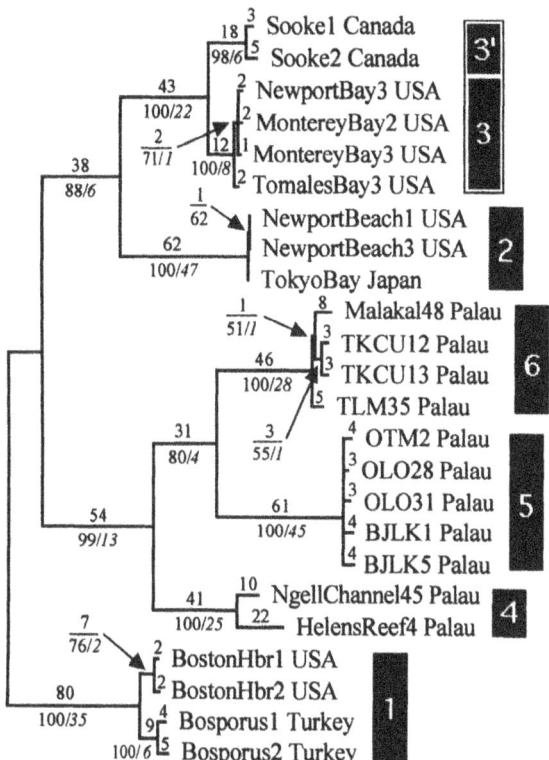

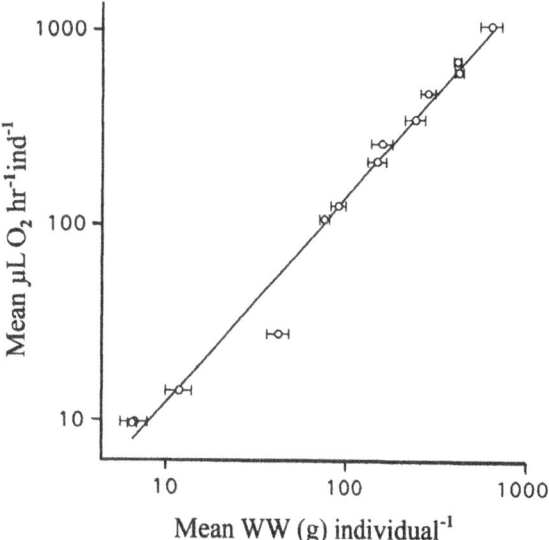

*Figure 2.* Unrooted phylogenetic tree showing the clades and evolutionary relationships supported by phylogenetic reconstructions completed for this study. Branch-lengths, the number of nucleotide changes or 'mutations' between adjacent nodes, from the 8 equally most-parsimonious trees recovered by unweighted parsimony analyses are shown above each branch. (Branch-lengths of all branches in clade 5, and the basal branches in clade 5/6 and 6 were variable, the average branch-lengths are shown). Bootstrap values (the percentage, if >50%, of 1000 bootstrap replicates in which each branch occurred, plain text) and Bremer support values (the increase in tree length, i.e. deviation from parsimony, required for a branch to be absent from the tree, italics) appear below branches. Clade numbers correspond to six clades recognized by Dawson & Jacobs (2001; also see Fig. 1), as follows: 1, Atlantic Ocean/Black Sea; 2, Californian Province; 3, Oregonian Province; 4, Lagoon; 5, Marine Lake; 6, Cove.

*Figure 3.* Mean wet weight (WW, g ind$^{-1}$) *versus* respiration rate (RR; $\mu$l O$_2$ h$^{-1}$ ind$^{-1}$, 30 °C) for *Aurelia* from BJLK. Error bars are included for wet weight because each point comprises multiple individuals enclosed in respirometry chambers. (RR = 1.084WW$^{1.059}$, $R^2$ = 0.982, $p < 0.001$).

when third positions were down-weighted or from ITS1 data alone. Variation within and between maximum parsimony analyses was limited to the number of shortest trees (2–8), the consistency index (0.7189–0.7341) and other such indices (e.g. retention index = 0.9063–0.9105), and differences of up to 2 steps in branch-lengths, depending on the weighting scheme employed. All analyses recovered at least five clades of *Aurelia* in the Pacific Ocean, a sixth in the Atlantic Ocean and Black Sea, and suggested shallower subdivision of several of these clades (#s 1, 3, & 4; Fig. 2). Sequences from *Aurelia* collected in Monterey Bay, California, are essentially indistinguishable from

others in Northern California and Oregon; they fall in the Oregonian biogeographic province. In contrast, sequences from *Aurelia* collected in Tokyo Bay, Japan, are related closely to Californian *Aurelia* and, thus, belong to a clade with members from two distant biogeographic regions. The sequences from Black Sea *Aurelia* also link two distant biogeographic regions, on opposite sides of the North Atlantic Ocean.

### Ecological studies in Palau, Micronesia

#### Respiration rates

Mean respiration rates at 30 °C regressed on mean wet weights demonstrated $\mu$L O$_2$ h$^{-1}$ ind.$^{-1}$= 1.084 WW$^{1.059}$ (WW measured in grams; $R^2$= 0.982; $p < 0.0005$; Fig. 3).

#### Daily ration

The mean digestion time ($\pm$ 95% confidence interval, CI) of BJLK copepods was 0.71 h ($\pm$0.05) and bivalve veligers was 2.3 h ($\pm$0.58). The SDR ranged from 11.4% to 1.3% in 1996 (2.9–23.9 cm *Aurelia*) and from 9.9% to 0.3% in 1997 (3.5–29 cm *Aurelia*). The log$_{10}$ of the SDR (log SDR) was related to the diameter (d) of *Aurelia* according to the equations log SDR$_{1996}$ = log 0.954–0.026d or log SDR$_{1997}$ = log 0.431–0.015d depending on the year of observation (significant negative slope, $p < 0.0005$ for

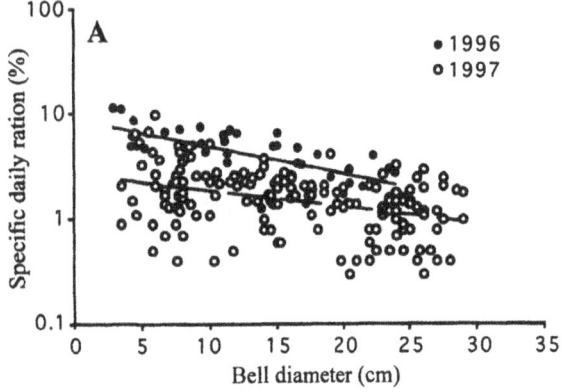

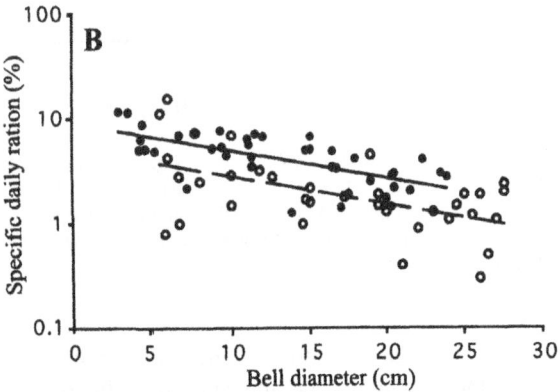

*Figure 4.* Specific daily ration (SDR,% of carbon content ingested per day) *versus* bell diameter (d) of *Aurelia* in BJLK in 1996 (black circles, solid line) and 1997 (white circles, dashed line). (A). 1996, log[SDR] = log[0.954 – 0.026d] ($R^2$ = 0.45, $p < 0.001$). 1997, log[SDR] = log[0.431 – 0.015d] ($R^2$ = 0.151, $p < 0.001$). (B). 0800 hours only. Although the slopes of the two regressions are equal (ANCOVA, $F_{(1,71)}$ = 0.001; $p$ = 0.997), the SDR is significantly greater in 1996 than 1997 (ANCOVA, $F_{(1,72)}$ = 17.006, $p <$ 0.001).

both years; Fig. 4a). Comparing similar times between years (0800–0900 h only), the slopes of the 1996 and 1997 regression equations were the same ($F_{(1,71)}$ = 0.001, $p$ = 0.997) although the daily ration was significantly greater in 1996 ($F_{(1,72)}$ = 17.006, $p <$ 0.0005; Fig. 4b).

### Population dynamics and growth rates

On all dates sampled, the population of *Aurelia* in BJLK consisted of a wide size range of individuals both sexually immature and mature. The population size varied from $1.2 \times 10^5$ to $1.6 \times 10^6$, and mean $\pm$ 95% CI density from 0.17 $\pm$0.04 medusa m$^{-3}$ to 2.7$\pm$0.40 medusa m$^{-3}$. On several occasions between 1994 and 1998, most notably 1996–97 and 1998, obvious peaks in the size-frequency distributions distinguished relatively discrete and large strobilation

events. Comparisons of successive size-frequency distributions indicated that diameters increased by 0.2–1.1 cm wk$^{-1}$ (mean = 0.7 cm wk$^{-1}$; Fig. 5) or 0.02–0.20 g C wk$^{-1}$. The dissipation of cohorts over time indicated inter-individual variation in growth rates.

### Pulse rates

During 1-min observation periods, *Aurelia* medusae swam either continuously or discontinuously, the latter consisted of series of pulses interrupted by quiescence. Continuous pulse rates ranged from 75 to 14 pulses min$^{-1}$ and discontinuous pulse rates ranged from 20 to 4 pulses min$^{-1}$in 2–25 cm medusae, respectively, between 30 and 31 °C. Medusae that pulsed continuously pulsed significantly more frequently than those that pulsed discontinuously in both BJLK (ANCOVA, $F_{(1,35)}$ = 24.163, $p < 0.001$) and OTM (ANCOVA, $F_{(1,58)}$ = 19.035, $p < 0.001$) but there was no significant difference in pulse rates between BJLK and OTM *Aurelia* that pulsed continuously (ANCOVA, $F_{(1,12)}$ = 2.363, $p$ = 0.15) nor between BJLK and OTM *Aurelia* that pulsed discontinuously (ANCOVA, $F_{(1,81)}$ = 0.134, $p$ = 0.715). All individuals observed for longer than one minute swam discontinuously, pulsing between 42% and 84% of the time (mean $\pm$ sd, 69$\pm$13%, n = 7). Pulse rates were correlated inversely with the bell diameter of medusae (Fig. 6).

### Discussion

Geographically concordant patterns of molecular variation in *Aurelia* have now been described by three independent studies (Zubkoff & Linn, 1975; Greenberg et al., 1996; Dawson & Jacobs, 2001; also this study; Fig. 1). In the northern Atlantic and Pacific oceans alone, at least six clades of *Aurelia* are distinguishable. These clades can be recognized by their geographic locations, habitats and unique genetic constitutions (Figs 1 and 2; Zubkoff & Linn, 1975; Greenberg et al., 1996), differences that suggest long histories of geographically or ecologically structured division and divergence of taxa (Avise et al., 1987; Lynch, 1989; Futuyma, 1998; Dawson & Jacobs, 2001). Whether allopatric or sympatric, the division and divergence of populations is the beginning and, if sufficiently long-lasting, can be the end of speciation (Ridley, 1993; Futuyma, 1998; but see Grant & Grant, 1998). Comparisons of COI and ITS1 sequence differences in *Aurelia* with species-level sequence differences in other taxa suggest each clade of *Aurelia* constitutes a distinct

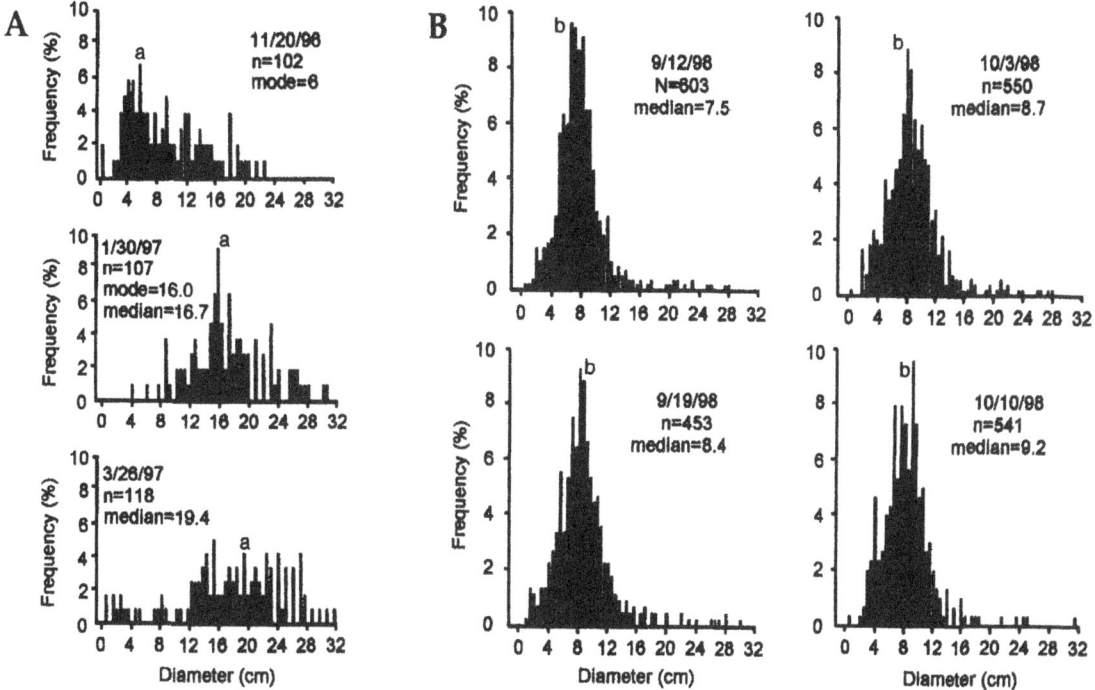

*Figure 5.* Size-frequency distributions of *Aurelia* in BJLK between 1996 and 1998. Tracing of modes or medians (a, b) between consecutive sampling times allows estimation of growth rates of (A). 1.1 and 0.4 cm week$^{-1}$ and (B). 0.9, 0.2 and 0.5 cm w$^{-1}$. (See also Table 1 & Martin, 1999.)

species (Dawson & Jacobs, 2001). Recognizing these clades as species also is supported by the coincidence of geographic or habitat separation with probable reproductive and physiological isolation (Dawson & Jacobs, 2001). Moreover, there are discernible morphological differences between molecularly identified clades of *Aurelia*, and these clades coincide approximately with previous morphological and geographic descriptions of *A. flavidula* Péron & Lesueur (western North Atlantic), *A. labiata* Chamisso & Eysenhardt (Pacific North America), *A. colpota* Brandt (Indian Ocean to Pacific) *A. hyalina* Brandt (North Pacific), and *A. maldivensis* Bigelow (Maldive Islands; Mayer, 1910; Kramp, 1961). Thus, the majority of data available at this time suggest strongly that descriptions of *Aurelia* as a bi- or tri-typic genus in which most populations belong to one almost ubiquitous generalist, *A. aurita*, are biogeographically and systematically overly conservative.

The existence of cryptic species of *Aurelia* implies classical taxonomic studies of this genus lack resolution. Kramp (1965, 1968) noted that there were few morphological characters that would distinguish reliably the different varieties of *Aurelia*. He attributed this to high levels of intra-specific variation and a lack of good species (Kramp, 1965, 1968). Kramp (1968) recognized only two species of *Aurelia*: *A. limbata* Brandt and *A. aurita*. However, Greenberg et al. (1996) found that only two of 12 morphological characters (manubrium length and canal structure) distinguished reliably two distinct genetic varieties of *Aurelia*, which suggests there is a dearth of good characters and not necessarily a dearth of good species. Indeed, molecular studies have demonstrated that morphological homoplasy is rampant among marine organisms (Knowlton, 1993; Potter et al., 1997; see also Gosliner & Ghiselin, 1984).

Phenotypic variation, such as that encountered by Kramp (1965, 1968) and Greenberg et al. (1996), is a double-edged sword. If there is not enough or there is too much and it is not distributed discretely among monophyletic taxa, variation will confound the recovery of accurate phylogenies (Yang, 1998). However, in light of a robust phylogeny, phenotypes can reveal much about the evolution of taxa. For example, convergent evolution often indicates intense selective pressure (e.g. for eyes on an illuminated planet [Dennett, 1995:128,134] and transparency in epipelagic plankton [McFall-Ngai, 1990; Hamner, 1995]), as does rapid evolution (Rooney & Zhang, 1999), whereas high levels of polymorphism or phenotypic plasticity suggest selectively neutral traits or selection for variation (Takahata, 1993; Futuyma, 1998:384–390; Rutherford & Lindquist, 1998). The power of molecular analyses, therefore, is not necessarily in their utility for reconstructing robust phylogenies as much as in the ability of robust phylogenies to explain patterns of geographic variation.

*Geographic variation. . .*

Laboratory experiments demonstrate that the respiration rates of many marine invertebrates more than double for every 10 °C increase in temperature (Arai, 1997). Similarly, medusae transferred to higher temperatures show greatly increased pulse rates (Berstadt et al., 1995; see also Arai, 1997: 127–130). Such changes in rate with temperature also occur naturally and, indeed, the most common patterns of geographic variation are clines effected by latitudinal changes in temperature (e.g. Briggs, 1974, 1995: 208–211; Unwin & Corbet, 1984; Nybakken, 1988:61; Stevens, 1989; Huntley & Lopez, 1992; Cockrell et al., 1993; Ridley, 1993; Van Voorhies, 1996; Sanz, 1999; but see Colwell & Lees, 2000). Variability in ambient temperature also affects life-history traits and the more variable temperate climate is more demanding physically and evolutionarily than is the relatively benign tropical climate (Dobzhansky, 1950; Briggs, 1974; Heinze et al., 1998; Schultz et al., 1998). These observations suggest two simple, testable, hypotheses: ($H_1$) that metabolic processes and their effects will be more rapid in tropical than in temperate organisms, and ($H_2$) that population dynamics will be more seasonally variable in temperate regions than in the tropics.

*. . . and ecological adaptation*

Annual life-cycles predominate in temperate *Aurelia*. Medusae generally are strobilated during spring and grow rapidly through summer, become reproductive by fall and senescent in late-autumn or early-winter (Rasmussen, 1973; Hamner & Jenssen, 1974; Möller, 1980a; Van Der Veer & Oorthuysen, 1985; Papathanassiou et al., 1987; Olesen et al., 1994; Costello & Mathieu, 1995; Schneider & Behrends, 1998). However, local variations exist. In Southampton Water, the cycle occurs between January and June (Lucas & Williams, 1994). In the Black Sea, strobilation between November and May produces two cohorts, the first matures in the spring and the second in the summer

(Lebedeva & Shushkina, 1991). Elsewhere, generations of medusae overlap between years and they may (Hamner & Jenssen, 1974; Miyake et al., 1997) or may not (Yasuda, 1971; Omori et al., 1995; Lucas, 1996) co-exist with reproductive medusae of the subsequent generation.

Cohorts of medusae are obvious in temperate populations of *Aurelia* in which strobilation is discrete and, like the senescence of medusae, typically annual (Möller, 1980a; Papathanassiou et al., 1987; Lucas & Williams, 1994; Olesen et al., 1994; Ishii & Båmstedt, 1998; Schneider & Behrends, 1998). Cohorts also are obvious in populations in which medusae persist for more than one season (Yasuda, 1968; Hamner & Jenssen, 1974; Omori et al., 1995; Miyake et al., 1997), strobilation occurs twice per year (Rasmussen, 1973; Hernroth & Gröndahl, 1985; Ledebeva & Shushkina, 1991), or strobilation occurs for many months of the year (Möller, 1980a; Lucas, 1996). Particularly in univoltine populations, the strobilation of strong cohorts can rapidly increase the density of medusae from effectively zero up to many tens (Lucas & Williams, 1994; Lucas, 1996; Nielson et al., 1997; Ishii & Båmstedt, 1998) or hundreds per cubic meter (Olesen et al., 1994). The most dense 'blooms' occur in fully- or semi-enclosed environments (Ishii & Båmstedt, 1998; see also Youngbluth & Båmstedt, 2001) perhaps due to reduced advection, dilution, and predation compared to more open habitats (Ishii & Båmstedt, 1998).

In contrast, tropical marine lake populations of *Aurelia* perennially consist of abundant medusae of all sizes and rarely contain strong cohorts. Here, medusae are strobilated and senesce throughout each year. In addition, the maximum densities of marine lake *Aurelia* populations are low and the maximum bell diameters of medusae high (>30 cm) relative to other enclosed systems (Table 1; Hamner et al., 1982; Ishii & Båmstedt, 1998; Martin, 1999).

The different characteristics of *Aurelia* populations in open, enclosed, and marine lake ecosystems suggest important effects of the environment on population dynamics. It is tempting to attribute perennial population structure and continual strobilation, which so far are recorded only from marine lake *Aurelia*, to the relatively constant tropical environment (cf. temperate regions). For example, marine lake *Aurelia* might be adapted to year-round food availability because more diffuse strobilation dissipates intra-cohort competition (see Bridle & Jiggins, 2000), allowing medusae to grow larger and thus more fecund (see

Table 1. Growth rates of *Aurelia* at different latitudes and temperatures. Additional information, such as population density, prey abundance, and bell diameter of medusae are provided as these also might influence growth rates. nd, no data. [1]"very low" (Ishii & Båmstedt, 1998)

| Growth (cm wk$^{-1}$) | Density (No. m$^{-3}$) | Location | Lat. (°N) | T (°C) | Prey (mg C m$^{-3}$) | Bell (cm) | Source |
|---|---|---|---|---|---|---|---|
| 0.4 | 0.20–0.23 | BJLK – 1996-1997 | 7 | 30 | 10–16 | 16–19 | This study |
| 1.1 | 0.17–0.20 | BJLK – 1997 | 7 | 30 | 10–27 | 6–16 | This study |
| 0.2–0.9 | 2.5 | BJLK – 1998 | 7 | 30 | 16 | 7–9 | This study |
| 0.25–0.7 | 3.1 | Urazoko Bay, Japan | 35 | 11–24 | nd | 7–19 | Yasuda (1969, 1971) |
| 0.9 | < 0.05–0.35 | Tokyo Bay, Japan | 35 | 17–23 | 134 | 12–17 | Omori et al. (1995) |
| 0.4 | ~7 | Horsea Lake, England | 51 | 16±3 | ≤1 | 4–5 | Lucas (1996) |
| 3.4 | 0.5–2 | Southampton, England | 51 | ~12 | 11–32 | 3–7 | Lucas & Williams (1994) |
| 4.5 | < 0.05–0.2 | Western Wadden Sea | 53 | nd | 40–60 | 6–14 | Van Der Veer & Oorthuysen (1985) |
| 4 | 0.09 | Kiel Bight – Western Baltic | 54 | ~15 | nd | 5–10 | Möller (1980b) |
| <0.1–1.25 | 50–125 | Kertinge Nor, Denmark | 58 | ~22 | 20–60 | 5–15 | Olesen et al. (1994) |
| 0.5 | < 20–60 | Kertinge Nor, Denmark | 58 | nd | 20 | 5–7 | Nielson et al. (1997) |
| <1.5, max 2.6 | 5–20 | Vågsbøpollen, Norway | 60 | 12–15 | v. low[1] | 5–8 | Ishii & Båmstedt (1998) |

Schneider, 1988; Lucas, 1996; Hansson, 1997). The potential for strong cohorts, or 'blooms', then may be reduced further by inter-cohort competition from larger medusae, which have a disproportionately high predation effect per unit bell diameter (Martin, 1999). Thus, the reduced 'peakiness' of density distributions of marine lake populations might be attributable in some way to reduced seasonality. However, protracted life-histories, spread throughout many months and even between years, also are apparent in temperate populations of *Aurelia* in enclosed habitats (Hamner & Jenssen, 1974; Möller, 1980a; Lucas, 1996), suggesting other contributing factors, including perhaps relaxed selection for synchronized strobilation and maturation because adults are rarely advected away (Martin, 1999; see also Schneider, 1988). In addition, pulses of strobilation do occur in BJLK (e.g. Fig. 5) and may be associated with changes in weather patterns (Martin, 1999), suggesting tropical populations may be adapted to respond to smaller climatic variations. Indeed, *Aurelia* in partially-open coves in Palau occur in distinct, approximately annual, cohorts (Dawson & Martin, pers. obs.) and the same is true of *Mastigias* (Dawson & Martin, pers. obs.). Thus, contrary to $H_2$ posited above, seasonality is obvious in tropical biota despite only modest seasonality in tropical weather (Dobzhansky, 1950).

Additional comparisons also reveal similarities between tropical and temperate *Aurelia* that are contrary to the latitudinal clines observed in other plankton (Huntley & Lopez, 1992). For example, the short-term effect of temperature change on respiration rate observed *in vitro* (Arai, 1997) is not realized when populations are studied across a range of latitudes (Table 2). Similarly, despite numerous factors that might influence daily ration (Table 3), growth rate (Table 1), and pulse rate (Fig. 7; see also Mayer, 1914) these metrics are not correlated to latitude. Digestion rates of copepods also are similar across a wide range of pelagic cnidarians and temperatures (2–4 h; Purcell, 1997), and faster rates may be attributable to small prey size as opposed to different rates of digestion *per se* (Martin, 1999; Ishii & Tanaka, 2001). Thus, contrary to $H_1$ posited above, the striking feature of these comparisons is the relative similarity in rates across a wide range of latitudes and temperatures.

Similarities in physiological indicators such as respiration, digestion, pulse and growth rates across many degrees of latitude and centigrade suggest *Aurelia* exhibit temperature compensation (Hochachka & Somero, 1984; Cossins & Bowler, 1987; but see Clarke, 1993). In light of the molecular variation in *Aurelia* (Figs 1 and 2), it seems likely that *Aurelia* exhibits countergradient variation (Martin, 1999), i.e. the non-random variation of genotypes across a latitudinal gradient such that genetic change counterbalances environmental change (Conover & Schultz, 1995). For example, increases in the volume and surface density of mitochondrial clusters in cold-water fish compensate for declining metabolic rates attributable to lower temperatures (Johnston et al., 1998). In addition, some variants of the mitochondrial gene COI

*Table 2.* Respiration rates in *Aurelia* at different latitudes and temperatures. T, Temperature °C; WW, wet weight; DW, dry weight; nd, no data. *Cited in Larson, 1987. **Cited in Arai, 1997

| T°C | $\mu$L O$_2$ h$^{-1}$ mg WW$^{-1}$ | $\mu$L O$_2$ h$^{-1}$ mg DW$^{-1}$ | Location | Latitude °N | Source |
|---|---|---|---|---|---|
| 30 | 0.0003–0.002 | 0.01–0.06 | Palau | 7 | This study |
| 22 | nd | 0.1 | Red Sea | 24 | Mergner & Svoboda (1977)* |
| 20 | 0.003–0.011 | nd | Black Sea | 48 | Kuzmicheva (1980)** |
| 20 | nd | 0.014 | Black Sea | 48 | Pavlova (1968)** |
| 10–21 | 0.002–0.007 | 0.09–0.36 | Black Sea | 48 | Yakovleva (1964)** |
| 16 | 0.003–0.005 | nd | Baltic Sea | 56 | Thill (1937)** |
| 10–15 | nd | 0.14–0.24 | Vancouver Is., N.E. Pacific | 44 | Larson (1987) |
| 12–14 | 0.002–0.004 | nd | Baltic Sea | 56 | Kerstan (1977)** |

*Table 3.* Specific daily ration (%) achieved by *Aurelia* at different latitudes and temperatures. Percent daily ration calculated from gut contents of medusae and carbon contents of medusae and prey. *T*, Temperature °C; $^{\tau}$ range of means; $^{\tau\,\delta}$ mean ± standard deviation

| T °C | Diameter (cm) | % Daily ration | Location | Latitude °N | Source |
|---|---|---|---|---|---|
| 30 | 3–29 | 1.1–6.0 | Palau | 7 | This study |
| 16–28 | 18–21 | 0.58–5.6$^{\tau}$ | Japan | 35 | Ishii & Tanaka (2001) |
| 10 | 11 | 6–8 | Black Sea | 48 | Shushkina & Musyeva (1983) |
| 10 | 8–10 | 1–2 | Black Sea | 48 | Shushkina & Arnautov (1985) |
| 9.5 | 3.5–29 | 1.8±5.7$^{\tau\,\delta}$ | Norway | 60 | Martinussen & Båmstedt (1995) |

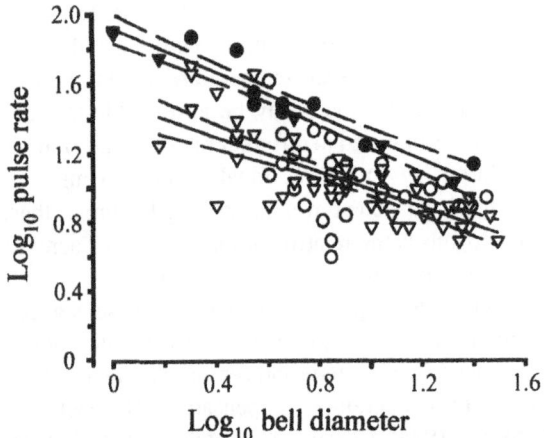

*Figure 6.* Pulse rate (pulses min$^{-1}$) *versus* bell diameter (cm) for individual *Aurelia* in BJLK (circles) and OTM (triangles), Palau. Closed symbols indicate continuous pulsing and open symbols indicate discontinuous pulsing. Pulse rate decreases significantly ($p \leq 0.019$) with increasing bell diameter in all datasets. Shown are regressions (with 95% CI, dashed lines) for all *Aurelia* pulsing continuously ($y = 1.918 - 0.623x$, $R^2 = 0.913$; $p < 0.001$) and those pulsing discontinuously ($y = 1.508 - 0.511x$, $R^2 = 0.476$; $p < 0.001$).

genotype and latitudinal clines in phenotype, as well as those between genotypic, phenotypic, and environmental variation, are well established in *Drosophila* (e.g. Dobzhansky, 1950; Rutherford & Lindquist, 1998; Azevedo et al., 1998; van't Land et al., 1999).

*Geographic variation and exotic species*

Current molecular data show that genetically very closely related *Aurelia* exist in Tokyo Bay, San Francisco Bay, and Southern California (clades 2, ii; Figs 1 and 2). The disjunct geographic distribution and genetic homogeneity of this clade, compared with the distributions and apparently greater genetic heterogeneity of other clades of *Aurelia*, suggest this is an introduced, or "exotic", species (Figs 1 and 2; also see Greenberg et al., 1996; Bastrop et al., 1998; Stepien et al., 1998). The phylogenetic affinity of the 'exotic' *Aurelia* for northeastern Pacific *Aurelia* (clade 3/c/i; Figs 1 and 2) suggests the populations in San Francisco Bay and Tokyo Bay were introduced from southern California. Morphologically and molecularly anomolous 'exotic' *Aurelia* first were observed in Foster City, San Francisco Bay, in 1988 (Greenberg et al., 1996). However, it is not known whether *Aurelia*

are associated with high growth rates in the shrimp *Penaeus vannamei* (Benzie, 1998). Links between

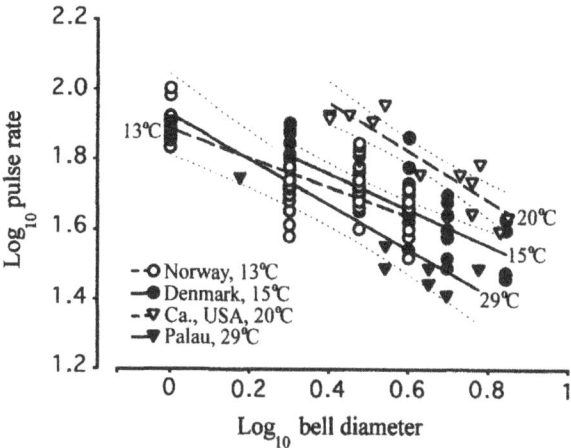

*Figure 7.* $\log_{10}$ pulse rate *versus* $\log_{10}$ bell diameter for populations of *Aurelia* from four different latitudes. All regressions show a significant ($p < 0.001$) decrease in pulse rate with increasing bell diameter. *Norway* (Berstadt et al., 1995), 60°N, the *in situ* temperature at which *Aurelia* were collected was 11.2–12.0°C and the laboratory experiment was conducted at 13 °C, $y = 1.883 - 0.408x$, $R^2 = 0.605$. *Denmark* (Olesen, 1995), 55 °N, *in situ* and laboratory temperatures were 15 °C, $y = 1.958 - 0.499x$, $R^2 = 0.525$. Ca., U.S.A. (Long Beach, California; this study), 34 °N, measurements made *in situ* at 20 °C, $y = 2.237 - 0.687x$, $R^2 = 0.822$. *Palau* (BJLK and OTM; this study), 7 °N, measurements *in situ* 29 °C, $y = 1.925 - 0.635x$, $R^2 = 0.8$. Dashed lines indicate 95% confidence intervals on regressions; only the highest and lowest confidence intervals for Norway, Denmark, and Palau *Aurelia* are shown to simplify the diagram.

that were present in Japan in 1891 and in Tokyo Bay in 1915 were endemic (Kishinouye, 1891, and Hirasaka, 1915, cited in Mills, 2001). Similarly, it remains to be demonstrated that modern southern California *Aurelia* are endemic; this depends on greater geographic sampling revealing that southern California *Aurelia* have both greater molecular heterogeneity and a larger continuous range than the introduced populations in Tokyo Bay and San Francisco Bay (e.g. Bastrop et al., 1998; Slade & Moritz, 1998). Presumably, southern California *Aurelia* also would maintain its current close, if not sister taxon, relationship to Oregonian *Aurelia*. The data describing Black Sea *Aurelia* also are provocative. According to historical records, *Aurelia* have been present in the Black Sea since at least the 1950s (Mutlu et al., 1994), but whether they are native or introduced remains to be determined, as is also the case for Boston Harbor *Aurelia*.

The absence of 'exotic' *Aurelia* from open coastal waters in central and northern California (Greenberg et al., 1996) suggests that somehow the species is limited to San Francisco Bay and southern California. In light of the evidence for local thermal adaptation

(Tables 1–3, Fig. 7), it seems likely that, among other things, water temperature may be limiting the range of 'exotic' *Aurelia*. Open coastal California waters are, on average, about 5 °C cooler than those in San Francisco Bay and Southern California (USGS, 1999; NOAA, 1999). Japanese bay waters also are considerably warmer than those of central California (Yasuda, 1969; Miyake et al., 1997). Consequently, environmental and genetic variation together may limit the spread of 'exotic' *Aurelia* (see also Bastrop et al., 1998). Introductions of *Aurelia* may be inhibited across lines of latitude or, more specifically, isotherms. For example, despite frequent shipping between Boston and Japan (MPA, 1999) these populations have remained discrete.

By contrast, introductions along lines of latitude or across small temperature changes might be facilitated if selective advantages are conferred on the 'exotic' relative to the native species. For example, translocation to warmer waters should increase pulse rates and, consequently, feeding rates in 'exotic' *Aurelia*, at least initially. The apparently high pulse rates of southern California *Aurelia* (Fig. 7), therefore, may be consistent with their introduction from slightly cooler waters and might be maintained by selection or an inability to acclimate due to the suite of isozymes available (e.g. see Zubkoff & Linn, 1975; Greenberg et al., 1996). More importantly, such effects might be manifested as large changes in population or ecosystem dynamics including 'blooms' of introduced species (see Shushkina & Musayeva, 1990; Zaitsev, 1992; Mutlu et al., 1994). Tracing the routes of introductions with molecular markers and understanding the genetic contributions to successful introductions are, therefore, potentially valuable areas of research.

*Implications for other species*

Many scyphozoan species, as defined by morphological criteria, have large geographic ranges. Hypothetically, distant dispersal across an homogeneous, continuous, ocean environment has maintained often pan-oceanic populations. However, molecular analyses of *Aurelia*, the most intensively studied jellyfish, have revealed considerable morphological homoplasy, genetic variation and multiple sibling species (Kramp, 1965, 1968; Zubkoff & Linn, 1975; Greenberg et al., 1996; Dawson & Jacobs, 2001; this study). This suggests that other, morphologically simple and widely distributed, but less well-studied, medusae also may require molecular systematic revision. Cryptic species most likely will be found among cosmopolitan, pan-oceanic, and disjunct taxa such as, *Atolla vanhoeffeni* Russell, *Atolla wyvillei* Haeckel, *Cyanea capillata* Linnaeus, *Periphylla periphylla* Péron & Lesueur, *Stygiomedusa gigantea* Browne, and *Stomolophus meleagris* L. Agassiz (see Wrobel & Mills, 1998). Similarly, molecular tools offer an unprecedented opportunity to distinguish endemic and introduced populations of scyphozoans in the absence of distinguishing morphological characters and limited information about original distributions (e.g. Bastrop et al., 1998).

However, the temptation to generalize across taxa should be tempered. Although morphological similarity may mask considerable molecular difference, morphological divergence does not necessarily imply molecular divergence. Preliminary analyses of *Mastigias* from Palau indicate considerable morphological, behavioral and ecological diversity but essentially no molecular differences in either COI or ITS1 (Hamner & Hauri, 1981; Muscatine & Marian, 1982; McCloskey et al., 1994; Dawson et al., 2001; Dawson, unpubl. data). Based on geographical data, the adaptive radiation of these populations likely has taken as little as 10 000–20 000 years (Hamner & Hauri, 1981; Hamner & Hamner, 1998; Dawson, unpubl. data), too little time for significant molecular differences to accrue in all but the fastest evolving DNA sequences. Thus, although there is considerable potential for cryptic species in scyphozoans, each taxon must be assayed independently.

*Closing remarks*

Morphological simplicity and homoplasy have confounded the traditional systematics of *Aurelia*, which in turn have confounded the interpretation of ecological data describing *Aurelia*. Consequently, *Aurelia aurita* has been described as a nearly cosmopolitan ecological generalist. However, new molecular and ecological data indicate *Aurelia aurita* actually is a species-complex comprised of numerous locally adapted species. Although local adaptation may limit the ability of *Aurelia* to exploit some habitats, care must be taken to avoid equating adaptation with extreme specialization or lack of variation (see Tables 1, 2 and 3; Fig. 7). At least one species of *Aurelia* has invaded at least two exotic locations and must, therefore, have a phenotype suitable for at least slightly different situations or the potential to express different phenotypes, i.e. adaptive polymorphism. Notably, the evolution of adaptive polymorphism is facilitated by overlapping generations (Ellner & Hairston, 1994; Sasaki & Ellner, 1997; Sasaki & de Jong, 1999) and so may be common in scyphozoans, such as *Aurelia*, that have bipartite life-histories. Moreover, adaptive polymorphism permits single species to be successful over broad geographic ranges (Dobzhansky, 1950). Adaptive polymorphism, therefore, may have contributed much to the evolutionary success and diversification of *Aurelia*.

Brown (1995) espouses 'macroecology' in which inter-population variation is related to both the biotic and abiotic environments across large geographic scales. Similarly, Miyake et al. (1997) encouraged future research on population attributes of *Aurelia* and physical features of the environment in the belief that inter-population variation is due to phenotypic plasticity and local adaptation. Here, we have presented data that suggest local phenotypic variation also reflects underlying genetic differences. Similar data often are collected routinely during ecological studies of scyphozoans and, in lieu of standardized data collected specifically for the purpose, we encourage their consideration in a phylogenetic context. Eventually, despite complicating effects such as morphological homoplasy, ecological similarity and introduced species, the biology of *Aurelia* should make sense in the light of evolution.

**Acknowledgements**

We thank those who provided specimens – the Coral Reef Research Foundation, Jack Costello, Haruto Ishii, Ahmet Kideys, Polly Rankin, West Wind Sealab Supplies, Dave Wrobel – or otherwise made this work possible or better by providing information, thoughtful discussion and logistical support – Phil Alderslade, anonymous reviewers, Keith Bayha, Paul Cornelius, the Coral Reef Research Foundation, Lisa Gershwin, Monty Graham, Haruto Ishii, David Jacobs, Winkler Maech, Claudia Mills, Makoto Omori, Kylie Pitt, Jenny Purcell, Kevin Raskoff, Jamie Seymour, John Starmer, Masaya Toyokawa, Francis Toribiong, and all at Fish 'n Fins. Special thanks are due to Bill and Peggy Hamner and to Lori and Pat Colin who have supported us in many ways throughout our graduate careers. Our work in Palau has been possible due to the support of Palau National Government and Koror State Government. This work was funded by the University of California, Los Angeles (UCLA), the Department of Organismic Biology, Ecology, and

Evolution at UCLA, the International Women's Fishing Association, the American Museum of Natural History (Lerner-Grey Award), and the British Schools and Universities Foundation.

# References

Aksnes, D. L., C. B. Miller, M. D. Ohman & S. N. Wood, 1997. Estimation techniques used in studies of copepod population dynamics – a review of underlying assumptions. Sarsia 82: 279–295.

Arai, M. N., 1997. A Functional Biology of Scyphozoa. Chapman & Hall, London: 316 pp.

Avise, J. C., J. Arnold, R. M. Ball, E. Bermingham, T. Lamb, J. E. Neigel, C. A. Reeb & N. C. Saunders, 1987. Intraspecific phylogeography: the mitochondrial DNA bridge between population genetics and systematics. Ann. Rev. Ecol. Syst. 18: 489–522.

Azevedo, R. B. R., A. C. James, J. McCabe & L. Partridge, 1998. Latitudinal variation of wing:thorax size ratio and wing-aspect ratio in *Drosophila melanogaster*. Evolution 52: 1353–1362.

Bastrop, R., K. Jurss & C. Sturmbauer, 1998. Cryptic species in a marine polychaete and their independent introduction from North America to Europe. Mol. Biol. Evol. 15: 97–103.

Behrends, G. & G. Schneider, 1995. Impact of *Aurelia aurita* medusae (Cnidaria, Scyphozoa) on the standing stock and community composition of mesozooplankton in the Kiel Bight (western Baltic Sea). Mar. Ecol. Prog. Ser. 127: 39–45.

Benzie, J. A. H., 1998. Penaeid genetics and biotechnology. Aquaculture 164: 23–47.

Berstad, V., U. Båmstedt & M. B. Martinussen, 1995. Distribution and swimming of the jellyfishes *Aurelia aurita* and *Cyanea capillata*. In Skjoldal, H. R., C. Hopkins, K. E. Erikstad & H. P. Leinaas (eds), Ecology of Fjords and Coastal Waters. Elsevier, Amsterdam: 257–271.

Bridle, J. R. & C. D. Jiggins, 2000. Adaptive dynamics: is speciation too easy? Trends. Ecol. Evol. 15: 225–226.

Briggs, J. C., 1974. Marine Zoogeography. McGraw-Hill, New York: 475 pp.

Briggs, J. C., 1995. Global Biogeography. Elsevier, Amsterdam: 454 pp.

Brown, J. H., 1995. Macroecology. University of Chicago Press, Chicago. 269 pp.

Clarke, A., 1993. Seasonal acclimatization and latitudinal compensation in metabolism: do they exist? Funct. Ecol. 7: 139–149.

Cockrell, B. J., S. B. Malcolm & L. P. Brower, 1993. Time, temperature, and latitudinal constraints on the annual recolonization of eastern North America by the monarch butterfly. In Malcolm, S. B. & M. P. Zalucki (eds), Biology and Conservation of the Monarch Butterfly. Nat. Hist. Mus. Los Angeles County Sci. Ser. 38: 233–251.

Colin, P. L. & C. Arneson, 1995. Tropical Pacific Invertebrates. Coral Reef Research Foundation, Coral Reef Press, Beverly Hills: 296 pp.

Colwell, R. K. & D. C. Lees, 2000. The mid-domain effect: geometric constraints on the geography of species richness. Trends Ecol. Evol. 15: 70–76.

Conover, D. O. & E. T. Schultz, 1995. Phenotypic similarity and the evolutionary significance of countergradient variation. Trends Ecol. Evol. 10: 248–252.

Cossins, A. R. & K. Bowler, 1987. Temperature Biology of Animals. Chapman & Hall, London: 339 pp.

Costello, J. H. & H. W. Mathieu, 1995. Seasonal abundance of medusae in Eel Pond, Massachusetts, U.S.A., during 1990–1991. J. Plankton Res. 17: 199–204.

Dawson, M. N & D. K. Jacobs, 2001. Molecular evidence for cryptic species of *Aurelia aurita* (Cnidaria, Scyphozoa). Biol. Bull. 200: 92–96.

Dawson, M. N, L. E. Martin & L. K. Penland, 2001. Jellyfish swarms, tourists and the Christ-child. Hydrobiologia 451 (Dev. Hydrobiol. 155): 131–144.

Dawson, M. N, K. A. Raskoff & D. K. Jacobs, 1998. Preservation of marine invertebrate tissues for DNA analyses. Mol. mar. Biol. Biotechnol. 7: 145–152.

Dennett, D. C., 1995. Darwin's Dangerous Idea: Evolution and the Meaning of Life. Touchstone, New York: 586 pp.

Devaney, D. M. & L. G. Eldredge, 1977. Reef and shore fauna of Hawaii. B.P. Bishop Mus. spec. Publ. 64(1).

Dobzhansky, T., 1950. Evolution in the tropics. Am. Sci. 38: 209–221.

Ellner, S. & N. G. Hairston, 1994. Role of overlapping generations in maintaining genetic variation in a fluctuating environment. Am. Nat. 143: 403–417.

Futuyma, D. J., 1998. Evolutionary Biology, 3rd edn. Sinauer, Sunderland: 763 pp.

Gosliner, T. M. & M. T. Ghiselin, 1984. Parallel evolution in opisthobranch gastropods and its implications for phylogenetic methodology. Syst. Zool. 33: 255–274.

Grant, P. R. & B. R. Grant, 1998. Speciation and hybridization of birds on islands. In Grant, R. R. (ed), Evolution on Islands. Oxford University Press, Oxford: 142–162.

Greenberg, N., R. L. Garthwaite & D. C. Potts, 1996. Allozyme and morphological evidence for a newly introduced species of *Aurelia* in San Francisco Bay, California. Mar. Biol. 125: 401–410.

Hamner, W. M., 1995. Predation, cover, and convergent evolution in epipelagic oceans. Mar. Freshwat. Behav. Physiol. 26: 71–89.

Hamner, W. M. & P. P. Hamner, 1998. Stratified marine lakes of Palau (Western Caroline Islands). Physical Geogr. 19: 175–220.

Hamner, W. M. & I. R. Hauri, 1981. Long-distance horizontal migrations of zooplankton (Scyphomedusae: *Mastigias*). Limnol. Oceanogr. 26: 414–423.

Hamner, W. M. & R. M. Jenssen, 1974. Growth, degrowth, and irreversible cell differentiation in *Aurelia aurita*. Am. Zool. 14: 833–849.

Hamner, W. M. & D. Schneider, 1986. Regularly spaced rows of medusae in the Bering Sea: role of Langmuir circulation. Limnol. Oceanogr. 31: 171–177.

Hamner, W. M., R. W. Gilmer & P. P. Hamner, 1982. The physical, chemical, and biological characteristics of a stratified, saline, sulfide lake in Palau. Limnol. Oceangr. 27: 896–909.

Hamner, W. M., P. P. Hamner & S. W. Strand, 1994. Sun-compass migration by *Aurelia aurita* (Scyphozoa): population retention and reproduction in Saanich Inlet, British Columbia. Mar. Biol. 119: 347–356.

Hansson, L. J., 1997. Effect of temperature on growth rate of *Aurelia aurita* (Cnidaria, Scyphozoa) from Gullmarsfjorden, Sweden. Mar. Ecol. Prog. Ser. 161: 145–153.

Heinze, J., S. Foitzik, V. E. Kipyatkov & E. B. Lopatina, 1998. Latitudinal variation in cold hardiness and body size in the boreal ant species *Leptothorax acervorum* (Hymenoptera: Formicidae). Entomol. Generalis 22: 305–312.

Hernroth, L. & F. Gröndahl, 1985. On the biology of *Aurelia aurita* (L.): 2. Major factors regulating the occurrence of ephyrae and young medusae in the Gullmar Fjord, Western, Sweden. Bull. mar. Sci. 37: 567–576.

Hochachka, P. W. & G. N. Somero, 1984. Biochemical Adaptation. Princeton University Press, Princeton: 537 pp.

Huntley, M. E. & M. D. G. Lopez, 1992. Temperature-dependent production of marine copepods: a global synthesis. Am. Nat. 140: 201–242.

Ishii, H. & U. Båmstedt, 1998. Food regulation of growth and maturation in a natural population of Aurelia aurita (L.). J. Plankton Res. 20: 805–816.

Ishii, H. & F. Tanaka, 2001. Food and feeding of Aurelia aurita in Tokyo Bay with an analysis of stomach contents and a measurement of digestion times. Hydrobiologia 451 (Dev. Hydrobiol. 155): 311–320.

Johnston, I. A., J. Calvo, H. Guderley, D. Fernandez & L. Palmer, 1998. Latitudinal variation in the abundance and oxidative capacities of muscle mitochondria in perciform fishes. J. exp. Biol. 201: 1–12.

Kimmerer, W. J., 1983. Direct measurement of the production:biomass ratio of the subtropical calanoid copepod Acrocalanus inermis. J. Plankton Res. 5: 1–14.

Kingsford, M. J., E. Wolanski & J. H. Choat, 1991. Influence of tidally induced fronts and Langmuir circulations on distribution and movements of presettlement fishes around a coral reef. Mar. Biol. 109: 167–180.

Kluge, A. G., 1998. Total evidence or taxonomic congruence: Cladistics or consensus classification. Cladistics 14: 151–158.

Knowlton, N., 1993. Sibling species in the sea. Ann. Rev. Ecol. Syst. 24: 189–216.

Kramp, P. L., 1961. Synopsis of the medusae of the world. J. Mar. biol. Assoc. U.K. 40: 337–342.

Kramp, P. L., 1965. Some medusae (mainly scyphomedusae) from Australian coastal waters. Trans. r. Soc. South Aust. 89: 257–278.

Kramp, P. L., 1968. The scyphomedusae collected by the Galathea expedition 1950–52. Vidensk. Meddr. Dansk Naturh. Foren. 31: 67–98.

Larson, R. J., 1987. Respiration and carbon turnover rates of medusae from the NE Pacific. Comp. Biochem. Physiol. 87A: 93–100.

Larson, R. J., 1992. Riding Langmuir circulations and swimming in circles: a novel form of clustering behavior by the scyphomedusa Linuche unguiculata. Mar. Biol. 112: 229–235.

Lebedeva, L. P. & E. A. Shushkina, 1991. Evaluation of population characteristics of the medusa Aurelia aurita in the Black Sea. Oceanology 31: 314–319.

Lucas, C. H., 1996. Population dynamics of Aurelia aurita (Scyphozoa) from an isolated brackish lake, with particular reference to sexual reproduction. J. Plankton Res. 18: 987–1007.

Lucas, C. H., 2001. Reproduction and life history strategies of the common jellyfish, Aurelia aurita, in relation to its ambient environment. Hydrobiologia 451 (Dev. Hydrobiol. 155): 229–246.

Lucas, C. H. & J. A. Williams, 1994. Population dynamics of the scyphomedusa Aurelia aurita in Southampton Water. J. Plankton Res. 16: 879–895.

Lynch, J. D., 1989. The gauge of speciation: on the frequencies of modes of speciation. In Otte, D. & J. A. Endler (eds), Speciation and its Consequences. Sinauer, Sunderland: 527–553.

Martin, L. E., 1999. The population biology and ecology of Aurelia sp. (Scyphozoa: Semaeostomeae) in a tropical meromictic marine lake in Palau, Micronesia. Ph.D. thesis, University of California, Los Angeles: 250 pp.

Martinussen, M. B. & Båmstedt, U., 1995. Diet, estimated daily food ration and predator impact by the scyphozoan jellyfishes Aurelia aurita and Cyanea capillata. In Skjoldal, H. R., C. Hop-

kins, K. E. Eirkstad & H. P. Leinaas (eds), Ecology of Fjords and Coastal Waters. Elsevier, Amsterdam: 127–145.

Mayer A. G., 1910. Medusae of the World, III: the Scyphomedusae. Carnegie Inst. Washington Publ. 109: 619–630.

Mayer A. G., 1914. The effects of temperature upon tropical marine animals. Carnegie Inst. Washington Publ. 183: 55–83.

McCloskey, L. R., L. Muscatine & F. P. Wilkerson, 1994. Daily photosynthesis, respiration, and carbon budgets in a tropical marine jellyfish (Mastigias sp.). Mar. Biol. 119: 13–22.

McFall-Ngai, M. J., 1990. Crypsis in the pelagic environment. Amer. Zool. 30: 175–188.

Mianzan, H. W. & P. F. S. Cornelius, 1999. Cubomedusae and scyphomedusae. In Boltovskoy, D. (ed.), South Atlantic Zooplankton, I. Backhuys, Leiden: 513–559.

Mills, C. E., 2001. Jellyfish blooms: are populations increasing globally in response to changing ocean conditions? Hydrobiologia 451 (Dev. Hydrobiol. 155): 55–68.

Miyake, H., K. Iwao & Y. Kakinuma, 1997. Life history and environment of Aurelia aurita. South Pacific Study 17: 273–285.

Möller, H., 1980a. Population dynamics of Aurelia aurita medusae in Kiel Bight, Germany (FRG). Mar. Biol. 60: 123–128.

Möller, H., 1980b. Scyphomedusae as predators and food competitors of larval fish. Meeresforsch 28: 90–100.

MPA, 1999. Massachusetts Port Authority. http://www.massport.com/about/publi.html.

Muscatine, L. & R. E. Marian, 1982. Dissolved inorganic nitrogen flux in symbiotic and nonsymbiotic medusae. Limnol. Oceanogr. 27: 910–917.

Mutlu, E., F. Bingel, A. C. Gücü, V. V. Melnikov, U. Niermann, N. A. Ostr & V. E. Zaika, 1994. Distribution of the new invader Mnemiopsis sp. and the resident Aurelia aurita and Pleurobrachia pileus populations in the Black Sea in the years 1991–1993. ICES J. mar. Sci. 51: 407–421.

Naumov, D. V., 1961. Scyphomedusae of the seas of the USSR. Opredeliteli po Faune SSSR 75: 1–98. (In Russian).

Nielson, A. S., A. W. Pederson & H. U. Riisgård, 1997. Implications of density driven currents for interaction between jellyfish (Aurelia aurita) and zooplankton in a Danish fjord. Sarsia 82: 297–305.

NOAA, 1999. Coastwatch: ocean color sst. http://coastwatch.noaa.gov/COASTWATCH/

Nybakken, J. W., 1988. Marine Biology: an Ecological Approach, 2nd edn. Harper & Row, Cambridge, 514 pp.

Olesen, N. J., K. Frandsen & H. U. Riisgård, 1994. Population dynamics, growth and energetics of jellyfish Aurelia aurita in a shallow fjord. Mar. Ecol. Prog. Ser. 105: 9–18.

Omori, M., H. Ishii & A. Fujinaga, 1995. Life history strategy of Aurelia aurita (Cnidaria, Scyphozoa) and its impact on the zooplankton community of Tokyo Bay. ICES J. Mar. Sci. 52: 597–603.

Papathanassiou, E., P. Panayotidis & K. Anagnostaki, 1987. Notes on the biology and ecology of the jellyfish Aurelia aurita Lam. in Elefsis Bay (Saronikos Gulf, Greece). Mar. Ecol. 8: 49–58.

Potter, D., T. C. Lajeunesse, G. W. Saunders & R. A. Anderson, 1997. Convergent evolution masks extensive biodiversity among marine coccoid picoplankton. Biodiv. Conserv. 6: 99–107.

Purcell, J. E., 1997. Pelagic cnidarians and ctenophores as predators: Selective predation, feeding rates and effects on prey populations. Ann. Inst. Oceanogr. 73: 125–137.

Purcell, J. E., E. D. Brown, K. D. E. Stokesbury, L. H. Haldorson & T. C. Shirley, 2000. Aggregations of the jellyfish Aurelia labiata: abundance, distribution, association with age-0 walleye pollock, and behaviors promoting aggregation in Prince William Sound, Alaska, U.S.A. Mar. Ecol. Prog. Ser. 195: 145–158.

Rasmussen, E., 1973. Systematics and ecology of the Isefjord marine fauna (Denmark) with a survey of the eelgrass (*Zostera*) vegetation and its communities. Ophelia 11: 1–46.

Ridley, M., 1993. Evolution. Blackwell, Boston: 670 pp.

Rooney, A. P. & J. Zhang, 1999. Rapid evolution of a primate sperm protein: Relaxation of functional constraint or positive Darwinian selection? Mol. Biol. Evol. 16: 706–710.

Russell, F. S., 1970. The Medusae of the British Isles. II Pelagic Scyphozoa with a Supplement to the First Volume on Hydromedusae. Cambridge University Press, Cambridge: 284 pp.

Rutherford, S. L. & S. Lindquist, 1998. Hsp90 as a capacitor for morphological evolution. Nature 396: 336–342.

Sabitini, M. & T. Kiorbe, 1994. Egg production, growth and development of the cyclopoid copepod *Oithona similis*. J. Plankton Res. 16: 1329–1351.

Sanz, J. J., 1999. Does daylength explain the latitudinal variation in clutch size of pied flycatchers *Ficedula hypoleuca*? Ibis 141: 100–108.

Sasaki, A. & G. De Jong, 1999. Density dependence and unpredictable selection in a heterogeneous environment: compromise and polymorphism in the ESS reaction norm. Evolution 53: 1329–1342.

Sasaki, A. & S. Ellner, 1997. Quantitative genetic variance maintained by fluctuating selection with overlapping generations: variance components and covariances. Evolution 51: 682–696.

Schneider, G., 1988. Larvae production of the common jellyfish *Aurelia aurita* in the Western Baltic 1982-1984. Kieler Meeresforsch. Sonderh. 6: 295–300.

Schneider, G. & G. Behrends, 1998. Top-down control in a neritic plankton system by *Aurelia aurita* medusae – a summary. Ophelia 48: 71–82.

Schultz, E. T, D. O. Conover & A. Ehtisham, 1998. The dead of winter: size-dependent variation and genetic differences in seasonal mortality among Atlantic silverside (Atherinidae: *Menidia menidia*) from different latitudes. Can. J. Fish. aquatic Sci. 55: 1149–1157.

Shick, J. M., 1973. Effects of salinity and starvation on the uptake and utilization of dissolved glycine by *Aurelia aurita* polyps. Biol. Bull. 144: 172–179.

Shick, J. M., 1975. Uptake and utilization of dissolved glycine by *Aurelia aurita* scyphistomae: temperature effects on the uptake process; nutritional role of dissolved amino acids. Biol. Bull. 148: 117–140.

Shushkina, E. A. & E. I. Musayeva, 1983. The role of jellyfish in the energy system of Black Sea plankton communities. Oceanology 23: 92–96.

Shushkina, E. A. & E. I. Musayeva, 1990. Structure of planktic community of the Black Sea epipelagic zone and its variation caused by invasion of a new ctenophore species. Oceanology 30: 225–228.

Shushkina, E. A. & G. N. Arnautov, 1985. Quantitative distribution of the medusa *Aurelia* and its role in the Black Sea ecosystem. Oceanology 25: 102–106.

Slade, R. W. & C. Moritz, 1998. Phylogeography of *Bufo marinus* from its natural and introduced ranges. Proc. r. Soc. Lond. B 265: 769–777.

Sorenson, M. D., 1996. TreeRot. University of Michigan, Ann Arbor.

Steel, R. G. D. & J. H. Torrie, 1980. Principles and Procedures of Statistics, A Biometrical Approach, 2nd edn. McGraw-Hill, Auckland: 633 pp.

Stepien, C. A., A. K. Dillon & M. D. Chandler, 1998. Genetic identity, phylogeography and systematics of ruffe *Gymnocephalus* in the North American Great Lakes and Eurasia. J. Great Lakes Res. 24: 361–378.

Stevens, G. C., 1989. The latitudinal gradient in geographical range: how so many species co-exist in the tropics. Am. Nat. 133: 240–256.

Stoecker, D. K., A. E. Michaels & L. H. Davis, 1987. Grazing by the jellyfish, *Aurelia aurita*, on microzooplankton. J. Plankton Res. 9: 901–915.

Swofford, D. L., 1993. PAUP: Phylogenetic Analysis Using Parsimony, Version 3.1. Computer program distributed by the Illinois Natural History Survey, Champaign, Illinois.

Swofford, D. L., 2000. PAUP: Phylogenetic Analysis Using Parsimony, v.4.0b3a. Smithsonian Institution and Sinauer Associates, Sunderland.

Takahata, N., 1993. Relaxed natural selection in human populations during the Pleistocene. Japan. J. Genet. 68: 539–547.

Unwin, D. M. & S. A. Corbet. 1984. Wingbeat frequency, temperature and body size in bees and flies. Physiol. Entomol. 9: 115–121.

USGS, 2000. Water quality of San Francisco Bay – temperature time serieshttp://sfbay.wr.usgs.gov/access/wqdata /overview/examp/charts/temp.html

Uye, S., 1994. Replacement of large copepods by small ones with eutrophication of embayments: cause and consequence. Hydrobiologia 292/293: 513–519.

Van Der Veer, H. W. & W. Oorthuysen, 1985. Abundance, growth and food demand of the scyphomedusa *Aurelia aurita* in the western Wadden Sea. Neth. J. Sea Res. 19: 38–44.

Van Voorhies, W. A., 1996. Bergmann size clines: a simple explanation for their occurrence in ectotherms. Evolution 50: 1259–1264.

Van't Land, J., P. Van Putten, B. Zwaan, A. Kamping & W. Van Delden, 1999. Latitudinal variation in wild populations of *Drosophila melanogaster*: heritabilities and reaction norms. J. Evol. Biol. 12: 222–232.

Watanabe, T. & H. Ishii, 2001. *In situ* estimation of ephyrae liberated from polyps of *Aurelia aurita* using settling plates in Tokyo Bay, Japan. Hydrobiologia 451 (Dev. Hydrobiol. 155): 247–258.

Wilkinson, L., M. Hill, J. P. Welna & G. K. Birkenbeuel, 1992. SYSTAT for Windows: statistics, version 5 edition. SYSTAT Inc., Evanston: 750 pp.

Wrobel, D. & C. Mills, 1998. Pacific Coast Pelagic Invertebrates, a Guide to the Common Gelatinous Animals. Sea Challengers and Monterey Bay Aquarium, Monterey: 108 pp.

Yang, Z., 1998. On the best evolutionary rate for phylogenetic analysis. Syst. Biol. 47: 125–133.

Yasuda, T., 1968. Ecological studies on the jelly-fish, *Aurelia aurita*, in Urazoko Bay, Fukuii Prefecture – II. Occurrence pattern of ephyrae. Bull. Jap. Soc. Sci. Fish. 34: 983–987.

Yasuda, T., 1969. Ecological studies on the jellyfish, *Aurelia aurita*, in Urazoko Bay, Fukui Prefecture – IV. Occurrence pattern of the medusa. Bull. Jap. Soc. Sci. Fish. 35: 1–6.

Yasuda, T., 1971. Ecological studies on the jelly-fish, *Aurelia aurita*, in Urazoko Bay, Fukuii Prefecture – IV. Monthly change in bell-length composition and breeding season. Bull. Jap. Soc. Sci. Fish. 37: 364–370.

Youngbluth, M. & U. Båmstedt, 2001. Distribution, abundance, behavior and metabolism of *Periphylla periphylla*, a mesopelagic coronate medusa in a Norwegian Fjord. Hydrobiologia 451 (Dev. Hydrobiol. 155): 321–333.

Zaitsev, Y. P., 1992. Recent changes in the trophic structure of the Black Sea. Fish. Oceanogr. 1: 180–189.

Zubkoff, P. L. & A. L. Linn, 1975. Isozymes of *Aurelia* aurita scyphistomae obtained from different geographical locations. In Markert, C. L. (ed.), Isozymes IV: Genetics and Evolution. Academic Press, New York: 915–930.

*Hydrobiologia* **451**: 275–286, 2001.
© 2001 *Kluwer Academic Publishers.*

# Observations on the distribution and relative abundance of the scyphomedusan *Chrysaora hysoscella* (Linné, 1766) and the hydrozoan *Aequorea aequorea* (Forskål, 1775) in the northern Benguela ecosystem

Conrad Sparks[1,2], Emmanuelle Buecher[2,3], Andrew S. Brierley[4], Bjørn Erik Axelsen[5], Helen Boyer[6] & Mark J. Gibbons[2]

[1]*Faculty of Applied Sciences, Cape Technikon, Box 652, Cape Town, 8000, South Africa*
[2]*Zoology Department, University of the Western Cape, Private Bag X17, Bellville, 7535, South Africa*
[3]*Laboratoire d'Océanographie Biologique et Ecologie du Plancton Marin, CNRS/UPMC 7076. Station Zoologique, B.P. 28, 06234 Villefranche-sur-Mer, France*
[4]*British Antarctic Survey, Biological Sciences Division, High Cross, Madingley Road, Cambridge, CB3 0ET, U.K.*
[5]*Institute if Marine Research, P.O. Box 1870 Nordnes, N-5817 Bergen, Norway*
[6]*National Marine Information and Research Centre, P.O. Box 912, Swakopmund, Namibia*

*Key words:* jellyfish, cross-shelf distribution, Namibia, spatial-partitioning, visual observations, trawl survey, Ecopath

## Abstract

Observations on the abundance of medusae at the surface were conducted in the northern Benguela ecosystem, over the period August 1997–June 1998. The results suggest that *Chrysaora hysoscella* is found inshore, whereas *Aequorea aequorea* tends to be found offshore. Although these relative observations are subject to bias caused by seasonal changes in the survey area, they are generally supported by the results of correlation analyses, and by the results of a more quantitative, cross-shelf trawl survey. Both species of medusae display marked patchiness, and can be very abundant. They appear to have mostly non-overlapping patterns of distribution in the upper layers of the water column, and so are able exert a consistent predation pressure across the width of the continental shelf. The estimates of biomass obtained are used as input variables to existing models of energy flow within the ecosystem.

## Introduction

The Benguela ecosystem is one of the major upwelling areas of the world, and extends from northern Namibia (∼17 °S), to south of Cape Point in South Africa (34 °S). The Luderitz upwelling cell (∼27 °S) represents a natural internal boundary to the system, and the areas to either side are referred to as the northern and southern Benguela (Shannon, 1985). The waters of the northern Benguela are less markedly seasonal than they are in the south (Shannon, op cit.; Estrada & Marrasé, 1987), and support a higher biomass owing to the less dynamic oceanography (Brown et al., 1991).

Although the dominant large medusae currently observed in the northern Benguela ecosystem are the scyphomedusan *Chrysaora hysoscella* (Linné, 1766)

and the hydromedusan *Aequorea aequorea* (Forskål, 1775), the detailed plankton studies that were conducted in the region during the 1950s (Hart & Currie, 1960) and 1960s (Stander & De Decker, 1969) did not report the presence of either species. *C. hysoscella* and *A. aequorea* were first noted in the early 1970s by King & O'Toole (1973), Cram & Visser (1973) and Schülein (1974) during their investigations on commercially important fishes. These studies did not examine either *C. hysoscella* or *A. aequorea* in any detail, however, and the first quantitative studies were only forthcoming some 15 years later (Venter, 1988; Fearon et al., 1992). Although both these latter studies reported on the abundance of large medusae in the region, they were based either on incidental statistics collected from the fishing industry (Venter, op. cit.), or from small-mouth Bongo nets (Fearon et al., op cit.). Both data sets are therefore, biased – the former by the problems attendant with fisheries dependent data (e.g. selective operational area, marked seasonality and poor records), and the latter by inadequate sampling. As a consequence they can only provide data of a relative nature. The most rigorous study of medusae in the northern Benguela was conducted by Pagès (1991), and it focused on small (principally hydro-) medusae that were collected by bongo nets. Despite the disparate methods that have hitherto been employed to study medusae in the region, they all agree on the dominant species present and on their general pattern of distribution.

The apparent increase in large medusae that has been observed in the northern Benguela ecosystem has been observed elsewhere in the world (Mills, 2001). Brodeur et al. (1999) documented a substantial increase in gelatinous zooplankton in the Bering Sea over the last 20 years, and Graham (2001) noted an increase in the abundance of *Chrysaora quinquecirrha* and *Aurelia aurita* in the Gulf of Mexico. A number of explanations have been put forward to explain these increases including climate change (Mills, 2001), eutrophication (Arai, 2001) and overfishing (Purcell & Arai, 2001).

Medusae are carnivorous, and can have negative impacts on zooplankton populations (e.g., Behrends & Schneider, 1995; Feigenbaum & Kelly, 1984; Purcell 1992) and fish eggs and larvae (reviewed in Purcell & Arai, 2001) when they occur at high densities. They are thought to have an important, if unquantified, role in the structure of pelagic ecosystems (Hernroth & Gröndahl, 1983). In the Benguela ecosystem, large medusae spoil and reduce fish catches, and are thought to interfere with the hydroacoustic assessment of pelagic fish stocks (Brierley et al., 2001; D. Boyer, NatMIRC, Swakopmund, pers. comm.). Although medusae are clearly important from an ecosystem and commercial point of view, the general decline in funds for ocean-going research in southern Africa has meant that studies on species of no direct commercial value have become limited. Research efforts directed towards medusae have, therefore, been of low priority, and have been conducted mainly opportunistically. The studies presented here are of such a nature, and deal largely with cross-shelf and alongshore patterns of distribution determined from observations of medusae at the surface that were conducted aboard cruises of opportunity.

## Materials and methods

Observations of *C. hysoscella* and *A. aequorea* were made from the bow of the RV *Welwitschia* between August 1997 and June 1998. The data were collected from six cruises in the area between 17 °S and 28 °S. Observations were concentrated between the Cunene River (17° 30' S) and Conception Bay (24° S) (Fig. 1a). All observations ($n = 410$) were made during the day between 07:00 and 18:00. Medusae were identified and counted within a band 5 m wide to the front of the bow, whilst the ship steamed at 10 knots. Counts were carried out for 10 minutes every hour, subject to visibility and sea-state, and these were converted to densities using knowledge of the distance steamed. When dense patches of medusae were observed within any sampling period, counts were estimated from smaller areas and then multiplied up by total patch area. No effort was made to examine patchiness within any sampling period, and so estimated density was based on the total counts per sampling period. Information on latitude, longitude, time and sea-surface temperature were recorded with each observation (sea-surface salinity was not recorded). Correlation analyses between these environmental variables and surface abundance were conducted using Statistica software. Only those data that were collected when the sea was flat and the visibility was good were used in the analyses.

Additional information on the cross-shelf distribution of large medusae was provided by a trawl survey conducted between 31 August and 6 September 1999 aboard the RV *Dr. Fridtjof Nansen*. The survey transect was situated off central Namibia (22° S), and extended between 12° 42' E (water depth 430 m) and 13° 47' E (102 m) (Fig. 1a). A total of 66 trawls were conducted using a pelagic Åkra trawl fitted with two nets of similar design. The smaller net was fitted with a Multisampler (Skeide et al., 1997) that enabled three separate samples to be collected from a single trawl, and the larger net was fitted with balloon floats that enabled surface (between 19 and 31 m) trawls to be made. The towing time varied between 2 and 15 min. The nets were towed at a speed of 3 knots. The nets had circular mouth openings of 12 m diameter, and they were fitted with a mesh that reduced in size from 400 mm diameter at the mouth, to 36 mm diameter at the cod-end. Few of the tows sampled the entire water column and most (78%) sampled within the upper 50 m only. Two trawls fished at depths greater than 200 m. CTD casts were made with a Seabird SBE 9+ probe

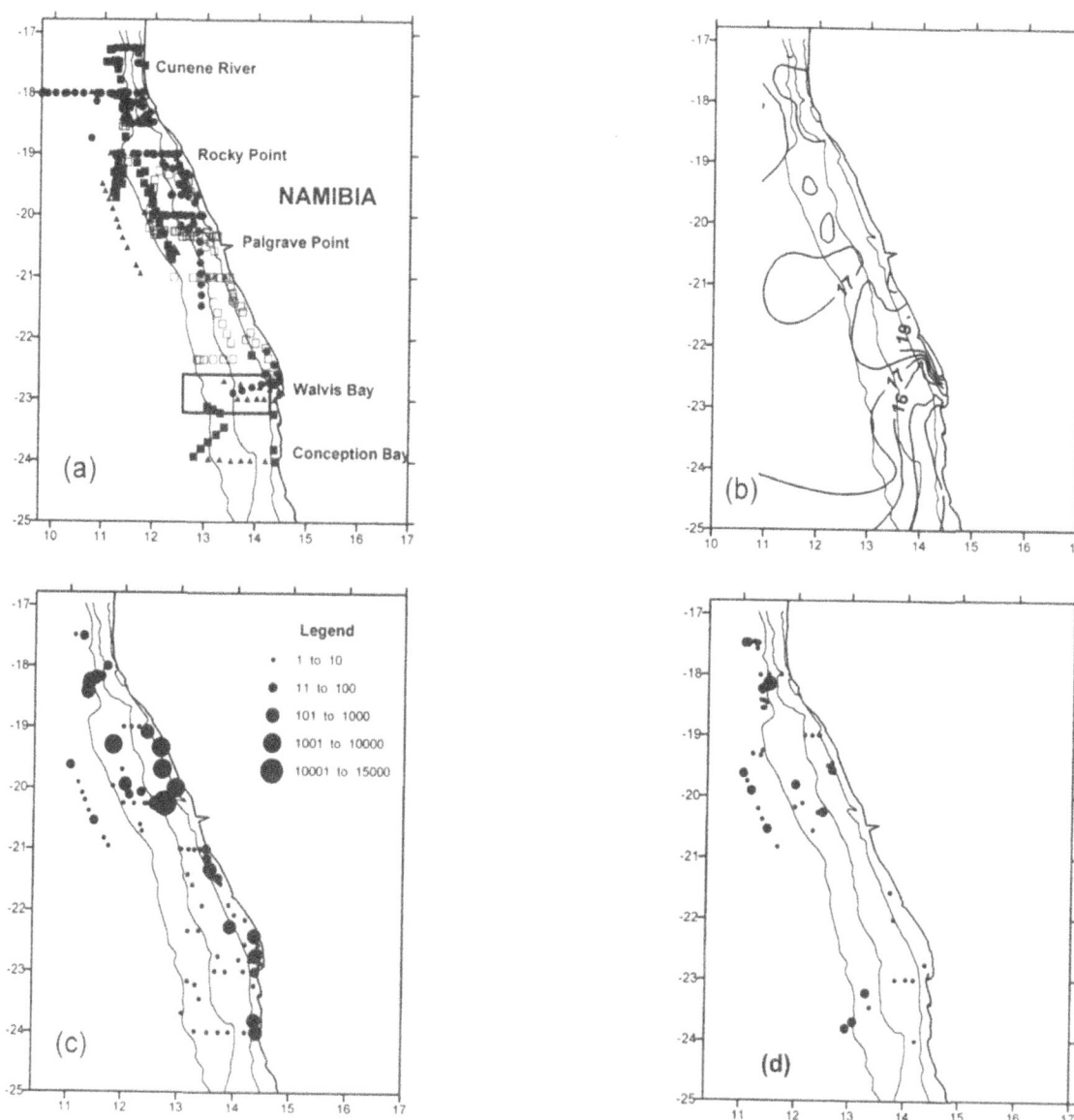

*Figure 1.* Study area across the Namibian shelf showing location of observation sites (▲ - spring, □ – summer, ● – autumn, ■ - winter), and the general area of the trawl survey conducted in August/September 1999 (blocked) (a). Isotherms of mean sea surface temperature (b), and abundance (numbers per 10 minute sampling period) of *Chrysaora hysoscella* (c) and *Aequorea aequorea* (d) between August 1997 and July 1998 (data pooled). Isobaths (100 m, 200 m and 500 m) indicated in each panel.

(SBE 11+ deck unit) to near bottom at six locations along the transect. Station spacings corresponded to approximately 50 m increments in bottom depth.

## Results

A total of six surveys were conducted in the area between Conception Point (24° S) and the Cunene River (17° S) (Fig. 1a), during the period August 1997 and July 1998. The study area varied between the surveys (Fig. 1a), as did the number of observations ($N = 133$, autumn; $N = 104$, winter; $N = 89$, spring; $N = 84$, summer). Sea surface temperature (SST) varied between 10 and 22 °C during the study period, and the modal SST of the samples was 17 °C (Fig. 2a). The mean SST of the survey area over the course of the study period was 17.4 °C, but this varied on a seasonal basis (18.68 °C , summer; 17.64 °C, autumn; 16.97 °C, winter; 17.37 °C, spring). SST was significantly affected by both latitude and longitude (stepwise mul-

278

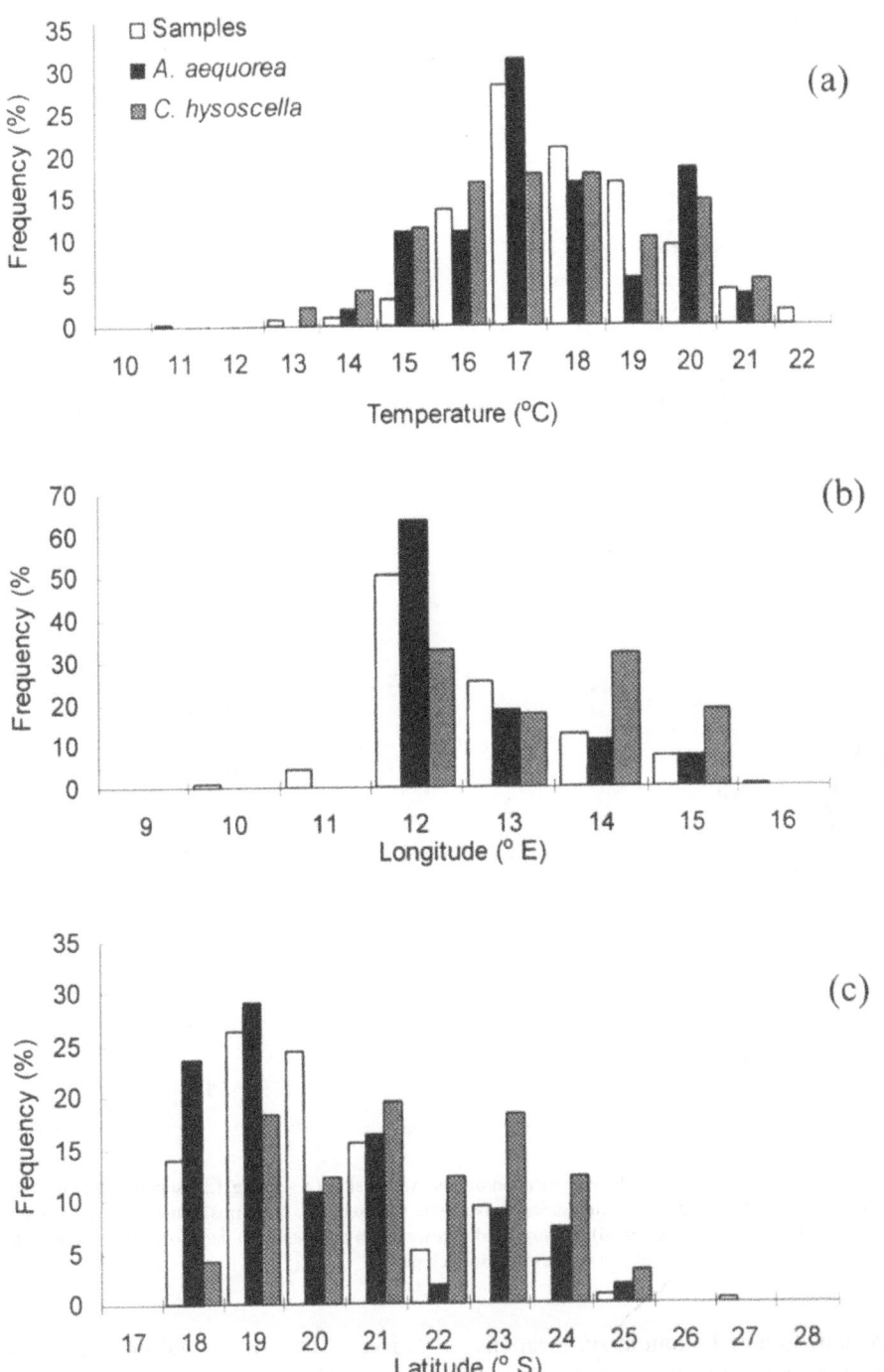

*Figure 2.* Distribution of *Chrysaora hysoscella* and *Aequorea aequorea* observations, within classes of (a) temperature, (b) longitude, and (c) latitude. All data pooled.

tiple regression analysis, $p < 0.0001$), and warmer water was encountered offshore and to the north.

Over the course of the year, *C. hysoscella* was recorded throughout the entire observation area (Fig. 1c) ($n = 97$). The species was characterised by gener-ally low abundance (<10 medusae per observation) throughout the observation area but high (>100 medu-sae per observation) densities were recorded close to shore. Highest densities (>1000 medusae per obser-vation) were observed inshore, between Rocky Point

(19° S) and Palgrave Point (20° 30′ S). The distribution of *A. aequorea* (*n* = 55) was sparser than *C. hysoscella*, particularly in the southern regions of the observation area between 22° S and 24° S. *A. aequorea* was most abundant between Palgrave Point and the Cunene River (Fig. 1d).

Although neither species of medusae occurred in samples collected at the extremes of SST observed, *C. hysoscella* tended to be found more frequently at temperatures ≤16 °C (31%) than *A. aequorea* (24%) (Fig. 2a). Temperatures within this range were found in the inshore and southern waters of the study area (Fig. 1b). It would appear from the distribution of *A. aequorea* in relation to longitudinal classes, that this species occurred further offshore than *C. hysoscella* (Fig. 2b). Longitude has been used as a proxy for depth (in the absence of these data), due to the near linear coastline and narrow continental shelf within the study area. 79% of the observations of *A. aequorea* occurred north of 22 °S, as opposed to only 53% of those of *C. hysoscella* (Fig. 2c).

An analysis of the correlations between abundance and the various environmental parameters (time of observation, SST, latitude and longitude), revealed some conflicting results. When the entire data set was analysed, the only significant (*p* < 0.05) correlation was between *C. hysoscella* and observational time. An examination of the scatter-plot between observational abundance and time (Fig. 3) suggested that the numbers of *C. hysoscella* at the surface were greatest at dawn and dusk. This implied the population underwent some sort of diel vertical migration, and required that, for other trends, the data set be analysed between 08:00 and 17:00.

Analysis of the time-filtered data set was undertaken on a cruise-by-cruise basis, because each cruise was conducted at a different time of the year, had a different survey grid and collected a different number of samples (Fig. 1a). Generally, *C. hysoscella* increased in abundance to the south[1], and it was more abundant in nearshore waters (Table 1). The abundance of *C. hysoscella* was inversely correlated with SST in summer, and positively correlated with SST in winter. The abundance of *A. aequorea* showed fewer significant

---

[1] Although these results may appear to contrast with the data illustrated in Fig. 1b (wherein higher abundances of *C. hysoscella* were observed in the northern, than southern, part of the study area), all data were pooled in the construction of this figure. The results of the correlation analyses between abundance and latitude were derived from seasonal analyses, of smaller areal data sets (Figure 1a). The high abundances observed at 20S were noted on a survey where this latitiude was at the southern limit of the survey grid.

*Table 1.* Significant (*p*) correlation coefficients (*r*) between the observed abundance (m$^{-2}$) of *Chrysaora hysoscella* and *Aequorea aequorea* and measured environmental variables from August 1997 to July 1998. The months for the seasons are: Winter (June and July); Spring (October); Summer (November and December) and Autumn (March and April)

| Season | Variable | Species | *r* | *P* | *N* |
|---|---|---|---|---|---|
| Autumn | Longitude | *C. hysoscella* | 0.22 | 0.01 | 131 |
| Autumn | Latitude | *C. hysoscella* | 0.21 | 0.02 | 131 |
| Spring | Longitude | *A. aequorea* | −0.30 | 0.05 | 43 |
| Summer | Temperature | *C. hysoscella* | 0.34 | 0.04 | 36 |
| Winter | Latitude | *C. hysoscella* | 0.51 | 0.00 | 72 |
| Winter | Temperature | *C. hysoscella* | −0.50 | 0.00 | 72 |
| Winter | Latitude | *A. aequorea* | −0.26 | 0.03 | 72 |
| Winter | Longitude | *C. hysoscella* | 0.70 | 0.00 | 72 |

trends within the different seasons (Table 1). However, the abundance of *A. aequorea* was correlated negatively with both latitude and longitude, which suggests that it was more common in the north and offshore. It should be noted that the correlation coefficients were low in most cases, implying that other factors were also important in determining the abundance and distribution of both species.

The results of the trawl survey conducted during the cruise in Autumn 1999 showed that higher densities of *C. hysoscella* were found in the inner and middle regions of the shelf (<150 m), whereas *A. aequorea* was more abundant in the deeper (>150 m) offshore waters (Fig. 4). This cross-shelf pattern in the distribution of *C. hysoscella* and *A. aequorea* may reflect the strong temperature and salinity gradients (Fig. 5) that were associated with the sharp shelf-break front (between 13 and 13.1 E). The shelf-break front represented the western limit of large catches of *C. hysoscella* (>100 kg per min). Inshore of this front the water column was not well-mixed. There was a weak, sub-surface frontal feature around 13.4 E, and this, broadly speaking, coincided with the transition between catches dominated by *C. hysoscella* and *A. aequorea*. Both species were patchy in their distribution, and although most trawls were of relatively low density, some were of very high density (Fig. 6a,b). These results generally support the findings of the visual observations (Fig. 6c,d).

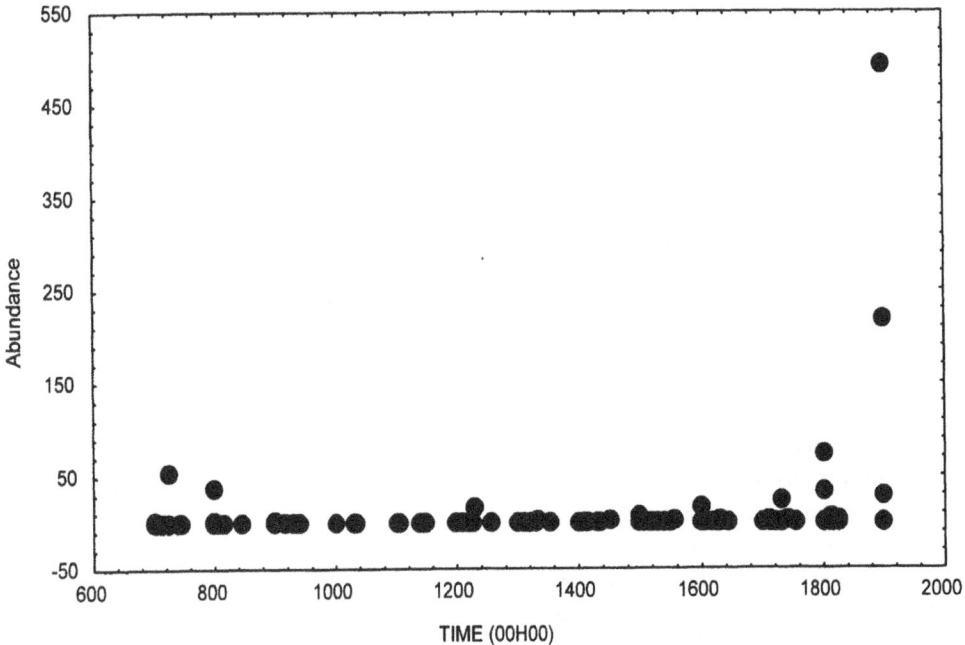

*Figure 3.* Density of *Chrysaora hysoscella* (m$^{-2}$) observed at the surface versus time of day. Only non-zero data shown; all data pooled.

## Discussion

Information on the distribution and abundance of medusae that has been derived from surface observations is subject to a number of criticisms. Observations are dependent upon sea-state and visibility, reflect observer bias, and are also influenced by diel vertical migration (DVM). Despite the fact that all these variables were standardised here, our results should be interpreted as measures of relative abundance. Therefore, concordance with other data sets need not be strict. Having said that, the relative cross-shelf distribution of the two species was similar in both data sets, which lends some credibility to the use of surface observations in determining the distribution and relative abundance of large medusae. The use of surface observations in determining the relative abundance and distribution of medusae is not new, and Purcell et al. (2000) used a modified approach to estimate the density of *Aurelia labiata* aggregations in Prince William Sound.

The inferences on DVM of *C. hysoscella* made here are supported, in part, by the results of an hydroacoustic survey (Brierley et al., 2001). Those authors noted that the sound scattering layer of *C. hysoscella* (at 25 m depth) was clearly detectable during the day, but that it was less distinct at night, probably as individuals migrated upwards in the wa-

ter column, returning to depth the following day. A number of other authors have noted that species of the genus *Chrysaora* display DVM (e.g., Pagès, 1991; Schuyler & Sullivan, 1997), although few have been able to correlate the behaviour with the environment.

There was tendency within any seasonal set of samples, for *C. hysoscella* to occur inshore and in the south (Table 1). These results are consistent with the cross-shelf distribution patterns recorded by Pagès & Gili (1991), and Fearon et al. (1992). Medusae of the genus *Chrysaora* are known to exhibit seasonal patterns of abundance (Kramp, 1961). Although the life cycle of *C. hysoscella* within the region remains unknown, it has been postulated that the ephyrae released by benthic scyphistomae in the north are carried south by the inshore undercurrent (Fearon et al., op cit.). Juvenile medusae are then thought to mature on their southward journey, which results in their higher observed numbers there, than in the north[2]. The variable response of *C. hysoscella* to temperature (Table 1) may reflect seasonality in hydrography. Inshore water temperatures during summer tend to be lower than in winter, because of upwelling, which is seasonal at these latitudes (Shannon, 1985). *C. hysoscella* might

---

[2] Unfortunately, the data sets collected here did not allow us to confirm the seasonal changes in distribution postulated by Fearon et al. (op cit.), owing to the lack of conformity in sampling area, and a lack of full-shelf coverage.

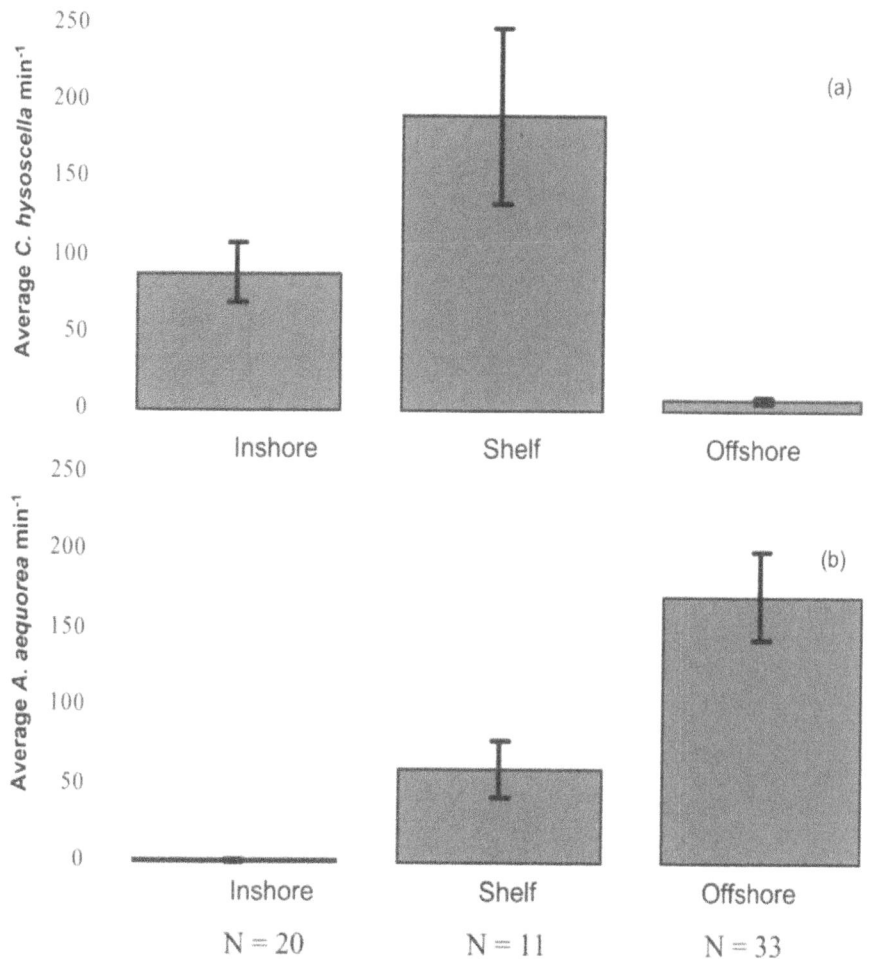

*Figure 4.* The mean (± S. E.) catch (kg min⁻¹) of (a) *Chrysaora hysoscella* and (b) *Aequorea aequorea* versus distance from shores. Number of trawls indicated; data from the cruise conducted in August/September 1999.

be considered a cold water species, but newly up-welled water is low in food and requires some level of maturation before populations of zooplankton prey begin to develop (Denman & Powell, 1984).

Although *A. aequorea* tended to be found offshore and in the north (Fig. 3), this was poorly supported by the correlation analyses, and can be attributed to their relatively low abundance. Pagès (1991) regarded *A. aequorea* as a shelf species, and found that it was particularly abundant in a broad strip separating coastal from oceanic waters. Fearon et al. (1992) reported that *A. aequorea* was common in the region between 20° S and 21° S (off Palgrave Point), but that it decreased in the areas north of 19 °S and south of 24° 30′ S. Our understanding of the biology of this species in the region is poor and precludes detailed comment.

The data for both species were characterised by great horizontal patchiness (Fig. 6). A number of factors influence the horizontal distribution of medusae, including physical processes, and behavioural responses to the prevailing environment (Graham et al., 2001). Langmuir convection cells result in regularly spaced (micro- and mesoscale) patches of plankton at the surface (Boero, 1991; Larson, 1992), and have been evoked to explain patchiness in the distribution of red tides off the South African west coast (A. Boyd, M&CM, Cape Town; unpublished data). Elsewhere in the world, medusae have been shown to be aggregated at upwelling fronts (Graham, 1994) and in areas of flow discontinuity (Purcell et al., 2000), and may accumulate in areas influenced by wind (Axiak et al., 1991), as well as eddies, currents and tides (Arai, 1992a). Medusae show behavioural responses to their food environment (Bailey & Batty, 1983; Arai, 1992b), which may allow them to aggregate in patches of high food density.

282

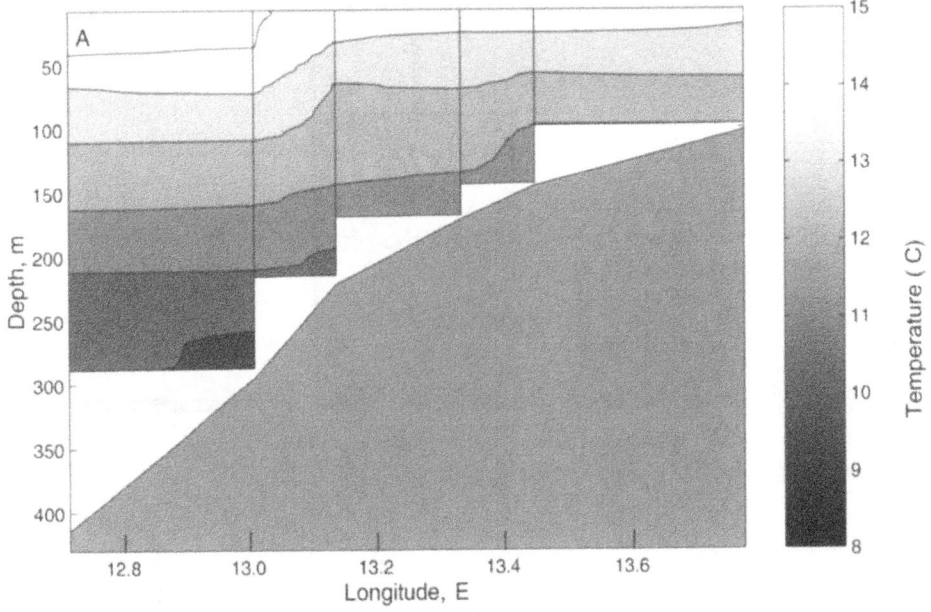

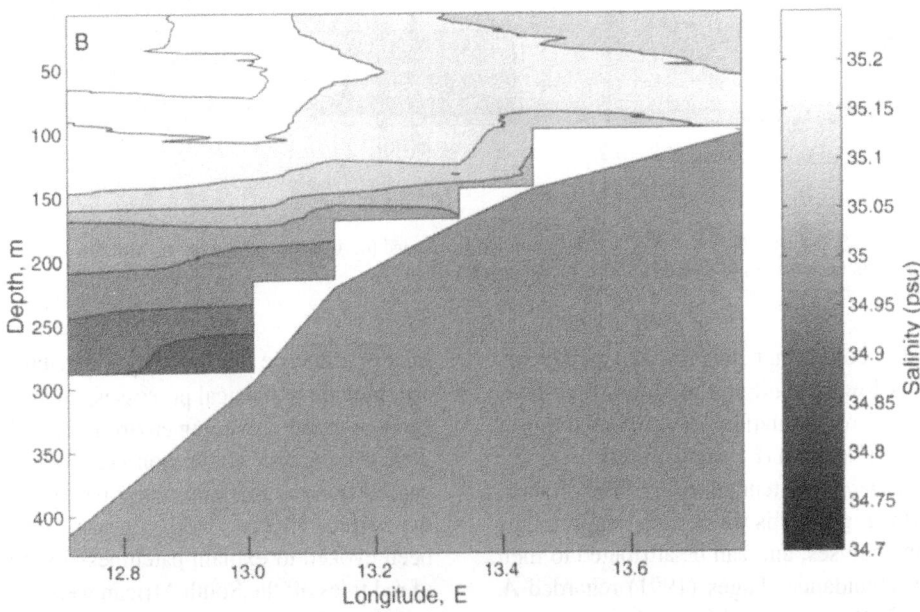

*Figure 5.* Vertical sections of temperature (a) and salinity (b) across the Namibian shelf at 22° S (12° 42′ E to 13° 47′ E), in August/September 1999. The data were collected during the trawl survey and were de-spiked prior to contouring.

One of the most significant results of this study is the apparent segregation in space between *C. hysoscella* and *A. aequorea*. This finding was common to both the observational, and the trawl data sets (Fig. 7), and suggests that the one species only occurred at high numbers when the other was rare. In his analysis of cnidarian assemblages in the region, Pagès (1991) as-

signed the two species to different water masses (*A. aequorea* – shelf, *C. hysoscella* – coastal-shelf) but did not comment further. A negative relationship between the abundance of the *C. hysoscella* and *A. aequorea* could be interpreted in terms of predation, because both genera are known to include pelagic coelenterates in their diet (Russell, 1970; Purcell, 1991). How-

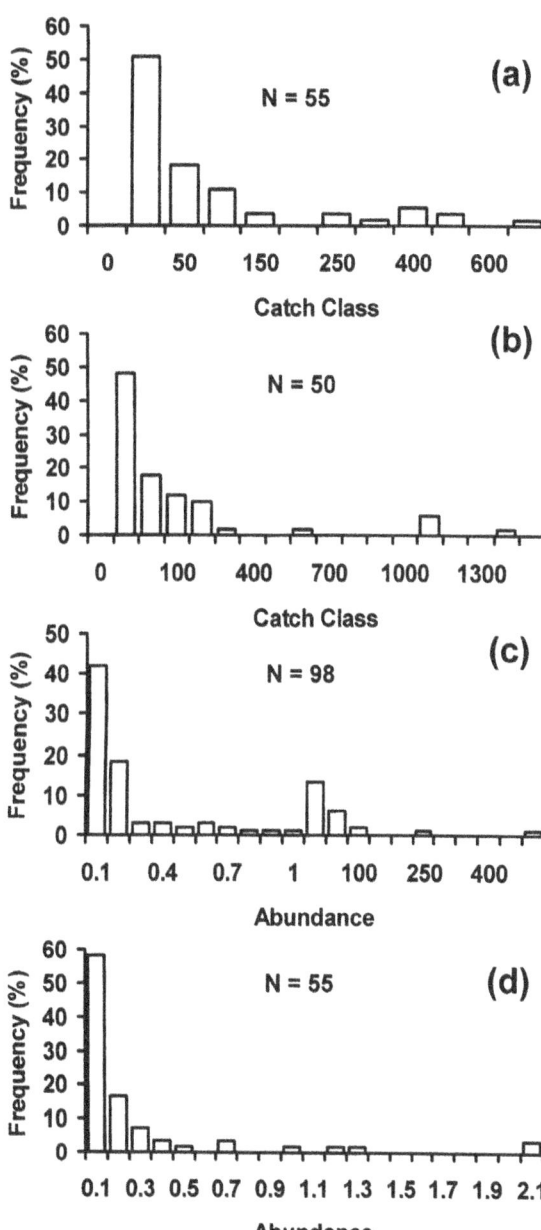

*Figure 6.* Frequencies (%) of observational density and catch abundance of *Chrysaora hysoscella* (a,c) and *Aequorea aequorea* (b,d). Trawl data (as kg min$^{-1}$, a,b), observational data (numbers 1000 m$^{-2}$, c,d).

ever, the absence of a negative relationship in this instance could imply some sort of spatial partitioning. Although this conclusion is based on data collected in the upper layers of the water column, and does not take cognisance of any deep-water populations, it is nevertheless valid over that depth range. Should there be deeper-living populations of either species, then this partitioning has an additional vertical component.

The 'partitioning' of horizontal and vertical space has been observed in other zooplankton from the Benguela ecosystem (Gibbons & Hutchings, 1996). Barange (1990) noted cross-shelf changes in the euphausiid assemblage in the northern Benguela, whilst Gibbons (1994) observed cross-shelf and vertical segregation of the dominant chaetognaths. It has been suggested that these examples of spatial partitioning might reflect competition (or the 'ghost of competition past'), and a similar explanation can be invoked here, as species of both genera eat a variety of hard- and soft-bodied zooplankton (Purcell, 1991; Purcell & Cowan, 1995; Purcell & Sturdevant, 2001).

This explanation would require that both species be able to identify some sort of environmental feature that would act as a boundary to their centres of preferred distribution. The front may represent just such a feature, because the catches of medusae in the trawl survey were not related to its position directly, but rather it acted as a boundary between the two species. This interpretation of the results implies that there is some biological structure to the assemblage of large medusae in the northern Benguela ecosystem, as has been postulated for cnidarians in the Mediterranean Sea (Buecher & Gibbons, 1999), and for other zooplankton assemblages in the southern Benguela ecosystem (Gibbons et al., 1999).

Whatever the underlying cause for the spatial partitioning of the waters of the northern Benguela region, it results in the distribution of large medusae across the regional shelf. A spatially persistent level of predation pressure is, therefore, applied across the shelf, the impact of which will depend upon the rate processes and biomass estimates of predator and prey. The trawl data derived from the *Dr. Fridtjof Nansen* survey represent the first estimates of large medusae abundance for the region, and therefore have value as input parameters for models of ecosystem functioning. A simple mass-balance model (Ecopath – Christensen & Pauly, 1992) has been used locally to explore ecosystem functioning (Jarre-Teichmann et al., 1998; Shannon & Jarre-Teichmann, 1999), and the results suggest that medusae play a negligible role in the system. Although the estimate of biomass employed in the model ($\sim$5 × $10^6$ tons) was little more than a guess (L. Shannon, M&CM, Cape Town, pers. comm.), the refined estimate generated here[3] differs little from it. It is estimated from the mean catch data that a total of 4.9 million

---

[3] Using mean catch data and assuming that both species are homogeneously distributed across the entire 179 000 km$^2$ of the northern Benguela.

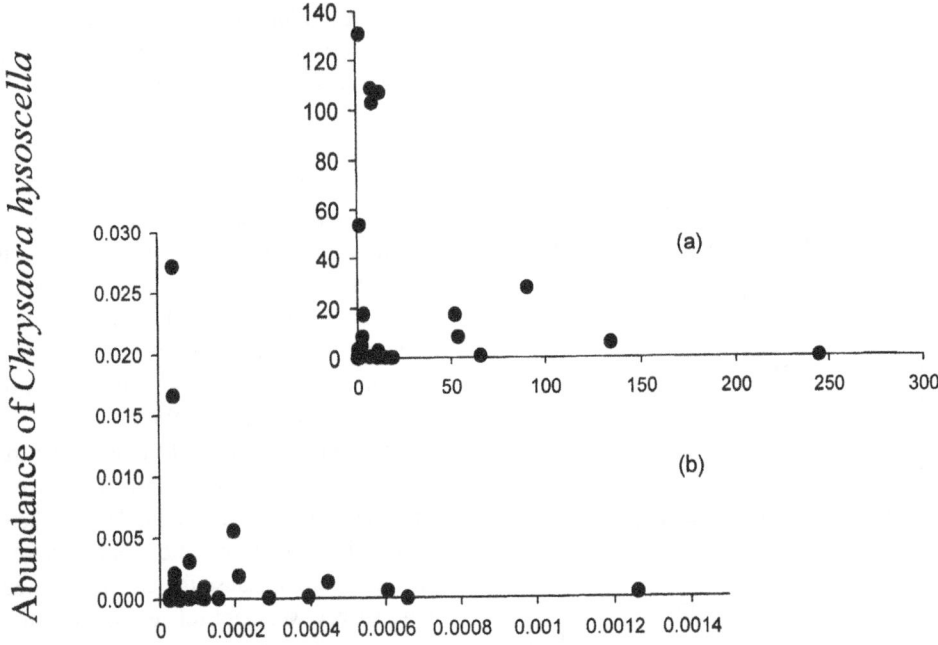

**Abundance of *Aequorea aequorea***

*Figure 7.* The relationship between the abundance of *Chrysaora hysoscella* and *Aequorea aequorea*, in the northern Benguela. Data from trawl surveys conducted in August/September 1999 (a) – as kg min$^{-1}$, and observations (b) – as numbers m$^{-2}$. Only non-zero data shown; all data pooled.

tons of medusae may occur in the region ($1.8 \times 10^6$ tons *C. hysoscella*, $3.1 \times 10^6$ tons *A. aequorea*). It should be realised that these data were collected over a one-week period during winter. Consequently, the estimate should be viewed with caution. However, if the maximum densities of medusae are used as input parameters ($13.1 \times 10^6$ tons *C. hysoscella* and $22.3 \times 10^6$ tons *A. aequorea*), the model becomes unbalanced and it is clear that medusae play a very significant role within the system (L. Shannon, M&CM, Cape Town, pers. comm.). An expansion of the model, as it pertains to medusae, falls outside the scope of this paper. Rather, we believe that the model needs to be re-visited, and greater attention needs to be paid to the estimates of production and ingestion used, as well as to the biomass values employed. We would ultimately anticipate that the significance of medusae will lie somewhere between the two extremes, though this is likely to vary on a seasonal basis.

The persistent decline in financial support for research into species of no commercial value in southern Africa, means that alternative methods for collecting data on these species will become more important. These alternative methods include beach-surveys of stranded specimens, the use of offshore drilling-platforms to collect samples and to make daily observations. They also include participation on cruises (research, fishing, etc.) of opportunity, and analyses of commercial fishing returns. Such methods were employed by Axiak et al. (1991), Benović (1991) and Carli (1991) for medusae in the Mediterranean Sea, and they proved to be relatively successful, despite their limited scope. Although the observational data collected here on cruises of opportunity were of a relative nature, they were supported (in part) by the results of a dedicated cruise, which suggests that the technique can be more widely applied.

### Acknowledgements

We would like to thank the Namibian Ministry of Fisheries and Marine Resources (NMFMR) for allowing C. Sparks to participate in cruises on the RV *Welwitschia,* and for providing logistic support in Swakopmund. The captains and crew of the RVs *Welwitschia* and '*Dr Fridtjof Nansen*' are thanked for their help in looking for and catching medusae. Dr Mark Brandon and Liz Hawker (BAS) provided advice and assistance with analysis of oceanographic data. De Beers Marine (Pty) Ltd, the NRF and the France-South Africa Science and Technology Exchange Agreement provided financial support for C. Sparks and E. Buecher. The Royal Society (London)-NRF provided funds to support the work of A. Brierley, who was given permission to participate by BAS. We thank the BENEFIT Committee for allowing ship's time, and NORAD for providing the RV *Dr. Fridtjof Nansen*. Grateful thanks are owed to Dr Jenny Purcell and the anonymous referees for greatly improving the manuscript.

# References

Arai, M. N., 1992a. Active and passive factors affecting aggregations of hydromedusae: a review. Sci. mar. 56: 99–108.

Arai, M. N., 1992b. Attraction of *Aurelia* and *Aeguorea* to prey. Hydrobiologia 216/217: 363–366.

Arai, M. N., 2001. Pelagic coelenterates and eutrophication: a review. Hydrobiologia 451 (Dev. Hydrobiol. 155): 69–87.

Axiak, V., C. Galea & P. J. Schrembri, 1991. Coastal aggregations of the jellyfish *Pelagia noctiluca* (Scyphozoa) in Maltese coastal waters during 1980–1986. Proceedings of the 2nd Workshop on Jellyfish Blooms in the Mediterranean Sea. UNEP, Athens. Map tech. Rep. Ser. 47: 32–40.

Bailey, K. M. & R. S. Batty, 1983. Laboratory study of predation by *Aurelia aurita* on larval herring (*Clupea harengus*): experimental observations compared with model predictions. Mar. Biol. 72: 295–301

Barange, M., 1990. Vertical migration and habitat partitioning of six euphausiid species in the northern Benguela upwelling system. J. Plankton Res. 12: 1223–1237.

Behrends, G. & G. Schneider, 1995. Impact of *Aurelia aurita* medusae (Cnidaria, Scyphozoa) on the standing stock and community composition of mesozooplankton in the Kiel Bight (western Baltic Sea). Mar. Ecol. Prog. Ser. 127: 39–45.

Benović, A. F., 1991. The aspect of jellyfish distribution in the Adriatic. Proceedings of the 2nd Workshop on Jellyfish Blooms in the Mediterranean Sea. UNEP, Athens. Map tech. Rep. Ser. 47: 41–50.

Boero, F., 1991. Contribution to the understanding of blooms in the marine environment. Proceedings of the 2nd Workshop on Jellyfish Blooms in the Mediterranean Sea. UNEP, Athens. Map tech. Rep. Ser. 47: 72–76.

Brierley, A. S., B. E. Axelsen, E. Buecher, C. A. J. Sparks, H. Boyer & M. J. Gibbons, (2001). Acoustic observations of jellyfish in the Namibian Benguela. Mar. Ecol. Prog. Ser. 210: 55–66.

Brodeur, R. D., C. E. Mills, J. E. Overland, G. E. Walters & J. D. Schumacher, 1999. Evidence for a substantial increase in gelatinous zooplankton in the Bering Sea, with possible links to climate change. Fish. Oceanogr. 8: 296–306.

Brown, P. C., S. J. Painting & K. L. Cochrane, 1991. Estimates of phytoplankton and bacterial biomass and production in the northern and southern Benguela ecosystems. S. Afr. J. mar. Sci. 11: 537–564.

Buecher, E. & M. J. Gibbons, 1999. Temporal persistence in the vertical structure of the assemblage of planktonic medusae in the NW Mediterranean Sea. Mar. Ecol. Prog. Ser. 189: 105–115.

Carli, A., 1991. Macroplanktonic jellyfish in the Ligurian Sea (1984–1986). Monitoring and biological characteristics. Proceedings of the 2nd Workshop on Jellyfish Blooms in the Mediterranean Sea. UNEP, Athens. Map tech. Rep. Ser. 47: 77–81.

Christensen, V. & D. Pauly, 1992. ECOPATH II – a software for balancing steady-state ecosystem models and calculating network characteristics. Ecol. Mod. 61: 169–185.

Cram, D. L. & G. A. Visser, 1973. SWA pilchard stock shows first signs of recovery (Summary of results of Phase III of the Cape Cross programme). S. Afr. Ship. News Fishg. Ind. Rev. 28: 56–63.

Denman, K. L. & T. M. Powell, 1984. Effects of physical processes on planktonic ecosystems in the coastal ocean. In Barnes, M. (ed.), Oceanogr. mar. biol. Ann. Rev. 22: 125–168.

Estrada, M. & C. Marrasé, 1987. Phytoplankton biomass and productivity off the Namibian coast. S. Afr. J. mar. Sci. 5: 347–356.

Fearon, J. J., A. J. Boyd & F. H. Schülein, 1992. Views on the biomass and distribution of *Chrysaora hysoscella* (Linné, 1766) and *Aequorea aequorea* (Forskål., 1775) off Namibia, 1982–1989. Sci. Mar. 56: 75–85.

Feigenbaum, D. L. & M. Kelly, 1984. Changes in the lower Chesapeake Bay food chain in the presence of the sea nettle *Chrysaora quinquecirrha* (Scyphomedusae). mar. Ecol. Prog. Ser. 19: 39–47.

Gibbons, M. J., 1994. Diel vertical migration and feeding of *Sagitta friderici* and *Sagitta tasmanica* in the southern Benguela upwelling region, with a comment on the structure of the guild of primary carnivores. Mar. Ecol. Prog. Ser. 225–240.

Gibbons, M. J., N. Gugushe, A. J. Boyd, L. J. Shannon & B. A. Mitchell-Innes, 1999. Changes in the composition of the non-copepod zooplankton assemblage in St Helena Bay (southern Benguela ecosystem) during a six day drogue study. Mar. Ecol. Prog. Ser. 180: 111–120.

Gibbons, M. J. & L. Hutchings, 1996. Zooplankton diversity and community structure around southern Africa, with special attention to the Benguela upwelling system. S. Afr. J. Sci. 92: 63–76.

Graham, W. M., 1994. The physical oceanography and ecology of upwelling shadows. PhD thesis, University of California, Santa Cruz.

Graham, W. M., 2001. Numerical increases and distributional shifts of *Chrysaora quinquecirrha* (Desor) and *Aurelia aurita* (Linné) (Cnidaria: Scyphozoa) in the northern Gulf of Mexico. Hydrobiologia 451 (Dev. Hydrobiol. 155): 97–111.

Graham W. M., Pagès, F., & W. M. Hamner. 2001. A physical context for gelatinous zooplankton aggregations: a review. Hydrobiologia 451 (Dev. Hydrobiol. 155): 199–212.

Hart, T. J. & R. I. Currie, 1960. The Benguela Current. Discovery Rep. 31: 123–298.

Hernroth, L. & F. Gröndahl, 1983. On the biology of *Aurelia aurita* (L.) 1. Release and growth of *Aurelia aurita* (L.) ephyrae in the Gullmar Fjord, western Sweden, 1982–83. Ophelia 22 (2): 189–199.

Jarre-Teichmann, A., L. J. Shannon, C. L. Moloney & P. A. Wickens, 1998. Comparing trophic flows in the southern Benguela to those in other upwelling ecosystems. S. Afr. J. mar. Sci. 19: 391–414.

King, D. P. F. & M. J. O'Toole, 1973. A preliminary report on the findings of the South West African pelagic egg and larval surveys. SFRI Internal Rep. Cape Cross Progr. Phase III.

Kramp, P. L., 1961. Synopsis of medusae of the world. J. mar. biol. Ass. U.K. 40: 1–469.

Larson, R. J., 1992. Riding Langmuir circulations and swimming in circles: a novel form of clustering behavior by the scyphomedusae *Linuche unguiculata*. Mar. Biol. 112: 229–235.

Mills, C. E., 2001. Jellyfish blooms: are populations increasing globally in response to changing ocean conditions? Hydrobiologia 451 (Dev. Hydrobiol. 155): 55–68.

Pagès, F., 1991. Ecología sistemática de los Cnidarios planktónicos de la corriente de Benguela (Atlántico Sudoriental). Ph.D. thesis, Univ. Barcelona, 466 pp.

Pagès, F. & J. M. Gili, 1991. Effects of large scale advective processes on gelatinous zooplankton populations in the northern Benguela ecosystem. Mar. Ecol. Prog. Ser. 75: 205–215.

Purcell, J. E., 1991. Predation by *Aequorea victoria* on other species of potentially competing hydrozoans. Mar. Ecol. Prog. Ser. 72: 255–260

Purcell, J. E., 1992. Effects of predation by the scyphozoan *Chrysaora quinquecirrha* on zooplankton populations in Chesapeake Bay, U.S.A. Mar. Ecol. Prog. Ser. 87: 65–76.

Purcell, J. E. & M. N. Arai, 2001. Interactions of pelagic cnidarians and ctenophores with fish: a review. Hydrobiologia 451 (Dev. Hydrobiol. 155): 27–44.

Purcell, J. E. & J. H. Cowan Jr, 1995. Predation by the scyphomedusan *Chrysaora quinquecirrha* on *Mnemiopsis leidyi* ctenophores. Mar. Ecol. Prog. Ser. 129: 63–70.

Purcell, J. E. & M. V. Sturdevant, 2001. Prey selection and dietary overlap among zooplanktivorous jellyfish and juvenile fishes in Prince William Sound, Alask. Mar. Ecol. Prog. Ser. 210: 67–83.

Purcell, J. E., E. D. Brown, K. D. E. Stokesbury, L. H. Haldorson & T. C. Shirley, 2000. Aggregations of the jellyfish *Aurelia labiata*: abundance, distribution, association with age-0 walleye pollock and behaviors promoting aggregation in Prince William Sound, Alaska, U.S.A. Mar. Ecol. Prog. Ser. 195: 145–158

Russell, F. S., 1970. The Medusae of the British Isles. II. Pelagic Scyphozoa with a Supplement to the First Volume on Hydromedusae. Cambridge University Press, Cambridge: 284 pp.

Schülein, F., 1974. A review of the SWA pelagic fish stocks in 1973. SFRI Internal Rep.; Cape Cross Progr. Phase IV: 3 pp.

Schuyler, Q. & B. K. Sullivan, 1997. Light responses and diel migration of the scyphomedusae *Chrysaora quinquecirrha* in mesocosms. J. Plankton Res. 19: 1417–1428.

Shannon, L. J. & A. Jarre-Teichmann, 1999. A model of trophic flows in the northern Benguela upwelling system during the 1980s. S. Afr. J. mar. Sci. 21: 349–366.

Shannon, L. V., 1985. The Benguela ecosystem. 1. Evolution of the Benguela, physical features and processes. In Barnes, M. (ed.), Oceanography and Marine Biology. University Press, Aberdeen: 105–182.

Skeide, R., A. Engås & C.W. West, 1997. Multisampler – a new tool for use in sampling trawls. In Shleinik, V. & M. Zaferman (eds), Seventh IMR-PINRO Symposium, Murmansk: 65–76.

Stander, G. H. & A. H. B. De Decker, 1969. Some physical and biological aspects of an oceanographic anomaly off South West Africa in 1963. Investl Rep. Div. Sea Fish. S. Africa 81: 1–46.

Venter, G. E., 1988. Occurrence of jellyfish on the west coast off South West Africa/Namibia. Rep. S. Afr. Natn Scient. Progms 157: 56–6.

*Hydrobiologia* **451**: 287–294, 2001.
© 2001 *Kluwer Academic Publishers.*

# A novel cilia-based feature within the food grooves of the ctenophore *Mnemiopsis mccradyi* Mayer

Anthony G. Moss[1], Rebecca C. Rapoza[2] & Lisa Muellner[1]

[1]*Dept. Biological Sciences, 131 Cary Hall, Auburn University, Auburn, AL 36849, U.S.A.*
[2]*Biology Dept., Woods Hole Oceanographic Institution, Woods Hole, MA 02543, U.S.A.*
*E-mail: mossant@mail.auburn.edu*

*Key words:* food groove, plankton, feeding, tentacle, compound cilium, actin, rod bacteria

## Abstract

We describe a novel compound ciliary structure (g-cilium) from the food groove of the lobate ctenophore *Mnemiopsis mccradyi*. G-cilia are small, flat compound ciliary organelles that are oriented with their tips pointing toward the mouth. Typically three to four rows of g-cilia line the inner surface of the tentacular groove, which together with the transport groove, make up the food groove. G-cilium cells are ~11.4 $\mu$m long and ~4.2 $\mu$m wide at the g-cilium base. The g-cilium itself is ~3.4 $\mu$m long and tapers to a flat, sharp tip. G-cilia are not motile but are surrounded by many hundreds of smaller, actively motile cilia that beat with orally-directed effective strokes. G-cilia contain ~50 conventional '9+2' cilia embedded in a fibrous core that arises from the cell body. In addition, g-cilia contain mitochondria, thousands of small membrane-bounded vesicles and rod bacteria. G-cilia basal bodies are anchored by large, strongly-banded rootlets that extend approximately the entire length of the cell. G-cilia may have organizational, sensory and/or secretory function within the feeding apparatus. Their placement strongly suggests that they play critical roles in feeding. They may enhance the efficiency of prey capture and so contribute to *M. mccradyi's* well-known voracious appetite. By enhancing prey capture they probably play a critical role in the capacity of this organism to follow prey dynamics, so contributing to dense blooms in mid-late summer in coastal regions.

## Introduction

Ctenophores are common coastal and oceanic gelatinous plankters that are well-known to have exploited cilia to form a variety of sensory and motor apparati (Tamm, 1982). Ctenophore comb plates are giant ciliary paddles used for both locomotion and the generation of feeding currents (Tamm & Moss, 1985; Matsumoto & Hamner, 1988; Barlow & Sleigh, 1993; Waggett & Costello, 1999). Beroid ctenophore macrocilia, formed by a single membrane surrounding up to several hundred axonemes, are used to capture and bite off pieces of prey (Horridge, 1965; Swanberg, 1974). Ctenophores depend on cilia for both geotactic orientation (Lowe, 1997) and transport of materials in the digestive system (Bumann & Puls, 1997).

The lobate ctenophores have evolved complex feeding mechanisms that allow them to efficiently capture and digest food. Waggett & Costello (1999) demonstrated that *Mnemiopsis leidyi* A. Agassiz employs two distinct feeding mechanisms. Adult copepods tend to swim into and be captured on the inner surface of the oral lobes. By contrast, copepod nauplii are entrained on feeding currents generated by auricular comb plates and are captured on the tentilla (tentacle side branches), which lie adjacent to the structures we describe here. Thus, the system and putative sensory g-cilia described herein may be specializations for the capture of smaller crustacean zooplankters, allowing expansion of the useful range of prey available to *Mnemiopsis* spp., and so extension of the active growth stage of this animal to periods dominated by smaller prey.

We have recently begun a detailed examination of the behavior and underlying microanatomy of feeding in lobate ctenophores, with particular emphasis on the common New World Atlantic ctenophores *Mnemiopsis leidyi* and *Mnemiopsis mccradyi* Mayr. Close examination of the food groove reveals previously undescribed details common to both species. Although the food groove has long been known to exist and to be important to the capture of prey (Chun, 1880; Reeve & Walter, 1978), the microanatomical elements and microscale behavior of this feeding apparatus have not yet been elucidated. Exploration of food groove microanatomy constitutes an essential step toward understanding the capacity of both species to undergo explosive growth to form dense blooms during favorable conditions (as reported by Purcell et al., 2001; Sullivan et al., 2001). Since lobate ctenophores are critically important predators in the coastal marine environment (Reeve & Walter, 1978; Monteleone & Duguay, 1988; Vinogradov et al., 1989; Cowan & Houde, 1993), knowledge of their feeding mechanisms will enable us to understand their impact on coastal ecology with greater precision.

Here, we describe a novel and intriguing cilia-based structure of the lobate feeding apparatus, specifically as it occurs in the Gulf ctenophore, *Mnemiopsis mccradyi*. These structures were noted in two other ctenophore species by Chun (1880), who gave them the name 'gemshornförmigen cilien' (goathorn-shaped cilia), but no further investigation has been made since that time. We use the abbreviated term 'g-cilium' to denote this feature, describe g-cilium organization and microanatomy and show that this organelle contains an unusually massive fibrous cytoskeleton surrounding immotile cilia. Our findings not only describe a new and fundamental ciliary-based cellular organelle, but in addition, have implications for understanding prey capture in *Mnemiopsis* spp., and furthermore should help coastal marine scientists to understand why *Mnemiopsis* spp. are capable of rapid population growth leading to the production of massive blooms in mid-late summer.

## Materials and methods

### Ctenophores

*Mnemiopsis mccradyi* were collected from Gulf of Mexico sites by dipping as described previously (Estes et al., 1997). Specimens were maintained in plankton kreisels or large jars for less than 2 days prior to fixation and embedding for electron microscopy. Only obviously healthy animals, displaying no morphological damage, were used for ultrastructural examination.

Live specimens were examined for overall microscopical anatomy with a high power dissecting microscope. Food grooves were microsurgically dissected and held in wet mounts for brightfield, phase contrast and differential interference contrast microscopy.

### Light microscopy

Light photomicrographs were taken on Technical Pan rated at ISO 100 (type 2415, Eastman Kodak, Rochester, NY) or Kodak Gold (ISO 200) print film with a dedicated microscopy camera (model PM10-AD, Olympus, NJ) attached either to a compound microscope (model BHS, Olympus) or dissecting microscope (model SZ11, Olympus). Digitally-enhanced video images were taken with a high-grade video camera (model VE1000, Dage-MTI, Indianapolis, IN). Images were captured with a computer digitizing board (Flashpoint model 3030, Integral Technologies, Indianapolis, IN) controlled by image-processing software (Image-Pro, Media Cybernetics, Bethesda, MD).

### Electron microscopy

Microsurgically-dissected specimens for electron microscopy were quickly fixed as described by Tamm & Tamm (1981) and Estes et al. (1997). Fixed tissues were dehydrated through an ethanol series into propylene oxide, infiltrated and embedded in Durcupan epoxy resin (Fluka Chemie, Buchs, Switzerland). Quarter-micron sections for examination by light microscopy were cut with glass knives on an Ultracut E (Reichert, Vienna, Austria); 50–70 nm thin sections were cut on the same instrument by diamond knife (Microstar, Houston, TX, U.S.A.) for electron microscopy.

Thick-sections were stained with borate-buffered 1% Toluidine Blue O (Sigma Chemical Co.) for examination of overall food groove morphology. Thin sections were stained in saturated methanolic uranyl acetate and 1% lead citrate and examined at 60 KeV in a transmission electron microscope (model 301, Philips Electron Optics, Amsterdam, The Netherlands).

All measurements are given as mean±standard error.

## Results

### G-cilium location and orientation

Lobate ctenophores have four food grooves (one per quadrant) that extend from the base of the auricles

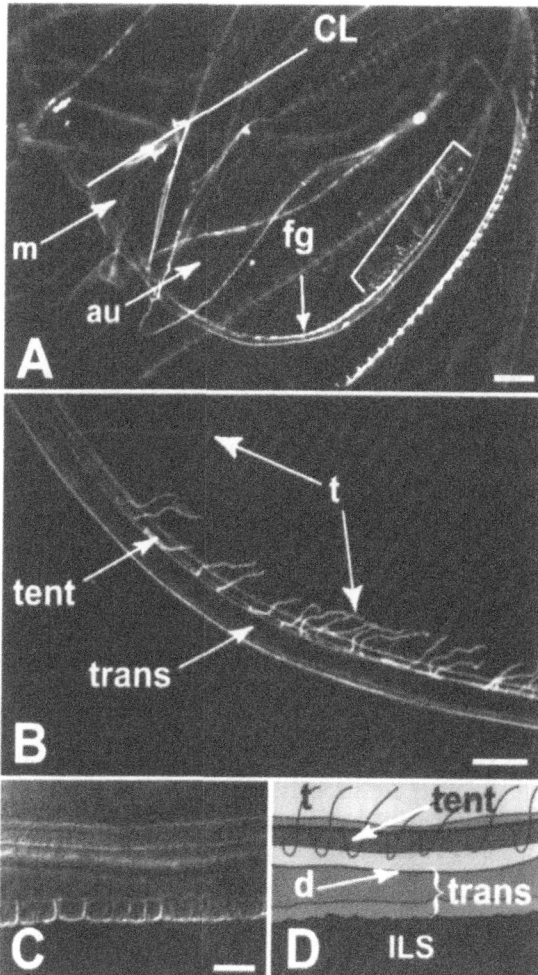

*Figure 1. Mnemiopsis mccradyi* food groove morphology. (**A**) Partial view of intact animal. Arrows indicate the food groove complex. The oral-aboral axis is indicated by the **CL**. **Bracket** indicates extended tentilla. **m**, mouth; **fg**, food groove; **au**, a fingerlike auricle. Scale, 5 mm. (**B**) Food groove anatomy at higher magnification. **trans**, transport groove; **tent**, tentacular groove; **t**, tentilla. Scale, 1 mm. (**C, D**) Transport and tentacular groove ('tent') at higher magnification (C) and interpretive diagram (D). The granular appearance of the tentacular groove (dark banded region indicated by 'tent' in D) indicates the tentacular groove bearing the g-cilia. The curly bracket indicates the width of the transport groove; 'd' indicates the deepest extent of the transport groove. In images prepared for differential interference microscopy and covered with a coverslip, the tentilla were always snapped off, and so none show in C. In order to to help orient the reader, in D tentilla (*t*) are shown diagrammatically, extending outward from the tentacular groove. The tentilla originate inside the tentacular groove, near the field of g-cilia. The thick line at the base of the tentilla indicates the attached tentacle from which the tentilla arise. Note the wrinkled appearance of the edge of the transport groove in C. The surface behind the wrinkled edge of the transport groove is the inner lobe surface (ILS). Darkfield images by dissecting microscope (A, B); successive images are all oriented as closely as possible in the same direction. Scale in C: 100 μm.

to the mouth and deliver prey for ingestion following lobe or tentillar capture. The position and overall morphology of the food groove of *M. mccradyi* is depicted in Figure 1. In the following discussion the terms 'medial' and 'lateral' are used with reference to the oral-aboral axis, or body centerline (see Fig. 1A). Each food groove is composed of two distinct microanatomical elements, a food transport groove and a shallower tentacular groove located medial to the transport groove (Fig. 1B, C, D). Near the base of the medial wall of the tentacular groove (Fig. 2A, B), we find 3–4 rows of g-cilia of similar size and overall design (Fig. 2C). An additional 1–2 rows of dagger-like g-cilia are positioned at the edge of the tentacular groove and are just visible under high power and ideal conditions in the dissecting microscope. A high magnification DIC image of one of the latter g-cilia is shown in Figure 2D.

Tentilla arise as lateral attachments to the tentacle proper which lies within the tentacular groove and is attached along its length. In intact, freely-swimming animals and under normal feeding conditions, tentilla are drawn out away from the body surface by the auricle-generated feeding current. The overall organization and functionality of the food groove will be described elsewhere and is outside of the scope of this report.

*G-cilia fine structure*

Seventy-one mid-row g-cilia were examined with regard to their dimensions using differential interference contrast light microscopy. Each triangular g-cilium has a broad base (4.2 μm±0.09), yet is only 3.4 μm (±0.09) long. G-cilia are ~1 μm thick at their base by light microscopy; this is difficult to estimate in living preparations because of the juxtaposition of nearby epithelial cells. Electron microscopy reveals that the cell tapers from the g-cilium base to a blunt end ~8 μm (±0.19 μm) distant. The g-cilium cell extends through the entire epithelial thickness (Fig. 3A).

Each g-cilium contains approximately 50 conventional 9+2 cilia. They do not display the 3–8 interciliary doublet bridges – also known as 'compartmenting lamellae' (Afzelius, 1961), which are thought to hold the individual cilia together (Dentler, 1981) – nor do they have the asymmetric organization with respect to the midfilament as previously reported for comb plate cilia (Afzelius, 1961). Each individual cilium within a g-cilum is wrapped in its own ciliary membrane, unlike the macrocilia of *Beroë* spp. (Horridge, 1965).

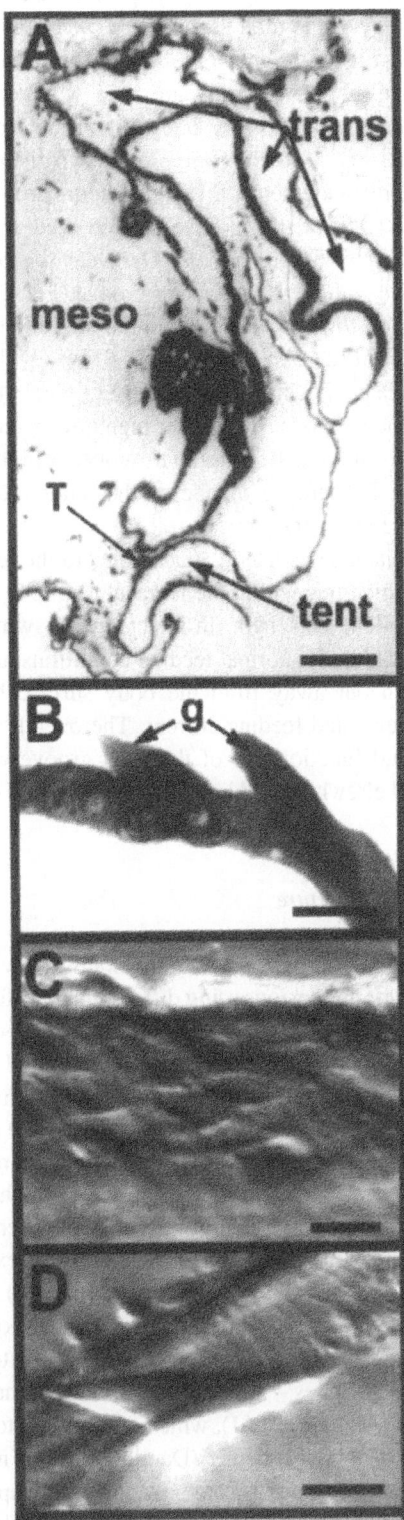

Figure 2. →

Individual cilia are very straight throughout the distal two thirds of the g-cilium.

Each g-cilum contains a dense fibrous matrix that arises from the cell body (Fig. 3A). Cilia contained within the g-cilium are embedded in the lateral ramifications of the matrix. High magnification TEM reveals that the ciliary membranes are continuous with the matrix membrane (not shown), demonstrating that the cilia and the matrix arise from the same cell. The longest cilia extend the full length of the g-cilium and the tips are exposed to sea water (Fig. 3C).

The matrix contains thousands of small clear membrane-bounded vesicles (Fig. 3C). Vesicles are found in rows throughout the g-cilium, often clearly aligned in the direction of the matrix fibers (Fig. 3C, D). The matrix also contains mitochondria (Fig. 3D).

### Ciliary rootlets

Individual cilia are anchored within the cell body by massive striated rootlets (Fig. 3A; apparent in Fig. 2B also). Rootlets entirely surround the ciliary kinetosomes and appear to extend to the oralmost end of the cell. Rootlets are strongly striated with a periodicity of ~40 nm ($39.3\pm3.3$, $n=40$) and can be up to ~400 nm wide ($185\pm79$, $n=40$). They display a distinct pattern of cross-linkages in transverse thin section (not shown).

### G-cilium bacteria

In many g-cilia, we observed rod bacteria embedded within the matrix, usually but not always aligned with the g-cilium axis (Fig. 4). Such bacteria are typically 0.25 $\mu$m dia. and are often several microns long. We did not observe flagella on the bacteria.

*Figure 2.* Microanatomy of the food groove and g-cilia. (**A**) Transverse thick section of the food groove region. Note the g-cilia on the wall of the tentacular groove. **T**, attached tentacle; **meso**, mesoglea; otherwise legend as in Figure 1. Scale, 500 $\mu$m. (**B**) Higher magnification of the tentacular groove medial wall, showing g-cilia. **g**, g-cilium. Scale: 10 $\mu$m. A, B: brightfield microscopy, stained with Toluidine Blue O. (**C**) Living g-cilia in 3–4 staggered rows deep within the tentacular groove, as described in text. Scale: 10 $\mu$m. (**D**) G-cilium from one of the edge-positioned rows. Note somewhat different morphology from the g-cilia seen in C. Scale: 5 $\mu$m. C, D: Differential interference microscopy; g-cilia are aligned as they would be in the lower magnification images of Figure 1.

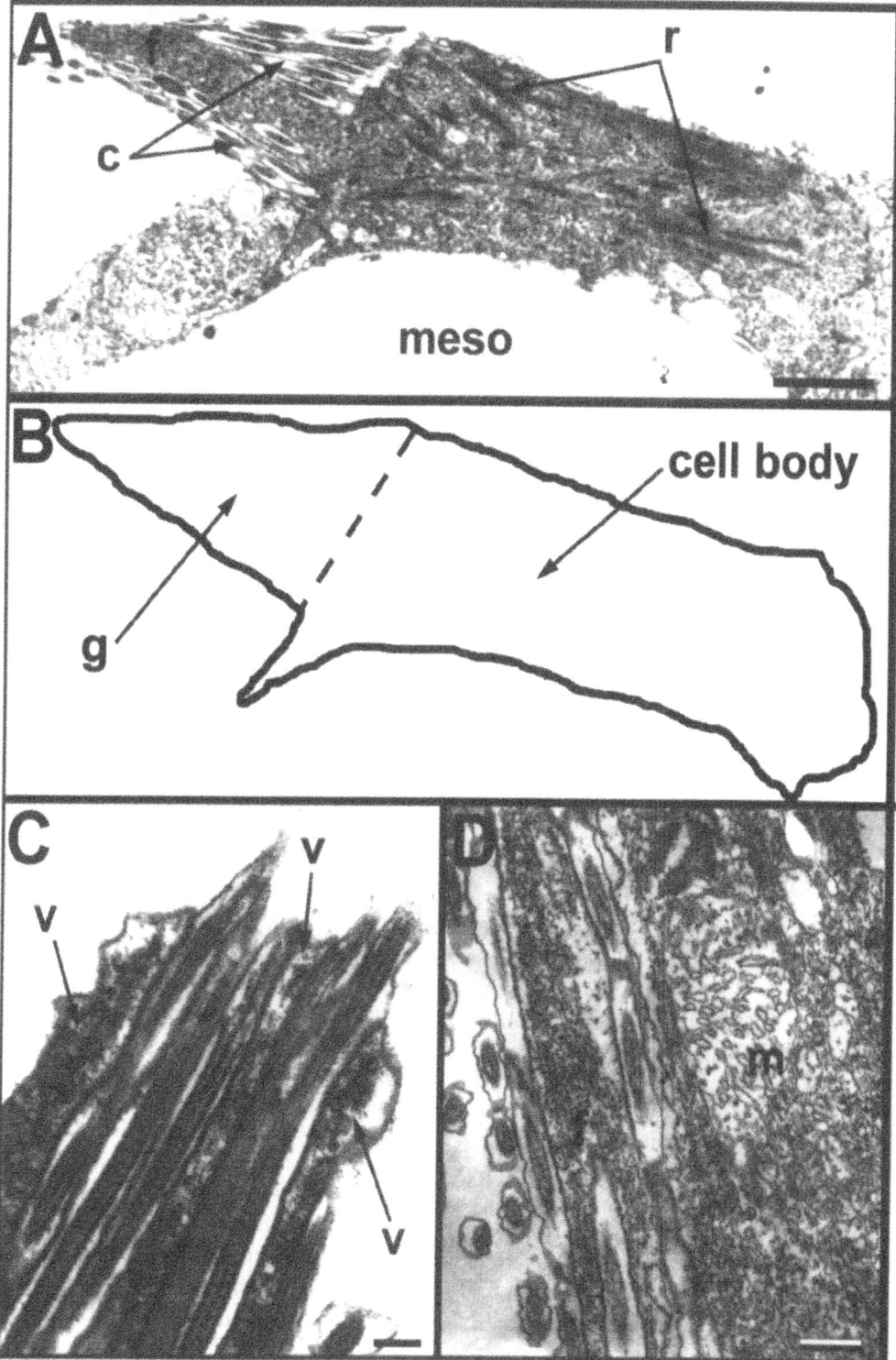

*Figure 3.* Electron microscopy of g-cilia. (**A**) Longitudinal section of g-cilium. **c**, constituent '9+2' cilia. **f**, fibrous core; **r**, ciliary rootlets; **meso**, mesoglea. Scale, 1 μm. (**B**) Interpretive diagram of Figure 3A. **g**, g-cilium 'tooth'. (**C**) Detail of the tip of a g-cilium. The cilia are sectioned longitudinally – indicating that they are very straight and strongly aligned along the length of the g-cilium. The constituent cilia are separated by fingerlike protrusions of the cytoplasm bounded by the plasma membrane. Strings of aligned, small, uniform vesicles (**v**) are apparent within the matrix near the tips of the protruding processes. The cilia protrude beyond the tip of the g-cilium (see also A). Scale: 0.25 μm. (**D**) Detail showing the fibrous nature of the matrix (**f**) and a mitochondrion (**m**). The cytoplasm contains many small vesicles. Conventional cilia to the left border of the image arise from neighboring ciliated epithelial cells. Scale: 0.5 μm.

292

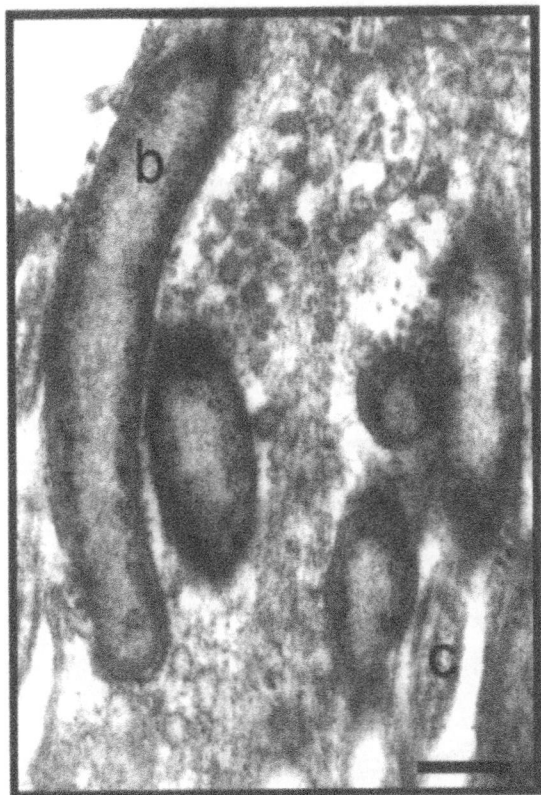

*Figure 4.* G-cilia bacteria (b). Longitudinal and transverse section through five bacteria embedded in the g-cilia matrix and accompanied by numerous vesicles. c, glancing section of a cilium. Scale: 0.2 μm.

## Discussion

### *Possible roles for the g-cilia in lobate feeding activity*

Tentacular groove g-cilia are located just medial to the transport groove, which quickly translocates prey items to the oral cavity of this animal (Rapoza & Moss, unpubl. results). We have observed that prey can be tethered by a single sticky tentillum. Tentilla, which arise from the cavity of the tentacular groove, are thus well placed to pull prey close to the transport groove.

The role of the g-cilia in this process, however, is unclear. Within the tentacular groove, tentilla lie over the bed of g-cilia, which suggests that g-cilia may play a role in tentilla organization or spacing. This is an attractive concept because the tentilla, which secrete a sticky mucilage (Bargmann et al., 1972), can become very entangled. During feeding, tentilla are drawn out and kept suspended by currents generated by the auricular comb plates. Strong tentillar contractions, elicited by prey contact, draw prey to the tentacular groove opening for transfer to the transport

groove. As tentilla contract into the tentacular groove, the comb-like g-cilia may facilitate prey transfer to the transport groove and provide a mechanism for maintaining tentillar spacing as they are resuspended for prey capture. G-cilia are surrounded by numerous individual cilia that continuously beat in an oral direction; these cilia may function to push contracted tentilla out of the tentacular groove for resuspension. Observation of the fine movements of the tentacular groove during prey capture is very difficult because it is small and extremely contractile. Nevertheless, we intend to examine this possibility in subsequent studies using a modified preparation that will allow higher resolution observation of the living food groove during feeding.

G-cilia contain mitochondria, which suggests that they are metabolically active. Our observation of many small vesicles in conjunction with the presence of mitochondria leads us to conclude that g-cilia contain an energetically-expensive intracellular transport system.

We have observed captured prey to become suddenly quieter when pulled near the food groove. Prey probably become entangled in a growing mass of mucus in the vicinity, but it is also possible that prey are immobilized by a locally-secreted toxin. The tentacular groove is ideally located to release an immobilizing agent upon captured prey prior to transfer to the transport groove.

G-cilia may serve as toxin delivery systems to anaesthetize prey, and the numerous g-cilium vesicles could carry a secretory substance. We have never observed omega-bodies (thereby indicating vesicle/plasma membrane fusion) in g-cilia; however, our fixation technique is unlikely to capture such a rapid process. The vesicles could also have other roles. For instance, they might transport membrane components to the plasma membrane and/or structural proteins to the g-cilium matrix. We plan to determine whether g-cilia might have a toxic effect on prey by collaboration with an expert toxicologist.

Lobate g-cilia may serve as vibration sensors. G-cilia bear some resemblance to presumed mechanosensory cells initially described by Hernandez-Nicaise (1974). Tamm & Tamm (1991) examined the ultrastructure of such cells, which combine a stiff actin-containing peg with a single cilium. To the best of our knowledge, mechanosensory function has not been firmly established for actin peg cells.

The g-cilium fibrous core almost certainly provides physical strength and immobilizes the embedded cilia. Furthermore, the rootlets that anchor the basal bodies of this structure are peculiarly massive and heavily striated. We conclude that the g-cilium cell is probably stiff.

It is well-known that mechanosensory cells are stiff. For instance, vertebrate hair cells are known to be very stiff and exquisitely sensitive to lateral displacement (for a review see Howard et al., 1988). Invertebrate and protistan vibration-sensing organelles are nearly always based upon the eukaryotic '9+2' cilium (Weiderhold, 1976; Stommel et al., 1980; Machemer & Machemer-Röhnisch, 1984) and the extensive rootlet organization of the mechanosensory hair cells of the nudibranch statocyst (Kuzerian et al., 1981) has been strongly implicated in that cell's mechanosensory function. The well-developed rootlet system of the g-cilium cell could similarly play an important role in mechanosensation.

A vibration-sensing function for the g-cilia would furthermore be developmentally and physiologically sound in that the sensory structures would be distributed along the full length of the tentillar prey capture apparatus. G-cilia could thus provide feedback information about the presence of ensnared prey, indicated by prey movement or by contraction of tentilla within the tentacular groove. Preliminary evidence by Moss suggests that the tentacular groove is indeed sensitive to mechanical vibration. Experiments are currently underway to more closely characterize the sensory characteristics of this feeding apparatus, and to determine whether the mechanosensation arises as a result of sensory activity within the g-cilium. If indeed the g-cilia are sensory cells, then they must have connection with the nervous system of the animal either via. chemical or electrical synapses; to date, we have not observed this, however we will more closely examine this possibility by thin-section TEM analysis.

G-cilia of the lobate *Mnemiopsis mccradyi* are unique among ciliary structures and unique among the already-broad complement of ciliary effector systems in ctenophores. The closest morphological relatives are the actin-peg bearing cells of Hernandez-Nicaise (1974) and Tamm & Tamm (1991) mentioned previously, which are liberally distributed over the surface of the animal. By contrast, g-cilia are distinctly different ultrastructurally and are in a clearly defined specialized food-capture organ. To date, we have observed g-cilia to be in only the tentacular groove. We conclude that the ctenophore g-cilium is a unique cilia-based organelle whose position and microanatomy strongly suggest roles important to prey capture. Potential roles for the g-cilium include organization

of the tentilla, secretion of a prey-disabling toxin, or detection of prey by mechanoreception.

Lobate ctenophores are notoriously voracious and are often observed to occur in prodigious numbers in many coastal locations (Purcell et al., 2001; Sullivan et al., 2001). Waggett & Costello (1999) have reported that there appears to be a distinct partitioning of capture effort in different locations of the ctenophore, with smaller prey more likely to be captured by tentilla (which originate only in the tentacular groove). However the g-cilium acts, it is very likely to be involved in enhancement of prey-capture capability. Enhancement of prey capture in any manner could give this predator a critical trophic advantage and would probably play an important role in the explosive growth characteristics of *Mnemiopsis* spp. in coastal waters.

## Acknowledgements

We wish to thank Dr Laurence Madin for the generous use of his laboratory facilities, and Ms Maria Tovio-Kinnucan for expert assistance with the electron microscopy. We also wish to thank Dr John Waterbury for stimulating discussions. Supported by grants to: A. G. M. (National Institutes of Health, Dept. Zoology and Wildlife Sciences, Auburn University); L. M. (The Program for Undergraduate Excellence and the Howard Hughes Future Life Sciences Scholars Program, Auburn University); R. R. (The Seaver Institute and Woods Hole Oceanographic Institution Sea Grant Project No. R/B-143 under NOAA Sea Grant No. NA86RG0075). The U. S. Government is authorized to produce and distribute reprints for governmental purposes notwithstanding any copyright notation that may appear hereon.

## References

Afzelius, B. A., 1961. The fine structure of the cilia from ctenophore swimming-plates. J. Biophys. Biochem. Cytol. 9: 383–394.

Bargmann, W., K. Jacob & A. Rast, 1972. Über Tentakel und Colloblasten der Ctenophore *Pleurobrachia pileus*. Z. Zellforsch. 123: 121–152.

Barlow, D. & M. Sleigh, 1993. Water propulsion speeds and power output by comb plates of the ctenophore *Pleurobrachia pileus* under different conditions. J. exp. Biol. 183: 149–163.

Bumann, D. & G. Puls, 1997. The ctenophore *Mnemiopsis leidyi* has a flow-through system for digestion with three consecutive phases of extracellular digestion. Physiol. Zool. 70: 1–6.

Chun, C., 1880. Die Ctenophoran des Golfes von Neapel und der angrenzenden Meeres-Abschnitte. Flora und Fauna des Golfes von Neapel, Vol. 1, Engelmann, Leipzig: 1–311.

Cowan, J. H. & E. D. Houde, 1993. Relative predation potentials of scyphomedusae, ctenophores and planktivorous fish on ichthyoplankton in Chesapeake Bay. Mar. Ecol. Prog. Ser. 95: 55–65.

Dentler, W. L., 1981. Microtubule-membrane interactions in ctenophore swimming plate cilia. Tissue Cell 13: 197–208.

Estes, A. M., B. S. Reynolds & A. G. Moss, 1997. *Trichodina ctenophorii* n. sp., a novel symbiont of ctenophores of the northern coast of the Gulf of Mexico. J. Euk. Microbiol. 44: 420–426.

Hernandez-Nicaise, M-L., 1974. Ultrastructural evidence for a sensory-motor neuron in Ctenophora. Tissue Cell 6: 43–47.

Horridge, G. A., 1965. Macrocilia with numerous shafts from the lips of the ctenophore *Beroë*. Proc. roy. Soc. Lond., 162B: 351–364.

Howard, J., W. M. Roberts & A. J. Hudspeth, 1988. Mechanoelectrical transduction by hair cells. Ann. Rev. Biophys. Biophys. Chem. 17: 99–124.

Kuzirian, A. M., D. L. Alkon & L. G. Harris, 1981. An infraciliary network in statocyst hair cells. J. Neurocytol. 10: 497–514.

Lowe, B. T., 1997. Studies on calcium regulation of motile, mechanoresponsive statocyst cilia and locomotory cilia in ctenophores. Ph.D. Thesis, Boston University: 180 pp.

Machemer, H. & S. Machemer-Röhnisch, 1984. Mechanical and electric correlates of mechanoreceptor activation of the ciliated tail in *Paramecium*. J. Comp. Physiol. 154A: 273–278.

Matsumoto, G. I. & W. H. Hamner, 1988. Modes of water manipulation by the lobate ctenophore *Leucothea* sp. Mar. Biol. 97: 551–558.

Monteleone, D. M. & L. E. Duguay, 1988. Laboratory studies of predation by the ctenophore *Mnemiopsis leidyi* on the early stages in the life history of the bay anchovy, *Anchoa mitchilli*. J. Plankton Res. 10: 359–372.

Purcell, J. E., T. A. Shiganova, M. B. Decker & E. D. Houde, 2001. The ctenophore *Mnemiopsis* in native and exotic habitats: U.S. estuaries *versus* the Black Sea basin. Hydrobiologia 451 (Dev. Hydrobiol. 155): 145–175.

Reeve, M. R. & M. A. Walter, 1978. Nutritional ecology of ctenophores – a review of recent research. Adv. mar. Biol. 15: 249–287.

Stommel, E. W., R. E. Stephens & D. L. Alkon, 1980. Motile statocyst cilia transmit rather than directly transduce mechanical stimuli. J. Cell Biol. 87: 652–662.

Sullivan, B. K., D. Van Keuren & M. Clancy, 2001. Timing and size of blooms of the ctenophore *Mnemiopsis leidyi* in relation to temperature in Narragansett Bay, RI. Hydrobiologia, 451 (Dev. Hydrobiol. 155): 113–120.

Swanberg, N., 1974. The feeding behavior of *Beroe ovata*. Mar. Biol. 24: 69–76.

Tamm, S. L., 1982. Ctenophora. In Shelton, G. A. B. (ed.), Electrical Conduction and Behaviour in 'Simple' Invertebrates. Clarendon Press, Oxford: 266–358.

Tamm, S. L. & A. G. Moss, 1985. Unilateral ciliary reversal and motor responses during prey capture by the ctenophore *Pleurobrachia pileus*. J. exp. Biol. 114: 443–461.

Tamm, S. L. & S. Tamm, 1981. Ciliary reversal without rotation of axonemal structures in ctenophore comb plates. J. Cell Biol. 89: 495–509.

Tamm, S. & S. L. Tamm, 1991. Actin pegs and ultrastructure of presumed sensory receptors of *Beroë* (Ctenophora). Cell Tissue Res. 264: 151–159.

Vinogradov, M. Y., E. A. Shushkina, E. I. Musayeva & P. Yu. Sorokin, 1989. A newly acclimated species in the Black Sea: The ctenophore *Mnemiopsis leidyi* (Ctenophora: Lobata). Oceanology 29: 220–224.

Waggett, R. & J. H. Costello, 1999. Capture mechanisms used by the lobate ctenophore, *Mnemiopsis leidyi*, preying on the copepod *Acartia tonsa*. J. Plankton Res. 21: 2037–2052.

Weiderhold, M. L., 1976. Mechanosensory transduction in 'sensory' and 'motile' cilia. Ann. Rev. Biophys. Bioeng. 5: 39–62.

*Hydrobiologia* **451**: 295–304, 2001.
© 2001 *Kluwer Academic Publishers.*

# Protistan epibionts of the ctenophore *Mnemiopsis mccradyi* Mayer

Anthony G. Moss, Anne M. Estes, Lisa A. Muellner & Darrell D. Morgan
*Biological Sciences, 131 Cary Hall, Auburn University, Auburn, AL 36849, U.S.A.*
*E-mail: mossant@mail.auburn.edu*

*Key words:* comb jellies, comb plate, marine parasites, amoebae, ciliates, dinoflagellates, coastal health

## Abstract

*Mnemiopsis mccradyi*, a common coastal ctenophore, was observed to bear two distinct, exclusive assemblages of protistan epibionts. The mobiline peritrich, *Trichodina ctenophorii* (Estes et al., 1997), and small *Flabellula*-like gymnamoebae inhabited only the surface of the comb plates. By contrast, small *Vexillifera*-like gymnamoebae and large *Protoodinium*-like dinoflagellates were found on the ectoderm. The relationship of the epimicrobial protists with their host varied from possible mutualism (vexilliferids) to commensalism (trichodinids) to parasitism (flabellulids and protoodinids). Trichodinids may benefit from comb plate attachment by enhanced food capture. Although they did not obviously impair comb plate beating, they did distort the surface and appear to produce fissures in the comb plate surface, which could provide inroads for more severe comb plate damage by amoebae. By contrast, when flabellulid amoebae occurred in very high surface densities (up to $\sim$5000 mm$^{-2}$), they clearly damaged comb plates by eroding the surface. Where flabellulid pseudopodia invaded the comb plate, we observed local degradation of comb plate cilia, as evidenced by central pair disorientation and plasma membrane perturbation and overt phagocytosis of comb plate cilia. Ectodermal vexilliferids, which occurred at much lower densities, did not appear to have any degradative impact on the ctenophore. By contrast, clusters of ectodermal protoodinids were found in localized depressions most likely caused by invasive phagocytosis. The impact of the protistan assemblages on ctenophore populations is unclear, but under conditions of severe infestation they might depress ctenophore population density.

## Introduction

Ctenophores commonly harbor parasites, although they are usually located in the mesogloea (Crowell, 1976; Stunkard, 1980; Yip, 1984; Purcell and Arai, 2001). Like many marine jellies, ctenophores have been observed to bear ectoparasitic amphipods (Harbison et al., 1977; Harbison & Madin, 1979). Mills & McLean (1991) described an ectodermal dinoflagellate parasite of several jellies from the Pacific Northwest. To date, however, no microbial assemblages have been reported to be associated specifically with ctenophore comb plates, or to have specific distributions upon different regions of the animal.

Trichodinids are mobiline peritrich ciliates found on a variety of hosts (Uzmann & Stickney, 1954; Lom & Hoffman, 1964), usually as parasites of freshwater fish (Lom, 1970; Arthur & Margolis, 1984; Urawa & Awakura, 1994). They are found as ecto-commensals of marine teleosts (Arthur & Margolis, 1984; Poynton & Lom, 1989), although many inhabit the urogenital tracts of amphibians (Lom, 1958) and fish (Van As & Basson, 1996). Some are known to cause severe infestations of hatchery fish, particularly salmonids (Arthur & Margolis, 1984). Several trichodinids are endo- and ectoparasites of invertebrates (Uzmann & Stickney, 1954; Lom, 1958; Lom & Haldar, 1976), including *Hydra* (in James-Clark, 1866).

Marine gymnamoebae (Subclass Gymnamoebia, *sensu* Page, 1988) exhibit very broad species diversity and yet their habits continue to be poorly understood. Gymnamoebae are relatively rare as free-floating forms in the water column, but are common on marine surfaces, especially those of dead plants and animals (Sieburth, 1984). They are also known to be a major biotic component of benthic sediments (Butler & Rogerson, 1995) where they can occur at densities of up to several thousand cm$^{-3}$. Despite showing modest bacterial consumption rates, their abundance implies that they may be important benthic grazers (Butler & Rogerson, 1997).

Less is known about microbial assemblages on the surfaces of free-swimming gelatinous zooplankton. To the best of our knowledge, other than our previous publication (Estes et al., 1997) and that of Mills & Mclean (1991), no reports describe ctenophore epimicrobial infestations. We describe here protistan epibionts that display considerable specificity with regard to their position of residence, either solely on the giant ciliary comb plates or solely the ectoderm. Where possible, we describe the effects that the epibionts have on the host.

In coastal regions of the Gulf of Mexico, the ctenophore *Mnemiopsis mccradyi* Mayer, is accepted to be a key predator in the coastal marine food web (Reeve et al., 1978; Edmiston, 1979). Indeed, *M. mccradyi* and *M. leidyi* A. Agassiz, the Atlantic coast species, both form dense blooms in the mid-late summer. *Mnemiopsis* spp. are profilic, with each individual releasing thousands of eggs during spawning; furthermore, they are very efficient predators with digestive efficiencies >70% and being capable of consuming copepod zooplankton at >1000% of their body carbon in 24 h (Reeve et al., 1978). By virtue of sheer numbers and capacity for feeding, *Mnemiopsis* spp. can exert significant control over zooplankton assemblages, and so indirectly, the phytoplankton assemblages of the coastal zone.

We describe here an assemblage of protists commonly found on *Mnemiopsis mccradyi* along the coastal Gulf of Mexico. Our observations suggest that some members of this group – a small flabellulid gymnamoebae and a sessile protoodinid dinoflagellate – have the potential to impact the general health of the ctenophore. Although heavily-infested *M. mccradyi* appeared to be fundamentally healthy, it is conceivable that the protists could impair ctenophore feeding, and so, possibly fecundity, which in turn, would impact the ability of this organism to develop dense blooms. The protist assemblage described here could therefore have important impact on the health of the coastal zone indirectly through its effects on *Mnemiopsis mccradyi*.

## Materials and methods

### Collection of specimens

Ctenophores were gently collected by dipping from the surface waters of Apalachicola Bay, Florida or Mobile Bay at Dauphin Island, Alabama throughout the year. They were sometimes examined in the field with a dissecting microscope for trichodinids. The amoebae described here were not reliably visible with a dissecting microscope under field conditions. Ctenophores were held in natural sea water in glass or plastic jars in coolers for transport back to the laboratory. Upon arrival in the lab, ctenophores were placed in jars containing sea water from the collection site, or a planktonkreisel apparatus with a recirculating sea water system. Ctenophores were fed *Artemia salina* nauplii (San Francisco Bay Brand, Inc., Newark, CA). All ctenophores were carefully selected for quality and only those in excellent physical condition as determined by overall inspection were used for subsequent observations. In some cases, we were able to perform close examination by light microscopy and additionally, perform fixation for scanning electron microscopy, within 1–2 h of capture. In all cases, specimens were prepared for light or electron microscopy within a day of capture.

### Light microscopy

Immobilized ctenophores were examined with a dissecting microscope capable of high power magnification and equipped with a video camera and time-lapse or conventional S-VHS video tape recorder. Higher magnification differential interference contrast (DIC) views of dissected comb plates and comb rows were generated and digitally-enhanced (=video enhanced DIC; VE-DIC) as described previously (Estes et al., 1997).

### Electron microscopy

Comb plates and associated microbes were dissected free of the ctenophore with fine iridectomy scissors. Specimens for scanning electron microscopy (SEM) were prepared and viewed as described by Estes et al. (1997).

Samples for transmission electron microscopy (TEM) were fixed as described in Estes et al. (1997), postfixed with 1% aqueous osmium tetroxide, rinsed once in 100 mM sodium cacodylate (pH 7.4) and *en bloc* stained in 1% aqueous uranyl acetate at 4 °

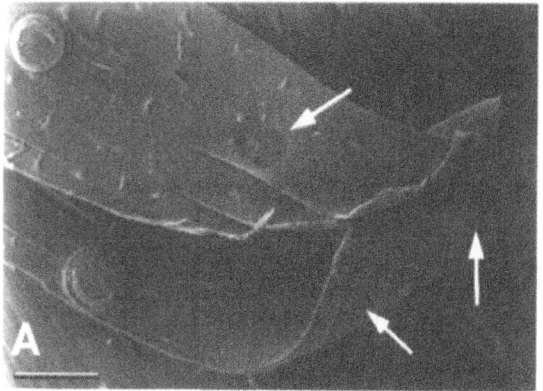

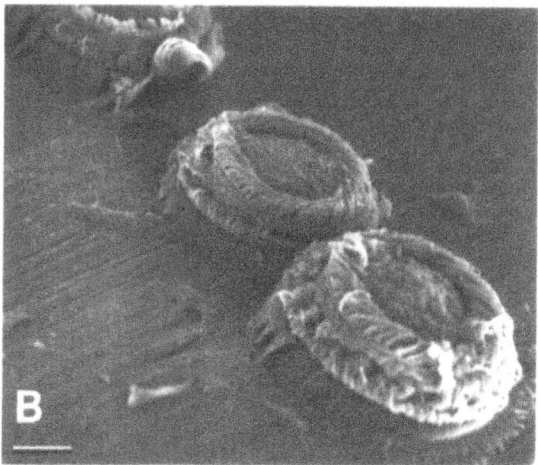

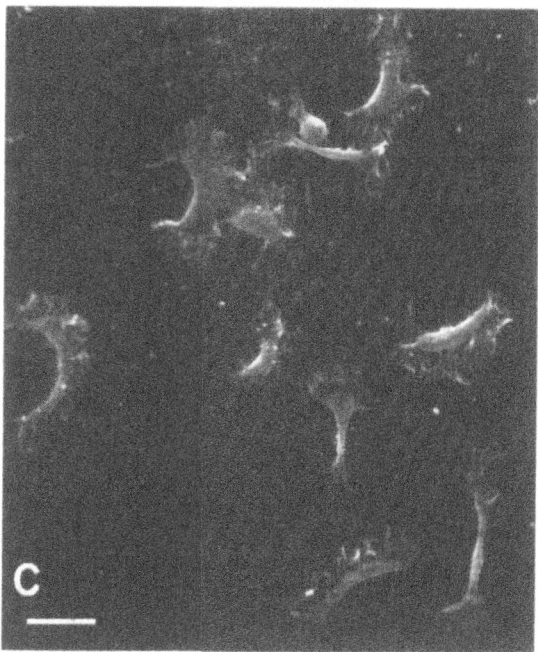

*Figure 1.*

C overnight. Stained specimens were washed briefly with distilled water, dehydrated as in Estes et al. (1997), subsequently infiltrated and flat-embedded in Durcupan ACM epoxy resin (Fluka Chemie AG, Buchs, Switzerland), which was polymerized at 60 ° C for 2–5 days. Seventy-nanometer thin sections were cut with a diamond knife. Thin sections were stained with 1% aqueous uranyl acetate and Reynold's lead citrate, and examined at 60 kV in a transmission electron microscope.

All measurements are presented as mean±standard error of the mean.

## Results

### Comb plate assemblage

Trichodina ctenophorii

We confirmed our previous observation (Estes et al., 1997) that *M. mccradyi* frequently bear *Trichodina ctenophorii* upon comb plates. We sometimes observed large numbers of *T. ctenophorii* attached at the time of collection, as determined by examination of freshly-collected ctenophores with a portable dissecting microscope. *T. ctenophorii* scrupulously attached only to the subsagittal, subtentacular and auricular plates and were not found on any other location. The trichodinids resided preferentially on the aboral side (Fig. 1). We examined video images of 193 subsagittal and subtentacular comb plates on 21 animals. On average 12.2±0.6 trichodinids were attached per comb plate. The maximum observed density was 55.7±1.2 mm$^{-2}$. Trichodinids were rare on ctenophores caught from November to April.

### Flabellula-*like gymnamoebae*

Examination of comb plates by VE-DIC and SEM revealed small amoebae attached to both the oral and aboral sides of the comb plates (Fig. 1A–C). Amoebae were found on subtentacular and subsagittal locomotory comb plates as well as the auricular comb

*Figure 1.* The assemblage of protozoan epibionts on comb plates of the ctenophore, *Mnemiopsis mccradyi*. Scanning electron microscopy. (**A**) The ciliate, *Trichodina ctenophorii*, and numerous flabellulid amoebae attached to comb plates. Arrows indicate dimples in the comb plates produced *T. ctenophorii* attached to the reverse side. Scale: 50 μm. (**B**) *T. ctenophorii* accompanied by several amoebae. Scale: 10 μm. (**C**) Flabellulid amoebae at higher magnification. The ends of the comb plate cilia show as a jagged pattern on the comb plate. Scale: 5 μm.

298

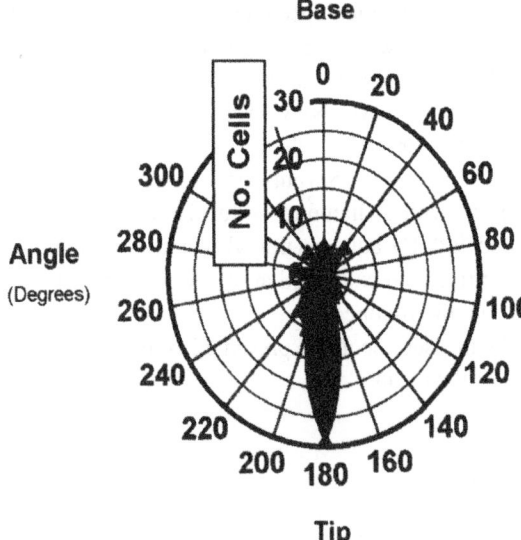

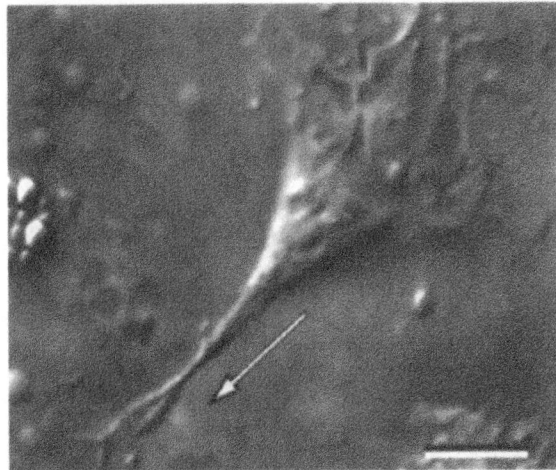

*Figure 3.* Actively crawling (arrow shows direction) ectodermal vexilliferid amoeba. VE-DIC Scale: 10 μm.

*Figure 2.* Polar plot of locomotory vectors of 100 flabellulid amoebae on the comb plate surface of the ctenophore, *Mnemiopsis mccradyi*, during a 30 s interval.

plates, which generate feeding currents. While occasionally there were larger individuals, the majority of comb plate amoebae had a maximum width of ~15 μm. Comb plate amoebae advanced smoothly, without eruptive cytoplasmic activity, typically displaying a broad anterior hyaline zone (Fig. 1C). Most locomotory cells were crescent-shaped with a distinctly blunt posterior and showed little evidence of a uroid (Fig. 1C). Free-floating forms were spherical with small, extremely active non-tapered pseudopodia (not shown). We tentatively identify comb plate gymnamoebae as a species of *Flabellula* based on information from previously published keys (Page, 1976, 1988). Amoebae were found over nearly the entire length of the comb plate but appeared to avoid the region of the most extreme basal bend.

We examined flabellulid movement on excised comb rows in the compound microscope, using VE-DIC combined with time-lapse video. Flabellulids crawled at ~10±0.3 μm min⁻¹ (N=100) along the length of the comb plate cilia. They crawled alternately distally and proximally, usually moving as a group. Over one 30 s interval, flabellulid amoebae were observed to move nearly *en masse* toward the comb plate tip (Fig. 2).

We determined the density of amoeboid infestation by examination of SEM images. Only amoebae on the aboral sides of the comb plates were counted because only those sides could be reliably observed. The number of amoebae varied greatly in our sample, but on average, plates bore 2726±395 mm⁻² (N=15 comb plates). The greatest number of amoebae occurred during the summer, when densities could exceed 5000 mm⁻². Comb plates bore amoebae year-round from the summer of 1993 to the summer of 1999.

### Ectodermal assemblage

#### Ectodermal vexilliferid amoebae

Time-lapse video of the interplate ectoderm revealed small but very different gymnamoebae. Ectodermal amoebae would slowly protrude one of several long, fingerlike subpseudopodia, which elongated and expanded while it advanced. The cell body would subsequently transfer into the newly attached and broadened pseudopodium. We recognize these amoebae as a genus very similar to *Vexillifera* based upon their distinctive locomotory behaviour and morphology (Fig. 3). We therefore refer to them as vexilliferids. Vexilliferids occurred at 20-fold lower density compared with the flabellulids at ~400 mm⁻². We were unable to detect the ecotodermal vexilliferid amoebae except by close study of low-magnification time-lapse video.

#### Protoodinid dinoflagellates

We observed large (30–100 μm diameter), distinctly green dinoflagellates attached to the ectodermal surface (Fig. 4A). The cells were easily identified as dinoflagellates based upon the presence of condensed chromosomes in specialized nuclei, which are termed dinokarya. The cingulum and sulcus, which are cell wall grooves that house the transverse and longitudinal flagella, respectively, were clearly visible (Fig.

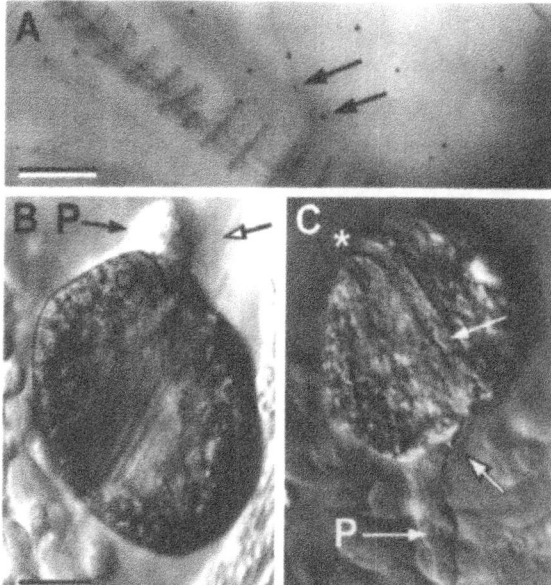

*Figure 4. Protoodinium*-like dinoflagellates on the ectoderm of the ctenophore, *Mnemiopsis mccradyi*. (**A**) Survey view. Arrows indicate two of 13 visible dinoflagellates. Scale: 1 mm. (**B, C.**) Dinoflagellates attached to the ctenophore interplate ectoderm. In B, open arrow indicates peduncular fimbrae. In C, open arrow indicates longitudinal flagellum, asterisk indicates cingulum, and white arrow indicates transverse flagellum. In both B and C, P refers to the peduncle. Scale: 20 μm.

4B, C). The transverse flagella, which were always present, were clearly motile, while the longitudinal flagella, when present, rarely beat in a rhythmic manner. Smaller cells appeared to be pelliculate, while larger cells did not appear to have obvious thecal plates in VE-DIC. The well-developed peduncle bore many fine processes that spread out from the peduncular stalk, and which we identify as distal rhizoids (Fig. 4B). Dinoflagellate overall morphology suggested that this was a mixotrophic dinoflagellate very similar to *Protoodinium chattoni* (Hovasse, 1935; Cachon & Cachon, 1971; Fensome et al., 1993).

The putative protoodinids were non-specific with regard to attachment site and could be found virtually anywhere on the surface of the host (Fig. 4A). Dinoflagellates occurred unpredictably on ctenophores throughout the year but were, like the rest of the epibionts, most common during the warm months.

*Effects of microbial attachment*

*T. ctenophorii* distorted comb plates by virtue of their attachment, particularly if attachment occured near the tip (cf. Fig. 1A), suggesting that they applied negative pressure. We observed displaced comb plate cilia

in scars left by previously attached trichodinids (not shown). Despite their ability to distort the plate and produce minor scars of the surface, trichodinids did not appear to have any effect on comb plate motility. We directly observed *T. ctenophorii* capture numerous small free-swimming flagellates with the adoral ciliature and ingest them (not shown). We conclude that *T. ctenophorii* does not feed upon comb plate cilia.

By contrast, comb plate flabellulids fed continually upon the comb plate cilia. This was evident while using high magnification, high resolution VE-DIC to examine flabellulids attached to comb plates. Phagocytic activity was routinely observed in actively crawling amoebae (not shown). TEM confirmed that flabellulids degrade and engulf comb plate ciliary components (Fig. 5A). Partially or entirely degraded ciliary membranes, outer doublets and/or central pair microtubules were evident beneath the amoebae (Fig. 5B), suggesting that the flabellulids utilize comb plates as a nutritional source.

Both SEM (not shown) and TEM showed that flabellulids can invade comb plates (Fig. 6A). Invading amoebae phagocytosed comb plate cilia and produced long-range disturbance of the normally precise ciliary packing (cf. Fig. 6A, B, D). The 3–8 doublet microtubule interciliary bridges (Afzelius, 1961; Tamm, 1982) were lost and the central pair became disoriented. The central pair midfilament, which is distinctive of ctenophore comb plates (Fig. 5B; Tamm & Tamm, 1981), remained intact (Fig. 6C). The ciliary membranes appeared to become detached from the 9+2 axonemal structure, and appeared 'loose'. Such effects suggest that the invading amoebae might release degradative enzymes.

Ectodermal vexilliferids were never observed to occur in high numbers, and we saw no evidence of any degradative effect on the ectoderm. Even though we closely examined ectodermal vexilliferids by high-resolution, high magnification VE-DIC, we never observed overt phagocytic activity, even though we were able to easily and closely observe fine details of amoeboid locomotory behaviour.

We did not not directly observe transport processes in the peduncle of the dinoflagellates. It is well known that protoodinids protrude an invasive amoeboid process through the peduncle (Hovasse, 1935). Ctenophores that bore many attached dinoflagellates displayed localized collapse of the mesogloea and ectoderm (not shown), suggesting that dinoflagellates feed upon the ctenophore.

300

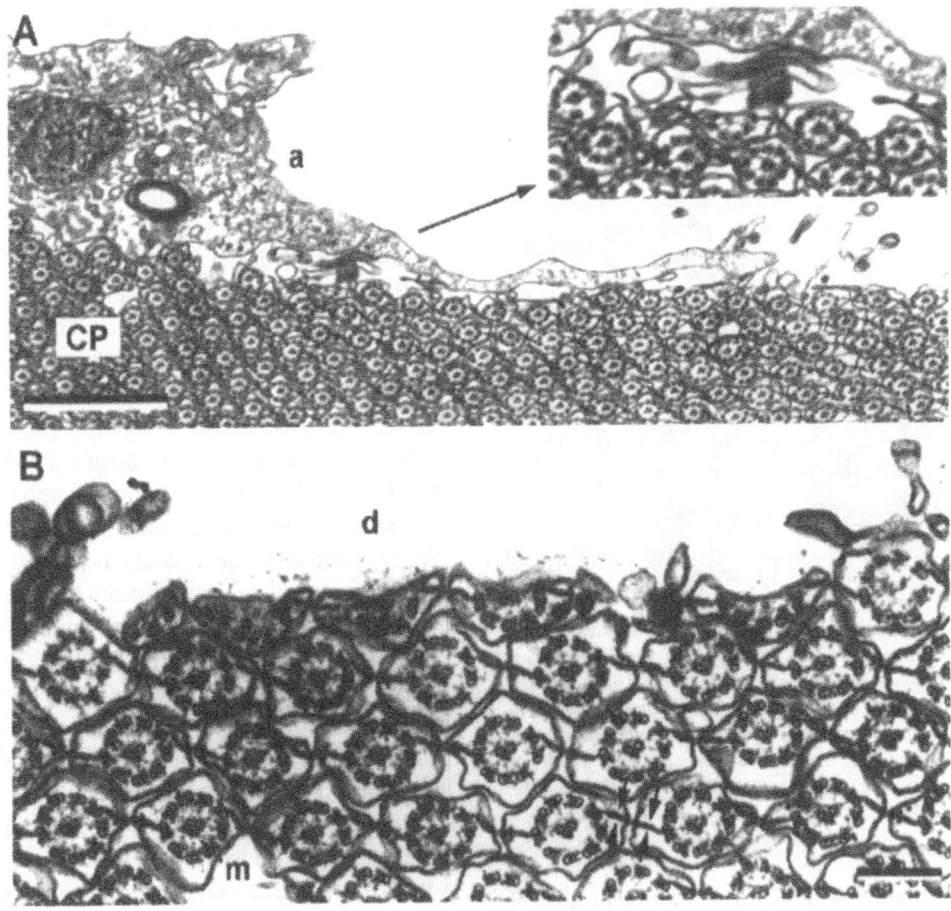

*Figure 5.* Flabellulid-dependent degradation of the comb plate of the ctenophore, *Mnemiopsis mccradyi*. (**A**) Flabellulid (a) crawling upon the comb plate (CP). Insert expands a region undergoing degradation; note loss of outer doublet microtubules in the comb plate cilia. Scale: 1 $\mu$m. (**B**) Degraded region (d) of the comb plate after passage of a flabellulid amoeba. Opposing arrows indicate the 3–8 outer doublet interciliary bridges. m indicates the central pair midfilament, a marker of ctenophore comb plate cilia. Scale: 0.25 $\mu$m.

## Discussion

### Specific, limited protistan assemblages inhabit healthy Mnemiopsis mccradyi

Ctenophores are hosts for a variety of parasites, including trematodes (Stunkard, 1980; Koie, 1991; Martorelli, 2001; Purcell & Arai, 2001), amphipods (Harbison et al., 1977), protists (Kinne, 1990) and anthozoa (Crowell, 1976). Most parasites are found in the mesoglea, gut or canals.

The protistan epibionts described here occupied distinctly nonoverlapping anatomical domains of *Mnemiopsis mccradyi*, a common coastal ctenophore of the Gulf of Mexico (Edmiston, 1979). Trichodinids and flabellulid amoebae occupied solely the comb plates, while vexilliferid amoebae and protoodinid dinoflagellates attached only to the ectoderm.

The trichodinids and flabellulids live in a unique niche, the giant ciliary comb plate. Comb plates are powerful water-propulsive paddles comprised of hundreds of thousands of very long cilia, and so it might seem unusual for organisms to select such an attachment location.

Barlow et al. (1993) showed that comb plates generate turbulent eddies during beating, due to the relatively high Reynold's number ($\sim$9 near the tip), which results in very effective water exchange near the comb plate surface. Vigorous water exchange would bring small suspended particles such as planktonic nanoflagellates in close proximity to the comb plate surface, where they could be preyed upon by *Trichodina ctenophorii*. We propose that *T. ctenophorii* and the flabellulids reside on the comb plates because of enhanced food availability.

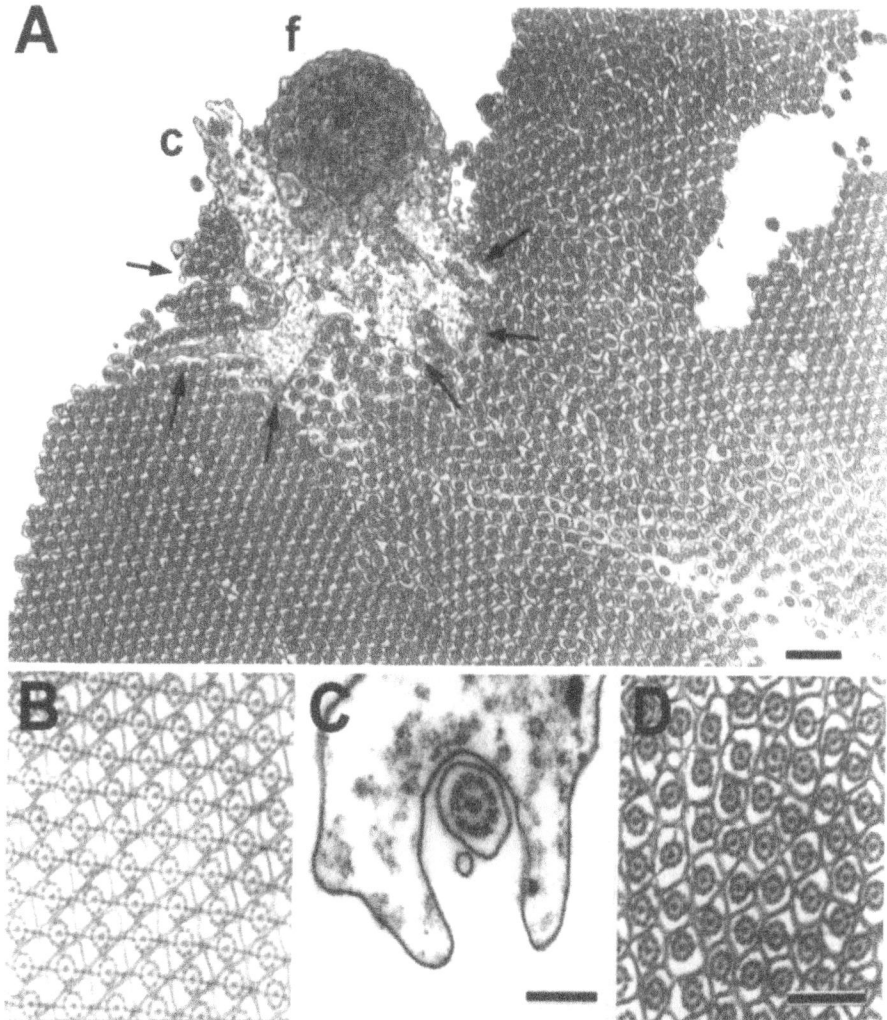

*Figure 6.* Invasion of a comb plate of the ctenophore, *Mnemiopsis mccradyi* by a flabellulid amoeba. (**A**) Survey view. Small arrows indicate invasive subpseudopodia. f: flabellulid; c, region expanded in C. Scale: 1.0 μm. (**B**) Normal comb plate organization, from lower left of A. Scale: 0.5 μm. (**C**) Phagocytosis of a comb plate cilium. Note loss of 3–8 bridges. Scale: 0.25 μm. (**D**) Disorganized comb plate structure, from right of the invading amoeba. Note partly detached ciliary membranes, disoriented ciliary 3–8 bridges and central pair complexes. Scale: 0.5 μm.

To date, the only other epicommensal protist associated with ctenophores has been described by Mills & McLean (1991) for ctenophores from Friday Harbor, Washington State, U.S.A. The sessile, stalked dinoflagellate *Oodinium* (probably *jordanii* sp.) was observed to attach near the comb rows of *Beroë abyssicola* Mortensen, *Bolinopsis infundibulium* (O. F. Müller), *Euplokamis dunlapae* Mills, *Dryodora glandiformis* (Mertens) and *Pleurobrachia bachei* L. Agassiz. *Oodinium* sp. also was seen on Friday Harbor cnidarians and chaetognaths. Although not specific with regard to host species preference, *Oodinium* sp. is scrupulously specific in its attachment to the ectoderm. We similarly observe *Protoodinium*-like

dinoflagellates on ectodermal locations on *M. mccradyi*. *Vexillifera*-like gymnamoebae also inhabit the ectoderm.

We tentatively identify the comb plate amoebae as flabellulids after Bovee & Sawyer (1979) and Page (1976). Although the comb plate amoebae resemble vannellids or platyamoebae, we did not observe any evidence of extracellular pentagonal glycostyles or the characteristic hexagonal arrangements of filamentous material in thin sections of the plasma membrane (Page, 1988). We feel that our fixation technique falls within the range of methods described by Page & Blakely (1979) to reveal extracellular glycostyles as described by Page (1986, 1987).

## Protist/host relationships

The precise relationship between any of these protists and the host is incompletely characterized and therefore uncertain. *Trichodina ctenophorii* may best be classified as a commensal, since it does not appear to have much adverse effect on the host although it clearly derives benefit by attachment to the comb plate. We suspect that it uses the comb plate as a feeding platform to capture flagellate prey under normal conditions since we have seen such behaviour under the compound microscope. Trichodinid attachment could lead to further consequences. Minor comb plate surface damage resulting from trichodinid attachment could provide inroads for flabellulid amoebae, which we show here to ingest comb plate cilia (Figs. 5 and 6), and which we, therefore, classify as at least a facultative, if not obligate, parasite. We have not observed any bacteria on the surface of the comb plates and it is possible that the flabellulids clean the surface of the comb plates by phagocytosing any bacteria they encounter, although we have never observed such behaviour.

The ectodermal vexilliferid amoebae could conceivably be mutualistic with the ctenophore. The ctenophore almost certainly provides an advantage to the attached vexilliferids, perhaps as a prey capture surface, much as we suggest for the comb plate-trichodinid relationship indicated above. We have not observed vexilliferids to ingest any part of the ctenophore, and the ctenophore ectoderm is remarkably clean of bacteria or other small eukaryotic epimicrobial colonists. The vexilliferid amoebae may in effect clean the ectoderm, if they normally ingest bacteria or other small food particles.

We propose the protoodinids to be mixotrophic parasites. All were deep green and so we presume that they were photosynthetically active, however, they almost certainly also receive nutrient support from the ctenophore host through an invasive amoeboid process that protrudes from the peduncle into the mesoglea, as described for *Protoodinium chattoni*-infested *Podocoryne* (Cachon & Cachon, 1971). Our observation of locally collapsed ectoderm and mesogloea under attached dinoflagellates supports this hypothesis.

## Importance for the coastal marine environment

Trichodinids are typically fish parasites (Lom & Hoffman, 1964; Wellborn, 1967; Lom & Haldar, 1976; Arthur & Lom, 1984; Van As & Basson, 1996). Intense trichodinid infestations can cause fish kills, and outbreaks are common in areas with high fish densities and sluggish water flow (Arthur & Margolis, 1984). We have not yet attempted to elucidate the host range of *Trichodina ctenophorii*. However, we have observed predation of ctenophores by fish, as mentioned in Estes et al. (1997) and reported elsewhere (Purcell & Arai, 2001). It is possible, although we think unlikely, that *T. ctenophorii* could be transferred to and subsequently infest a fish host. Our observations show that *T. ctenophorii* is very specific in its recognition and attachment to *Mnemiopsis mccradyi* comb plates. The specificity of attachment is underscored by our observations that *Beroë ovata* Brugière, a ctenophore predator that feeds upon *Mnemiopsis* spp. (Swanberg, 1974), rarely bears *T. ctenophorii* (in Estes et al., 1997).

It is unclear whether mnemiopsid epimicrobial populations are related to the environmental conditions. Elevated marine hydrocarbons have been demonstrated to exacerbate piscine trichodinid infestations (Khan, 1990), as do dioxin-enriched paper mill effluents (Khan et al., 1993). The Alabama coast is the site of many gas drilling platforms, and several pulp mills operate in the Mobile River and Apalachicola River watershed that drain into the coastal Gulf region covered in this study; in addition, coastal Gulf of Mexico experiences very heavy shipping and recreational boat traffic. *Trichodina ctenophorii* populations could be stimulated by the presence of these sources of pollution. Additionally, all of the protists could be stimulated or nutritionally supported either directly or indirectly (via increased marine microbial populations in general) by the generally elevated nutrient loading associated with human habitation.

Coastal salinity during intense blooms of *Mnemiopsis mccradyi* is typically no greater than 29 ppt. Usually salinity at our Apalachicola Bay/St. George Sound collection site is between 21 and 29 ppt. By contrast, salinity at the Mobile site can be as low as 3 ppt and yet still supports heavy blooms of ctenophores. During the summer of 2000, the U. S. Southeast experienced one of the most severe droughts on record, resulting in reduced riverine runoff into the Northern Gulf. We directly observed that salinity at our Apalachicola and Mobile Bay sites was routinely above 35 ppt and only a few *Mnemiopsis mccradyi* and *Beroë ovata* were collected. Examination of the comb plates of these animals by SEM revealed no amoebae or trichodinids, and dinoflagellates also were not observed. Because of the isolated nature of these observations we cannot draw any strong inferences, however, it is quite possible that protistan infestations are supported by low to moderate coastal salinity. We cannot distinguish this from the possibility that lower salinity conditions could lead to intense ctenophore blooms, which in turn could lead to high epibiont infestation rates.

The impact of the protistan epibionts on coastal ctenophore populations is equally uncertain. As noted above, ctenophore populations can become very dense in the Gulf (Edmiston, 1979) but the same is true for locations elsewhere (e.g., Purcell et al., 2001), and one of us (AGM) has routinely collected *Mnemiopsis leidyi* in the Woods Hole harbor at extraordinary concentrations. Despite many years of collections at sites along the east coast by the same author, such populations of ctenophore epimicrobes have not been observed. Therefore, the Gulf population of epimicrobial organisms may be confined to the Gulf region.

Ctenophores that were heavily infested with *T. ctenophorii* rarely appeared to be in poor health, as indicated by malformation of the body, however, comb plates that are heavily infested with flabellulid amoebae sometimes appear quite ragged, which we interpret to result from amoeboid phagocytic activity. If amoebae actively destroy the comb plates, then they might have a severely detrimental effect on the infested individual's ability to capture food and escape predation (Kreps et al., 1997). Similarly, if parasitic oodinids become very dense we would expect them to have a negative impact on fecundity in this animal.

*Mnemiopsis leidyi* has had severe impacts on the food webs of U. S. estuaries and the Black Sea, (Shushkina & Musayeva, 1990; Purcell et al., 2001), especially on fish stocks and associated food sources. Reduced numbers of *Mnemiopsis* spp. would reduce the predation pressure on copepod populations, which in turn would increase the cropping effect of copepods on the phytoplankton. Reduced predation due to epibionts on predators such as *Beroë* (Finenko et al., 2001; Shiganova et al., 2001) on copepods could have significant stimulatory impact on zooplanktivorous fish that directly depend on crustacean zooplankton. Alternatively, that could allow undesirable growth of stinging jellyfish. *Mnemiopsis* spp. population are dramatically reduced by predation by *Beroë* spp. (Shiganova et al., 2001). Thus, circumstances that have significant impacts on the ctenophore population could have great, and possibly difficult to predict, impacts on coastal health. We plan to examine the spaciotemporal occurrence of the ctenophore epibiotic protists in an effort to elucidate the effects, if any, that these organisms might have on coastal *Mnemiopsis* populations.

## Acknowledgements

C. Sundermann, B. Estridge; R. Dute, M. Miller and M. Toivio-Kinnucan provided expert insight and aid with ultrastructural analysis. E. Bovee, R. O. Anderson and A. Rogerson generously provided guidance regarding gymnamoeba identification; similarly D. W. Coats provided indispensible insight with the dinoflagellate identification. Two anonymous reviewers aided greatly by their insightful remarks. R. Henry, C. Bailey, A. Hitt, S. Kempf, S. Smith and T. Wakefield aided with animal collection. L. Edmiston of the Apalachicola National Marine Sanctuary provided near-shore laboratory facilities to study freshly-caught animals. Supported by the Auburn Univ. Howard Hughes Future Life Sciences Scholars Program, the NIH and the Auburn University Office of the Vice President for Research.

## References

Afzelius, B., 1961. The fine structure of the cilia from ctenophore swimming plates. J. Biophys. Biochem. Cytol. 9: 383–394.

Arthur, J. R. & J. Lom, 1984. Trichodinid Protozoa (Ciliophora: Peritrichida) from freshwater fishes of Rybinsk Reservior, USSR. Protozoology 31: 82–91.

Arthur, J. R. & L. Margolis, 1984. *Trichodina truttae* Müller, 1937 (Ciliophora: Peritrichida), a common pathogenic ectoparasite of cultured juvenile salmonid fishes in British Columbia: Redescription and examination by scanning electron microscopy. Can J. Zool. 62: 1842–1848.

Barlow, D., M. A. Sleigh & R. J. White, 1993. Water flows around the comb plates of the ctenophore *Pleurobrachia* plotted by computer: a model system for studying propulsion by antiplectic metachronism. J. exp. Biol. 177: 113–128.

Bovee, E. C. & T. K. Sawyer, 1979. Marine flora and fauna of the Northeastern United States. Protozoa: Sarcodina: Amoebae. NOAA Technical Report. National Marine Fisheries Service Circular 419. 1–56.

Butler, H. & A. Rogerson, 1995. Temporal and spacial abundance of naked amoebae (Gymnamoebae) in marine benthic sediment of the Clyde Sea area, Scotland. J. Euk. Microbiol. 42: 724–730.

Butler, H. & A. Rogerson, 1997. Consumption rates of six species of marine benthic naked amoebae (Gymnamoebia) from sediments in the Clyde Sea area. J. mar. biol. Assoc. U. K. 77: 989–997.

Cachon, J. & M. Cachon, 1971. *Protoodinium chattoni* Hovasse. Manifestations ultrastructurales des repports entre le Péridinien et la Médusa-hôte: fixation, phagocytos. Arch. Protistenk. 113: 293–305.

Crowell, S., 1976. An edwardsiid larva parasitic in *Mnemiopsis*. In Mackie, G. O. (ed.), Coelenterate Ecology and Behaviour, Plenum Publishing Corp., New York: 247–250.

304

Estes, A. M., B. S. Reynolds & A. G. Moss, 1997. *Trichodina ctenophorii* n. sp., a novel symbiont of ctenophores of the northern coast of the Gulf of Mexico. J. Euk. Microbiol. 44: 420–426.

Edmiston, L., 1979. The Zooplankton of the Apalachicola Bay System. Dissertation, Florida State University, Tallahassee, Florida: 104 pp.

Fensome, R. A., F. J. R. Taylor, G. Norris, W. A. S. Sarjeant, D. I. Wharton & G. L. Williams, 1993. A classification of living and fossil dinoflagellates. Micropaleontol. Press spec. Publ. No. 7: 351 pp.

Finenko, G. A., B. E. Anninsky, A. A. Romanova, G. I. Abolmasova & A. E. Kideys, 2001. Chemical composition, respiration and feeding rates of the new alien ctenophore, *Beroe ovata*, in the Black Sea. Hydrobiologia 451 (Dev. Hydrobiol. 155): 177–186.

Harbison, G. R. & L. P. Madin, 1979. Diving – A new view of plankton biology. Oceanus 22: 21–27.

Harbison, G. R., D. C. Biggs & L. P. Madin, 1977. The associations of Amphipoda Hyperiidea with gelatinous zooplankton – II. Associations with cnidarians, ctenophora and radiolaria. Deep Sea Res. 24: 465–485.

Hovasse, R., 1935. Deux Péridiniens parasites convergents: *Oodinium poucheti* (Lemm.), *Protoodiium chattoni*. Gen. Nov. sp. Nov. Bull. Biol. Fr. Belg. 69: 59–86.

James-Clark, H., 1866. On the anatomy and physiology of the vorticellidan parasite (*Trichodina pediculus*, Ehr) of *Hydra*. Ann. Mag. N. Hist. 17: 401–425.

Khan, R. A., 1990. Parasitism in marine fish after chronic exposure to petroleum hydrocarbons in the laboratory and to the Exxon Valdez oil spill. Bull. envir. Contam. Toxicol. 44: 759–763.

Khan, R. A., D. E. Barker, K. Williams-Ryan & R. G. Hooper, 1993. Influence of crude oil and pulp and paper mill effluent on mixed infections of *Trichodina cottidarium* and *T. saintjohnsi* (Ciliophora) parasitizing *Myoxiocephalus octodecemspinosus* and *M. scorpius*. Can. J. Zool. 72: 247–251.

Kinne, 0., 1990. Diseases of Marine Animals. Vol. III, Biologische Anstalt Helgoland, Hamburg, Germany: 450 pp.

Koie, M., 1991. Aspects of the morphology and life cycle of *Lecithocladium excisum* (Digenia: *Hemiuridae*) – a parasite of *Scomber*. Int. J. Parasitol. 21: 597–602.

Kreps, T. A., J. E. Purcell & K. B. Heidelberg, 1997. Escape of the ctenophore *Mnemiopsis leidyi* from the scyphomedusa predator *Chrysaora quinquecirrha*. Mar. Biol. 128: 441–446.

Lom, J., 1958. A contribution to the systematics and morphology of endoparasitic trichodinids from amphibians, with a proposal of uniform specific characteristics. J. Protozool. 5: 251–263.

Lom, J., 1970. Observations on trichodinid ciliates from freshwater fishes. Arch. Protistenk. Bd. 112: 153–177.

Lom, J. & D. P. Haldar, 1976. Observations on trichodinids endocommensal in fishes. Trans. am. Micro. Soc. 95: 527–541.

Lom, J. & G. L. Hoffman, 1964. Geographic distribution of some species of trichodinids (Ciliata: Peritricha) parasitic on fishes. J. Parasitol. 50: 30–35.

Martorelli, S. R., 2001. Digenea parasites of jellyfishes and ctenophores of the southern Atlantic. Hydrobiologia, this volume.

Mills, C. E. & N. McLean, 1991. Ectoparasitism by a dinoflagellate (*Dinoflagellata*: *Oodinidae*) on 5 ctenophores and a hydromedusae (Cnidaria). Dis. aquat. Org. 10: 211–216.

Page, F. C., 1976. An illustrated key to freshwater and soil amoebae. Freshwater Biological Association, Ambleside, Cumbria: 155 pp.

Page, F. C., 1986. The genera and possible relationships of the family amoebidae, with special attention to comparative ultrastructure. Protistologica 22: 301–316.

Page, F. C., 1987. The classification of 'naked' amoebae (Phylum Rhizopoda). Arch. Protistenk. 133: 199–217.

Page, F. C., 1988. A New Key to Freshwater and Soil Gymnamoebae. Freshwater Biological Association, Ambleside, Cumbria, U.K.: 122 pp.

Page, F. C. & S. M. Blakely, 1979. Cell surface structures: a taxonomic character in the Thecamoebae (Protozoa: Gymnamoebia). Zool. J. linn. Soc. 66: 113–135.

Poynton, S. L. & J. Lom, 1989. Some ectoparastic trichodinids from Atlantic cod, *Gadus morhua* L., with a description of *Trichodina cooperi* n. sp. Can. J. Zool. 67: 1793–1800.

Purcell, J. E. & M. N. Arai, 2001. Interactions of pelagic cnidarians and ctenophores with fish: a review. Hydrobiologia 451 (Dev. Hydrobiol. 155): 27–44.

Purcell, J. E., T. A. Shiganova, M. B. Decker & E. D. Houde, 2001. The ctenophore *Mnemiopsis* in native and exotic habitats: U.S. estuaries *versus* the Black Sea basin. Hydrobiologia 451 (Dev. Hydrobiol. 155): 145–175.

Reeve, M. R., M. A. Walter & T. Ikeda, 1978. Laboratory studies of ingestion and food utilization in lobate and tentaculate ctenophores. Limnol. Oceanogr. 23: 740–751.

Shushkina, E. A. & E. I. Musayeva, 1990. Structure of planktonic community of the Black Sea epipelagic zone and its variation caused by invasion of a new ctenophore species. Oceanology 30: 225–228.

Shiganova, T. A., Yu. V. Bulgakova, S. P. Volovik, Z. A. Mirzoyan & S. I Dudkin, 2001. The new invader *Beroe ovata* Mayer, 1912 and its effect on the ecosystem in the northeastern Black Sea. Hydrobiologia 451 (Dev. Hydrobiol. 155): 187–197.

Sieburth, J. M., 1984. Protozoan bacterivory in pelagic marine waters. In Hobbie, J. E. & P. J. le B. Williams (eds), Heterotrophic Activity in the Sea, Plenum Publishing Corp., N.Y.: 569 pp.

Stunkard, H. W., 1980. The morphology, life cycle, and taxonomic relations *Lepocreadium areolatum* (Trematoda: Digenea). Biol. Bull. 158: 154–163.

Swanberg, N., 1974. The feeding behaviour of *Beroë ovata*. Mar. Biol. 24: 69–76.

Tamm, S. L., 1982. Ctenophores. In Shelton, G. A. B. (ed.), Electrical Conduction and Behaviour in 'Simple' Invertebrates. Oxford University Press, London: 261–358.

Tamm, S. L. & S. Tamm, 1981. Ciliary reversal without rotation of axonemal structures in ctenophore comb plates. J. Cell Biol. 89: 495–509.

Urawa, S. & T. Awakura, 1994. Protozoan diseases of freshwater fishes in Hokkaido. Sci. Rep. Hokkaido Fish Hatchery 48: 47–58.

Uzmann, J. R. & A. P. Stickney, 1954. *Trichodina myicola* n. sp., a peritrichous ciliate from the marine bivalve *Mya arenaria* L. J. Protozool. 1: 149–155.

Van As, J. G. & L. Basson, 1996. An endosymbiotic trichodinid, *Trichodina rhinobatae* sp. n. (Ciliophora: Peritrichia) found in the lesser guitarfish, *Rhinobatos annulatus* Smith, 1841 (Rajiformes: Rhinobatidae) from the South African Coast. Acta Protozool. 35: 61–67.

Wellborn, T. L., 1967. *Trichodina* (Ciliata: Urceolariidae) of freshwater fishes of the southeastern United States. J. Protozool. 14: 399–412.

Yip, S. Y., 1984. Parasites of *Pleurobrachia pileus* Müller (Ctenophora) from Galway Bay, Western Ireland. J. Plankton Res. 6: 107–122.

*Hydrobiologia* **451:** 305–310, 2001.
© 2001 *Kluwer Academic Publishers.*

# Digenea parasites of jellyfish and ctenophores of the southern Atlantic

Sergio R. Martorelli
*CEPAVE- (CONICET), Centro de Estudios Parasitologicos y Vectores 2 Nro. 584 (1900) La Plata, Argentina*
*E-mail martorelli@usa.net*

*Key words:* Argentine Sea, Digenea, metacercariae, zooplankton

## Abstract

Parasites of planktonic cnidarians and ctenophores in the southern Atlantic Ocean are little known. The aim of this study was to describe three new metacercariae from jellyfish and ctenophores, and assess the importance of the gelatinous zooplankton as intermediate hosts in the life history of digeneans. During examination of zooplankton in Argentine Sea for digeneans that mature in fishes, two species of jellyfish (*Phialidium* sp. and *Liriope tetraphylla* Chamiso & Eysenhardt, 1821), and one ctenophore (*Mnemiopsis mccradyi* Mayer, 1900) were analyzed for parasites. The samples were obtained in Mar del Plata. Three metacercariae belonging to Faustulidae, Lepocreadiidae and Hemiuridae are described. The prevalence (percent of hosts infected) varied from 1.4–30% and the range of intensity (number of individuals of a parasite species in a single infected host) was from 1 to 30 for the different metacercariae. Given the important position of free-swimming cnidarians and ctenophores in the marine food web, and the great number of fishes that have been found with these organisms in their digestive tracts, their importance in the life histories of digeneans should not be underrated.

## Introduction

Digeneans are abundant parasites of vertebrates and invertebrates in the marine environment. They usually have (with exceptions) three host life histories, with two intermediate hosts and one definitive host, a vertebrate that harbors the adults. The metacercaria infects the second intermediate host and is an important resting stage that allows the parasite to survive until it can reach the definitive host. The gelatinous zooplankton could act as hosts of different metacercariae.

Numerous metacercariae have been reported, mainly in the North Hemisphere. Lebour (1916) reported the metacercaria of *Opechona bacillaris* (Molin, 1859) in *Cosmetira pilosella* Fobres, 1898, *Obelia* sp., *Leuckartiara octona* (Fleming, 1823), and *Pleurobrachia pileus* (Fabricius, 1780). Dollfus (1963) reported all the well-known metacercariae at that time, and stated that it was unknown if jellyfish are intermediate hosts for digeneans or only are accidental hosts, since the metacercariae were always unencysted. In a revision of digeneans found in planktonic organisms,

Rebec (1965) concluded that in some cases, planktonic invertebrates could be occasional or facultative hosts for digeneans. Stunkard (1967, 1969, 1974, 1978a, b, 1980a, b) described different unencysted metacercariae in jellyfish and ctenophores and considered these as paratenic or reservoir hosts. For *Neopechona pyriforme* (Linton, 1900), however, the same author mentioned that metacercariae from jellyfish were immediately infective, and in this case, the jellyfish could be true intermediate hosts. Reimer et al. (1971) reported three metacercariae in plankton samples of the North Sea, and mentioned for the metacercaria of *O. bacillaris* that all the planktonic host found could be considered as second intermediate hosts. Køie (1975, 1979, 1983, 1985a, b) reported several life histories of digeneans that use jellyfish and ctenophores as second intermediate hosts. This author also mentioned that these invertebrates are true intermediate hosts for *O. bacillaris* and *Lepidapedon rachion* (Cobbold, 1858), (Lepocreadiidae). In this parasite, the cercariae actively penetrated the mesoglea of the jellyfish. The same author also mentions that hemiurid metacercaria of *Derogenes varicus* (Müller, 1784), *Brachiphalus crenatus* Rudolphi, 1802, and *Hemiurus communis* Odnher, 1905, which use copepods as intermediate hosts, also could use ctenophores and chaetognaths as transport hosts. Køie (op. cit.) also comments that metacercariae found in jellyfish and ctenophores did not need to encyst, to gain protection against the host's response. Lauchner (1980a, b) compiled all known parasites from jellyfish and ctenophores, and explained that these invertebrates served as intermediate or paratenic host for helminthes. More recently, Matsumoto et al. (1997) reported high prevalence (49%) and intensity (up to 100) for an unencysted metacercaria hemiuridae in the manubrium of the deep-sea trachymedusa *Benthocodon pedunculata* (Bigelow, 1913) in California.

Digenean parasites of planktonic organisms in the Southern Atlantic Ocean are still little known. Only three metacercariae have been reported in the Argentine Sea: *Opechona* sp. in *Olindias sambaquiensis* Delle Chiaje, 1841 (in Martorelli, 1991); *Monaschus filiformis* (Rudolphi, 1819) in *Phialidium* sp., *Liriope tetraphylla* Chamizo & Eysenhardt, 1821, *Eucheilota ventricularis* McCrady, 1857 and *Aglauropsis kawari* Moreira, 1972 (in Girola, et al., 1992; Martorelli & Cremonte, 1998); and an encysted lepocreadioid metacercaria in *Phialidium* sp., *L. tetraphylla* and *Mnemiopsis mccradyi* (in Martorelli, 1996).

During examination of gelatinous zooplankton for digeneans that mature in fishes, three unidentified metacercariae were found in ctenophores and jellyfish. The specific goals of the present research were to identify and describe these metacercariae, to analyze the prevalence and intensity, and to evaluate the importance of the gelatinous plankton as intermediate host for digeneans.

## Materials and methods

Jellyfish and ctenophore samples were collected in Mar del Plata port (38° 00′ S and 57° 30′ W), Buenos Aires, Argentina. One hundred specimens of *Mnemiopsis mccradyi*, 70 of *Phialidium* sp. and 20 *Liriope tetraphylla* were collected by plankton net between 1996 and 1998, transferred to containers with marine water, transported to the laboratory, and observed alive under a binocular microscope for parasites. Some infected jellyfish were fixed in 5% formalin in sea water. Parasites were observed *'in vivo'* in the host jelly. Prevalence (percent of hosts infected), and intensity

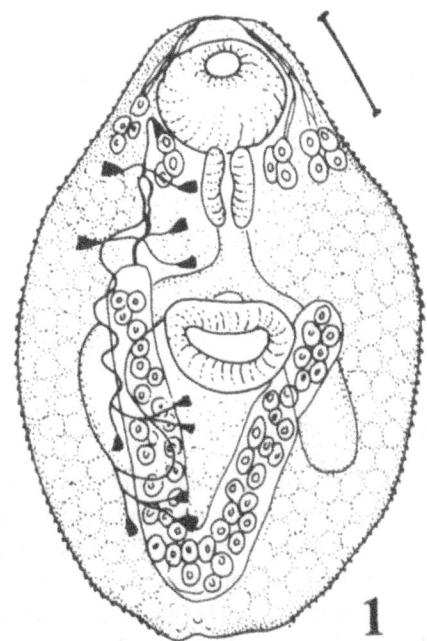

*Figure 1. Bacciger* sp. metacercaria ventral view. scale bar 60 μm.

(number of individuals of a parasite species in a single infected host) was defined according to Margolis et al. (1982) and Bush et al. (1997).

Live metacercariae used for morphological studies and identification were removed from the bodies of the hosts with needles. The metacercariae were fixed in AFA (alcohol-formalin-acetic acid), stained in hydrochloric carmine, cleared in creosote and mounted in natural Canada balsam. All dimensions are in micrometers, and means reported followed by the range in parentheses. The material studied was deposited in the Invertebrate Collections (Helminths) of La Plata Natural Sciences Museum, La Plata, Argentina.

## Results

In this study, three different unencysted metacercariae belonging to Faustulidae, Lepocreadiidae and Hemiuridae families were found.

### Metacercaria of *Bacciger* sp. (Fig. 1)

*Description:* (Based on 10 mounted and several living specimens, measurements taken from 10 mounted specimens): Body small, spinose 179 (175–191) long by 118 (95–134) wide, at acetabulum level. Oral sucker subterminal 35 (27–40) long by 36 (34–39) wide. Seven pairs of cephalic glands in four groups. Ventral sucker equatorial 33 (27–39) long by 38 (32–42) wide. Sucker width ratio 1: 0.94. Pharynx 25 (16–28) long; esophagus short; intestinal ceca reaching the posterior end of acetabulum. Excretory vesicle V-shape with arms extending to cecal bifurcations, filled with rounded concretions. Flame cell formula 2[(3+3)+(3+3)]=24.

*Comments:* Bray (1988) remarked that the presence of the spinous tegument is the principal difference between Lintoninae and Bacigerinae. According to these criteria, the metacercariae studied here, belong to the latter subfamily because they have a spinous tegument. This subfamily was recently included in the newly named family Faustulidae (Hall et al., 1999). The general morphology of the new metacercaria resembles the metacercariae of *Bacciger bacciger* (Rudolphi, 1819) and *Pseudobacciger harengula* (Yamaguti, 1938), reported by Palombi (1934) and Kim & Chun (1984), respectively, and probably could be included in this genus. In the same area where the infected jellyfish and ctenophores were collected, a *Bacciger*-like cercaria was reported released from the clam, *Tagelus plebeius*, by Cremonte (1999). At present, only two life histories have been reported for the genus *Bacciger,* by Palombi (1934) and Kim & Chun (1984), and both species used crustaceans as the second intermediate hosts. In spite of that, Wardle (1983) reported, without descriptions, the presence of Fellodistomid metacercariae closely related to *Bacciger* in *Beroe ovata* and *Mnemiopsis mccradyi* from Galveston Bay in the Gulf of Mexico.

### Taxonomic summary

*Hosts: Phialidium* sp. and *Mnemiopsis mccradyi.*

*Location:* Mesoglea near the circular canal in jellyfish and infundibulum in ctenophores.

*Prevalence:* 13% *(Phialidium* sp.), 2% *(M. mccradyi).*

*Intensity:* 1–13 *(Phialidium* sp.), 1–2 *(M. mccradyi).*

### Metacercaria of *Opechona* sp. (Fig. 2)

*Description:* (Based on 10 mounted specimens): Body oval, spinose 464 (397–552) long by 151 (136–161) wide at acetabular level. Oral sucker subterminal 71 (68–77) long by 62 (59–68) wide, with 5 ventral

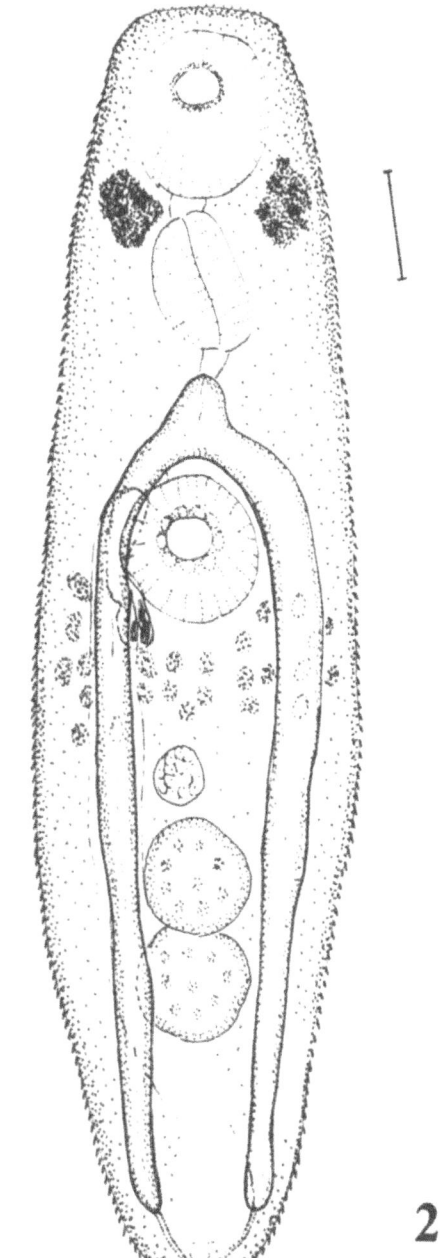

Figure 2. *Opechona* sp. metacercaria, ventral view scale bar 60 μm.

papillae. Acetabulum 58 (52–61) long by 57 (50–64) wide, with nine papillae near the middle. Prepharynx small, only visible in compressed specimens; pharynx 57 (59–64) long by 37 (32–45) wide; esophagus smaller than the pseudo-esophagus; intestinal ceca reaching the posterior end of body. Uroproct present. Eyespot pigment present, in some specimens dispersed in the forebody. Excretory formula not observed; excretory vesicle I-shaped, extending to the posterior end of acetabulum, and containing many concretions. Primordia of the testes, ovary, seminal vesicle and vitelline glands were present.

*Comments:* The morphology of these metacercariae resembled that of lepocreadiid digeneans. This group has been reported encysted in polychaetes, gastropods, bivalves, echinoids and fishes, and unencysted in medusae, ctenophores and molluscs (Martin, 1945; Dollfus, 1963; Stunkard, 1969; Køie, 1975; Lauckner, 1980a, b; Bray, 1988). The new metacercaria is close to metacercaria of *Opechona* sp. reported for the same area, in the jellyfish, *Olindias sambaquiensis*, and related to a cercaria released by *Buccinanops moliniferus* (Valenciennes, 1834) (Gastropoda, Nassariidae) (in Martorelli, 1991). The described metacercaria is similar in morphology to this cercaria. Usually lepocradiid metacercariae reported in ctenophores and medusae have been described unencysted in the host mesoglea and are very similar to the cercaria stage. At present, only one lepocreadiid metacercaria has been reported encysted in jellyfish and comb jellies by Martorelli (1996). In this metacercaria, the cyst was partly attached to the host mesoglea and was probably produced by the parasite.

In the same area where the new metacercaria was found, Cremonte & Sardella (1997) reported adults of *Opechona* sp. in *Scomber japonicus* Houttuyn, 1782. These authors assigned the high prevalence found to a greater ingestion of gelatinous zooplankton.

**Taxonomic summary**

*Hosts: Phialidium* sp., *Liriope tetraphylla* and *Mnemiopsis maccradyi*

*Location:* Mesoglea near the circular canal in jellyfish and infundibulum in ctenophores.

*Prevalence:* 17% (*Phialidium* sp.), 5% (*L. tetraphylla*), 30% (*M. mccradyi*).

*Intensity:* 1–7 (*Phialidium* sp.), 1–2 (*L. terraphylla*), 2–20 (*M. mccradyi*).

**Metacercaria Hemiuridae (Fig. 3)**

*Description:* (Based on 5 mounted specimens): Body long without ecsoma, and smooth surface, 411 (319–503) long by 113 (79–131) wide. Pharinx 28 (22–33) long; short esophagus, and intestinal ceca reaching the posterior end of body. Oral sucker 46 (37–57) by 50 (37–59). Ventral sucker very large, 96

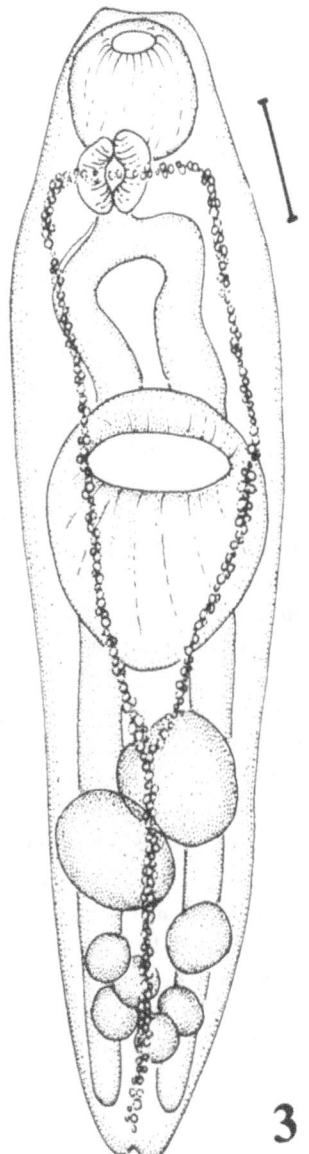

*Figure 3.* Hemiuridae metacercariae, ventral view scale bar 60 μm.

(79–103) by 90 (63–98). Excretory vesicle Y-shape with arms joined at pharyngeal level. Primordia of the testes, ovaries and vitelline glands were present.

*Comments:* The morphology of these metacercariae resembled that in the Hemiuridae family. Presently, several metacercariae of this family have been reported in pelagic cnidarians and ctenophores: *Hemiurus communis* in *Pleurobranchia pileus* by Yip (1984); *Brachyphallus crenatus* Rudolphi, 1802 in *P. pileus* by Køie (1983); *Aponurus* sp. in *Aequorea pensilis* (Haeckel) by Rao (1958); and *Lecithocladium excisum* Rudolphi, 1819 and *Derogenes varicus* in *P. pileus* by Køie (1983). By the absence of the ecsoma, the present metacercaria differs from those of genera *Hemiurus*, *Brachyphallus* and *Lecithocladium*, and probably could be included in the genus *Derogenes* or *Aponurus*. *Derogenes varicus* has a wide distribution and was found in almost 100 species of marine fishes (Marcogliese & Price, 1997). The metacercariae of *D. varicus* have been reported in many zooplankton species (Marcogliese 1995), and as progenetic metacercariae in *P. pileus* (in Køie, 1979; Lauckner 1980a). Adults of *D. varicus* also were reported in fishes from the Argentine Sea (Gaevskaya et al., 1985; Sardella & Timi, 1996). Experimetal studies will be necessary for the definitive classification of this metacercaria.

## Taxonomic summary

*Hosts: Phialidium* sp. and *Mnemiopsis mccradyi.*

*Location:* Mesoglea near the circular canal in jellyfish and infundibulum in ctenophores.

*Prevalence:* 1.4% *(Phialidium* sp.), 2% *(M. mccradyi).*

*Intensity:* 1–2 *(Phialidium* sp. and *M. mccradyi).*

## Conclusions

This is the first report of hemiurid and faustulid metacercariae in gelatinous organisms from the South Atlantic Ocean.

The presence of metacercariae in jellyfish and ctenophores has been reported in different oceans of the world and is probably an indirect indication of the use of these organisms as food by the fishes in which the adult digeneans are present. Works like Arai (1988), Ates (1988) and Purcell & Arai (2001) reinforce the importance of the gelatinous zooplankton in the pelagic food webs. In the same area as the present study, Mianzan et al. (1996) found 20 species of fishes feeding on ctenophores and gelatinous zooplankton. According to Lauchner (1980a), the importance of pelagic cnidarians in marine food webs is reflected in their use as intermediate or paratenic hosts by many helminthes. As hosts in the life cycles of marine digeneans, gelatinous zooplankton could play an important role in the distribution of the metacercaria stage.

In general, the prevalence of digeneans in zooplankton species is very low, and this fact contrasts with the high percentages of infections by digeneans found in fishes (Marcogliese 1995). The relatively heavy prevalence found in the present study for the metacercariae of *Bacciger* sp. and *Opechona* sp. seems to be a good indication that jellyfish and ctenophores are true intermediate host for these parasites. On the other hand, the low prevalence of hemiurid metacercariae supports the idea that gelatinous zooplankton act as transport host for this parasite (Køie, 1979).

According to Marcogliese (1995), the role of jellyfish and ctenophores as intermediate host for digeneans is still unclear, and the future elucidation of life cycles, mainly of the Fellodistomidae, Faustulidae, Lepocreadiidae and Hemiuridae families will help to clarify the actual situation.

## References

Arai, M. N., 1988. Interactions of fish and pelagic coelenterates. Can. J. Zool. 66: 1913–1927.

Ates, R. M., 1988. Medusivorous fishes, a review. Zool. Med. Leiden 62: 29–42.

Bray, R. A., 1988. Discussion of the subfamily Baccigerinae Yamaguti, 1958 (Digenea) and the constitution of the family Fellodistomidae Nicoll, 1909. Syst. Parasitol. 11: 97–112.

Bush, A. O., K. D. Lafferty, J. M. Lotz & A.W. Shostak, 1997. Parasitology meets ecology on its own terms: Margolis et al. (revisited). J. Parasitol. 83: 575–83.

Cremonte, F., 1999. Estudio parasitologico de bivalvos que habitan ambientes marinos mixohalinos en Argentina. Thesis of La Plata National University, La Plata, Argentina: 260 pp.

Cremonte, F. & N. H. Sardella, 1997. The parasite fauna of *Scomber japonicus* Houttuyn, 1782 (Pisces: Scombridae) in two zones of the Argentine sea. Fish. Res. 31: 1–9.

Dollfus, R., 1963. Liste des coelenteres marins, paleartiques et indiens ou ont ete trouves des trematodes digenetiques. Bull. Inst. Peches. mar. Maroc. 9–10: 33–57.

Gaevskaya, A. V., A. A. Kovaliova & G. N. Rodjuk, 1985. Parasitofauna of the fishes of the Falkland-Patagonian Region. In Hargis, W. Jr. (ed.), Parasitology and Pathology of Marine Organisms of the World Ocean. NOAA Technical Report. NMFS: 25 pp.

Girola, C. V., S. R. Martorelli & N. H. Sardella, 1992. Sobre la presencia de estadios larvales de *Monascus filiformis* (Digenea: Fellodistomidae) en hidromedusas del Atlántico Sur. Contribución al conocimiento de su ciclo de vida. Rev. Chil. Hist. Nat. 65: 409–415.

Hall, K. A., T. H. Cribb & S. C. Barker, 1999. V4 region of samm subunit rDNA indicates polyphyly of the Fellodistomidae (Digenea) which is supported by morphology and life-cycle data. Syst. Parasitol. 43: 81–92.

Kim, Y. G. & S. K. Chun, 1984. Studies on the lif history of *Bacciger harengulae*. Bull. Korean Fish. Soc. 17: 449–470.

Køie, M., 1975. On the morphology and life-history of *Opechona bacillaris* (Molin, 1859) Looss, 1907 (Trematoda, Lepocreadiidae). Ophelia 13: 63–86.

Køie, M., 1979. On the morphology and life-history of *Derogenes varicus* (Müller, 1784) Looss, 1901 (Trematoda, Hemiuridae). Z. Parasitenkd. 59: 67–78.

Køie, M., 1983. Digenetic trematodes from *Limanda limanda* (L.) (Osteichthyes, Pleuronectidae) from Danish and adjacent waters, with special reference to their life-histories. Ophelia 22: 201–228.

Køie, M., 1985a. The surface topography and life-cycles of digenetic trematodes in *Limanda limanda* (L.) and *Gadus morhua* L. Marin Biol Lab: 20 pp.

Køie, M., 1985b. On the morphology and life-history of *Lepidapedon elongatum* (Lebour, 1908) Nicoll, 1910 (Trematoda, Lepocreadiidae). Ophelia 24: 135–153.

Lauchner, G., 1980a. Diseases of Cnidaria. In Kinne, O. (ed.), Diseases of Marine Animals Vol. 1. General Aspects. Protozoa to Gastropoda. John Wiley & Sons, New York: 167–237.

Lauchner, G., 1980b. Diseases of Ctenophora. In Kinne, O. (ed.), Diseases of Marine Animals Vol. 1. General Aspects. Protozoa to Gastropoda. John Wiley & Sons, New York: 239–253.

Lebour, M. V., 1916. Medusae as hosts for larval trematodes. J. mar. biol. Assoc. U.K. 11: 57–59.

Marcogliese, D. J., 1995. The role of zooplankton in the transmission of helminth parasites to fish. Rev. fish. Biol. Fish. 5: 336–371.

Marcogliese, D. J. & J. Price, 1997. Marine trematode: *Derogenes varicus*. Global Biodivers. 7: 16 pp.

Margolis L., G. W. Esch, J. C. Holmes, A. M. Kuris & G. A. Schad, 1982. The use of ecological terms in parasitology (Report of an ad hoc committee of the American Society of Parasitology) J. Parasitol. 68: 131–33.

Martin, W. E., 1945. Two new species of marine cercariae. Trans. am. microsc. soc. 64: 203–212.

Martorelli, S. R., 1991. Primera cita de una cercaria tricocerca parásita de *Dorsanum moniliferum* (Mollusca, Buccinidae) para el Atlántico Sudoccidental. Aportes al conocimiento de su ciclo de vida. Neotrópica 37: 57–65.

Martorelli, S. R., 1996. First record of encysted metacercariae in jellyfish of Southern Atlantic. J. Parasitol. 82: 352–53.

Martorelli, S. R. & F. Cremonte, 1998. A proposed three-host life-history of *Monascus filiformis* (Rudolphi, 1819) (Digenea: Fellodistomidae) in the Southwest Atlantic. Can. J. Zool. 76: 1198–1203.

Matsumoto, G. I., Ch. Baxter & E. H. Chen, 1997. Observations of deep-sea trachymedusa *Benthocodon pedunculata*. Invert. Biol. 116: 17–25.

Mianzan, H. W., N. Mari, B. Prenski & F. Sanchez, 1996. Fish predation on neritic ctenophores from Argentine continental shelf: a neglected food resource? Fish. Res. 27: 69–79.

Palombi, A., 1934. *Bacciger bacciger* (Rud.). Trematode digen-

etico: Fam. Steringophoridae Odhner. Anatomia, sistematica e biologia. Pub. Staz. Zool. di Napoli 13: 438–478.

Purcell, J. E. & M. N. Arai, 2001. Interactions of pelagic cnidarians and ctenophores with fish: a review. Hydrobiologia 451 (Dev. Hydrobiol. 155): 27–44.

Rao, K. H., 1958. Hemiurid larvae (Trematoda) in the medusa *Aequorea pensilis* (Haeckel) from the Bay of Bengal. Ann. Mag. nat. Hist. 1: 702–704.

Rebec, J., 1965. Considerations sur la place des trematodes dans le zooplancton marin. Ann. Fac. Sci. Marseilles 38: 61–84.

Reimer, L. W., H. Berger, B. Hewer, H. Lainka, I. Rosenthal & I. Scharnweber, 1971. The distribution of larvae of helminths in planktonic animals of the North Sea. Parasitologiya 5: 542–550.

Sardella, N. & J. T. Timi, 1996. Parasite communities of *Merluccius hubbsi* from Argentinian-Uruguayan common fishing zone. Fish. Res. 27: 81–88.

Stunkard, H. W., 1967. The life-cycle and developmental stages of a digenetic trematode whose unencysted stages occur in medusae. Biol. Bull. 133: 488.

Stunkard, H.W., 1969. The morphology and life-history of *Neopechona pyriforme* (Linton, 1900) n. gen., n. comb., (Trematoda: Lepocreadiidae). Biol. Bull. 136: 96–113.

Stunkard, H. W., 1974. New intermediate hosts of the digenetic trematodes, *Monorcheides cumingiae* (Martin, 1938) and *Neopechona pyriforme* (Linton, 1900). J. Parasitol. 60: 858.

Stunkard, H. W., 1978a. The life-cycle and taxonomic relations of *Lintonium vibex* (Linton, 1900) Stunkard and Nigrelli, 1930 (Trematoda: Fellodistomidae). Biol. Bull. 155: 383–94.

Stunkard, H. W., 1978b. Metacercariae of digenetic trematodes from ctenophores and medusae in the Woods Hole, Massachusetts area. Biol. Bull. 155: 467–68.

Stunkard, H. W., 1980a. The morphology, life-history, and taxonomic relations of *Lepocreadium areolatum* (Linton, 1900) Stunkard, 1969 (Trematoda: Digenea). Biol. Bull. 158: 154–163.

Stunkard,. H. W., 1980b. Successive hosts and developmental stages in the life history of *Neopechona cablei* sp. n. (Trematoda: Lepocreadiidae). J. Parasitol. 66: 636–641.

Wardle, W. J., 1983. Two new non-ocellate trichocercous cercariae (Digenea: Fellodistomidae) from estuarine bivalved molluscs in Galveston Bay, Texas. Cont. mar. Sci. 26: 15–22.

Yip, S. Y., 1984. Parasites of *Pleurobranchia pileus* Müller, 1776 (Ctenophora), from Galway Bay, western Ireland. J. Plankton Res. 6: 107–121.

*Hydrobiologia* **451**: 311–320, 2001.
© 2001 *Kluwer Academic Publishers.*

311

# Food and feeding of *Aurelia aurita* in Tokyo Bay with an analysis of stomach contents and a measurement of digestion times

Haruto Ishii & Fusako Tanaka
*Tokyo University of Fisheries, 4-5-7 Kounan, Minato-ku, Tokyo, 108-8477 Japan*
*E-mail: ishii@tokyo-u-fish.ac.jp*

*Key words:* Aurelia aurita, digestion time, daily ration, stomach contents, jellyfish, Tokyo Bay

## Abstract

*In situ* seasonal variations in stomach contents of *Aurelia aurita* (L.) in Tokyo Bay, Japan, were analyzed. Copepods, such as *Oithona davisae* Ferrari & Orsi were the predominant food items of *A. aurita* from June to November. The mean digestion time measured in incubation experiments was 0.95 h. Daily rations calculated using stomach content data and digestion times were 2.2–21.8 mg C ind$^{-1}$ corresponding to 0.58–5.56% of body carbon. The ingestion rate increased significantly with an increase in medusa size, although no significant relationship was found between medusa size and carbon specific daily ration. The zooplankton community in Tokyo Bay is characterized by the significant dominance of *O. davisae* and it is assumed that the prosperity of *A. aurita* is caused by the high abundance of the *O. davisae* population. It is suggested that a food chain comprised of microflagellates, cyclopoid copepods *O. davisae*, and *A. aurita* is the most significant one in Tokyo Bay and only a small portion of production is transferred to fish.

## Introduction

In recent years, gelatinous predators and scyphomedusae, in particular, have received scientific interest all over the world with respect to their role in the food web (e.g. Yasuda, 1971; Olesen et al., 1994; Schneider & Behrends, 1994; Arai, 1997; Purcell, 1997; Purcell et al. , 2000). These population studies observed high-density aggregations of *Aurelia aurita* medusae in summer. In addition to these studies, dense aggregations (150 ind m$^{-2}$) of *A. aurita* were also recorded in Saanich Inlet, Canada, during summer (Hamner et al., 1994). In Tokyo Bay, particularly in summer, mass occurrences of *A. aurita* are often observed (Toyokawa et al., 1997) and periodically cause serious problems as they clog power plant intakes and fishing nets.

*Aurelia aurita* is usually a predominant scyphomedusan species in coastal waters and some studies exist on their predation impact on prey populations (e.g. Lindahl & Hernroth, 1983; Schneider & Behrends, 1994; Martinussen & Båmstedt, 1995; Olesen, 1995; Omori et al., 1995; Lucas et al., 1997; Schneider & Behrends, 1998). For example, Lindahl & Hernroth

(1983) suggested that the outbreak of *A. aurita* in Gullmar Fjord, Sweden, in summer regulates the structure of pelagic ecosystems. Lucas et al. (1997) estimated the potential impact of *A. aurita* on the mesozooplankton communities in Southampton Water and Horsea Lake, England, from their secondary production. To estimate the feeding impact on prey populations, the *in situ* ingestion rate of medusae is highly desirable (Purcell, 1997). Olesen (1995) investigated the predation impact of *A. aurita* on zooplankton in Kertinge Nor, Denmark, using clearance rate data measured by laboratory incubation experiments. Previously, only Matsakis & Conover (1991) and Martinussen & Båmstedt (1995) estimate the predation impact of *A. aurita* on zooplankton prey based on their *in situ* ingestion rates with an analysis of stomach contents and measurement of digestion times.

The present study documents the ingestion rate and carbon specific daily ration of *A. aurita* based on field observations of stomach contents and laboratory incubation experiments on digestion time. We also discuss the relationship between the medusae population and the zooplankton community in Tokyo Bay.

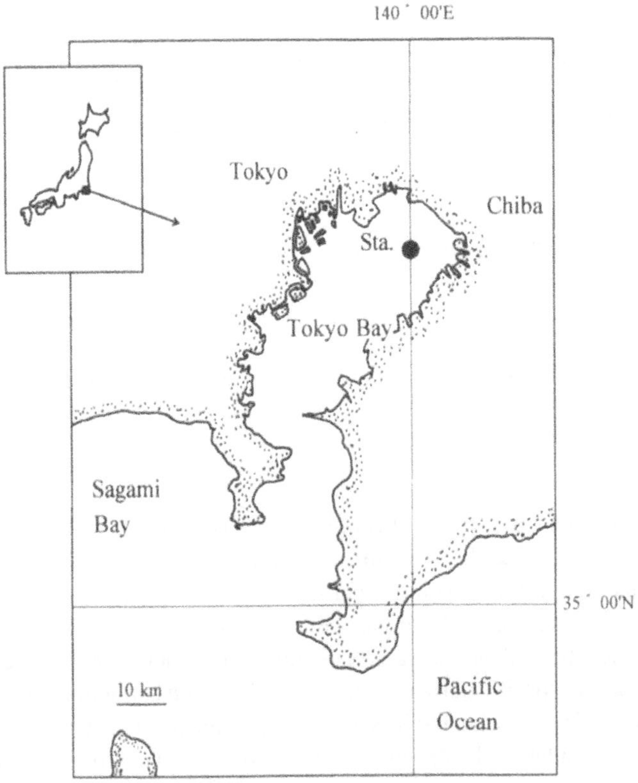

*Figure 1.* Monthly sampling station of *Aurelia aurita* in Tokyo Bay, Japan.

## Materials and methods

### Sampling

Sampling of *A. aurita* medusae was conducted in daylight hours once or twice a month between June 10 and December 2, 1997, aboard the "T. S. Hiyodori", Tokyo University of Fisheries, at a sampling area in the innermost part of Tokyo Bay (Fig. 1). Medusae were occasionally scooped with a hand net of 10 mm in mesh size from surface aggregations. Specimens for stomach content analyses and the incubation experiments were carefully handled with a vinyl sheet hand net so as to get undamaged specimens. The specimens used in incubation experiments were immediately transferred with a plastic bowl into a 20 l bucket. Surface water temperature was simultaneously measured during sampling. Bell diameter, sex and the presence of planula larvae of medusae were immediately determined after sampling. Measurements of the bell diameter, to the nearest 1.0 mm, were made by placing specimens on a scale with the exumbrellar side down to flatten out the bell. Immediately after collection, the gastric pouches of each medusa were opened using a scalpel and the stomach contents

were removed using a pipette. The stomach content of each medusa was preserved in 5% buffered formalin. Medusae for incubation experiments were carefully transferred to the laboratory in buckets and incubated at 22 °C, similar to the mean ambient water temperature in the medusae sampling area during summer and autumn. They were kept in 1 μm filtered seawater until the start of the incubation experiments.

Zooplankton prey for the incubation experiments was collected in the same area with a 100 μm conical mesh net, in oblique hauls from 3 m depth. They were also incubated in 1 μm filtered seawater at 22 °C for 2–3 days until the experiments started.

### Weight analyses

Some of the medusae were brought to the laboratory for dry weight and carbon analyses. Specimens were rinsed with an isotonic ammonium formate to remove external salts. Whole specimens were dried at 60 °C for 1 week and weighed. Parts of a dried medusa were homogenized, re-dried and re-weighed, and their carbon contents were analyzed using a CHN Corder (Yanako MT-3).

## Stomach content analyses

The preserved stomach contents, excluding medusae eggs and planula larvae, were examined using a dissection microscope. Prey items were identified, counted and their sizes were measured. On the basis of literature (e.g. Hirota, 1981; Martinussen & Båmstedt, 1995), our own data and the size of the prey items, we estimated the individual dry weight and carbon content of each prey species. For the prey species where the carbon content was not found in the literature, estimations were made on the basis of size and taxonomy. Summed carbon contents of prey items in each stomach were used to calculate the ingestion rate and daily ration of *A. aurita*.

## Digestion experiments

For the measurement of the digestion time of *A. aurita*, medusae having a bell diameter between 12.1 and 22.3 cm (*N*=18; average 16.6 cm) were used. Experiments were run in 2 l transparent plastic bowls kept at 22 °C, with one medusa per bowl. Sampled zooplankton prey were added near the mouth of the medusae. Cyclopoid copepods, *Oithona davisae*, formed 64% of the average gross composition of the prey population, and calanoid copepods and nauplius larvae of cirripede *Balanus* spp. were also observed. Each fed medusa was continuously observed until the unrecognizable prey was egested. Digestion time was defined as the time elapsed between ingestion and egestion.

To follow the digestion process, medusae having a bell diameter between 17.9 and 20.7 cm (*N*=8; average 19.3 cm) were used. Experiments were run in 20 l buckets kept at 22 °C. One medusa was incubated in each bucket. *Oithona davisae* sorted from the live zooplankton samples was used as prey. After prey ingestion, the medusae were carefully transported to the other buckets filled with 1 μm filtered seawater. Every 5 min after prey ingestion, each medusa was individually taken out of the water and the condition of the prey in the gastric pouches was observed under a dissection microscope. The experiment continued until the prey were unrecognizable.

After the digestion experiments, the bell diameters and dry weights of all specimens were measured by the methods as mentioned above.

## Ingestion rate and daily ration

The ingestion rate (I) and carbon specific daily ration (DR) of *A. aurita* was calculated using the following

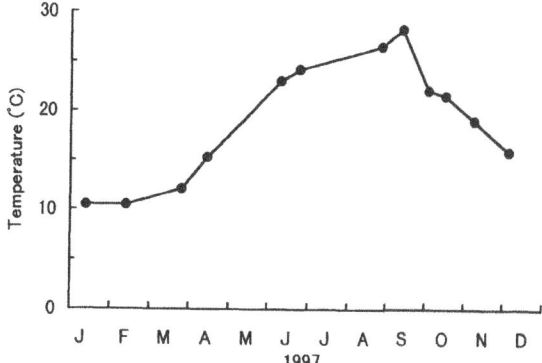

*Figure 2.* Monthly changes of surface water temperature at the sampling station in Tokyo Bay in 1997.

equations:

$$I = CC \cdot DT^{-1} \cdot 24h,$$

$$DR = I \cdot CW^{-1} \cdot 100,$$

where $I$ is the ingestion rate (mg C ind$^{-1}$ d$^{-1}$), CC is the summed carbon content in the stomach (mg C ind$^{-1}$), DT is the digestion time (h), DR is the carbon specific daily ration (%) and CW is the body carbon weight (mg C ind$^{-1}$).

## Results

### Temperature

Typically, the inner part of Tokyo Bay is well mixed during all the months except from June to September, when stratification occurs (Nomura, 1993). Monthly changes in surface water temperature at the sampling station from January 14 to December 2, 1997, are shown in Figure 2. The maximum surface temperature was 28.2 °C on September 11 and the minimum was 10.5 °C on 14 January and 13 February.

### Size distribution

Monthly size frequency distributions of the bell diameter of medusae are shown in Figure 3. In Tokyo Bay, ephyra larvae appear in March and young medusae in April (Omori et al., 1995). More than 50% of medusae had bell diameters of over 20 cm by 10 June, and the mode became 21 cm by 26 August. The growth was lowered and the size range of individuals in the monthly samples was narrower in August. Reduction of the bell diameter occurred in December and the mode became 16 cm.

314

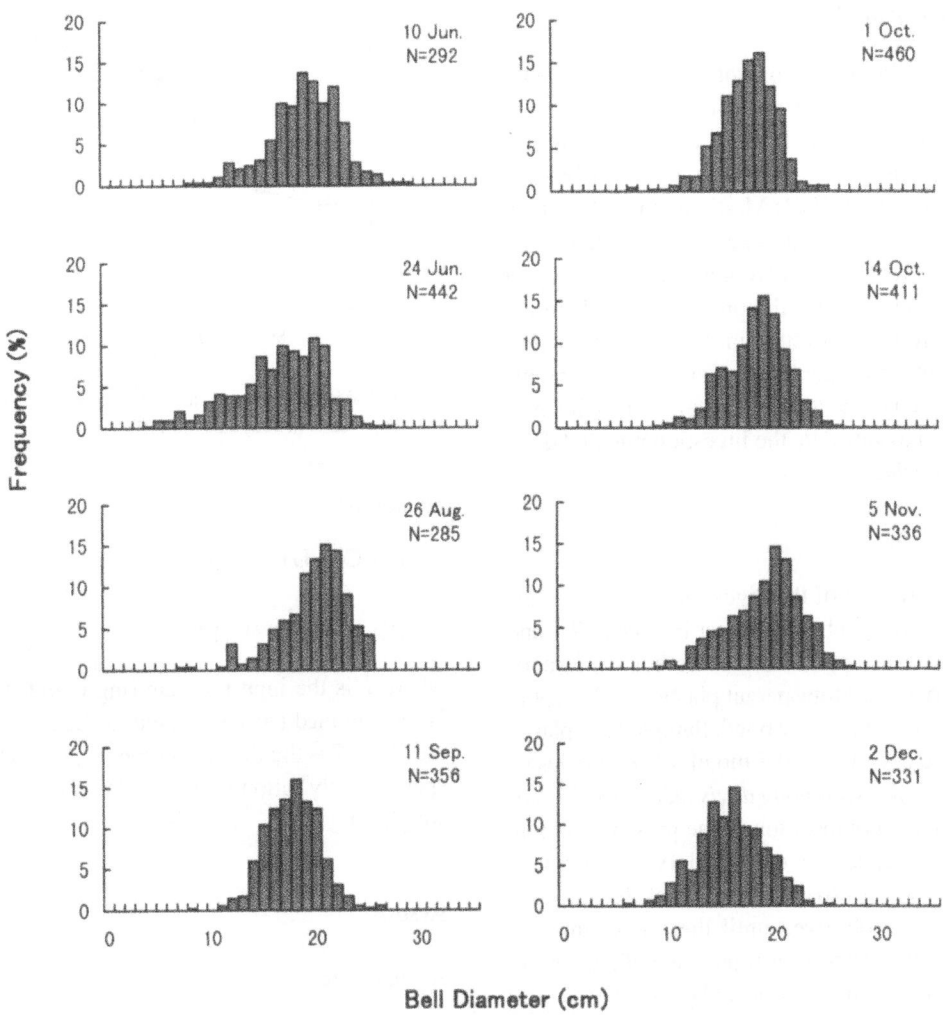

*Figure 3.* Relative frequency distributions of the bell diameter of *Aurelia aurita* in 1997.

## Stomach content analyses

Figure 4 presents the percentage (average of 10 specimens) of different prey observed in the stomachs of *A. aurita* on each sampling date. Except for 2 December when bivalve larvae dominated numerically (78%), cyclopoid copepods, *O. davisae*, were always dominant throughout the sampling period. The percentage frequency of *O. davisae* in the stomach contents ranged between 3% (2 December) and 94% (1 October).

## Digestion experiments

Digestion experiments performed on 18 individuals of *A. aurita* using mixed zooplankton showed a non-significant decrease in digestion time with increasing medusa diameter (Fig. 5). The mean digestion time for

all experiments was 0.95±0.25 h. This value was used for the estimation of their ingestion rate.

Observation of the digestive process of *O. davisae* gave the following information. After 10 min of digestion, the carapace of copepods were damaged, and material started to leak out after 35 min. After 40 min, most of the tissues were digested, and the body became partly transparent. By 45 min, the carapaces of the copepods were destroyed and the separation of prosomes and urosomes had occurred, and afterward they were egested as faeces.

## Ingestion rate and daily ration

The relationships between the bell diameter and the ingestion rate (carbon basis) and the carbon specific daily ration are shown in Figures 6 and 7, respectively. The ingestion rate increased significantly with

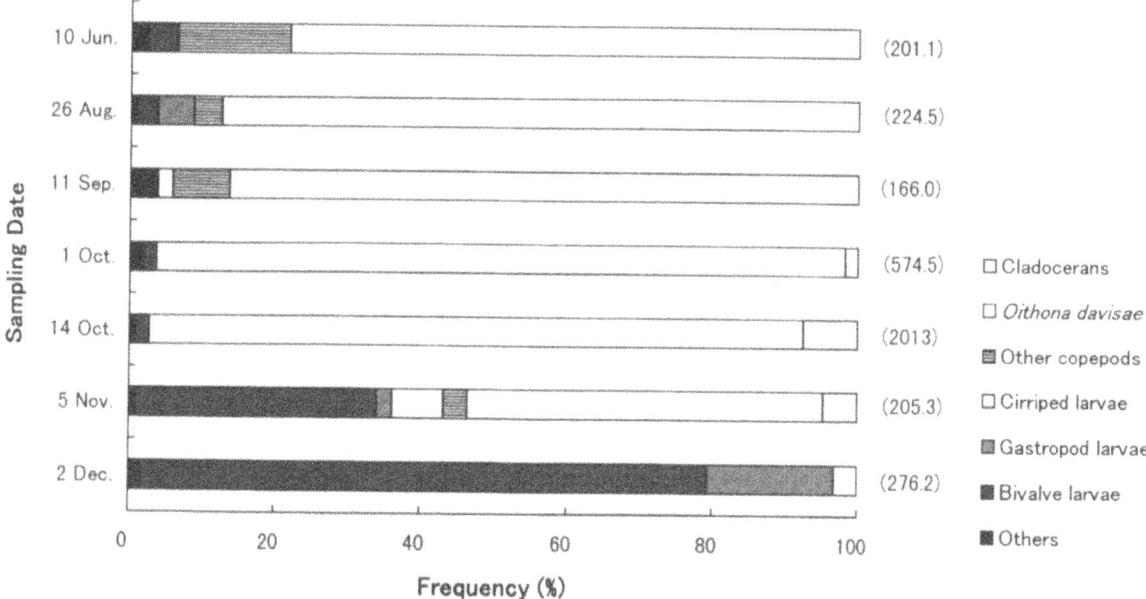

*Figure 4.* Relative frequency of prey as numbers in the stomach of *Aurelia aurita* in 1997. The average number of prey items in a stomach on each date is indicated in parentheses. Ten medusae were examined on each date.

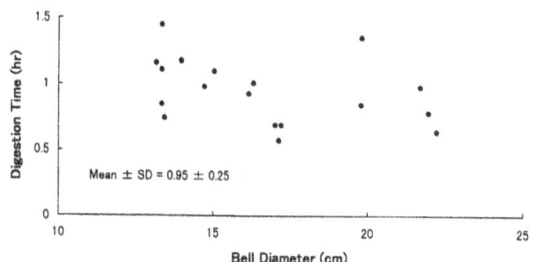

*Figure 5.* Relationship between the bell diameter and digestion time of *Aurelia aurita*.

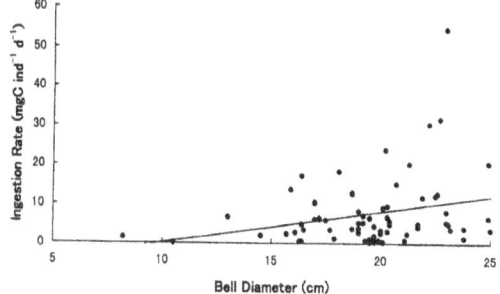

*Figure 6.* Relationship between the bell diameter and ingestion rate of *Aurelia aurita* estimated from the stomach contents and digestion time.

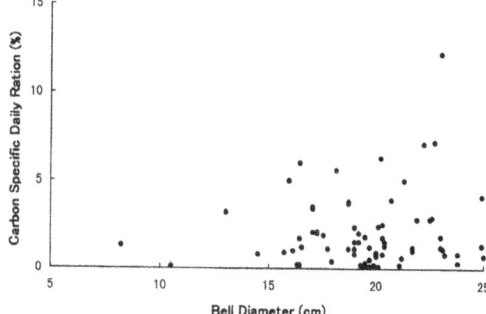

*Figure 7.* Relationship between the bell diameter and carbon specific daily ration of *Aurelia aurita* expressed in percentage.

The ingestion rate and carbon specific daily ration of *A. aurita* on each sampling date calculated from the stomach contents and the digestion time is expressed in Table 1, together with the diameters and carbon weights of medusae. The highest ingestion rate and carbon specific daily ration were 21.83 mg C ind$^{-1}$ d$^{-1}$ and 5.56%, respectively, on 14 October. No seasonal trends were detected in ingestion rate or carbon specific daily ration.

## Discussion

Although medusae sizes within the monthly size frequency distributions varied, seasonal occurrence patterns were clear and medusae showed continuous

an increase in medusae size ($r^2 = 0.076$; $p < 0.05$), however, there was no correlation between the size of medusae and the carbon specific daily ration ($r^2 = 0.017$). The average carbon specific daily ration of all medusae was $2.06 \pm 2.13\%$.

*Table 1.* Ingestion rate and carbon specific daily ration of *Aurelia aurita* estimated from the stomach contents and digestion time. $N=10$ on each date

| Sampling date | Temperature (°C) | Bell diameter (cm) | Carbon weight (mg ind$^{-1}$) | Ingestion rate (mg C ind$^{-1}$ d$^{-1}$) | Carbon specific daily ration (%) |
|---|---|---|---|---|---|
| 10 Jun. | 23.0 | 19.2±4.3 | 355±96.5 | 6.73±3.16 | 1.90±0.83 |
| 26 Aug. | 26.5 | 20.1±3.0 | 377±73.4 | 2.25±1.56 | 0.58±0.34 |
| 11 Sep. | 28.2 | 18.3±1.4 | 330±33.4 | 3.66±2.36 | 1.12±0.73 |
| 1 Oct. | 22.0 | 19.4±2.7 | 358±66.4 | 5.61±2.89 | 1.57±0.70 |
| 14 Oct. | 21.5 | 20.1±3.8 | 376±92.3 | 21.83±14.07 | 5.56±2.78 |
| 5 Nov. | 19.0 | 21.2±2.0 | 402±50.8 | 4.33±3.49 | 1.05±0.78 |
| 2 Dec. | 15.9 | 17.8±3.1 | 320±69.6 | 8.82±8.27 | 2.62±2.39 |

growth until August. The pattern after autumn in this study was in contrast to the investigation by Omori et al. (1995), and was more similar to the observations of Ishii et al. (1995). During 1990–1992, Omori et al. (1995) showed two groups of the *A. aurita* population after October. One grew continuously until the following spring and another reduced in bell diameter and disappeared by winter. On the other hand, the *A. aurita* population in 1993 simply decreased in bell diameter from summer to autumn (Ishii et al., 1995). Ishii et al. (1995) suggested that medusae decreased in size during autumn and winter after releasing their gametes into the sea, simultaneously lowering their metabolic activity. The size distribution pattern in this study also indicates the decrease of the late reproducing population. The females probably released their gametes into the sea before November in 1997, and thereafter decreased in size. Möller (1980a) also found that the bell diameter reduced by 13–18% from August to September after sexual reproduction. Hansson (1997) suggests that the cessation in growth of *A. aurita* medusae after reproduction period is not caused by food limitation. As discussed below, the zooplankton abundance, especially *O. davisae*, is enough to sustain the daily ration of medusae in Tokyo Bay. In laboratory incubation of medusae, we often observed a medusa having relatively long oral arms under food shortage conditions, however, this type of medusae was not collected at all even in a period of size reduction, like December. Further experimental studies are needed to elucidate the effects of the prey concentration on growth rate of medusae after reproduction, however, we consider that the reduction in medusae size is independent of food availability.

*Aurelia aurita* is known to be an opportunistic tactile predator, feeding on zooplankton of a broad size and taxonomic range. Matsakis & Conover (1991) found that the food of *A. aurita* in Bedford Basin, Canada, consisted of copepods, fish larvae, eggs and a significant number of the co-occurring cnidarian *Rathkea octopunctata* (M. Sars). In Kertinge Nor, Denmark, Olesen et al. (1994) found that the stomach of *A. aurita* contained tintinids, gastropod larvae, harpacticoid copepods, polychaete larvae and sometimes rotifers as a major part of their food. Möller (1980b) found that herring larvae were a major part of food of *A. aurita* in Kiel Fjord, Germany. Martinussen & Båmstedt (1995) found 42 different prey species in *A. aurita* by stomach analyses. The present study also found a broad taxonomic range of prey in the stomach of *A. aurita*, and a number of prey co-occurred with *O. davisae* throughout the sampling period. The typical composition of small zooplankton (>100 $\mu$m of body length) in Tokyo Bay was studied by Anakubo & Murano (1991) and was always dominated by *O. davisae*, except in April and May (Fig. 8). Therefore, it is considered that the stomach contents of medusae during the sampling period reflect the zooplankton composition in ambient seawater. Martinussen & Båmstedt (1995) also described the feeding behavior of *A. aurita* according to the prey composition in ambient seawater.

The digestion time is an important factor in the calculation of daily ingestion rate from stomach contents. We have used a single value, 0.95 h, for *A. aurita* irrespective of the medusa size. This average digestion time was rather low when compared with previous studies (Table 2), however, for *Aurelia* sp. in Big Jellyfish Lake, Palau, Dawson & Martin (2001) observed lower digestion time (0.71 h) in experiments with copepods as a prey. These results indicate different digestion times for different prey items, and we believe

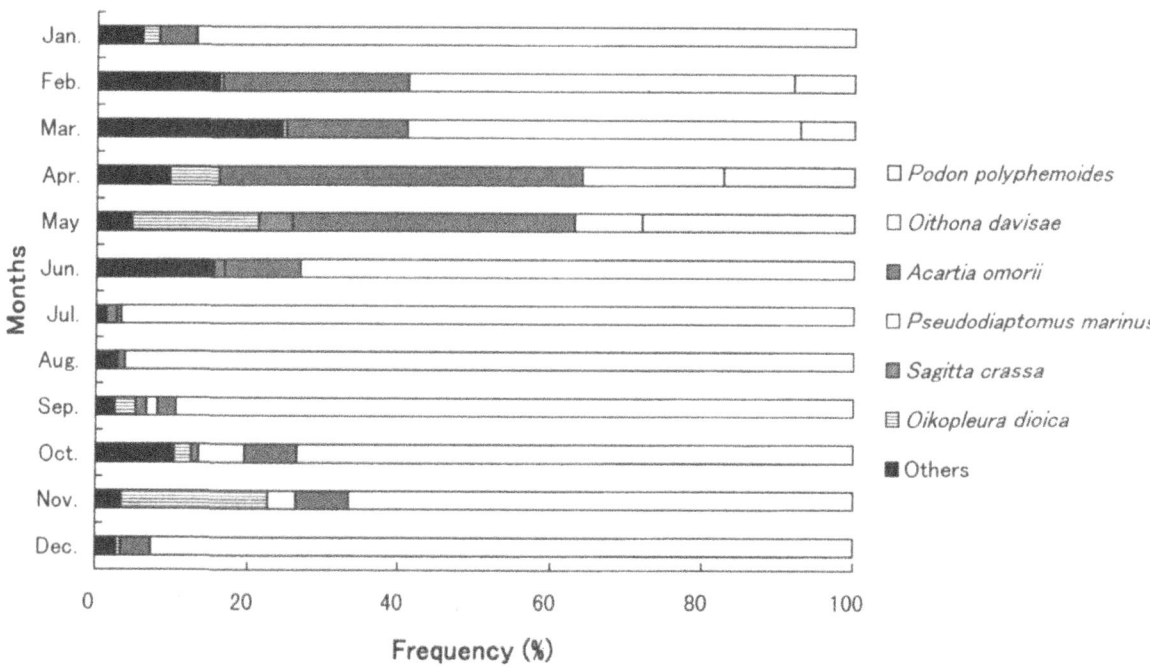

*Figure 8.* Seasonal variation of species composition of zooplankton in Tokyo Bay except microzooplankton (redrawn from Anakubo & Murano, 1991).

*Table 2.* Review of digestion time and ingestion rate estimated from the stomach contents of *Aurelia aurita*

| Area | Bell diameter (cm) | Prey | Temperature (°C) | Digestion time (h) | Ingestion rate (mgC ind⁻¹ d⁻¹) | Reference |
|---|---|---|---|---|---|---|
| Kiel Fjord, Germany | 0.6–2.5 | Herring larvae | 10–12 | 5 | – | Möller (1980b) |
| Kiel Fjord, Germany | 1.8–2.4 | Herring larvae | 22 | 3.8 | – | Heeger & Möller (1987) |
| Bedford Basin, Canada | 8 | Copepods, eggs and *Rathkea octopunctata* | 4 | 3.85 | 0.5 | matsakis & Conover (1991) |
| Kiel Bight, Germany | 16–26 | Mixed zooplankton | – | 4 | 8–15 | Schneider & Behrends (1994) |
| Narragansett Bay, U.S.A. | 3 | Copepods | 7 | 3.5 | – | Sullivan et al. (1994) |
| " | " | Larval fish | " | 2.3 | – | " |
| Raunefjorden, Norway | 3.5 | *Calanus finmarchicus* | 9.5 | 2.83 | – | Martinussen & Båmstedt (1995) |
| " | 13.89 | Mixed zooplankton | – | – | 1.6 | " |
| Raunefjorden, Norway | 0.35–1.45 | *Artemia* nauplii | 9.5 | 1.1–1.9 | – | Martinusen & Båmstedt (1999) |
| " | 0.45–1.35 | *Pseudocalanus elongatus* | " | 3.7 | – | " |
| " | 0.87–1.3 | *Temora longicornis* | " | 3.2 | – | " |
| " | 0.43–5.4 | *Calanus finmarchicus* | " | 2.3–5.3 | – | " |
| " | 2–7.5 | *Clupea harengus* | " | 3.3–4.1 | – | " |
| Tokyo Bay, Japan | 16.6 | Mixed zooplankton | 22 | 0.95 | – | Present study |
| " | 8.2–25.0 | " | 15.9–28.2 | – | 2.2–21.8 | " |
| " | 20.0 | *Perinereis nuntia* | 22 | 10.42 | – | Ishii (unpublished) |

that prey size is one of the crucial factors affecting digestion time. Martinussen & Båmstedt (1999) showed by incubation experiments with five different prey of various sizes that the digestion time of *A. aurita* medusae was proportionally related to the prey number and size. They observed that digestion of herring larvae (ca. 11 mm) was 3.8 times slower than for *Artemia* nauplii (ca. 0.5 mm). *Oithona davisae*, mainly used as a food for medusae in our experiment, is a very tiny copepod having only 0.3 mm prosome length in the copepodite VI stage, and it would be digested more rapidly than other larger prey. Digestion of a polychaete, *Perinereis nuntia* (Savigny) (ca. 30 mm), by *A. aurita* medusae required 10.42 h at the same temperature (Table 2).

Predator size also can affect digestion rates. Martinussen & Båmstedt (1999) found a weak tendency of decreased digestion time with an increased size of medusa using a calanoid copepod, *Calanus finmarchicus* (Gunner), as prey, and emphasized that the relative sizes of the prey and predator was the important factor for the estimation of digestion time. The size of medusae in Tokyo Bay is relatively large compared with other areas (Table 2). Considering the size of prey (mainly *O. davisae*) and predators, we think the digestion time of 0.95 h is reasonable for *A. aurita* medusae in Tokyo Bay.

Temperature also affects digestion times of medusae (Purcell, 1997). Heeger & Möller (1987) found that the digestion time of herring larvae decreased with increasing temperature. Our digestion experiments were conducted only at 22 °C, while the *in situ* temperature of ambient seawater varied from 15.9 to 28.2 °C during the medusae sampling period. The daily rations of medusae during summer was markedly lower during autumn, however, considering the effect of higher water temperature, higher daily rations would also be observed during the summer.

Our ingestion rates from field sampled *A. aurita* were almost within the range of results of previous studies with *in situ* observations of the stomach contents (Table 2). We compared our data with those of Schneider & Behrends (1994) and Martinussen & Båmstedt (1995) because copepods were the main food and the observed bell diameters were similar. The mean observed predation rate of Martinussen & Båmstedt (1995) was 1.6 mg C ind$^{-1}$ d$^{-1}$, which is lower than Schneider & Behrends (1994) and that of the present study. Our results were more comparable to those of Schneider & Behrends (1994), which showed

a predation rate of 8–15 mg C ind$^{-1}$ d$^{-1}$ and up to 25 mg C ind$^{-1}$ d$^{-1}$.

Laboratory experiments supplying various food items to estimate the ingestion rate of *A. aurita* medusae are restricted to a few studies (Båmstedt, 1990; Båmstedt et al., 1994; Olesen et al., 1994). Ingestion rates in laboratory experiments are high in comparison with *in situ* rates. Båmstedt (1990) estimated the daily ration in the range of 5–40% using mixed zooplankton, while the estimations of Båmstedt et al. (1994) exceeded 250% for large (>1 mm) zooplankton. However, these results were mainly extrapolations from incubation experiments with a balance method using very small medusae (e.g. 5.9 cm; Båmstedt et al., 1994), which makes it difficult to compare with results from field studies. There are no incubation experiments to estimate ingestion rate using large medusae (> 20 cm), similar to those that frequently emerge in Tokyo Bay. Hansson (1997) and Purcell (1997) suggested that feeding and growth are reduced in laboratory incubation experiments because medusae are often subjected to a variety of stress factors, particularly confinement in small containers. To know the ingestion rate or daily ration of large medusae, *in situ* observation of stomach contents may be the most suitable way.

The present study revealed that the most significant prey of *A. aurita* medusae was *O. davisae* throughout the sampling period. Long-term fluctuations of the zooplankton community and abundance in Tokyo Bay based on the results of a 10-year investigation have been compiled (Nomura, 1993). According to his study, *O. davisae* (copepodids) occurred throughout the year (average 657 000 ind m$^{-3}$), and the highest abundance (maximum 2 520 000 ind m$^{-3}$) was observed in summer. Based on the data of Hirota (1981), the average dry weight of *O. davisae*, 0.70 μg, can be converted to its average carbon weight, 0.34 μg. If we calculate their production from the data of dry weight and temperature according to Ikeda & Motoda (1978), the production rate of *O. davisae* would be 0.18 μg C ind$^{-1}$ d$^{-1}$, and the average production rate of *O. davisae* population would be 118 mg C m$^{-3}$ d$^{-1}$. Even if *A. aurita* totally depends on *O. davisae* as a prey, considering the average abundance of *A. aurita* medusae in the recent study in Tokyo Bay (0.32 ind m$^{-3}$; Omori et al., 1995) and the ingestion rate (2–22 mg C ind$^{-1}$ d$^{-1}$), the production rate of *O. davisae* is enough to sustain the daily ration of medusae. However, we could not estimate their feeding impact on the zooplankton population in dense aggregations of medusae. Toyokawa et al. (1997) sampled dense aggregations of *A. aurita* medusae up to 13.4 ind. m$^{-3}$. Although it is not elucidated whether medusae are food limited in dense aggregations, we believe that the zooplankton production is enough to sustain the food demands of medusae at average abundance in Tokyo Bay.

In Tokyo Bay, the abundance of zooplankton as a prey of *A. aurita* greatly increased after the post-World War II period because of eutrophication (Nomura, 1993). Since the 1960s, the zooplankton community has been dominated by microzooplankton and relatively small copepods, *O. davisae*, instead of the relatively large mesozooplankton, such as *Acartia omorii* Bradford or *Paracalanus parvus* (Claus) (Anakubo & Murano, 1991; Nomura, 1993). The prosperity of *O. davisae* is believed to be caused by the increase of microflagellates instead of diatoms (Yamaguchi & Aruga, 1988), since *O. davisae* is not a diatom feeder but a microflagellate feeder (Uchima, 1988). The transfer of primary production via. *O. davisae* results in interesting situations in Tokyo Bay. Prey visibility is an important difference in the foraging of medusae and fish as zooplankton predators (Eiane et al., 1999). Eiane et al. (1999) investigated two fjords, one that is dominated by fish and has a lower biomass of large mesozooplankton and greater water clarity than the second 'jellyfish' fjord. They suggested that visual predators like fish exert a selective pressure against larger zooplankton, and the prey size is not critical for non-visual predators such as jellyfish. In Tokyo Bay, we assumed that *O. davisae* and other microzooplankton would be too small for visual predators such as planktivorous fish, which ingest relatively large copepods. The biomass of large copepods has been decreasing since the 1960s (Nomura, 1993), and only a small portion of the primary production is transferred to fish production in Tokyo Bay. Although prey size is important for planktivorous fish, it is not for *A. aurita*, which are opportunistic tactile predators. Therefore, an increase in the abundance of *O. davisae* may have contributed to increases in medusa biomass relative to fish biomass. Medusae in Tokyo Bay are usually larger than in other areas throughout year (Omori et al., 1995), and Ishii & Båmstedt (1998) showed experimentally that the abundance of prey contributes to the success of medusae growth and maintenance of the population. For *O. davisae*, there is no dominant predator in Tokyo Bay except for *A. aurita* medusae (Nomura, personal information). We propose that the food chain of microflagellates, *O. davisae*, and *A. aur-* *ita* medusae is the most significant one in the pelagic biological transformation of Tokyo Bay.

## Acknowledgements

Assistance given by the captains and crews of the T. S. 'Hiyodori' and by members at the Ecology Laboratory of Tokyo University of Fisheries is greatly appreciated. We are grateful to Dr U. Båmstedt for the comments on the manuscript and Mr R. Vassallo-Agius for correcting the manuscript. We thank two anonymous referees for their constructive criticisms.

## References

Anakubo, T. & M. Murano, 1991. Seasonal variation of zooplankton in Tokyo Bay. J. Tokyo Univ. Fish. 78: 145–165.

Arai, M. N., 1997. A Functional Biology of the Scyphozoa. Chapman & Hall, London: 316 pp.

Båmstedt, U., 1990. Trophodynamics of the scyphomedusae *Aurelia aurita*. Predation rate in relation to abundance, size and type of prey organism. J. Plankton Res. 12: 215–229.

Båmstedt, U., M. B. Martinussen & S. Matsakis, 1994. Trophodynamics of the two scyphozoan jellyfishes, *Aurelia aurita* and *Cyanea capillata*, in western Norway. ICES J. mar. Sci. 51: 369–382.

Dawson, M. N. & L. E. Martin, 2001. Geographic variation and ecological adaptation in *Aurelia* (Scyphozoa, Semaeostomeae): some implications from molecular phylogenetics. Hydrobiologia 451 (Dev. Hydrobiol. 155): 259–273.

Eiane, K., D. L. Aksnes, E. Bagøien & S. Kaartvedt, 1999. Fish or jellies – a question of visibility? Limnol. Oceanogr. 44: 1352–1357.

Hamner, W. M., P. P. Hamner & S. W. Strand, 1994. Sun-compass migration by *Aurelia aurita* (Scyphozoa): population retention and reproduction in Saanich Inlet, British Columbia. Mar. Biol. 119: 347–356.

Hansson, L. J., 1997. Effect of temperature on growth rate of *Aurelia aurita* (Cnidaria, Scyphozoa) from Gullmarsfjorden, Sweden. Mar. Ecol. Prog. Ser. 161: 145–153.

Heeger, T. & H. Möller, 1987. Ultrastructural observations on prey capture and digestion in the scyphomedusa *Aurelia aurita*. Mar. Biol. 96: 391–400.

Hirota, R., 1981. Dry weight and chemical composition of the important zooplankton in the Setonaikai (Inland Sea of Japan). Bull. Plankton Soc. Japan 28: 19–24.

Ikeda, T. & S. Motoda, 1978. Estimated zooplankton production and their ammonia excretion in the Kuroshio and adjacent seas. Fish. Bull. U.S. 76: 357–367.

Ishii, H. & U. Båmstedt, 1998. Food regulation of growth and maturation in a natural population of *Aurelia aurita* (L.). J. Plankton Res. 20: 805–816.

Ishii, H., S. Tadokoro, H. Yamanaka & M. Omori, 1995. Population dynamics of the jellyfish, *Aurelia aurita*, in Tokyo Bay in 1993 with determination of ATP-related compounds. Bull. Plankton Soc. Japan 42: 171–176.

Lindahl, O. & L. Hernroth, 1983. Phyto-zooplankton community in coastal waters of Western Sweden – an ecosystem off balance? Mar. Ecol. Prog. Ser. 10: 119–126.

320

Lucas, C. H., A. G. Hirst & J. A. Williams, 1997. Plankton dynamics and *Aurelia aurita* production in two contrasting ecosystems: comparisons and consequences. Estuar. coast. shelf Sci. 45: 209–219.

Martinussen, M. B. & U. Båmstedt, 1995. Diet, estimated daily food ration and predator impact by the scyphozoan jellyfishes *Aurelia aurita* and *Cyanea capillata*. In Skjoldal, H. R., C. Hopkins, K. E. Erikstad & H. P. Leinaas (eds), Ecology of Fjords and Coastal Waters. Elsevier Science Publishers B.V., Amsterdam: 127–145.

Martinussen, M. B. & U. Båmstedt, 1999. Nutritional ecology of gelatinous planktonic predators. Digestion rate in relation to type and amount of prey. J. exp. mar. Biol. Ecol. 232: 61–84.

Matsakis, S. & R. J. Conover, 1991. Abundance and feeding of medusae and their potential impact as predators on other zooplankton in Bedford Basin (Nova Scotia, Canada) during spring. Can. J. Fish. aquat. Sci. 48: 1419–1430.

Möller, H., 1980a. Population dynamics of *Aurelia aurita* medusae in Kiel Bight, Germany (FRG). Mar. Biol. 60: 123–128.

Möller, H., 1980b. Scyphomedusae as predators and food competitors of larval fish. Meeresforschung 28: 90–100.

Nomura, H., 1993. Community structure and succession in zooplankton in Tokyo Bay. Ph. D. Thesis, Tokyo University of Fisheries: 82 pp.

Olesen, N. J., 1995. Clearance potential of jellyfish *Aurelia aurita*, and predation impact on zooplankton in a shallow cove. Mar. Ecol. Prog. Ser. 124: 63–72.

Olesen, N. J., K. Frandsen & H. U. Riisgård, 1994. Population dynamics, growth and energetics of jellyfish *Aurelia aurita* in a shallow fjord. Mar. Ecol. Prog. Ser. 105: 9–18.

Omori, M., H. Ishii & A. Fujinaga, 1995. Life history strategy of *Aurelia aurita* (Cnidaria, Scyphomedusae) and its impact on the zooplankton community of Tokyo Bay. ICES J. mar. Sci. 52: 597–603.

Purcell, J. E., 1997. Pelagic cnidarians and ctenophores as predators; selective predation, feeding rates and effects on prey populations. Ann. Inst. Oceanogr., Paris 73: 125–137.

Purcell, J. E., E. D. Brown, K. D. E. Stokesbury, L. H. Haldorson & T. C. Shirley, 2000. Aggregations of the jellyfish *Aurelia labiata*: abundance, distribution, association with age-0 walleye pollock, and behaviors promoting aggregation in Prince William Sound, Alaska, U.S.A. Mar. Ecol. Prog. Ser. 195: 145–158.

Schneider, G. & G. Behrends, 1994. Population dynamics and the trophic role of *Aurelia aurita* medusae in the Kiel Bight and western Baltic. ICES J. mar. Sci. 51: 359–367.

Schneider, G. & G. Behrends, 1998. Top-down control in a neritic plankton system by *Aurelia aurita* medusae – a summary. Ophelia 48: 71–82.

Sullivan, B. K., J. R. Garcia & G. Klein-MacPhee, 1994. Prey selection by the scyphomedusan predator *Aurelia aurita*. Mar. Biol. 121: 335–341.

Toyokawa, M., T. Inagaki & M. Terazaki, 1997. Distribution of *Aurelia aurita* (Linnaeus, 1758) in Tokyo Bay; observations with echosounder and plankton net. Proc. 6th Int. Conf. Coelenterate Biol. 1995: 483–490.

Uchima, M., 1988. Gut content analysis of neritic copepods *Acartia omorii* and *Oithona davisae* by a new method. Mar. Ecol. Prog. Ser. 48: 93–97.

Yamaguchi, Y. & Y. Aruga, 1988. Transition of primary production in Tokyo Bay. Bull. Coast. Oceanogr. 25: 87–95.

Yasuda, T., 1971. Ecological studies on the jelly-fish, *Aurelia aurita* in Urazoko Bay, Fukui Prefecture-IV. Monthly change in the bell-length composition and breeding season. Bull. Japanese Soc. Sci. Fish. 37: 364–370.

*Hydrobiologia* **451**: 321–333, 2001.
© 2001 *Kluwer Academic Publishers.*

# Distribution, abundance, behavior and metabolism of *Periphylla periphylla*, a mesopelagic coronate medusa in a Norwegian fjord

Marsh J. Youngbluth[1] & Ulf Båmstedt[2]

[1]*5600 U.S. 1, North, Harbor Branch Oceanographic Institution, Fort Pierce, FL. 34946 U.S.A.*
*E-mail: youngbluth@hboi.edu*
[2]*Department of Fisheries and Marine Biology, University of Bergen, P.O. Box 7800, N-5020, Bergen, Norway*
*E-mail: ulf.baamstedt@ifm.uib.no*

*Key words:* jellyfish, vertical migration, swimming, nematocysts, net feeding, respiration, ROV

## Abstract

The distribution, behavior and metabolism of the mesopelagic jellyfish, *Periphylla periphylla* (Péron & Lesueur), were investigated in Lurefjorden, Norway. Field studies, conducted in 1998–1999 with plankton nets and a remotely operated vehicle, indicated that 80-90% of the dense (up to 2.5 m$^{-3}$) population migrated 200–400 m vertically each day throughout the year. *In situ* observations with red light revealed that swimming rates and feeding activity varied with age and time of day. Detection of turbulence and contact with surfaces caused this medusa to conceal one or all of its tentacles in the stomach or to shed nematocyst-laden tissue from the tentacles. Stomachs of medusae collected with nets were often full of prey entangled with the sloughed tissue. Stomachs of medusae captured individually with ROV samplers were empty or contained only a few prey in their stomachs (typically, 1–4 copepods *Calanus* spp. or chaetognaths *Eukrohnia hamata* Möbius per medusa). Low rates (0.4–5.6 $\mu$l O$_2$ mg C$^{-1}$ h$^{-1}$) of oxygen consumption of *P. periphylla* suggested that this species was sustained by relatively few (1–34) prey d$^{-1}$.

## Introduction

There is a growing awareness that invasions of gelatinous zooplankton can be as ecologically damaging as oil spills. For example, the comb jelly, *Mnemiopsis leidyi* A. Agassiz, was introduced into the Black Sea from ballast water about 1987 and now constitutes up to 95% of the zooplankton biomass in the Black Sea. The collapse of the Black Sea fisheries, worth 250 million US$ per year, has been directly attributed to this ctenophore (Volovik et al., 1993; Purcell et al., 2001). Apart from their roles as competitors and predators in marine communities (e.g. Alldredge, 1984; Purcell & Arai, 2001), swarms of soft-bodied zooplankton are known to foul fishing trawls (Rogers et al., 1978), clog seawater inlets of electrical power stations (Hay et al., 1990) and damage salmon farming (Båmstedt et al., 1998).

The degree to which gelatinous zooplankton can be regulatory components of marine food webs is difficult to assess (Mills, 1995). Their bodies are fragile and easily damaged by traditional sampling with plankton nets. Consequently, reports on their natural history as well as quantitative accounts of the abundance, feeding and metabolism of these animals are rare, especially for mesopelagic species (e.g. Youngbluth et al., 1990; Thuesen & Childress, 1994; Robison et al., 1998). The deep fjords of Norway, particularly those with shallow sills, are ideal marine environments for studies of various midwater fauna (Eiane et al., 1998). The normally calm sea states allow access to deep (up to 1200 m) environments throughout the year, and therefore facilitate investigations of processes that influence zooplankton abundance and recruitment over diel and seasonal scales. Physical exchange of water is restricted primarily to the layer above the sill depth, and consequently, the animal communities that live in deep fjord basins tend to remain undisturbed by advective forces for prolonged periods.

Gelatinous zooplankton, especially those living below the euphotic zone, are difficult to study in the open ocean. The papers that have mentioned the meso-

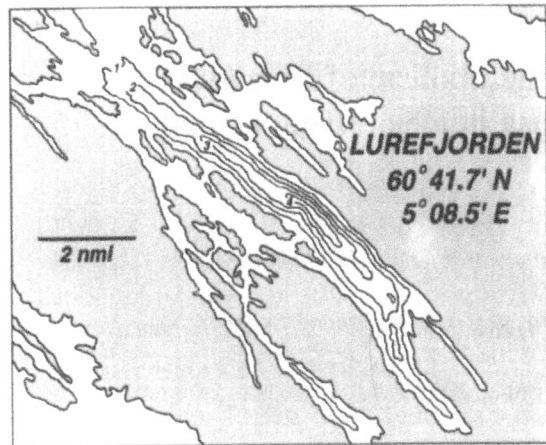

*Figure 1.* Geographic location of Lurefjorden, Norway. Numbers (1–4) designate 100 m depth intervals. The study site was within the innermost (= 400 m) contour.

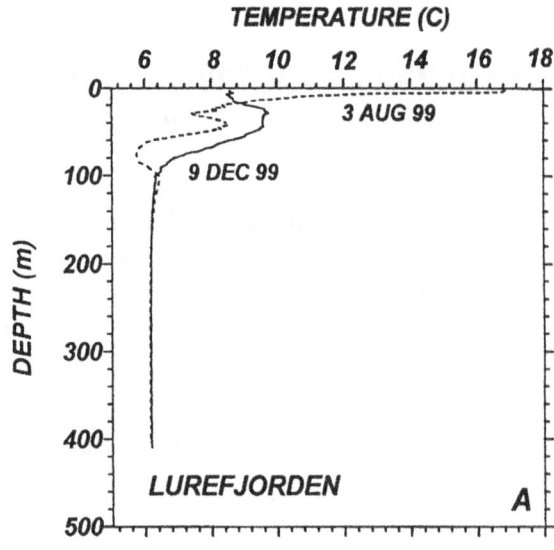

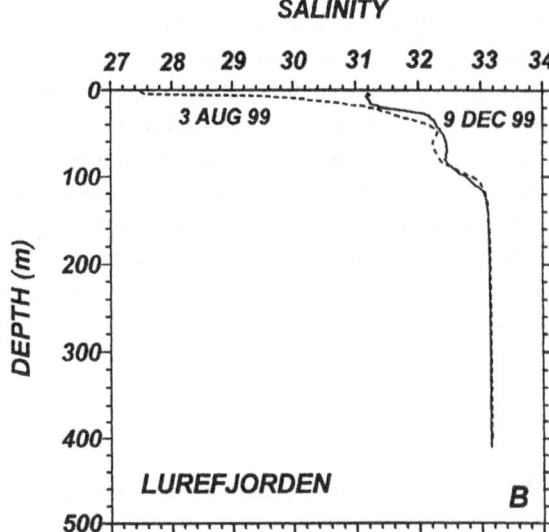

*Figure 2.* Vertical profiles of temperature (a) and salinity (b) in Lurefjorden during winter and summer.

pelagic coronate scyphomedusa, *Periphylla periphylla* (Péron and Lesueur), are typical. Collections with plankton nets have shown that this deep water species is cosmopolitan, but rarely numerous (Larson, 1986, 1990; Larson et al., 1991; Pagès & Kurbjeweit, 1994; Dalpadado et al., 1998). However, during the last century, dense populations have been reported to occur in Norwegian fjords (Fosså, 1992; Jens Nejstgaard, pers. comm.). Lurefjorden, 50 km north of Bergen, is one such fjord, and investigations there have indicated that the jellyfish population is several orders of magnitude greater than in the open ocean. Ongoing studies since 1992 have defined the life cycle of this species (Jarms et al., 1999) and have dealt with its population dynamics and growth. In addition, Lurefjorden provides an opportunity to study an ecosystem continuously dominated by a jellyfish as the top predator (Eiane et al., 1999). This condition may eventually provide insight about how physical and biological factors interact to maintain such a situation. Since jellyfishes, in contrast to fishes, usually are considered as useless production for the human population, such knowledge is of more than purely academic interest. Furthermore, the use of a remotely operated vehicle (ROV) since 1999 has allowed more detailed studies, both at the population and the individual levels. In this paper, we present new observations about the relative abundance, vertical distribution, diel migration, feeding behavior and metabolic demand of *P. periphylla*.

## Materials and methods

### Description of the environment

Lurefjorden is a relatively isolated fjord located along the western coast of Norway (Fig. 1). The seaward entrance is about 200 m wide and 20 m deep; the broadest section along its 20 km extent is nearly 2 km. The narrow entrance and shallow sill limit the exchange of water between the fjord and its surrounding areas. The greatest depth in the main basin is 440 m. The total volume of the fjord is about 14 km$^3$, of which 0.29 km$^3$ is 400 m or deeper, 1.76 km$^3$ is 300 m or deeper, 4.2 km$^3$ is 200 m or deeper (Fig. 1). Stratification begins in May, reaches a maximum in July and becomes weak from December through April. The largest vertical gradients appear in the upper 50 m. Below this depth, temperature and salinity remain rather constant throughout the year, ca. $6.1\pm0.1$ °C and $33.2\pm0.02$, respectively (Fig. 2a, b). Oxygen concentrations corresponding to 50–70% saturation persist throughout the year below 150 m (Fosså, 1992). There are no rivers entering the fjord, and the water exchange is mainly driven by the tidal currents and local weather conditions. Internal waves generated by the semidiurnal tide are probably of some importance for the vertical mixing of the basin water (Golmen et al., 1998). The basin water is mainly dominated by coastal water, indicating that the sill is too shallow for Atlantic water to intrude (Eiane et al., 1998). Weather conditions, such as prolonged periods with southerly winds, force outflowing Skagerrak water along the coast and into the fjords below the surface water (Aure & Saetre, 1981). Northerly winds and thereby outward Ekman transport of the surface water probably also affect water renewal in the fjords.

*Abundance, distribution and behavior*

Abundance patterns, diel vertical migrations and feeding activities were documented in February 1998, August 1999 and December 1999 with several time-series trials. Plankton nets and the ROV *Aglantha* were deployed for these field studies. Two kinds of net gear were towed vertically to collect medusae. A 2 m diameter net (3 mm mesh) was hauled in a continuously open position throughout the entire water column; a closing WP-3 net (1 m diameter, 500 $\mu$m mesh) was pulled through 100 m intervals from 400 m depth. Physical and physiological damages to the medusae were mitigated by hauling the nets slowly (6 m min$^{-1}$) and attaching a large (= 20 l) plastic bag at the cod end.

Medusae also were observed, enumerated and captured with the ROV. Estimates of abundance were quantified by counting the number of individuals encountered in the field of view of a Sony Hi8 video camera, set at minimum focal length. Initial measurements defined the angle of view and maximum distance of visibility for a small (= 2 cm) coronal diameter (CD) and a large (= 10 cm CD) medusa. Two 500 W halogen lights and two HID gas arc lights (comparable effect = 1000 W each) illuminated the field of view. A scale was mounted 0.9 m from the lens, the view of this scale in water indicated the angle

of view ($\alpha$) was 36.5 degrees. A 'visibility distance' was measured in the dark at the surface at night by suspending a medusa from a string in front of the video camera. Each medusa was moved away from the lens until the medusa disappeared. The small medusa had a visibility distance of 2.65 m; the larger one 4.3 m. Simple geometric relationships [$\tan(\alpha/2) = (X/2)/0.9$, where $X$ is the width of the view] gave the visibility area for these two sizes of medusae, i.e. 2.31 and 6.09 m$^2$, respectively. The volume covered was determined by multiplying this area by the number of meters moved vertically. The descent rate of the ROV was constant throughout a given dive. Each profile of the 440 m water column was completed within 30–35 min. For small medusae, the total volume sampled was ca. 1000 m$^3$, whereas large medusae were counted in ca. 2700 m$^3$. The depth of occurrence of every medusa was recorded and abundance was calculated per 100 m$^3$ in 10 m vertical intervals.

A problem with this enumeration method was the simplification of defining medusae as either 'small' or 'large' and using one of the two visibility constants in the calculations. In reality, the size of a medusae observed varied from 2 to 17 cm CD. The visibility distance, therefore, also varies over a continuous range. In the future, a stereometric video system will be used to allow an observer to more easily categorize medusae into several size groups. Post-processing of stereo frames with image analysis software will provide more precise size information.

Medusae were collected individually within any one of four static samplers that were attached to the ROV. These samplers (Tietze & Clarke, 1986) were clear acrylic tubes (1.3 cm thick × 16.5 cm ID, 6.5 l volume). Each tube was sealed by a pair of lids that simultaneously rotate horizontally across the open ends (Youngbluth, 1984). This closing mechanism minimized the production of turbulence in the water inside the chambers during the capture process.

Quantitative assessments of predation were based on medusae collected in nets and medusae captured individually in the ROV samplers. Immediately after capture, each medusa was carefully removed from a given collecting device and dissected. When medusae were handled gently, prey remained inside the stomach. All prey in the stomach were identified and counted at 10–50× magnification with a dissecting microscope. Subjective appraisals of the degree of digestion, i.e. fresh or partially digested, were also made.

Diel studies of behaviors (i.e. resting, swimming, and apparent feeding) of *Periphylla periphylla* were conducted with the ROV. All of these observations were made in red light (660 nm) and were recorded on SVHS videotape with a low light B/W camera. The visibility distance for a 10 cm CD medusa was 2.35 m under these light conditions.

During the behavioral studies, sizes of individual medusae were determined *in situ* with a paired-beam laser (Tusting & Davis, 1993) that was mounted on the ROV. This scaling device (1.5 cm distance between beams) consisted of two small, 635 nm wavelength, 10 mw solid-state modules contained in a single housing. The two bright, red-orange light spots projected by the laser were easily detected and recorded clearly on videotape. The output beams were parallel to within 0.1 milliradian and provided a minimum spot size of ca. 1 mm at a range of 1–3 m.

*Biomass and chemical composition*

Medusae were sized [CD and coronal height ±1 mm] and weighed [wet weight (WW) ±5 g] immediately after capture. Subsequently, these medusae were placed in pre-weighed containers, frozen and processed for mass [dry weight (DW) ± 0.001 g] and chemical content (percent carbon and nitrogen of dry weight) with standard procedures (Parsons et al., 1985).

The chemical content of copepods (*Calanus* spp., mostly *Calanus glacialis* Jashnov and chaetognaths *Eukrohnia hamata* Möbius) were also determined. Individuals of each taxon (six replicate sets of 500 copepods and 50 chaetognaths, respectively) were selected, washed briefly with distilled water, placed on aluminum foil, frozen and processed for dry weight, carbon and nitrogen as noted above.

*Metabolic demand*

*Periphylla periphylla* were collected in Lurefjorden during day trips with the R/V Hans Brattstrom. All captures were with the 2 m net that was hauled vertically from 400 m to the surface. Jellyfishes were carefully transferred from the plastic cod-end bag to 10 l buckets containing seawater from the depth of capture. These containers were shielded thereafter from daylight and transported to the field station at Espegrend. Medusae were maintained in the dark at 6–8 °C (= natural, ambient condition) in a temperature/light-controlled laboratory. Medusae obtained with ROV samplers were not used for metabolic studies because

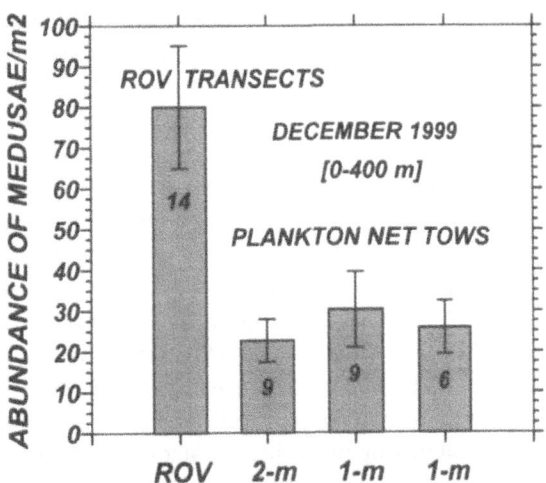

*Figure 3.* Comparisons of total mean abundance m$^{-2}$ of *Periphylla periphylla* (0–400 m): medusae counted during ROV transects versus medusae capture in net tows. Error bars are 95% CL. The number of vertical transects/tows is shown inside each bar and 2 m and 1 m designate the mouth diameter (m) of the nets. The last bar contains data from 6 sets of four, 100 m vertical tows (400–300 m, 300–200 m, 200–100 m and 100–0 m) with the 1 m net rigged to close.

temperature and light could not be controlled during the extended cruises aboard the support ship R/V Håkon Mosby.

Oxygen consumption rates of medusae were measured with an Orion (Model 840) system after 12–24 h. This period allowed medusae to clear any food in their stomachs and allowed for selection of healthy individuals. Measurements of oxygen concentrations were discontinuous, i.e. instrument readings were taken at the beginning and end of incubations, which ranged from 9 to 13 h. Seawater for these experiments was collected in Lurefjorden at 350 m with a rosette of 12 l Niskin water bottles. The volume of the respiration chambers varied from 1 to 5 l. A single medusa was placed in each chamber and incubations were conducted in the dark. Observations of medusae and readings of instruments were made under red light. A typical trial consisted of 8–10 chambers plus two controls (chambers without medusae). Medusae were measured (= CD) after removing them from the respirometers.

The accuracy of the Orion instrument was verified with standard Winkler titration procedures. Antibiotics were not added to the water in the incubation chambers. Respiration chambers were cleaned with 1 N HCl and thoroughly rinsed before each trial.

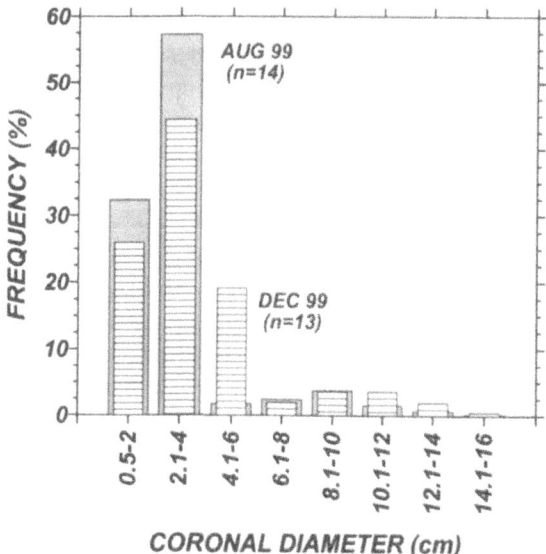

*Figure 4.* Percent frequency of standing stocks of medusae in several size (cm) classes captured in WP-3, 1 m diameter net (500 μm mesh) in summer (solid bars) and winter (lined bars) (*n* = number of tows).

## Results

### Abundance

On average, visual assessments of the abundance of *Periphylla periphylla* from the ROV deployments were three times greater that estimates based on medusae collected in net tows on a m$^{-2}$ basis (Fig. 3). Averaged data are presented as mean value ± standard error. The difference in abundance between ROV and net tows suggested that medusae were capable of detecting and avoiding the physical disturbance caused by slowly towed nets. The rapid (twice the haul speed of plankton nets) descent rate of the ROV, the larger (2–7 times) volume surveyed, and the lack of disturbance by the vehicle to the water column in the field of view probably acted in concert to improve abundance estimates. At least 90% of the visually recorded medusae were defined as 'small' (see 'Methods'). The actual size distribution, based on net-caught medusae, confirmed the strong numerical dominance (ca. 90% of all medusae collected) of small (0.5–6 cm) individuals (Fig. 4).

### Vertical migration

Vertical transects with the ROV indicated that *Periphylla periphylla* migrated upward each night (Fig. 5a, b). Typically, the vast majority (80–90%) of the medusae moved from the deep basin to a depth zone below

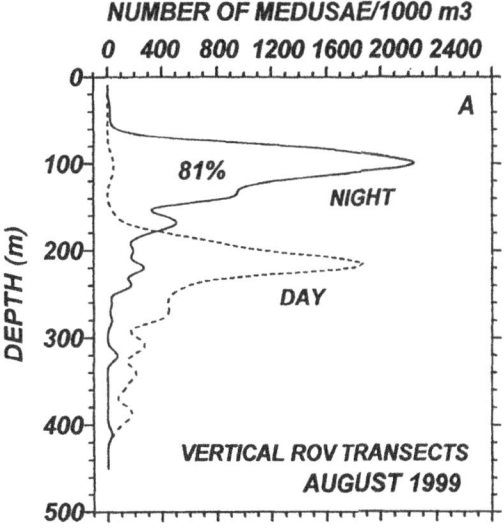

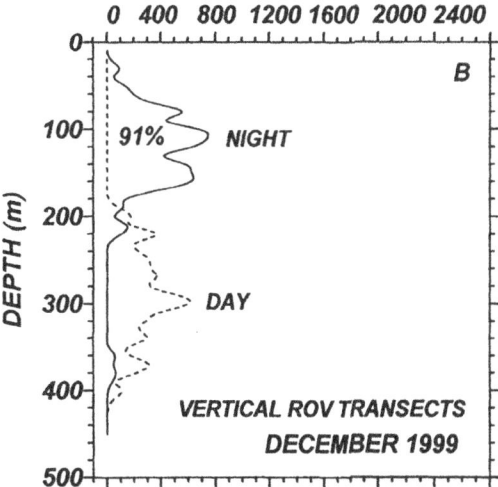

*Figure 5.* Vertical distribution and densities (no. 1000 m$^{-3}$) of *Periphylla periphylla* during day (1000–1200) and night (2300–0100) periods in summer (a) and winter (b) based on medusae counted during vertical ROV transects.

the thermocline, a distance ranging from 100 to 400 m. At dawn, medusae moved downward and dispersed within the water column below 200 m, mostly from 200 to 350 m. During the daylight hours, most of the medusae resided in narrow vertical intervals (ca. 60 m) during August and broader intervals (ca. 100 m) in December. Throughout the year, most of the population was observed in the low (6–8 °C) temperature water below the thermocline.

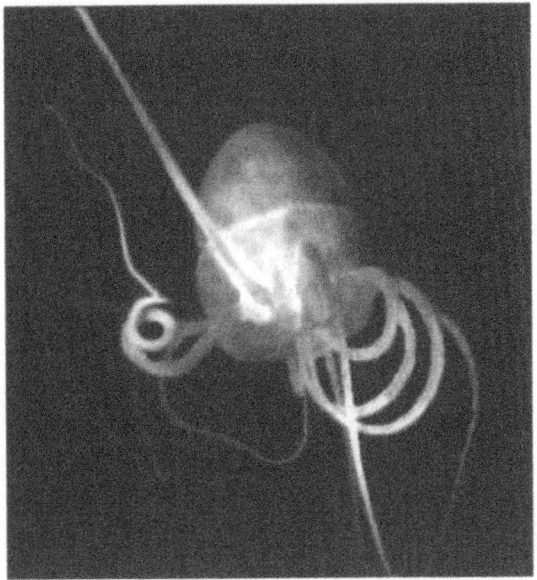

*Figure 6.* The variety of postures exhibited by the tentacles of *Periphylla periphylla*.

## Swimming and apparent feeding activity

The natural behaviors of medusae were observed from the ROV by using red light, because when exposed to incandescent illumination or daylight, *Periphylla periphylla* was always strongly negatively phototactic (see also Larson et al., 1991), most notably the large individuals. Medusa swimming speed under red light was typically 1 m min$^{-1}$, measured as the change in vertical position over at least 5 min ($n = 25$ individuals). When disturbed by white light or turbulence from the ROV, medusae swam much faster, up to 10 m min$^{-1}$ for prolonged (5–20 min) periods ($n = 15$ individuals). In most cases, medusae were initially disoriented in white light, swimming rapidly in any direction for a few minutes, and then swam downward toward the seafloor. Even upon reaching the seabed, medusae would continue swimming downward, repeatedly impacting the sediment. Some individuals swam downward for over 100 m after only a few seconds of exposure to white light. This response was never observed when medusae were viewed with red light.

*In situ* observations of *Periphylla periphylla* also revealed that swimming behaviors were apparently linked to feeding and differed with age, depth and time. During daytime, the larger (up to 17 cm CD) medusae were distributed mostly from 200 to 350 m where they performed dive and drift activities. These jellyfishes adopted an exumbrellar side down posture with all 12 tentacles positioned aborally alongside the umbrella. They repeatedly swam slowly downward for 10 m or so, and then drifted upward for similar vertical distances, occasionally making rapid upward sweeps with individual tentacles. At night, medusae were often upright with their tentacles folded upward along the outer surface of the conical dome or flared at right angles from the coronal margin. When viewed in the upper 150 m where copepods were conspicuous, one or more tentacles were often quickly arched away from the dome and downward toward the mouth. As a tentacle approached the mouth, the distal half shortened, not by contraction but by coiling . The coiled portion of a tentacle was then inserted into the gastric cavity. This rapid concealment of tentacles persisted for a few seconds to several minutes. From one to all 12 of the tentacles were held in the gastric cavity. The variety of tentacle postures is depicted in Figure 6.

By contrast, small medusae (up to 6 cm CD) were always active. They swam haphazardly in no particular direction, coiling and inserting tentacles into the stomach throughout a day. They remained mostly in a 25 m vertical zone just below the thermocline where copepods (*Calanus* spp.) were numerous.

The bursts of tentacle movements were assumed to be feeding bouts. Unfortunately, video records lacked resolution sufficient to identify prey attached to the tentacles. The placement of tentacles into the gastric cavity must serve as a defensive reflex to protect tentacles from damage. When a medusa experienced turbulence or sudden exposure to bright white light in the field or the laboratory, some or all of the tentacles were simultaneously curled, coiled and pushed into the stomach.

## Prey consumption

The stomachs of medusae ($n = 325$) collected with nets contained varying amounts of food. In most cases, the guts of the medusae from a given net tow contained copepods (*Calanus* spp. and *Euchaeta norvegica* Boeck, up to 200 individuals per medusa), chaetognaths (*Eukrohnia hamata* and *Sagitta elegans* Verril, up to 55 individuals per medusa), and ostracods (*Conchoecia* spp., up to 50 individuals per medusa). These prey were often entangled in a pale yellow matrix of what appeared to be mucus. Krill *Meganyctiphanes norvegica* M. Sars and small *Periphylla periphylla* occasionally appeared in the stomachs. The number of prey in the stomach varied with the size of a medusa as well as the depth of capture and time of

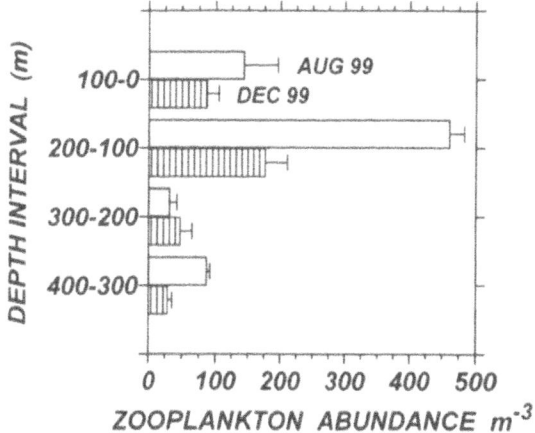

*Figure 7.* Total abundance of calanoid copepods (*Calanus* spp. and *Euchaeta norvegica*) and chaetognaths (*Eukhronia hamata*) collected at night (2300–0100) in 100 m intervals with vertical tows of a 1 m diameter (500 μm mesh) closing nets in summer and winter.

day. In general, the larger the medusa the greater the number of prey in its stomach. On average, more prey were found in the stomachs during the evening hours in shallow water.

It was surprising, then, that only 10% of the medusae (*n* = 41) collected individually in the ROV samplers had prey in their stomachs. The number of prey per medusa was also much smaller, i.e. 1–3 copepods or a single chaetognath, and 2–4 ostracods regardless of the size of the medusa, the depth of capture or the time of day. Such conflicting results suggest strongly that medusae obtained with nets had captured their prey from the zooplankton contained in the cod-end bag. The diets and foraging trends based on net tows therefore probably only reflected the vertical distribution, abundance and diversity of prey caught in the nets (Fig. 7). A small portion of the prey in some medusae, collected at night in the upper 100 m or in the lower 100 m, was partially digested, which suggests that at least a few individuals had fed prior to entering a towed net.

Further evidence supports the conclusion that medusae captured prey in the cod end of the net. Medusae collected individually in static chambers sloughed portions of the epidermal tissue from their tentacles within 15–30 s after contact with the walls of the sampler. The exfoliated tissue was pale yellow and clumped into aggregates. Microscopic examination of these aggregates revealed a tangled mass of nematocysts, discharged and undischarged. Once this behavior was noticed, the masses of such material, always seen in the cod end bags, were carefully inspected.

In all cases, what were assumed to be mucous secretions expelled by the medusae were in fact masses of nematocysts. Immobile copepods and chaetognaths were often entangled in this material. Collectively, these observations indicated that medusae 'fed' in the cod end. When confined in the relatively small volume of the cod end bag, it appears that globs of the immobilized prey were easily packed into the stomach, not with the tentacles but rather, via umbrellar movements. In the net tows that were open continuously from 440 m to the surface, medusae spent up to 70 min with the prey. The corresponding time for the closing-net collections ranged from 70 to 17 min.

Attempts to induce prey capture in the laboratory by pushing prey against the tentacles were unsuccessful. In part, this lack of response probably resulted because the mature nematocysts had been shed from the tentacles during collection. Examinations of the outer surfaces of the tentacles revealed closely-packed, ellipsoid nematocysts (predominately microbasic mastigophores, $20–25 \times 40–55$ μm). The inner surfaces of the stomach contained numerous, but smaller (18 μm), ovoid nematocysts (isorhizas). Both types of nematocysts are known to penetrate the bodies of zooplankton prey (Purcell & Mills, 1988).

*Biomass and chemical composition*

Regression equations, developed from measurements of the biomass and chemical content of *Periphylla periphylla*, are presented in Table 1. Dry weight averaged 3.24±0.2% WW, ranging from 2.0 to 3.9% (*n* = 18). Carbon averaged 19.6±0.5% DW, varying from 12.1 to 30.8% (*n* = 18). Nitrogen averaged 2.9±0.2% DW, ranging from 1.9 to 4.4% (*n* = 18). The body mass of the main prey of the medusa, copepods *Calanus* spp. and chaetognaths *Eukrohnia hamata*, averaged 0.234±0.01 and 0.252±0.01 mg C per individual (57.4 and 44.9% of DW, respectively).

*Metabolism*

During confinement in the experimental chambers, medusae were calm. They alternately displayed slow swim and sink behaviors, floated motionless and rested on the bottom of the chamber. Oxygen levels, in controls and experiments, were >70% of saturation. Decreases in the oxygen content of water in the controls were insignificant ($P < 0.05$; paired *t*-test). Differences in the decline of oxygen content between controls and experimental chambers were significant in every case ($P < 0.01$; *t*-test).

*Table 1. Periphylla periphylla.* Relationships between coronal diameter (CD in cm), wet weight (WW in g), dry weight (DW in g), carbon (C in mg) and nitrogen (N in mg) are expressed as linear regressions of natural logarithmic transformations. ($r^2$) is the coefficient of determination. ($n$ indicates the number of individuals measured for each category

| Y | (Range) | X | (Range) | Linear regression | $r^2$ | ($n$) |
|----|----|----|----|----|----|----|
| WW | (7–1125 g) | CD | (2.5–14.9) cm) | $Y = -0.877+2.922 * X$ | 0.98 | (75) |
| DW | (0.01–25.48 g) | CD | (1.4–14 cm) | $Y = -4.728+3.123 * X$ | 0.98 | (28) |
| DW | (0.38–25.48 g) | WW | (8.5–918 g) | $Y = -2.943+0.899 * X$ | 0.98 | (18) |
| C | (80.42–4719.72 mg) | DW | (0.38–25.48 g) | $Y = 5.373+0.924 * X$ | 0.97 | (18) |
| N | (11.64–702.22 mg) | DW | (0.38–25.48 g) | $Y = 3.396+0.971 * X$ | 0.98 | (18) |

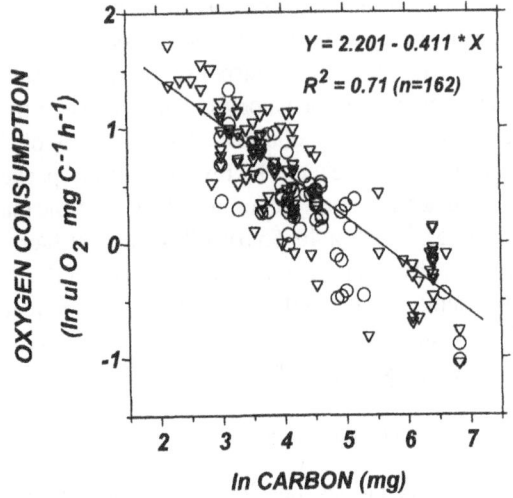

$$Y = 2.201 - 0.411 * X$$
$$R^2 = 0.71 (n=162)$$

*Figure 8.* Logarithmic (ln)-scale plot of oxygen consumption rates ($\mu$l O$_2$ mg C$^{-1}$ h$^{-1}$) of small to medium-sized (1.7–7.5 cm CD) *Periphylla periphylla* as a function of carbon weight (mg). Open triangles (March 1999); open circles (November, 1999).

Small to medium-sized (1.5–7.5 cm CD, 1.4–150 g WW) medusae had weight specific respiration rates that averaged 0.3±0.01 ($n = 162$), ranging from 0.1 to 0.7 $\mu$l O$_2$ mg DW$^{-1}$ h$^{-1}$. Weight-specific oxygen consumption decreased with increasing weight (Fig. 8), as observed for other medusae (Thuesen & Childress, 1994). Data from March, 1999 ($n = 107$) and November, 1999 ($n = 55$) were pooled because there was no significant difference between these two data sets (ANCOVA, $P < 0.05$). When expressed on a carbon-specific basis, respiration rates averaged 1.8±0.07 and ranged from 0.4 to 5.6 $\mu$l O$_2$ mg C$^{-1}$ h$^{-1}$. Daily carbon turnover rates averaged 1.7±0.08% of total body carbon and ranged from 0.4 to 5.8% (RQ = 0.8). The number of copepods and chaetognaths that would fulfill the daily metabolic requirements of small medusae averaged 7.8±0.6 and 7.2±0.5 prey, with ranges of

1–34 and 1–31 prey, respectively. An assimilation efficiency of 90% was assumed (Conover, 1978; Purcell, 1983).

## Discussion

### Abundance

In this study, collections with 1 and 2 m diameter nets towed vertically throughout the 440 m water column, indicated that the abundance (hundreds to thousands of individuals 1000 m$^{-3}$) of *Periphylla periphylla* was consistently 2–3 orders of magnitude more numerous than reported by Fosså (1992). Visual observations of medusae from ROV deployments indicated densities that were 3 times higher than those estimated from medusae captured with nets. Abundance estimates of *P. periphylla* from net tows in open ocean environments are usually <1 individual 1000 m$^{-3}$ (Pagès & Gili, 1992; Pagès & Kurbjeweit, 1994; Pagès et al., 1996; Dalpadado et al., 1998). The disparity in estimates of abundance between our results and those of Fosså (1992) is not surprising since in that study the density of medusae was based on individuals collected with a Harstad trawl. The rapid tow speed (3 kt) of that gear probably forced a substantial number of small (<5 cm CD) medusae through the broad (mostly 10 cm) mesh.

A greater number of medusae than previously reported would also be expected in 1999 if the population in Lurefjorden has grown in size. Interviews with local fishermen have indicated that since the 1970s there have been large variations in the occurrence of jellyfish there. Young stages (1–4 cm CD) were collected in the 1980s (Johannessen, 1980, pers. comm.). Fosså (1992, pers. comm.) reported only a few small (2–4 cm CD) medusae in late 1991, and in early 1992, sampling with Juday and MOCNESS net gear

(333 $\mu$m mesh) failed to collect small individuals. Jarms et al. (1999) remarked that larval recruitment of medusae did not occur until late in 1993. During this study (1998–1999), small medusae (< 5 cm CD) were numerically predominant. A sampling program from 1992 to 1997 gave abundance values integrated for the whole water column of 3–11 large medusae 1000 m$^{-3}$ (Båmstedt et al., unpublished). In June 1996, a scanning sonar (Simrad Mesotech, 675 kHz) mounted an ROV (Båmstedt, unpublished data) indicated 10–20 large medusae 1000 m$^{-3}$. In this study, stocks of large individuals were 17±3 (n = 4) for February 1998, and 8±2 (n = 14) and 9±2 (n = 9) for August and December, 1999, respectively. Thus, it seems reasonable to assume that the main part of the population increase since the early nineties is due to a pronounced, but variable recruitment of small medusae.

Physical and biological factors, acting together, probably favor the retention and proliferation of *Periphylla periphylla* in Lurefjorden. First of all, the movement of water into and out of the fjord is restricted to the upper 20 m by the shallow sill depth. The narrow, 200 m-wide entrance further limits the exchange rate. The water column is stratified during most of the year and temperature and salinity in the deep water are very stable. Young and old medusae tend to remain within this aphotic, 6–8 °C environment throughout their lifetime. The medusa undergoes direct development and fertilized eggs are released in the deep water where they remain for several months as free-floating and non-motile organisms through several non-feeding stages (Jarms et al., 1999). Eggs have been collected from June to October and larvae are present throughout the year (Jarms et al., 1999). Such information suggests that spawning is continuous. It is, therefore, puzzling that medusae smaller than 5 cm CD were not captured in net hauls during the period from 1991 to 1993 (Fosså, 1992; Jarms et al., 1999). A reasonable explanation for the absence young medusae would be that water in the fjord, at the depth where eggs and the early developmental stages occur, was renewed before larvae developed the ability to swim.

Once the young medusae develop rhopalia and tentacles, they begin a lifetime of diel vertical migration, swimming hundreds of meters upward and then downward each night. Upward excursions usually terminate at the base of the mixed layer. *Periphylla periphylla* is negatively phototactic and responds to the onset of daylight by returning to deep water. Thus, once inside the fjord, this species will tend to breed and grow in the deep basin. Observations since

1991 have indicated that between October and March, large medusae may appear at the surface during the night, except during periods of rough weather or when heavy rainfall causes a lens of brackish water to form at the surface. The relative importance of such surface swarmings to recruitment and survival and the mechanisms underlying this recurrent behavior require further study. Previous investigations in open ocean environments have suggested that *P. periphylla* may perform diel migrations (Pagès & Gili, 1992; Pagès et al., 1996).

If the advective loss is low, there are at least two other factors that would allow for the high population density within the fjord, i.e. low mortality and high longevity. Unpublished data from 1991 to 1997 (Båmstedt et al.) indicate that *P. periphylla* grows slowly, that mortality of the large individuals is low, and that longevity is very high, perhaps 10–30 years. These data are presently being prepared for publication. If, for example, the mortality is low and average longevity is 10 years, the intrinsic rate of natural increase does not have to be high to maintain a large population.

The perennial life cycle of *Periphylla periphylla* in Lurefjorden is quite different from other medusae that commonly occur in the coastal waters of Northwestern Europe. In those regions, there is an annual spring/summer outburst of newly recruited medusae (e.g. *Aurelia aurita* (Linnaeus) and *Cyanea capillata* (Linnaeus) that quickly grow to large size (Möller, 1980; Hernroth & Grøndal, 1983; Van Der Veer & Oorthuysen, 1985; Berstad et al., 1995; Lucas, 2001). The former species is also heavily predated by the latter species (Båmstedt et al., 1994; Hansson, 1997). Total mortality of both species usually occurs in the autumn shortly after they spawn.

*Periphylla periphylla* has no obvious competitors or predators in Lurefjorden. Other gelatinous zooplankton, such as ctenophores, siphonophores and medusae, are uncommon (pers. obs.). Fishes are rare in Lurefjorden, but quite numerous in nearby fjords where relatively few or no *P. periphylla* are present (Eiane et al., 1999). Disfigurement of a small portion of the medusae collected in net tows suggested that there may be predators. Dome mesoglea on a few medusae was damaged but healing. Parts or all of a given tentacle were missing on other individuals. In these cases, it is possible that a hyperiid amphipod *Hyperoche medusarum* (Kröyer), which was symbiotic with 5% (n = 1105) of the medusae examined, inflicted such mutilations. All medusae hosting an amphipod were intact at the time of collection, but this amphipod will consume any part of a medusa in the laboratory. Other hyperiid amphipods (*Cyllopus magellanicus* Dana and juvenile *Themisto guadichaudii* Guérin) have been reported to associate with *P. periphylla* (Pagès et al., 1996). Pycnoconids were described as ectoparasitic with *P. periphylla* because these crustaceans appeared to have fed on the tentacles (Child & Harbison, 1986).

Measurement of individual and population growth of *Periphylla periphylla* remains a topic for future investigations. Size class information for the 4-month period from August and December, 1999 showed a pronounced decrease in the 2.1–4 cm group and a strong increase in the 4.1–6 cm size class (Fig. 4). By using mid-point values for each class (3 and 5 cm CD, respectively) and converting CD to wet weight, it appears that small individuals increased in weight by 9 g WW per month. If constant, this rate is three times greater than reported by Jarms et al. (1999). That rate, however, was based on extrapolation of data obtained in 1992 and again in 1997. More frequent sampling is needed to define population growth rates.

## Behavior

The most exciting results from *in situ* observations were the behavioral responses of the medusae. As might be expected, these medusae, which live in no light or dim light regimes, are characteristically negatively phototactic. The flight reactions exhibited by *Periphylla periphylla* when exposed to white light confirm that future studies of the *in situ* behavior of deep water pelagic fauna should be conducted under red light fields. When viewed in red light, swimming behaviors differed with age and depth during the course of a day. Observations of the mode and manner of tentacle movements suggested that, during the day, the larger medusae cruised through narrow depth intervals feeding as a hunter. At night, these animals migrated into zones where prey are more numerous and behaved more like ambush predators, waiting for prey to swim into the tentacle field. Like many midwater animals, large *Periphylla periphylla*, can remain immobile at depth during the daytime, hanging motionless for minutes to hours (e.g. Barham, 1966; Youngbluth et al., 1988; Robison, 1999). Is it possible that the older medusae feed on small sinking particles? Probably not. When immobile, the umbrella is spread wide, but the flaccid lips of the stomach are held closed rather than open. The tentacles hang downward, surrounding the dome. Does this posture of nematocyst-laden tentacles provide the umbrella with protection from predators? Larson (1979) reported similar postures and sweeping behaviors by tentacles of the small coronate medusa, *Nausithoe punctata* Kölliker. More *in situ* observations are needed to determine the stimulus-response repertoire.

## Net Feeding and predation

The relatively high number of prey found in the stomachs of *Periphylla periphylla* collected in the nets suggested that this medusa was a voracious predator. In contrast to these observations, the stomachs of medusae sampled individually from the ROV rarely contained prey. Fosså (1992) also reported small numbers of prey consumed per medusa, i.e. 1–6 prey (the same copepods and an unknown decapod species) in 60% of the medusae that were captured individually at the surface at night. The major difference in prey consumption between net and individually-captured medusae indicated that medusae 'fed' in the cod end of the net. This conclusion was strengthened with the fact that prey found in the stomachs were usually entangled in nematocyst-laden tissue (see 'Results'). Other studies (Maas, 1897: Larson, 1979), which have reported crustacean (copepods and decapods) and fish remains in the stomachs of *P. periphylla*, are based on medusae captured in nets. The results of this study suggested that all such net-derived, trophic data are probably unreliable accounts of the predatory habits of this species.

## Metabolism

Measurements of oxygen consumption indicated that *Periphylla periphylla* survives on an extremely low energy intake. The rates were consistent with previous estimates for other scyphomedusae (Larson, 1987). Interestingly, metabolic data for *P. periphylla* reported by Thuesen & Childress (1994) were 6 times lower than determined in this investigation. The discrepancy may reflect variability between populations or decompression effects (Bailey et al., 1994). In any event, estimates of oxygen consumed by *P. periphylla* in this study remain preliminary. The relatively small experimental chambers may have restricted the normal activity patterns of small to medium medusae and thereby underestimated their usual rates of respiration. Determinations of respiration rates for large medusae were unsuccessful for at least two reasons. Individuals (10 cm or larger) were always physically injured when captured with plankton nets. Medusae collected with ROV samplers were not obviously damaged and survived, on one occasion, for a few days in shipboard aquaria. However, within 12–24 h, these medusae digested the mesoglea at the apex of the stomach. Gastric filaments extruded from this apical hole, which became progressively larger with time. This phenomenon occurred in smaller medusae maintained ashore but usually appeared only after several days. Another unexpected complication was the shedding of epidermal tissue from the surface of the tentacles. Small medusae exhibited less sloughing than large medusae, presumably because their tentacles are smaller and did not contact the sides of a container as frequently. Such stress, which invariably results from turbulence and abrasion during capture procedures, may be mitigated in the future by using larger ROV samplers to collect large medusae. However, even with carefully handling, seemingly robust individuals become moribund under laboratory conditions within a few days. Perhaps, the best way of obtaining large medusae for live experimentation will to be to capture them at the surface in the winter months. Alternatively, measurements of metabolism could be performed *in situ* to reduce the stresses associated with handling procedures (Youngbluth et al., 1988; Ikeda et al., 2000) and decompression effects (Bailey et al., 1994).

The daily metabolic demand of 1–6% of body carbon indicated that on average there should be 2–6 times more prey in the stomach at any one time, even allowing for rapid (3–4 h) digestion rates. The observations of low prey consumption also contrasted with the fact that >80% of the population performed extensive (100–400 m) diel vertical migrations throughout the year. Why do the medusae migrate daily? There was no evidence that they aggregate and spawn. Dispersal by the slow surface currents is probably minimal. In spite of the low metabolic rates and low prey consumption, there are at least 35 million *Periphylla periphylla* in Lurefjorden, assuming 80 medusae m$^{-2}$ in the area (4.2 10$^{-6}$ m$^2$) encompassed by the 200 m depth contour (see Fig. 1). Even if individual and population growth rates of medusae are slow, these gelatinous predators should eventually effect the standing stocks of the prey populations. Unfortunately, the abundance data available lack sufficient resolution to detect interspecific interactions. At present, both predator and prey are numerous. However, estimates of carbon turnover suggest that *P. periphylla* should consume the yearly production of zooplankton

(principally copepods *Calanus* spp.). The average zooplankton biomass in Lurefjorden is 7.5±1.7 g C m$^{-2}$ (monthly range = 1–17, Båmstedt unpublished data, 80 $\mu$m mesh net samples, 0–400 m vertical tows). By assuming that the annual P/B ratio is directly related to body mass (Aksnes & Magnesen, 1983) and using a value of 0.1 (Mauchline, 1998), the annual secondary production of carbon would be 75 g C m$^{-2}$. By using 80 medusae m$^{-2}$, a daily carbon requirement based solely on oxygen consumption reported in this paper (carbon demand per day expressed according to carbon weight using the arithmetic regression

$$Y = 0.6531 + 0.0061(X),$$

where $X$ is the carbon weight (mg) of a given medusa), a distribution weighted for small medusae (80% of population in 5, 1 cm size classes comprising 16 individuals each) with large medusae (20% in 10, 1 cm of 2 individuals each), a very conservative estimate of daily carbon consumption would be 0.57 g C m$^{-2}$ d$^{-1}$ or 208 g C m$^{-2}$ yr$^{-1}$, a value 2.8 times greater than the annual secondary production. Clearly, long-term investigations of the ecology and behavior of *P. periphylla* are needed to understand its role in fjord ecosystems.

## Acknowledgements

The assistance of scientists Per Flood, Gaby Gorsky, Gerhard Jarms, Monica Martinussen, Jennifer Purcell, Henry Tiemann, students Erling Heggoy, Dawn Murray, Carlos Montero, Shawn Osborn, Eva Visauta, ROV pilots Trond Bolstad, Jan Bryn, Bjorn Øystein, Kåre Saue, Vidar Saue, the crews of the R/V Hans Brattstrom (Frank Berland, Tore Heggheim, Kolbjørn Øyjordsbakken) and the R/V Håkon Mosby, and staff at the Marine Biological Station, Espegrend (Agnes Aadnesen, Halvdan Gjertsen) is greatly appreciated. The study was supported in part by a Guest Scientist fellowship awarded by the University of Bergen and a grant from the Division of International Programs, National Science Foundation (INT-9903467).

## References

Alldredge, A. L., 1984. The quantitative significance of gelatinous zooplankton as pelagic consumers. In Fasham, M. J. R. (ed.), Flows of Energy and Materials in Marine Ecosystems: Theory and Practice. Plenum New York: 407–433.

Aksnes, D. L. & T. Magnesen, 1983. Distribution, development and production of *Calanus finmarchicus* (Gunnerus) in Lindåspollene, western Norway, 1979. Sarsia 68: 195–208.

Aure, J. & R. Sætre, 1981. Wind effects on the Skagerrak outflow. In Sætre, R. & M. Mork (eds), The Norwegian Coastal Current. University of Bergen: 263–293.

Bailey, T. G., J. J. Torres, M. J. Youngbluth & G. P. Owen, 1994. Effect of decompression on mesopelagic gelatinous zooplankton: A comparison of *in situ* and shipboard measurements of metabolism. Mar. Ecol. Prog. Ser. 113: 13–27.

Båmstedt, U., M. B. Martinussen & S. Matsakis, 1994. Trophodynamics of the two scyphozoan jellyfishes, *Aurelia aurita* and *Cyanea capillata*, in western Norway. ICES J. Mar. Sci. 51: 369–382.

Båmstedt, U., J. H. Fosså, M. B. Martinussen & A. Fosshagen, 1998. Mass occurrence of the physonect siphonophore *Apolemia uvaria* (Lesueur) in Norwegian waters. Sarsia 83: 79–85.

Barham, E. G., 1966. Deep scattering layer migration and composition: observations from a diving saucer. Science 151: 1399–1403.

Berstad, V., U. Båmstedt & M. B. Martinussen, 1995. Distribution and swimming of the jellyfishes *Aurelia aurita* and *Cyanea capillata*. In Skjoldal, H. R., C. Hopkins, K. E. Erikstad & H. P. Leinaas (eds), Ecology of Fjords and Coastal Waters. Elsevier Science, London: 257–271.

Child, C. A. & G. R. Harbison, 1986. A parasitic association between a pycnogonid and a scyphomedusa in midwater. J. mar. biol. Ass. U.K. 66: 113–117.

Conover, R. J., 1978. Transformation of organic matter. In Kinne, O. (ed.), Marine Ecology IV, Dynamics. Wiley, Chichester: 221–499.

Dalpadado, P., B. Ellertsen, W. Melle & H. R. Skjoldal, 1998. Summer distribution patterns and biomass estimates of macrozooplankton and micronekton in the Nordic seas. Sarsia 83: 103–116.

Eiane, K., D. L. Aksnes & M. D. Ohman, 1998. Advection and zooplankton fitness. Sarsia 83: 87–93.

Eiane, K., D. L. Aksnes, E. Bagoeien & S. Kaartvedt, 1999. Fish or jellies – a question of visibility? Limnol. Oceanogr. 44: 1352–1357.

Fosså, J. H., 1992. Mass occurrence of *Periphylla periphylla* (Scyphozoa, Coronatae) in a Norwegian fjord. Sarsia 77: 237–251.

Golmen, L., H. Svendsen, A. Bakke & J. Molvaer, 1998. Strong tide-induced vertical mixing in a deep fjord with a shallow sill. Oceanography 11: 1–5.

Hansson, L. J., 1997. Effect of temperature on growth rate of *Aurelia aurita* (Cnidaria, Scyphozoa) from Gullmarsfjorden, Sweden. Mar. Ecol. Prog. Ser. 161: 145–153.

Hay, S. J., J. R. G. Hislop & A. M. Shanks, 1990. North Sea scyphomedusae: summer distribution, estimated biomass and significance particularly for 0-group gadoid fish. Neth. J. Sea Res. 25: 113–130.

Hernroth, L. & F. Grøndal, 1983. On the biology of *Aurelia aurita*. I. Release and growth of *Aurelia aurita* (L.) ephyrae in the Gullmar Fjord, western Sweden, 1982–83. Ophelia 22: 189–199.

Ikeda, T., J. J. Torres, S. Hernández-León & S. P. Geiger, 2000. Metabolism. In Harris, R., P. Wiebe, J. Lenz, H. R. Skjoldal & M. Huntley (eds), ICES Zooplankton Methodology Manual. Academic Press, New York: 455–532.

Jarms, G., U. Båmstedt, H. Tiemann, M. B. Martinussen & J. H. Fosså, 1999. The holopelagic life cycle of the deep-sea medusa *Periphylla periphylla* (Scyphozoa, Coronata). Sarsia 84: 55–65.

Johannessen, P., 1980. Resipientundersøkelse av enkelte fjordavsnitt I Lindås kommune med hovedvekt lagt på bunnforhold og bunndyr. Institutt for Marin-biologi, Universitetet I Bergen: 39 pp.

Larson, R. J., 1979. Feeding in coronate medusae (Class Scyphozoa, Order Coronatae). Mar. Behav. Physiol. 6: 123–129.

Larson, R. J., 1986. Pelagic scyphomedusae (Scyphozoa: Coronatae and Semaeostomeae) of the southern Ocean. In Kornicker, L. (ed.), Biology of the Antarctic Seas. XVI. Ant. Res. Ser. 41: 59–165.

Larson, R. J., 1987. Respiration and carbon turnover rates of medusae from the NE Pacific. Comp. Biochem. Physiol. 87A: 93–100.

Larson, R. J.,1990. Scyphomedusae and cubomedusae from the Eastern Pacific. Bull. mar. Sci. 47: 546–556.

Larson, R. J., C. E. Mills & G. R. Harbison, 1991. Western Atlantic midwater hydrozoan and scyphozoan medusae: *in situ* studies using manned submersibles. Hydrobiologia. 216/217: 311–317.

Lucas, C. H., 2001. Reproduction and life history strategies of the common jellyfish *Aurelia aurita*, in relation to its ambient environment. Hydrobiologia 451 (Dev. Hydrobiol. 155): 229–246.

Maas, O., 1897. Reports on an exploration off the west coasts of Mexico, Central and South America, and off the Galapagos Islands. XXXI. Die Medusen. Mem. Mus. Comp. Zool. Harv. 23: 9–92.

Mauchline, J., 1998. The biology of calanoid copepods. Adv. mar. Biol. 33: 1–710.

Mills, C. E., 1995. Medusae, siphonophores and ctenophores as planktivorous predators in changing global ecosystems. ICES J. mar. Sci. 52: 575–581.

Möller, H., 1980. Population dynamics of *Aurelia aurita* medusae in Kiel Bight, Germany (FRG). Mar. Biol. 60: 123–128.

Pagès, F. & J. –M. Gili, 1992. Influence of the thermocline on the vertical migration of medusae during a 48 h sampling period. S. –Afr. Tydskr. Dierk. 27: 50–59.

Pagès, F. & F. Kurbjeweit, 1994. Vertical distribution and abundance of mesoplanktonic medusae and siphonophores from the Weddell Sea, Antarctica. Polar Biol. 14: 243–251.

Pagès, F., M. G. White & P. G. Rodhouse, 1996. Abundance of gelatinous carnivores in the nekton community of the Antarctic polar frontal zone in summer 1994. Mar. Ecol. Prog. Ser. 141: 139–147.

Parsons, T. R., Y. Maita & C. M. Lalli, 1985. A Manual of Chemical and Biological Methods for Seawater Analysis. Pergamon Press, New York.

Purcell, J. E., 1983. Digestion rates and assimilation efficiencies of siphonophores fed zooplankton prey. Mar. Biol. 73: 257–261.

Purcell, J. E. & C. E. Mills, 1988. The correlation between nematocyst types and diets in pelagic Hydrozoa. In Hessinger, D. & H. Lenhoff (eds), The Biology of Nematocysts. Academic Press, New York: 463–485.

Purcell, J. E. & M. N. Arai, 2001. Interactions of pelagic cnidarians and ctenophores with fish: a review. Hydrobiologia 451 (Dev. Hydrobiol. 155): 27–44.

Purcell, J. E., T. A. Shiganova, M. B. Decker & E. D. Houde, 2001. The ctenophore *Mnemiopsis* in native and exotic habitats: U.S. estuaries versus the Black Sea basin. Hydrobiologia, this volume.

Robison, B. H., 1999. Shape change behavior by mesopelagic animals. Mar. Fresh. Behav. Physiol. 32: 17–25.

Robison, B. H., K. R. Reisenbichler, R. Sherlock, J. M. B. Silguero & F. P. Chavez, 1998. Seasonal abundance of *Nanomia bijuga* in Monterey Bay. Deep-Sea Res. II, 45: 1741–1751.

Rogers, C. A., D. C. Biggs & R. A. Cooper, 1978. Aggregation of the siphonophore *Nanomia cara* in the Gulf of Maine. Fish. Bull. 76: 281–284.

Thuesen, E. V. & J. J. Childress, 1994. Oxygen consumption rates and metabolic enzyme activities of oceanic California medusae in relation to body size and habitat depth. Biol Bull. 187: 84–98.

Tietze, R. C. & A. M. Clark, 1986. Remotely operated tools for undersea vehicles. In McGuiness, T. (ed.), Current Practices and New Technology in Ocean Engineering, Am. Soc. Mech. Engin. 11: 219–223.

Tusting, R. F. & D. L. Davis, 1993. Laser systems and structured illumination for quantitative undersea imaging. Mar. Technol. Soc. J. 26: 5–12.

Van Der Veer, H. M. & W. Oorthuysen, 1985. Abundance, growth and food demand of the scyphomedusa *Aurelia aurita* in the western Wedden Sea. Neth. J. Sea Res. 19: 38–44.

Volovik, S. P., Z. A. Myrzoyan & G. S. Volovik, 1993. *Mnemiopsis leidyi* in the Azov Sea: biology, population dynamics, impact to the ecosystem and fisheries. ICES-CM-1993/L:69: 11 pp.

Youngbluth, M. J., 1984. Manned submersibles and sophisticated instrumentation: Tools for oceanographic research. In Proceedings of SUBTECH 1983 Symposium, Society of underwater technology, London: 335–344.

Youngbluth, M. J., T. G. Bailey & C. A. Jacoby, 1990. Biological explorations in the mid-ocean realm: foods webs, particle flux and technological advancements. In Lin, Y. C. & K. K. Shida (eds), Man in the Sea, Volume II, Best Publishing, San Pedro: 191–208.

Youngbluth, M. J., P. Kremer, T. G. Bailey & C. A. Jacoby, 1988. Chemical composition, metabolic rates and feeding behavior of the midwater ctenophore *Bathocyroe fosteri*. Mar. Biol. 98: 87–94.

The manufacturer's authorised representative in the EU is Springer
Nature Customer Service Centre GmbH, Europaplatz 3, 69115 Heidelberg,
Germany. If you have any concerns regarding our products, please
contact ProductSafety@springernature.com

Printed and bound by CPI Group (UK) Ltd, Croydon, CR0 4YY
23/04/2026
02095658-0001